ENERGY SCIENCE AND APPLIED TECHNOLOGY

PROCEEDINGS OF THE 2ND INTERNATIONAL CONFERENCE ON ENERGY SCIENCE AND APPLIED TECHNOLOGY, (ESAT 2015), WUHAN, CHINA, 28–30 AUGUST 2015

Energy Science and Applied Technology

Editor

Zhigang Fang
Wuhan University of Technology, Wuhan, Hubei, China

CRC Press is an imprint of the
Taylor & Francis Group, an **informa** business
A BALKEMA BOOK

Published by:
CRC Press/Balkema
P.O. Box 447, 2300 AK Leiden, The Netherlands
e-mail: Pub.NL@taylorandfrancis.com
www.crcpress.com – www.taylorandfrancis.com

First issued in paperback 2020

© 2016 by Taylor & Francis Group, LLC
CRC Press/Balkema is an imprint of the Taylor & Francis Group, an informa business

No claim to original U.S. Government works

Typeset by V Publishing Solutions Pvt Ltd., Chennai, India

ISBN 13: 978-0-367-73763-4 (pbk)
ISBN 13: 978-1-138-02833-3 (hbk)

Visit the Taylor & Francis Web site at
http://www.taylorandfrancis.com

and the CRC Press Web site at
http://www.crcpress.com

Energy Science and Applied Technology – Fang (Ed.)
© *2016 Taylor & Francis Group, London, ISBN 978-1-138-02833-3*

Table of contents

Environmental engineering and sustainable development

Electrical and electronic technology, power system engineering

Mechanical, manufacturing, process engineering

Applied and computational mathematics

Methods and algorithms optimization

Preface

Dear Distinguished Delegates and Guests,

The organizing Committee of ESAT 2015 warmly welcomes you to join the 2015 International Conference on Energy Science and Applied Technology (ESAT 2015: www.ESATconf.org/), held on August 28–30, 2015 in Wuhan, China.

The aim of ESAT 2015 is to provide a platform for researchers, engineers, and academicians, as well as industrial professionals, to present their research results and development activities in energy science and engineering and its applied technology. The themes which will be in this proceeding would be: Technologies in Geology, Mining, Oil and Gas; Renewable Energy, Bio-Energy and Cell Technologies; Energy Transfer and Conversion, Materials and Chemical Technologies; Environmental Engineering and Sustainable Development; Electrical and Electronic Technology, Power System Engineering; Mechanical, Manufacturing, Process Engineering; Control and Automation; Communications and Applied Information Technologies; Applied and Computational Mathematics; Methods and Algorithms Optimization; Network Technology and Application; System Test, Diagnosis, Detection and Monitoring; Recognition, Video and Image Processing. It provides opportunities for the delegates to exchange new ideas and application experiences, to establish business or research relations and to find global partners for future collaboration.

Hopefully, all participants and other interested readers benefit scientifically from the proceedings.

With our warmest regards,
Zhigang Fang
Conference Organizing Chair,
Wuhan, China

Energy Science and Applied Technology – Fang (Ed.)
© 2016 Taylor & Francis Group, London, ISBN 978-1-138-02833-3

Committees

SCIENTIFIC COMMITTEE

Chairman

Dr. Z.G. Fang, *Wuhan University of Technology, China*

Co-Chairmen

Dr. C. Yang, *Wuhan University of Technology, China*
Dr. Y.F. Zhao, *Wuhan University of Technology, China*
Prof. G.J. Fu, *Northeast Petroleum University, China*

Members

Dr. Z.G. Fang, *Wuhan University of Technology, China*
Prof. J.J. Xu, *Northeast Petroleum University, China*
Dr. J. Hu, *Ohio State University, USA*
Dr. W. Zhong, *New York State University at Stony Brook, USA*
Prof. H. Davis, *Boya Century Publishing Ltd., Hong Kong*
Associate Prof. D. Fang, *Wuhan University, China*
Prof. G.J. Fu, *Northeast Petroleum University, China*
Prof. P. Wang, *Guangxi College of Education, China*
Associate Prof. H. Chen, *Shanghai University of Engineering Science, China*
Dr. C. Yang, *Wuhan University of Technology, China*
Dr. Y.F. Zhao, *Wuhan University of Technology, China*
Dr. X.H. Deng, *Wuhan University of Technology, China*
Dr. J. Luo, *Wuhan University of Technology, China*
Dr. G.L. Liu, *Wuhan University of Technology, China*
Dr. H.H. You, *Zhicheng Conference Services Ltd., China*

Organizing Committee

Dr. Z.G. Fang, *Wuhan University of Technology, China*
Associate Prof. L. Liu, *Wuhan University of Technology, China*
Dr. C. Yang, *Wuhan University of Technology, China*
Dr. Y.F. Zhao, *Wuhan University of Technology, China*
Dr. J. Luo, *Wuhan University of Technology, China*
Dr. X.H. Deng, *Wuhan University of Technology, China*
Dr. G.L. Liu, *Wuhan University of Technology, China*
Dr. X. Xiao, *Wuhan University of Technology, China*
Dr. C. Zhang, *Wuhan University of Technology, China*
Associate Prof. D. Fang, *Wuhan University, China*
Dr. H.H. You, *Zhicheng Conference Services Ltd., China*
Prof. Y. Ma, *Zhicheng Conference Services Ltd., China*
Dr. J. Shi, *Zhicheng Times Culture Development Co. Ltd., China*
Dr. L. Xu, *Zhicheng Times Culture Development Co. Ltd., China*

Energy Science and Applied Technology – Fang (Ed.)
© 2016 Taylor & Francis Group, London, ISBN 978-1-138-02833-3

Sponsors

Northeast Petroleum University
Boya Century Publishing Ltd.
Research Center of Engineering and Science (RCES)
Asian Union of Information Technology
Zhicheng Times Culture Development Co. Ltd.

Technologies in geology, mining,
oil and gas exploration and exploitation of deposits

Energy Science and Applied Technology – Fang (Ed.)
© *2016 Taylor & Francis Group, London, ISBN 978-1-138-02833-3*

Reinforcing polyamide 1212 via graphene oxide

Ziqing Cai, Xiaoyu Meng, Yingshuai Han, Qiong Zhou & Lishan Cui
New College of Science, China University of Petroleum, Beijing, China
Beijing Key Laboratory of Failure, China University of Petroleum, Beijing, China

ABSTRACT: In this paper, Graphene Oxide (GO) was synthesized from graphite powder by Hummers' method. The results indicated that the GO single layer was successfully synthesized. The polyamide 1212 (PA1212)/GO composites were prepared by a two-step melt-compounding method. First, GO concentrates were prepared via solution coagulation. Then the resulting product was melt-compounded with a PA1212 matrix. The mechanical properties of the PA1212/GO composites were improved with the addition of 0.3 wt. % GO. The GO nanosheets promoted the crystallization process according to DSC results.

Keywords: Polyamide 1212 composites; Graphene oxide; Mechanical properties; Crystallization properties

1 INTRODUCTION

Graphene has attracted tremendous amount of attentions recently due to its remarkable properties, such as high mechanical properties, superior thermal conductivity, large specific surface area and excellent electronic transport properties (Kim 2010, Stankovich 2006, Zhao 2010).

Polyamide 1212 (PA1212) is an important engineering plastic in the family of even-even series polyamide and it has superior physical-chemical properties (Liu 2003). Nevertheless, PA1212 has not enough good mechanical properties at room temperature, which greatly restrict its application in mechanics, automobile, aerospace and petroleum fields (Peng 2001). In order to improve the mechanical properties, composites of PA1212/ nanofiller are of particular interest (Tibbetts 2007, Coleman 2006, Potts 2011). Graphene is widely used in polymer blends to improve the mechanical properties, thermal stability and electric conductivity based on its excellent properties (Potts 2011). A number of papers have been published on the reinforcement of mechanical properties in polyamide/graphene composites by in situ polymerization, solution mixing and melt-compounding (Zheng 2012, Yan 2012, Cai 2015). Zhen et al. (Xu 2010). The mechanical properties of prepared NG fibers by melt spinning process were reinforced significantly by 0.1 wt. % graphene. Rafiq et al. (Rafiq 2010) fabricated functionalized graphene/nylon 12 composites were by melt compounding. The results revealed that the incorporation of 0.6 wt. % of the functionalized graphene could improve ultimate tensile strength, elongation, impact strength and toughness significantly.

In this work, GO was prepared by Hummers' method (Zhao 2010, Marcano 2010, Wang 2008). GO was characterized by several techniques, such as XRD, TEM and AFM. The PA1212/GO composite was prepared by the one-step and two-step compounding process to compare the effect of the two methods. The mechanical and crystallization properties of pure PA1212 and the composites were investigated.

2 EXPERIMENTAL SECTION

2.1 Materials

Copy Potassium persulfate ($K_2S_2O_8$) was purchased from Tianjin Guangfu Technology Development Co. Ltd. Phosphorus pentoxide (P_2O_5) was purchased from Tianjin Chemical Reagent Factory. Concentrated sulfuric acid (H_2SO_4, Purity > 98%), potassium permanganate ($KMnO_4$), hydrogen peroxide (H_2O_2, Purity > 30%) were both supplied by Beijing Chemical Factory. Natural graphite (0.5~1.0 μm, Purity > 90%) was purchased from Qingdao Jinrilai Graphite Co. Ltd. PA1212 pellets were purchased from Shandong Dongcheng Engineering Plastic Co. Ethanol Soluble polyamide (ES-1) was purchased from Shanghai Zhenwei Composites Co.

2.2 Synthesis of Graphene Oxide (GO) and PA1212/GO composite preparation

Pre-oxidation process. Natural flake graphite (1 g) was added into a mixture of concentrated H_2SO_4 (5 mL), $K_2S_2O_8$ (1.6 g) and P_2O_5 (1.6 g). The mixture solution was then heated to 80°C and stirred for 3h.

The black solid product obtained was cooled to room temperature and then diluted with deionized water until the pH value of the water reached 7. The resulting product was vacuum-dried for 24 h at room temperature.

Oxidation process. The pre-oxide graphite was added into concentrated H_2SO_4 (23 mL) and stirred in an ice bath. Then $KMnO_4$ (3 g) was added into the solution and stirred uniformly. The solution was kept at 35°C and stirred for 2h. Then the solution was diluted with deionized water (115 mL) slowly. After 15 min, H_2O_2 (3 mL) solution was added to terminate this reaction. The color of mixture changed from dark brown to brilliant yellow. After the solution was cooled to room temperature, the mixture was centrifuged and dialyzed. Replace dialysis water two times a day until dialysis water was neutral. The product of graphene oxide was thus obtained after drying by freeze drier.

Nanocomposite preparation. Two-step method Firstly, GO (0.15 g) was added into water (100 mL) and the solution was dispersed by ultrasonic (100 w) treatment for 10 min. Then, the prepared Ethanol soluble polyamide (ES-1) (2 g) solution (10 wt.%) was mixed with GO solution under mechanical stirring. The pre-mixed product was dried in vacuum oven at 80°C for 12 h. The composite containing 0.3 wt.% GO were prepared by melt blending in a Brabender mixer for 12 min at 205°C with a rotating speed of 40 rpm. Plate specimens were prepared by compression molding at 210°C for 10 min. The drawing dumbbell specimens were cut with the dimensions of 1.2 mm in thickness, 5 mm in width, and 30 mm in length.

One-step method. Prepared GO powders were used in melt blending. A composite containing 0.3 wt.% GO was prepared by melt-blending in a Brabender mixer directly.

2.3 *Performance testing and structural characterization*

The mechanical property testing was carried out on the Electronic Universal Tensile Testing Machine. The tensile strength, yield strength and elongation at break were measured at the room temperature with a speed of 50 mm/min.

The Wide-Angle X-ray Diffraction (WAXD) experiments were conducted on a D8 Focus diffractometer at room temperature, with a wavelength of 0.154 nm. The X-ray diffractograms were collected in the 2θ range of 5~45° at the scan speed of 2 °/min.

Thermal analysis of the samples was conducted on a Differential Scanning Calorimetry (DSC 204 F1, NETZSCH) under argon atmosphere. For non-isothermal melt crystallization, the samples were heated from 25°C to 220°C at a heating rate of 10°C/min, holding for 2 min to remove thermal

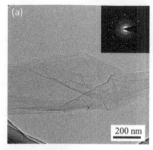

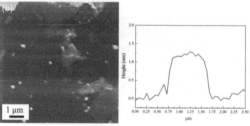

Figure 1. TEM image and select area electronic diffraction of GO layers (a), AFM image of single layer GO (b).

history, and then cooled down to 25°C. The melting, crystallization points (T_m, T_c) and crystallinity (X_c) were obtained.

The morphology of GO and the composite was characterized by optical microscope (DM2500P), SEM instrument (Quanta 200F), TEM instrument (FEI F20) and a typical tapping-mode Atomic Force Microscope (AFM) measurement (Nanoman VS).

3 RESULTS AND DISCUSSION

3.1 *Morphology of graphene oxide and the composites*

Figure 1a shows a general view of GO nanosheets, clearly illustrating the flake-like shape of GO. Multilayer GO nanosheets corrugate together with sizes in the range of dozens to several hundreds of nanometers. Selected Area Electron Diffraction (SAED, inset in Figure 1a) further confirms the ordered graphite lattices are clearly visible. It indicates that the graphene nanosheets are ordered crystal structure to some extent. As presented in AFM image (Figure 1b), the size of single-layer GO nanosheet is about 2~3 μm. The measured thickness of GO nanosheet is about 1 nm, which is consistent with previous reports (Zhao 2010).

Figure 2 shows WAXD patterns of GO and natural graphite. The WAXD pattern of natural graphite is presented for comparison with GO. The diffraction peak of GO at 2θ = 9.30° is corresponding to the 002 interplanar spacing of 0.95 nm. The 002 diffraction peak for natural graphite at

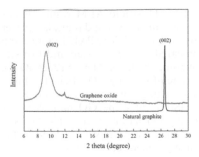

Figure 2. The WAXD analysis of GO and natural graphite.

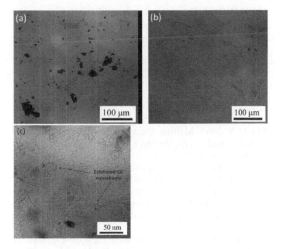

Figure 3. Optical micrographs of the PA1212/GO composites prepared via one- and two-step melt compounding: (a) 0.3wt. % GO/PA1212 (one-step method), (b) 0.3 wt. % GO/PA1212 (two-step method). The TEM image (c) of 0.3 wt. % PA1212/GO composite (two-step method).

$2\theta = 26.60°$ with an interlayer distance of 0.34 nm disappeared in GO sample. After oxidation of natural graphite, the interlayer distance has been increased from 0.34 nm to 0.95 nm.

Figure 3 shows the overall dispersion state of PA1212/GO composites prepared via one- and two-step methods. For the 0.3 wt. % GO/PA1212 composite prepared via one-step method, GO powders were not well-dispersed in the PA1212 matrix because of the strong interaction between graphene layers. The GO aggregates were observed in the matrix (as shown in Figure 3a). For GO/PA1212 composite prepared via two-step method, GO was uniformly dispersed in the PA1212 matrix (Figure 3b). This favorable dispersion state also can be observed in Figure 3c, where the GO nanosheets were efficiently exfoliated. When ultrasonic-treated GO was dissolved in an ethanol-soluble polyamide

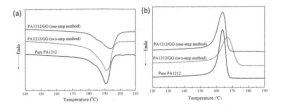

Figure 4. DSC scans of pure PA1212 and 0.3 wt.% PA1212/GO composites: (a) melting curves, (b) crystallization curves.

Table 1. DSC analysis of pure PA1212 and 0.3 wt.% PA1212/GO composites.

Samples	T_m(°C)	$T_{c,o}$(°C)	$T_{c,p}$(°C)	X_c(%)
Pure PA1212	190.9	166.8	163.2	12.60
PA1212/GO (two-step method)	190.4	173.7	168.5	12.49
PA1212/GO (one-step method)	193.7	167.2	163.8	11.27

(ES-1) solution, exfoliated GO sheets were wrapped with ES-1 polymer chains, which have good compatibility with the PA1212 matrix. During melt-compounding, wrapped GO nanosheets were effectively dispersed by shear force in the PA1212 matrix. And it is important for reinforcement of the PA1212/GO composite.

3.2 Thermal analysis of pure PA1212 and the composites

The effects of GO on crystallization behavior of PA1212/GO composite prepared by two different methods were analyzed via non-isothermal DSC experiments as reported in Figure 4 and Table 1.

$$X_c = \frac{\Delta H_m \times 100}{\Delta H_m^\circ} \tag{1}$$

where ΔH_m and ΔH_m° are regarded as the enthalpy of fusion of samples and the equilibrium melting enthalpy, respectively. The value of ΔH_m° for PA1212 is 292.2 J/g (Ren 2004).

As shown in Figure 4 and Table 1, for two-step method, with addition of 0.3 wt.% GO, the crystallization temperature $T_{c,p}$ and $T_{c,o}$ increases by about 5°C and 7°C respectively. It indicates that GO nanosheets promote the crystallization process. Meanwhile, the melting point (T_m) and crystallinity of the composite change slightly. For one-step method, the melting point of the composite increases obviously possibly due to the perfect crystalline induced by the GO nanosheets. However, the crystallization temperatures increase slightly which can be attributed to the bad dispersion of GO in the matrix.

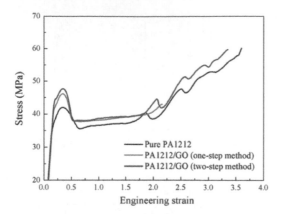

Figure 5. The stress-strain curves of pure PA1212 and 0.3 wt. % PA1212/GO composites stretched at room temperature.

Table 2. Mechanical properties of pure PA1212 and PA1212/GO composites.

Samples	Tensile strength (MPa)	Yield strength (MPa)	Young's modulus (MPa)	Elongation at break (%)
Pure PA1212	60 ± 1.3	43 ± 1.5	251 ± 30.9	373 ± 7.9
PA1212/GO (one-step method)	47 ± 3.5	46 ± 0.6	263 ± 26.3	231 ± 15.7
PA1212/GO (two-step method)	60 ± 0.8	47 ± 1.1	291 ± 22.1	335 ± 26.3

3.3 Mechanical properties of pure PA1212 and the composites

As shown in Figure 6 and Table 2, compared with pure PA1212, the yield strength and Young's modulus of the PA1212/GO composite (two-step method) are improved from 43 and 251 MPa to 47 and 291 MPa, respectively. And the elongation at break of the composite decreases slightly. For the GO/PA1212 composite prepared via one-step method, its yield strength improved from 43 to 46 MPa, and its tensile strength and elongation at break decreased significantly because of structural defects (caused by GO aggregates, Figure 3a). Thereby, the mechanical performance of the composites prepared via two-step method was better than that of obtained via one-step method. The favorable diapersion of GO in PA1212 matrix is crucial to reinforcement of the composites (Dzenis 2008).

4 CONCLUSIONS

The Graphene Oxide (GO) was synthesized from natural graphite powder by Hummers' method.

From SEM, TEM and AFM results, it indicated that the GO single layer and multi layers were successfully synthesized. The PA1212/GO composites were prepared by the one-step and two-step melt compounding methods. The PA1212/GO composite can be reinforced by GO efficiently through the two-step melting compounding process. According to DSC results, the GO nanosheets accelerated the crystallization process. The two-step melt compounding is a promising method to fabricate polyamide/graphene composites.

REFERENCES

Cai Z.Q. 2015. Reinforcing polyamide 1212 with graphene oxide via a two-step melt compounding process. Composites Part A, 69:115–123.
Coleman J.N. 2006. Small but strong: A review of the mechanical properties of carbon nanotube–polymer composites, Carbon, 44(9):1624–1652.
Dzenis Y. 2008. Structural Nanocomposites. Science, 319(5682):419–420.
Kim H. 2010. Graphene/Polymer Nanocomposites. Macromolecules, 43(16):6515–6530.
Liu M.Y. 2003. Melting behaviors, isothermal and non-isothermal crystallization kinetics of nylon 1212, Polymer 44(8):2537–2545.
Marcano D.C. 2010. Improved Synthesis of Graphene Oxide. Nano, 4(8):4806–4814.
Peng Z.H. & Shi Z.P. 2001. Manual of the plastics industry: Chemical Industry Press, 7.
Potts J.R. 2011. Graphene-based polymer nanocomposites. Polymer, 52(1):5–25.
Rafiq R. 2010. Increasing the toughness of nylon 12 by the incorporation of functionalized grapheme Carbon, 48(15):4309–4314.
Ren M.Q. 2004. Crystallization kinetics and morphology of nylon 1212 Polymer, 45(10):3511–3518.
Stankovich S. 2006. Graphene-based composite materials. Nature, 442(7100):282–286.
Tibbetts G.G. 2007. A review of the fabrication and properties of vapor-grown carbon nanofiber/polymer composites. Composites Science and Technology, 67(7):1709–1718.
Wang G.X. 2008. Facile Synthesis and Characterization of Graphene Nanosheets. J. Phys. Chem. C, 112(22):8192–8195.
Xu Z. & Gao C. 2010. In situ Polymerization Approach to Graphene-Reinforced Nylon-6 Composites. Macromolecules, 43(16):6716–6723.
Yan D. 2012. Improved Electrical Conductivity of Polyamide 12/Graphene Nanocomposites with Maleated Polyethylene-Octene Rubber Prepared by Melt Compounding. Applied Materials and Interfaces, 4(9):4740–4745.
Zhao X. 2010. Enhanced Mechanical Properties of Graphene-Based Poly (vinyl alcohol) Composites Macromolecules, 43(22):9411–9416.
Zheng D. 2012. In situ thermal reduction of graphene oxide for high electrical conductivity and low percolation threshold in polyamide 6 nanocomposites. Composites Science and Technology, 72(2):284–286.

Energy Science and Applied Technology – Fang (Ed.)
© 2016 Taylor & Francis Group, London, ISBN 978-1-138-02833-3

Analysis of horizontal seepage prevention based on three-dimensional finite element method

Liting Qiu, Li Zhang & Renjie Zhou
The College of Water Conservancy and Hydropower Engineering, Hohai University, Nanjing, China

Zhenzhong Shen
State Key Laboratory of Hydrology, Water Resources and Hydraulic Engineering, Hohai University, Nanjing, China

Lei Yang
Large Dam Safety Supervision Center, National Energy Administration, Hangzhou, China

ABSTRACT: Shangmo Reservoir is located on the Jinjiahe River in the west of Tianshui in Gansu Province, China. It is a loam core, sandy gravel dam with a maximum height of 50.0 m. The *in situ* drilling and geological prospects show that the rock masses under the dam and reservoir area have a good permeability and the deep riverbed fault fracture zones are difficult to cut off completely by the impervious curtain. Thus, a seepage scheme namely "dam surface geomembrane reservoir bottom geomembrane bank concrete protection (gunite)" is applied for this project. Based on the establishment and calculation of the three-dimensional finite element seepage model, the analysis and comparison among three different horizontal seepage control measures was made. The results show that, in comparison with other schemes, the whole reservoir basin anti-seepage scheme have distinct advantages such as seepage-control effectiveness and operational stability. Besides, the seepage gradient of each structure is also within an allowable range. Therefore, the whole reservoir basin anti-seepage scheme is recommended for the Shangmo Reservoir Project. The achievement and experience of this horizontal seepage prevention design should be taken into consideration for other similar projects.

Keywords: Shangmo Reservoir, Saturated–Unsaturated Unsteady Seepage, FEM, Horizontal Seepage Prevention

1 INTRODUCTION

The Shangmo Reservoir is located on the Jinjiahe River, which is a tributary of the Xihe River, in the west of Tianshui. It is a level IV minor (1) project for the purpose of water supply and flood control. The mean annual runoff at the site is 0.355 m³/s. Its main buildings include the dam, the spillway on the right bank, and the water delivery and flood discharge tunnel on the left bank. The rock masses under the dam and reservoir area have a good permeability and the deep riverbed fault fracture zones are difficult to cut off completely by the impervious curtain. Thus, a horizontal seepage control measure is an appropriate choice for the seepage prevention. A seepage scheme namely "dam surface geomembrane reservoir bottom geomembrane bank concrete protection (gunite)" for seepage prevention is set up. Specifically, based on a previously completed project, we cancel the grout curtain and the loam core wall, and then fill the dam body directly. The geomembrane is set up on the upstream slope of the dam and the bottom of the reservoir (the riverbed surface). Meanwhile, concrete slabs are set up to the check flood level on both sides of the river. Combining the above measures to establish an overall anti-seepage system namely "upstream slope reservoir bottom + reservoir surround".

2 STABLE SEEPAGE FIELD CALCULATION

The three-dimensional, non-steady, saturated–unsaturated seepage finite element analysis program CNPM3D is applied to establish the three-dimensional finite element seepage model and to calculate and analyze the seepage field of the key project during the operating period to demonstrate the seepage behavior of the main dam and the subsidiary buildings.

Finite Element Model. The three-dimensional model is established based on the real size of the main structures. The coordinate system and the boundary of the calculation domain for the finite element model are taken as follows: the model geodetic coordinates (X, Y) = (3823010.51,18532505.61) are taken as the origin of the coordinates. In the X-axis, the positive direction is from upstream to downstream; in the Y-axis, the positive direction is from the right abutment to the left abutment; in the Z-axis, the positive direction is from bottom to top. The scope of the FEM model is required to be 150 m beyond the edge of the bank until the midline of the gully in the right downstream mountain, 1500 m beyond the dam heel along the channel direction, 250 m beyond the dam toe and 150 m lower than the bottom of the grout curtain. The discretized three-dimensional finite element mesh calculation model is shown in Figure 1 and Figure 2. The finite element mesh of the geomembrane on the reservoir bottom and upstream slope is shown in Figure 3. The finite element mesh of the concrete slab surrounding the reservoir is shown in Figure 4.

Calculation Conditions and Seepage Coefficient. The seepage coefficients of the dam body materials and the dam foundation rock masses are listed in Table 1. The calculation conditions are as follows: in all the conditions, the water level takes the normal water level condition of 1632.27 m in the upstream, and the corresponding downstream water level is the riverbed elevation.

Condition 1: set up the geomembrane on the reservoir bottom and the concrete slab around the reservoir up to the tail of the backwater area;

Condition 2: lay the geomembrane on the reservoir bottom and the concrete slab around the reservoir up to 3/4 of the backwater area, about 1000 m beyond the dam heel;

Condition 3: lay the geomembrane on the reservoir bottom up to the tail of the backwater area and the concrete slab around the reservoir up to 3/4 of the backwater area, about 1000 m beyond the dam heel.

Figure 1. Three-dimensional finite element mesh calculation model.

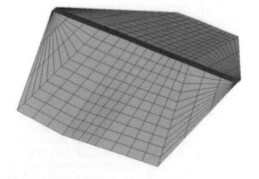

Figure 2. Three-dimensional finite element mesh calculation model of the dam body.

Figure 3. Finite element mesh of the geomembrane on the reservoir bottom and upstream slope.

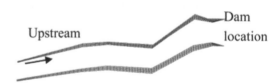

Figure 4. Finite element mesh of the concrete slab around the reservoir.

Table 1. Seepage coefficients of the dam body materials and the dam foundation rock masses.

Partition	Seepage coefficient (cm/s)
Strong permeable 100 Lu ≤ q	2.0×10^{-3}
Medium permeable 10 Lu ≤ q < 100 Lu	3.0×10^{-4}
Weakly permeable 5 Lu ≤ q < 10 Lu	5.0×10^{-5}
Relatively impermeable q < 5 Lu	2.0×10^{-5}
Fault fracture zones	1.5×10^{-2}
Dam body	2.0×10^{-1}
Geomembrane	1.0×10^{-9}
Concrete slab	1.0×10^{-7}

3 CALCULATION RESULTS: COMPARISON AND ANALYSIS OF EACH SCHEME

Seepage Field Analysis of The Reservoir Area. As shown in Figure 5 and Figure 6, the underground water level of the dam surroundings and the dam abutment are both below the normal water level, and the water potential decreases evenly from upstream to downstream. At the dam area, the underground water level of the left bank is higher than the level of the right bank, which is only a little higher than the bed elevation. Due to the barrier property of the geomembrane and the concrete slab, the saturated surface has a sudden drop at the place of the geomembrane and the concrete slab. Therefore, the impact of reservoir storage on the groundwater level is very small. Meanwhile, the design of the anti-seepage system is to some extent good: the geomembrane and the concrete slab bring down the water head obviously, although its seepage gradient is relatively large; the seepage gradient in the rock masses of both banks is homogeneous and small; the saturated surface in the dam body is only 3 m higher than the natural water surface. Finally, the underground water potential before and after the reservoir filling is in fact the same.

Comparison of Seepage Discharge. The whole reservoir basin anti-seepage scheme has the best seepage-control effectiveness, its total seepage discharge being only 0.007 m³/s, much less than the mean annual runoff of 0.355 m³/s, meeting the requirements of stored water and reservoir operation. The seepage discharge of the 3/4 reservoir basin anti-seepage scheme is 0.285 m³/s and that of the whole reservoir bottom and 3/4 reservoir surround anti-seepage scheme is 0.210 m³/s, both being close to the mean annual runoff of 0.355 m³/s. Considering the localized seepage caused by construction problems, the real seepage discharge will increase and even seriously affect the project benefits of the reservoir. Therefore, the whole reservoir basin anti-seepage scheme is an ideal scheme for seepage control. Based on the calculation of the recommended scheme, the seepage discharge of the dead water level is 0.005 m³/s, and both the design water level and the check flood level are 0.008 m³/s, much less than the mean annual runoff of 0.355 m³/s, meeting the requirements of the reservoir operation.

Comparison of Seepage Gradient. The maximum average seepage gradient of the anti-seepage system (geomembrane and concrete slab) of each scheme is in fact the same, as shown in Table 2. Meanwhile, the differences between the exit gradients in the rock-fill area and in the rock masses are very small.

Stability of Anti-seepage Structures. In the whole reservoir basin anti-seepage scheme namely "dam surface geomembrane + reservoir bottom geomembrane + bank concrete protection (gunite)", the geomembrane and the concrete slab bring down the water head obviously, and the underground water potential before and after the reservoir filling is in fact the same. However, in other partial reservoir basin anti-seepage schemes, the underground water level is higher, which leads

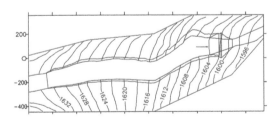

Figure 5. Groundwater contours of the dam under run-time condition (unit: m).

Figure 6. Equipotential line distributions at the maximum cross-section of the dam (unit: m).

Table 2. Maximum average seepage gradient of the anti-seepage system and important rock masses.

Material partition	Maximum seepage gradient	Location	Allowable range
Geomembrane	322.70	Border of the dam heel and the geomembrane on the reservoir bottom	/
Concrete slab	161.35	Border of the dam heel and the geomembrane on the reservoir bottom	200
Rock-fill area	0.0317	Exit of the left downstream slope	>1
Downstream slope	0.263	Exit of the left downstream slope	>1

to a higher uplift pressure as well. Therefore, the stability of anti-seepage structures in a partial reservoir basin anti-seepage scheme is worse than that of the whole reservoir basin anti-seepage scheme.

4 CONCLUSION

Compared with other schemes, the whole reservoir basin anti-seepage scheme has distinct advantages such as the effective control of seepage discharge and operational stability of anti-seepage structures. Meanwhile, in this scheme, the seepage gradient of each structure is within an allowable range. It is suggested that the whole reservoir basin anti-seepage scheme should be recommended as the seepage prevention design for security and economic benefits.

REFERENCES

Chapuis, R.P. and Aubertin, M.: Canadian Geotechnical Journal, Vol. 38(2001) No. 6, p.1321.
Chen, S.K., Liu, S.W., Guo, L.X., Yan, J. and Xie, Z.Q.: Journal of Basic Science and Engineering, Vol. 20(2012) No. 4, p. 612.
Jiang, F., Mi, Y.N. and Zhang, R.: Journal of Water Resources and Architectural Engineering, Vol. 4 (2006) No. 4, p. 94.
Jury, W.A., Wang, Z. and Tuli, A.: Vadose Zone Journal, Vol. 2(2003) No.1, p. 61.
Liu, B., Shen, Z.Z., Zong, Y. and Zhang, Q.: Water Resources and Power, Vol. 26 (2008) No. 4, p. 67.
Shen, Z.Z. and Mao, C.M.: Journal of Hohai University, Vol. 22(1994) No. 5, p. 75.
van Genuchten, M. Th.: Soil Science Society of America Journal, Vol. 44(1980) No. 5, p. 892.
Wang, Z., Wu, Q.J., Wu, L., et al: Journal of hydrology, Vol. 231(2000), p. 265.

Renewable energy, bio-energy and cell technologies

Energy Science and Applied Technology – Fang (Ed.)
© 2016 Taylor & Francis Group, London, ISBN 978-1-138-02833-3

New digital graphics tablet based on solar energy and cloud storage

Zitong Wang
Beijing University of Posts and Telecommunications, Beijing, China

ABSTRACT: The digital graphics tablet is a kind of computer input device, which provides a digital board for the users to paint and design images. The present digital graphics tablet has limitations such as low battery time and insufficient storage space. To fill this gap, we envisioned a new digital graphics tablet that has solar energy panels to collect solar energy for transforming it into electricity as a double source and has cloud storage to expand storage for the users. The working details of the graphics tablet is presented in the paper, and some issues that need to be improved are discussed. Further research should devote more attention to the implementation of the new graphics tablet.

Keywords: Solar energy; Cloud storage

1 INTRODUCTION

Nowadays, the digital graphics tablet is a common device in both colleges and companies. The digital graphics tablet is a kind of computer input device that mainly orients designers, fine art-related students and teachers, advertising companies, design studios, Flash vector animators and some art lovers. Most of the paintings related to design and characters in movies are accomplished by drawing on the digital board and by post-processing on the computer. The present digital graphics tablet can well meet all the requirements, but with limitations such as the low battery time and insufficient storage space, which even cause a serious impact on the users' experience. To remedy this limitation of low battery time, Dreyer proposed the use of solar panels to collect solar energy for transforming it into electricity to be used by the tablet. In his paper, the solar panel is the core part of the solar energy system, which plays a big role in the output of the direct current in the battery after the transformation of sunlight energy into electrical energy. The present solar energy panel is normally used by silicon panels, which are inefficient with high cost. While the search for more efficient but less expensive solar panels is ongoing, we can take the advantage of David Bradley's paper. In his paper, a US team has found that stacking perovskites on to a conventional silicon solar cell can boost efficiency. Michael McGehee of Stanford University and colleagues point out that the dominant photovoltaic technology based on silicon got stuck at 25% efficiency about 15 years ago. However, a way to break through to higher

efficiencies might be to create tandem devices that exploit silicon together with a second inexpensive photovoltaic material. They performed two experiments and suggested that currently, a layered perovskite cell might be the solution for the best solar panels: "Since most, if not all, of the layers in a perovskite cell can be deposited from solution, it might be possible to upgrade conventional solar cells into higher-performing tandems with little increase in cost,". For insufficient storage space, the graphics tablet can provide cloud storage for its users to both upload and download their works. Cloud computing plays a major role in the business over the Internet today, and cloud storage is one of the services provided in cloud computing which has been increasing in popularity. Leesakul and his colleagues stated that "The main advantage of using cloud storage from the customers' point of view is that customers can reduce their expenditure in purchasing and maintaining storage infrastructure while only paying for the amount of storage requested, which can be scaled-up and down upon demand". Some datasets may be frequently accessed or updated by multiple users at the same time, while others may need the high level of redundancy for reliability requirement. Therefore, it is crucial to support this dynamic feature in cloud storage. However current approaches are mostly focused on a static scheme, which limits their full applicability in the dynamic characteristic of data in cloud storage.

In their paper, Leesakul, Townend and Xu proposed a dynamic deduplication scheme for cloud storage, aiming to improve storage efficiency and maintaining redundancy for fault tolerance. The

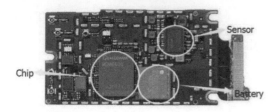

Figure 1. Current models.

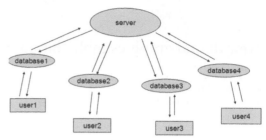

Figure 3. Cloud storage server.

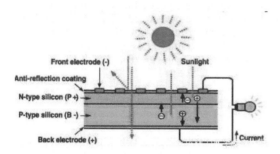

Figure 2. Solar panels arrayed in the system process.

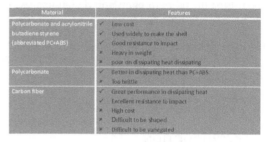

Figure 4. Most widely used shell materials.

more specific details are organized as follows. Section 2 introduces the function modules of the product. Section 3 provides some discussions. Section 4 concludes this paper.

2 PRODUCT FUNCTION MODULE

2.1 Pressure pen and touch panel

Our product is aimed at the medium-end and high-end customers. Our current model feature has 1024/2048 levels of pressure sensitivity and a resolution of 3048/4000/5080 lines per inch (1000 lines/cm). Most of the models have a 5.8 × 3.6 in (14.7 × 9.2 cm) active surface area (see Figure 1).

2.2 Solar energy

In the digital graphics tablet, the battery is incorporated within the body of the pad. There are solar panels arrayed spreading across the frame of the device, as well as a built-in Li-ion storage cell. This double-source system can effectively prolong the battery time of the digital tablet. Under a good weather condition, the solar array system will receive and save energy into the photovoltaic array until it is fully charged, as shown in Figure 2. Once the Li-ion storage cell is used up, the system will immediately switch to make use of the storage power in the photovoltaic array.

2.3 Cloud storage

2.4 Other function module

Concerning the selection of shell materials, the tablet shell can provide better heat dissipating performance, while it also functions well with respect to weight and cost. Moreover, this kind of shell material can be shaped and variegated, and thus we can meet each individual's demand. The unique design of the changeable shell can also add colour to the existing tablet market. The most widely used shell materials today are shown in Figure 4. As a result, we choose magnesium alloy for our tablet's changeable cover.

3 DISCUSSION

The paper presents a new digital graphics tablet with new features enabling longer battery time with solar energy and more sufficient storage space with cloud storage. Previous papers proposed that some silicon panels was used to collect solar energy for transformed it into electricity. We take advantage of David Bradley's paper for more efficient but less expensive solar panels. They show a perovskite cell of 12.7% efficiency layered on to a relatively low-quality silicon cell efficiency of 11.4%. The two layers then worked synergistically to give a total efficiency of 17% for the tandem device. Obviously, such a drastic improvement in efficiency has the potential to redefine the commercial viability of low-quality silicon in the future. It can never be denied that there are limitations of solar energy use, such as instability, the short battery life during cloudy days and the heavy burden of the double-source system.

Cloud storage is another new feature, where storage resources in the cloud can be used by the users to access the data easily at anytime, anywhere, through any network device connected to the cloud. Indeed, cloud storage can expand insufficient storage space. However, there are some security issues related to cloud storage, as reported in Yujuan Tan's paper, such as personal privacy: the administrator can actually directly view and delete uploaded files of the users from the server platform. Data safety is also a big problem, such as hacking and data loss due to data synchronization delay. However, cloud storage will become a trend in the development of storage in the future. With the inevitable development of the cloud storage technology, all kinds of search, application technology and the combination of cloud storage applications need improvement with respect to safety, portability and data access.

4 CONCLUSION

This paper introduced a new digital graphics tablet that overcomes the traditional graphics tablet's limitation such as low battery time and insufficient storage space. In our design, we use solar energy as our tablet double-source system to expand the battery time, and cloud storage is used to provide more extra storage by a built-in wireless card linking to the Internet, which could boost the market and reduce the current high cost of the graphics tablet.

This paper presents no experimental results. The main objective of this paper is to provide envisioned design and ideas for relative companies and manufactures. The next step of this research is to manufacture prototypes and solve implementation issues.

REFERENCES

Bradley D. Perovskite promise for solar energy [J]. Materials Today, 2015, 18(3): 124–125.

Bailie C D, Christoforo M G, Mailoa J P, et al. Semi-transparent perovskite solar cells for tandems with silicon and CIGS[J]. Energy & Environmental Science, 2015.

Bradley D. Perovskite promise for solar energy [J]. Materials Today, 2015, 18(3): 124–125.

Cui H, Meng Z, Xiao C, et al. Microstructure and properties of plasma remelted AZ91D magnesium alloy [J]. Transactions of Nonferrous Metals Society of China, 2015, 25(1): 30–35

Dreyer P, Morales-Masis M, Ballif C, et al. Copper and Trans parent-Conductor Reflectarray Elements on Thin-Film Solar Cell Panels[J]. 2013.

Leesakul W, Townend P, Xu J. Dynamic Data Deduplication in Cloud Storage[C]//Service Oriented System Engineering (SOSE), 2014 IEEE 8th International Symposium on. IEEE, 2014: 320–325.

Tan Y, Jiang H, Sha E H M, et al. SAFE: A Source Deduplication Framework for Efficient Cloud Backup Services [J]. Journal of Signal Processing Systems, 2013, 72(3): 209–228.

Energy Science and Applied Technology – Fang (Ed.)
© 2016 Taylor & Francis Group, London, ISBN 978-1-138-02833-3

SOC estimation of lithium-ion battery using adaptive extended Kalman filter based on maximum likelihood criterion

Xuanju Dang, Kai Xu, Hui Jiang, Xiangwen Zhang, Xiru Wu & Li Yan
School of Electronic Engineering and Automation, Guilin University of Electronic Technology, Guilin, China

ABSTRACT: A method for estimating the State Of Charge (SOC) is presented, which combines the backward difference equation model of the power battery with the adaptive extended Kalman filter based on the maximum likelihood criterion algorithm (MLC-Based AEKF). The discrete model of the Thevenin equivalent circuit for the power battery is built by the backward difference method. Compared with the bilinear transform equation model, the proposed method has the advantages of a simple structure and low computational complexity. The forgetting factor recursive least squares algorithm (FFRLS) is used to identify the parameters of the model. The MLC-Based AEKF is applied to realize an online SOC estimation under the unknown interference noise environments. The simulation experiment results verified the effectiveness of the SOC estimation method; that is, the SOC estimation average error is less than 0.15% and its maximum error is less than 0.5% in the Dynamic Stress Test (DST) condition.

Keywords: Lithium-ion battery; Thevenin model; Least squares; Adaptive Extended Kalman filter based on Maximum Likelihood Criterion algorithm (MLC-Based AEKF); State Of Charge (SOC)

1 INTRODUCTION

In the electric vehicle, an accurate prediction of SOC, also called the remaining power of the battery, is a premise and a key to effectually running of the Battery Management System (BMS). In recent years, the lithium-ion battery has become the main candidate for the power battery of the electric vehicle. Compared with the conventional lead-acid battery or the nickel metal hydride battery, the lithium-ion battery has the advantages of high energy density, no memory effect, long cycle life and environment friendly.

The SOC estimation of the battery includes two main components: parameter identification and SOC estimation. (1) The methods for parameter identification usually include the genetic algorithm, the least squares algorithm, the dual extended Kalman filter. (2) The SOC estimation methods usually include the open-circuit voltage method, the current integral method, the Kalman filter method, the extended Kalman filter method, the unscented Kalman filter method and the adaptive extended Kalman filter method.

The different combinations of the two parts construct different kinds of SOC estimation methods: a combination of the least squares algorithm and the unscented Kalman filter; a combination of the least squares algorithm and the open-circuit

voltage method; a combination of the genetic algorithm and the open-circuit voltage method; a combination of the least squares algorithm and the extended Kalman filter. It has been shown that the dual extended Kalman filter simultaneously estimates SOC and the internal parameters of the lithium-ion battery.

The parameters of the backward difference equation model are obtained by the FFRLS. Meanwhile, the MLC-Based AEKF is applied for the online SOC estimation. This combination has the advantages of a simple structure, low computational complexity and high accuracy of SOC estimation.

2 MODEL FOR POWER BATTERY

For the BMS, the main battery models include the thermal model, the equivalent circuit model and the electrochemical model. Compared with other battery models, the equivalent circuit model intuitively describes the relationship between the current and the voltage to easily analyze and identify the model parameters. The equivalent circuit model includes the Rint model, the Thevenin model, the PNGV model and the second-order RC model. Among these, the Thevenin model is the most widely used the equivalent model at present. It can be utilized to

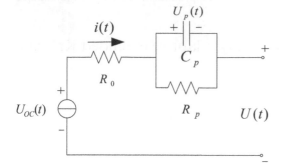

Figure 1. Thevenin model.

describe the dynamic and static characteristics of the battery, and is of relatively simple structure to easily identify the parameters with high accuracy.

The Thevenin model is shown in Figure 1, where $U_{oc}(t)$ is the open-circuit voltage; R_0 is the Ohmic resistance; R_p is the polarization resistance; C_p is the polarization capacitance; RC is the polarization response of the battery; $U_p(t)$ is the polarization voltage across C_p; $i(t)$ is the current flowing through R_0; and $U(t)$ is the battery terminal voltage.

According to Kirchhoff's voltage law and Kirchhoff's current law, the mathematical expressions of the Thevenin model are as follows:

$$\frac{dU_p(t)}{dt} = -\frac{U_p(t)}{R_pC_p} + \frac{i(t)}{C_p} \tag{1}$$

$$U(t) = U_{oc}(t) - R_0 i(t) - U_p(t) \tag{2}$$

3 MODEL PARAMETER IDENTIFICATION

3.1 Discretization of the model

There are many discrete methods for the aforementioned models, which include the forward difference method, the backward difference method, the bilinear transform method and the constant impulse response method. The backward difference method with a simple structure and low computational complexity is selected in this paper.

The discretization form of Equations (1) and (2) is given by

$$U(k) - U_{oc}(k) = a[U(k-1) - U_{oc}(k-1)]$$
$$+ bI(k) + cI(k-1) \tag{3}$$

where a, b and c are the discrete model parameters.

The charge/discharge process of the battery is a relatively slow process, and the open-circuit voltage is relatively stable in a short period of time, which is given by

$$\frac{dU_{oc}(t)}{dt} = \frac{U_{oc}(k) - U_{oc}(k-1)}{T} \approx 0 \tag{4}$$

$$\Delta U_{oc}(k) = U_{oc}(k) - U_{oc}(k-1) \approx 0 \tag{5}$$

Substituting Equation (5) into Equation (3), we obtain

$$U(k) = aU(k-1) + bI(k)$$
$$+ cI(k-1) + (1-a)U_{oc}(k) \tag{6}$$

The parameters of the battery backward difference model can be obtained as follows:

$$R_0 = \frac{c}{a} \tag{7}$$

$$R_p = \frac{-ab - c}{a(1-a)} \tag{8}$$

$$C_p = \frac{Ta^2}{-ab - c} \tag{9}$$

where T is the sampling interval.

The characteristics of the battery backward difference model are that the open-circuit voltage is not related to the history data of the voltage, and is easy to estimate.

3.2 Parameter identification

During the running of the electric vehicle, the power battery parameters are slowly changed. The conventional recursive least squares method will lead to the "data saturation" phenomenon in the continuous iterative updating data. The FFRLS is used to identify parameters of the backward difference equation model, in which the "data saturation" can be eliminated by strengthening the weight of the latest data and reducing the impact of historical data in the FFRLS.

The process of parameter estimation in the FFRLS is realized as follows:

$$y(k) = \phi^T(k)\theta + e(k) \tag{10}$$

$$e(k) = y(k) - \phi^T(k)\hat{\theta}(k-1) \tag{11}$$

$$\hat{\theta}(k) = \hat{\theta}(k-1) + K(k)e(k) \tag{12}$$

$$K(k) = \frac{P(k-1)\phi(k)}{\lambda + \phi^T(k)P(k-1)\phi(k)} \tag{13}$$

$$P(k) = [I - K(k)\phi^T(k)]P(k-1) \tag{14}$$

where $\hat{\theta}(k)$ is the estimate of the parameter vector θ; $e(k)$ is the predicted error of $y(k)$; $K(k)$ is the algorithm gain; $P(k)$ is the covariance matrix; and the constant λ is the forgetting factor.

The least squares form of the backward difference equation model Equation (6) can be written as

$$y(k) = \phi^T(k)\theta \tag{15}$$

$$\phi(k) = [U(k-1), I(k), I(k-1), 1]^T \tag{16}$$

$$\theta = [a, b, c, (1-a)U_{OC}(k)]^T \tag{17}$$

The values of a, b and c can be obtained by the FFRLS. From Equations (7), (8) and (9), the parameter values of the models R_0, R_p, C_p and U_{oc} can be obtained.

3.3 Identification results

The Dynamic Stress Test (DST) is a typical dynamic driving cycle, which is often used to evaluate the performance of the vehicle in the experiment.

The lithium-ion battery with a nominal voltage of 24V and a nominal capacity of 20 Ah is used for this experiment. One cycle lasted 360 seconds in the DST condition. Before the experiment, the battery module is fully charged at a constant current rate, i.e. SOC = 1. The experiment begins after standing for a long time. The battery is almost empty after a total of 98 cycles.

The current time distribution of the DST condition is shown in Figure 2.

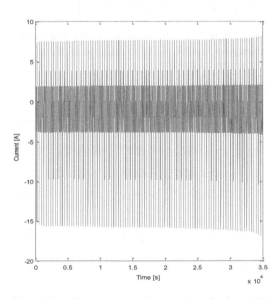

Figure 2. Current time distribution of the DST condition.

The FFRLS is used for the online identification of the Thevenin model. Figure 3 shows the change in the coefficients a, b, c and U_{oc} $(1-a)$.

3.4 Validation and comparative analysis

The simulation model under the environment of MATLAB/Simulink is shown in Figure 4.

In references, Equations (1) and (2) are discretized by the bilinear transform method. The backward difference method is applied in this paper. Using the two discrete methods, the mathematical expressions of calculated capacitors and resistors are different. The results of the different discrete methods in the DST condition are given in Table 1.

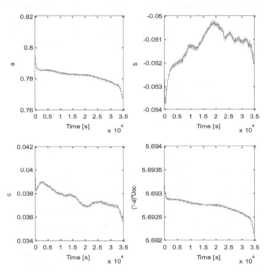

Figure 3. Identification results of the model parameters.

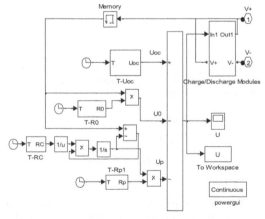

Figure 4. Simulation model of the battery.

19

Table 1. Under the condition of the DST.

Discrete method	Maximum error	Average error	Maximum error rate	Average error rate
Backward difference method	0.1576 V	0.0125 V	0.6567%	0.0521%
Bilinear transform method	0.1579 V	0.0119 V	0.6579%	0.0496%

Table 1 shows that the value of the terminal voltage error is almost the same in a small range for the backward difference equation model and the bilinear transform method. Under the condition of the same results or the consistent accuracy, the backward difference method with a simple structure and low computational complexity is selected in this paper.

4 SOC ESTIMATION BASED ON MLC-BASED AEKF

4.1 MLC-Based AEKF

In a real system, the dynamic model parameters and noise statistical characteristics of the system are unknown. The satisfactory results cannot be obtained by the EKF algorithm for estimating SOC. The MCL-Based AEKF is used to improve the accuracy of the filter by estimating unknown noise and adaptively describing the noise statistical properties or the filter gain matrix.

The SOC and the terminal voltage of the capacitor C_p are selected as the state variables, namely $X_k = [SOC_k \ U_{p,k}]^T$, the state equation and measurement equation of the system, which can be described as follows:

$$\begin{cases} X_k = A_{k|k-1}X_{k-1} + B_{k-1}i_{k-1} + w_{k-1} \\ Y_k = U_{oc}(SOC_k) - R_0 i_k - U_{p,k} + v_k \end{cases} \tag{18}$$

where $U_{oc}(SOC_k)$ represents the non-linear relationship between the open-circuit voltage U_{oc} and the SOC as follows:

$$U_{oc}(SOC_k) = k_1 SOC_k{}^8 + k_2 SOC_k{}^7 + k_3 SOC_k{}^6$$
$$+ k_4 SOC_k{}^5 + k_5 SOC_k{}^4 + k_6 SOC_k{}^3$$
$$+ k_7 SOC_k{}^2 + k_8 SOC_k + k_9 \tag{19}$$

The nine coefficients k_1–k_9 were calculated by the least squares algorithm, in which the open-circuit voltage U_{oc} is obtained by the online identification, and the SOC is obtained by the experiment.

The steps for SOC estimation are illustrated as follows:

(1) State prediction

$$\hat{X}_{k|k-1} = A_{k|k-1}\hat{X}_{k-1|k-1} + B_{k-1}i_{k-1} \tag{20}$$

$$A_{k|k-1} = \begin{pmatrix} 1 & 0 \\ 0 & 1 - \dfrac{T}{R_p C_p} \end{pmatrix} \tag{21}$$

$$B_{k-1} = \left[-\dfrac{\eta T}{Q_N} \quad \dfrac{T}{C_p} \right]^T \tag{22}$$

where Q_N and η are the power battery nominal capacity and the columbic efficiency, respectively.

(2) Error variance matrix prediction

$$P_{k|k-1} = A_{k|k-1}P_{k-1|k-1}A_{k|k-1}^T + Q_{k-1} \tag{23}$$

where Q_k is the covariance of the system process noise w_k.

(3) Filter gain matrix

$$K_k = P_{k|k-1}H_k^T(H_k P_{k|k-1}H_k^T + R_k)^{-1} \tag{24}$$

$$H_k = \left[\dfrac{dU_{oc}(SOC_k)}{dSOC_k}\bigg|_{SOC_k = \hat{SOC}_{k|k-1}} \quad -1 \right] \tag{25}$$

where $\hat{SOC}_{k|k-1}$ is the estimate of the SOC at the moment k, and R_k is the covariance of the system measurement noise V_k.

(4) Updating state

$$\hat{X}_{k|k} = \hat{X}_{k|k-1} + K_k(Y_{m|k} - \hat{Y}_k) \tag{26}$$

$$\hat{U}_{OC,k|k-1} = k_1\hat{SOC}_{k|k-1}^8 + k_2\hat{SOC}_{k|k-1}^7 + k_3\hat{SOC}_{k|k-1}^6$$
$$+ k_4\hat{SOC}_{k|k-1}^5 + k_5\hat{SOC}_{k|k-1}^4 + k_6\hat{SOC}_{k|k-1}^3$$
$$+ k_7\hat{SOC}_{k|k-1}^2 + k_8\hat{SOC}_{k|k-1} + k_9 \tag{27}$$

$$U_{p,k} = (1 - \dfrac{T}{R_p C_p})U_{p,k-1} + \dfrac{T}{C_p}i_{k-1} \tag{28}$$

$$\hat{Y}_k = \hat{U}_{OC,k|k-1} - R_0 i_k - U_{p,k} \tag{29}$$

where $Y_{m|k}$ is the measure value of the terminal voltage of the battery at the moment k.

20

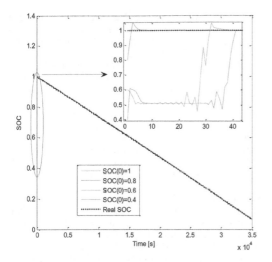

Figure 5(a). SOC estimation curves.

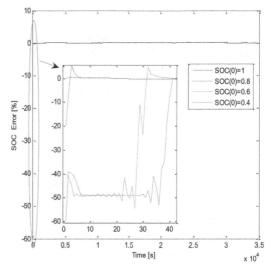

Figure 5(b). SOC error curves.

(5) Updating error variance matrix

$$P_{k|k} = (I - K_k H_k) P_{k|k-1} \qquad (30)$$

Table 2. Statistical results of the SOC error.

SOC initial	SOC maximum error	SOC average error
1	0.4290%	0.0853%
0.8	0.4290%	0.0863%
0.6	0.4366%	0.1245%
0.4	0.4525%	0.1375%

(6) Updating noise covariance

$$\hat{U}_{OC,k|k} = k_1 \hat{SOC}_{k|k}^8 + k_2 \hat{SOC}_{k|k}^7 + k_3 \hat{SOC}_{k|k}^6$$

$$+ k_4 \hat{SOC}_{k|k}^5 + k_5 \hat{SOC}_{k|k}^4 + k_6 \hat{SOC}_{k|k}^3$$

$$+ k_7 \hat{SOC}_{k|k}^2 + k_8 \hat{SOC}_{k|k} + k_9 \qquad (31)$$

$$\tilde{Y}_k = \hat{U}_{OC,k|k} - R_0 i_k - U_{p,k} \qquad (32)$$

$$\mu_k = Y_{m|k} - \tilde{Y}_k \qquad (33)$$

$$F_k = \sum_{n=k-L+1}^{k} \frac{\mu_n \mu_n^T}{L} \qquad (34)$$

$$R_k = F_k + H_k P_{k|k} H_k^T \qquad (35)$$

$$Q_k = K_k F_k K_k^T \qquad (36)$$

where $\hat{SOC}_{k|k}$ is the optimal estimate of the SOC at the moment k, setting the adaptive window L as 20 in this paper.

4.2 SOC estimation results

The experiment for SOC estimation is conducted on a high-precision ARBIN platform. The actual value of the SOC is obtained by the definition of SOC. The algorithm validation test under the condition of the DST is shown in Figure 2.

The SOC estimation includes the parameter identification of the backward difference model

by the FFRLS, and estimation of the SOC by the MLC-Based AEKF.

The SOC estimation results are plotted in Figure 5, where the initial state of the estimator is initialized to 0.4, 0.6, 0.8 and 1.

From Figure 5(b), the SOC can quickly converge to the true value at the different initial SOC. After the converging of the SOC to the true value, the results of the SOC maximum error and SOC average error are all relatively small, as shown in Table 2.

5 CONCLUSION

Due to the behavior of the slow charge/discharge process of the battery, i.e. the open-circuit voltage being stable in a relatively short time, the discrete model obtained by the backward difference method for the battery is of a simple structure and low computational complexity. The combined FFRLS and the MLC-Based AEKF is applied for estimating SOC. In the DST condition, the SOC estimation average error is less than 0.15% and its maximum error is less than 0.5%.

ACKNOWLEDGEMENTS

This work was supported by the National Nature Science Foundation Project (61263013), the Scientific Research Project of Guangxi Education Department (201101ZD007) and the Guangxi Experiment Center of Information Science, and the Guilin University of Electronic Technology Project (20130110).

REFERENCES

Amardeep Singh, Afshin Izadian, Sohel Anwar. 2014. Model based condition monitoring in lithium-ion batteries, Journal of Power Sources, 268:459–468.

Chiang YH, Sean WY, Ke JC. 2011. Online estimation of internal resistance and opencircuit voltage of lithium-ion batteries in electric vehicles, Journal of Power Sources, 196: 3921–3932.

Dai Haifeng, Sun Zechang, Wei Xuezhe. 2009. Estimation of Internal States of Power Lithium-ion Batteries Used on Electric Vehicles by Dual Extended Kalman Filter, Journal of Mechanical Engineering, 45(6):95–101.

Dang Xuanju, Chen Bo, and Jiang Hui. 2013. Ekf-Based a Novel SOC Estimation Algorithm of Lithium-ion Battery, J. Sensors and Transducers, 23(7), 137–143.

Dave Andre, Christian Appel, Thomas Soczka-Guth, Dirk Uwe Sauer. 2013. Advanced mathematical methods of SOC and SOH estimation for lithium-ion batteries, Journal of Power Sources, 224:20–27.

H.W. He, X.W. Zhang, R. Xiong, and Y.L. Xu. 2012. Online model-based estimation of state-of-charge and open-circuit voltage of lithium-ion batteries in electric vehicles, Energy, 39(1):310–318.

Hu C, Youn BD, Chung J. 2012. A multiscale framework with extended kalman filter for lithium-ion battery SOC and capacity estimation, Appl Energy, 92:694–704.

J. Kim, J. Shin, C. Chun and B.H. Cho, 2012. Stable Configuration of a Li-Ion Series Battery Pack Based on a Screening Process for Improved Voltage/SOC Balancing, IEEE Transactions on Power Electronics, 27(1): 252–292.

Johnson V H. 2002. Battery performance models in ADVISOR, J Power Sources, 110 (2):321–329.

Kaiser R. 2007. Optimized battery-management system to improve storage lifetime in renewable energy systems, Journal of Power Sources, 168(1):58–65.

Liu Jiang, Shi Yikai, Yuan Xiaoqing, et al. 2013. SOC Estimation of Lithium Battery Based on RLS and EKF, Measurement & Control Technology, 32(8):123–125.

Ng KS, Moo CS, Chen YP, Hsieh YC. 2009. Enhanced coulomb counting method for estimating state-of-charge and state-of-health of lithium-ion batteries, Appl Energy, 86:1506–1511.

Roscher MA, Sauer DU. 2011. Dynamic electric behavior and open-circuit-voltage modeling of LiFePO4-based lithium-ion secondary batteries, Journal of Power Sources, 196:331–336.

Santhanagopalan S, White RE. 2010. State of charge estimation using an unscented filter for high power lithium ion cells, J Energy, 34:152–163.

Spagnol P, Rossi S, Savaresi SM. IEEE. 2011. Kalman Filter SoC estimation for Li-Ion batteries, 2011 IEEE international conference on control applications, 587–592.

Sun Fengchun, Hu Xiaosong, Zou Yuan and Li Siguang.2011. Adaptive unscented Kalman filtering for state of charge estimation of a lithium-ion battery for electric vehicles, Energy, 36(5):3531–3540.

Wang J P, Guo J G, Ding L. 2009. An adaptive Kalman filtering based state of charge combined estimator for electric vehicle battery pack, Energy Conversion and Management, 50(12): 3182–3186.

Wladislaw Waag, Christian Fleischer, Dirk Uwe Sauer. 2014. Critical review of the methods for monitoring of lithium-ion batteries in electric and hybrid vehicles, Journal of Power Sources, 258: 321–339.

Xiong Rui, He Hongwen, and Ding Ying. 2011. Study on Identification approach of Dynamic Model Parameters for Lithium-ion Batteries Used in Hybrid Electric Vehicles. Power Electronics, 45(4):100–102.

Xiong Rui, Sun Fengchun, He Hongwen. 2012. State-of-charge estimation of Lithium-ion batteries in electric vehicles based on an adaptive extended Kalman filter, High Technology Letters, 22(2):198–204.

Y. Hu, S. Yurkovich, Y. Gyezennec et al. 2009. A technique for dynamic battery model identification in automotive applications using linear parameter varying structures, Control Engineering Practice, 17: 1190–1201.

Yue Xiaokui, Yuan Jianping. 2005. An Adaptive Kalman filtering Algorithm Based on Maximum-Likelihood Criterion, Journal of Northwestern Polytechnical University, 23(4):469–474.

Zhang C P, Zhang C N, Sharkh S M. 2010. Estimation of real-time peak power capability of a traction battery pack used in an HEV, Power and Energy Engineering Conference (APPEEC), 2010 Asia-Pacific, 1–6.

Zhang Yanqin, Guo Kai, Liu Hanyu, Wu Bin, Li Yixia. 2013. Research on identification of dynamic model parameters for power batteries, Chinese Battery Industry, 18(1):29–32.

Zhao Kai, Zhu Liming. 2013. Experimental Study on SOC Estimation of Power Battery Based on Unscented Kalman Filter Method, Chinese Journal of Automotive Engineering, 3(5):332–337.

Energy Science and Applied Technology – Fang (Ed.)
© 2016 Taylor & Francis Group, London, ISBN 978-1-138-02833-3

SOC estimation strategy based on online feed-forward compensation for lithium-ion battery

Xuanju Dang, Yan Mo, Hui Jiang, Xiru Wu & Xiangwen Zhang
School of Electronic Engineering and Automation, Guilin University of Electronic Technology, Guangxi, China

ABSTRACT: Accurately estimating the internal parameters is very important for electric vehicles, which can not only guarantee that the State Of Charge (SOC) of the power battery is maintained within a reasonable range of the operation to avoid the risk of the overcharge or the deep discharge of the power battery, but also both improve the efficiency and extend the life of the battery. In this paper, the static model is obtained by transforming the first-order RC equivalent circuit model in the frequency domain, which is only associated with the voltage and the current. The adaptive Kalman filter is applied to implement the online identification of the parameters and Open-Circuit Voltage (OCV) in the static model. The OCV-SOC model is established by the neural network. The SOC is obtained by the OCV-SOC neural network model with an estimation error of approximately 8%. The current feed-forward method is introduced to compensate the SOC for improving the estimation precision. The simulation experimental results show that this scheme has a higher precision and a simpler structure. The proposed method provides an approach for estimating the SOC by the OCV, which takes fully into account the hysteresis characteristics of the OCV-SOC model during the dynamic condition in the long charge and discharge process of the battery for electric vehicles.

Keywords: State of charge; Adaptive Kalman filter; OCV-SOC; Feed-forward compensation

1 INTRODUCTION

Electric vehicles have the advantages of non-pollution, high efficiency and a comfortable driving environment over traditional fossil-fuel vehicles. Lithium-ion batteries are commonly used as the power source for electric vehicles since they have a high efficiency, a high charge and discharge rate, a low self-discharge rate and no memory effect (Du J, Liu Z, Wang Y. 2014)

State Of Charge (SOC) is an important parameter of the Battery Management System (BMS). However, due to the internal electrochemical process of the battery, the SOC cannot be directly obtained by the sensor measurements but can only be indirectly estimated by using the algorithm based on the established model (Chen X, Shen W, Cao Z, et al. 2014).

Currently, SOC estimation methods can be divided into direct and indirect ways. The direct estimation method is the Ampere-hour method with the accumulated error problem (Ng, K. S., C. S. Moo, Y. P. Chen, and Y. C. Hsieh. 2009). Conversely, in the indirect estimation method, the SOC is estimated by the internal relationships between the SOC and the battery parameters (Xing Y, He W, Pecht M, et al. 2014, Fendri D, Chaabene M. 2012,

He H, Xiong R, Guo H. 2012), such as the relationship between the SOC and the Open-Circuit Voltage (OCV). Generally, the OCV is proportionally declined as the energy is consumed, which is widely used for SOC estimation. However, it is difficult to obtain the OCV online since it needs a long time for the battery to attain a stable condition. The other indirect estimation methods include the state estimation methods of the battery model, such as the recursive Kalman filter, the extended Kalman filter (Sepasi S, Ghorbani R, Liaw B Y. 2014a,b Sepasi S, Ghorbani R, Liaw B Y. 2014) and the unscented Kalman filter (Partovibakhsh M, Liu G. 2014).

The information about the noise statistical characteristics is a precondition of applying the Kalman filter. Moreover, the determination of the statistical characteristics for the system process noise and measurement noise has a great significance for the stability of the Kalman filter and the state estimation. In the presence of the noise statistical characteristics dramatically fluctuating with the actual operation condition, the Kalman filter cannot adaptively adjust, which leads to the decline in filtering accuracy or even divergence. In order to overcome the aforementioned problems, the adaptive Kalman filter is introduced (Partovibakhsh M, Liu G. 2014, Xiong R, He H, Sun F, et al. 2013) in this paper.

In this paper, the static equation is obtained by transforming the first-order RC equivalent circuit model in the frequency domain, which is only associated with the current and voltage parameters. Besides, the static equation of a simple structure is not related to the historical data of the resistance and capacitance. Compared with the state equation of the state estimation, the static equation without consideration of the initial value problem can improve the identification speed.

The static equation can be employed for the adaptive Kalman filter to identify the model parameters online, such as the internal resistance and OCV, and then the OCV-SOC is modeled by using the neural network. The SOC is obtained by the OCV-SOC neural network model with an estimation error of about 8%. In order to improve the accuracy, the current feed-forward compensation is introduced to compensate the SOC. The simulation experimental results demonstrate the effectiveness of the proposed method.

For estimating the SOC by the OCV, the hysteresis characteristics exhibited in the long charge and discharge process of the battery are considered, and the battery characteristics can be fully described.

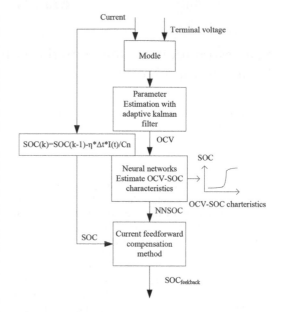

Figure 1. Flow of the proposed SOC estimation method of the online current feed-forward compensation.

2 BATTERY MODEL AND PARAMETER IDENTIFICATION

A flow of the proposed SOC estimation method of the online current feed-forward compensation is shown in Figure.1. On the basis of transforming the first-order RC equivalent circuit model in the frequency domain, the OCV is obtained by the adaptive Kalman filter, and then the OCV-SOC model is established by the neural network. The SOC obtained by the OCV-SOC neural network model is compensated online based on the current feed-forward method.

2.1 Battery model

In order to obtain a reliable SOC estimation of the battery, an accurate battery model is needed. At present, electric vehicle battery models mainly include the electrochemical model (Corno M, Bhatt N, Savaresi S M, et al., N. A. Windarko, J. Choi and G. B. Chung, 2015), the neural network model (Liu Z, Wang Y, Du J, et al. 2012, Kang L W, Zhao X, Ma J. 2014, He W, Williard N, Chen C, et al. 2014), the impedance model (Xu J, Mi C C, Cao B, et al. 2013) and the equivalent circuit model (H. Chaoui and P. Sicard, 2011, Bing D, Yantao T, Changjiu Z. 2014).

In this paper, the battery model is shown in Figure 2, where IL is the load current, in which

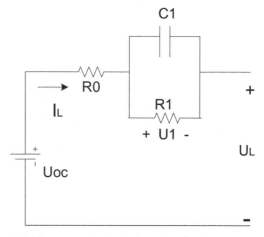

Figure. 2. First-order RC equivalent circuit model.

the discharge process is positive and the charge process is negative; UL is the terminal voltage; UOC is the OCV; and R0, R1, C1 are the internal resistance, equivalent polarization resistance and equivalent polarization capacitance, respectively. These elements constitute an RC network that describes the electricity transmission effects and dynamic voltage performances (Sun F, Xiong R, He H. 2014).

According to the literature (He H, Zhang X, Xiong R, et al. 2012), the first-order RC model in the frequency domain is given in Eq. (1):

$$U_L(s) - U_{oc}(s) = -I_L(s)(R_0 + \frac{R_1}{1 + R_1 C_1 s}) \quad (1)$$

By defining EL(s) = UL(s)-UOC(s), the transfer function $G(s)$ of Eq. (1) can be written as follows:

$$G(s) = \frac{E_L(s)}{I_L(s)} = -R_0 - \frac{R_1}{1 + R_1 C_1 s} = -\frac{R_0 + R_1 + R_0 R_1 C_1 s}{1 + R_1 C_1 s} \quad (2)$$

A bilinear transformation method is given by

$$s = \frac{2(1 - z^{-1})}{T(1 + z^{-1})} \quad (3)$$

By discretizing Eq. (2), a_1, a_2, a_3 can be obtained as follows:

$$a_1 = -\frac{T - 2R_1 C_1}{T + 2R_1 C_1},$$

$$a_2 = -\frac{R_0 T + R_1 T + 2R_0 R_1 C_1}{T + 2R_1 C_1},$$

$$a_3 = -\frac{R_0 T + R_1 T - 2R_0 R_1 C_1}{T + 2R_1 C_1}$$

where $T = 1$. Then, R_0 can be solved according to the combined equation of a_1, a_2 and a_3 as follows:

$$R_0 = \frac{a_3 - a_2}{1 + a_1} \quad (4)$$

$$R_1 = \frac{2(a_3 + a_1 a_2)}{a_1^2 - 1} \quad (5)$$

$$C_1 = \frac{-(a_1 + 1)^2}{4(a_3 + a_1 a_2)} \quad (6)$$

By discretization, Eq. (1) can be rewritten as

$$E_L(k) = a_1 E_L(k-1) + a_2 I_L(k) + a_3 I_L(k-1) \quad (7)$$

$$\Delta Uoc(k) = Uoc(k) - Uoc(k-1) \approx 0 \quad (8)$$

Then, Eq. (7) is given by

$$U_L(k) = (1 - a_1)Uoc(k) + a_1 U_L(k-1)$$
$$+ a_2 I_L(k) + a_3 I_L(k-1) \quad (9)$$

Let $\varphi_1(k) = [1 \; U_L(k-1) \; I_L(k) \; I_L(k-1)]$, $\theta_1(k) = [(1 - a_1)Uoc(k) \; a_1 \; a_2 \; a_3]^T$ and $y_k = U_L(k)$, then

$$y_k = \varphi_1(k)\theta_1(k) \quad (10)$$

In the online application, UL (k) and IL (k) are measurable parameters and the vector θ_1 can be identified by the adaptive Kalman filter algorithm according to Eq. (10). This static equation, which needs no historical data of the voltage across RC and does not have the initial value problem, can accelerate the identification speed for the state and parameter.

2.2 Model parameter identification

The model parameters $\theta = [(1 - a_1)Uoc \; a_1 \; a_2 \; a_3]^T$ of Eq. (10) are directly identified by the adaptive Kalman filter algorithm.

According to the literature (Partovibakhsh M, Liu G. 2014, Xiong R, He H, Sun F, et al. (2013), the process of identification parameters by the adaptive Kalman filter algorithm is summarized as follow as:

$$\theta_{k|k-1} = \theta_{k-1|k-1} \quad (11)$$

$$Y_k = H_k \theta_{k|k-1} \quad (12)$$

$$P_{k|k-1} = P_{k-1|k-1} + Q_{k-1} \quad (13)$$

$$K_k = P_{k|k-1}H^T_k(H_k P_{k|k-1}H_k^T + R_k)^{-1} \quad (14)$$

$$H_k = [1 \; U_{L,k-1} \; I_{L,k} \; I_{L,k-1}] \quad (15)$$

$$\theta_{k|k} = \theta_{k|k-1} + K_k(Y_{m|k} - Y_k) \quad (16)$$

$$P_{k|k} = (I - K_k H_k)P_{k|k-1} \quad (17)$$

By adjusting the measurement noise variance matrix R and the process noise variance matrix Q, the following equations can be obtained

$$\tilde{Y}_k = H_k \theta_{k|k} \quad (18)$$

$$u_k = Y_{m|k} - \tilde{Y}_k \quad (19)$$

$$F_k \approx \frac{\sum_{n=k-L+1}^{k} u_n u_n^T}{L} \quad (20)$$

$$R_k = F_k + H_k P_{k|k}H_k^T \quad (21)$$

$$Q_k = K_k F_k K_k^T \quad (22)$$

Identifying a_1, a_2, a_3 then gives the following equations:

$$R_0 = \frac{a_3 - a_2}{1 + a_1}, R_1 = \frac{2(a_3 + a_1 a_2)}{a_1^2 - 1},$$

$$C_1 = \frac{-(a_1 + 1)^2}{4(a_3 + a_1 a_2)}, Uoc = \frac{\theta(1,1)}{(1 - a_1)}$$

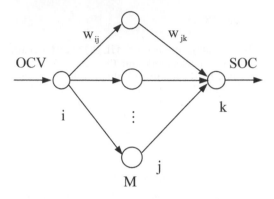

Figure 3.   BP neural network structure.

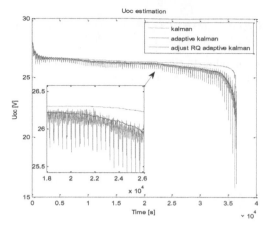

Figure 4.   Identification of the OCV under the DST operation.

3   RELATIONSHIP OF THE OCV-SOC IS DESCRIBED BY THE NEURAL NETWORK

The BP network is a multilayer feed-forward network (He H, Zhang X, Xiong R, et al. 2012). Since the curve of the relationship of the OCV-SOC is similar to the S-type excitation function of the BP network neuron, the BP network is selected to describe the mapping between the OCV and the SOC for the purpose of accelerating the learning speed of the neural network.

The structure with an M-node hidden layer of the BP network is shown in Figure 3, where i is the input layer; j is the hidden layer; and o is the output layer (Liu jinkun. Interlligent control).

The OCV-SOC is described by using the neural network. The SOC is obtained from the OCV-SOC neural network model, by which the complex non-smooth hysteresis characteristics between the OCV and the SOC can be conveniently described. Nevertheless, the calculation of the Jacobian matrix for the state functions, measurement functions and the complex hysteresis relationship of the OCV-SOC model must be introduced to apply the extended Kalman filter, and the differentiable problem of the complex non-smooth hysteresis of the OCV-SOC is inevitable.

However, the SOC estimation by the OCV-SOC and modeling the look-up table approach can effectively avoid the problem.

4   SIMULATION AND EXPERIMENT

In order to verify the feasibility and accuracy of the method for SOC estimation, a 20 Ah/ 24 V power lithium battery pack is selected for two kinds of typical working condition test in order to demonstrate the feasibility of the proposed algorithm for SOC estimation.

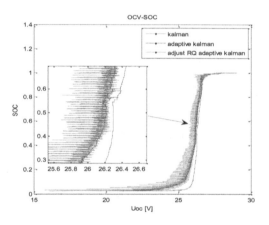

Figure 5.   Relationship between the OCV and the SOC.

The DST working condition lasts 360 seconds (United States Advanced Battery Consortium.1996). The experiment includes two initial conditions: (1) the battery is fully charged ($SOC = 1$) and (2) the battery rests for a certain period of time to attain a stable condition. The battery is subjected to 102 DST cycles until it is emptied.

According to the literature [8], the regulatory factors δ and ε ($\delta = 0.8$, ($\varepsilon = 1.3$) are used to adjust the process noise covariance Q and the measurement noise covariance R, respectively, in which δ and ε determine the ability to inhibit the noise.

The Kalman filter model, the adaptive Kalman filter model and the adaptive Kalman filter model with the adjustment factor are used to identify the OCV and to estimate the relationship between the OCV and the SOC, as shown in Figure 4 and Figure 5, respectively.

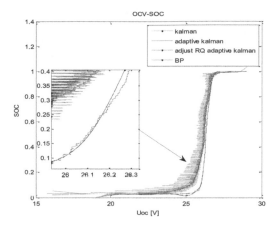

Figure 6. Modeling and testing.

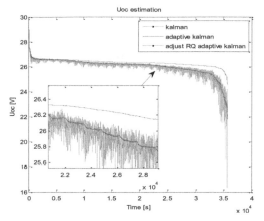

Figure 7. Identification of the Uoc under the FUDS operation.

As ureshown in Figure 4, the identification results of the Uoc show large differences between the three curves, because the real Uoc cannot be obtained in practice but only its approximate estimation value can be obtained. Adjusting the regulatory factors δ and ε by the final SOC result can compensate the estimation error of the Uoc.

In this paper, the DST working condition, the FUDS working condition and BP neural network are employed for the training and testing samples, the actual data and modeling, respectively. In the BP neural network, M is the implied node (let M = 10), losing is defined as the S-type function of the hidden layer, and purelin is the output linear function. The Mean Square Error (MSE) of the BP neural network model of the OCV-SOC is about 8.08e−04. The modeling and testing of the OCV-SOC neural network is shown in Figure 6.

The OCV-SOC model is established by the BP neural network, and then tested as shown in Figure 6. The neural networks fails to fit the relationship between the OCV and the SOC in the range of the SOC < 0.05 and SOC > 0.9. In order to avoid overcharge or deep discharge in the real battery application, the range of the SOC is only 0.1–0.9. Therefore, the proposed OCV-SOC model can meet the practical requirement.

The FUDS working condition lasts for 1373seconds (United States Advanced Battery Consortium.1996). Under the FUDS operation, the experiment with the initial condition of $SOC = 1$ is executed for 27 cycles until the battery is emptied.

The Kalman filter model, the adaptive Kalman filtering model and the adaptive Kalman filter model with the adjustment factor are used to identify the Uoc, as shown in Figure 7.

The SOC estimation results and errors under the FUDS operation are shown in Figure 8 and Figure 9, respectively.

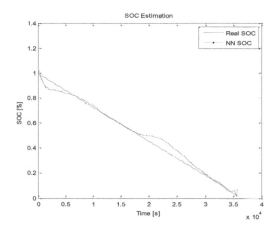

Figure 8. SOC estimation result.

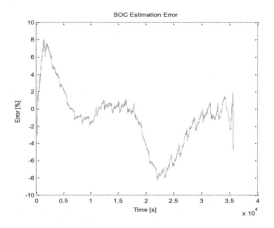

Figure 9. SOC estimation error.

27

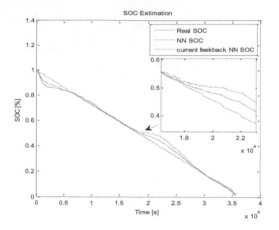

Figure 10. SOC estimation result of online feed-forward compensation.

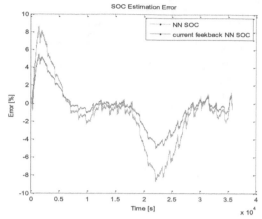

Figure 11. SOC estimation error of online feed-forward compensation.

The SOC is obtained by the neural network model of the OCV-SOC. Due to the lack of the feedback compensation for the OCV-SOC characteristics, the estimation error is probably about 8%, as shown in Figure 9. In order to improve the estimation accuracy, the feed-forward method is introduced to compensate the SOC online. The feed-forward method is summarized as follow:

$$SOC_k = SOC_{k-1} - \frac{\eta_{k-1}\Delta t}{C_n} I_{k-1} \qquad (23)$$

where the Coulomb efficiency $\eta_{k-1} = 1$; $\Delta t = 1$; and C_n is the nominal capacity of the battery. The initial value of the SOC is obtained by the neural network identification as follows:

$$SOC_{feedback,k} = \alpha NNSOC_k + \beta SOC_k \qquad (24)$$

where NNSOCk is the SOC obtained by the OCV-SOC neural network model; SOC feedback, k is the SOC of online feed-forward compensation; $\alpha = 0.6$; and $\beta = 0.6$.

The online feed-forward compensation method can be used to estimate the SOC and the SOC errors, as shown in Figure 10 and Figure 11, respectively.

5 CONCLUSION

The static equation is obtained by transforming the RC network equivalent circuit model in the frequency domain, which is only associated with the current and voltage parameters. Besides, compared with the state equation which is used to estimate the SOC, the static equation with a simple structure is not related to the history data of the resistance, capacitance and the initial value of the state, accelerating the identification speed.

The adaptive Kalman filter is used to implement the online identification of the model parameters in the static equation, such as the internal resistance and the OCV. The OCV-SOC model is described by the neural network, and the SOC is obtained via the OCV-SOC neural network model. The current feed-forward compensation method is introduced to effectively improve the SOC estimation accuracy.

The time of the discharge current under two working conditions is so short that the hysteresis characteristics fail to obviously show the corresponding curves, but the hysteresis characteristics are inevitable for the energy recovery of the car downhill, as shown in this paper. The SOC is obtained by the OCV, which can be used to consider the dynamic hysteresis characteristics in the case of the dynamic charge and discharge process. Therefore, the proposed SOC estimation method will further improve the accuracy of SOC estimation.

ACKNOWLEDGEMENTS

This work was supported by the National Nature Science Foundation Project (61263013), the Scientific Research Project of Guangxi Education Department (201101ZD007) and the Guangxi Experiment Center of Information Science, Guilin University of Electronic Technology Project (20130110).

REFERENCES

Bing D, Yantao T, Changjiu Z. 2014. One Estimating Method of the State of Charge of Power Battery for Electronic Vehicle[C]//Measuring Technology and Mechatronics Automation (ICMTMA), 2014 Sixth International Conference on. IEEE: 439–442.

Chen X, Shen W, Cao Z, et al. 2014. A novel approach for state of charge estimation based on adaptive switching gain sliding mode observer in electric vehicles [J]. Journal of Power Sources, 246: 667–678.

Corno M, Bhatt N, Savaresi S M, et al. Electrochemical Model-Based State of Charge Estimation for Li-Ion Cells[J].

Du J, Liu Z, Wang Y. 2014.State of charge estimation for Li-ion battery based on model from extreme learning machine [J]. Control Engineering Practice, 26: 11–19.

Fendri D, Chaabene M. 2012. Dynamic model to follow the state of charge of a lead-acid battery connected to photovoltaic panel [J]. Energy Conversion and Management, 64: 587–593.

H. Chaoui and P. Sicard, 2011. "Accurate State of Charge (SOC) Estimation for Batteries using a Reduced-order Observer," IEEE International Conference on Industrial Technology (ICIT), pp. 39–43, March.

He H, Xiong R, Guo H. 2012. Online estimation of model parameters and state-of-charge of LiFePO4 batteries in electric vehicles [J]. Applied Energy, 89(1): 413–420.

He H, Zhang X, Xiong R, et al. 2012. Online model-based estimation of state of charge and open-circuit voltage of lithium-ion batteries in electric vehicles [J]. Energy, 39(1): 310–318.

He W, Williard N, Chen C, et al. 2014. State of charge estimation for Li-ion batteries using neural network modeling and unscented Kalman filter-based error cancellation [J]. International Journal of Electrical Power & Energy Systems, 62: 783–791.

Kang L W, Zhao X, Ma J. 2014. A new neural network model for the state-of-charge estimation in the battery degradation process [J]. Applied Energy, 121: 20–27.

Liu jinkun. 2011. Interlligent control [M]. Beijing: Publishing House of Electronics Industry.

Liu Z, Wang Y, Du J, et al. 2012. RBF network-aided adaptive unscented kalman filter for lithium-ion battery SOC estimation in electric vehicles[C]//Industrial Electronics and Applications (ICIEA), 2012 7th IEEE Conference on. IEEE, 1673–1677.

N.A. Windarko, J. Choi and G.B. Chung, 2011. "SOC estimation of LiPB batteries using Extended Kalman Filter based on high accuracy electrical model," 8th International Conference on Power Electronics—ECCE Asia, pp. 2015–2022, May, 30–June, 3.

Ng, K.S., C.S. Moo, Y.P. Chen, and Y.C. Hsieh, 2009. "Enhanced coulomb counting method for estimating state-of-charge and state-of-health of lithium-ion batteries," Applied Energy, vol. 86, no. 9, pp. 1506–1511, September.

Partovibakhsh M, Liu G. 2014. An adaptive unscented Kalman filtering approach for online estimation of model parameters and state-of-charge of lithium-ion batteries for autonomous mobile robots [J].

Sepasi S, Ghorbani R, Liaw B Y. 2014. A novel on-board state-of-charge estimation method for aged Li-ion batteries based on model adaptive extended Kalman filter [J]. Journal of Power Sources, 245: 337–344.

Sepasi S, Ghorbani R, Liaw B Y. 2014. Improved extended Kalman filter for state of charge estimation of battery pack [J]. Journal of Power Sources, 255: 368–376.

Sun F, Xiong R, He H. 2014. Estimation of state-of-charge and state-of-power capability of lithium-ion battery considering varying health conditions [J]. Journal of Power Sources, 259: 166–176.

United States Advanced Battery Consortium. 1996. Electric vehicle battery test procedures manual Revision 2 Published, January.

Xing Y, He W, Pecht M, et al. 2014. State of charge estimation of lithium-ion batteries using the open-circuit voltage at various ambient temperatures [J]. Applied Energy, 113: 106–115.

Xiong R, He H, Sun F, et al. 2013. Evaluation on state of charge estimation of batteries with adaptive extended Kalman filter by experiment approach [J]. Vehicular Technology, IEEE Transactions on, 62(1): 108–117.

Xu J, Mi C C, Cao B, et al. 2013. A new method to estimate the state of charge of lithium-ion batteries based on the battery impedance model [J]. Journal of Power Sources, 233: 277–284.

Energy Science and Applied Technology – Fang (Ed.)
© 2016 Taylor & Francis Group, London, ISBN 978-1-138-02833-3

A kind of distributed seawater desalination generation system

Liming Liu
Department of North China Electric Power University, Baoding, Hebei Province, China

ABSTRACT: This paper presents a kind of a distributed seawater desalination generation system based on a solar collector that gathers solar radiation using a concentrating solar cooker in order to desalinate seawater by distillation. The system uses four quadrant photosensitive sensors to obtain the position of the sun, which is controlled by a microcontroller. We seek to always achieve the largest solar radiation absorption rate of the apparatus. At the same time, we seek to make full use of solar energy to achieve self-powered supply in the whole system using a Sterling generator and a solar panel battery for power generation control. The system not only saves the consumption of non-renewable energy and reduces the cost of desalination, but also reduces the pollution of the environment. The system adapts to the trend of developing low-carbon economy and environmental protection. It has a good research value and prospects.

Keywords: Solar Collector, Seawater Desalination, Microcontroller, Sterling generator

1 INTRODUCTION

Water is the source of life. With the rapid growth of the population and fast urbanization, the demand for fresh water is increasing, leading to the lack of the water source. The contradiction of water demand is serious. Therefore, it is a prime and strategic matter to develop and utilize water sources. However, the existing seawater desalination technologies exhibit features of complex process, high energy consumption, large investment, low profits and serious secondary pollution. Due to these weaknesses, it is hard to be further developed. Owing to the decrease in non-renewable energy such as coal and petroleum, energy problem has gradually become a bottleneck, restricting the development of economy in the international society. In fact, more and more countries have carried out the "Project Sunshine" that explores solar energy resources and regards it as a new power of economic growth. With the advantages of low operation cost, simple equipment maintenance, no noise and no pollution emission, the limitless solar energy is of great importance to many institutions and people. At present, the solar energy is mainly used for power generation and preparation of heating water. However, because the solar radiation on the surface of the earth fluctuates day and night, it cannot satisfy the requirement when the light is weak. When the light is strong, the collection of the solar energy is often not fully utilized, resulting in tremendous waste. Consequently, it is still a difficult problem for researchers to improve the utilization of the solar energy. The purpose of the project aims at the defects of the existing technology, providing a solar seawater desalination and power generation device, in order to reduce the cost of desalination and the pollution of the environment.

2 SYSTEM DESIGN

2.1 Basic principles

The system itself is self-powered, i.e. generates power through their own power generation equipment to meet the power required by the entire system in a four quadrant photosensitive sensor pursuit and device steering. It does not need external power support (as shown in Figure 1).

The system can realize automatic control, with intelligence, independently complete solar cooker steering, distillation vessel replenishment, water distilling, distilled water collection, condensing steam power and brine collection function.

2.2 Design proposal

The system consists of three units: solar collector unit, seawater desalination unit and photovoltaic power generation unit. The solar collector system achieves the function of solar thermal radiation collection, seawater heating and solar light tracking. The photovoltaic power generation system is responsible not only for the conversion of solar

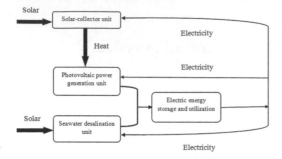

Figure 1. Diagram of the system energy cycle.

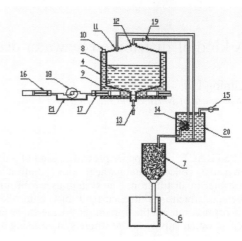

Figure 2. Distillation tank.

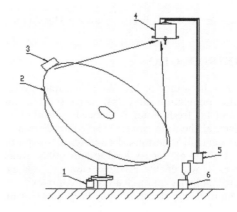

Figure 3. Sealed casing.

energy, but also for the storage and comprehensive utilization of the electric energy in the entire system.

Solar Collector Unit. The major components of the solar collector unit are the concentrating solar cooker, gear, motor and optical tracking sensor. Its main functions are as follows:

Solar thermal radiation collection will be completed by the concentrating solar cooker, gathering the larger area of the sunlight into the bottom of the pot so that the temperature rises to a higher level.

Sunlight tracking is based on four quadrant photosensitive sensors (Zhong Yufang, 2013) to obtain the position of the sun and transmit the messages to the microcontroller that controls motor rotation so that the solar cooker faces the sun.

Sea water Desalination Unit. The sea water desalination unit is mainly composed of a distillation tank, a condenser, a fresh water bucket, an activated carbon tank (Jiang Wenxin, 2007), a medium oil and brine discharge electromagnetic valve, a water pump, a steam solenoid valve and a Sterling generator (Zhou Shouming, 2013).

Seawater desalination part: Figure 2 shows the distillation tank positioned on the focus of the solar cooker and is connected to the solar cooker through the bracket. The top is arranged with a water inlet and a steam outlet. At the bottom, there is a brine discharge outlet with an electromagnetic valve. Steam condensation is accomplished by the condenser pipe. The condenser includes a sealed casing and the condenser pipe is installed in the interior. The sealed casing inlet connects the water inlet of the water pipe, with the outlet connecting the inlet of the distillation tank (Figure 3). The condenser uses the seawater for desalination, which can achieve the purpose of preheating the seawater. One end of the condenser pipe is connected to the steam outlet of the distillation tank, while the other end is connected to the freshwater drum, in order to complete the desalination of the seawater.

The power generation part: Figure 2 shows the side wall and the bottom plate of the distillation

tank arranged in two layers, forming a two-layer bottom oil cavity filled with medium oil. A Sterling generator is installed on the outer side wall of the distillation tank, with a total of four motor settings. They are evenly distributed around the retort. The heating pipe of the Sterling generator threads into the medium oil through the generator installation hole on the outer side wall of the distillation tank. It completes the power generating function through the heating pipe of the Sterling generator,

Photovoltaic Power Generation Unit. This unit consists of the solar panel battery, the charger and other parts, to complete the generation, storage, power supplement of the whole system and other functions. Considering that the steam power may not be enough to supply the entire system, we intro-

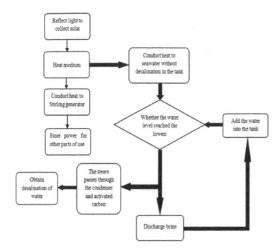

Figure 4.   Flowchart of system implementation.

duce the photovoltaic power generation system. On the one hand, the photovoltaic power generation system makes up the lack of steam power. On the other hand, photovoltaic power generation system plays a significant role in integrating the electric energy and comprehensive utilization, making a closer collaboration between the system units.

The description of each label list is as follows:

1. Motor
2. Solar cooker
3. Optical tracking sensor
4. Distillation tank
5. Condenser (Jiang Xiang, 2002)
6. Fresh water bucket
7. Activated carbon tank
8. Seawater
9. Medium oil
10. Oil change mouth
11. Water inlet
12. Steam outlet
13. Brine discharge electromagnetic valve
14. Condenser pipe
15. Water pump
16. Cooling pipe
17. Heating tube
18. Gear
19. Steam solenoid valve
20. Containment shell
21. Air channel

3   PROCEDURE OF OPERATION

Sunlight tracking is based on four quadrant photosensitive sensors, in order to obtain the position of the sun and transmit the messages to the microcontroller that controls motor rotation so that the solar cooker faces the sun.

Retort replenishment of the distillation tank is controlled by the microcontroller. The distillation tank has a water level-detecting sensor that feedbacks the signal to the microcontroller, and the water pump is controlled to hydrate.

Solar cookers first gather sunlight for heating medium oil (boiling point of 250DEG C; Xia Liping, 1997), which, in turn, heats the water in the distillation tank and the internal gas of the Sterling generator. The total amount of oil reduces in a period of time after work because medium oil is easily volatilized. At the moment, we need to add medium oil to the oil chamber through the oil change mouth manually.

When the water level in the distillation tank reaches its limit, the brine discharge electromagnetic valve is opened and the distillation tank brine is discharged.

At the same time, the photovoltaic power generation unit completes the absorption of the solar energy by the solar panel battery. Under the control of the microcontroller, electricity can be produced from the solar energy by boosting the voltage stabilizer stored in the battery to meet the lack of power.

The device adopts the method of distillation for seawater desalination, which produces steam directly through the condenser pipe. Cooling water is used to desalt seawater, playing its role in preheating the seawater.

When the steam solenoid valve is opened, the solar energy is used for power generation and desalination of seawater simultaneously. When it is closed, the solar energy is fully used for power generation (including all Sterling engine power and solar panels for electricity).

4   CONCLUSION

The system achieve the basic automation control. The design of the microcontroller and the use of four quadrant photosensitive sensors solve the problem of utilizing complete energy from the sun. In addition, the design of the distillation vessel is exquisite and discrete. It can realize the efficient heating of seawater, the collection of condensed steam, and prevent water overflow within the vessels.

This system using the renewable solar energy as the main energy source, which is based on the intention of solving the prominent social problem, not only saves the consumption of non-renewable energy and reduces the cost of desalination, but also reduces the pollution of the environment. The system adapts to the trend of developing low-carbon economy and environmental protection.

The solar energy heating, seawater desalination and solar power are integrated into one system, called the design of automatic control. The reasonable distribution of the consumption of energy can improve the comprehensive utilization of solar energy and reduce the waste of energy. Based on the automatic control and the building block design of the microcontroller, the system can expand the scale of production, as the product is convenient for installation and maintenance, as well as has good prospects for development.

REFERENCES

Jiang Wenxin, Zhang Wei, Chang Qigang, Zhang Huaixu, Fu Lijun, Ying Weiqi, Li Meining, Yuyuan, Enhanced activated carbon adsorption process for advanced treatment of biotreated coking plant wastewater, Environmental Pollution & Control, 2007.

Jiang Xiang, Zhu Dongsheng, Present Situation of Condenser and Development of Evaporative Condenser, Refrigeration, 2002.

Xia Liping, Huang Ping. Conduct Heat Oil is One Excellent Intermediate Heat Transfer Medium. Petro-Chemical Equipment Technology. 1997.

Zhong Yufang, Wang Huifen. Ye Jianfeng, Ye Xing, Wu Mingguang, Analysis of Solar Azimuth Tracking Device of PV System. Acta Energise Solaris Sinica. 2013.

Zhou Shouming, Wu Hongxing. Xiao Chong, Li Liyi, Technology Review of Free Piston Sterling Linear Generator System. Micromotors, 2013.

Energy transfer and conversion, materials and chemical technologies

Energy Science and Applied Technology – Fang (Ed.)
© 2016 Taylor & Francis Group, London, ISBN 978-1-138-02833-3

Separation of flavonoids from kudzu root extract using poly (vinylidene fluoride) ultrafiltration membranes

Hai Xiang, Hui Wang & Lu Zeng
College of Life Science, Zhejiang Chinese Medical University, Zhejiang, China

Songxue Wang & Xiuzhen Wei
College of Biological and Environmental Engineering, Zhejiang University of Technology, Zhejiang, China

ABSTRACT: Poly (vinylidene fluoride) (PVDF) ultrafiltration (UF) membranes with different molecular weight cut-off (MWCO) were used to separate the flavonoids from kudzu root extract via a cross-flow filtration system. The effects of operation pressure, operation temperature and flavonoids concentration on the permeate flux and the loss rate were investigated. The morphology of PVDF UF membranes before and after the separation process was characterized by Scanning Electron Microscopy (SEM). An obvious layer is formed on the PVDF-30 membrane surface. The experiment results indicated that the PVDF UF membrane with relative smaller MWCO is more suitable for separation the flavonoids from kudzu root extract.

Keywords: flavonoids, kudzu root extract, separation, ultrafiltration membranes

1 INTRODUCTION

Kudzu root (Gegen in Chinese) is the dried root of a perennial climber, Pueraria lobate (Willd). It is also one of the most widely used traditional Chinese medicine herbs. It has been commonly used for the relief of headaches, as well as other minor aches and pains. Many studies have revealed that kudzu root has anti-aging, antihypertensive, antioxidant and anti-diabetic effects. For example, kudzu root extract can decrease LDL cholesterol in patients with coronary heart disease. The major effective constituent of kudzu root extract is flavonoids. And various separation methods have been applied to further enhance the production of flavonoids content from the kudzu root crude, such as Super Critical Fluid Extraction (SFE), organic solvent extraction, ultrasonic extraction technology, and microporous-resin adsorption. Among them, microporous-resin adsorption and organic solvent extraction have been widely used as effective separation techniques for the extraction of the kudzu root recently. However, organic solvent extraction and microporous-resin adsorption have some inherent shortcomings. Organic solvent extraction and adsorption processes are tedious and some hazardous materials are inevitably kept in the final products. What's more, some active ingredients are easily lost. SFE is one of the feasible and effective technologies. However, a wide range of applications have limited due to expensive equipment's and complicated operation process.

Membrane separation technology has attracted great attention in recent years. Membrane technologies have been established as very effective and promising technologies for separation and purification processes in dealing with biochemical, pharmaceuticals and foods. Compared with the above-mentioned traditional separation methods, ultrafiltration (UF) can be a feasible alternative due to various advantages, including high mechanical strength, high selectivity and low energy consumption. Commercially available poly(vinylidene fluoride) (PVDF) UF membranes have received great attention relying on its outstanding properties such as no phase change, high mechanical strength, corrosion resistance, thermal stability, chemical stability etc.. PVDF UF membrane has been widely used in wastewater treatment and medicine field. However, the serious membrane fouling is a major impediment to the use of membrane separation technology in this field. On the other hand, the operation parameters and the character of the membranes also affect the separation of the biochemical.

In this study, PVDF UF membranes were used to separate the flavonoids from the kudzu root extract. The chemical composition of kudzu root extract is very complex, containing inorganic salts, amino acids, organic acids, alkaloids, flavonoids, ketones, sapiens, steroidal and terpene compounds, proteins, polysaccharides, starch, cellulose, colloidal and other polymer materials. Flavonoids was used to evaluate the effect of ultrafiltration due to flavonoids is the main effective constituent of kudzu root. The effects of operation parameters and MWCO of PVDF on the loss rate of flavonoids were studied in detail. At the same time, membrane fouling was also characterized by using Scanning Electron Microscopy (SEM).

2 EXPERIMENTAL

2.1 *Materials*

Poly (vinylidene fluoride) was purchased from Hangzhou Joel Membrane Separation Technology Co., Ltd. The molecular weight cut off of PVDF membranes was 30000 and 50000, which denoted as PVDF-30 and PVDF−50. Kudzu root was purchased from Zhejiang Chinese Medical University, Chinese Herbal Medicine Company. Ethanol (AR grade) was purchased from Ling Feng Chemical Reagent Co. Ltd., Shanghai, China.

2.2 *Methods*

Preparation of kudzu root extract. A certain amount of kudzu root was crushed using plants pulverize (40 mesh). Then boiling water was added into the solution. boiling water of ten times was first added into for 1.5 hours, then boiling water of eight times was added into for 1 hour, finally boiling water of six times was added into for 30 minutes. Liquid was collected by filtration. The decoction was merged and cooled for use.

Ultrafiltration. A cross-flow filtration system was designed to evaluate the separation performance of membranes. The UF membrane was precompacted at 0.15 MPa about 30 min, then the pressure was reduced to the operating pressure of 0.1 MPa and the water flux (J_w) was measured every 5 minutes. All the water flux measurements were conducted at room temperature. The water flux was calculated using the following equation (1):

$$J_w = \frac{V}{S\Delta t} \tag{1}$$

Where V (L) was the volume of permeated water, S (m^2) was the membrane area and Δt (h) was the permeation time.

Loss rate. The flavonoids concentration in the permeation, feed and retentive solution was measured and loss rate was calculated according to the amount difference in the three solutions. The

concentration of flavonoids was measured by UV-visible spectrophotometer (752 UV).

2.3 *Morphology of PVDF membranes*

Morphology of PVDF membranes before and after separation the flavonoids from the kudzu root extract was characterized by using a scanning electron microscopy (SEM, JSM-5900 LV). All the membrane samples were sputtered with a thin gold layer for SEM imaging.

3 RESULTS AND DISCUSSION

3.1 *Effect of operation pressure*

Effect of operation pressure on the permeate flux. On the premise of same operation temperature and flavonoids concentration, the effect of operation pressure on permeate flux was studied which was presented in Fig.1. It shows that the permeate flux increased proportional with the operation pressure increased. The permeate flux of PVDF-50 membrane increased from 21 to 89 L/m²·h as operation pressure increased from 0.05 to 1.50 MPa. The permeate flux of PVDF-30 membrane increased from 20 to 40 L/m²·h as the operation pressure increased from 0.05 to 1.50 MPa. However, the permeate flux decline gradually and presented stable finally with the operation time prolonging. This phenomenon can be explained as following: Concentration polarization and membrane fouling was more serious with the proceeding of ultrafiltration, and some macromolecules deposited onto membrane surface forming sedimentary layer that was become denser gradually. When the sedimentary layer exceeds a critical value, the permeation rate will not change obviously.

Effect of operation pressure on the loss rate. The effect of operation pressure on the loss rate was shown in Fig. 2. As shown in Fig. 2, when the operation pressure increased, the loss rate increased. The loss rate of PVDF-50 was relative higher than that of PVDF-30 due to the pore size of PVDF-50 was relative larger than that of PVDF-30. The results indicated that PVDF UF membrane with

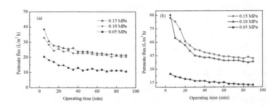

Figure 1. Effect of operating pressure on the permeate flux (a) PVDF-30, (b) PVDF-50.

smaller MWCO can effectively reserve flavonoids under our experiment condition.

3.2 *Effect of operation temperature*

Effect of operation temperature on the permeate flux. The effect of operation temperature on the permeate flux was investigated on the premise of keeping operation pressure and flavonoids concentration constant. As shown in Fig. 3, the permeate flux declined with the operation temperature increase. The temperature has a dual influence on the permeate flux. On the one hand, the viscosity of the raw material drops and the diffusion coefficient increases with the rise of the temperature, which lead to the increase of the permeate flux. On the other hand, the hydrolysis of hydrolysable material will accelerate and the protein in the raw solution will precipitate which result in more serious membrane fouling and the reduction of permeate flux.

Effect of operation temperature on the loss rate. Fig. 4 shows the effect of operation temperature on the loss rate. For PVDF-50, the higher of the operation temperature was, the larger of the loss rate would be. The phenomenon was caused by the more serious membrane fouling. However, for PVDF-30, the loss rate did not change obviously with the change of operation temperature which is consistent with the results of the permeate flux.

3.3 *Effect of flavonoids concentration*

Effect of flavonoids concentration on the permeate flux. In this study, the operation temperature and pressure are kept constant. The effect of flavonoids on the permeate flux was investigated on the above-mentioned conditions and results were shown in Fig. 5. The permeate flux declined with the flavonoids of kudzu concentration increase. The phenomenon is caused by more active substance in the raw solution would deposit onto the membrane surface with the increasing of the flavonoids concentration which lead to the decline of permeate flux. For PVDF-50, the permeate flux declined more rapidly than that of PVDF-30. At the initial ultrafiltration stage, some composition of the solution would be rejected by the membrane and form a cake type of layer on the membrane surface. PVDF-50 has relative larger pore than that of PVDF-30, it is relatively easy for the constituents in the raw solution to enter into the membrane pores leading to the clog of the pores.

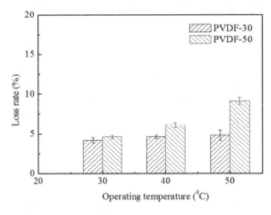

Figure 4. Effect of operation temperature on the loss rate.

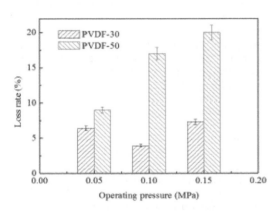

Figure 2. Effect of operation pressure on the loss rate.

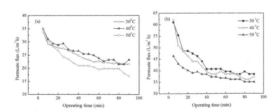

Figure 3. Effect of operation temperature on the permeate flux (a) PVDF-30, (b) PVDF-50.

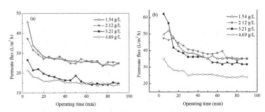

Figure 5. Effect of flavonoids concentration on the permeate flux (a) PVDF-30, (b) PVDF-50.

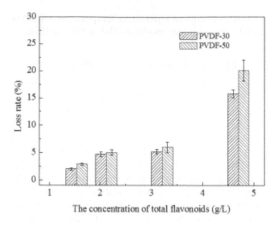

Figure 6. Effect of the total flavonoids concentration on the loss rate.

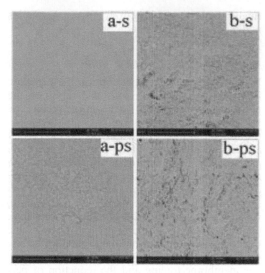

Figure 7. SEM images of PVDF membrane.

Effect of flavonoids concentration on the loss rate. Fig. 6 shows the effect of flavonoids concentration on loss rate. The loss rate increased with flavonoids concentration increased due to more active substance in the raw solution would deposit onto the membrane surface with the increasing of the flavonoids concentration. The highest loss rate is 15.8% and 20.1% for PVDF-30 and PVDF-50, respectively. It can be also found that the loss rate increased with the increasing of the MWCO of PVDF membranes. The phenomenon is attributed to the fact that more substance would penetrate into or through the membrane due to the pores are relative larger for PVDF-50. All the results is consistent with that of permeate flux. The experimental results indicated that PVDF-30 is more suitable for separation the flavonoids from the kudzu root extract.

3.4 Morphology of PVDF membrane

The morphology of the cleaned and fouled membrane is examined using SEM. As shown in Fig. 7, the surface of PVDF-30 is smooth and relative dense, while the surface of PVDF-50 is relative looser and presented some pores. For the fouled PVDF-30 membrane, a cake layer and some particle substance showed on the surface. This morphology confirmed the declined of the permeate flux was resulted from the fouling of the membrane. For the fouled PVDF-50, no obvious cake layer or particle substance is presented on the surface. The results indicated that the decline of the permeate flux was due to some active molecules penetrated into and clogged the pores of the membranes.

a-s: the cleaned membrane of PVDF-30, b-s: the cleaned Membrane of PVDF-50, a-ps: the fouled membrane of PVDF-30, b-ps: the fouled membrane of PVDF-50.

4 CONCLUSION

PVDF UF membranes with different MWCO were used to separate the flavonoids from the kudzu root extract. The permeate flux and loss rate were affected by the operation pressure, operation temperature, and the flavonoids concentration. The permeate flux increased with the operation pressure increased. The permeate flux decreased gradually with increasing of operation temperature and the concentration of flavonoids. The loss rate increased with increasing of operation pressure, operation temperature and the concentration of flavonoids. For PVDF-30, after fouling, an obvious cake type of layer formed on the membrane surface which led to the permeate flux decline. For PVDF-50, the permeate flux decline was attributed to the clog of the pores. Under our experimental conditions, PVDF-30 is more suitable for the concentration of kudzu root extract.

ACKNOWLEDGEMENTS

The authors gratefully acknowledge financial support from the Science Technology Department of Zhejiang Province (Grant No. 2014C33214). And the authors also sincerely thank the State Administration of Traditional Chinese Medicine of Zhejiang Province (Grant No. 2013ZQ006) and the Natural Science Foundation from Zhejiang Province (Grant No. LY13H280009).

REFERENCES

Busch, J., Cruse1, A. & Marquardt, W., 2007, modelling submerged hollow-fibre membrane filtration for wastewater treatment, J. Member. Sci., 288, p. 94–111.

Chiu, K.L., Cheng, Y.C., Chen, J.H., Chang, C.J. & Yang, P.W., 2002 Supercritical fluids extraction of Ginkgo ides and flavonoids, J. Superscript. Fluids. 24, p. 77–87.

El-Bourawi, M.S., Ding, Z., Ma, R. & Khayet, M., 2006, A framework for better understanding membrane distillation separation process, J. Member. Sci., 285, p. 4–29.

Keung, W.M. & Vallee, B.L., Kudzu root, 1998: an ancient Chinese source of modern antidipsotropic agents, Phytochemical. 47, p. 499–506.

Krzysztof K., Marek, G. & Antonin, W.M., 2009, Membrane processes used for separation of effluents from wire productions, Chem. Pap., 63, p. 205–211.

Liang, H., Pan, W.H. & Zhang, W.F., 1999, the extract technique of flavonoid of Ginkgo balboa L. leaves, J. Plant. Retour. Environ. 8, p. 12–17.

Liu, Y. & Wei, S.L., 2013, Optimization of ultrasonic extraction of phenolic compounds from Euryale faro seed shells using response surface methodology, Ind. Crop. Prod., 49, p. 837–848.

Liang, H., Gong, W., Chen, J. & Li, G., 2008, cleaning of fouled ultrafiltration (UF) membrane by algae during reservoir water treatment, Desalination, 220, p. 267–272.

Laroche, G., Marois, Y., Guidoin, R., King, M.W., Martin, L., How, T & Douville, Y., 1995, Polyvinylidene fluoride (PVDF) as a biomaterial: from polymeric raw material to monofilament vascular suture, J. Biomed. Mater. Res., 29, p. 1525–1536.

Meng, X.R., Zhang, N., Wang, X.D., Wang, L., Huang, D.X. & Miao, R., 2014, Dye effluent treatment using PVDF UF membranes with different properties, Desalin. Water Treat. 52, p. 5068–5075.

McGregor, M. R., 2007, Pueraria lobate (Kudzu root) hangover remedies and acetaldehyde-associated neoplasm risk, Alcohol, 41, p. 469–78.

Miao, R., Wang, L., Feng, L., Liu, Z.W. & Lv, Y.T., 2014, Understanding PVDF ultrafiltration membrane fouling behaviour through model solutions and secondary wastewater effluent, Desalin. Water Treat. 52, p. 5061–5067.

Pagliero, C., Ochoa, N., Marches J. & Matteo, M., 2001, Degumming of crude soybean oil by ultrafiltration using polymeric membranes, J. Am. Oil Chem. Soc., 78, p. 793–796.

Pinto, C.G., Laespada, M.E.F., Pavo'n, J.L.P. & Cordero, B.M., 1999, Analytical applications of separation techniques through membranes, Lab. Auto. Inf. Manage., 34, p. 115–130.

Pierre, L.C., Vicki, C. & Tony, A.G., 2006, Fane Fouling in membrane bioreactors used in wastewater treatment, J. Member. Sci., 284, pp. 17–53.

Tam, W.Y., Chook, P., Qiao, M., Chan, L.T, Chan, T. & Poon, Y.K., 2009, the efficacy and tolerability of adjunctive alternative herbal medicine (Salvia miltiorrhiza and Pueraria lobate) on vascular function and structure in coronary patients, J. Alter. Complex. Med., 15, p. 415–21.

Wang, Z., Wu, Z., Yin, X. & Tian, L, 2008. And Membrane fouling in a submerged membrane bioreactor (MBR) under sub-critical flux operation: Membrane foul ant and gel layer characterization, J. Member. Sci., 325, p. 238–244.

Wang, X.J., Lu, X.J., Xu, W.Y., Zhu, J.C. & Wang, L.G., Preparation and Characterization of PVDF Modified Ultrafiltration Membrane for Purification of Hg in Water, Key Eng. Mater., 575–576, p. 265–269, 2013.

Wong, K.H., Li, G.Q., Li, K.M., 2011, Razmovski-Naumovski, V. & Chan, K., Kudzu root: Traditional uses and potential medicinal benefits in diabetes and cardiovascular diseases, J. Ethnopharmacol., 134, p. 584–607.

Xu, M.C., Shi, Z.Q., Shi, R.F., Liu, J.X., Lu, Y.L. & He, B.L., 2000, Synthesis of the adsorbent based on microporous copolymer MA–DVB beads and its application in purification for the extracts from Ginkgo balboa leaves, React. Funct. Polyp. 43, p. 297–304.

Xu, Z.H., Li, L., Wu, F.W, Tan, S.J. & Zhang, Z.B., 2005, the application of the modified PVDF ultrafiltration membranes in further purification of Ginkgo biloba extraction, J. Member. Sci., 255, p. 125–131.

Yi, X.S., Yu, S.L., Shi, W.X., Wang, S., Jin, L.M., Sun, N., Ma, C. & Sun, L.P., 2013, Separation of oil/water emulsion using nano-particle (TlO2/AI2O3) modified PVDF ultrafiltration membranes and evaluation of fouling mechanism, Water Sci. Technol., 67, p. 477–484.

Yang, C., Xu, Y.R. & Yao, W.X., 2002 and Extraction of pharmaceutical components from Ginkgo balboa leaves using supercritical carbon dioxide, J. Agric. Food Chem., 50, p. 846–849.

Energy Science and Applied Technology – Fang (Ed.)
© 2016 Taylor & Francis Group, London, ISBN 978-1-138-02833-3

Preparation and characterization of Cr_2O_3–TiO_2–Co_2O_3–ZnO green pigment

L. Chen, X.L. Weng & L.J. Deng
National Engineering Research Center of Electromagnetic Radiation Control Materials, University of Electronic Science and Technology of China, Chengdu, China

L. Yuan
Center for Advanced Materials and Energy, Xihua University, Sichuan, China

ABSTRACT: A new green pigment with a high near-infrared reflectance based on a Cr_2O_3–TiO_2–Co_2O_3–ZnO composition was synthesized by the solid state reaction method. The phase structure, near-infrared reflectance and color parameters were characterized by XRD, SEM, UV–vis and colorimeter, respectively. The effect of the zinc oxide content in the original mixture on spectral reflectance and color was investigated. The results show that the reflectance of the pigment increases with the increasing content of ZnO. In addition, the reflectance increases rapidly from a low level at the red edge, ranging from 680 to 750 nm. In addition, a blue shift of the red edge was observed by increasing the content of ZnO.

Keywords: Pigment; Solid state reaction; Near-infrared reflection

1 INTRODUCTION

In recent years, Near-Infrared (NIR) reflective pigments have been receiving significant attention for their wide applications in energy-saving coating and military camouflage (Jeevanandam. P et al. 2007). The reflective spectral characteristic of a pigment is an important parameter for determining its properties. In the visible region (400–700 nm), pigments selectively absorb the visible light and reflect the remaining light. The wavelength of reflection light has effects on the hue of pigments. In addition, in the infrared region (700–2500 nm), the reflection characteristic is concerned with the solar heat. These radiations on absorption result in heating up of the surface (Bendiganavale. A. K. et al. 2008). Infrared reflective inorganic pigments are mainly metal oxides. And they are primarily useful in two major applications: one is reducing heat build-up on the surface of the building as cool roofing materials (Han. A, 2013). Since the energy of the near-infrared region accounts for half of the whole solar energy (Ferrari. C et al. 2013). Cool roofs with high Near-Infrared (NIR) reflectance can reflect a significant fraction (52%) of the solar energy that arrives as NIR radiation (Zou. J et al. 2014, Li. Y. Q et al. 2014). The other application is military camouflage. Although the NIR region of the spectrum covers the wavelength range from 700 to 2000 nm, the current camouflage requirement is focused on the wavelength range of 700–1200 nm. In this region, objects can be "seen" by reflection (Rubeziene. V, et al. 2008). The high infrared reflective pigments have the ability to reflect electromagnetic waves in such a way that they give the same signature to the observed object as the signature of the surrounding environment, both in the visible and near-infrared spectra (Zhou. Y. X, 2013).

Thongkanluang et al. synthesized a new pigment by mixing cheap precursors of Cr_2O_3 (host component), TiO2, Al_2O_3 and V_2O_5. It was found that the pigment has a NIR reflectance of 82.8% (Li. P, 2008, Thongkanluang. T et al. 2011). Zhang jie et al. prepared Ti-doped Cr_2O_3 pigments by the solid-phase synthesis process and the NIR reflectance of doped Cr_2O_3 increased significantly. The pigments have a maximum NIR reflectance when the content of TiO_2 is 1.07 wt% (Zhang. J et al. 2010). In the military field, USA Ferro Company invented a commercial chromium cobalt oxide pigment, "Ferro camouflage pigment". The reflectance of the pigment increased sharply at near 700 nm and the reflectance beyond 60%, which had a similar reflection spectral curve to that of chlorophyll. Hu et al. modified the pigment of the Al_2O_3-CoO system as a green camouflage pigment by using ferric oxide, and the result proved the efficiency and availability of this scheme (Hu. Z. H et al. 2010).

In this paper, a new green pigment based on a Cr_2O_3–TiO_2–Co_2O_3–ZnO composition was synthesized and the optical property of these pigments was studied. In addition, depending on the application of the pigment in military camouflage, the effect of the ZnO content on NIR reflection performance is discussed.

2 EXPERIMENTAL METHOD

2.1 *Powder preparation*

A green pigment based on a Cr_2O_3–TiO_2–Co_2O_3–ZnO composition was designed according to the assumption that Zn^{2+} replaces Co^{2+} in tetrahedral sites. The resulting nominal compositions are $Co_{0.4-x}M_xCr_{1.58}Ti_{0.02}O_4$. The starting materials used for the synthesis were commercially obtained powders: TiO_2, Cr_2O_3, ZnO and Co_2O_3. All the samples were synthesized via the solid state reaction. The nominal compositions of the samples are given in Table 1.

2.2 *Characterization of techniques*

To identify the crystallization phase of the crystallites, phase analysis of the samples was performed by X-ray powder diffraction using Ni-filtered Cu-Kα radiation ($\lambda = 0.154$ nm). Data were collected by steps scanning over a 2θ range from $20°$ to $80°$ with a step size of $0.02°$ and a counting time of 5 s at each step. Scanning electron micrographs of the samples were recorded on a Scanning Electron Microscope (SEM) JEOL JSM-7600F model, with an acceleration voltage of 15 kV.

The pigment powders were compressed in a mold to obtain the sample in a form of a thin disk, with a diameter of 3 cm and a thickness of 4 mm. The UV–vis spectroscopy study of the samples was carried out in a Perkin-Elmer Lambda 750 spectrophotometer in the 350–2500 nm range using barium sulfate as a reference.

The color properties are described in terms of the CIE $L*$ $a*$ $b*$ system (1976). The value $a*$ (the red–green axis) and $b*$ (the yellow–blue axis) indicate the color hue. The value $L*$ represents the brightness or darkness of the color in relation to the neutral gray scale. In the $L*$ $a*$ $b*$ system, it is described by numbers from zero (black) to hundred (white).

Table 1. Nominal compositions of the samples.

Sample	x	Cr_2O_3	Co_2O_3	ZnO	TiO_2
S1	0.05	7.5	1.84	0.25	0.1
S2	0.1	7.5	1.58	0.51	0.1
S3	0.15	7.5	1.31	0.76	0.1
S4	0.2	7.5	1.05	1.01	0.1
S5	0.25	7.5	0.79	1.27	0.1
S6	0.3	7.5	0.52	1.52	0.1
S7	0.4	7.5	0	2.03	0.1

3 RESULTS AND DISCUSSION

3.1 *Crystal structure of pigment powder*

Figure 1 shows the X-ray diffraction patterns for $Co_{0.4-x}M_xCr_{1.58}Ti_{0.02}O_4$ powder with different doping contents of ZnO, where x = 0.05, 0.10, 0.15, 0.20, 0.25, 0.30 and 0.40, respectively. The characteristic peaks of $CoCr_2O_4$ (JCPDS 01–1122), $ZnCr_2O_4$ (JCPDS 87–0082) and Cr_2O_3 (JCPDS 38–1479) are distinguished in the synthesized samples. It is observed that the peak positions of the spinel phase shift towards lower 2θ values with the introduction of Zn^{2+}. $ZnCr_2O_4$ and $CoCr_2O_4$ have a similar characteristic peak of XRD patterns. Thus, when the quantity of ZnO increases, the 2θ peak position shifts closer to the $ZnCr_2O_4$ structure. In addition, the peaks of TiO_2 do not appear in all samples due to the entry of Ti^{4+} into the lattice replacing Cr^{3+}. This behavior is expected as both Cr^{3+} (0.061 nm) and Ti^{4+} (0.061 nm) have a similar ionic radius.

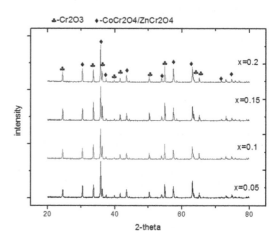

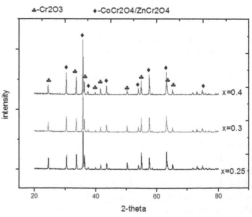

Figure 1. XRD patterns of pigment powder samples.

3.2 NIR reflectance characterization

The UV–vis absorbance spectra of the samples, ranging from 400 to 2500 nm, is shown in Figure 2. In the reflectance spectra, low reflectance values indicate high absorption in that wavelength region. The pigments have low reflectance values in the visible region. The reflection peak is observed at near 530 nm that can be contributed to the absorption by Cr^{3+} at 470 nm and Co^{2+} at 600 nm, respectively. These absorption bands can be associated with the spin-allowed transition of Cr^{3+} in the octahedral coordination, $^4A_{2g} \rightarrow ^2T_{2g}$ at 470 nm (Eliziário. A. S et al. 2011). The band appearing at 590–600 nm results most probably from a superposition of two bands assigned to the $^4A_{2g} \rightarrow ^4T_{2g}$ transition characteristic of Cr^{3+} and the $^4A_{1g} \rightarrow ^4T_{1g}(P)$ of Co^{2+} transition of the octahedral high spin ion (Mindrua. I et al. 2014). In addition, the reflectance of the samples, x = 0–0.35, decreases with the wavelength increasing in the range of 1100–1600 nm. It can be attributed to the characteristic absorption of Co^{2+} in tetrahedral interstices (Llusar. M et al. 2001).

The reflection spectra of the samples with different amounts of Zn^{2+} in the range of 400–650 nm are shown in Figure 3. The reflectance spectra of the samples increase from x = 0 to 0.4. All the optical absorption detected in the visible region can be attributed to the intra—or inter-atomic transitions involving the Cr^{3+} cation and Co^{2+} cation, because electronic transitions involving Zn^{2+} orbitals are known to occur only at the origin of charge transfers in the UV region (Pailhe. N et al.2008). As mentioned previously, these absorption bands at 470 nm and 590 nm can be associated with the spin-allowed transition of Cr^{3+} in the octahedral coordination. Regarding Co^{2+} in a tetrahedral ligand field, the three spin-allowed transitions are as follows: $^4A_2(F) \rightarrow ^4T_2(F)$, $^4A_2(F) \rightarrow ^4T_1(F)$ and $^4A_2(F) \rightarrow ^4T_1(P)$. The third one is present in the visible region, usually as a triple band around 540 nm, 590 nm and 640 nm, which can be attributed to a Jahn–Teller distortion of the tetrahedral structure (Pailhe. N et al.2008). Therefore, when ZnO (x) is added from a value of 0 to 0.4 to substitute the same quantity of Co_2O_3, the characteristic absorption decreased gradually.

The results of the reflectance measurements on the pigment in the region 600–1100 nm are depicted in Figure 4. Their corresponding NIR reflectance spectra in the wavelength range of 780–1100 nm for all pigment powder samples increase with increasing x. Pigment S8 gives a maximum reflectance of 87.4% compared with a minimum reflectance of 49.1% for S1. In addition, the reflectance increases rapidly from a low level at the red edge,

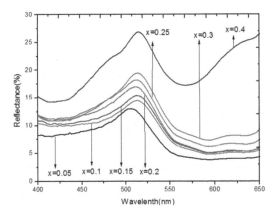

Figure 3. Reflectance spectra of the samples ranging from 400 to 650 nm.

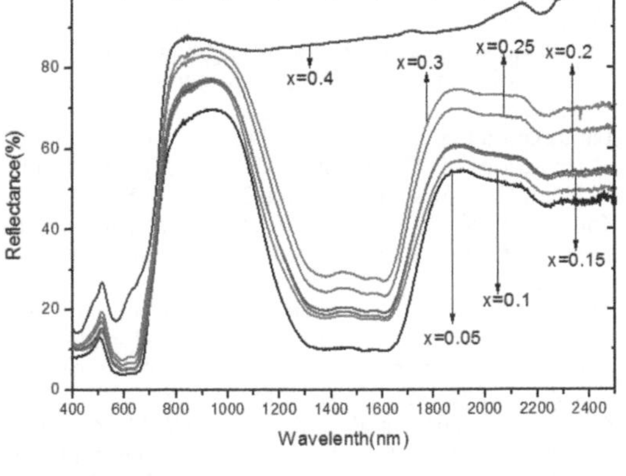

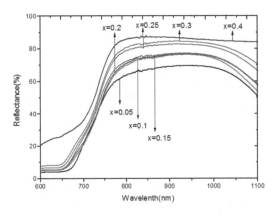

Figure 2. Reflectance spectra of the samples ranging from 400 to 2500 nm.

Figure 4. Reflectance spectra of the samples ranging from 600 to 1100 nm.

Table 2. L*, a* b* color coordinate data of the samples.

Sample	x	L*	a*	b*
1	0.05	29.47	−17.58	−7.02
2	0.1	30.16	−16.63	−5.06
3	0.15	31.80	−17.34	−4.97
4	0.2	32.63	−19.73	−4.24
5	0.25	34.76	−17.62	−3.56
6	0.3	35.05	−15.63	−0.22
7	0.4	44.68	−3.33	5.58

ranging from 680 to 750 nm. Compared with all the samples, a blue shift was also observed in the $Co_{0.4-x}Zn_xCr_{1.58}Ti_{0.02}O_4$ system with the increasing values of x. It may be attributed that Zn^{2+} replaces Co^{2+} in the tetrahedral site. Zn^{2+} have a higher ionic radius, with the bond length of Co-O in the tetrahedral site being changed.

3.3 Color coordinates

The L* a* b* color coordinate data of the samples are summarized in Table 2. It can be seen that the value of L* increased continuously from 27.26 to 44.68 as the ZnO contents (x) increased from 0 to 0.4, corresponding to brightening of the pigment. The change in the color of the pigment is due to the phase component, which leads to the charge in the crystal field around the chromophore ions. S2-S7 produced more green color hues (a* values around -17). Zn^{2+} substituting Co^{2+} in composition results in an increase in the b* value from -5.46 to 8.25, which shows the loss of yellow hue. It can be attributed to the decreasing absorption of Co^{2+} in tetrahedral interstices at 590 nm.

4 CONCLUSION

Green pigments were prepared by the solid-state reaction from Cr_2O_3 and the mixtures of TiO_2, Co_2O_3 and ZnO. The XRD patterns indicate that the composition consists of the corundum phase and the spinel phase. The effect of the zinc oxide content in the original mixture on the spectral reflectance and color was evaluated. The results show that the reflectance of the pigment increases with the increasing content of ZnO (x). Pigment S8 gives a maximum reflectance of 87.4% compared with a minimum reflectance of 49.1% for S1. In addition, a blue shift of the red edge was observed by increasing the values of x. In addition, the influence of ZnO content on color performance was discussed. The value of L* increased continuously

from 27.26 to 44.68 as the ZnO contents increased from 0 to 0.4. The value of a* changes insignificantly with varying x. The decreasing value of b* is related to the decrease in the absorption by Co^{2+} around 590 nm.

REFERENCES

Bendiganavale. A. K. et al. 2008, Infrared Reflective Inorganic Pigments, Recent patents on chemical engineering: 1, 67–79.

Eliziário. A. S et al. 2011, Black and green pigments based on chromium–cobalt spinels, Materials chemistry and physics: 129, 619–624.

Ferrari. C et al. 2013, Design of ceramic tiles with high solar reflectance through the development of a functional engobe, Ceramics international: 39, 9583–9590.

Han. A, 2013, Crystal structure, chromatic and near-infrared reflective properties of iron doped YMnO3 compounds as colored cool pigments, Dyes and pigments: 99, 527–530

Hu. Z. H et al. 2010, research on new green camouflage pigments, Advance materials research: 282–283, 666–669

Jeevanandam.P et al. 2007, near infrared reflectance properties of metal oxide nanoparticles, Journal of physical chemistry C: 111, 1912–1918

Li. P, 2008, a green process to prepare chromic oxide green pigment, Environmental science and technology: 42, 7231–7235.

Llusar. M et al. 2001, Colour analysis of some colbalt-based blue pigments, Journal of the European ceramic society: 21, 1121–1130.

Li. Y. Q et al. 2014, highly solar radiation reflective Cr_2O_3–3TiO2 orange Nano pigment prepared by a polymer-pyrolysis method, ACS sustainable chemistry and engineering: 2, 318–321.

Mindrua. I et al. 2014, Cobalt chromite obtained by thermal decomposition of oxalate coordination compounds, Ceramics international: 40, 15249–15258

Pailhe. N et al. 2008, Correlation between structural features and vis–NIR spectra of α-Fe2O3 hematite and AFe2O4 spinel oxides (A = Mg, Zn), Journal of solid state chemistry: 181, 1040–1047.

Rubeziene. V, et al. 2008, Evaluation of camouflage effectiveness of printed fabrics in visible and near infrared radiation spectral ranges, Materials science: 14, 361–365.

Thongkanluang. T et al. 2011, Preparation and characterization of Cr_2O_3–TiO2–Al2O3–V2O5 green pigment, Ceramics international: 37, 543–548

Zhang.J et al. 2010, near infrared reflectance of the doped Cr_2O_3 pigment, Journal of inorganic materials: 25, 1303–1306.

Zhou. Y. X et al. 2013, Influence of composition on near infrared reflectance properties of M-doped Cr_2O_3 (M = Ti, V) green pigments, journal of the ceramic society of Japan: 122: 311–316.

Zou. J et al. 2014, highly dispersed (Cr, Sb)-co-doped rutile pigments of cool color with high near-infrared reflectance, Dyes and pigments: 109: 113–119.

Energy Science and Applied Technology – Fang (Ed.)
© 2016 Taylor & Francis Group, London, ISBN 978-1-138-02833-3

The development of tensioner for automated fiber placement

Zhao Jin, Zhibin Wang, Hongchun Su, Chun Yuan & Maliya
Chongqing Communication College, Chongqing, China

ABSTRACT: Tensioner can precisely generate and control the tension of fibers, which is one of the important modules for automated fiber placement machine. This paper performs the development periods of tensioner by summing up the development of tensioner in China and other countries. In addition, this paper also aims at giving a perspective in such area in the future.

Keywords: Automated fiber placement, tensioner, development

1 INTRODUCTION

Composite materials are widely used in the aviation field because of the advantages such as high specific modulus, high specific strength, corrosion resistance, excellent fatigue resistance which traditional materials don't have. Carbon fiber reinforced resin matrix composites are most widely used whose utilization rate is about 80% or more (Sheng Lin, 2009). Fiber placement technology which is an important process method in composite molding process, has been widely used. Fiber placement machine is the key equipment of the process, which is the key support in improving the automation of processing the composite materials in aviation industry.

Tension is an important parameter which plays a significant role in fiber placement. The tension size, the tension instability and the tension level change of each layer of the fiber bundles determine the final quality of the product (Chunxiang Wang et al, 2010. Qian Fan et al, 1988). Therefore, the effective control of tesion has important practical significance so as to manufacture high-strength, high-performance composite material components.

2 THE DEVELOPMENT PERIODS OF TENSIONER

The tensioner for fiber placement has roughly experienced three development periods which are mechanical tensioner, electric controlled tensioner and computer controlled tensioner (M. Lossie et al, 1994). The instances of each periods and the characteristics of which will be summarized In the following contents.

The mechanical tensioner achieves the goal of controlling the tension by the mechanical structure.

Yuanmin Zhang proposed an early compensative tensioner the type of which force balanced is presented (Yuanmin Zhang, 1991). The braking torque is generated by the friction of the brake band and the brake wheel, while the adjustment of which is realized by regulating the brake spring deformation. When the decreases of the yarn ball radius bring about the increases of the tension in the course of work, the brake spring deformation will decrease so as to keep the tension constant.

Ryan P. Emerson invents a tensioner including a frame which rotatably supports ten axles arranged in two vertical columns of five axles each (Ryan P. Emerson et al, 2008). Each axle includes a drum on one end for engaging the fiber and a wheel on the opposite end for coupling a belt. The fiber is drawn from the supply spool under a relatively small tensile force threading the number of the drums that can be choosen while the drag torque comes into being on the belt which is controlled by a magnetic particle brake.

Alexander Hamlyn & Yvan Hardy describe that a tensioner is provided in order to exert a tensile stress on the fibers coming from the balls thereby restricting the take-up tension of the fibers at the application roller(Alexander Hamlyn, et al 2009). When fiber is conveyed from said storing means to the application head, ech fiber comes into contact with the cylinders by its two principal surfaces, over substantially identical lengths for each of its principal surfaces. Then the tension comes into being which can be controlled by the frictional force of the cylinders on the fiber.

3 THE ELECTRIC CONTROLLED TENSIONER

The electric controlled tensioner commonly uses the strain-gauged force sensor to detect the real time fiber tension which afterwards feedbacks to the controller that is the analog electronic system composed of the electronic components. The controller then makes the tension keep in a suitable range.

P. Plante shows an electric controlled tensioner is recommended which is comprised of tension sensor, tension controller, electromagnetic brake and fiber rollers (P. Plante, 1982). The real time tension which is detected by the tension sensor will feedback to the tension controller. The controller then compares the feedback value with the set value which will be adjusted so as to export the control signal which is enlarged to drive the yarn ball motors or electromagnetic clutch keeping the tension in a suitable range.

Lloyd G. Miller's invention comprises a support frame, a motor, a dancer arm, a low-friction pneumatic cylinder, a controller and associated circuitry (Lloyd G. Miller, et al 2002). The tension on each fiber is controlled by monitoring its respective dancer arm's position, as being sensed by the angular-position sensor co-located with the dancer arm at its pivot point. When the diversification of tension changes the position of the dancer arm, the microcontroller controls the stepper motor so as to feed or not feed fiber in order to maintain the dancer arm in a preselected position, usually its centre position.

Georges J. Cahuzac et al introduced a tensioner which comprises a roller for controlling the pivoting arm of the poppet driven by the rod of a pneumatic jack, the arm being elastically brought back by a spring (Georges J. Cahuzac, et al, 1997). In order to enable the roving to be reeled under a high tension, the jack is fed with compressed air under an adjustable pressure. It presses the roller onto the arm of the poppet so that the potentiometer gives the order to the motor to brake and no longer pull until the torque due to the tension of the roving balances the torque exerted by the jack on the poppet.

The fiber bundles of electric controlled tensioner can go back to the yarn easily and the tension can be regulated automatically, of which the control accuracy is higher than the mechanical tension, but the tension of the fibers is subject to the interference of the external environment.

The computer controlled tensioner By the appearance of the high performance microprocessor with the development of electronic technology, the computer controlled tensioner comes into being.

P. D. Mathur describes the entire process of control system designed for a prototype tape spooler that is used for conducting research on the dynamics of high-speed and low-tension tape transport. The system is similar to the tensioner for AFP.The tape speed and tension are regulated by controlling the torque applied to the supply and the take-up reels(Priyadarshee D. Mathur, et al, 1998). Adaptive tension ripple cancellation and fault detection and compensation are demonstrated.

Shengle Ren, et al introduces a new tensioner. The system use PLC singlechip as the main controller, low friction air cylinder and AC frequency conversion motor as the executing element. In the system, the change of fiber tension is detected by wobbly lever and angle sensor which is used to replace the radius check device and the tension sensor (Shengle Ren, et al, 2007).

Xingjian Chen et al proposed a novel Proportional-Integral-Derivative (PID) self-adaptive controller by fuzzy adaptive mechanism for the tension control of filament winding system is presented using motion controller (PMAC) as core and DC torque motor as actuator. PMAC collect tension data and control the motor with tuned PID parameters which is tuned by upper computer. A comparative study shows that the overshoot and the fluctuation of tension decrease considerably (Xingjian Chen, et al 2007).

Because of the high calculation speed, high precision and high reliability of the computer, the complexity of hardware circuit of the electronic control system reduces to make the system simplify and more reliable.

4 A NEW KIND OF TENSIONER

Drawing on the experience of the invention and research before, a new kind of tensioner for fiber placement is designed. This tensioner uses AC servo motor and the cylinder as actuators, with the displacement measuring encoder as feedback devices. The mechanical structure is shown in Figure 1.

When the tensioner works, the tow is unwound from the spool. The traction effect of the traction

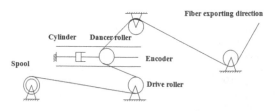

Figure 1.   The mechanical picture of the tensioner.

side generates tension while the servo motor provides resistance to unwound roller, so that the fibers will be pulled. Under thrust of the cylinder, the tension could be set by adjusting cylinder pressure so as to keep the dancer roller on a equilibrium position. When the fiber moves steadily the dancer roller will be on the balanced position. When the tension changes due to the perturbations, the dancer roller will swing at the same time so that the encoder can gather the signal as the feedback signal for the controller. Then the tensioner will adjust the rotational speed of the servo motor in order to achieve purpose of keeping the tension constant.

As a result of the modeling analysis, the tensioner can achieve the function that fiber placement machine needs.

5 CONCLUSIONS

1. Tensioner has gradually developed from mechanical tensioner without control to computer controlled tensioner of closed-loop control with feedback device.
2. With the rapid development of fiber placement technology, it is foreseeable that the development of the tensioner in the future will tend to be digital, integrated, modular, miniaturized, dedicated and high-precision to adapt the diverse needs of tensioner.

REFERENCES

Alexander Hamlyn, Yvan Hardy 2009. Fiber Application Machine, United States Pantent 0229760.
Chunxiang Wang, Yongzhang Wang, Hua Lu, 2010. A Research of the Precise Tension Control System and Control Precision. Chinese Journal of Scientific Instrument, vol. 21, pp. 407–409.
Georges J. Cahuzac et al. Machine for the Simultaneous Laying down and Winding of a Plurality of Individual Fiber Rovings, United States Pantent 5645677. 1997.
Lloyd G. Miller, Keith G. Shupe. Position-controlled tensioner system, United States Pantent 6491773. 2002.
M. Lossie, H. V. Brussel. Design Principles in Filament winding. Composites Manufacturing. vol. 5, 1994, pp. 5–13.
P. Plante. Applying Electromagnetic Cluthes and Brakes. Instruments & Control System. vol. 4, 1982, pp. 31–33.
Priyadarshee D. Mathur, William C. Messner. Controller Development for a Prototype High-Speed Low-Tension Tape Transport. IEEE Transactions on Control Systems Technology. vol. 6, 1998, pp. 534–542.
Qian Fan, Guowei Li. A Research on the Tension Control Effected in a Microcomputer-Controlled Filament Winding Machine. Journal of Harbin Institute of Technology, vol. 4, 1988, pp. 89–95.
Ryan P. Emerson, Robert P. Kaste. Fiber Tensioning Device, United States Patent 0272221. 2008.
Sheng Lin. ATL/AFP-The key Machine for Manufacturing of Modern Large Airplane. WMEM, vol. 4, 2010, pp. 99–104.
Shengle Ren, et al. The research of new tension system controlled by pneumatic-electricity. Machinery Design & Manufacture, vol. 6, 2007, pp. 4–5.
Xingjian Chen et al. Fuzzy adaptive PI D tension control for filament winding. Manufacturing Automation, vol. 29, 2007, pp. 62–64.
Yuanmin Zhang, 1991. The FRP mechanism and equipment. China Building Industry Press, pp. 347–348.

Energy Science and Applied Technology – Fang (Ed.)
© 2016 Taylor & Francis Group, London, ISBN 978-1-138-02833-3

Erbium-doped Na$_2$O-ZnO-MoO$_3$-TeO$_2$ glasses

Jia Li, S.Q. Man, Q.R. Zhao & Ning Zhang
School of Physics and Electronic Information, Yunnan Normal University, Kunming, China

ABSTRACT: Er^{3+}-doped tellurite glasses (Na$_2$O-ZnO-MoO$_3$-TeO$_2$ or NZMT) were fabricated and characterized under certain conditions. The J-O spectral line intensity parameters Ω ($\Omega_2 = 4.88 \times 10^{-20}$ cm^2, $\Omega_4 = 1.18 \times 10^{-20}$ cm^2, $\Omega_6 = 0.86 \times 10^{-20}$ cm^2) and some other significant properties were calculated by using the Judd–Ofelt theory and the experimental oscillator strengths. The fluorescence branching ratio and the spontaneous transition probability were measured. The emission cross section of this glass sample was acquired by Judd–Ofelt and McCumber theories. At about 1.53-µm wavelength window, the glass samples had the emission peak of the $^4I_{13/2} \rightarrow {}^4I_{15/2}$ transition. The FWHM bandwidth was ~82 nm. The lifetime of the $^4I_{13/2}$ level was 4.79 ms. The emission cross section was estimated to be about 8.51×10^{-21} cm^2. The results demonstrated that the NZMT glass is an excellent host material for applications in fiber amplification.

Keywords: Judd–Ofelt theory, Rare-earth ion, Optical transitions, Fluorescence spectrum, Tellurite glasses

1 INTRODUCTION

Rare-earth Er^{3+}-doped tellurite glasses have been studied for use as laser work materials and fiber amplifiers (S.Q. Man, 2006). With the rapid development of information processing and other data transmission services, the long-distance optical fiber transmission system needs greater communication capacity and much larger bandwidth. The erbium-doped silica fiber amplifier is unable to meet the development requirements of a high-speed and large-capacity communication transmission. In order to get a new kind of fiber amplifier with a wide bandwidth and a flat gain, it is necessary to find new active fiber materials.

Recently, phosphate glass, fluoride glass and tellurite glass have been studied as excellent host materials. The gain spectrum of tellurite glass is much wider than that of existing EDFAs, and the full width at half-maximum of this new kind of tellurite glass is ~82 nm. In addition, this kind of glass was characterized by a wide infrared transmission area, a high coefficient of photoelectric coupling, a good anticorrosion property, low phonon energy, and a high refractive index. So, it was considered to be an ideal host material.

In this study, we discussed some properties of erbium-doped NZMT glasses with a molar composition of 10%Na$_2$O-10% ZnO-10%MoO$_3$_70%TeO$_2$. The experimental oscillator strengths, the spontaneous transition probability, the fluorescence branching ratio and other parameters were found by using the J-O theory (S.T. Tanabe, 1992). The stimulated emission cross section was measured by using some of the aforementioned theories. All results demonstrated that the NZMT glasses are an outstanding host material for application in fiber amplification.

2 EXPERIMENTS

Erbium—doped NZMT glass samples were prepared from anhydrous sodium carbonate (Na$_2$CO$_3$), zinc oxide (ZnO), molybdenum oxide (MoO$_3$) and tellurium oxide (TeO$_2$) powders. All these powders were obtained from Chinese domestic chemical company. The glass samples examined in this paper had similar compositions: 10 Na$_2$O-10ZnO-10MoO$_3$_70TeO$_2$. In addition, 1wt% Er^{3+} was added to all these glass samples. At first, we placed these powders in a platinum crucible and released some undesired gas at 400 degrees Celsius for 30 minutes. Then, they were melted at about 750 degrees Celsius. Finally, the glass samples were placed in an annealing resistance furnace at about 250 degrees Celsius for more than 8 hours.

The density of this sample was 5.31 g/cm^3. By using the Metricon 2010 prism coupler technique at three wavelengths, the refractive indices were obtained. The refractive indices of the NZMT glasses were 1.950 and 2.011 at wavelengths of 633 nm and 1550 nm, respectively.

Using the Cary 5000 double-bean spectrophotometer, the absorption spectra were included between 350 nm and 1700 nm. Using the SPEX 500M monochromatic, we detected the fluorescence spectra. This fluorescence lifetime of the ground state of the erbium ion was calculated by using a 980 nm laser diode light pulse and an InGaAs photo detector by a semiconductor 980 nm laser.

3 RESULTS AND DISCUSSION

By using the J–O theory, we analyzed the radiative transition (G.S. Ofelt, 1962). S_{ed}, which is the oscillator strength of the electric dipole intensity of spectral lines, was expressed as follows:

$$S_{ed}(J;J') = \sum_{t=2,4,6} \Omega_t \left\langle \alpha SL, J \| U^{(t)} \| \alpha' S' L', J' \right\rangle^2 \quad (1)$$

The absorption spectrum of the new kind of NZMT glasses doped with 1wt% Er^{3+} is shown in Figure 1. The peaks are labeled in this figure.

Ω_t(t = 2, 4, 6) are the Judd–Ofelt parameters; J is the initial level; and J' is the final state. The value of the element of matrix ($<\|U^{(t)}\|>$) depends on the difference of the rare-earth ion itself. It is entirely unrelated to the host materials.

The experimental oscillator strengths f_{exp} include two aspects: the electric dipole and magnetic dipole absorption transitions:

$$f_{exp} = \frac{2303mc^2}{N_A \pi e^2} \int \varepsilon(v)dv = 4.318 \times 10^{-9} \int \varepsilon(v)dv \quad (2)$$

where m is the mass of the electron and $\varepsilon(v)$ is the molar absorption coefficient.

As a result of the electric dipole and magnetic dipole transitions, the integral and the absorption coefficient is proportional to the experimental oscillator strength. So, the expression is given by (W.T. Carnall, 1968)

$$f_{md} = nf', \quad (3)$$

$$f(J,J') = \frac{8\pi^2 mc(n^2+2)^2}{3h\bar{\lambda}(2J+1)9n} S_{ed}(J;J'), \quad (4)$$

where n is the refractive index. The J–O intensity parameters Ω_t were calculated as $\Omega_t = 4.88 \times 10^{-20}$ cm^2, 1.18×10^{-20} cm^2 and 0.86×10^{-20} cm^2. These values were closer to those reported by McDougall et al (J. McDougall, 1996).

Table 1 shows the absorption S_{ed} for each transition given in Figure 1, as well as the J–O intensity parameters of this new kind of glass sample. According to Jørgensen et al., the value of Ω_t associated with the material properties of the NZMT glass (C. K. Jorgensen, 1983).

The fluorescence spectrum of the $^4I_{13/2} \rightarrow {}^4I_{15/2}$ transition for the Na_2O-ZnO-MoO_3-TeO_2 glass is shown in Figure 2. The bandwidth of the $^4I_{13/2} \rightarrow {}^4I_{15/2}$ transition is ~82 nm. The value of a silica-based glass is much narrower than this.

Some important radiative properties were found by using Ω_t. The spontaneous transitions probability, A, can be obtained by using the following equation (R. R. Jacobs, 1976):

$$A[S,L]J : (S',L')J'$$

$$= \frac{64\pi^4 e^2 n}{3h(2J+1)\bar{\lambda}^3} \left[\frac{(n^2+2)^2}{9} \right]$$

$$\sum_{t=2,4,6} \Omega_t \left\langle (S,L)J \| U^{(t)} \| (S',L')J' \right\rangle^2 \quad (5)$$

Table 1. Measured and calculated oscillator strengths, the electric-dipole line strengths Sed for some transitions, and the Judd–Ofelt intensity parameters of Er3+-doped tellurite glasses.

Absorption	Energy (cm^{-1})	f_{exp} (10^{-6})	f_{cal} (10^{-6})	f_{md} (10^{-6})	S_{ed} (10^{-20})
$^4I_{15/2} \rightarrow {}^4I_{13/2}$	6532	1.8415	1.2202	0.5938	1.4976
$^4I_{15/2} \rightarrow {}^4I_{11/2}$	10256	0.6057	0.6385		0.4533
$^4I_{15/2} \rightarrow {}^4I_{9/2}$	12500	0.3345	0.3529		0.2019
$^4I_{15/2} \rightarrow {}^4F_{9/2}$	15314	2.1579	2.1450		1.0351
$^4I_{15/2} \rightarrow {}^4S_{3/2}$	18349	0.4609	0.4918		0.1782
$^4I_{15/2} \rightarrow {}^2H_{11/2}$	19157	10.9180	11.0309		4.0019
$^4I_{15/2} \rightarrow {}^2F_{7/2}$	20492	2.0153	2.1164		0.6782
$^4I_{15/2} \rightarrow {}^2F_{5/2,3/2}$	22124	0.8626	0.9899		0.2626
$^4I_{15/2} \rightarrow {}^2H_{9/2}$	24570	0.7374	0.8205		0.1944
$^4I_{15/2} \rightarrow {}^4G_{11/2}$	26316	21.9116	21.7777		5.2350
Ω_2 (10^{-20} cm^2)			4.88		
Ω_4 (10^{-20} cm^2)			1.18		
Ω_6 (10^{-20} cm^2)			0.86		
Root mean square deviation (10^{-6})			0.090		

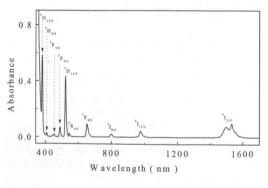

Figure 1. Absorption spectra of the Er3+-doped tellurite glass.

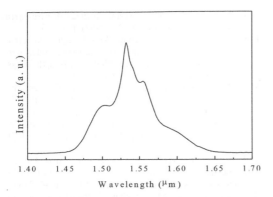

Figure 2. Fluorescence spectra of the 1 wt% Er3± doped tellurite glass.

Table 2. Predicted spontaneous emission probabilities of Er3± doped tellurite glasses.

Transition	Energy (cm⁻¹)	Aed (⁻¹)	A_{md} (s⁻¹)	A_{total} (s⁻¹)	β_{ij}	τ_{rad} (ms)
$^4I_{13/2} \rightarrow {}^4I_{15/2}$	6525	149	64	213	1	4.69
$^4I_{11/2} \rightarrow {}^4I_{13/2}$	3640	22	14	36	0.14	3.94
$^4I_{15/2}$	10163	218		218	0.86	
$^4I_{9/2} \rightarrow {}^4I_{11/2}$	2252	2	2	4	0.02	4.29
$^4I_{13/2}$	5882	57		57	0.24	
$^4I_{15/2}$	12121	172		172	0.74	
$^4F_{9/2} \rightarrow {}^4I_{9/2}$	2880	6	5	11	<0.01	0.41
$^4I_{11/2}$	5110	95	11	106	0.04	
$^4I_{13/2}$	8750	107		107	0.04	
$^4I_{15/2}$	15240	2203		2203	0.91	
$^4S_{3/2} \rightarrow {}^4F_{9/2}$	3125	1		1	0	0.37
$^4I_{9/2}$	6080	87		87	0.03	
$^4I_{11/2}$	8240	52		52	0.02	
$^4I_{13/2}$	11870	681		681	0.25	
$^4I_{15/2}$	18315	1901		1901	0.70	

The value of the matrix elements ($\langle\langle \|U^{(t)}\| \rangle\rangle$) was obtained from erbium ions doped in lanthanum fluoride (LaF_3) (M. J. Weber, 1968).

Because the methods of computation of spectral line intensity between spontaneous radiation and stimulated radiation transition strength are similar, the fluorescence branching ratio (β) and the radiative lifetime (τ_{rad}) are expressed as

$$\beta[(S,L)J \cdot (S',L')J']$$
$$= \frac{A[(S,L)J : (S',L')J']}{\sum\limits_{S',L',J'} A[(S,L)J : (S',L')J']} S_{ed}(J;J'), \quad (6)$$

$$\tau_{rad} = \left\{ \sum\limits_{S',L',J'} A[(S,L)J : (S',L')J'] \right\}^{-1} = A_{total}^{-1} \quad (7)$$

Table 2 shows the spontaneous transition probabilities, the branching ratios and the calculated lifetimes of optical transitions in Er^{3+}-doped tellurite glasses.

For a three-level gain system such as Er^{3+} at 1500 nm, the stimulated emission and the absorption cross sections play important roles in weighing the performance of a host material. Cross sections quantify the ability of an ion to absorb and emit light. The emission cross section from the absorption spectrum is measured by using the McCumber theory (D. E. Mc Cumber, 1964).

In the theory of McCumber, the absorption cross section spectrum $\sigma_a(v)$ is related to the stimulated emission cross section spectrum $\sigma_e(v)$ for transitions between the same two state levels by

$$\sigma_a(v) = \sigma_e(v)\exp\left[(hv - \varepsilon)/kT \right], \quad (8)$$

Where k is Boltzmann's constant; T is the temperature; and ε is the energy that varies directly

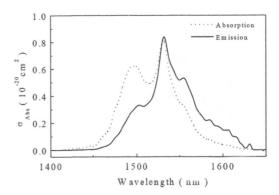

Figure 3. Measured absorption and calculated emission cross sections of the Er3± doped tellurite glass.

with temperature (W. J. Miniscalco, 1991). Figure 3 shows the calculated emission cross section measured by using the McCumber theory.

4 CONCLUSION

The Er^{3+}-doped NZMT glass was frabricated and characterized under some certain conditions. The J–O spectral line intensity parameters $\Omega_t = 4.88 \times 10^{-20}$ cm², 1.18×10^{-20} cm², 0.86×10^{-20} cm² (t = 2, 4, 6) and some other significant properties were calculated by using the J-O theory. The spontaneous transition probability, the fluorescence branching ratio and the emission cross section of this glass sample were obtained by the Judd–Ofelt and McCumber theories. The FWHM of the emission from the first excited state to the ground state at about the 1.55μm wavelength window was ~82 nm. The value of erbium-doped Al/P silicate glasses was about 43.3 nm, which was much narrower than the value of this new kind of NZMT glasses. The fluorescence lifetime of the $^4I_{13/2}$ state was 4.79 ms, and the quantum efficiency was ~100%. The emission cross section was calculated as about 8.5×10^{-21} cm^2. It was larger than that of the Al/P silicate glasses (5.7×10^{-21} cm^2) and the silicate glasses (7.27×10^{-21} cm^2). The results of this study indicate that this new kind of Er^{3+}-doped tellurite glass is an outstanding host material for application in fiber amplification.

ACKNOWLEDGMENT

This work was supported by the National Natural Science Foundation of China (NSFC) (Grant No. 21171072 and Grant No. 21361028).

REFERENCES

C.K. Jorgensen and R. Reisfeld. Judd-Ofelt parameters and chemical bonding[J]: J. Less-Common Metals, Vol. 93 (1983), p.107–110.

D.E. McCumber. Theory of phonon-terminated optical masers[J]: Phys. Rev, Vol. 134 (1964), p.A299.

G.S. Ofelt. Intensities of crystal spectra of rare-earth ions[J]: J. Chem. Phys, Vol. 37 (1962), p.511–515.

J. McDougall, D.B. Hollis and M.J.P. Payne. Spectroscopic properties of Er3+ in fluorozirconate, germanate, tellurite and phosphate glasses[J]: Phys. Chem. Glasses, Vol. 37 (1996), p.73–79.

M.J. Weber. Spontaneous emission probabilities and quantum efficiencies for excited states of Pr^{3+} in LaF$_3$[J]: J. Chem. Phys, Vol. 48 (1968), p.4774–4779.

M.J. Weber. Probabilities for Radiative and Nonradiative Decay of Er^{3+} in LaF^{3+}[J]: Phys. Rev, Vol. 157 (1967), p.262–276.

M.J. Weber, J.D. Myers, and D.H. Blackburn. Optical properties of Nd^{3+} in tellurite and phosphotellurite glasses[J]: J. Appl. Phys, Vol. 52 (1981), p.2944–2949.

R.R. Jacobs and M.J. Weber. Dependence of the $^4F_{3/2} \rightarrow ^4I_{11/2}$ induced-emission cross section for Nd^{3+} on glass composition[J]: J. Quantum Electron, Vol. 12 (1976), p.103–111.

S.Q. Man, H.L. Zhang, Y.L. Liu, J.X. Meng. Energy transfer in Pr^{3+}/Yb^{3+} codoped tellurite glasses[J]: J. Optical Materials, Vol. 45 (2006), p.334–337

S.T. Tanabe, T. Ohyagi, N. Soga. Compositional dependence of Judd-Ofelt parameters of Er^{3+} ions in alkali-metal borate glasses[J]: J. Phts Rev B, Vol. 46 (1992) No.6, p.3305–3310.

W.T. Carnall, P.R. Fields, K. Ragnak. Electronic energy levels in the Trivalent Lanthanide Aquo ions. i. Pr^{3+}, Nd^{3+}, Pm^{3+}, Sm^{3+}, Dy^{3+}, Ho^{3+}, Er^{3+}, and Tm^{3+}[J]: J. Chem. Phys, Vol. 49 (1968), p.4424–4442.

W.J. Miniscalco, R.S. Quimby. General procedure for the analysis of Er^{3+} cross sections[J]: Opt. Lett, Vol. 16 (1991), p.258–261.

Energy Science and Applied Technology – Fang (Ed.)
© 2016 Taylor & Francis Group, London, ISBN 978-1-138-02833-3

Energy transfer of the Airy pulse during nonlinear propagation in the presence of high-order effects

Nanjing Mei & Xiquan Fu
College of Computer Science and Electronic Engineering, Hunan University, Changsha, China

ABSTRACT: It is important to study the energy transfer of Finite Energy Airy Pulse (FEAP) under the action of the Third-Order Dispersion (TOD), self-steepening (SS) and Raman effects. It is demonstrated that the influences of TOD and Raman slow down the main lobe energy of FEAP transfer to the secondary lobe during the FEAP propagation in nonlinear media. During the input low power Airy pulse, the SS effect has a little impact on the energy transfer of the FEAP. The SS effect deaccelerates the movement toward the trailing-edge of the FEAP with the increase in initial power. Meanwhile, the main lobe will form a soliton in the regime of high pump power.

Keywords: Airy pulses; energy transfer; third-order dispersion; Raman effects; self-steepening

1 INTRODUCTION

Recently, a new class of non-diffracting pulses in free space has been observed experimentally, which is known as Airy pulses. Airy pulses can resist dispersion and parabolic trajectory propagation in the linear and nonlinear media. Airy pulses under the condition of the interference or partial shade has the property to heal itself (Berry M. V. & Balazs. N. L. 1979; Siviloglou G. A. & Demetrios N. C. 2007; Besieris I. M. & Shaarawi A. M. 2007). These unusual features bring about tremendous potential applications of the FEAB in many areas. However, the Airy pulse contains infinite energy, which is physically unrealizable in reality. Experiments have found that the Airy function has a decaying coefficient. To generate the Airy pulse, a common experimental method is used that imposes a cubic spectral phase on an input Gaussian pulse. Recently, the propagation dynamics of temporal Airy pulses have been numerically studied in the dispersive medium (Wang S. et al. 2014). These Airy pulses can only maintain their profiles over several dispersive distances. Properties of temporal Airy pulses have been studied from the linear to nonlinear regimes. Since then, the unique features of Airy pulses have been used for many applications such as supercontinuum (Ament C. et al. 2011) and plasmonic routing (Liu W. et al. 2011; Zhang P. et al. 2011). In the nonlinear conditions, the dynamic of Airy pulses under the action of dispersion and nonlinearity has been extensively researched

(Fattal Y. et al. 2011; Hu Y. et al. 2013). Under the strong nonlinearity conditions, one of the most important features is induced (solitary wave). The formation of the soliton is due to the balance of the dispersion and nonlinear Self-Phase Modulation (SPM).

When the input pulse width is less than 100 femtoseconds, the higher-order effects should be taken into consideration to accurately express the pulse propagation in nonlinear media. Some research related to the spread of the pulse with these higher-order effects has been made. The third-order dispersion effect on pulse propagation has been studied (Besieris I. M. & Shaarawi A. M. 2008; Driben R. et al. 2013). For ultrashort Airy pulses, it is necessary to consider all the higher-order effects, because even small third-order dispersion and Raman parameters can cause an obvious change in the energy transfer of the lobes of the Airy pulses (Agrawal G. P. 2001; Bourkoff E. et al 1987). In this paper, we are interested in the energy transfer between the lobes of Airy pulses during nonlinear propagation under the individual influence of TOD, Raman and self-steepening effects.

This paper is divided into four parts. In Section 2, we introduce the propagation model that includes the high-order dispersion and high-order nonlinear Schrödinger equation. Airy pulses are analyzed, respectively, in the high-order dispersion and high order nonlinear effect under the action of energy transfer described in Section 3. The last part provides a summary of all the above-mentioned results and a conclusion.

2 THEORETICAL PROPAGATION MODEL

During the FEAP propagation in the dispersive-nonlinear medium, the evolution of the pulse is governed by the simplified Nonlinear Schrödinger Equation (NLSE).

$$\frac{\partial A}{\partial z} + i\frac{\beta_2}{2}\frac{\partial^2 A}{\partial \tau^2} - \frac{\beta_3}{6}\frac{\partial^3 A}{\partial \tau^3}$$
$$= i\gamma\left[|A|^2 A + \frac{i}{\omega_0}\frac{\partial}{\partial \tau}\left(|A|^2 A\right) - \tau_R A\frac{\partial |A|^2}{\partial \tau}\right], \quad (1)$$

where A is the amplitude of the field; z is the pulse propagation distance; τ is the delay time having the following relationship with the group velocity ($\tau = t - z/v_g$), where t is the time of pulse and v_g is the group velocity; β_2 is the Group Velocity Dispersion (GVD); β_3 is the third-order dispersion coefficient; τ_R is the decay time of Raman effects; ω_0^{-1} is the self-steeping; and γ is the nonlinear coefficient. In order to achieve the dimensionless processing, the normalized equation for NLSE can be written as

$$\frac{\partial U}{\partial Z} + i\frac{\text{sgn}(\beta_2)}{2}\frac{\partial^2 U}{\partial T^2} - \text{sgn}(\beta_3)\frac{\delta_3}{6}\frac{\partial^3 U}{\partial T^3}$$
$$= iN^2\left[s\frac{\partial}{\partial T}\left(|U|^2 U\right) + iT_R U\frac{\partial |U|^2}{\partial T}\right] + iN^2|U|^2 U, \quad (2)$$

where $\delta_3 = L_{D2}/L_{D3}$, $N^2 = L_{D2}/L_{NL}$, $s = \left(\omega_0 \tau_p\right)^{-1}$ and $T_R = \tau_R/\tau_p$.

The input Airy pulse is expressed as

$$A(Z = 0, T) = \text{Ai}(T)\exp(aT) \quad (3)$$

where a is the decay coefficient. The Airy pulses have a finite energy in Eq. (3). Based on Perseval's theorem, the total energy of the FEAP is given by

$$\int_{-\infty}^{\infty} ds\,|A(T, Z = 0)|^2 = \sqrt{\frac{1}{8\pi a}}\exp\left(\frac{2a^3}{3}\right). \quad (4)$$

The energy is normalized by the total input pulse energy in Eq. (4). Since the main lobe and the secondary lobe of the FEAP carry almost the pulse's energy, we only consider the energy of the main lobe and the secondary lobe to study energy transfer of the FEAP. The main lobe relative energy (S_1) and the secondary relative energy (S_2) are defined as (c0, c1 and c2; see Figure 1):

$$S_1 = \frac{\int_{c_1}^{c_0}|A(Z, T)|^2 ds}{\int_{-\infty}^{\infty}|A(Z = 0, T)|^2 ds}, \quad S_2 = \frac{\int_{c_2}^{c_1}|A(Z, T)|^2 ds}{\int_{-\infty}^{\infty}|A(Z = 0, T)|^2 ds} \quad (5)$$

3 NUMERICAL SIMULATION

3.1 The energy transfer of Airy pulse propagation in the presence of TOD

Now we consider the TOD effects while neglecting the nonlinear effects. In linear propagation, we only consider the anomalous dispersion ($|\beta_2| = T_0 = \gamma = 1, \text{sgn}(\beta_2) = -1$), where the energy of the main lobe reduces rapidly with increasing propagation distance, leading to a worst performance of FEAP transmission in the optical fiber. After a certain distance, the GVD radiation intensity is a Gaussian distribution. Figure 2 (a) and (b) shows the temporal and the main lobe relative energy dynamic evolutions of the FEAP, with $a = 0.1$. As it can be seen, during the FEAP propagation in the presence of both GVD and TOD, when GVD is positive, the accelerating energy of the main lobe decreases as the propagation distance increases. If the third-order dispersion is negative, the energy

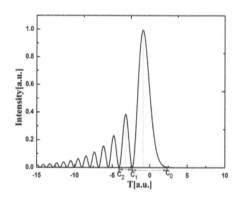

Figure 1.   Defining the initial position for the lobes of the Airy pulse to calculate the relative energy when a = 0.1.

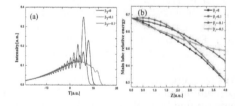

Figure 2.   (a) Intensity distributions of the FEAP as a function of time at distance $Z = 6$, (b) the main lobe energy evolution as a function of the propagation distance with different TOD parameters: $\beta_3 = 0$ (Black line); $\beta_3 = 0.1$ (Red line); $\beta_3 = -0.1$ (Blue line); $\beta_3 = -0.5$ (Green line), respectively.

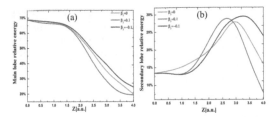

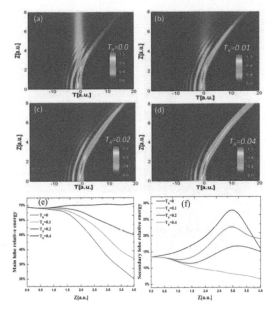

Figure 3. (a) Relative energy of the main lobe and (b) relative energy of the secondary lobe as a function of the propagation distance with different TOD values: $\beta_3 = 0$ (Green line); $\beta_3 = 0.1$ (Red line); $\beta_3 = -0.1$ (Blue line).

evolutions of Airy pulses are different, the energy of the main lobe decreases slowly with the decrease in the third-order dispersion value, and the FEAP keep their profiles over several dispersive distances with a larger TOD parameter.

Next, we investigate the influence of TOD during the pulse propagation in nonlinear media, and discuss the energy evolution between the lobes of pulse during propagation in the presence of dispersion and nonlinearity. In order to investigate the effects of both TOD and nonlinearity on its evolution, we varied the TOD parameter with a low initial launched power. It is well known that soliton formation due to the combination of GVD and SPM effects, which the energy transfer from the main lobe to the next lobe when propagating in the weak nonlinear regime, is larger than that in linear conditions. Figure 3 shows pulse evolution examples for different values. From Figure 3, it can be found that the positive TOD parameter accelerates the main lobe energy transfer to the secondary lobe. If the TOD parameter is negative, it can be seen that under the effect of TOD, the FEAP retains its shape for a longer distance. From Figure 3, it can be seen that the stronger the value of TOD is, the lower the amount of the main lobe energy will be. For the same propagation distance, the corresponding secondary lobe energy increases first and then decreases at different values of TOD. We found that TOD can delay the main lobe energy flow to the secondary lobe and the pulse can propagate for a longer distance. The larger value of TOD can retain the pulse profile better than the small value of TOD.

3.2 *The energy transfer of Airy pulse during nonlinear propagation in the presence of Raman effects*

We know that pulse Raman effects have an important influence on pulse propagation. It was first observed by Mitschke et al. in 1986. To study the effects of Raman during nonlinear propagation of the FEAP, we only add the Raman effects term during nonlinear propagation. It is well known that the energy of Airy

Figure 4. Temporal evolution of the FEAP in the nonlinear propagation medium with different T_R values: (a) $T_R = 0.0$; (b) $T_R = 0.1$; (c) $T_R = 0.2$; and (d) $T_R = 0.4$. (e) Relative energy of the main lobe. (f) Relative energy of the secondary lobe as a function of the propagation distance with different T_R values.

pulse during nonlinear propagation is separated into two parts: one is used to form the soliton pulse and the other one continues to accelerate. Figure 4(a–d) shows the time evolution of Airy pulse with different T_R values during the nonlinear propagation. The Raman effects play a dominant role in propagation as the T_R values increase. It is interesting to find from Figure 4 (e, f) that the influence of Raman effects prevents the energy of the main lobe transfer to the next lobe, even change the direction of energy transfer between the lobes of the pulse in the nonlinear conditions. With increasing T_R values, most of the energy is transferred to the main lobe, and the main lobe can be propagated for a longer distance.

3.3 *The energy transfer of Airy pulse in the presence of self-steepening*

In the previous section, we analyzed the pulse energy transfer distribution under the TOD and Raman effects, respectively. We continue to explore the energy evolution of the FEAP in the presence of both Raman effects and self-steepening. We set $T_R = 0.1$ and $N = 2$, the TOD effect is negligible because the wavelength of the pulse is not near the zero-dispersion wavelength. However, in this case, Raman and self-steepening terms should not be neglected, it is necessary to take into account at

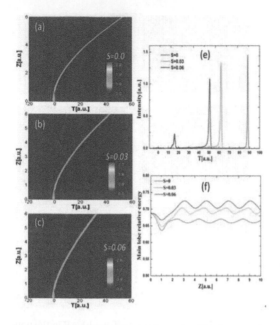

Figure 5. (a–c) Evolution of intensity distributions in different SS values, (e) output pulse intensity at the propagation distance Z = 8, and evolution of the FEAP in time for S = 0 (red line); S = 0.03 (green line); S = 0.06 (blue line), respectively. (f) Main lobe relative energy of the FEAP with different self-steepening (SS) values.

the same time. Meanwhile, at a low launched power, the influence of self-steepening for the pulse energy transfer is not obvious, thus we set the SS parameter within the scope of 0–0.06 at a high launched power. Figure 5(a–c) shows the intensity distributions of the Airy pulse under different SS parameters. We can note that the SS effect deaccelerates the main lobe shift toward the trailing side of the pulse. The larger the SS values, the greater the pulse width of the main lobe. The corresponding secondary lobe has the same behavior change under the condition of different SS values. As shown in Figure 5(e, f), the energy of the main lobe during propagation has periodic oscillation, and can propagate for a longer distance, similar to the behavior of the soliton. The main lobe keeps its energy, and will not flow to the secondary lobe of the pulse. We found that the self-steepening effect is contrary to the Raman effects: the influence of self-steepening reduces the degree of the main lobe deviation pulse, and at the same time transfers part of the pulse energy to the main lobe.

4 CONCLUSIONS

We numerically and theoretically study the rule of Airy pulse energy transfer in the nonlinear medium by using the nonlinear Schrödinger equation. The paper proposed the Airy pulses solution for NLSE including high-order dispersion and high-order nonlinearity. The positive TOD accelerates the energy of the main lobe transfer to the secondary lobe, the negative TOD can delay the main lobe energy flow to the secondary lobe and the pulse can propagate for a longer distance. The Raman effects prevent the action of the main lobe energy flow to the secondary lobe, most energy can transferred to the main lobe, and the main lobe carries most of the energy of pulse with increasing T_R values. Finally, we show the energy evolution of the FEAP under both Raman and self-steepening effects. We found that the self-steepening effect is contrary to the Raman effects The SS effect deaccelerates the main lobe shift toward the back edge of the pulse, and the main lobe evolved into a soliton. We believe that these results will be useful for many potential applications of the FEAP.

REFERENCES

Ament, C. Polynkin, P. &Moloney, J. V. 2011. Su percontinuum generation with femtosecond self-healing airy pulses. Phys. Rev. Lett. 107: 243901.
Agrawal, G. P. 2001. Nonlinear Fiber Optics, Optics and Photonics. Academic Press, London.
Berry,M.V.& Balazs, N. L. 1979. Nonspreading wave packets. Am. J. Phys. 47: 264–26.
Besieris, I. M. & Shaarawi, A. M. 2007. A note on an accelerating finite energy Airy beam. Opt. Lett. 32: 2447–2449.
Besieris, I. M. & Shaarawi, A. M. 2008. Accelerating airy wave packets in the presence of quadratic and cubic dispersion. Phys. Rev. E: 78.
Bourkoff, E. Zhao, W. & Joseph, R. I. 1987. Evolution of femtosecond pulses in single-mode fibers having higher-order nonlinearity and dispersion, Opt. Lett. 12: 272–274.
Driben, R. 2013. Inversion and tight focusing of Airy pulses under the action of third-order dispersion. Opt. Lett. 38:2499.
Fattal, Y. Rudnick, A. & Marom, D. M. 2011. Soliton shedding from airy pulses in Kerr media. Opt. Express 19: 17298–17307.
Hu, Y. 2013. Spectrum to distance mapping via nonlinear Airy pulses. Opt. Lett. 38: 380–382.
Liu, W. Neshev, D. N. Shadrivov, I. V. Miroshnichenko, A. E. & Kivshar, Y. S. 2011. Plasmonic Airy beam manipulation in linear optical potentials. Opt. Lett. 36: 1164–1166.
Siviloglou, G. A. & Demetrios, N. C. 2007. Accelerating finite energy Airy beams. Opt. Lett. 32: 979–981.
Wang, S. Fan, D. Bai, X. & Zeng, X. 2014. Propagation dynamics of Airy pulses in optical fibers with periodic dispersion modulation. Phys. Rev. A 89: 023802.
Zhang, P. Wang, S. & Zhang, X. 2011. Plasmonic Airy beams with dynamically controlled trajectories. Opt. Lett. 36: 3191–3193.

Energy Science and Applied Technology – Fang (Ed.)
© 2016 Taylor & Francis Group, London, ISBN 978-1-138-02833-3

Simulation and experimental validation of mass loss induced by vacuum ultraviolet irradiation on polymer materials

Y. Wang
Lanzhou Hongrui Ht Mechanical and Electrical Equipment Co. Ltd., Lanzhou, China

X. Guo, X.R. Wang, S.S. Yang & X.J. Wang
Science and Technology on Vacuum Technology and Physics Laboratory, Lanzhou Institute of Space Technology Physics, Lanzhou, China

ABSTRACT: Outgassing of polymer materials induced by Vacuum Ultraviolet (VUV) causes molecular contamination. Based on diffusion theory and radiation chemistry, simulation of mass loss induced by VUV irradiation on polymer materials is studied. The mathematical formulation between mass of volatile in the material and irradiation time is obtained by using the variable separation approach. Mass loss for the Polyimide (PI) or Poly-Ethylene Terephthalate (PET) film obtained from experiments and parameters in the model is calculated through curve fit in order to validate the applicability of the model. The results show that the model of mass loss is applicable to the PI or PET film and can be used for predicting mass loss of polymer materials induced by VUV irradiation.

Keywords: Vacuum Ultraviolet (VUV); Polymer materials; Mass Loss; Molecular contamination; Diffusion theory

1 INTRODUCTION

Outgassing of polymer materials induced by space irradiations could cause molecular contamination (Laikhtman et al., 2009). Contamination of sensitive surfaces of optical systems and cells of solar batteries is a complex physical and chemical process, which contains three principle courses including outgassing of volatile components from materials, transport and deposition on sensitive surfaces. The effects of molecular contamination not only relate to polymer materials used for the spacecraft, but also relate to the dimension of vehicles and location of sensitive surfaces, as well as the influence by the temperature and irradiation. (Bertrand, 1995) Simulation on the effect of contamination is based on a series of physical models. However, existing models that do not take into account the irradiation can forecast the effects of contamination to a certain extent, which leads to differences in the orbit (Hall, 2000; Wang, 1989, 1994; Khassanchine, 2004). Consequently, elaboration of prediction models describing the outgassing processes of polymer materials is a present-day problem to predict the contamination of sensitive surfaces.

VUV, which can be strongly absorbed by the polymer material, will decompose the film material and increase mass loss in these materials, hence causing contaminations and ultraviolet-enhanced contaminations, thereby influencing the optical, chemical and physical performances of sensitive surfaces of optical systems and cells of solar batteries (Brinza et al., 1991; Keith, 2007; Dever, 2005). Although research on the performance of polymer materials caused by VUV irradiation is mainly focused on the performance of common materials, such as polyimide and silicone rubber, simulations or predictions of mass loss induced by irradiation are scarcely studied. The model of mass loss induced by VUV irradiation on polymeric materials is studied based on the physical and chemical mechanisms between ultraviolet irradiation and materials. Thus, experiments of VUV radiation on Polyimide (PI) and Polyethylene Terephthalate (PET) are carried out to validate the applicability of the model.

2 SIMULATION OF MASS LOSS

Several principal postulates applied in the model of outgassing at thermal-vacuum action are utilized in order to describe mathematically the influence of VUV on physical and chemical processes that occur in the material and on its surface. The change in the concentration of volatile components in the material applied to a hermetically enclosed substrate is stipulated by the following processes: desorption from the surface at the material–vacuum boundary; photodecomposition reactions of the material; and diffusion resulting from the aforementioned processes.

This model is based on the following assumptions:

1. Thickness of the polymer film is significantly less than other linear dimensions so that it can be treated as a one-dimensional massive plate.
2. Temperature of the materials is fixed.
3. Coefficients of diffusion desorption and VUV decomposition only depend on time.
4. Some components of the outgassing process can be produced as a result of destruction of other ones; desorption are not considered in the model, that is, all molecules diffused to the surface volatilize to vacuum.
5. Volatile components in the materials are involved only in the first-order reaction.
6. Outgassing occurs only through the material–vacuum boundary.
7. Coefficients used in the model are considered as effective coefficients.

Under these assumptions, as shown in Figure 1, the change in the concentration of volatile components C(x, t) in the film material being in vacuum under exposure to VUV radiation can be described by the following partial differential equations:

$$
\begin{cases}
\dfrac{\partial C}{\partial t} = D\dfrac{\partial^2 C}{\partial x^2} - kC + Ae^{-\alpha(l-x)} & 0 < x < l, t > 0 \\
C(0,t) = C(l,t) = 0 & t > 0 \\
C(x,0) = C_0 & 0 \le x \le 1
\end{cases} \quad (1)
$$

where l is the thickness of the material; k is the coefficient of photodecomposition reactions;

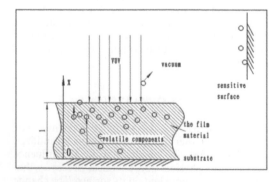

Figure 1. Simulation of outgassing induced by VUV irradiation on the polymer film material.

$S(x,t) = Aexp[-\alpha(l - x)]$ is the source function of volatile components in the material, in which A is a parameter that depends on the composition of the material and the UV source, α is the effective coefficient of linear reduction of VUV radiation; and C_0 is the initial concentration of volatile components.

Based on the partial differential equation theory, Eq. (1) is calculated by using the variable separation approach:

$$
C = \sum_{n=1}^{\infty} \Big\{ \frac{4C_0}{n\pi} e^{-\left(\frac{n\pi}{l}\right)^2 Dt} + \frac{2Al^2}{[(\alpha l)^2 + (n\pi)^2]Dn\pi}\Big[1 - e^{-\left(\frac{n\pi}{l}\right)^2 Dt}\Big] + \frac{4C_0}{n\pi} e^{-\left(k + \frac{Dn^2\pi^2}{l^2}\right)t} \Big\} \sin\frac{n\pi}{l}x \quad n = 1,3,5... \quad (2)
$$

The outgassing rate of volatiles from the unit of the surface can be determined by Fick`s second law:

$$
F(x,t) = -\frac{D}{l} \sum_{n=1}^{\infty} \Big\{ 4C_0 e^{-\left(\frac{n\pi}{l}\right)^2 Dt} + \frac{2Al^2}{(\alpha l)^2 + (n\pi)^2}\Big[1 - e^{-\left(\frac{n\pi}{l}\right)^2 Dt}\Big] + 4C_0 e^{-\left(k + \frac{Dn^2\pi^2}{l^2}\right)t} \Big\} \cos\frac{n\pi}{l}x \quad n = 1,3,5... \quad (3)
$$

For a one-dimensional massive plate, the outgassing rate is given by

$$
F(l,t) = \frac{4C_0 D}{l} \sum_{n=1}^{\infty} \Big[e^{-\left(\frac{n\pi}{l}\right)^2 Dt} + e^{-\left(k + \frac{Dn^2\pi^2}{l^2}\right)t}\Big]
$$
$$
- \sum_{n=1}^{\infty} \frac{2Al}{(\alpha l)^2 + (n\pi)^2} e^{-\left(\frac{n\pi}{l}\right)^2 Dt}
$$
$$
+ \sum_{n=1}^{\infty} \frac{2Al}{(\alpha l)^2 + (n\pi)^2} \quad n = 1,3,5... \quad (4)
$$

Then, the mass loss of the film $W(T,t)$ is obtained by integrating $F(l,t)$ from 0 to t:

$$
W(T,t) = \frac{l^2}{\pi^2 D}\Big[\frac{4DC_0}{l} - b\Big]\Big[1 - e^{-\left(\frac{\pi}{l}\right)^2 Dt}\Big] + \frac{4DC_0 l}{kl^2 + \pi^2 D}\Big[1 - e^{[-k - (\frac{\pi}{l})^2 D]t}\Big] + bt \quad (5)
$$

where

$$
b = \sum_{n=1}^{\infty} \frac{2Al}{(\alpha l)^2 + (n\pi)^2}
$$

3 MATHEMATICAL FIT AND VALIDATION OF THE MASS LOSS MODEL

3.1 *Mathematical fit of the mass loss model*

To determine the mass loss model for the special film material, experiments of PI and PET irradiated by VUV are processed in the VUV irradiation simulation experimental equipment, respectively. In addition, the Total Mass Loss (TML) is tested in the contamination condensation effect equipment (CCEE) (Wang, 2002). The outgassing data obtained from the experiment are listed in Table 1 and Figure 2.

As can be seen from Figure 2, the mass loss caused by VUV radiation on PI or PET increases sharply at the initial time, whereas it enhances mildly in the latter.

Utilizing the experimental data, coefficients in Eq. (6) are fitted. The result is summarized in Table 2, and comparisons between the fit line and the experimental data are shown in Figure 3 and Figure 4.

When introducing the coefficients listed in Table 2 to Eq. (6), the mathematical formula for mass loss caused by VUV irradiation on PI or PET can be represented as Eq. (7) and Eq. (8), respectively:

$$W(T,t) = \frac{l^2}{\pi^2 D}\left[\frac{4DC_0}{l} - b\right]\left[1 - e^{-\left(\frac{\pi}{l}\right)^2 Dt}\right]$$
$$+ \frac{4DC_0 l}{kl^2 + \pi^2 D}\left[1 - e^{[-k-(\frac{\pi}{l})^2 D]t}\right] + bt$$
$$= 2.66\times10^{-5}\times(1 - \exp(-4.3\times10^{-4}t))$$
$$+ 2.5\times10^{-15}\times(1 - \exp(-4.6\times10^{6}t))$$
$$+ 1\times10^{-10}t \qquad (7)$$

Table 1. Outgassing data of VUV irradiation.

Irradiation time	PI TML	PET TML
ESH	(g/cm2)	(g/cm²)
500	4.24E-6	8.74E-7
1000	9.47E-6	1.03E-6
2200	1.66E-5	1.19E-6
3000	1.92E-5	1.26E-6
4400	2.23E-5	1.28E-6
5000	2.31E-5	1.29E-6

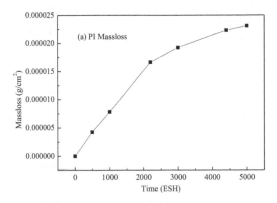

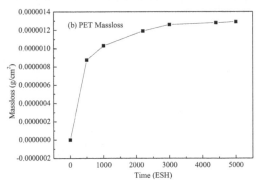

Figure 2. Mass loss of PI or PET induced by VUV irradiation.

Table 2. Fit coefficients in the mass loss equations of PI or PET.

Coefficients	C0 g/cm³	k 1/h	D cm²/h	b g/(cm²·h)	R2
PI fit data	1.6E-2	2.2E-6	1.7E-10	1.0E-15	0.99
PET fit data	1.3E-3	7.0E-11	3.2E-10	7.9E-12	0.98

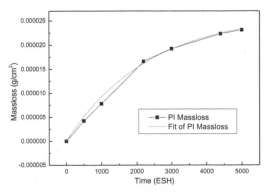

Figure 3. Comparison between fit line and experiment for PI mass loss.

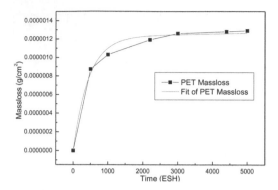

Figure 4. Comparison between fit line and experiment for PET mass loss.

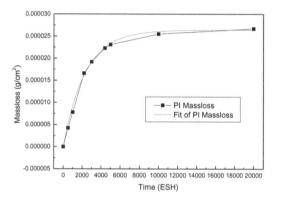

Figure 5. Comparison between extrapolative line and experiment for PI mass loss.

$$W(T,t) = \frac{l^2}{\pi^2 D}\left[\frac{4DC_0}{l} - b\right]\left[1 - e^{-\left(\frac{\pi}{l}\right)^2 Dt}\right]$$

$$+ \frac{4DC_0 l}{kl^2 + \pi^2 D}\left[1 - e^{[-k-(\frac{\pi}{l})^2 D]t}\right] + bt$$

$$= -2.74 \times 10^{-9} \times (1 - \exp(2.26 \times 10^{-2}t))$$

$$+ 6.03 \times 10^{-7} \times (1 - \exp(-2.26 \times 10^{-2}t))$$

$$+ 7.9 \times 10^{-12}t$$

(8)

3.2 Validation of the mass loss model

To validate the application of Eq. (7) and Eq. (8) for a longer irradiation time, the irradiation time is extended to 200 hours (=20000 ESH). The simulation and experiment data are compared in Figure 5 and Figure 6 and in Table 3. As shown from these figures and table, deviation factors between simulations and experiments are less than 2.5%, which

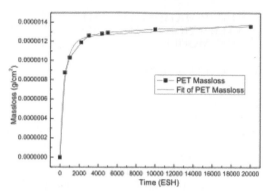

Figure 6. Comparison between extrapolative line and experiment for PET mass loss.

Table 3. Comparison between mass loss simulation data and experiment data.

Irradiation time (ESH)	10000	20000
PI mass loss experiment data (g·cm-2)	2.55E-5	2.67E-5
PI mass loss simulation data (g·cm-2)	2.61E-5	2.64E-5
Deviation factors (%)	2.35	−1.12
PET mass loss experiment data (g·cm-2)	1.33E-6	1.36E-6
PET mass loss simulation data (g·cm-2)	1.30E-6	1.38E-6
Deviation factors (%)	−2.26	1.47

testify that the simulation is valid for mass loss of PI or PET irradiated by VUV.

4 CONCLUSIONS

(1) Based on chemical and physical processes influenced by VUV irradiation on the polymer film material, simulation of mass loss is studied. The mathematical function about mass loss and irradiation time is acquired by analysis and solving the model.

(2) The mass change in the PI or PET film is studied by using VUV irradiation equipment and contamination condensation effect equipment. When the mass of the sample is introduced into the function, coefficients in the model are calculated.

(3) The mass loss model for PI or PET is validated by longer irradiation time experiments. The results show that the mass loss model is applicable for the PI or PET film material and it can be used for predicting mass loss of polymer materials induced by VUV irradiation.

REFERENCES

Brinza D E, Stiegman A E, Staszak P R,et al.1991. Vacuum Ultraviolet (VUV) Radiation Induced Degradation of Fluorinated Ethylene Propylene (FEP) Teflon aboard the Long Duration Exposure Facility (LDEF). LDEF-69 Months in Space-First Post-Retrieval Symposium, 817–829.

Bertrand W T. 1995. Effects of Spacecraft Material Outgassing on Optical System in the Vacuum Ultraviolet. ADA296935.

Dever, Joyce A, Yan L. 2005. Vacuum Ultraviolet Radiation Effects on DC93–500 Silicone Film Studied. NASA, 20050217221.

Hall D F, Arnold G S, Simpson T R, etal. 2000. Progress on Spacecraft Contamination Model Development. SPIE 4096:138–156.

Keith C, Albyn. 2007. Outgassing Measurements Combined with Vacuum Ultraviolet Illumination of the Deposited Materials. Journal of Spacecraft and Rochets, 44(2):102–108.

Khassanchine R H, Grigorevskiy A V, Galygin A N. 2004. Simulation of Outgassing Processes in Spacecraft Coatings Induced by Thermal Vacuum Influence [J]. Journal of Spacecraft and Rochets, 41(3):384–388.

Laikhtman A, Gouzman, I, VerkerR, et al. 2009. Contamination Produced by Vacuum Outgassing of Kapton Acrylic Adhesive Tape. Journal of Spacecraft and Rochets, 46(2): 236–240.

Wang X R. 1989. Mathematical Analysis and Experimental Verification on Spacecraft Materials Mass Loss Process. Chinese Space Science and Technology, 3: 8–20.

Wang X R, Fan C Z.1994. A Diffusion Model upon Spacecraft Materials Acceleration Outgassing Process. Journal of Astronautics, (1): 55–59.

Wang X R, Ma W J. 2002. A Simulation Equipment Used for Determining the Outgassing Contamination Condensation Effect on Cryogenic Sensitive Surface in Space. Journal of Astronautics, 23(3): 68–71.

Energy Science and Applied Technology – Fang (Ed.)
© 2016 Taylor & Francis Group, London, ISBN 978-1-138-02833-3

Solar irradiance interval prediction based on set pair analysis theory

C. Du & Z.Q. Liu

North China Electric Power University, Beijing, China

ABSTRACT: Output of Photovoltaic (PV) power plants always fluctuates obviously. Interval prediction of the output is necessary for a scheduled planning and system operation. The basis of output interval prediction is irradiance interval prediction. Based on the scientific division of meteorological data range, this paper proposes a set pair analysis method to construct irradiance interval. First, we normalized the history data. Then, we selected similar days. Thereafter, we constructed pairs and calculated the Identical Discrepancy Contrary (IDC) distance. The results reveal that the method proposed in this paper is effective.

Keywords: Irradiance interval prediction; Set pair analysis

1 INTRODUCTION

Studies have shown that the output of photovoltaic power plants always fluctuates obviously by the influence of meteorological factors such as solar radiation, temperature, wind speed, clouds and shadows. When large-scale photovoltaic power plants are connected to the grid, the difficulty of grid scheduling will increase greatly and have an impact on the grid stable operation. Therefore, an accurate prediction of output is the key to the mass exploitation of photovoltaic power plants. Power prediction can be divided into deterministic prediction and interval prediction. The methods of deterministic prediction include the direct prediction method, the neural network method and the support vector machine method. Interval forecast estimates the fluctuation range of power. It contains more information and has more value to the schedule of the power system operation and regulation plan.

Because the PV output is affected by the meteorological factors, the PV output prediction can be based on numerical weather prediction. The principle and experiments of power generation show that the meteorological factor solar irradiance has the greatest impact on the PV output. Therefore, by constructing the solar irradiance range, we can predict the future periods of PV output. However, our solar radiation observation site can only provide measured records, as solar radiation forecasting analysis has not yet been carried out and there is no solar irradiance fluctuation range for reference. This paper proposes a set pair analysis method based on historical data. By choosing similar days to the future, we can get irradiance fluctuation that can assist in the photovoltaic output range forecast. The results reveal that the method proposed in this paper is effective.

2 SET PAIR THEORY

Set Pair Analysis (SPA) is a new system analysis method proposed by the scholar Zhao Keqin in the National System Theory Conference in 1989. The so-called set pair refers to the pair of two collections that have some relevance. The basic idea of the set pair analysis is that under some kind of problem in the background, the Identical Discrepancy Contrary (IDC) of the collections under the background can be measured. Then, systematic evaluation, simulation, prediction, control and research can be conducted. The IDC expression of set pair can be described as follows:

$$u = a + bi + cj \qquad (1)$$

where a is the same extent to the two sets of the problem; b is the uncertainty extent to the two sets of the problem; c is the antithetic extent to the two sets of the problem; i is the different degree mark symbol, and also as a factor participating in operations. Its value is usually between [0,1]; and j is the antithetic degree mark symbol, and also as a factor participating in operations. Its value is -1.

a, b and c satisfy the normalization condition. This means they obey the relation $a + b + c = 1$, and therefore expression (1) is often abbreviated as $u' = a + bi$. u is called the connection degree; strictly speaking, it should be a function of the analysis, but it can be a number in the operation. The set pair analysis links the certainty and uncertainty between the set pairs. It converts the uncertainty between ranges into a concrete mathematical tool and does not need any hypothesis for prediction error distribution. So, this method is more reliable.

3 PREDICTION STEP

3.1 Data preprocessing

The original historical data have some bad data because of equipment and signal transmission problems. First, we screen the data. Then, we eliminate roughness of the historical data by excluding the data that have a big error or serious discrepancies from experiments through observations. Then, we normalize the selected data. The expression is as follows:

$$x = \frac{x' - x'_{min}}{x'_{max} - x'_{min}} \quad (2)$$

where x is the original data; x'_{min} is the minimum value of certain factors (e.g. solar irradiance, temperature and wind speed); x'_{max} is the maximum value; and x' is the normalized value. Its value is between [0,1].

3.2 Selection of similar days

We use Pearson's correlation coefficient method to calculate the degree of correlation between other meteorological factors (e.g. temperature and wind speed) and solar irradiance. We then set two variables x, y. Each set of samples is expressed as (x_i, y_i) $(i = 1,2,...,n)$. Pearson's correlation coefficient is given by

$$\rho = \frac{\sum_{i=1}^{n}(x_i - \bar{x})(y_i - \bar{y})}{\sqrt{\sum_{i=1}^{n}(x_i - \bar{x})^2}\sqrt{\sum_{i=1}^{n}(y_i - \bar{y})^2}} \quad (3)$$

where $\bar{x} = \frac{1}{n}\sum_{i=1}^{n}x_i, \bar{y} = \frac{1}{n}\sum_{i=1}^{n}y_i$.

The absolute value of Pearson's correlation coefficient is closer to 1, which is the higher degree of the linear relationship between two variables. From these meteorological factors, we select factors that have a strong correlation with solar irradiance as solar irradiance influence factors. We use Gray Relational Analysis (GRA) to determine the similar day of the day to be predicted. The influence factor data sequence of the day to be predicted is called

the generating sequence. It can be expressed as follows: $X_j = (x_j(1), x_j(2),....x_j(n))$, where $i = 1,2,...,m$, is the number of influence factors. The influence factor data sequence of historical days is called the subsequence. It can be expressed as follows: $X_j = (x_j(1), x_j(2),....x_j(n))$, where $j = 1,2,...,M$, is the number of historical days. At time t, the gray correlation coefficient of X_i and X_j is given by

$$r_{ij}(t) = \frac{\Delta_{min} + \beta\Delta_{max}}{\Delta_{ij}(t) + \beta\Delta_{max}} \quad (4)$$

where $\Delta_{ij}(t) = |x_i(t) - x_j(t)|$, $\Delta_{max} = max|x_i(t) - x_j(t)|$, are the maximum of the absolute difference between subsequence factors and generating subsequence factors. $\Delta_{min} = min|x_i(t) - x_j(t)|$ is the minimum of the absolute difference between subsequence factors and generating subsequence factors. β is the resolution factor and its value is usually 0.5. β denotes the indirect impact the system or subfactors made to the correlation. Expression (4) shows the correlation of the influence factors of the jth historical day and the day to be predicted at the time t.

The correlation between the ith influence factor of the jth day and the ith influence factor of the day to be predicted is given by

$$r_{ij} = \frac{1}{n}\sum_{t=1}^{n}r_{ij}(t) \quad (5)$$

The correlation between the jth historical day and the day to be predicted is given by

$$r_j = \sum_{i=1}^{m}\rho_i r_{ij} \quad (6)$$

where ρ_i $(i = 1,2,...,m)$ is Pearson's correlation coefficient between this influence factor and solar irradiance. The greater the correlation value is, the more similar the two-day weather conditions will be.

3.3 Set pair analysis

3.3.1 Establishing pairs of solar irradiance and the strongest influence factors

The stronger the correlation between individual pair is, the more accurate the results will be. To reduce the error that the weak correlation factors made to the results, we select only the largest influence factor with Pearson's correlation coefficient of solar irradiance as considerations, which is called "the strongest factors."

We use the strongest factor data of the similar day from a small to a large order from a new sequence F, $F = \{f_1, f_2, ..., f_k\}$, where k is the number of this sequence. After considering the error distribution and the prediction precision synthetically, the strongest factors affecting the gear is set. The gear value can be expressed as: $f_m(m = 1,2,...,k')$, where k' is the number of gear. After rearranging solar irradiance values corresponding to the factors, a sequence of G is formed. G will be divided into the following collections according to the corresponding solar irradiance: $G = \{G_1, G_2, ..., G_{k区}\}$, $G_m(m = 1,2,...,k')$ is the mth collection of the solar irradiance sequence. The maximum and minimum range of this sequence is expressed as: $[g_{m-}, g_{m+}]$.

3.3.2 Establishing pairs of the strongest influence factors of similar days and the strongest influence factors of the day to be predicted

We use the self-comparison method to form the set pair, which is the collection of the strongest influence factors of similar days and itself, and then we get the set pair whose correlation is equal to 1. According to expression (1), the IDC correlation between collection F and itself is established as:

$$u_m = a_m + b_m i + c_m j = f_m + b_m i + c_m j \qquad (7)$$

where f_m is the factor gear and $b_m = c_m = 0$.

Similarly, we form the set pair of the strongest factor collection H of the day to be predicted and itself. $H = \{h_1, h_2, ..., h_{k*}\}$, where k^* is the number of factor data of the day to be predicted. The correlation of the set pair is 1. The identical part is all data in the collection and does not exhibit a different part and an opposite part. The IDC correlation is given by

$$u_{m*} = a_{m*} + b_{m*} i + c_{m*} j = h_{m*} + b_{m*} i + c_{m*} j$$
$$(m^* = 1,2,...,k^*) \qquad (8)$$

where h_{m*} is the data of the m^*th factor and $b_{m*} = c_{m*} = 0$.

3.3.3 Calculation of the IDC distance between the strongest influence factor of similar days and the strongest influence factor of the day to be predicted

The IDC distance of two set pairs can be calculated as follows:

$$d = \sqrt{(a_m - a_{m*})^2 + (b_m - b_{m*})^2 + (c_m - c_{m*})^2} \qquad (9)$$

We calculate the IDC distance between the correlation of the day to be predicted and the correlation of all similar days. Then, according to the closeness principle, we choose the nearest IDC distance. This is the distance corresponding to the selected gear. We set this gear as f_m, and then classify the

factor into this gear. All the corresponding solar irradiance $[g_{m-}, g_{m+}]$ is the solar irradiance range of this predicted point. Finally, the prediction of solar irradiance range is realized.

4 SIMULATION RESULTS

In this paper, the data measured as samples are acquired from a PV plant in southwest China, including PV output data and meteorological data. The meteorological data include solar irradiance, temperature and wind speed. All data were recorded every 15 minute. The solar irradiance of September 15 was chosen to be predicted, and August 14 to September 14 as the chosen range for similar days.

First, we normalize the solar irradiance, temperature and wind speed data. We calculate Pearson's correlation coefficient between irradiance and temperature, between irradiance and wind speed. The results are given in Table 1.

It is generally regarded as highly correlated when the coefficient is greater than 0.8, and relatively highly correlated when the number is between 0.5 and 0.8, and weakly correlated when the number is less than 0.5. As shown in Table 1, temperature has a stronger correlation than wind speed, and the coefficient between temperature and irradiance is greater than 0.5. So, based on the set pair analysis theory, we set temperature and irradiance as a set pair and calculate irradiance range depending on temperature numerical analysis.

We determine the similar data using the gray relational analysis.

We then choose the biggest 7 days to be the similar day0

We select the forecast period from 07:00 to 15:45.

Table 1. Pearson's correlation coefficient.

	Temperature	Wind speed
Coefficient between irradiance	0.6567	0.4924

Table 2. Biggest gray correlation coefficient.

	Aug.29	Sep.1	Aug. 16	Aug.18
Gray correlation coefficient	0.8961	0.8615	0.8553	0.8316

	Aug. 23	Aug. 24	Aug.17
Gray correlation coefficient	0.8080	0.7987	0.7566

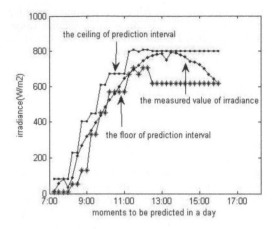

Figure 1. Predicted irradiance interval and measured value.

According to the set pair analysis method mentioned above, we select temperature as the strongest factor. We arrange temperature data in ascending order, and then divide them by 0.1, getting the irradiance collection divided by temperature gear. The minimum and maximum values in each collection constitute the illumination value interval of this gear.

We establish the IDC connection between similar days' temperature and itself. Similarly, we establish the identical discrepancy contrary between future days' temperature and itself. We then calculate the IDC distance between the similar days and the day to be predicted and choose the shortest distance, with the corresponding gear shift being the right temperature gear, thereby getting the irradiance interval.

According to the set pair analysis method, we select temperature as "the strongest factors". We arrange temperature of similar days in ascending order and consider 0.1 as a portion for division. Then, we get each thermal gear and corresponding solar irradiance collection. The minimum and maximum values of each set constitute the illumination value interval of this gear.

The predicted result is shown in Figure 1.

As shown in Figure 1, there are 32 groups that lie between this interval, with an accuracy rate of 88.89%. This test based on evenly spaced gear can be applied widely. If we divide non-equidistantly, the result value will be more accurate. The results reveal that the method proposed in this paper is effective.

5 CONCLUSIONS

1. Based on the scientific division of meteorological data range, the accuracy of historical data and the applicability of the set pair analysis theory, this paper proposes a new idea for solar irradiance prediction using the set pair analysis theory.
2. We use Gray Relational Analysis (GRA) to choose the similar day that has similar meteorological conditions to the day to be predicted and build a strong small-sized correlation sample.
3. We establish the set pair analysis model of similar days and the day to be predicted, build the IDC correlation, calculate the IDC distance, and get future predictions of solar irradiance range.

REFERENCES

Aznarte J L,Girard R,Kariniotakis G,et al.2008.Short Term Forecasting of Photovoltaic Power Production. DB2:*Fore-casting Functions with focus to PV prediction for Microgrids*,2008.

Bai Yongqing.2011. Preliminary research on prediction of hourly solar radiation based on WRF model output statistics. *Transactions of Atmospheric Sciences*, 2011,34 (3):363–369.

Cao S H,Cao J C.2005.Forecast of solar irradiance using recurrent neural networks combined with wavelet analysis.*Applied Ther-mal Engineering*,2005,25(2–3):161–172.

Chen C S,Duan S X,Cai T,et al2.011.Online 24-h solar power forecasting based on weather type classification using artificial neural network.*Solar Energy*,2011,85(11):2856–2870

Hiyama T.1997.Neural network based estimation of maximum power generation from PV module using environmental information.*IEEE Transactions on Energy Conversion*,1997,12(3):241–247.

Li Ran, Li Guangmin. 2008. Based on support vector machine regression PV output forecast. *Electric Power, 2008,41(2):74–78*

Liu Sifeng, Guo Tianbang,Dang Yaoguo.1999. Grey System Theory and Its Applications (Second Edition). *Beijing: Science Press*, 1999, 1–18.

Luo Yufeng.2011. PV principles and processes. *BeiJing, Central Radio and TV University Press*, 2011,2–3.

Ren hang, Ye lin.2010. Operation Characteristics of PV System Under the Influence of Environmental Factors. *Transactions of China Electrotechnical Society*. 2010,(10):158~165

Xu Jing.2.11. Design and implementation of Solar photovoltaic forecast website system *Water Resources and Power*, 2011,29(12):193–195.

Yona A2008..Application of neural network to 24-hour-ahead generating power forecasting for PV system. IEEE Power and Energy Society General Meeting-Conversion and Delivery of Electrical Energy in the 21st Century.

Zhang Xueli.. 2012.Analysis of Influencing Factors of Output Power of Photovoltaic Power Plant. *Power System and Clean Energy*.2012, 28(5): 76–81.

Zhao Keqin.1996. Set pair analysis theory, a new method of uncertainty theory and application. *Journal of systems engineering*.1996, 14(1): 18–23.

Energy Science and Applied Technology – Fang (Ed.)
© 2016 Taylor & Francis Group, London, ISBN 978-1-138-02833-3

H₂O₂ electrocatalytic reduction on Pd/SiC catalyst

C.P. Han, Y.R. Bao, L.M. Sun & Z.R. Liu

College of Chemistry and Chemical Engineering, Inner Mongolia University for the Nationalities, Tongliao, China

ABSTRACT: The Pd/SiC catalyst was prepared by the chemical reduction method. The composition and structure of the catalyst were characterized using X-ray diffraction and SEM analysis. The electrocatalytic activity and stability of the catalyst for the H_2O_2 reduction were studied using cyclic voltammetry and chronoamperometry. The results showed that the performance of the Pd/SiC catalyst for the H_2O_2 reduction reaction is higher than that of the Pd/C catalyst.

Keywords: Pd/SiC; Hydrogen peroxide electroreduction; Fuel cell

1 INTRODUCTION

It is well known that the H_2O_2 fuel cell can convert chemical energy into electrical energy by using H_2O_2 as the oxidizing agent, which is an amazing generation device with high energy density, excellent environment friendly, easy storability and fine security. Due to the above-mentioned fascinating properties, the H_2O_2 fuel cell can be used for the power supply in versatile applications, such as underwater navigation of unmanned underwater vehicle, communication field, and oil and gas equipment.

As the critical section of the redox reaction in the fuel cell, the catalyst plays an important role in operating the fuel cell. Therefore, the catalyst has attracted considerable attention on account of its high catalytic activity. In particular, the pore structure, surface structure and composition of the carrier will have an impact on the activity of the catalyst. At present, there are some materials that are often used as the carriers of H_2O_2 electroreduction catalysts, such as carbon materials [5–8] (e.g. carbon powder, carbon nanotube, carbon paper and carbon fiber), conducting ceramic materials [5–8] (e.g. ITO, WO_3, TiO_2, TiB_2, SiC, CeO_2, WC, CeO_2-ZrO_2 and Ti_4O_7), graphite [12] and foamed nickel [13]. However, carbon materials show lack of stability because they are easily corroded in an acid environment. By contrast, ceramic materials have good thermostability, mechanical strength, hardness, thermal conductivity, electrical conductivity, corrosion resistance and property because of which they hardly adsorb electrolytes. Therefore, improving the electric catalytic properties of the catalyst has important implications for promoting the performance of the full cell.

2 EXPERIMENTAL SECTION

2.1 *Materials and characterization*

Vulcan XC-72 activated carbon was obtained from Cabot Inc., USA, as well as SiC, NaOH, $NaBH_4$ and H_2O_2 (30%) (analytical grade).

The phase composition and phase structure of the as-synthesized products were examined by X-Ray Diffraction (XRD) using a Shimadzu XRD-6000 X-ray diffractometer with Cu Ka radiation ($\lambda = 0.154056$ nm, scanning angle range 20–90°, scanning speed 5°/min and scanning resolution 0.01°). The morphologies and microstructure analysis were characterized with a field emission Scanning Electron Microscope (SEM) (JSM-6480, Hitachi).

2.2 *Carrier pretreatment*

Pretreatment of silicon carbide carrier. In the typical synthesis, 2 g silicon carbide (SiC) powder was added in excess to hydrochloric acid and soaked for 12 hours. The product was collected by suction filtration and washed with distilled water. Then, the product was placed in an oven at 50°C for a long period of time, followed by thermal treatment at 600°C for 3 hours in a muffle furnace, and cooled and stored at room temperature.

Pretreatment of carbon carrier. 2 g carbon powder was added in excess to nitric acid and stirred for 2 hours at 80°C. The product was collected by suction filtration and washed with distilled water. Thereafter, the product was placed in an oven at 50°C for a long period of time, followed by dissolving in excess ammonium hydroxide and stirring at 60°C for 4 hours. Then, the products were washed with distilled water and placed in an oven for 12 hours.

Preparation of Pd/SiC and Pd/C catalysts. For the typical synthesis, 60 mg of the prepared SiC powder was added to a 6.3 ml mixed solution containing $PdCl_2$ and HCl (0.023 mol/L). The pH of the solution was adjusted to 8–9. Then, the solution was treated with ultrasonic vibration after stirring for 4 hours at 70°C. Excess $NaBH_4$ solution (0.048 mol/L) was added dropwise during ultrasonic vibration. The solution was stirred continuously for another 1 hour until the reduction reaction was complete. The resulting mixture was washed constantly with distilled water until Cl^- disappeared completely. Thereafter, the Pd/SiC catalyst (20 wt%) was prepared after drying in a vacuum oven for 12 hours at 65°C. The preparation method of the Pd/C catalyst was the same as that of the Pd/SiC catalyst.

Preparation of the electrode. A glassy carbon electrode was polished like the mirror surface and then washed and air-dried. The turbid liquid composed of 5 mg catalyst, and 2 ml ultrapure water was treated with ultrasonic vibration for 30 min. 15 μL of the turbid liquid was transferred to the surface of the glassy carbon basement, and air-dried.

Electrochemical test. An electrochemical test was carried out by using the AUTOLAB (PARSTAT2273 PARSTAT2273) electrochemical workstation. The test employs a traditional three-electrode system using the prepared electrode as the working electrode, high-purity carbon rod as the auxiliary electrode and saturated calomel electrode as the reference electrode.

The first test method was cyclic voltammetry, which used 0.1 mol/L sulfuric acid solution as the electrolyte, 1.2~-0.3V as the scanning range of electric potential and 50 mV/s as the scanning speed. The second method was the polarization curve, which used mixed solution containing 0.1 mol/L sulfuric acid and 0.5 mol/L hydrogen peroxide as electrolytes, 50 mV/s as the scanning speed and 0.6 ~ –0.3V as the voltage range.

3 RESULTS AND DISCUSSION

3.1 *Characterization of the catalysts*

The structure of Pd/SiC and Pd/C is confirmed by XRD. As shown in the XRD spectra (Figure 1), the peaks of the two catalysts are mainly in accord with standard spectra, indicating that their structures are typical face-centered cubic (corresponding to the (111), (200), (220) and (311) crystal face diffraction peaks of Pd). According to the Scherrer equation (d = 0. 89 λ / Bcosθ, where d being the grain size; λ, the diffraction wavelength; and B, the integrated intensity of half-width of X-ray diffraction), the grains of Pd/SiC are significantly smaller

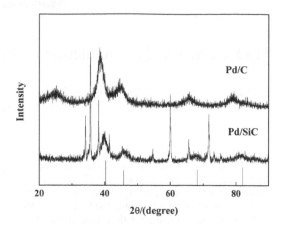

Figure 1. XRD patterns of Pd/SiC and Pd/C catalysts.

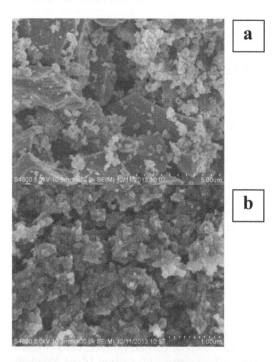

Figure 2. SEM images of (a) Pd/SiC and (b) Pd/C catalysts.

than those of Pd/C, indicating that the former has a greater specific surface area.

The morphology of the pretreated Pd/SiC and Pd/C catalysts is also investigated. The SEM images in Figure 2a and Figure 2b clearly show that Pd is loaded onto the surface of two kinds of catalysts. In addition, the agglomeration of Pd particles appears during the load.

Figure 3 shows the cyclic voltammetry curves of the two catalysts in the 0.1 mol/L H_2SO_4 solution.

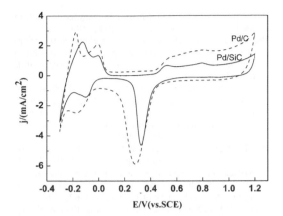

Figure 3. Cyclic voltammetry curves of Pd/SiC and Pd/C catalysts in the 0.1 mol/L H2SO4 solution.

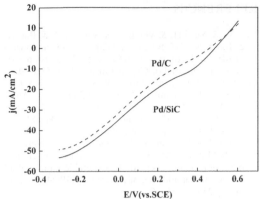

Figure 4. Potential-current patterns of Pd/SiC and Pd/C electrodes (in the mixture containing 0.1 mol/L H_2SO_4 and 0.5 mol/L H_2O_2).

All curves show the characteristic peaks of Pd, and two pairs of redox peaks at -0.3~1.2 V are assigned to the absorption–desorption process of the hydrogen atom on the surface of Pd. The electrochemical specific surface area of Pd/SiC is found to be larger than that of Pd/C when comparing the electric current density of the two curves.

3.2 Electrocatalytic property of the reduction of H2O2 catalyzed by Pd-related catalysts

The reductions of H_2O_2 catalyzed by Pd/SiC and Pd/C catalysts were carried out in a mixed solution containing 0.5 mol/L H_2O_2 and 0.1 mol/L H_2SO_4, with the scanning speed of 50 mV/s. As shown in the potential-current patterns (Figure 4), the onset potentials of reductions of H_2O_2 catalyzed by Pd/SiC and Pd/C catalysts are identical, that is 0.6 V approximately. Under the same electric potential ($E = 0.2$ V), the current density of reductions catalyzed by Pd/SiC and Pd/C catalysts vary considerably, which are 12.14 mA/cm^2 (Pd/C) and 20.16 mA/cm^2 (Pd/SiC), respectively.

Time-current curves are obtained by placing Pd/SiC and Pd/C electrodes in the mixture containing 0.1 mol/L H_2SO_4 and 0.5 mol/L H_2O_2, as shown in Figure 5. It is observed that both electrodes have good stability in a particular situation (E = 0.2 V, temperature 20°C and scanning time 1800s). The current density of reduction catalyzed by the Pd/SiC electrode is approximately 19 mA/cm^2, exceeding 6 mA/cm^2 than the Pd/C electrode.

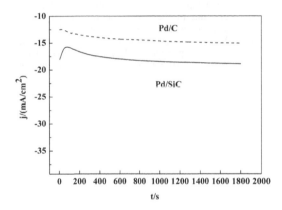

Figure 5. Time-current curves of Pd/SiC and Pd/C electrodes (in the mixture containing 0.1 mol/L H_2SO_4 and 0.5 mol/L H_2O_2).

that the catalyst has a face-centered cubic structure, and Pd particles are distributed on the surface of SiC, which exhibit certain agglomeration. The performance of the catalyst that catalyzed the H_2O_2 reduction reaction indicated that the Pd/SiC catalyst possesses better catalytic activity than the traditional Pd/C catalyst.

ACKNOWLEDGEMENTS

This work was financially supported by the National Natural Science Foundation of China (21003070, 21463017) and the Natural Science Foundation of Inner Mongolia, China (2012MS0208).

4 CONCLUSION

In conclusion, the Pd/SiC catalyst was prepared by the chemical reduction method. It was found

REFERENCES

Dai, W. & Yu, J.H. & Wang, Y. & Song, Y.Z. & Bai. H. & Jiang, N. 2015. Single crystalline 3C–SiC nanowires grown on the diamond surface with the assistance of graphene. Journal of Crystal Growth 420(15):6–10.

Gudarzi, D. & Ratchananusorn, W. & Turunen, I. & Heinonen, M. & Salmi, T. Promotional effects of Au in Pd–Au bimetallic catalysts supported on Activated Carbon Cloth (ACC) for direct synthesis of H2O2 from H2 and O2.Catalysis Today 248(15):58–68.

Ishihara, T. & Nakashima, R. & Ooishi, Y. & Hagiwara, H. & Matsuka, M. 2015. H2O2 synthesis by selective oxidation of H2 over Pd–Au bimetallic nano colloid catalyst under addition of NaBr and H3PO4.Catalysis Today 248(15):35–39.

Li, S.Z. & Zhang, W. & Chen, F.X. & Chen, R. 2015. One-pot hydrothermal synthesis of Pd/Fe3O4 Nano composite in HEPES buffer solution and catalytic activity for Suzuki reaction. Materials Research Bulletin 66:186–191.

Padovano, E. & Badini, C. & Celasco, E. & Biamino, S. & Pavese, M. & Fino, P. 2015. Oxidation behavior of ZrB2/SiC laminates: Effect of composition on microstructure and mechanical strength. Journal of the European Ceramic Society 35(6):1699–1714.

Patil, N.M. & Bhanage, B.M. Fe@Pd/C: An efficient magnetically separable catalyst for direct reductive amination of carbonyl compounds using environment friendly molecular hydrogen in aqueous reaction medium. Catalysis Today 247(1):182–189.

Prosviryakov, A.S. 2015. SiC content effect on the properties of Cu–SiC composites produced by mechanical alloying. Journal of Alloys and Compounds 632(25):707–710

Riyapan, S. & Boonyongmaneerat, Y. & Mekasuwandumrong, O. & Praserthdam, P. & Panpranot, J. 2015. Effect of surface Ti3+ on the sol–gel derived TiO2 in the selective acetylene hydrogenation on Pd/TiO2 catalysts. Catalysis Today 245(1):134–138

Wu, H.D. & Zhang, H. & Chen, S. & Fu, D.F. Flow stress behavior and processing map of extruded 7075 Al/SiC particle reinforced composite prepared by spray deposition during hot compression. Transactions of Nonferrous Metals Society of China 25(3): 692–698

Zhang, Q.F. & Xu, W. & Li, X.N. & Jiang, D.H. & Xiang, Y.Z. & Wang, J.G. & Cen, J. & Romano, S. & Ni, J. 2015. Catalytic hydrogenation of sulfur-containing nitrobenzene over Pd/C catalysts: In situ sulfidation of Pd/C for the preparation of PdxSy catalysts. Applied Catalysis a: General 497(5):17–21.

Zhao, Y. & Xia, H.Y. & Tang, R. & Shi, Z.Q. & Yang, J.F & Wang, J.P. 2015. A low cost preparation of C/SiC composites by infiltrating molten Si into gel casted pure porous carbon preform. Ceramics International 41(5):6478–6487.

Energy Science and Applied Technology – Fang (Ed.)
© 2016 Taylor & Francis Group, London, ISBN 978-1-138-02833-3

Investigation on characteristic of high-frequency impulse water chain formation process under surfactant systems

Yuxin Fan, Bin Li, Zhiqian Sun & Zhenbo Wang
Department of Chemical Engineering, China University of Petroleum, China

ABSTRACT: The impact mechanism of surfactants on water chain formation in a high-frequency and high-voltage pulse electric field is investigated using a digital microscope system. The results show that surfactants adsorbed on the oil–water interfacial membrane make an increment, with the oil–water interfacial tension dropping sharply, membrane intensity making a remarkable decrement, and coalescence of water droplets taking place increasingly easily with increasing surfactant concentrations. As a result, average particle diameters are enlarged. In contrast, water chain length increases with increasing surfactant concentrations. This can be attributed to the sharp drop in interfacial energy, the steric hindrance domino effect and the compression of the thickness of the electric double layer domino effect. For the impact of the surfactant molecular structure on membrane intensity and Maranon convection caused by surfactants in the continuous phase, the emulsion whose water chain length and particle diameter take the first place is Span-80, while OP-10 takes the second place, and Tween-80 takes the last place.

Keywords: high-frequency impulse; W/O emulsions; surfactant; water chain length; particle diameter

1 INTRODUCTION

Crude oil exploited from oil reservoir is always combined with water and mineral salt for exploitation methods. In recent years, water content of oilfield produced fluid that sometimes reached up to 90%. The worst part is that crude oil contains indigenous surfactants such as resin, asphaltene and paraffin that result in stable W/O emulsions. Moreover, not only the water content of crude oil has a sharp rise, but also the physical property of crude oil becomes more complex due to the recent oilfield development and widely used polymer flooding. As a result, oil–water separation turns out to be even more difficult (Sun Jiang-bo, 2004). Several techniques (R. A. Mohammed 1994, D. Sun 1999, M. Goto 1989, R. A. Mohammed 1993, Hirato T. 1991), such as chemical emulsification, gravity or centrifugal settling, filtration, ultrasonic method, microwave heating method, magnetic treatment, heating treatment and electrostatic emulsification, are used in order to achieve oil–water separation. Among these techniques, one of the most effective and utilized method in the industry today is electrostatic emulsification. Furthermore, pulse electrostatic emulsification proposed by Bails in the 1980s gains favorable emulsification results, and attracts extensive attention (Zhang Liming 2007). Concerning heavy oil, some problems such as electric field collision, low efficiency and sub-standard dewatering occur in pulse electrostatic coalescence instruments. Moreover, influencing factors of pulse electrostatic coalescence are the lack of deep and comprehensive analysis (Jin Youhai 2010). Single factor analysis of electric field strength, electric field frequency and duty ratio on water chain formation is adopted in this article. The results achieved will provide a theoretical foundation for further exploitation of efficient and coherent pulse electrostatic coalescence equipment.

2 EXPERIMENTAL INSTRUMENTATION AND METHODS

In this micro-experiment, the dispersion phase was distilled water. The continuous phase was conduction oil of good light-admitting quality. The solution was oscillated for 1200 times. The density of conduction oil was 856.0 kg·m-3, and the apparent viscosity was 89.6 mPa·s at room temperature. The initial moisture content was 4%. The operating temperature was 20oC. The high-frequency and high-voltage rectangular wave pulse dehydration power supply adopted has a continuously adjustable voltage, frequency and duty ratio. The range of the voltage was 1.2 ~ 7.0 kV, the frequency was 1.8 ~ 6.3 kHz, and the duty ratio was 0.1 ~ 0.875. The experimental cell was a double-insulated plate, whose dimensions were 85 mm × 11 mm × 11 mm.

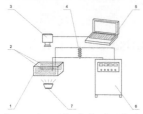

1- experimental cell; 2- electrical plate; 3- digital camera; 4-
transformer; 5- image and video data processing system;
6- high-frequency and high-voltage power-supply system;
7- light source

Figure 1.   Experimental set-up of the microscopic test.

The cell was made with Perspex. The electrodes
were made with polished brass plates. The form
of the electrode was planar. The upper part of
the cell was open. Electric field intensity can be
changed by regulating the voltage in the cell. The
micro-imaging system was composed of a PH50-
DB310U digital biomicroscope, computer, and
PHMIA2008 micrograph reprocessing software.
The experimental set-up of the microscopic test is
shown in Figure 1.

In order to investigate the effect of surfactant
types on water chain formation, OP-10, Tween-80
and Span-80 were adopted to carry out the micro-
experiment. Moreover, in order to investigate
the effect of surfactant concentrations on water
chain formation, the concentrations used were
40 mg·L-1, 200 mg·L-1 and 1000 mg·L-1. The
experimental cells were set up. The objective lens
was focused. The digital camera was turned on
and connected to a computer. The electric field
strength was adjusted at 1.86 kV·cm-1, the electric field frequency at 3 kHz, and the duty ratio
at 50%. Experimental phenomena were observed,
and experimental images were recorded. Experimental data were processed by micrograph reprocessing software.

3   RESULTS AND DISCUSSION

During processing the experimental data, the water
chain whose length is close to the average length of
the overall experiment is selected as the characteristic water chain under the same condition. This
is because the selected water chain can reflect the
general characteristics more appropriately to minimize random error.

3.1   Effect of surfactant concentrations

Figure 2 to Figure 7 show the variations in the
average particle diameter and water chain length
under different concentration conditions of OP-10,

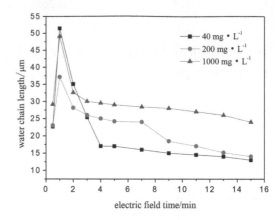

Figure 2.   Variation in the average particle diameter in
different concentration conditions of OP-10.

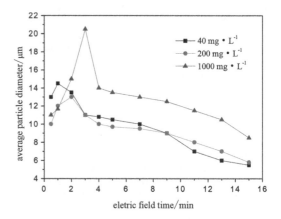

Figure 3.   Variation in the water chain length in different concentration conditions of OP-10.

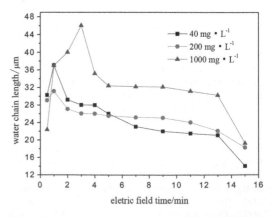

Figure 4.   Variation in the average particle diameter in
different concentration conditions of Span-80.

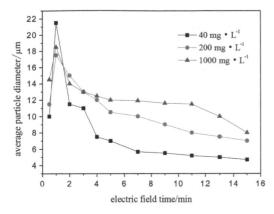

Figure 5. Variation in the water chain length in different concentration conditions of Span-80.

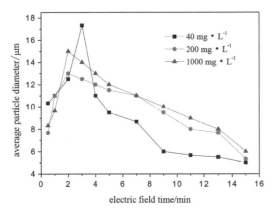

Figure 6. Variation in the average particle diameter in different concentration conditions of Tween-80.

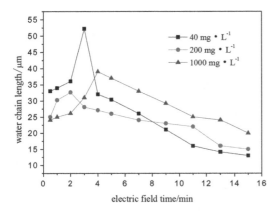

Figure 7. Variation in the water chain length in different concentration conditions of Tween-80.

Span-80 and Tween-80 in a time-dependent electric field. As shown in these figures, oil–water emulsions mix evenly and stabilize at an early stage. The conductivities and permittivities, the electric field force on droplets and the process of water chain formation, coalescence and sedimentation differ in different surfactant systems. Different electric field forces on emulsions lead to different variation trends in the initial average particle diameter and initial water chain length. In a time-dependent electric field, fine aqueous droplets coalesce into large droplets and sedimentation takes place. Then, a continuous aqueous layer evolves. The conductivity and permittivity of water are higher than that of emulsions, which results in the inconspicuous impact of different conductivities and permittivities of emulsions on the process of water chain formation, coalescence and sedimentation in different surfactant systems. As a result, both the average particle diameter and the water chain length first make an increment and then a decrement in a time-dependent electric field. This is because fine droplets arrange water chains and coalesce into large droplets under the effect of the dipole coalescence force in a time-dependent electric field at an early stage, which leads to the increment of the particle diameter and water chain length. At a later stage, large droplets fall under the effect of oil–water density difference and gravity in a time-dependent electric field, which leads to the decrement of the particle diameter and water chain length. Surfactants adsorbed on the oil–water interfacial membrane increase with increasing surfactant concentrations. The oil–water interfacial tension decreases sharply. The oil–water interfacial membrane strength drops rapidly (Ding Xuancai 1990), thereby coalescence taking place easily. Thus, the average particle diameter makes an increment.

In contrast, interfacial energy reduces sharply with increasing surfactant concentrations. The process of bounce declines remarkably after the rupture of interfacial films. The arrangement of surfactants on oil–water interfacial films becomes compact with increasing concentrations. Long chains and branched chains of molecules lead to the steric hindrance domino effect, which hinders coalescence. When two droplets get closer, compression of the thickness of the electric double layer domino effect is enhanced with increasing concentrations. Thus, the electrostatic repulsive force between droplets increases, which hinders coalescence. As a result, the water chain length increases with the increment of the particle diameter and droplet quantities. Interestingly, when the concentrations of Tween-80 increase from 200 mg·L-1 to 1000 mg·L-1, the critical micelle concentration is reached. Residual surfactants separate from the oil–water interface and form a micelle. Interfacial intensity and interfacial energy tend to be stable. Therefore, the average particle diameter of the Tween-80 system makes a negligible increment from 200 mg·L-1 to 1000 mg·L-1.

3.2 *Effect of surfactant types*

Figure 8 shows two kinds of ideal surfactant systems. Surfactant molecules entirely dissolve in the continuous phase in ideal system 1, while surfactant molecules entirely dissolve in the dispersed phase in ideal system 2 (Sun Zhiqian 2013). The HLB values of OP-10, Tween-80 and Span-80 are 14.5, 15.0 and 4.3, respectively. The properties of OP-10 and Tween-80 are similar to those system 2, while the property of Span-80 is similar to that of system 1.

Figure 9 to Figure 11 show the comprehensive analysis of different surfactant types, and the effect of surfactant types on the average particle diameter and water chain length, respectively. The molecular weight of Span-80 and Tween-80 is about 429 and that of OP-10 is about 647. The intensity of new interfacial membranes absorbed with surfactants increases with increasing molecular weight. Aqueous droplet coalescence is therefore difficult. However, the molecular structure of OP-10 is simpler than that of Span-80 and Tween-80. The steric hindrance effect of OP-10 is therefore weaker than that of Span-80 and Tween-80, which results in easy coalescence in the OP-10 system. The results show that the impact of molecular structure is higher than the impact of molecular weight on interfacial membrane intensity. Therefore, the average particle diameter and the water chain length of the OP-10 system is the largest, and that of the Span-80 and Tween-80 systems is similar at an early stage.

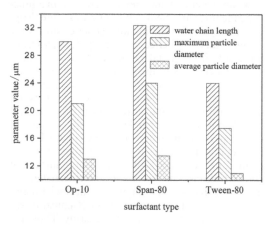

Figure 8.   Two surfactant systems.

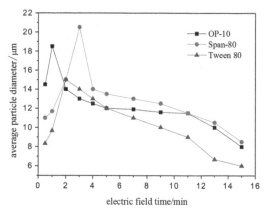

Figure 10.   Impact of surfactant types on the average particle diameter.

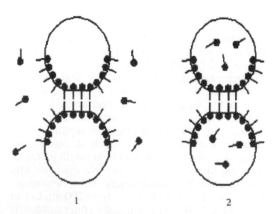

Figure 9.   Comprehensive analysis of different surfactant types.

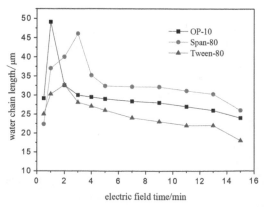

Figure 11.   Impact of surfactant types on the water chain length.

In contrast, Span-80 is an oil-soluble surfactant. When Span-80 dissolves in the continuous phase, surfactant molecules permeate through aqueous droplets under the flow of the continuous phase. Then, surface diffusion of surfactants takes place. This results in Marangoni convection. Coalescence and sedimentation is therefore hindered. Droplets tend to form water chains. As depicted above, sedimentation of large droplets is hindered in the Span-80 system, which accounts for the maximum average particle diameter and water chain length. The HLB value of OP-10 is slightly lower than that of Tween-80. In other words, the oil solubility of OP-10 is slightly higher than that of Tween-80. Therefore, the trend of water chain formation in the OP-10 system is more obvious than that in the Tween-80 system. The increment of the particle diameter and droplet quantities leads to increasing water chain length. In industrial applications, oil–water emulsions usually contain indigenous surfactants such as resin and asphaltene whose combined mechanisms with OP-10, Span-80 and Tween-80 need further investigation.

4 CONCLUSIONS

1. Surfactants adsorbed on the oil–water interfacial membrane make an increment, with the oil–water interfacial tension dropping sharply, membrane intensity making a remarkable decrement, and coalescence of water droplets taking place increasingly easily with increasing surfactant concentrations. As a result, average particle diameters are enlarged. In contrast, the water chain length increases with increasing surfactant concentrations. This is attributed to a sharp drop in interfacial energy, the steric hindrance domino effect and the compression of the thickness of the electric double layer domino effect.
2. For the impact of the surfactant molecular structure on membrane intensity and Marangoni convection caused by surfactants in the continuous phase, the emulsion whose water chain length and particle diameter take the first place is Span-80, while OP-10 takes the second place, and Tween-80 takes the last place.

ACKNOWLEDGMENTS

This work was supported by the National Natural Science Foundation of China under Grant No. 21406267, the Natural Science Foundation of Shandong Provincial under Grant No. ZR2014BL029, and the Graduate Innovation Project Foundation of China University of Petroleum under Grant No. YCX2014019.

REFERENCES

Sun, D.,Jong, S.C., Duan, X.D. & Zhou, D. 1999, Emulsification of water-in-oil emulsion by wetting coalescence materials in stirred—and—packed columns. Colloids and Surfaces A, 150, pp.69–75.
Ding Xuan-cai, Xie Fu-quan & Ding Cui, 1990, a study on swelling properties of emulsion liquid membrane. Membrane Science and Technology, 10(2), pp. 1–25.
Hirato T., Koyama K., & Tanaka T., 1991, Emulsification of water-in-oil emulsion by an electrostatic coalescence method. Mater Trans JIM, 32(3), pp. 257–263.
Jin You-hai, Hu Jia-ning & Sun Zhi-qian, 2010, Experimental study of factors acting on the performance of high-voltage and high-frequency pulse electrostatic dewatering process. Journal of Chemical Engineering of Chinese Universities, 24(6), pp. 917–922.
Goto, M., Irie, J., Kondo, K. & Nakashio, F. 1989, Electrical emulsification of w/o emulsion by continuous tubular coalesce. Journal of Chemical Engineering of Japan, 22(4), pp. 401–406.
Mohammed, R.A., Baile, A.I., Luckham, P.F. & Taylor, S.E., 1994, Dewatering of crude oil emulsions. Colloids and Surfaces A, 83 (3), pp. 261–271.
Mohammed, R.A., Baile, A.I., Luckham, P.F. & Taylor, S.E., 1993, Dewatering of crude oil emulsions. Colloids and Surfaces A, 80 (2&3), pp. 223–235.
Sun Zhi-qian, Jin You-hai & Wang Zhen-bo, 2013, Investigation on characteristic of high frequency impulse electrostatic coalescing process under surfactant systems. Speciality Petrochemicals, 30(5), pp. 49–53.
Sun Jiang-bo, 2004, Treatment technique of polymer production stream in feiyantan oil field. Geology and Harvest Rate of Oil Field, 11(4), pp. 56–57.
Zhang Li-ming, He Li-min & Ma Hua-wei, 2007, Coalescence characteristics of droplets in insulated compact electric demulsified. Journal of China University of Petroleum, 31(6), pp. 82–86.

Energy Science and Applied Technology – Fang (Ed.)
© 2016 Taylor & Francis Group, London, ISBN 978-1-138-02833-3

The optimization of ultrasonic-assisted extraction technology of rosmarinic acid from *Prunella vulgaris* L. by the response surface method

Tian Long & Yiceng Lou
School of Chemistry, Chemical Engineering and Life Science, Wuhan University of Technology, Wuhan, China

ABSTRACT: The objective of this paper was to analyze and optimize ultrasonic-assisted extraction technology of *Prunella vulgaris* L. by using the response surface method and the establishment of the Box–Behnken design. We determined the extraction of rosmarinic acid using the response surface method to optimize the process parameters of three factors such as the solid–liquid ratio, ethanol concentration and extraction time on the basis of a single-factor experiment. The optimization of the extraction conditions of *Prunella vulgaris* L. was as follows: the solvent–solid ratio was 28, ethanol concentration was 74% and the extraction time was 55 min. Under the condition of the process parameters, the rosmarinic acid extraction rate can reach 0.5231%. In conclusion, the optimization of extraction technology by the response surface method has a high efficiency, and can be performed within the continuous analysis. It is suitable for the extraction of rosmarinic acid.

Keywords: *Prunella vulgaris* L., ultrasonic-assisted extraction, rosmarinic acid, response surface method

1 INTRODUCTION

Prunella vulgaris L., called the headdress flower, is the dry ear of Lamiaceae herbs. It can be used in the clinical treatment of red eyes and sore, goiter, resistance to high blood pressure, fall hematic fat, etc. (H. Yang, 2007).

This paper studies the surface optimization method (RSM), which can not only establish a continuous variable, but can also form a response surface. The factors affecting the extraction process are evaluated, and their interaction can make up for the inadequacy of traditional methods (H.Y. Zhu, 2011).

2 EXPERIMENTAL

Equipment and materials: Waters 1525 Binary HPLC pump (Waters, USA); Waters 2998 photodiode array detector (Waters, USA); methanol (chromatographic level, Tedia Company, USA); reagents (analytic grade) for sample preparation; liquid water; rosmarinic acid reference substance (China's Food and Drug Verification Research Institute, 111871–201203). *Prunella vulgaris* L. was purchased from Anhui Bozhou Herbs Company (R. Guo, 2006).

The chromatographic conditions were as follows: ZORBAX-C18 (250 mm × 4.6 mm, 5 microns) with a detection wavelength of 330 nm. The standard curve was as follows: place a standard stock solution of 1.0, 2.0, 4.0, 6.0, 8.0, 10.0 ml in a 10 ml volumetric flask; use the peak area as the ordinate, with rosmarinic acid concentration as the abscissa; and draw the standard curve. The regression equation for rosmarinic acid was as follows: $Y = 38002X + 41731$, $r = 0.9999$.

The procedure for the determination of rosmarinic acid extraction is as follows: determine the filtrate volume (v); precisely measure a certain amount of the filtrate for dilution; record the dilution factor (n) in accordance with the method of preparation of the standard curve, according to the construction of the regression equation for the determination of rosmarinic acid in the filtrate quality concentration (c) (Z.Z. Wang, 2007):

$$\text{Rosmarinic acid extraction } (\%) = \frac{c \times n \times v}{m} \times 100\%.$$

3 RESULTS AND DISCUSSION

3.1 *Single-factor test results and analysis*

The study of the ratio of liquid to material: *Prunella vulgaris* L. powder is made into six balls. The results showed that when the material to liquid ratio is 1:30, the extraction yield reaches the highest, and then gently moves towards a downward trend. At 1:50, the rosmarinic acid extraction yield is basically no longer affected by the material to liquid ratio.

The study of the ethanol volume fraction: the results showed that when other conditions are kept constant and the volume fraction of the alcohol extract conditions is changed, the rosmarinic acid extraction rate increases rapidly, when extracting ethanol volume fraction of 70% up to a maximum, when the ethanol to continue to increase the volume fraction of rosmarinic acid extraction yield declines.

Ultrasonic time factor investigation: The results show that when the ultrasonic extraction of 50 min, the extraction yield reaches the maximum, then extends the ultrasonic extraction time, the extraction yield of rosmarinic acid will no longer increase slightly down instead (L. He, 2010).

Optimization design of the response surface method experiment: select the material to liquid ratio (g/ml, the X_1), the ethanol volume fraction (%, X_2), and the three factors of ultrasonic time (min, X_3) response surface analysis to determine the main factors influencing the rosmarinic acid extraction yield. The experimental design factors and levels are listed in Table1.

3.2 The response surface method to optimize the ultrasonic extraction rosmarinic acid test

In order to optimize the *Prunella vulgaris* L. rosmarinic acid extraction process conditions according to the principle of the Box–Behnken test design, a total of 17 site design, analysis of 12 points, five

Table 1. Test factor level and coding.

Level	Factors		
	X1(g/mL)	X2(%)	X3(min)
−1	1:20	60	40
0	1:30	70	50
+1	1:40	80	60

Table 2. Test results of rosmarinic acid response surface optimization ultrasonic extraction from *Prunella vulgaris* L.

Serial no.	X1 (g/ml)	X2 (%)	X3 (min)	Rosmarinic acid extraction yield (%)
1	0	1	1	0.4932
2	0	0	0	0.5177
3	−1	0	−1	0.3523
4	1	1	0	0.4094
5	1	−1	0	0.3321
6	−1	0	1	0.4762
7	0	−1	1	0.4353
8	0	0	0	0.5144
9	−1	−1	0	0.3484
10	0	0	0	0.5171
11	0	−1	−1	0.3214
12	1	0	−1	0.3973
13	−1	1	0	0.4843
14	0	0	0	0.5133
15	0	1	−1	0.4969
16	1	0	1	0.4561
17	0	0	0	0.5134

Table 3. Significant regression equation of regression coefficients.

Sources of variance	Sum of squares	Freedom	Mean square	F	Prob > F	Significance
Model	0.079	9	0.008829	25.64	0.0002	**
X_1	0.0005495	1	0.0005495	1.60	0.2470	
X_2	0.25	1	0.025	72.41	<0.0001	**
X_3	0.011	1	0.011	31.14	0.0008	**
$X_1 X_2$	0.0008585	1	0.0008585	2.49	0.1583	
$X_1 X_3$	0.001060	1	0.001060	3.08	0.1228	
$X_2 X_3$	0.003457	1	0.003457	10.04	0.0157	*
X_1^2	0.020	1	0.020	58.10	0.0001	**
X_2^2	0.012	1	0.012	33.96	0.0006	**
X_3^2	0.002798	1	0.002798	8.13	0.0247	*
Residual	0.002410	7	0.0003443			
Loss of quasi item	0.002393	3	0.0007976	183.92	<0.0001	**
Error	0.00001735	4	0.000004337			
Sum	0.082	16				
Correction factor	$R^2 = 0.9706$	Radj = 0.9327				

Note: '*' denotes significance at the 0.05 level; "**" denotes significance at the 0.01 level.

zeros for estimating error, each precision, respectively, taking 1.00 g *Prunella vulgaris* L. are considered to determine the extraction of rosmarinic acid content and to calculate its extraction yield. The results are given in Table 2.

The establishment of the model: using the data in Table 2, the quadratic multinomial regression model equation can be obtained as follows: $Y = 0.51518 - 8.28750 \times 10^{-3} X_1 + 0.055825 X_2 + 0.036613 X_3 0.014650 X_1 X_2 0.016275 X_1 X_3 0.029400 X_2 X_3 \quad 0.068928 X_1^2 0.052703 X_2^2 0.025777 X_3^2$.

Analysis of variance:

Multiple regression analysis of variance is given in Table 3.

The reliability of the model from the variance analysis and the related coefficient examined: the p value <0.01 suggests that the chosen quadratic polynomial model by the tests is highly significant. Moreover, the quadratic term in the equation $X_1^2 X_2^2$ is remarkably influenced, the interaction between $X_2 X_3$ and the secondary item X_3^2 is significant, the decision of the equation coefficient is 97.06%, showing that 97.06% of the response value is derived from the selected variables. The regression equation can well describe the true relationship between the response value and various factors, which can use the model instead of real sites with the result of the experiment being analyzed and forecast.

3.3 *Response surface analysis and optimization*

According to the quadratic regression equation response to the three-dimensional curve, two factors of the interaction between the corresponding response values (rosmarinic acid extraction yield Y) can be obtained. The results are shown in Figure 1.

The ethanol volume fraction (X_2) (X_3) is much greater than the extraction time and the solid–liquid ratio (X_1), the ethanol volume fraction of extraction used in rosmarinic acid extraction has an extremely significant effect $(p < 0.0001)$. Among these three factors, the ethanol volume fraction and the extraction time and their interaction have a significantly greater impact on the response curve of the contour than the solid–liquid ratio and the extraction time, and the interaction between the solid–liquid ratio and the ethanol volume fraction, i.e. $X_2, X_3 > X_1, X_3 > X_1, X_2$, with the ethanol volume fraction and the extraction time of rosmarinic acid extraction yield having a significant effect $(p < 0.05)$.

4 CONCLUSIONS

The results showed that the best extraction technology conditions for *Prunella vulgaris* L. extraction were as follows: solvent–solid ratio 28;

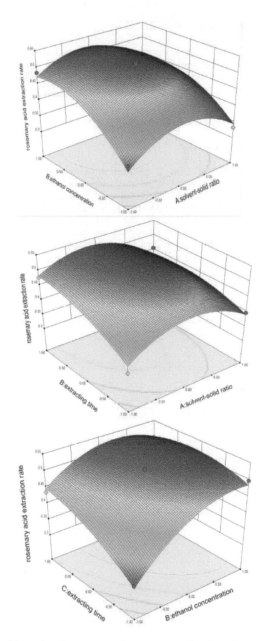

Figure 1. Response surface contours of various factors for rosmarinic acid extraction.

ethanol concentration 74%; and extraction time 55 min. Under these conditions, the average yield of the extraction rate of rosmarinic acid in three validation experiments was 0.5231%. The method for making effective use of *Prunella vulgaris* L. resources provide a certain theoretical reference to improve its efficiency in the industrial application (M.K. Akalin, 2013).

ACKNOWLEDGEMENTS

This work was supported by the National Natural Science Foundation of China under Grant No. 81173509 and Public Science and Technology Research Projects of Hubei Province under Grant No. 20121 g0186.

REFERENCES

H. Yang, W.Y. Guo, Y, Zhao, et al. Orthogonal experiment optimization selfheal extraction technology research. (Lishizhen Medicine and Materia Medica Research), Vol. 17(2007), NO. 10, p.2505–2506.

H.Y. Zhu, X.L. Peng, B.Feng, Orthogonal experiment optimization selfheal rosemary acid ultrasonic extraction technology. (Journal of Jilin institute of medicine), Vol. 32 (2011), NO.1, p.11–14.

L. He, G.W. Zhang, J. Yang, et al. The response surface method to optimize the ultrasonic extraction selfheal medium yellow ketone technology research (Journal of food industry science and technology), Vol.31 (2010), No. 9, p. 259–264.

M.K. Akalin, S. Karagoz, M. Akyuz, Application of response surface methodology to extract yields from stinging nettle under supercritical ethanol conditions(Journal of Supercritical Fluids), Vol. 84(2013),p.164–172.

R. Guo, J.J. Li, L. Guo, et al. Licorice slices was optimized by using response surface method of ultrasonic extraction technology of glycyrrhizin acid (Journal of northwest A&F university, natural science edition), Vol. 9,2006, p.188–191.

Z.Z. Wang, W.J. Wu, Wild chrysanthemum flower extraction technology of total flavonoids response surface design optimization (Lishizhen Medicine and Materia Medica Research), Vol. 19, (2007) No.3,p.648–650.

Environmental engineering and sustainable development

Energy Science and Applied Technology – Fang (Ed.)
© 2016 Taylor & Francis Group, London, ISBN 978-1-138-02833-3

Research on the hydrological effects of forest litter-fall

J.H. Zhang, X. Hou & Q.S. Shu
Chongqing Water Resources and Electric Engineering College, Chongqing, China
Chongqing Institute of Water Resources and Science, Chongqing, China

ABSTRACT: Litter-fall in the forest ecological system has a very important hydrological function. The storage capacity of litter-fall depends on its accumulation in the forest and its water-holding ability, which, in turn, depend on forest tree species, stand of the horizontal and vertical structures, the condition of litter-fall decomposition, and other factors. Litter-fall leads to redistribution of rainfall, which has more porosity than top soil, which is beneficial for retaining moisture, controlling surface evaporation, and promoting the effect of infiltration. Litter-fall facilitates the capture and infiltration of rainwater into lower soil layers. Litter-fall protects soil aggregates from the impact of raindrops, preventing the release of clay and silt particles from plugging soil pores. This plays an important role in soil and water conservation.

Keywords: Litter-fall; Hydrological; Soil and water conservation; Soil water

1 INTRODUCTION

Forest litter-fall is composed of dead plant material, such as leaves, bark, needles and twigs, that has fallen to the ground. This detritus or dead organic material and its constituent nutrients are added to the top layer of soil, commonly known as the litter layer or O horizon ("O" for "organic"). Litter-fall has attracted the attention of ecologists at length for the reasons that it is an instrumental factor in ecosystem dynamics, is indicative of ecological productivity, and may be useful in predicting regional nutrient cycling and soil fertility. Litter-fall is characterized as fresh, undecomposed and easily recognizable plant debris. This can be anything from leaves, cones, needles, twigs, bark, seeds, logs or reproductive organs (Table 1).

Items larger than 2 cm of diameter are referred to as coarse litter, while anything smaller is referred to as fine litter or litter. The type of litter-fall is most directly affected by ecosystem type. For example, leaf tissues account for about 70% of litter-fall in forests, but woody litter tends to increase with forest age. In grasslands, there is very little above-ground perennial tissue.

Forest litter layer leads to redistribution of rainfall, which has more porosity than top soil, that is beneficial for retaining moisture, controlling surface evaporation and promoting the effect of infiltration. Litter-fall facilitates the capture and infiltration of rainwater into lower soil layers. Litter-fall protects soil aggregates from the impact of raindrops, preventing the release of clay and silt particles from plugging soil pores. Litter-fall can reduce the impact of raindrops, reduce the soil structure damage caused by the raindrop splash and soil erosion, prevent soil splash erosion, hysteresis, runoff, production flow time, and increase soil resistance. Litter-fall is considered as the main function of water conservation in the forest ecosystem, which plays a very good role in soil and water conservation and water conservation. Among different forest types with various forest canopies, under the wood or ground cover layer, litter-fall's structure, composition, types, quantity and properties vary, which affect the forest hydrological function. Therefore, the study of the hydrological function of litter-fall on water cycle and water balance in the forest ecosystem is very useful and of great significance.

Table 1. Monthly variations of litter-fall in the *P. massoniana* forest.

Litter-fall component	Month				
	3	5	7	9	11
Needle	277	119	335	410	396
Twig	2.5	11	13	39	3
Inflorescence	26	35	86	6	5
Bark	13	18	33	22	3
Cone	15	8	6	3	6

2 LITTER-FALL'S WATER-HOLDING CAPACITY

Water-holding capacity of litter-fall is an important indicator of the function of hydrology. Most of the domestic research in China that focused on the problem has found that the water-holding capacity of litter-fall depends on its accumulation in the forest and its water-holding ability, which depends on the development of forest tree species, stand of the horizontal and vertical structures, and the condition of litter-fall decomposition. Forest hydrology researchers also found that forests such as litter-fall reserves have a significant effect on its water-holding capacity. Wang et al. studied the litter-fall characteristics of the Masson pine forest in south China, which is a widely distributed local tree species. Researchers found that the order of litter-fall components was as follows: needles > twigs > broad leaves > debris > bark and cones. The monthly proportion of dead needle to total litter-fall was almost the highest in the whole year and fluctuated from 32.8% to 95.7%. The monthly variation of broad leaves also presented a double-peak pattern, but the highest peak appeared during the period of April–May and during the period of November–December. The monthly proportion of litter-fall components such as green needles, bark and cones was always the lowest. The monthly proportion of debris reached to a maximum during the period of April–May and during the dry summer period of August, respectively.

Li and Lu investigated five kinds of broad-leaved forest litter-fall's water-holding characteristics in southwest China, which were as follows: *Quercus aquifolioides, Populus davidiana, Betula platyphylla, Rhododendron aganniphum* and *Salix cupularis.* The results showed that the range of broad-leaved forest litter reserves was 2240 t/hm²~11040 t/hm², in the following order: *Quercus aquifolioides > Populus davidiana > Betula platyphylla > Rhododendron aganniphum > Salix cupularis.* The water-holding ability was in following order: *Populus davidiana > Betula platyphylla > Quercus aquifolioides > Rhododendron aganniphum > Salix cupularis.* Litter-fall's water-holding capacity was closely related to the amount and soaking time. The order of litter-fall's water-holding rate was as follows: *Salix cupularis > Betula platyphylla > Rhododendron aganniphum > Populus davidiana > Quercus aquifolioides,* while the water absorption rate was related to the soaking time.

To understand the reserves and water-holding characteristics of litter-fall in *Larix principis-ruppre-chtii, Betula platyphylla* and *Populus davidiana,* Zhang and Ma conducted an experiment on different forest ages, different components and different proportions of litters in *Larix principis-rupprechtii* in the Yan Shan mountain. The results showed that (1) the reserves of six forest stands were 35.75, 45.50, 60.00, 65.94, 25.40 and 19. 39 t/ hm², respectively. (2) In each period, the water-holding rate of litter-fall of different ages in *Larix*

principis-rupprechtii plantations all exhibited the same trend in the order 16a > 23a > 34a > 42a; the different mixed modes of litter, the water-holding rate of litter of *Larix principis-rupprechtii + Betula platyphylla* was the highest. The water-holding rate increased in a logarithmic manner when the soak time was extended. (3) The water absorption rate reduced in turn among 16a, 23a, 34a and 42a. The water absorption rate of litter of *Larix principis-rupprechtii + Betula platyphylla* was the highest. The water absorption rate decreased in a power function manner when the soak time was extended.

3   LITTER-FALL PREVENTS RUNOFF VELOCITY

Litter-fall's rain entrapment quantity changes with the change in forest types, as revealed by the research conducted in different forest vegetation types in the Chinese Liao dong mountainous area. The research on the litter-fall's layer intercept rainfall behavior pointed out that the litter-fall's intercept rain quantity increased with the increase in rainfall, which had a significant linear relationship, that litter-fall's interception rate falling gradient was larger at low rainfall areas, then increased with forest precipitation and reduced gradually, finally reaching a certain value, and that litter-fall's intercept saturated, with the saturation of different forest types of litter-fall's interception rate being not the same. Forest litter-fall's rainfall interception ability was improved after reconstruction.

The famous American forest hydrologist Richard Lee suggested that the amount of rainfall interception of litter-fall depends on the water storage capacity of litter-fall. Gao found that litter-fall's water-holding ability following a period of precipitation was dependent on the standing crop of litter-fall and the average natural moisture content. The study also found that mixed bamboo leaf litter-fall showed stronger hydrological adjusting function than the pure bamboo forest.

The runoff flow velocity and flow rate in the forest surface are the mainspring of soil erosion, while flow velocity plays a greater role than runoff flow velocity, so delaying runoff and reducing runoff scouring soil are key factors that maintain water and soil. Forest litter-fall plays an important role in slow flow velocity. The research conducted in the Ding Hushan biosphere reserve found that litter-fall in three forest communities adjusted and blocked surface runoff, reducing runoff and flow velocity, increased the function of the surface roughness, and the thicker litter-fall layer had stronger water absorption and water retention ability, and the water conservation function of the forest was greater, which reduced the chance of surface runoff. The research of four kinds of forest litter-fall layer in the Ba dealing forest farm founded that litter-fall layer can obviously prevent the runoff, and delay the production flow time. Litter-fall could prevent the effect of flow velocity, mainly because litter-fall could absorb runoff and increase the surface roughness.

4 LITTER-FALL COULD IMPROVE SOIL STRUCTURE

Soil structure and permeability are the main factors that reflect soil and water conservation, while soil infiltration ability is also an important factor for the evaluation of the forest, soil and water conservation. Forest has a variety of effects on soil physical properties, but the most significant influence on 0~20 cm soil layers is forest litter-fall. Researchers found that by means of soil and water conservation, we must find an easy way to make more soil and surface runoff infiltration and penetration into the soil under gravity into the groundwater. The existence of litter, especially its lower decomposition layers, could increase soil organic matter, and improve soil structure that has a great effect on infiltration. A large number of experimental results showed that soil infiltration condition not only depends on forest appearance or stand density, but also depends on the characteristics of litter-fall.

The sand flow rate is an important indicator for measuring the degree of soil and water loss. Researchers found different sediment yields with litter-fall, which means litter-fall has a certain influence on sediment yield. There were obvious differences between bare land and litter-fall covered sediment production flow: with the increase in slope, the sediment and flow reduction effect of litter-fall became more obvious. Litter-fall could hold rainfall effectively, or block runoff, which ultimately reflected with respect to flow and sediment reduction of forest land, which effectively shows that litter-fall plays an important role in the process of soil and water conservation, preventing a large number of soil and water loss every year globally.

Litter-fall layer in the forest ecological system has a very important hydrological function, the storage capacity of litter-fall depends on its accumulation in the forest and its water-holding ability, which, in turn, depend on forest tree species, stand of the horizontal and vertical structures, the condition of litter-fall decomposition, and other factors. The current natural forest protection project and the project of returning farmland to forest adjust measures to local conditions to choose suitable afforestation tree species, in order to increase the ground and surface coverage and enhance its hydrological ecological function, reduce soil and water loss, which plays an important role in soil and water conservation.

ACKNOWLEDGEMENTS

Funding was provided by scientific research projects of the Chongqing municipal education committee <Research on the vegetation pattern of reservoir shore zone and its ecological effect> (KJ1403603); Talent introduction of scientific research project of Chongqing Water Resources and Electric Engineering College <Vegetation pattern of and their ecological effects on soil and water conservation> (KRC201403); The Major projects of Chongqing Water Bureau <Research and extension conservancy of water resources and water ecological restoration plants applied in water conservancy projects> (YSK201306). The Basic and Advanced Project of Chongqing <The mechanisms of soil macrospore flow and contaminant transport for wetland in the three gorges, China> (cstc2014jcyjA20009); the Science and Technology Projector Chongqing Education Commission <Spatial distribution and variation pattern of soil conservation effect for slope hedgerow systems in the three gorges, China> (KJ1403602).

REFERENCES

Curiel Yuste J, Janssens I. A, Carrara A, et al. 2004. Annual Q10 of soil respiration reflects plant phonological patterns as well as temperature sensitivity. Global Change Biology, 10 (2): 161–169.

Janssens, I.A, Dore S, Epron D, et al. 2003. Climatic influences on seasonal and spatial differences in soil CO2 efflux, in fluxes of carbon, water and energy of European forests. Berlin: Springer-Verlag.

Li, J, Lu, J. 2014. Research on the litter-fall water-holding characteristics in Sela Mountain. Sichuan Forestry Exploration and Design, (4):23–27.

Li, Z.H, YU, P.T, Wang, Y.H, et al. 2011. Characters of Litter-Fall in Damaged Pinus massoniana Forests and Its Responses to Environmental Factors in the Acid Rain Region of Chongqing, China. Forestry Science, 47(8):19–26.

Meentemeyer, V, Box, E.O, &Thompson R. 1982. World patterns and amounts of terrestrial plant litter production. Bioscience 32(2):125–128.

Wang, Y.H, & Pengtao, Y.U.2013. The litter-fall characteristics and their response to drought stress in the Masson pins forests damaged by acid rain at Chongqing, China. Act Ecologica Sonica 32(6): 1842–1851.

Zhang, Y, MA, C,M. 2014. Reserve and water-holding characteristics of litters in *Larix principis-rupprechtii* Plantation. Forest resources management, (6):63–68.

Energy Science and Applied Technology – Fang (Ed.)
© 2016 Taylor & Francis Group, London, ISBN 978-1-138-02833-3

Life cycle assessment on Battery Electric Vehicles

Yaowei Zhang, Yanfen Liao, Guicai Liu, Shanchao Hu & Xiaoqian Ma
Key Laboratory of Efficient and Clean Energy Utilization of Guangdong Higher Education Institutes, South China University of Technology, Guangzhou, China

Ru Yang & Junfei He
Guangzhou Institute of Energy Testing, Guangzhou, China

ABSTRACT: The method of life cycle assessment is carried out for Battery Electric Vehicles (BEVs) including production, use and recycling. After taking the weighted average, the coefficient of resource depletion is 0.39 mPR90 and the total environmental impact load is 14.63PET. BEVs save resources by 50.6% and reduce pollution by 5.2% compared with internal combustion engine vehicles. The main source of energy consumption and emissions of electric vehicles is in the utilization phase.

Keywords: Battery electric vehicle; life cycle assessment; resource consumption; environmental impact

1 INTRODUCTION

Nowadays, electric vehicles (EV) have become an important direction to develop the automobile industry (Walker and Zhang, 2014, Seixas et al., 2015, Nanaki and Koroneos, 2013, Feng and Figliozzi, 2012, Sharma et al., 2013, Zhou et al., 2013, Lucas et al., 2012, Ma et al., 2012). With energy and environment problems drawing much attention of public, electric vehicles have become a research focus and have another rapid development period. The China State Council has approved the "Energy Saving and New Energy Vehicle Industry Development Plan" to achieve leapfrog development of China's automobile industry. The plan states that the battery electric vehicle (BEV) will be the main strategic direction in China.

The battery electric vehicle industry has started quickly and developed rapidly (Zhou et al., 2013, Wu et al., 2015b, Shuhua, 2014, Huo et al., 2015, Lin et al., 2013). Having no emission while running because its driving power comes from the electric battery, battery electric vehicle is also called 'zero-emission vehicles'. However, it causes a lot of resource consumption and environmental pollution due to the production part that requires a plenty of electric power, consuming much fossil fuel. The present study (Nanaki and Koroneos, 2013, Feng and Figliozzi, 2012, Sharma et al., 2013, Lucas et al., 2012, Ma et al., 2012, Wu et al., 2015a) shows that electric vehicles have advantages of energy saving and environmental protection compared with traditional fuel vehicles; however, we can see only a few life cycle assessments on

electric vehicles. Therefore, it is necessary to make quantitative analysis of its whole life cycle to investigate the advantages and disadvantages.

2 RESEARCH GOAL AND SCOPE

2.1 *The definition of product system*

In this paper, a battery electric vehicle will be considered as a functional unit to value its goal and scope. This battery electric vehicle is powered by a 24 kWh lithium-ion battery, weighing (excluding batteries) 1735 kg and its power consumption is 28 per 100 km. The internal combustion engine vehicle weighs 1345 kg and its fuel consumption is 8 L per 100 km.

Studies are arranged into four categories: supply of raw materials, production of auto spare parts, assembly and transportation of vehicles, and recycling of materials from vehicle scrapping. The assessment of resource consumption contains three major fossil fuels, including coal, oil and natural gas. The assessment of environmental impact includes greenhouse gas emissions and standard emissions.

2.2 *System boundary*

The whole life cycle of vehicles is divided into 3 stages: production, use and recycling, including raw material supply, processing, transportation, electric production and emissions. The life cycle boundary is shown in Figure 1.

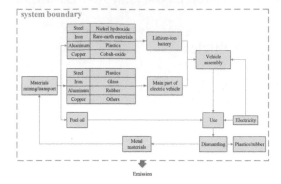

Figure 1. System boundary of the whole life cycle of vehicles

3 INVENTORY ANALYSIS

Inventory is the description of basic data in the process of life cycle assessment. Inventory analysis is the basis of life cycle assessment data and is an important intermediate step of life cycle assessment. Therefore, the integrity and accuracy of the data collection of the inventory play a decisive role in the results of inventory analysis during the entire LCA process.

3.1 Data sources

The inventory of automobile manufacturing was obtained from a full life-cycle model called GREET, which was developed by the US Department of Energy's Argonne National Laboratory (2010, Hertwich and Hammitt, 2001, Hertwich E G, 2001). Then, the Chinese Life Cycle Database (CLCD), the China Energy Statistical Yearbook and other native data were used to calculate the resource consumption, and atmospheric pollution and greenhouse gas emission. The fuel cycle from wells to wheels and the vehicle cycle through material recovery and vehicle disposal were considered in the GREET model in order to fully evaluate energy and emission impacts of advanced vehicle technologies and new transportation fuels. The GREET model was based on the database Argonne National Laboratory, and the corresponding data should be modified according to the native database when the model was employed outside the USA. It was relatively difficult to implement the local modification of data on the model directly. The acquisition phase and the transportation phase of raw material, production phase, utilization phase, and recycling treatment phase were included in the process of automobile manufacturing. Data of raw material inventory and the assembly phase used in this paper were obtained from the GREET 2014, and data of other phases were collected from China's domestic data.

Table 1. Battery consumables.

Composition/kg	BEV
$LiMn_2O_4$	206.3118
Graphite/Carbon	90.1842
Binder	15.4425
Copper	67.3293
Wrought Aluminum	117.9807
$LiPF_6$	11.1186
Ethylene Carbonate	32.7381
Dimethyl Carbonate	32.7381
Polypropylene	10.5009
Polyethylene	1.8531
Polyethylene Terephthalate	7.4124
Steel	8.6478
Thermal Insulation	1.8531
Coolant: Glycol	6.177
Electronic Parts	7.4124
Total	617.7

Table 2. Vehicle component consumables.

Composition/kg	BEV	ICEV
Total Weight	1117.449	1435
Steel	742.0487	952.9204
Iron	22.28305	28.61534
Aluminum	72.73811	93.40846
Copper	52.31784	67.18527
Plastics	135.3901	173.8646
Glasses	38.85929	49.90213
Rubber	19.63805	25.21869
Others	33.93469	43.57808

3.2 Acquisition phase of raw material

The composition of raw materials of the electric vehicle body was simplified into steel, iron, aluminum, copper, plastics, glasses and rubber in this paper. The simplification would ensure the accuracy of data used in the process of inventory analysis, and would relatively decrease the unnecessary workloads. Moreover, with the simplification of the variety of materials, it would be more intuitive and convenient to compare the resource consumption and environmental impact assessment (EIA) between different automobiles. According to the GREET 2014, the raw material composition of the lithium-ion battery and the vehicle body are listed in Table 1 and Table 2, respectively.

3.3 The manufacturing and assembly phase

The manufacturing and assembly phase of vehicles was primarily carried out in the assembly plant, whose resource consumption was almost electric power consumption. Therefore, only the secondary energy electric power was considered in the manufacturing and assembly phase of electric vehicle, and the primary energy consumption and atmospheric pollution and greenhouse gas emission were calculated and converted into electric power consumption on the basis of the local data in China. The manufacturing and assembly phase included the production of vehicle fluids and vehicle component, and component assembly.

3.4 *The transportation phase*

The transportation phase consisted of the distances from the mine, via raw material-producing plants, component-manufacturing plants, to vehicle body assembly, and the distances between collection points of waste vehicles and processing plants. According to statistics, the transportation distances were about 1500 km, and the emissions of engine are in accordance with the Euro III emissions standard using trucks. The energy consumption of the transportation phase is given in Table 4.

3.5 *The utilization phase*

In this investigation, the value of electric vehicles was set to 250000 km. The charging efficiency of the battery of BEV was 85%, and the fuel economy was 0.212 kW.h/km. The gasoline consumption of

Table 3. Energy consumption in the manufacturing and assembly phase.

Values /kW·h	BEV	ICEV
Total Power Consumption	7253.25	6890.34

Table 4. Energy consumption of the transportation phase.

Values/kg	BEV	ICEV
Diesel	137.7243	113.9103

per 100 km for ICEV was 9 L. In this paper, it was assumed that the emissions of electric vehicles are in accordance with the China V emissions standard.

3.6 *The recycling treatment phase*

The recycling treatment and utilization inside the LCA was considered. The life cycle of waste vehicles was employed in other products and other utilizations regarded as materials were neglected here. It was assumed that only the electric energy was consumed in the process of the recycling treatment, and the recovery rate was 30%. The energy consumption of the recycling treatment phase is given in Table 5.

3.7 *Comprehensive analysis*

The energy consumption and atmospheric pollution and greenhouse gas emission of three electric vehicles of BEV and ICEV at each phase in the full life cycle were calculated, and the results are summarized in Table 6 and Table 7, respectively.

As can be seen from Table 6 and Table 7, the main sources of the energy consumption and atmospheric pollution and greenhouse gas emission of BEV and ICEV were from the utilization phase, which accounted for more than 80%. Some data under the recycling treatment phase had negative values, which suggested that the process of the recycling treatment and utilization could decrease the energy consumption and atmospheric pollution and greenhouse gas emission in the full life cycle of BEV and ICEV.

Table 5. Energy consumption of the recycling phase.

	Unit energy consumption (kW·h /kg)	BEV	ICEV
Vehicle body	0.287	96.19	123.53
Battery	24.037	4453.21	0
Total		4549.41	123.53

Table 6. Energy consumption and emission list of the BEV full life cycle.

BEV	Energy consumption (kg/per vehicle)			Emissions (kg/per vehicle)						
	Natural gas	Coal equivalent	Crude oil	CO_2	CH_4	NO_x	SO_2	PM	CO	VOC
RM[a]	0.00	1120.50	157.08	8281.15	24.48	23.32	55.72	38.94	5.43	1.78
MT[a]	0.00	2792.50	139.54	6391.15	3.88	7.00	2.20	0.76	1.90	0.25
U[a]	0.00	24005.88	15.65	51316.13	33.30	40.72	17.77	6.01	8.73	0.00
RT[a]	0.00	1325.43	−46.04	1067.54	−5.04	−4.18	−15.49	−11.27	−1.03	−0.53
Total	0.00	29244.32	266.24	67055.97	56.62	66.86	60.21	34.45	15.04	1.49

[a]RM: Raw materials; MT: Manufacturing and transportation; U: Utilization; RT: Recycling treatment.

Table 7.   Energy consumption and emission list of the ICEV full life cycle.

HEV	Energy consumption (kg/per vehicle)			Emissions (kg/per vehicle)						
	Natural gas	Coal equivalent	Crude oil	CO_2	CH_4	NO_x	SO_2	PM	CO	VOC
RM[a]	0.00	1438.92	201.72	5299.35	10.95	21.31	40.88	40.59	4.62	0.92
MT[a]	0.00	2652.78	115.64	6019.55	3.69	6.37	2.07	0.71	1.70	0.20
U[a]	11.77	1944.00	20930.40	59940.00	27.54	16.52	0.00	1.24	275.40	0.00
RT[a]	0.00	−384.12	−60.48	−1488.14	−3.22	−6.31	−12.23	−12.17	−1.37	−0.28
Total	11.77	5651.58	21187.27	69770.76	38.96	37.89	30.72	30.38	280.35	0.85

[a]RM: Raw materials; MT: Manufacturing and transportation; U: Utilization; RT: Recycling treatment.

4   IMPACT ASSESSMENT

4.1   *The coefficient of resource depletion*

To provide a clear picture of how great the resource consumption values are and which of these values are the most important, the resource consumption for one electric vehicle was normalized and then weighted.

During normalization, the resource consumption (RC (j)) was divided by the corresponding normalization references. The normalization references for consumption of a non-renewable resource were found to be the duration, T (year) for the service defined in the functional unit, multiplied by the annual consumption of the resource in 1990, RR90 computed per person in the world. The normalized resource consumption, NR (j), was calculated as follows:

$$NR(j) = RC(j)/(T*RR(j)90) \qquad (1)$$

The normalized resource was expressed as mPEw90, which was the milli person equivalents (mPE) calculated as the source consumption for an average person in the world (W) 1990. The normalized resource consumption merely reflected the relative size of the consumption of various resources, but did not reflect the scarcity of these resources; therefore, a weighted analysis and resource depletion coefficient was calculated. The weighted resource consumption was expressed as mPR90, which was equal to the normalized resource consumption (NR(j)) multiplied by the corresponding weighting factor. The weighting factor depended on the number of years for which current consumption of the resource can continue before known reserves were exhausted. Therefore, the weighted resource consumption, WR (j), was calculated as follows:

$$WR(j) = NR(j)/(time) \qquad (2)$$

Table 8 shows the primary energy consumption per electric vehicle product.

Table 8 shows that the resource consumption (mPR90) of BEV and ICEV is 0.389 mPR90 and

Table 8.   System resource consumption before and after standardizing and weighting.

	Unit	BEV	ICEV
90's coal	kg/(a, person)	574	574
90's oil	kg/(a, person)	592	592
90's gas	kg/(a, person)	382	382
Coal consumption	kg	36944.32	5435.58
Oil consumption	kg	271.25	18861.67
Gas consumption	kg	0	10.46
Coal's Normalized resource consumption/ PEw_{90}	mPEw90	64.36	9.47
Oil's Normalized resource consumption/ PEw_{90}	mPEw90	0.46	31.86
Gas's Normalized resource consumption/ PEw_{90}	mPEw90	0	0.027
Coal's available supply time	a	170	170
Oil's available supply time	a	43	43
Gas's available supply time	a	60	60
Coal's Weighted resource consumption (mPR90)	mPR90	0.378	0.055
Oil's Weighted resource consumption (mPR90)	mPR90	0.01	0.74
Gas's Weighted resource consumption (mPR90)	mPR90	0	0.00045
Total	mPR90	0.389	0.797

0.797 mPR90. The full life cycle of 1 BEV energy consumption was much lower than 1 ICEV because the available supply period of crude oil was much shorter than that of coal. Because the energy

Table 9. Environmental impact potential value of the research system.

	Weighted environmental impact potential value	
	BEV	ICEV
Soot and ashes	1.23	1.02
Global warming	8.14	6.29
Acidification	2.47	1.16
Nutrient enrichment	1.29	0.58
Photochemical ozone formation	1.49	6.38
Total	14.63	15.44

consumption of BEV and ICEV in the process of using is in different ways, in the life cycles of BEV, the coal resource consumption accounted for the major part; For ICEV, oil consumption takes the largest proportion.

4.2 Environment impact assessment

The measures of environmental impacts fall into five general categories: global warming potential, acidification potential, eutrophication, dust and photochemical oxidation. After normalization and weighting, 5 environmental impact potentials were obtained. The analysis of the environmental impacts is given in Table 9.

Table 9 shows the weighted environmental impacts of BEV and ICEV being 14.63PET and 15.44PET, which means the BEV reduces pollution by 5.2% compared with the ICEV. In this paper, the way of power generation is assumed to be coal-fired powered; furthermore, the ICEV emission is in accordance with the China V emissions standard, thus achieving a small gap of environmental impact potential value between the BEV and the ICEV. In a variety of environmental impacts, the greatest impact is global warming, acidification, photochemical pollution being less, and eutrophication being the smallest. The BEV and ICEV in the life cycle mainly consume gas or electricity. Gasoline combustion and power generation process emissions of VOC, CO_2 and other harmful gases are the main factors that lead to photochemical pollution and global warming.

5 CONCLUSIONS

In this paper, based on the life cycle hypothesis, combined with the stage of raw material production, manufacture and assembly, using course, material recovery, the resource consumption and environmental effects of BEV and ICEV are analyzed in detail. The main results are as follows:

i. The resource consumption (mPR90) of the BEV and ICEV is 0.389 and 0.7971, respectively. The environment burden (PET2010) is 14.63 and 15.44, respectively.
ii. Resource depletion and environmental impact potential coefficient of the BEV are both smaller than those of the ICEV. In the life cycle of the BEV, coal accounts for the main part of energy consumption, while crude oil accounts for the main part in the life cycle of the ICEV. The main effect on the environment is global warming, and the reason is that both BEV and ICEV consume petrol and electricity mostly in the life cycle, which produces large amounts of VOC, CO_2 and other harmful gas.
iii. Apparently, the optimization of energy structure is one of the important measures that promote the energy saving and emissions reduction effect of a new energy vehicle. Therefore, we should pay more attention to the development of clean energy such as nuclear and wind power, and improve the energy efficiency of natural gas, coal and other energy. This will bring great economic and social benefits to the electric power and vehicle industry in China.

ACKNOWLEDGEMENTS

This work was supported by the Guangzhou Science and Technology Plan Projects No. 2014 J4100232, the Quality and Technology Supervision of Guangdong Province Science and Technology Projects No. 2013CJ01, the Guangdong Province Key Laboratory of Efficient and Clean Energy Utilization No.2013 A061401005, and the Key Laboratory of Efficient and Clean Energy Utilization of Guangdong Higher Education Institutes (KLB10004).

REFERENCES

2010. Carbon-emission analysis of vehicle and component manufacturing. Argonne National Laboratory (ANL).
Feng, W. & Figliozzi, M.A. 2012. Conventional vs Electric Commercial Vehicle Fleets: A Case Study of Economic and Technological Factors Affecting the Competitiveness of Electric Commercial Vehicles in the USA. Procedia—Social and Behavioral Sciences, 39, 702–711.
Hertwich E.G., H.J.K. 2001. A decision-analytic framework for impact assessment Part2: Midpoints, endpoints, and criteria for method development. International Journal of Life Cycle Assessment, 6, 265–272.
Hertwich, E.G. & Hammitt, J.K. 2001. A decision-analytic framework for impact assessment part I: LCA and decision analysis. The International Journal of Life Cycle Assessment, 6, 5–12.

Huo, H., Cai, H., Zhang, Q., Liu, F. & He, K. 2015. Life-cycle assessment of greenhouse gas and air emissions of electric vehicles: A comparison between China and the U.S. Atmospheric Environment, 108, 107–116.

Lin, C., Wu, T., Ou, X., Zhang, Q., Zhang, X. & Zhang, X. 2013. Life-cycle private costs of hybrid electric vehicles in the current Chinese market. Energy Policy, 55, 501–510.

Lucas, A., Alexandra Silva, C. & Costa Neto, R. 2012. Life cycle analysis of energy supply infrastructure for conventional and electric vehicles. Energy Policy, 41, 537–547.

Ma, H., Balthasar, F., Tait, N., Riera-Palou, X. & Harrison, A. 2012. A new comparison between the life cycle greenhouse gas emissions of battery electric vehicles and internal combustion vehicles. Energy Policy, 44, 160–173.

Nanaki, E.A. & Koroneos, C.J. 2013. Comparative economic and environmental analysis of conventional, hybrid and electric vehicles—the case study of Greece. Journal of Cleaner Production, 53, 261–266.

Seixas, J., Sim E.S., S., Dias, L., Kanudia, A., Fortes, P. & Gargiulo, M. 2015. Assessing the cost-effectiveness of electric vehicles in European countries using integrated modeling. Energy Policy, 80, 165–176.

Sharma, R., Manzie, C., Bessede, M., Crawford, R.H. & Brear, M.J. 2013. Conventional, hybrid and electric vehicles for Australian driving conditions. Part 2: Life cycle CO2-e emissions. Transportation Research Part C: Emerging Technologies, 28, 63–73.

Shuhua, L. 2014. Life Cycle Assessment and Environmental Benefits Analysis of Electric Vehicles. D, Jilin University.

Walker, P.D. & Zhang, N. 2014. Active damping of transient vibration in dual clutch transmission equipped powertrains: A comparison of conventional and hybrid electric vehicles. Mechanism and Machine Theory, 77, 1–12.

Wu, G., Inderbitzin, A. & Bening, C. 2015a. Total cost of ownership of electric vehicles compared to conventional vehicles: A probabilistic analysis and projection across market segments. Energy Policy, 80, 196–214.

Wu, Z., Ma, Q. & Li, C. 2015b. Performance investigation and analysis of market-oriented low-speed electric vehicles in China. Journal of Cleaner Production, 91, 305–312.

Zhou, G., Ou, X. & Zhang, X. 2013. Development of electric vehicles use in China: A study from the perspective of life-cycle energy consumption and greenhouse gas emissions. Energy Policy, 59, 875–884.

Energy Science and Applied Technology – Fang (Ed.)
© 2016 Taylor & Francis Group, London, ISBN 978-1-138-02833-3

Analysis and evaluation of the navigation risk of offshore wind project

Y.B. Guo, F.C. Jiang & H.B. Zou
School of Navigation, Wuhan University of Technology, Wuhan, China

ABSTRACT: In recent years, China's offshore wind power project developed very rapidly; however, the offshore wind farm construction and operation has been facing many problems. Its navigation safety issues during construction and operation of the nearby waters cannot be ignored. Research on offshore wind power project risk navigation has been comprehensive and thorough, but the study on many risk factors has many shortcomings. This paper analyzes the navigation features and risks of offshore wind power projects using the AHP and fuzzy comprehensive evaluation theory methods, to determine the offshore wind power project navigation risk factors for the development of offshore wind power project traffic risk assessment system, perform quantitative risk analysis, and ultimately obtain risk assessment results.

Keywords: Off-shore wind power project; Navigation risk; Analytic Hierarchy Process (AHP); Fuzzy Comprehensive Evaluation

1 INTRODUCTION

In this paper, we analyze the traffic characteristics and the risk of offshore wind power project using the analytic hierarchy process, the theory and method of fuzzy comprehensive evaluation, to determine the offshore wind power project navigation risk factor, the formulation of offshore wind power project navigation risk assessment system, carry out the risk quantitative analysis, and finally obtain the results for risk assessment. The risk evaluation results presented security recommendations, which provide the reference for the planning and construction of the offshore wind power project, with the risk being clearer, the location being more reasonable, and the navigation risk of offshore wind farm reducing.

2 NAVIGATION RISK CHARACTERISTICS

2.1 *Analysis of the characteristics of risk*

To fully understand the characteristics of navigation risk of the offshore wind power project, the project risk management personnel and the risk of the project itself have a very big significance. Especially, at the present stage, no design has been formulated for China's offshore wind power project site selection rules, thus we design a method based on the advantage of location with reference to the foreign offshore wind power project, with a correct understanding of the navigation risk and grasping the offshore wind power project.

2.2 *Risk assessment method*

Fuzzy comprehensive evaluation method is a kind of comprehensive evaluation based on the fuzzy mathematics method. The method is to evaluate the fuzzy concept of unified into a certain fuzzy sets, and then establish the appropriate membership functions based on the fuzzy mathematics, the membership degree of fuzzy theory, and the qualitative evaluation into quantitative evaluation. The method has the following advantages: the mathematical model is simple and easy to master; good comprehensive evaluation of multi-factor effects on complex multi-level problems; an excellent evaluation of other mathematical model is hard to replace method.

3 NAVIGATION RISK ASSESSMENT PROCESS

3.1 *Index system*

According to the risk of the offshore wind power project on the navigation area, the establishment of the off-shore wind power project navigation risk assessment hierarchy is shown in Figure 1.

3.2 *The weight of each risk factor level calculation*

The Analytic Hierarchy Process (AHP) in the system engineering theory is a good method for determining the weights of the risk factor levels. It can be of various factors in a complex problem divided

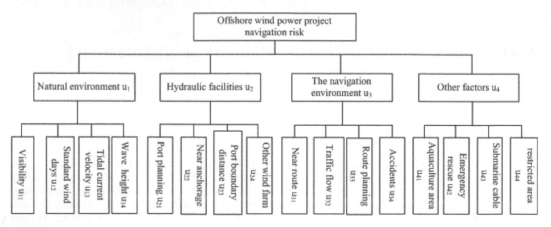

Figure 1. Navigation hierarchical structure chart of risk assessment.

Table 1. First level of risk factors of comparative judgment table.

	u_1	u_2	u_3	u_4
u_1	1	1/3	1/6	1/5
u_2	3	1	1/3	1
u_3	6	3	1	2
u_4	5	1	1/2	1

into levels is associated with and orderly, make a decision making method of orderly, multi-objective, multi-criteria evaluation method, is a combination of quantitative and qualitative analysis. The AHP divides a complex problem into various components, and these factors are grouped to form a hierarchy. To determine the relative importance of the factors in the hierarchy by two compared with the way. And then integrated the relevant personnel to judge, to determine the total ranking relative importance of alternatives.

Table 1 shows the comparative judgment for the relative important degree of two to the first layer of risk factors.

From the above table, we can get the judgment:

$$A = \begin{pmatrix} 1.0000 & 0.3333 & 0.1667 & 0.2000 \\ 3.0000 & 1.0000 & 0.3333 & 1.0000 \\ 6.0000 & 3.0000 & 1.0000 & 2.0000 \\ 5.0000 & 1.0000 & 0.5000 & 1.0000 \end{pmatrix}$$

$$A' = \begin{pmatrix} 0.0667 & 0.0625 & 0.0833 & 0.0476 \\ 0.2000 & 0.1875 & 0.1667 & 0.2381 \\ 0.4000 & 0.5625 & 0.5000 & 0.4762 \\ 0.3333 & 0.1875 & 0.2500 & 0.2381 \end{pmatrix}$$

According to the normalized matrix A', we can calculate the weights u1, u2, u3 and u4:

$$W_1 = \frac{M_1'}{\sum_{i=1}^{m} M_i'} = 0.0645 \qquad W_2 = \frac{M_2'}{\sum_{i=1}^{m} M_i'} = 0.1987$$

$$W_4 = \frac{M_4'}{\sum_{i=1}^{m} M_i'} = 0.2499 \qquad W_3 = \frac{M_3'}{\sum_{i=1}^{m} M_i'} = 0.4868$$

Then, the computation comparison judgment matrix a maximum characteristic root:

$$AW = \begin{pmatrix} 1.0000 & 0.3333 & 0.1667 & 0.2000 \\ 3.0000 & 1.0000 & 0.3333 & 1.0000 \\ 6.0000 & 3.0000 & 1.0000 & 2.0000 \\ 5.0000 & 1.0000 & 0.5000 & 1.0000 \end{pmatrix}$$

$$\bullet \begin{pmatrix} 0.0645 \\ 0.1987 \\ 0.4868 \\ 0.2499 \end{pmatrix} = \begin{pmatrix} 0.2618 \\ 0.8044 \\ 1.9697 \\ 1.0145 \end{pmatrix}$$

$$\lambda_{\max} = \sum_{i=1}^{4} \frac{(AW)_i}{nW_i} = 4.0533$$

Then, we test the greatest characteristic root, calculate the "consistency index" CI and "random consistency ratio" the value of CR.

$$a = \frac{\lambda_{\max} - n}{n - 1} = \frac{4.0533 - 4}{4 - 1} = 0.018$$

Corresponding to that four order judgment matrix RI value is 0.9, with the CR value being

$$CR = \frac{CI}{RI} = \frac{0.018}{0.9} = 0.02 < 0.10$$

Table 2. First level of risk factors of weight table.

Level 1	u_1	u_2	u_3	u_4
(W)	0.0646	0.1987	0.4868	0.2499

Table 3. Second level of risk factors of weight table.

u_{11}	u_{12}	u_{13}	u_{14}	u_{21}	u_{22}	u_{23}	u_{24}
0.4	0.15	0.34	0.08	0.26	0.52	0.12	0.0897

u_{31}	u_{32}	u_{33}	u_{34}	u_{41}	u_{42}	u_{43}	u_{44}
0.36	0.27	0.23	0.12	0.19	0.49	0.05	0.2491

So we can judge the consistency of the judgment matrix in the first level of risk factors. It can draw the factors of U1, U2, U3 weight table, as shown in Table 2.

By using the same method, we can calculate the weights of second levels of risk factors. The results of the calculation are given in Table 3.

3.3 Establishing the evaluation set

The establishment of the evaluation scale set. The evaluation scale set V represents a collection of reviews of the composition of the index factors. We can classify each factor risk into the bottom, low, high, in general, highly dangerous grade five, then the evaluation scale set V = (V1, V2,... V5) = (low, lower, general, higher, high) = (1, 2, 3, 4, 5).

Determining the degree of membership of each index. According to the risk analysis and the corresponding specification and related experts on the evaluation of offshore wind power project navigation, we summarize the established risk assessment standards of off-shore wind power project navigation, as shown in Tables 4–7. Based on the evaluation standard, we determine the risk of each evaluation index membership function, and finally determine the risk degree of membership.

4 CONCLUSION

In this paper, by the analysis of offshore wind power project navigation risk characteristics, by using the analytic hierarchy process and fuzzy comprehensive evaluation method, the theoretical basis is established by the offshore wind power project navigation risk analysis method and the fuzzy evaluation model, introducing in detail the evaluation method and steps of offshore wind power project navigation risk, and the evaluation

Table 4. Evaluation standard (a).

Risk Factors	Low	Lower	General	Higher	High
$u_{11} < 1$ km(d)	15 ↓	15 ~ 25	25 ~ 40	40 ~ 50	50 ↑
u_{12}(d)	30 ↓	30 ~ 60	60 ~ 100	100 ~ 150	150 ↑
u_{13}(m/s)	0.25	0.2 ~ 0.75	0.75 ~ 1.25	1.25 ~ 2	2↑
u_{14}(m)	0.3 ↓	0.3 ~ 0.5	0.5 ~ 1	1 ~ 1.5	1.5 ↑

Table 5. Evaluation standard (b).

Level Factors	Low	Lower	General	Higher	High
u_{21}	20 ↑	15 ~ 20	10 ~ 15	5 ~ 10	5 ↓
u_{22}	10 ↑	6 ~ 10	1 ~ 6	0.5 ~ 1	0.5 ↓
u_{23}	1000 ↑	500 ~ 1000	100 ~ 500	0 ~ 100	0 ↓
u_{24}	10 ↑	6 ~ 10	1 ~ 6	0.5 ~ 1	0.5↓

Table 6. Evaluation standard (c).

Level Factors	Low	Lower	General	Higher	High
u_{31}(km)	10 ↑	6 ~ 10	1 ~ 6	0.5 ~ 1	0.5 ↓
u_{32}(Trips/day)	20 ↓	20–50	50–100	100–300	300↑
u_{33}(km)	10 ↑	6 ~ 10	1 ~ 6	0.5 ~ 1	0.5 ↓
u_{34}(Play/year)	1 ↓	1 ~ 3	3 ~ 6	6 ~ 10	10 ↑

Table 7. Evaluation standard (d).

Level Factors	Low	Lower	General	Higher	High
u_{41}(km)	10 ↑	5 ~ 10	3 ~ 5	1 ~ 3	1 ↓
u_{42}(km)	10 ↓	10 ~ 20	20 ~ 30	30 ~ 40	50 ↑
u_{43}(km)	10 ↑	6 ~ 10	1 ~ 6	0.5 ~ 1	0.5 ↓
u_{44}(m)	5 ↓	6 ~ 9	10 ~ 15	15 ~ 20	20 ↑

method was applied to a practical case to case, on the off-shore wind field of navigation risk assessment, and obtained a good evaluation effect.

REFERENCES

Chang, J.N &Jiang, T.L. 2007. AHP to determine weights of factors, Journal of Wuhan University of Technology (Information & Management Engineering Edition), (1):29–32.
Deng, Y.C & Yu, Z. 2010. An evaluation method in wind farm siting of traffic conditions, East of China electric power. (2): 282–285.

Liu, K.Z & Zhang, J.F. 2010. Offshore wind farm to the study of marine radar detection performance, Journal of Wuhan University of Technology (Transportation Science & Engineering), (4): 28–32.

Peak. J. H, Lee. Y.W. 1993.Pricing Construction Risk-fuzzy Set Application, ASCE Journal of Construction Engineering and Management, 119(4): 201–269.

Ragnar, R.1998.Risk influence analysis: A methodology for identification and assessment of risk reduction strategies, Reliability Engineering & System Safety, 60: 153~164.

Sun, Y.S. 2008.Fuzzy network analysis in offshore wind power project risk appraisal application, Dalian University of Technology. 35~38.

Xu, S.B. 1988.The theory of analytic hierarchy process, Tianjin publishing house, 78~82.

Yun, L.P. 2008. Fuzzy comprehensive evaluation of, North China Electric Power University, 121~123.

Energy Science and Applied Technology – Fang (Ed.)
© *2016 Taylor & Francis Group, London, ISBN 978-1-138-02833-3*

A comparison research on pollution index between thermal power generation and solar power generation

Jiayu Huang, Xunzhe Wang, Xinyue Mao & Liugang Li
North China Electric Power University, Boading, China

ABSTRACT: This paper mainly analyzes the list and the disposal cost of the pollutant discharge of solar power generation and thermal power generation, aiming to understand the advantages and disadvantages of two types of power generation and their respective characteristics. This paper analyzes five parts of disposal costs of various pollutants including dust particles, sulfur dioxide, nitrogen oxide, effluent water and ash. In addition, this paper analyzes the solar power generation based on the concept of life cycle that is divided into eight life cycle processes, calculating the disposal cost of solar panels adequately in raw materials, production, use and recycling process. After analyzing as well as contrasting the disposal cost, this paper compares the environmental impact of the two generators and provides some basis for macroeconomic control and power planning.

Keywords: solar power generation, thermal power generation, life cycle, environmental indicators, disposal cost

1 INTRODUCTION

The electric power industry is mainly composed of thermal power generation in China, and the total capacity of power generation is growing rapidly with the development of economy. At the same time, it also brings huge pollution to the environment. In addition, from the point of view of resource storage, coal resources have been largely consumed, which is arousing attention worldwide. With the reform of electric power enterprises and protection of the environment, aspects concerning environmental protection and efficient way of power generation have led many experts and scholars to research.

As a new type of energy, solar energy arouses wide attention due to its inexhaustible advantages. However, the environment pollution by solar power has become very serious. Meanwhile, its pollution to the environment does not draw enough attention and there are hardly any data to show the pollution level. Another fact is that in America, solar modules rely on import instead of production in the domestic. With its operation process running only for two years, the United States has stopped the solar power generation because of the serious issue of pollution.

2 PRINCIPAL ANALYSIS

2.1 Thermal power generation

This paper divides the pollutants of thermal power generation into five parts, which include dust particles, sulfur dioxide, nitrogen oxide, effluent water and ash, according to the classification and treatment measures of power plant pollutants. This paper reports the result by, respectively, analyzing and calculating the pollutant treatment cost.

2.2 Solar power generation

For solar power generation, this paper calculates the disposal cost of solar panels for every life cycle based on the method of life cycle assessment. It includes silica mining, industrial silicon production, solar energy polysilicon production, polysilicon wafer production, solar cell production, solar energy battery component production, solar energy generating systems and waste recycling.

2.3 Comparison between thermal power generation and solar power generation

This paper calculates the cost of the pollutant discharge of the two types of power generation, and converts it into pollution treatment costs per watt of generating power for comparison. We do not take the cost of equipment into account. This paper considers the power supply cost, the material consumption cost, the adding cost, the artificial cost and the maintenance cost. Because some parts of the cost are relatively low, they are out of the scope of this paper.

3 DISPOSAL COST OF THERMAL POWER GENERATION

This paper considers the burning of fossil fuels as the main process because large amounts of pollutants are produced in this process of thermal power generation. This paper makes a detailed analysis of the disposal cost of the pollutant discharge of thermal power generation.

3.1 *Pollutant treatment costs*

The electrostatic dust removal costs. Electrostatic precipitator consumes large amounts of electricity during the operation process. Four electric fields of electric dust collector that match with a 600 MW set have an electricity consumption of about 2000 KW per hour.

The desulfurization cost. A coal-fired power plant desulfurization FDG flue gas system mainly includes the flue gas system, the absorption tower system, the limestone slurry preparation system, the gypsum dewatering system, the process water system, the cooling water system, the air compressed system, and the discharge system. The raw materials in the operation process of the desulfurization system are mainly limestone, water and electricity. The operation cost of an 845 MW coal-fired power plant desulfurization system is as follows: unit [10,000 yuan/year]

Item	Limestone	Water and electricity	Spares	Labor costs	total
Amount	344.36	164.39	300	250	7082.21

The denigration cost. An 845 MW set adopts Selective Catalytic Reduction (SCR) denigration technology. The reactor is arranged in the boiler economizer export and among air preheater. It is located in the boiler room. The denigration cost of running per year is as follows:

Item	Urea and catalyst	fuel and power consumption	steam and spare parts	Labor costs	Total
Amount	4054.7	2240	276	2240	8810.7

Theure effluent. water treatment cost. A power plant with a total installed capacity of 1000 MW is a condensing generating set. Waste water emissions are about 2.19 million tons in the factory area. The waste water treatment project runs well and is easy to maintain.

Total installed capacity	Waste water emission	treatment cost	Total
1000 (MW)	219 (ton/year)	2.85 (yuan/ton)	624.15

Ash handling costs. The pulverized coal ash and dust ash discharge processes are simple, because they are generally stored in the yard and then dumped. The processing cost can be ignored.

3.2 *Comprehensive cost analysis*

This paper estimates that annual utilization hours are 6000 hours and the electricity price is 0.48 yuan per kilowatt-hour. This papers calculates the disposal cost in 1 kwh of solar power generation. In conclusion, the calculation results of pollutant treatment are as follows: unit [yuan/ (kW·h))]

Electrostatic removal	0.001600
Electrostatic removal	0.001600
desulfurization	0.01397
denitration	0.01738
effluent water treatment	0.001040
ash treatmen	0
total	0.03399

Thiures paper. concludes that the disposal cost of pollutant discharge per watt is 0.03399 yuan/(kW·h).

4 DISPOSAL COST OF SOLAR POWER GENERATION

According to the life cycle process of the solar cell component, this paper analyzes each unit of data.

The source of data is mainly for actual production data of enterprise, which is representative in China of the results of all kinds of excellent thesis and literature data. This part of the paper does not include the details of the number of pollutants and its processing method involving each life cycle process of solar panels.

4.1 *Pollutant treatment costs*

Silica mining. The main pollutant of the silica mining process is dust. The dust produced during crushing and screening can be removed by using the method of Bag-type strainer. Silicon ore, which has an output of 9×10^7 kilogram per year, costs RMB 250,000 in a year. Mining is only a physical process, so the waste water does not cause too much pollution to the environment. This is not taken into the account.

Industrial silicon. The dust can be mostly removed by using the method of Bag-type strainer, an annual output of 7200 tons of industrial grade silicon enterprises cost 1.9 million per year, including electricity, wages and other costs. There is little water pollution, so this is not taken into consideration.

Solar energy polysilicon production. This process is actually a process of purification. A project with an output of 2000 tons of polysilicon mainly adopts the method of spraying to remove waste gas, which costs about 1.6 million each year in leaching operation. The processing of waste liquid reagents used mainly has 20% lime, 5% PAM and 0.5% PAC. In addition, the solution processing cost is 12.6583 million yuan per year.

Polysilicon production. During the production of polycrystalline silicon slice, the bag filter is used to remove the dust particles. Acid gases mainly adopt the alkali leaching process. An enterprise with about 1 GW can output 8300 tons of polycrystalline silicon slice each year. On the whole, the dust removal cost is RMB 55,000/year and the spray scrubber operation cost is RMB 7.8 million/year.

Pollutant treatment costs. In this process, there are various types of exhaust. As a result, different waste gases need to be bubbled into different towers, which include acid mist purification tower, organic waste gas purification tower and silage treatment tower. The operation cost of each tower for a 25 MW company is 100,000 yuan per year, 60,000 yuan per year and 80,000 yuan per year. The total cost is 240,000 yuan per year.

The waste water mainly includes alkaline waste water, acid and alkali washing waste water, and highly concentrated fluoride waste water. The waste water discharge amount of a 200 MW polysilicon solar panel factory in Xining is 750.6 m³/day. The total cost is 5.42555 million yuan/year, adding up the electric consumption, drug consumption and staff salary.

Solar cell component production. It mainly includes organic waste gases (TVOC) and welding gases. The waste gas cost is RMB 570,000 yuan per year for a 100 MW company. The waste liquid can be ignored because of their small amount.

Solar power generation and system waste recycling. Solar panels almost do not produce pollution in the operation process, so pollutant emission can be ignored.

Abandoned solar panels will be specially treated by recycling companies. There is also little pollution to the environment, so it can be ignored.

4.2 Comprehensive cost analysis

By calculation, when manufacturing 1 kilogram polycrystalline silicon of solar cell components, the dosage of each life cycle is as follows:

The life cycle process	Weight (kg)	Processing cost (yuan/kg)
Silica mining	5.0319	0.002778
Preparation of industrial silicon	1.8769	0.26389
Solar energy polysilicon production	1.2519	7.12915
Polycrystalline silicon production	1.0741	1.93133
Solar cell production	1.0204	10.80228
The production of solar battery components	1.0000	0.9814

To sum up, the total processing cost of pollutants in the solar power plant is around 23.5128 yuan per watt of generating capacity when 1 kilogram polycrystalline silicon of solar cell components is manufactured.

For the entire system of solar power generation, the generating capacity and the required amount of polysilicon directly refer to representative data:

Power generation Capacity (MW)	capacity of Polysilicon (kg)	total power generation (degrees)
1	3574.4688	2724.88125*10^4

In conclusion, the total disposal cost of pollutants in the solar power plant is around 3.0844×10^{-3} yuan/(kw·h) per watt of generating capacity.

5 COMPARISON OF THERMAL POWER GENERATION AND SOLAR POWER GENERATION

5.1 The disposal cost

This paper concludes that the disposal cost of the pollutant discharge of thermal power generation is 0.03399 yuan/ (kw·h), while solar power generation of pollutants processing cost is 3.0844×10^{-3} yuan/(kw·h). The cost of thermal power generation is nearly 10 times than that of the solar thermal power generation. In conclusion, solar power generation renders more environmental protection than thermal power generation.

Pollution treatment costs of thermal power generation mainly concentrate on the process of desulfurization and denigration, which account for more than 90% of the total cost. While, pollution treatment costs of solar power generation mainly concentrate on the process of solar energy poly-silicon production and solar cell production, which account for more than 80% of the total cost. So, these main parts of the disposal cost should be more focused in the improvement of both power generation technology and management of pollutants.

5.2 *Comparison of other characteristics*

The largest environmental impact of thermal power generation is gas materials such as smoke, sulfur dioxide and nitrogen oxide. While in solar power generation, the main pollutants are liquid materials such as silicon-based liquid and waste water with impurity. Therefore, the impact of thermal power generation on the environment mainly exists in the atmosphere, while it exists in the soil and rivers (the earth's surface) of solar power generation.

The pollutant types are relatively few and the handling method is relatively simple for thermal power generation. Also, the research of pollutant treatment is more thorough. By contrast, the pollutant types are diverse and the handling method is relatively difficult for solar power generation. There are not too many research papers on solar power generation.

6 CONCLUSION

There are both advantages and disadvantages of thermal power generation and solar energy power generation. The influence on the environment is demonstrated in different respects. In summary, solar power generation is still better than thermal power generation when considering the disposal cost of the pollutant discharge. The development prospect of solar power, which is a kind of clean energy, is very broad. This paper believes that the proportion of solar power generation will increase with the continuous improvement of technology.

REFERENCES

Department of Energy Statistics, 2011, National Bureau of Statistics, Peoples of China. China energy statistical yearbook 2011[M].Beijing: China Statistics Press: 1–5 (in Chinese).
Kordy, M.N., Badr, M.A. Abed, K.A. et al, 2002. Ibrahim Economical evaluation of elect realty generation considering externalities [J]. Renewable Energy, 25:317–328.
Minwon Park, In Keun Yu, 2004. A study on the optimal voltage for MPPT obtained by surface temperature of solar cell [C], Industrial Electronics Society, 2004IECON 2004. 30th Annual conference of IEEE: 2040–2045.
Wolfram Krewitt, Joachim Nitsch, 2003. The German Renewable Energy Sources Act an investment into the future pays off already today [J]. Renewable Energy, 28:533–542.
Yang Zhiping, Yang Yongping, 1988. Energy consumption and distribution of 1000 MW coal-fired power generating unit [J].Journal of North China Electric Power University, 2012, 39(1):76–80(in Chinese). [1] Rahman S, Chowdhury B H. Simulation of Photovoltaic Power Systems and Their Performance Prediction [J]. IEEE Transactions on Energy Conversion, 3(3): 440–446.

Electrical and electronic technology, power system engineering

Energy Science and Applied Technology – Fang (Ed.)
© 2016 Taylor & Francis Group, London, ISBN 978-1-138-02833-3

ADS-based design and simulation of power divider

Luchao Han, Jieqing Fan & Ce Zheng
*Department of Electrical and Electronic Engineering, North China Electric Power University,
Beijing, China*

ABSTRACT: A power divider is a very important passive component in modern microwave communication systems, radar systems and electronic countermeasure systems. It occupies a very important position. Each indicator will be a very important influence factor on the quality of the whole system. The main work of this paper is as follows. The development of microwave power divider and the present research status are introduced. The theory of microwave power divider is studied and analyzed such as the theory of three-port and four-port networks and the theory of Wilkinson power divider. Second, based on the theory of Wilkinson power divider and impedance converter, we designed a 2.4 GHz broadband power divider. The simulation curves are given on ADS.

Keywords: ADS; power divider; modern microwave

1 INTRODUCTION

In recent years, with the rapid development of wireless communication, the modern communication system is gaining heights to meet the transmitter's request. The power divider takes the transmitter's important part, playing the key role of the transmitter system's performance index. In the entire wireless communication system, the power divider is a very important part or passive component in modern microwave communication systems, radar systems and electronic countermeasure systems. It occupies a very important position. Microwave workers are very concerned about the indicator of the power divider, such as amplitude balance, insertion loss, and the isolation between the output ports, the consistency of the phase and volume. Each indicator will be a very important influence factor on the quality of the whole system.

The Wilkinson power divider is one of the conventional and fundamental components in microwave engineering and exists in many microwave circuits. Both distributed and lumped Wilkinson power dividers have been applied in microwave integrated circuits and monolithic microwave integrated circuits. Recently, extensive studies have been made to enhance the performances of the Wilkinson power divider, including size reduction by capacitive loading, folded circuitry and resonating structure. The power dividers discussed in this paper are focused on the appropriate index in simulation, such as type font, type size and spacing.

2 GETTING STARTED

The Wilkinson power divider has a good wideband and equipage characteristic. Its structure is shown in Figure 1. The Wilkinson power divider consists of transmission line whose length is quarter-wave. The characteristic impedances of the input line and the output line are both. The characteristic impedances of the branch line between the input port and the output port is Z. The demand for the power divider is that the high degree of isolation between the output ports prevents crosstalk between the individual channels. The input power is split equally between two output ports.

It is viable to use the odd–even mode analytical method to analyze this circuit. Specifically, the asymmetrical and symmetrical drive source is used, and the transfer characteristic of the circuit is analyzed. Then, we can synthesize them and finally obtain the transfer characteristic. First, we use characteristic impedance uniformly in all impedances, and obtain the circuit as shown in Figure 2.

The even-mode excitation of the circuit is:

The odd-mode excitation of the circuit is:

When the excitations are added, the effective excitation is. From this, we can get the parameter S. In the even-mode excitation, there is no current flow for the isolation resistance r/2. In other words, the input of the two transmission lines in port 1 is short. Form port 2, the impedance is.

As a quarter-wave transformer, the transmission line is matching port 2 in the even-mode excitation. Under these circumstances, as the impedance r/2 is open circuit so it has no effect. From the

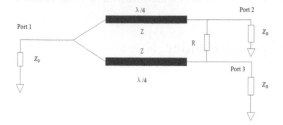

Figure 1. Wilkinson power divider.

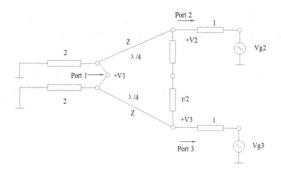

Figure 2. Wilkinson power divider circuit in the symmetric form.

transmission-line equation, we can get. Let x = 0 in port 1, then we can get x = −λ/4. The voltage in transmission line is.

So we can get:

In port 1, let us consider the resistance which is 2, the reflectance is:

In the same time.

As for the odd-mode excitation, so. IF r = 2, in the port 2 it is suited. When, any power is load in r/2. So, there is no power in port 1.

3   LAYOUT OF TEXT

Operating frequency. The structure of the power divider is bound up with operating frequency. Operating frequency is the precondition of the work. So, it is necessary to clearly define the operating frequency.

Power capability. The maximum operating frequency of the power divider cannot exceed power capability. Power capability is the precondition of the selection of the transmission line.

Return loss. RL is an important indicator to measure the matching degree of all the ports.

$$RL_1 = -20lg|S_{11}|$$
$$RL_2 = -20lg|S_{22}|$$
$$RL_3 = -20lg|S_{33}|$$

In general, RL is as far as possible more than 20dB, i.e. less than −20dB.

Isolation. When an output port has reflection, isolation can measure the disturbed condition, in which this reflection disturbs another port. It can be defined as. Generally, isolation is as far as possible more than 20db.

Insertion loss. Insertion loss is used to measure transmission performance objectives and recommendations of the power divider.

It can be defined as

$$IL_{21} = -20lg|S_{21}|$$
$$IL_{31} = -20lg|S_{31}|$$

4   PHOTOGRAPHS AND FIGURES

We set the index of the power divider as follows:

Operating frequency: 2.4 GHz
Return loss: $S_{11}, S_{22}, S_{33} > -4dB$
Insertion loss: $S_{21}, S_{31} > -4dB$
Isolation: $S_{23} < -2dB$

We set the substrate parameters as the TLines-Microstrip. In addition, we set the working central frequency similar to the operating frequency of 2.4 GHz. Finally, we set the input characteristic impedance as 50Ω, phase delay. The results are shown in Figure 3.

We set goal 1 optimization as follows:

Expr = "dB(S(1,1))"
SimInstanceName = "SP1"
LimitMin[1] = −20
Limitmax[1] = −15
Weight = 1
RangeVar = "freq"
RangeMin = 2.35GHz
RangeMax = 2.45GHz

We the set the other goal in the same way. The results are shown in Figure 4.

We choose components and parts to design the circuit diagram. The whole design results are shown in Figure 5.

We use "Simulation" and choose Rectangular Plot to insert four quadrangle maps. We then choose "S(1, 1), S(2, 2), S(2, 1), S(2, 3)" and make sure. Then, we can get the result for the simulatio0

We can find. that the return loss in S(1, 1) and S(2, 2) are −14.923 dB and −20.163 dB, which are less than the target −3.248 dB. Moreover, for the insertion loss, S(2, 1) is −14.923 dB, which meets the target that ranges from −3 dB to −6 dB. Finally, isolation in S(2, 3) is −20.633 dB, which also meets the target that is less than −20 dB.

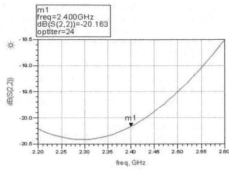

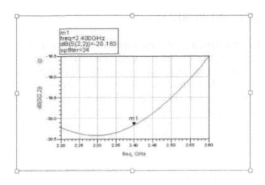

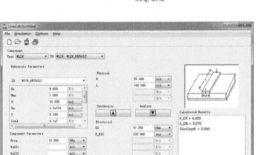

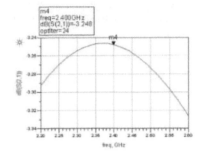

Figure 3. LineCalc.

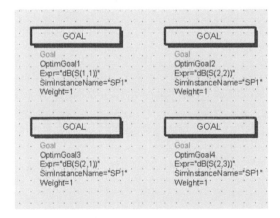

Figure 4. Goal.

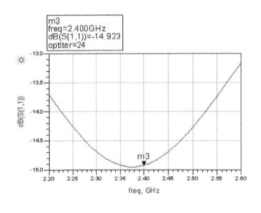

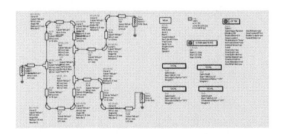

Figure 5. Principle diagram of the simulation.

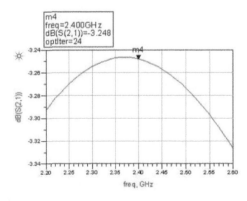

Figure 6. Simulation result.

5 CONCLUSION

In this article, we first analyze the principle and performance of the power divider. When comes to the structural analysis, we use the odd-even mode analytical method to discuss the circuit of the power divider. Then, a Wilkinson power divider working on 2.4 GHz was introduced. From these examples, we obtain the simulation result for the Wilkinson power divider on ADS.

REFERENCES

K.M. Shum, Q. Xue & C.H. Chan. Curved PBG cell and its applications. Asia–Pacific Microw. Conf., pp. 767–770, 2001.

K. Hettak, G.A. Morin & M.G. Stubbs. Compact MMIC CPW and asymmetric CPS branch-line couplers and Wilkinson dividers using shunt and series stub loading. IEEE Trans. Microw. Theory Tech., vol. 53, no. 5, pp. 1624–1635, 2005.

L. Chiu, T.Y. Yum, Q. Xue & C.H. Chan. The folded hybrid ring and its applications in balance devices. IEEE Eur. Microw. Conf., pp. 1–4, 2005.

L.H. Lu, P. Bhattacharya, L.P.B. Katehi & G.E. Ponchak. X-band and K-band lumped Wilkinson power dividers with a micromachined technology. IEEE MTT-S Int. Microw. Symp. Dig., pp. 287–290, 2000.

Energy Science and Applied Technology – Fang (Ed.)
© 2016 Taylor & Francis Group, London, ISBN 978-1-138-02833-3

Sensorless control of permanent magnet synchronous motor based on hybrid strategy position estimation

Chengyi Yu & Bo Tao

School of Mechanical Science and Engineering, HUST, Wuhan, Hubei, China

ABSTRACT: For the Permanent Magnet Synchronous Motor (PMSM) control, the Field Oriented Controllers (FOC) is the main method. For the sensorless motor, one of the problems is how to obtain the accurate position of sensor-less motor in foccontrol. In this paper, a method combing the high frequency signal injection method and the model current analysis method will be introduced for the salient pole PMSM. The estimation method can estimate position in the whole speed range, including zero speed. The simulations of this estimation method prove that this method can accurately estimate the position of the PMSM.

Keywords: PMSM; Position Estimation; Sensorless

1 INTRODUCTION

The PMSM has the characteristic of superior torque control, lower torque ripple, and in many cases, improved efficiency compared to traditional AC control techniques. In recent years, it has been widely used in the aviation, aerospace, electric vehicles and industrial control.

In PMSM vector volume control system, rotor position accuracy straightly affect the stability and accuracy of the system. Traditionally, the position is detected by a mechanical sensor mounted to the shaft of the motor. These sensors provide excellent angle feedback, but increase the cost of the system and drop the system reliability. Therefore, PMSM sensorless control technology has become an important research direction in the control theory.

Currently, sensorless control studies can be divided into two categories: high frequency signal injection and fundamental model analysis. In the fundamental model analysis, back Electro Motive Force (EMF) is always used to estimate the position. This method is simple and has been widely used. But the problem is that when the motor is running in low speed or even zero-speed, the current is too small, which makes it difficult to get the accurate position. The high frequency signal injection method is based on saliency detection motor through by injecting a specific high frequency voltage signal into the motor signal, and then measure the corresponding current and voltage to determine the position of the rotor. This method depends on the continuous high frequency excitation to show saliency, it has nothing to do with speed and is not sensitive to changes of

the motor parameters, but it also needs more complex signal processing and leads to bad dynamic performance. So this can be used in low speed but not suitable for high speed running. A combination of two techniques must be implemented to obtain a full-speed control.

This paper first introduces this two methods to estimate the rotor position and analysis their shortcoming, this two methods can be combined in a way to solve the problem, and can get accurate position of the rotor in whole speed running. An algorithm is used in the transition area to avoid the potential instability. Finally, the simulink simulations show that the result obtained by the new method can be used in practical application.

2 PMSM MODEL

The block of Field Oriented Controllers (FOC) is as Figure 1. In sensorless control system, this includes the position estimator. The system is a three-loop closed system, an external speed loop which is closed on the estimated value and two internal current loops for d-axis and q-axis. The estimator gets sample single of the stator through current sensors in the two phase.

The fundamental machine equation can be stated in two phase rotating coordinate system as follows:

$$\begin{bmatrix} V_d \\ V_q \end{bmatrix} = \begin{bmatrix} R + pL_d & -wL_q \\ wL_d & R + pL_q \end{bmatrix} \begin{bmatrix} i_d \\ i_q \end{bmatrix} + \begin{bmatrix} 0 \\ \omega\psi_f \end{bmatrix} \quad (1)$$

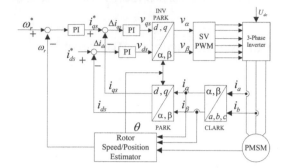

Figure 1.   Block Diagram of FOC.

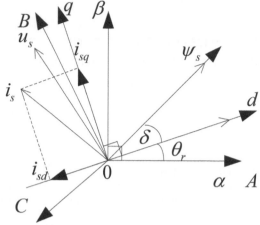

Figure 2.   Coordinate conversion.

Where, V_d, V_q is the d-q coordinate system voltage component, i_d, i_q is the current in d-q coordinate, L_d, L_q is the phase inductance in d-q coordinate, ψ_f is the permanent flux of the rotor, R is the stator resistance, p is the differential operator.

In sensorless motor system, rotor position and speed is unknown, so the estimated two phase rotating coordinate system (α – β) is introduced. In this coordinate system, the coordinate conversion is as follow:

Transformation matrix is:

$$T^e = \begin{bmatrix} \cos(\theta) & \sin(\theta) \\ -\sin(\theta) & \cos(\theta) \end{bmatrix} \qquad (2)$$

So the voltage equation under (α – β) coordinate is as follow:

$$\begin{bmatrix} V_\alpha \\ V_\beta \end{bmatrix} = T^\theta \begin{bmatrix} V_d \\ V_q \end{bmatrix} = R \begin{bmatrix} i_\alpha \\ i_\beta \end{bmatrix} + p \begin{bmatrix} \psi_\alpha \\ \psi_\beta \end{bmatrix} \qquad (3)$$

$$\begin{bmatrix} \psi_\alpha \\ \psi_\beta \end{bmatrix} = L_q \begin{bmatrix} i_\alpha \\ i_\beta \end{bmatrix} + \left[(L_d - L_q)i_d + \psi_r \right] \begin{bmatrix} \cos\theta \\ \sin\theta \end{bmatrix} \qquad (4)$$

$$\begin{bmatrix} \psi_\alpha - L_q i_\alpha \\ \psi_\beta - L_q i_\beta \end{bmatrix} = \left[(L_d - L_q)i_d + \psi_r \right] \begin{bmatrix} \cos\theta \\ \sin\theta \end{bmatrix} \qquad (5)$$

$$\cos\theta = \frac{\psi_\alpha - L_q i_\alpha}{(L_d - L_q)i_d + \psi_r}$$

$$\sin\theta = \frac{\psi_\beta - L_q i_\beta}{(L_d - L_q)i_d + \psi_r} \qquad (6)$$

In vector control system, $i_d = 0$. So the position can be calculated from (6).

In ((α – β) coordinate, the flux can be stated as (7),

$$\begin{cases} \psi_\alpha = \int (u_\alpha - R i_\alpha)dt \\ \psi_\beta = \int (u_\beta - R i_\beta)dt \end{cases} \qquad (7)$$

The position can be calculated as in (8),

$$\tan\theta = \frac{\int (u_\beta - R i_\beta)dt - L_q i_\beta}{\int (u_\alpha - R i_\alpha)dt - L_q i_\alpha}$$

$$\omega = \frac{d\theta}{dt} \qquad (8)$$

The i_α, i_β can be get from CLARK conversion through the three phase current.

$$\begin{bmatrix} i_\alpha \\ i_\beta \end{bmatrix} = \sqrt{\frac{2}{3}} \begin{bmatrix} 1 & -1/2 & -1/2 \\ 0 & \sqrt{3}/2 & -\sqrt{3}/2 \end{bmatrix} \begin{bmatrix} i_A \\ i_B \\ i_C \end{bmatrix} \qquad (9)$$

The advantage of this method in estimating the position and speed includes the conciseness and accuracy in tracking the position and speed. When the motor is running in low speed or even zero speed, the current and voltage is too small. Considering of external disturbances, it is difficult to get the accurate position and speed of PMSM. Also, in the case of the motor start up, the current is zero, this method is unsuitable to get the position and start the motor.

3   HIGH FREQUENCY INJECTION

High frequency injection to measure the magnetic irregularities is widely used to get the rotor position down to zero speed.

With the PWM voltage source inverter supplied, the balanced three-phrase high-frequency voltage signal can be injected in the stator winding of three phase symmetrical to generate position estimation error, as show in Figure 3.

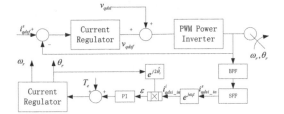

Figure 3.   The principle of high frequency injection.

Suppose the injected frequency is ω_i and $w_i \gg \omega_r$, then the injected high-frequency voltage can be stated as,

$$v_{qdsi}^s = \begin{bmatrix} v_{qsi}^s \\ v_{dsi}^s \end{bmatrix} = v_{si} \begin{bmatrix} \cos \omega_i t \\ -\sin \omega_i t \end{bmatrix} = v_{si} e^{j\omega_i t} \approx L_{qds}^s \frac{di_{qdsi}^s}{dt} \quad (10)$$

The injected high frequency signal is in the range of about 0.5~2 kHz which is much higher than the rated fundamental frequency, the impedance of the motor is dominated by the self—inductance. So the model can be simplified to (10).

In the stationary coordinate system, the stator inductance can be show as,

$$L_{qds}^s = \begin{bmatrix} L + \Delta L \cos(2\theta_r) & -\Delta L \sin(2\theta_r) \\ -\Delta L \sin(2\theta_r) & L - \Delta L \cos(2\theta_r) \end{bmatrix} \quad (11)$$

Where, $L = (L_d + L_q)/2$, $\Delta L = (L_q - L_d)/2$, θ_r is the position of rotor.

According to (10) and (11), the current of high frequency can be represented as,

$$i_{qdi} = i_{ip} e^{j(\theta_i(t) - \pi/2)} + i_{in} e^{j(2\theta_r - \theta_i(t) + \pi/2)} \quad (12)$$

The amplitudes of positive and negative sequence can be given as,

$$i_{ip} = \left[\frac{L}{L^2 - \Delta L^2}\right]\frac{v_i}{\omega_i}, i_{in} = \left[\frac{\Delta L}{L^2 - \Delta L^2}\right]\frac{v_i}{\omega_i} \quad (13)$$

From (13), it can be seen that only the negative sequence contains the information about position of the rotor θ_r, it needs to analysis the signal to get the accurate position.

To get the position information in the negative sequence, the Band Pass Filter (BPF) and the Synchronous Frame Filter (SFF) are employed to remove the positive sequence. Then use the position tracking observer to estimate the position and speed.

The vector angular error can be expressed as

$$\varepsilon = i_{qs\_in}^i \cos(2\hat{\theta}_r) - i_{ds\_in}^i \sin(2\hat{\theta}_r)$$

$$= \sin(2\hat{\theta}_r - 2\theta_r) \quad (14)$$

When the vector error is zero, the estimated position $\hat{\theta}_r$ is equal to the real position θ_r. So the motor speed is equal to the time derivative of the position.

This method introduces some other control problems, such as, using a lot of filters causes the rotor position and speed estimation lags from the real running, the dynamic performance is unsatisfactory, especially in high speed when the load has a sudden change. This angle measurement technique only works well at lower speed when the motor fundamental frequency does not interfere with the interrogation frequency.

4   HYBRID-CONTROL STRATEGY

To get the full speed control of the Permanent Magnet Synchronous Motor, a new Hybrid-Control strategy is introduced to avoid the drawback of high frequency injection and fundamental model. In this method, a new strategy is employed to achieve smooth transition of the two methods.

At the start up and low speed running of PMSM, high frequency injection signal is used to estimate the position and information. When the motor is running in high speed, the PMSM model is used to calculate the position and speed. Between the two speed status, there is an area to change the algorithm of the estimator. It should be changed smoothly to ensure the system stability. Here, ω_l is the low-speed border of the area, ω_h is the high-speed area border of the area.

The values of ω_h and ω_l is different with different motor parameters, it can be determined through experiments or simulations. The previous speed ω_{re}^3 is used here to determine the current weighting factor of this two methods.

$$\omega_{re} = \alpha \omega_{hfi} + (1 - \alpha) \omega_{fm} \quad (15)$$

Where, ω_{hfi} and ω_{hm} are the estimated speed through two thread, α is the weighting factor which can be calculated as follow,

$$\alpha = \begin{cases} 1 & \omega_{re} < \omega_l \\ \dfrac{1}{\omega_l - \omega_h}(\omega_{re} - \omega_h) & \omega_l < \omega_{re} < \omega_h \\ 0 & \omega_{re} > \omega_h \end{cases} \quad (16)$$

Where, ω_{re} is the speed of the previous step. The value of speed is used as absolute value because the system has to work in both positive and negative speed direction.

In order to reduce the adverse effects of high frequency signals injected to the system, when the speed is higher than the high-speed border ω_h, the injection of the high frequency signal should be gradually reduced until it reaches zero.

5 SIMULATION RESULTS

To verify the strategy introduced in this paper, some simulations are carried out by means of MATLAB/simulink. This simulations are based on the high frequency injection and fundamental model.

In low speed area, the high frequency signal in 2k HZ is injected into the system. When the motor runs in high speed, the high frequency signal disappears and the fundamental model estimator starts to work.

The simulation of the position estimation shows as in Figure 4. At the beginning of running, there is some delay of the estimator, then the estimator can estimate the position with satisfactory result. The error of the position is in a stable form. It means the error can be decreased through some compensation.

The simulation of the speed estimation shows as in Figure 6 and Figure 7. The estimated speed can track the actual speed quickly. In this simulation, the area border ω_h is 800 rpm and ω_l is 450 rpm, which is chosen by a series of simulations. The speed 600 rpm is chosen in the transition area where the two algorithms both work. At the start of the running, there is some deviation of speed under the condition of high frequency injection, but it will gradually track the correct speed value.

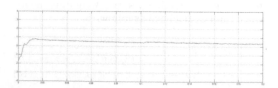

Figure 4. The simulation of position comparison.

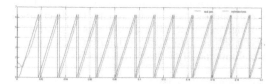

Figure 5. Estimated position error.

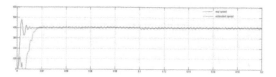

Figure 6. The simulation of speed comparison.

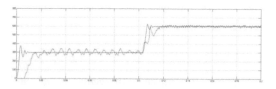

Figure 7. Actual and estimated speed response.

In low speed zones, there is a shock at the initial position of tracking, but the estimator did not lose track of the rotor and still work well with no motor appeared steps. In the transition area, there is no confusion, the system remains stable. When the speed has a sudden change, the system has a good response and track the actual speed.

6 CONCLUSION

In this paper, a hybrid control strategy is introduced to achieve the accurate position of PMSM in full speed range. The strategy is the combination of high frequency signal injection and motor fundamental model calculation. Simulation of the strategy shows the good result of this algorithm. The position and speed can be estimated in a wide speed range, even at zero-speed. The running performance of PMSM in whole speed has been improved based on this strategy and the transition area algorithm.

ACKNOWLEDGMENTS

This work is supported by the National Science and Technology Major Project under Grant 2014Z X04014101 and the National Fundamental Research Program of China under Grant 2013CB035803.

REFERENCES

Bisheimer, G. and Sonnaillon, M. 2010. Full speed range permanent magnet synchronous motor control without mechanical sensors. Electric Power Application. ET. p35–44.

Chan, T.F., Wang, W. and Borsje, P. Sensorless Permanent-magnet Synchronous Motor Drive Using a Reduced-order Rotor Flux Observer[J]. *Electric Power Applications, IET, 2011, 2(2), p.88–98.*

Eskola, M, Jussila, M, Tuusa, H.2004. Indirect matrix converter fed PMSM-sensorless control using carrier injection.*Power Electronics Specialists Conference. PSEC 04.*

French C, Acamley P. 1996. Control of permanent magnet motor drivers using a new position estimation technique. *IEEE Trans. 32(5).*

Qin Feng, He Yi-kang and Jia Hong-ping. 2007. Investigation of the Sensorless Control for PMSM Based on a Hybird Rotor Position Self sending Approach. *Proceedings of the CSEE.*

Qunjiang Wang, Yin Wei, Weidong Jiang. 2008. Simulation of Predictive Control Based ANN for PMSM servo system[J]. *Electric Drive, 38(1 0): S4-S7.*

Thiemann, P. and Mantala, C. PMSM Sensorless Rotor Position Detection for all Speed by Direct Flux Control. *Industrial Electronics(ISIE), 27–30, p673–678.*

Xie Ge, Lu, Kaiyuan. 2014. High bandwidth zero voltage injection method for sensorless control of PMSM. *Electrical Machines and Systems (ICEMS). 22–25. p3546–3552.*

Zhang Jian, Xu Zhenlin, Wen Xuhui. 2006. Study of a Novel Method of State-Estimation and Compensation for Sensorless PMSM Drive System. *Transactions of China Electrotechnical Society, 2006(01), p.7–12.*

Zhao, G., Liu, J. and Fen, J. 2010. The research of speed sensorless control of pmsm based on mras and high frequency signal injection method. *Computer, Mechanical, Control and Electronic Engineer(CMCC), 2010 International Conference on, vol, aug. p175–178.*

Energy Science and Applied Technology – Fang (Ed.)
© 2016 Taylor & Francis Group, London, ISBN 978-1-138-02833-3

Novel analysis and design of rectifier for wireless power transmission

Quanqi Zhang, Xiaodong Liang, Yongjian Xu, Xuming Lu & Hongzhou Tan
School of Information Science and Technology, Sun Yat-Sen University, Guangzhou, China
SYSU-CMU Shunde International Joint Research Institute, Foshan, China

ABSTRACT: This paper presents a novel theoretical analysis method for the current model of the Schottky diode rectifier. Varying of junction resistance and junction capacitance of diode is added to the model for better results. Through the analysis, a novel model of a rectifier system is constructed to figure out optimal value for a dc load and conversion efficiency. A maximal Power Conversion Efficiency (PCE) can be derived for each case as a function of the parameters of diode and dc load. The analysis is verified in a previous designed rectifier by the experimental results. Basing on this improved theoretical analysis method, the new single-diode half-wave rectifier for wireless power transmission at 5.8 GHz is reported. Measured results of the designed single-diode half-wave rectifier is presented, which is in general consistent with the theoretical analysis. Furthermore, com-pact design is achieved by placing the harmonic rejecting filters before rectifying element. Tapered transmission lines are applied for better impedance matching. A maximal PCE of 66.7% for the single-diode half-wave rectifier had been achieved at 5.8GHz, which agrees well with the theoretical prediction.

Keywords: current model, Power Conversion Efficiency (PCE), tapered transmission lines, harmonic rejecting filters

1 INTRODUCTION

In the past few years, researchers have tried to figure out how to optimize the rectenna for higher conversion efficiency through theoretical analysis. The relationship between input power, output voltage, output current and frequency, was investigated. Furthermore, closed-form equations were derived to express conversion efficiency as a function of diode parameters, In, Yoo and Chang conducted a theoretical investigation on 10 and 35 GHz rectenna and derived the closed-form equations which presented the relation between diode parameters and conversion efficiency. The calculated efficiencies agreed well with those simulated by LIBRA when input RF power was below 50 mW at 10 GHz and 35 GHz. Another theoretical analysis on Schottky diode were proposed and a class-F^{-1} rectifier with a conversion efficiency of 85% at 10 W based on microwave power amplifier was designed. In, the voltage swings across the rectifying element were assumed to be a sinusoidal waveform. In addition, the junction resistance and junction capacitance of diode were assumed to be constant.

Some enhancements have been implemented for rectifiers. Available 2.4/45/62 GHz rectifiers fabricated in 200/180/90/65 nm CMOS processes were presented in, respectively. Sub-rectifiers, automatic load control property, voltage sensing circuit were implemented in rectifiers. A dual-band compact rectifier for embedded system was presented and a chip CMOS rectifier was designed for powering RFIDs. Besides, a double Y-branch ballistic junction and a nonlinear model are proposed for the rectification of signals up to 94 GHz. The single diode design was more practical due to simplicity in structure and low cost.

In section 2, a novel theoretical analysis based on a current model is presented to show the relations between output voltage, PCE (Power Conversion Efficiency) and parameters of diode. The current from receiving antenna could be a sinusoidal waveform, which is different from previous works. The junction resistance and junction capacitance of diode were assumed to be constant in previous research, while they could be variable with voltage across the rectifying element in this paper. Rectifiers with single shunt-connected Schottky diodes is designed to verify the theory.

The analysis is compared with a previous design and verified in a new designs in section 3. The results of analysis are in general consistent with those of previous design in part 3.1. Additionally, the single-diode is proposed in part 3.2 and 3.3. Some enhancements will be applied to the new designs. Rejecting harmonics at both the input and output was common used in most of the reported work. In this paper, harmonic rejection and impedance matching network are placed in front of diode. Input filter and output higher harmonic rejecting filter could be removed, as they are not required for the suppression of higher order harmonics. Hence, a compact structure could be achieved.

2 THEORY ANALYSIS

In this part, a novel theoretical analysis for the current model of a Schottky diode rectifier will be presented. Varying of junction resistance and junction capacitance of diode is added to the model for better results.

The typical equivalent circuit of a Schottky diode is shown in Fig. 1 with its parameters in Table 1. R_S, C_p, R_j and V_{BV} primarily determine the maximum available PCE.

The structure of rectifier used in this paper is shown in Fig. 2. The waveform of voltages and current is showed in Fig. 3.

The current-voltage characteristic of a Schottky barrier diode is described by the following equation:

$$I_j = -I_S \left[\exp\left(\frac{-V_j}{knT}\right) - 1 \right] \qquad (1)$$

where I_j is the current flows through R_j and V_j is the voltage on C_j, in Fig. 2.

In addition, I_S and n may affect the PCE of rectifier. When diode is on, R_j can be derived as

$$R_j = \frac{knT}{I_S - I_j} \qquad (2)$$

When $V_j > 0$, junction capacitor C_j is proportional to $1/\sqrt{V_D + V_j}$. When $V_j < 0$, C_j is the sum of barrier capacitance C_B and diffusion capacitance C_D. For simplicity, assume C_j to be a constant when $V_j \in (-V_{on}, 0)$. When $V_j < -V_{on}$, R_j is very small and C_j

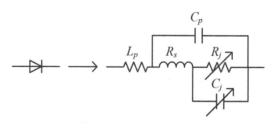

Figure 1. Equivalent circuit of a packaged Schottky diode.

Table 1. Parameters of Diode.

Parameter	Description
L_p	package inductance
C_p	Package capacitance
R_S	series resistance
C_j	Junction capacitance
R_j	Junction resistance
V_D	built-in voltage
V_{BV}	Peak inverse breakdown voltage
V_{j0}	Threshold voltage
I_{BV}	Maximum reverse leakage
I_S	saturation current
n	Ideality factor
C_D	Barrier capacitance
k	a factor that subjected to Boltzmann constant and the value of theelemental charge
T	temperature, K

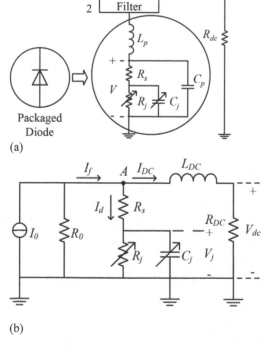

(a)

(b)

Figure 2. (a) Microwave rectifying circuit. Block 1 can be removed since impedance matching could be done in block 2. (b) Equivalent rectifying circuit of (a). The dc pass filter is simplified to a large inductance LDC.

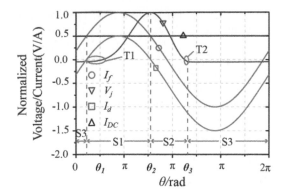

Figure 3. Current and voltage waveform modeled by Matlab in this paper, normalized to their peak values.

can be neglected. When $V_j \geq -V_{on}$, C_j can be written as:

$$C_j = \begin{cases} \dfrac{C_{j0}}{\sqrt{1+V_j/V_D}}, \text{when } V_j > 0 \\[3mm] \dfrac{C_{j0}}{\sqrt{1+V_j/V_D}} + C_D, \text{when } V_j \in [-V_{on},0] \end{cases} \quad (3)$$

Where C_{j0} is zero-bias junction capacitor. Fig. 2 shows the rectifying circuit and its equivalence. A novel theoretical analysis of a rectifier with shunt-connected diode will be presented based on following assumptions:

I_f is the driving current presented a sinusoidal waveform.

Output current I_{DC} is constant.

By applying Kirchhoff's current law to node A in Fig. 2 (b), the relation between current I_d, I_f and I_{DC} can be expressed as follows:

$$I_d = I_f - I_{DC} \quad (4)$$

When $V_j \leq -V_{on}$, diode is on and $I_j = I_d$. When $V_j > -V_{on}$, diode is off, and $I_j = 0$.

When diode is on, V_j can be derived from (1) as

$$V_j = -knT\, ln\left(-\dfrac{I_j}{I_s}+1\right) \quad (5)$$

I_f can be written as

$$I_f = I_m \sin \omega_0 t = I_m \sin \theta_t \quad (6)$$

where ω_0 is the fundamental frequency and $\theta_t = \omega_0 t$. Therefore

$$I_d = I_f - I_{DC} = I_m \sin \omega_0 t - I_{DC} = I_m \sin \theta_t - I_{DC} \quad (7)$$

When diode is off, I_d will charge or discharge C_j while C_j changes as (3) due to the changing of V_j. When diode is on, V_j is nearly constant and I_d will all flow through R_j.

Assume Q_j is quantity of charge on C_j. The relation between C_j and V_j can be written as

$$V_j = \dfrac{Q_j}{C_j} \quad (8)$$

Substituting (3) into (8) results in

$$V_j = \begin{cases} \dfrac{Q_j^2}{C_{j0}^2} \dfrac{1}{V_D} + \sqrt{\left(\dfrac{Q_j^2}{C_{j0}^2}\dfrac{1}{V_D}\right)^2 + \dfrac{4Q_j^2}{C_{j0}^2}} \\[4mm] Q_j\left(\dfrac{C_{j0}}{\sqrt{1-V_{on}/V_D}}+C_D\right)^{-1}, Q_j < 0 \left(V_j \in [-V_{on},0]\right) \end{cases}$$

$$(9)$$

When $V_j \leq -V_{on}$, V_j is given by (5). V_{on} should be chosen carefully to make sure that V_j is continuous. When diode is off, Q_j can be expressed as the integration of I_d.

$$Q_j = \int I_d dt \quad (10)$$

A mathematical modeling is implemented in Matlab to figure out the relation between V_j, I_f, I_d and I_{DC}. The results are shown in Fig. 3. In and, V_j was assumed to be a sinusoidal voltage and I_j was assumed to be a rectangular current when diode is on while V_j was assumed to be a rectangular voltage and I_j was assumed to be a zero when diode is off. Actually, the current that charges and discharges C_j periodically should be taken into consideration as it would result in additional power dissipation on R_S, which was neglected in and.

According to Fig. 3, the whole conduction period can be divided into three parts:

S1: $I_d > 0$ and diode is off. C_j is charged by I_d and V_j rises up to peak inverse voltage V_{BV} of diode. Because of the nonlinearity of C_j, V_j does not present a sinusoidal waveform.

S2: $I_d < 0$ and diode is off. C_j is discharged by I_d and V_j declined to $-V_{on}$. Because of the nonlinearity of C_j, V_j does not present a sinusoidal waveform.

S3: $I_d < 0$ and diode is on. V_j changes slightly according to (5). Charging or discharging current to C_j can be neglected.

According to, PCE will be highest when

$$\max (V_j) = V_{BV} \quad (11)$$

Assume $Q_j = Q_1$ when $V_j = -V_{on}$ and $Q_j = Q_2$ when $V_j = V_{BV}$. By using (9), the relation between Q_1 and Q_2 can be expressed as

$$\begin{cases} Q_2 - Q_1 = \int_{\theta_1/\omega_0}^{\theta_2/\omega_0} I_d dt = \int_{\theta_1}^{\theta_2} \dfrac{I_m \sin\theta_t - I_{DC}}{\omega_0} d\theta_t & (12a) \\[2ex] Q_1 - Q_2 = \int_{\theta_2/\omega_0}^{\theta_3/\omega_0} I_d dt = \int_{\theta_2}^{\theta_3} \dfrac{I_m \sin\theta_t - I_{DC}}{\omega_0} d\theta_t & (12b) \end{cases}$$

where Q_1, Q_2 are given by

$$\begin{cases} Q_1 = C_j V_j \big|_{V_j = -V_{on}} \\[1ex] Q_2 = C_j V_j \big|_{V_j = V_{BV}} \end{cases} \qquad (13)$$

Substitute (3) into (13) results in

$$\begin{cases} Q_1 = -\left(\dfrac{C_{j0}}{\sqrt{1 - V_{on}/V_D}} + C_D \right) V_{on} \\[3ex] Q_2 = \dfrac{C_{j0}}{\sqrt{1 + V_{BV}/V_D}} V_{BV} \end{cases} \qquad (14)$$

θ_1, θ_2 are solutions of equation

$$I_m \sin\theta_t - I_{DC} = 0 \qquad (15)$$

Solving (15), θ_1, θ_2 can be derived as

$$\begin{cases} \theta_1 = \arcsin \dfrac{I_{DC}}{I_m} \\[2ex] \theta_2 = \pi - \arcsin \dfrac{I_{DC}}{I_m} \end{cases} \qquad (16)$$

Combining (12), (14) and (16), I_m can be derived when I_{DC} is given or I_{DC} can be derived when I_m is given. Then, θ_3 can be derived by substituting I_m, I_{DC}, θ_1, Q_1 and Q_2 into (12).

During S1 and S2, Q_j can be expressed as follow:

$$Q_j = Q_1 + \int_{\theta_1/\omega_0}^{\theta/\omega_0} I_d dt$$

$$= -\left(\dfrac{C_{j0}}{\sqrt{1 + V_j/V_D}} + C_D \right) V_{on} + \int_{\theta_1/\omega_0}^{\theta/\omega_0} \dfrac{I_m \sin\theta_t - I_{DC}}{\omega_0} d\theta_t \qquad (17)$$

where $\theta_1 < \theta < \theta_3$. Hence, substituting Q_j into (9), V_j can be derived during S1 and S2. Besides, V_j is

given by substituting I_d into I_j in (5) during S3. The dc component of V_j is

$$V_j' = \frac{1}{T_0} \int_0^{T_0} V_j dt \big|_{T_0 = 2\pi/\omega_0} \qquad (18)$$

where V_j is given by (5) and (9). Thus, R_{DC} and output voltage V_{DC} is determined by

$$\begin{cases} R_{DC} = \dfrac{V_j'}{I_{DC}} - R_s \\[2ex] V_{DC} = V_j' - I_{DC} R_s \end{cases} \qquad (19)$$

Output power can be calculated by

$$P_{DC} = I_{DC}^2 R_{DC} \qquad (20)$$

Power dissipated by R_S can be calculated by

$$P_s = \frac{1}{T_0} \int_0^{T_0} I_d^2 R_s dt \big|_{T_0 = \frac{2\pi}{\omega_0}} = \frac{R_s}{2\pi}\left(I_m^2 + 2I_{DC}^2 \right) \qquad (21)$$

Since $I_j = 0$ during S1 and S2, power dissipated by R_j can be calculated by

$$P_j = \frac{1}{T_0} \int_{\theta_3/\omega_0}^{T_0 + \theta_1/\omega_0} I_d^2 R_j dt \big|_{T_0 = \frac{2\pi}{\omega_0}} \qquad (22)$$

Substituting (2) and $I_j = I_d$ into (22) results in

$$P_j = \frac{1}{T_0} \int_{\theta_3/\omega_0}^{T_0 + \theta_1/\omega_0} I_d^2 \frac{knT}{I_s - I_d} dt \big|_{T_0 = \frac{2\pi}{\omega_0}} \qquad (23)$$

Generally, the value of knT does not exceed 0.03V and I_s is of the order of 10^{-8} A. Since $|I_d| \gg I_S$ during most time of S3, it does not matter substituting with $-knTI_d$. So Equation (23) can be rewritten as

$$\begin{aligned} P_j &\approx \frac{1}{T_0} \int_{\theta_3/\omega_0}^{T_0 + \theta_1/\omega_0} -knTI_d dt \big|_{T_0 = \frac{2\pi}{\omega_0}} \\ &= \frac{knT}{2\pi}\left[I_m \left(\cos\theta_1 - \cos\theta_3 \right) + I_{DC} \left(2\pi + \theta_1 - \theta_3 \right) \right] \end{aligned} \qquad (24)$$

The rectifying efficiency is given by

$$\eta = \frac{P_{DC}}{P_{DC} + P_s + P_j} \qquad (25)$$

Finally, consider the impedance of the diode shown in Fig. 1. The fundamental frequency component $V(f_0)$ of V_j can be derived from the Fourier Transform of V_j. Then, the impedance can be approximately written as

$$Z_d = R_s + \frac{V(f_0)}{I_m} \qquad (26)$$

where $V(f_0)$ can be presented as

$$V(f_0) = |V(f_0)|e^{j\theta} \qquad (27)$$

So (26) can be rewritten as

$$Z_d = R_s + \frac{|V(f_0)|}{I_m}e^{j\theta} \qquad (28)$$

Considering package parameters of diode, the equivalent impedance of the diode in Fig. 1 is

$$Z_d' = Z_d \,\|\, \frac{1}{j\omega_0 C_p} + j\omega_0 L_p \qquad (29)$$

3   CIRCUIT DESIGN AND VERIFICATION

3.1   *Comparison with previous design*

In this part, the theory analysis will be verified in a previous designed rectifier. In (18), a Skyworks SMS7630 Schottky diode in the SC-79 package was used for a half-wave rectifier. The best PCE occurred at an input power of 6 dBm with a dc load of 1080 Ω in source-pull simulation. The best measured efficiency was 72.8% at 8 dBm with a dc load of 742 Ω. The result is listed in Table 2.

In this paper, the maximal conversion efficiencies were calculated for various dc loads. The results, including the corresponding input powers, are shown in Fig. 4. The maximal PCE is 86.61% considering no loss while 75.43% considering 0.6 dB matching loss. The optimal dc load is around 960 Ω, which is closer to the measured 742 Ω than the calculated 1080 Ω in. The best PCE is achieved between 6 dBm and 8 dBm, which is basically consistent with that in (18). The results are listed in Table 2 together with those in (18).

3.2   *Verification in single-diode half-wave rectifier*

A single-diode is proposed in this part. Some enhancements will be added to the designed.
Results of Mathematical Model
A Schottky-diode HSMS286B provided by Avago Technologies packaged in SOT-323 was usedin this paper.

Table 2.   The result in(18) and calculated results in this work at 2.45 GHz.

		Efficiency (%)	R_{DC} (Ω)	P_{in} (dBm)
(18)	Calculated	87 (max.)	1080	-
	Simulated	77.6 (max.)	1080	6
	Measured	72.8 (max.)	742	8
This work	Calculated (no loss)	86.56	1091	6.58
	Calculated (0.6dB loss)	75.39	1091	7.18
	Calculated (no loss)	86.61 (max.)	960	6.75
	Calculated (0.6dB loss)	75.43 (max.)	960	7.35
	Calculated (no loss)	86.39	731	7.58
	Calculated (0.6dB loss)	75.24	731	8.18

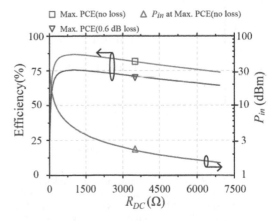

Figure 4.   Maximal PCE and corresponding input powers versus various dc load (RDC) for SMS7630 in this work.

To determine the parameters of rectifier, these steps should be followed:
Determine the operating frequency f_0, output current I_{DC} and threshold voltage V_{on}.
Obtain magnitude of input current I_m according to (12), (14) and (16).
Calculate Q_j by (17) and V_j according to (5) and (9).
Determine V_j, R_{DC}, V_{DC}, P_{DC}, P_S, P_j and η according to (18)–(25).
Obtain the impedance of diode Z_d' according to (26)–(29).
A mathematical model was implemented in Matlab to obtain all required data while an electromagnetic simulation was done in Agilent ADS.

119

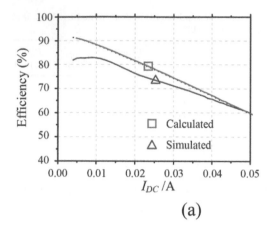

(a)

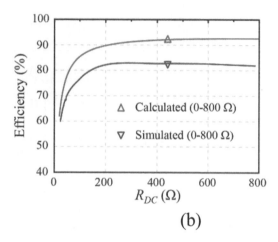

(b)

Figure 5. Results of mathematical modeling of HSMS286B compared to those of simulation: (a) Efficiency vs. IDC. (b) Efficiency vs. RDC (0–800 Ω), when Von = 0.3V and f0 = 5.8 GHz. No loss was taken into account.

The results are presented as efficiency vs. I_{DC} and efficiency vs. dc load (R_L) in Fig. 5. In the simulation, ideal lowpass filters were placed in front of the diode and load, respectively, while simple impedance matching was applied. The differences between efficiencies calculated and those simulated are not more than 10 percent, as described in Fig. 5(a), which are mainly caused by ideal assumptions in basic analysis and the matching circuit which does not present the optimal impedance in simulation. Fig. 5(b) presents PCEs over 92% from 400 Ω to 1400 Ω.

3.3 Parameters and structure setup

R_L = 470 Ω and f$_0$ = 5.8 GHz were selected for the new rectifier with a single shunt-connected diode.

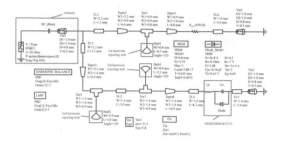

Figure 6. Half-wave rectifier with single shunt-connected diode.

According to (1)–(25), a theoretical efficiency of 92.2% is expected.

The structure of the designed rectifier is shown in Fig. 6. The antenna is presented by a power source and a dc block. Only the first three order harmonics were taken into consideration. The second ($2f_0$) and third ($3f_0$) order harmonics were terminated by the harmonics rejecting filter. Upon the harmonics rejecting filters placed in front of diode, the dcpass filter does not need to terminate higher order harmonics except the fundamental one and no input filter is required.

The results derived in basic analysis were used to design the matching network and calculate the maximal efficiency. A Harmonic Balance (HB) method was used to tune the sizes of micro-strip lines for maximal efficiency, minimal ripple on output and minimal higher order harmonics from diode to antenna. A large signal simulator (LSSP) was used to measure the input impedance. Tapered transmission lines are applied to get higher bandwidth. Matching network applied, the dc branch could present nearly a pure reactance, which means less power consumption on R_L,

And the diode branch could present an input impedance of 50 Ω at f_0. The diode branch should present nearly a pure reactance at $2f_0$ and $3f_0$ in order to reject the higher order harmonics. The dcpass filter should be designed firstly, followed by harmonic rejecting filter. The printed circuit board of the designed half-wave rectifier is shown in Fig. 7.

3.3 Simulation and experiment

An Arlon AD255C substrate with E_r = 2.55 and 0.8 mm thickness was selected for rectifiers.

Firstly, V_{on} was set to 0.3V and I_{DC} = 5.2 mA were determined by the mathematical modeling for R_L = 470 Ω, resulting in a maximal calculated efficiency of 92.2% with no loss, which were obtained by Matlab. With the help of ADS, the rectifier was

Figure 7. Photograph of the designed single-diode half-wave rectifier.

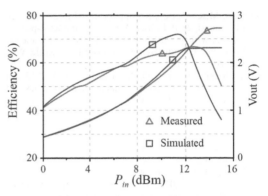

Figure 8. Measured and simulated RF-DC PCE and output voltage of the single-diode half-wave at 5.8 GHz.

optimized for maximal PCE at 5.8 GHz and all its parameters are shown in Fig. 6. The size of the designed rectifier is 7.8 cm × 3.2 cm. The PCE is defined as

$$\eta_0 = \frac{V_{DC}^2 / R_{DC}}{P_{in}} \tag{30}$$

where V_{DC} is output voltage and P_{in} is input power.

The simulated PCE and output dc voltage for V_{DC} at 5.8 GHz are plotted with those measured in Fig. 8, for an input power range from 0 dBm to 15.5 dBm. The highest simulated PCE is 72.2%, which was achieved at an input power of 11.5 dBm. The measured 66.7% achieved at 12.5 dBm (17.78 mW) is very close to the simulated one. Measured results show that the proposed rectifier can perform at an efficiency of more than 50% in the input power range from 3 to 15 mW. The non-loss calculated PCE for HSMS286B with a 470 Ω dc load at 5.8 GHz is 92.2%, which is very reasonable considering the loss about 0.6 dB inserted by matching network and only the second and third harmonics were rejected

4 CONCLUSION

This paper proposes a novel theoretical analysis for current model of a Schottky diode rectifier and experimental validation on a single-diode half-wave rectifier for wireless power transmission at 5.8 GHz. The theory for current model of rectifying element bases on a time domain analysis. More factors are

Used in this improve theoretical analysis to make the calculation results accurately. The varying of junction resistance and junction capacitance are first time taken into consideration to estimate the maximal PCE of given rectifier and the impedance of diode. Some parameters such as the optimal dc load and input power at a given frequency is derived as a function of the parameters of given diode. The experimental results show that good agreement is reached between theory and experiment. The results of the model agree well with those in (18).

The improved theoretically analysis proposed above is applied for designing rectifiers. A 66.7% efficiency was achieved at 12.5 dBm in the single-diode half-wave rectifier with 0.6 dB loss at 5.8 GHz. Although the diode used in this paper is not the best diode for input power levels in the mill watt range, intended for wireless power harvesting reported in (15), the maximal efficiency is close to those in (18), Placing the harmonic rejecting filters in front of rectifying element is beneficial to fabrication. Tapered transmission lines are applied for better impedance matching.

REFERENCES

B. Strassner & K. Chang. 2002. 5.8-GHz circularly polarized rectifying antenna for wireless microwave power transmission. Ieee Trans. Microw. Theory Tech. 50(8): 1870–1876.
E.B. Kaldjob & B. Geck. 2008. A low-cost 2.45/5.8 GHz Ism-band rectifier for embedded low-power systems in metallic objects. in Microwave Conference (GeMIC), Hamburg-Harburg, German, Mar. 2008, pp. 1–4.

H. Gao, M.K. Matters-Kammerer, D. Milosevic, A. van Roermund, & P. Baltus. A 62 GHz inductor-peaked rectifier with 7% efficiency. in IEEE Radio Frequency Integrated Circuits Symposium (RFIC), Seattle, WA, Jun. 2013, pp. 189–192.

H. Takhedmit, L. Cirio, O. Picon, & J.L.S. Luk. 2012. An accurate linear electrical model applied to a series and parallel 2.45 GHz dual-diode rectenna array. in 6th European Conference on Antennas and Propagation (EUCAP), Prague, Mar. 2012, pp. 2510–2513.

J.A.G. Akkermans, M.C. van Beurden, G.J.N. Doodeman & H.J. Visser. 2005. Analytical models for low-power rectenna design. IEEE Antennas Wireless Propag. Lett. 4: 187–190.

J. Guo & X. Zhu. 2012. An improved analytical model for RF-DC conversion efficiency in microwave rectifiers. in IEEE MTT-S International Microwave Symposium Digest, Montreal,QC,Canada, Jun. 2012, pp. 1–3.

J. Guo, H. Hong & X. Zhu. 2012. Automatic load control for highly efficient microwave rectifiers. in IEEE MTT-S International Microwave Workshop Series on Innovative Wireless Power Transmission: Technologies, Systems, and Applications (IMWS), Perugia, May 2012, pp. 171–174.

J. Hansen & K. Chang. 2011. Diode modeling for rectenna design. in IEEE International Symposium on Antennas and Propagation (APSURSI), Spokane, WA, Jul. 2011, pp. 1077–1080.

J. Li & T.C. Lee. 2014. 2.4-GHz high-efficiency adaptive power harvester. IEEE Trans. VLSI Syst. 22(2): 434–438

J. Lim, H. Cho, K. Cho & T. Park. 2009. High sensitive RF-DC rectifier and ultra low power dc sensing circuit for waking up wireless system. in Asia Pacific Microwave Conference, Singapore, Dec. 2009, pp. 237–240.

J.O. McSpadden, T. Yoo & K. Chang. 1992. Theoretical and experimental investigation of a rectenna element for microwave power transmission. IEEE Trans. Microw. Theory Tech. 40(12): 2359–2366

J. Zbitou, M. Latrach, & S. Toutain. 2006. Hybrid rectenna and monolithic integrated zero-bias microwave rectifier. IEEE Trans. Microw. Theory Tech. 54(1): 147–152

K.H. Chen, J.H. Lu & S.I. Liu. 2007. A 2.4GHz efficiency-enhanced rectifier for wireless telemetry. in IEEE Custom Intergrated Circuits Conference (CICC), San Jose, CA, Sep. 2007, pp. 555–558.

L. Bednarz, Rashmi, G. Farhi, B. Hackens, V. Bayot, I. Huynen, J.S. Galloo, Y. Roelens, S. Bollaert & A. Cappy. 2005. Theoretical and experimental characterization of Y-branch nano-junction rectifier up to 94 GHz. in European Microwave Conference, vol. 1, Oct. 2005, pp. 1–2.

M. Roberg, E. Falkenstein & Z. Popovic. High-efficiency harmonically-terminated rectifier for wireless powering applications. in IEEE MTT-S Microwave Symposium Digest, Montreal, QC, Canada, Jun. 2012, pp. 1–3.

S.M. Sze. 1981. Physics of Semiconductor Devices. New York: Wiley-Interscience.

S. Pellerano, J. Alvarado & Y. Palaskas. A mm-wave power harvesting RFID tag in 90 nm CMOS. IEEE J. Solid-State Circuits 45(8)L: 1627–1637

T. Reveyrand, I. Ramos, E.A. Falkenstein, & Z. Popovic, 2012. High-efficiency harmonically terminated diode and transistor rectifiers. IEEE Trans. Microw. Theory Tech. 60(12): 4043–4052

T. Umeda, H. Yoshida, S. Sekine, Y. Fujita, T. Suzuki, & S. Otaka. 2006. A 950-MHz Rectifier Circuit for Sensor Network Tags With 10-m Distance. IEEE Journal of Solid State Electronics 41(1): 35–41

E.B. Kaldjob & B. Geck. 2008. A low-cost 2.45/5.8 GHz ISM-band rectifier for embedded low-power systems in metallic objects. in Microwave Conference (GeMIC), Hamburg-Harburg, German, Mar. 2008, pp. 1–4.

W. Shockley. 1949. The theory of pn junctions in semiconductors and pn junction transistors. The Bell System Technical Journal 28(3): 435–489.

Energy Science and Applied Technology – Fang (Ed.)
© 2016 Taylor & Francis Group, London, ISBN 978-1-138-02833-3

The sensitive mapping of memories in microprocessor

Yuanfu Zhao, Chunqing Yu, Long Fan, Suge Yue, Maoxin Chen & Hongchao Zheng
IC Design, Department of Microelectronics, Beijing Technology Institute, Beijing, China

ABSTRACT: In this paper, the SEE (single event effects) sensitive mapping of different parts of device is explored in a 32-bit microprocessor with a five-stage instruction pipeline by the laser test. The comparison of the sensitive mapping and the layout of device indicates that they fit well with each other. The method in this paper is a meaningful way to validate the reinforcement effect and to provide guidance for improving reinforcement design by corresponding with the physical layout. The pulsed laser that can locate the sensitive sites is an extremely powerful and useful technique for SEE testing, and will provide invaluable information for characterizing the SEE in integrated circuits.

Keywords: SEE, cache, register, microprocessor, sensitive mapping

1 INTRODUCTION

Nowadays, with the reduced technology size of semiconductors, SEEs (Single Event Effects) of devices, which may cause a catastrophic system failure, are becoming more obvious. The chip must be estimated before used in space in order to determine whether it is suitable for space application or not. With the deeper exploring in space, more integrated circuits with a high performance and complex framework will be used in space missions. For complex circuits, there are a variety of functional units including registers, cache memory modules, combinational logic, sequential logic and controller registers.

Particle accelerator testing is the standard method used for characterizing the sensitivity of modern device technology to SEEs. However accelerator testing is often both expensive and not easily accessible, and the data obtained have provided only limited spatial and temporal information; therefore, other techniques that do not suffer from these limitations are needed. The pulsed laser is one such technique that has generated much interest in the SEE community. Pulsed laser is a fast and inexpensive alternative to the ion beam experiment for the general screening of large numbers of device types (J.S Melinger 1994). What is more, the laser system has the special ability to identify the physical location at which an upset occurs (R Jones 2000). In this paper, the SEEs of different parts are explored in a 32-bit microprocessor with a five-stage instruction pipeline by the laser test.

Here, a characteristic parameter "sensitive mapping" is defined as an indicator to distinguish the sensitivity of devices (Maoxin Chen, 2012). Generally, registers and cache memory modules are considered to be the most sensitive modules. However, in this paper, the sensitive mapping of register and cache modules are studied. What is more, the correlation of sensitive mapping and layout is studied, which offers significant guidance for improving reinforcement design. It represents the probability of the occurrence of a single event upset on the sensitive points. Combined with the circuit layout information to find the location of the sensitive node, the sensitive mapping provides direct guidance for radiation-hardened circuit design within the circuit area irradiated.

2 TEST DEVICE

The experimental circuit is a 32-bit embedded RISC microprocessor based on a SPARC V8 architecture, as shown in Figure 1. It can be used to meet a variety of aerospace missions in an onboard embedded real-time computer system. We just need to add memories and related peripheral circuits to form a complete single board computer system. The experimental circuit is fabricated in bulk silicon CMOS 180 nm technology, with a substrate thickness of 400 μm.

3 EXPERIMENT DETAILS

The major laser facility is shown in Figure 2. It generates an array of pulsed laser ionization tracking over a selected area of a die to simulate an SEE (Zhifeng Lei, 2011, Feng Guoqiang 2012).

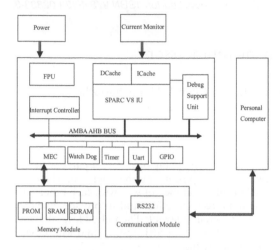

Figure 1. Architecture of the microprocessor.

Figure 2. Laser test facility.

With the advanced characteristics of pulsed laser whose injected position and injected energy can be precisely adjusted, different storage units are selected to irradiate. Sensitive mapping of cache and register is obtained, which indicates the distri-

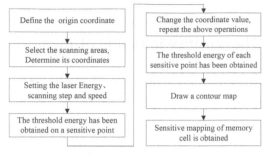

Figure 3. Flow chart of sensitive mapping.

bution characteristics of the circuit sensitive areas with different threshold of energy. The flowchart of the sensitive mapping is shown in Figure 3.

4 TEST RESULTS

The cache and registers in the microprocessor are selected for the experiment. A 10 μm × 10 μm area is selected as the experimental scanning area. The test step is 1 μm and the scanning speed is 1 μm/s, the total number of scanning points is 121, and the threshold energy of the SEU on each point is obtained. The results of the experiment are summarized as follows.

Figure 4 shows the sensitive mapping of a cache memory cell in the microprocessor. In Figure 4 the XY axes represent the XY plane coordinate, and different colors indicate different threshold energy of the laser. The gray part represents that the required laser energy to generate SEU is large. Figure 5 shows the layout of the cache memory cell.

Analyzing the correlation of the above-described cache sensitive mapping and the corresponding layout information, the size information in the X, Y direction on the sensitive circuit node is obtained in both the layout and sensitive mapping. The statistics is presented in Table 1.

According to the size information of the sensitive point given in Table 1, we can draw a conclusion that the sensitive mapping obtained by the laser test fits well with the layout of the cache memory cell. The sensitive node size almost matches the size of the drain of the device.

Similarly, the register sensitive mapping obtained is shown in Figure 6. The horizontal axis and the vertical axis in Figure 6 are the same as in Figure 4. Figure 7 is the layout of the register memory cell.

The size of the sensitive point given in Table 2 shows that the sensitive mapping obtained by the laser test almost fits well with the layout of the cache memory cell. In certain points, the sensitive node size in the sensitive mapping is larger

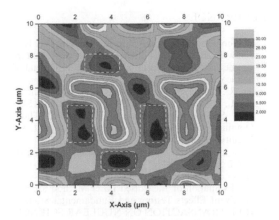

Figure 4. SEU sensitivity mapping of a cache memory cell.

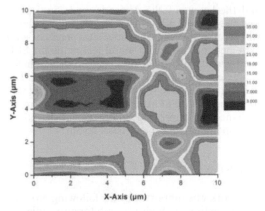

Figure 6. SEU sensitivity mapping of the register memory cell.

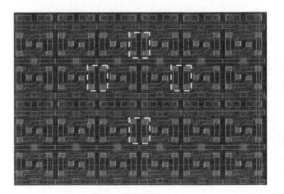

Figure 5. Layout of a cache memory cell.

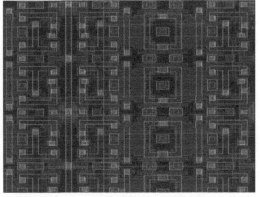

Figure 7. Layout of the register memory cell.

Table 1. Correspondence of sensitive node size between the layout and sensitive mapping in the cache memory cell.

Sensitive node size (um)	X direction	Y direction	Between unit in the X direction	Between unit in the Y direction
Layout	0.3–0.9	1.35–1.6	1.5–2.3	1.71–2.45
Sensitive mapping	1.1	1.8	2.1	2.5

Table 2. Correspondence of sensitive node size between the layout and sensitive mapping in the register memory cell.

Sensitive node size (um)	X direction	Y direction	Between unit in the X direction	Between unit in the Y direction
Layout	0.85–3.13	0.63–1.6	0.7–1.5	2.0–4.0
Sensitive mapping	4	2	4	4

than that in the layout. The possible reason may be the charge sharing of a single event effect in the surrounding areas of sensitive nodes. Besides, sensitive mapping of memory cells is in good agreement with the corresponding layout. The sensitive node size almost matches the size of the drain of the device.

5 CONCLUSION

In this paper, the SEE (single event effects) sensitive mapping of the cache and register memory cells of the device is explored in a 32-bit microprocessor with a five-stage instruction pipeline by the laser test. The correlation between the sensitive mapping and the layout of the device is made, indicating that they fit well with each other. The pulsed laser, which can position the sensitive sites precisely, is an extremely powerful and useful technique for SEE testing, and will provide invaluable information for characterizing the SEE in integrated circuits. The method proposed in this paper is a meaningful way to validate the reinforcement effect and to provide guidance for improving reinforcement design by corresponding with the physical layout.

ACKNOWLEDGEMENTS

This work was supported by the following projects: test evaluation methods research for anti-radiation performance of nanometer ICs and the SEE verification and assessment techniques of CVLSIC based on the laser micro beam.

REFERENCES

Feng Guoqiang, Shangguan Shipeng, Ma Yingqi & Han Jianwei, 2012. SEE characteristics of small feature size devices by using laser backside testing Journal of Semiconductors, Vol.33, and No.1 January.

Jones, R., Chugg, A. M., Jones, C. M. S. and Duncan, P. H., 2000. "Comparison Between SRAM SEE Cross-Section From Ion Beam Testing With Those Obtained Using A New Picosecond Pulsed Laser Facility" IEEE.

Maoxin Chen, 2012. "Single Event Upset Mapping of DICE SRAM Cells".

Melinger, J.S., Buchner, S. & McMorrow, D., 1994. "Critical Evaluation of the Pulsed Laser Method for Single Event Effects Testing and Fundamental Studies" IEEE TRANSACTION ON NUCLEAR SCIENCE, VOL.41, NO.6, DECEMBER.

Zhifeng Lei, 2011. Single Event Effects test for CMOS devices using 1064 nm pulsed laser [J]. IEEE 978-1-4577-1232-6.

Energy Science and Applied Technology – Fang (Ed.)
© 2016 Taylor & Francis Group, London, ISBN 978-1-138-02833-3

Portable refrigerator design

Z.S. Wang & W.X. Xu
Telecom and Business College, Qingdao Huanghai University, Qingdao, China

ABSTRACT: This project designs a portable refrigerator, which can keep the temperature constant at the set temperature range. A portable refrigerator uses temperature sensors, refrigeration equipment, and electronic temperature control and temperature control system. Under the normal operating conditions, it can control the temperature within a certain range. Its basic principle is to collect the refrigerator temperature by temperature probe, and then compare it with the temperature set by the user. When the collected temperature is higher than the set temperature, the thermostat starts the refrigeration system to work. When the collected temperature is less than or equal to the set temperature, the thermostat allows the refrigeration system to power down, the cooling system is stopped, so cycle operation, the temperature inside the refrigerator will remain essentially constant.

Keywords: refrigerator, refrigeration systems, temperature sense probe

1 INTRODUCTION

Refrigerator is a small cabinet or small room to keep food or other things cold. Refrigerator, also known as the ice bucket, evolved from ancient times "ice Kam" (ancient ice-loading containers). It has a clear function, not only to save the food, but also to distribute cool air and make the interior cool. It is an ancient human inventions, revealing to us one side of ancient life. In the mid-17th century, the "icebox" entered the American language word. With the use and development of ice, the refrigerator began to enter the Boston and Chicago home before 1880. As a new household equipment, the refrigerator, a precursor of the modern refrigerator, was invented. Because of the degree of concern growing for resources, energy-saving concept position is also growing in the minds of consumers. Since the late 1990s, energy-efficient refrigerators became popular. Because the refrigerator inhibiting bacterial reproduction through low temperatures cannot meet the consumer's increasing fresh demands, sterilization refrigerator emerged. After 2006, preservation technology abounded. The authentic preservation effect, zero fresh, soft fresh frozen, humidification preservation and thus new preservation technology emerged. Today, the refrigerator is the main tool for the modern home storage and refrigerated food. In China, 95% of urban households have a refrigerator. After 10 years of development, the refrigerator industry already has a high level of market segments. In a recent analysis of the retail market in refrigerator, it shows 97.63 percent of the market share occupied by Haier, Siemens, Meiling and other big brands.

Currently, the semiconductor refrigerator market has entered a mature development period. As the refrigerator company is growing, the semiconductor refrigerator market is gradually attaining specialization, being the scale of the situation, and this new rise adds comfort, warmth, protection, fashion and other functions to the people who live and travel, and it is being recognized by the society.

2 FEATURES

The portable refrigerator is small in size, light weight, easy to carry, and will make you feel relaxed. So, you can enjoy refreshing cold drinks, fruit and food in the hot summer, and you can always enjoy science and technology to bring you a comfortable and convenient life. The portable refrigerator is a electronic refrigeration equipment that can adjust the temperature. In the Intelligent mode, the temperature inside the box is maintained between 2 to 8 degrees Celsius. In the setting mode, the temperature inside the box can be set to anywhere between 2 and 25 degrees Celsius.

1. Lightweight and Portable
   The portable refrigerator has a mini body and small fine, configured lithium battery-powered systems and vehicle power supply connectors, and lightweight so you can enjoy the convenience of life.

2. Green Cooling

   It adopts the international advanced semiconductor electronic cooling and heating technology, directly realizes refrigeration function, has no freon, has low noise, no pollution and affords environmental protection, energy saving, lightweight and long service life.

3. Smart Temperature Control

   The portable refrigerator is equipped with smart settings consisting of the temperature control system, in which the temperature can be converted and the temperature can be regulated.

4. Smart Display

   The 1602 LCD module displays the refrigerator temperature and temperature set by the user, the dual temperature real-time, and the refrigeration temperature control.

3 OVERALL DESIGNS

The principle of the portable refrigerator is the use of hot and cold diversion, energy conduction. Because of its unique and creative research and development, the portable refrigerator can automatically protect from freezing, and is easy to carry and use, thereby having a very good response in the market. There are cabinet refrigeration systems, cooling fans, power supply systems, and a temperature display section. You can save the special conditions of drugs (such as insulin cold boxes, freezer insulin, and insulin small refrigerator). It is an electronic refrigeration equipment that has temperature-regulating functions. In the intelligent mode, the inside temperature is maintained between −8 and 8°C. In the setting mode, the temperature inside the box can be set anywhere between −8 and 25°C.

Its characteristics are as follows:

1. Small and easy to carry.
2. A wide power range (car: through the car's cigarette lighter to provide 12 V; home: through a small transformer to provide 12 V).
3. The unique hydraulic cooling system allows improved cooling efficiency.
4. Simple structure and easy maintenance.
5. No mechanical transmission parts, no wear, low noise, and long life.
6. No refrigerant refrigeration (only compression and absorption are required), greener and safer.
7. High efficiency, low power consumption (power consumption of only compression and absorption 1/3).

Whether you are traveling or in the office, you will find that the portable refrigerator will become your indispensable partner in life, work and travel.

So, you can always enjoy comfortable and convenient living afforded by science and technology.

4 OVERALL CIRCUIT

The overall system block diagram is shown in Figure 1.

5 COOLING SYSTEM

After power supply, the heat of the cold end is moved to the hot end, resulting in the decreased cold junction temperature and the increased hot end temperature, which is well known as the Peltier effect. This phenomenon was first developed in 1821 by a German scientist Thomas Seebeck, but he was doing the wrong inference, with no real insight into the underlying scientific principles. In 1834, a French watchmaker, Jean Peltier, also made a part-time study of this phenomenon only to find the real reason behind this phenomenon until modern times with the development of semiconductors have a practical application, that is the invention of [refrigeration] (Note that it is called the cooler, and not the semiconductor cooling device). The structure of semiconductor cooling is shown in Figure 2.

A semiconductor cooling device consists of a number of N-type and P-type semiconductors in mutual arrangement, and the N\P is connected by a general conductor to make a complete circuit, typically copper, aluminum or other metal conductors, and is finally clipped together by two ceramic sheets. Ceramic must have a good insulation and

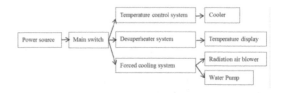

Figure 1.   Overall system block diagram.

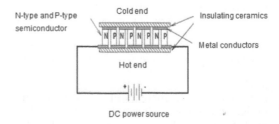

Figure 2.   Structure of semiconductor cooling.

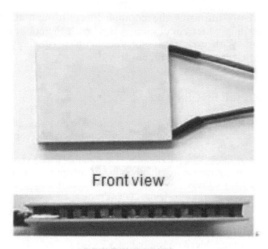

Front view

Lateral view

Figure 3. Semiconductor refrigeration appearance.

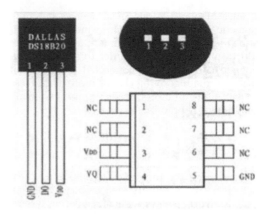

Figure 4. DS18B20.

thermal performance. The appearance looks like a sandwich, as shown in Figure 3:

The simple structure of the electronic refrigerator comprises the P-type semiconductor, the N-type semiconductors, as well as copper and copper wire connected into a circuit. Copper and lead are only used in electrical transmission. Loop power is provided by 12 V DC power. After switching current, the junction becomes cold (refrigerator inside) and another junction radiates (the radiator behind the refrigerator).

6 TEMPERATURE SENSORS DS18B20

Performance characteristics of DS18B20:

1. Singlet structure, only one signal line is needed to connect with the CPU.
2. Direct output serial data, not required for external components.
3. Power supplied directly through the signal line, no external power source. The range of power source voltage is 3.3V ~ 5V.
4. Temperature measurement accuracy is high and the temperature range is −55°C ~ + 125°C. In the range of −10°C ~ + 85°C, measurement accuracy is ± O.5°C.
5. Temperature resolution is high; when it uses 12-bit to convert, temperature resolution can reach 0.0625°C.
6. Digital conversion accuracy and conversion time can be controlled by a simple programming: 9-bit precision conversion time is 93.75 ms; 10-bit precision conversion time is 187.5 ms; 12-bit precision conversion time is 750 ms.

7. It has a non-volatile alarm settings, users can easily modify the value by programming.
8. It can identify whether the temperature of the chip DS18820 is collected beyond the upper or lower limit by the alarm search command.
Functional description of the foot:
DQ: digital signal input/output.
GND: power ground.
VDD: external power supply input (when the parasitic power wiring, this pin should be grounded)

The description of DS18B20 internal structure:
The main internal structure of DS18820 is 64 lithography ROM, temperature sensors, non-volatile temperature alarm triggers TH and TL, configuration registers and other components.

7 COOLING FAN

Fans generally use single-phase or shaded pole motors. The structure of a shaded pole electric motor is simple, easy to manufacture, easy to repair, and low in cost, but the starting torque is small, has poor overload, low efficiency, and can only be used in the small size of a desktop fan. A capacitor-run motor has a big power factor, good starting performance, high efficiency, strong overload capacity, and smooth operation, so most fans use the capacitor-run motor. Primary and secondary windings are in a 900 electrical angle to each other in space. After setting in series with a run capacitor, the secondary winding and the primary winding access the circuit in parallel. When the primary and secondary windings are passed into the single-phase AC power, the primary winding produces pulsating magnetic field, in the role of the capacitor, the secondary winding produces a phase advance of about 90 electrical degrees pulsating magnetic field than the primary winding. The pulsating magnetic field produces a rotating magnetic field to push the rotor in order to start running. The capacitor is always connected to the circuit when it is rotating, so this electric motor is called as the capacitor-run motor. In fact, it is a two-phase operation of the motor.

8 CONCLUSION

A portable refrigerator is mainly composed of heat exchangers, refrigeration systems, water circulation systems, smart displays, manual regulation system, and temperature sensor. It has cold switching function. Hot and cold conduct through aluminum inner with fan. It uses the DC fan to cool so that the working life can be up to 30,000 hours. Using the working life up to 180,000 hours of cooling chips, insulation can be connected to a power supply or 12 V car adapter AC 220 V.

Through several tests, it was concluded that the temperature display is precise, the range of temperature drop is wide, and has a good insulation effect, is small in size, easy to carry, and beneficial to be widely promoted. The portable refrigerator greatly improves the cooling effect through the cooling water circulation system with the cooling fan. With no compressor and freon, it is energy-efficient and environment friendly.

REFERENCES

Bai Bingxu, 2014. Refrigerators, air conditioners and maintenance of equipment principle, Beijing: People Post Press.

Chen Jiabin, 2008. Examples of commonly used electrical equipment troubleshooting. Zhengzhou: Henan Science Press.

Feng Yuqi, Lu Dao-Qing, 2009. Practical air conditioning, refrigeration equipment repair encyclopedia. Beijing: Electronic Industry Press.

Liu Jie, 2009. Mechatronics base and product design. Beijing: Metallurgical Industry Press.

Xu Shiyi, 2013. Emergency repair of household air conditioners refrigerators case. Wuhan: Hubei Science and Technology Press.

Yin Zhiqiang, 2010. Mechatronics System Design. Beijing: Machinery Industry Press.

Yang Xiangzhong, 2010. Refrigerator Repair Encyclopedia. Hangzhou: Zhejiang Science and Technology Press.

Zhang Jianmin, 2012. Mechatronics System Design. Beijing: Higher Education Press.

Energy Science and Applied Technology – Fang (Ed.)
© 2016 Taylor & Francis Group, London, ISBN 978-1-138-02833-3

Development of an electromagnetic microswitch based on non-silicon microfabrication

X.D. Miao
Shanghai University of Engineering Science, Shanghai, China

Y. Dong
Henan Polytechnic Institute, Henan, China

ABSTRACT: This paper developed an electromagnetic microswitch based on a two layer planar micro-coil, which consisted of a microcoil a supporter and a planar microspring. By inputting current into the microcoil, the microswitch is closed, and when the current is off, the device breaks up. Its non-silicon fabrication processes are also presented and then the prototype is obtained. The testing results showed that the device could be actuated at a voltage of 3–5V.

Keywords: MEMS; Electromagnetic; Microswitch; actuator

1 INTRODUCTION

With the development of the communication system, the tremendous need for a high performance switch increased the research of the micro switch. The MEMS technology enables the MEMS microswitch application in communication systems such as satellites, military tactical radio, and military phased array [H. Jaafar, 2014]. In past years, many types of MEMS microswitch were developed with different actuated mechanisms, including electrostatic [Jitendra Pal, 2015][T. Seki, 2003], electrothermal [S. Zhou,1999] [QIU J, 2005] and electromagnetic [Jeong Sam] effects. An electromagnetic microswitch could switch at low voltage and large displacement, adapt to the environment, and be compatible with integrated circuit. Recent research has found that most magnetic microswitches work at the air gap less than 100 um. This paper reports a new type of MEMS switch design, which is moved out of plane at a large air gap of 300 um. The device is developed by non-silicon micro fabrication and assembly process, which could be actuated under a voltage of 3~5V.

2 DESIGN AND PRINCIPLE

The design of the microswitch takes full use of the planar microstructure based on non-silicon surface microfabrication technology, which consists of a planar microcoil, a supporter and a planar microspring. All the structures, especially the microcoil, were designed as the planar structure, which is easily compatible with non-silicon surface microfabrication. The schematic diagram of the electromagnetic microswitch is shown in Figure 1. The key features of the structure include the following: (1) the planar microcoil is used to provide the electromagnetic force when inputting current into it; (2) the enclosed magnetic yoke in and around the planar microcoil is used to attract the magnetic flux, improving the magnetic field; (3) the planar microspring is used to provide the elastic force to move the device up to the break-up status; (4) the magnetic platform of the microspring is used to provide the magnetic force on the microspring to move it down to the close status; (5) the supporter is used to assemble the microspring and the microcoil.

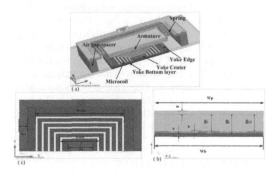

Figure 1. Schematic diagram of a new type of electromagnetic microswitch.

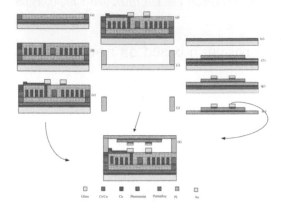

Figure2. Main microfabrication steps of the integrated microcoil, planar microspring and assembling process of the electromagnetic microswitch.

Initially, the microspring is at the flat status when no current is input into the microcoil. When the current is input into the microcoil, the electromagnetic force is generated in the magnetic circuit between the microcoil and the microspring. Since the magnetic force is larger than the elastic force, the microspring is attracted down until the device is closed. When the current is off, the electromagnetic force disappears, and then the elastic force is still large to move the microspring upward to the break-up status.

3 NON-SILICON MICROFARBRICATION

Among the three main parts including the microcoil, the supporter and the microspring were both fabricated by non-silicon surface microfabrication, as shown in Figure 2. The microcoil with an enclosed magnetic yoke was fabricated by the sputtering chromium/copper (Cr/Cu), spin-coating photoresistance, photolithography and electroplating copper to form the microcoil, and then electroplated to form an enclosed magnetic yoke (permalloy 80% Ni, 20% Fe) around the microcoil. The microspring was fabricated by spin-coating photoresistance as the sacrificial layer, and then by sputtering chromium/copper (Cr/Cu), spin-coating photoresistance, and photolithography with electroplating to form the microspring. Then, SU-8 was spin coated on the microspring to form the supporter. Finally, the microspring with the supporter was released from the assembly with the microcoil.

4 CHARACTERIZATION

The fabricated prototype is shown in Figure 3. The 3~5V voltage generated by the B&K 2706 power

Figure 3. Optical photograph of the assembled electromagnetic microswitch.

Figure 4. Diagrams of the characterization of the microswitch.

amplifier incorporated with a GW waveform generator (GFG-8016G) was input into the microcoil, and the output was connected with DC power supply (Agilent E3646 A Dual output) and a resistor. Both the input and output were monitored by the oscilloscope (Agilent MSO6034) to observe the response time.

As shown in Figure 4, the upper level 1 indicates the input of the microswitch, while the upper level 2 indicates the output. When the current is fed into the coil, the difference between the driving voltage and the switching voltage at the upper level indicates the response time at 4.96 ms.

5 CONCLUSION

A new type of electromagnetic microswitch based on a two layer planar microcoil, which consisted of a microcoil, a supporter and a planar microspring, is presented in this paper. By inputting the current into the microcoil, the microswitch is closed, and when the current is off, the device breaks up. Its non-silicon fabrication processes are also presented and then the prototype is obtained. The results showed that the device could be actuated at a voltage of 3~5V.

REFERENCES

Jaafar. K.H., et al. Comprehensive study on RF MEMS switch. Microsystem Technology, 2014, 20:2109–2121.

Jitendra Pal. Yong Zhu. Junwei Lu. RF MEMS switches for smart antennas. Microsystem Technology, 2015, 21:487–495.

Jeong Sam Han et al. Structural optimization of a large-displacement electromagnetic Lorentz force micro actuator for optical switching applications. Journal of Micromechanical and Microengineering, 20014, 14:1585–1596.

Qiu J., Lang J.H., slocum A.H., A bulk-micro machined bi-stable relay with U-shaped thermal actuators, Journal of Microelectromechanical Systems, 14 (2005) 1099–1109.

Zhou, S., Sun, X.-Q. and Carr, W. N., A Monolithic Variable Inductor Network Using Micro-relays with Combined Thermal and Electrostatic Actuation, J. Micromesh. Micron. 9(1999)45–50.

Seki, T., et al. Development of a large-force low-loss metal-contact RF MEMS switch. Sensors and Actuators A. 2003, 132:683–688.

Energy Science and Applied Technology – Fang (Ed.)
© 2016 Taylor & Francis Group, London, ISBN 978-1-138-02833-3

Research on two kinds of ultrasonic sensor array fixtures and design

Mingxi Zhao, Gege Tong & Xiaoxiao Wang
Hebei Provincial Key Laboratory of Power Transmission Equipment Security Defense, North China Electric Power University Baoding, China

ABSTRACT: Partial discharge can make physical and chemical properties of the insulating material to gradually change, and can affect electrical equipment of mechanical and electrical performance, leading to insulation failure. Therefore, the accurate positioning of partial discharge for the electrical equipment is vital. The partial discharge ultrasonic array location method is used to obtain the PD source location using the ultrasonic sensor array and related algorithm. The ultrasonic sensor is used to receive signals, so its poor fix will greatly affect the positioning result. Therefore, this article, respectively, designs two fixing devices for a nine yuan round ultrasonic sensor array and a nine yuan square ultrasonic sensor array, namely a nine yuan circular ultrasonic phased array of electrical equipment for partial discharge fault locating fixture and a nine yuan square of the electric equipment of ultrasonic phased array fixed partial discharge fault location devices.

Keywords: ultrasonic phased array, array detection, fixed device

1 INTRODUCTION

Partial Discharge (PD) is a discharge phenomenon that the applied field voltage generated in electrical equipment is enough to make the insulating part of the regional discharge, but not forming a fixed discharge passage in the discharge region (Li Yanqing, 2004). The faults in the majority of electrical equipment are caused by partial discharge. Partial discharge would make the physical and chemical properties of the insulating material to gradually change, decrease mechanical and electrical performances of electrical equipment, when the development is to a certain extent, it will cause damage to insulation materials. Therefore, accurate location of partial discharge in electrical equipment is necessary. The principle of the partial discharge ultrasonic array positioning method uses the ultrasonic sensor array to receive the target signal, to estimate the delay of each individual sensor array and to calculate the received signal of each single sensor array by an algorithm. Then, it can get the information on space PD source azimuth and elevation angle.

Currently, the market is lacking in the planar ultrasonic sensor array of the finished product. Mostly, by using the method of fixing a plurality of the ultrasonic array element in the electrical equipment of the outer wall, researchers realize the purpose of positioning the partial discharge ultrasonic array. Due to difficulty in disassembly, non-fixing of array element spacing and other reasons, the application of partial discharge ultrasonic

array location is limited. Therefore, it is necessary to design a simple and practical planar ultrasonic array sensor fixing device that can be applied to electrical equipment PD detection.

The device has the advantages of simple structure and easy to carry. It can not only ensure the consistency of the planar ultrasonic sensor array element spacing and the detection accuracy, but also has a low disassembly difficulty. So, it can greatly improve the efficiency of detection of partial discharge in electrical equipment.

Based on the above analysis, this paper mainly relates to the introduction of two kinds of partial discharge fault location of electrical equipment fixing device, namely a nine yuan circular ultrasonic phased array of electrical equipment for partial discharge fault locating fixture and a nine yuan square of the electric equipment of ultrasonic phased array fixed partial discharge fault location devices.

2 THE STRUCTURE OF ARRAY ELEMENTS

The structure of array elements on the nine yuan circular ultrasonic phased sensor includes piezoelectric element, sound absorption backing, acoustic matching layer, electrode lead, and outer covering, as shown in Figure 1.

1, sound absorption backing, 2, outer covering, 3, piezoelectric element, 4, acoustic matching layer, 5, electrode lead

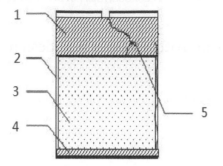

Figure 1. Graphical structure of the ultrasonic sensor array.

2.1 The piezoelectric element of ultrasonic phased sensor

By the piezoelectric effect of the piezoelectric element, the mechanical energy can be converted into electrical energy. As a material of the energy exchange, piezoelectric ceramic has higher piezoelectric constant and electromechanical coupling coefficient, and stable performance, which can withstand greater mechanical stress. Making up is also very convenient, so we often choose piezoelectric ceramics as the ultrasonic sensor of piezoelectric element (Li Jisheng, 2011).

In the production process of ultrasonic phased sensor, to ensure the consistency of the signals in the thickness and the width direction, the ultrasonic sensor selection of the size should consider the following aspects:

With respect to the sensitivity of the receiving area, the greater the receiving area the sensor element has, the greater value the equivalent capacitance is and the higher sensitivity it has. With respect to the production process, the greater the diameter of the sensor element is, the more simpler the making process will be. However, with respect to the acoustic coupling, a sensor element with smaller diameter is better. So, taking all these factors into account, the diameter of the sensor element is set to 7.8 mm.

In the condition of increasing the probe bandwidth on the sound absorption backing, the thickness finally is set to 13 mm. The sensor element is fixed at the tail end of part size: diameter 5.2 mm and length 13 mm.

2.2 The design of the sound absorbing backing

We aim at preventing the entry into the dorsal acoustic reflection in the lining on piezoelectric array and reducing the acoustic coupling between each array element. However, adding backing absorption will lead to the reduction of the receiving sensitivity. So,

considering the relationship between the bandwidth and the receiver sensitivity, we choose the back lining structure design that meets the requirements.

Backing material sensor in this paper is mainly used in tungsten powder and epoxy resin. Due to the moderate acoustic impedance of the two materials, they can guarantee the sensor probe receiving frequency bandwidth, and can meet the requirements of high sensitivity, but also can effectively reduce the coupling between the acoustic sensor array elements.

2.3 The design of the matching layer

When an acoustic wave propagates from one medium to another, the acoustic characteristic impedance mismatch will cause reflection, which will produce great attenuation. Therefore, the matching layer is used to balance the relationship between the receiving sensitivity of the transducer and its frequency band (Luo Yong-fen). As revealed by the analysis of the material, the more acoustic matching layers it has, the better it will be. However, the acoustic matching layer material will have an attenuation effect, and with the increasing number of layers, the making process will be more difficult. Therefore, combining the above consideration, we finally adopt two layers of sound matching: the first matching layer materials are mainly quartz crystal and the second matching layer materials are silicon rubber. The parameters of ultrasonic sensor array element are given in Table 1, while the physical map is shown in Figure 2.

3 A NINE YUAN CIRCULAR ULTRASONIC PHASED ARRAY OF ELECTRICAL EQUIPMENT FOR PARTIAL DISCHARGE FAULT LOCATING FIXTURE

3.1 Design scheme of the fixed device of a nine yuan circular ultrasonic sensor

(1) Structure design scheme. Cylindrical holes on the assembly of a nine yuan circular ultrasonic sensor are circular in shape, as shown in Figure 5. The left and right sides of the assembly are, respectively, equipped with a bar magnet slot.

The body size of the nine element circular ultrasonic sensor array assembly is as follows: length 50 mm; width 40 mm; assembly cylindrical hole

Table 1. Ultrasonic sensor array parameters.

Length	Head diameter	Tail diameter
13 mm	7.8 mm	5.2 mm

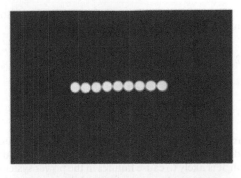

(a)　Front view.

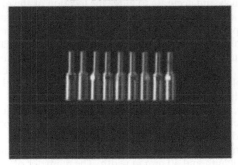

(b)　Side view.

Figure 2. Ultrasonic sensor array element physical map.

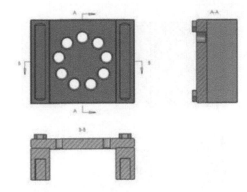

Figure 3. Nine yuan circular array ultrasound transducer fixtures in three views.

Figure 4. Nine yuan circular array ultrasound transducer assembly physical map.

diameter 5.2 mm; cylindrical hole depth 5 mm; adjacent to the center distance of the cylindrical hole 8 mm; radius of the ring surrounded by a 9 cylindrical hole center 11.5 mm.

(2) Material design. The assembly of the ultrasonic array sensor is mainly used for ultrasonic signal reception for the fixed electrical equipment PD sensor array. Considering that most electrical devices are in an open environment, the assembly materials selected for the ultrasonic array sensor should adapt not only to the change in the external environment, day and night temperature and humidity conditions, but also to the change in acid alkali and salt corrosion caused by the air and other pollution (Luo Yong-fen, 2011). In view of the above-mentioned conditions, the ultrasonic array sensor selects the aviation aluminum as the assembly material. Aviation aluminum is a kind of aluminum alloy, whose main chemical components are: aluminum, copper, zinc, magnesium and silicon. Because of its oil resistance property, it shows good performance when dissolved in acid, alkali, salt and other chemical reagents, and can also withstand to some extent when dissolved in an organic solvent. In addition, it has good mechanical properties and

mold ability, processed products smooth surface, good wear resistance, easy processing, secondary processing machinery, low quality, and easy to carry. Although its high temperature strength is low (Hu Ping, 2004), it is fully able to meet the requirements under normal environmental conditions. Because of its excellent comprehensive performance, it is more suitable for the ultrasonic array sensor as the assembly material.

3.2 *Real figure of the assembly of a nine yuan circular ultrasonic sensor*

3.3 *Using step of the fixed device of a nine yuan square ultrasonic sensor*

First, we fix each sensor to the device in turn. Then, we make the device coupled on the outer wall of the electrical equipment. We then use the magnet to adsorb on the outer wall and hold the fixture at the same time. Multi-channel synchronous data collector is adapted to give access to the signal transmission line. That is, we can collect real-time ultrasonic data of partial discharge. If we want to

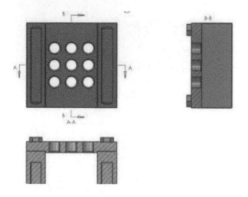

Figure 5. Nine yuan square array ultrasound transducer fixtures in three views.

Figure 6. Nine yuan square array ultrasound transducer assembly physical map.

change the detection position, we can just remove the device and put the coupling in other locations. The operation is simple and convenient.

4 A NINE YUAN SQUARE ULTRASONIC PHASED ARRAY OF ELECTRICAL EQUIPMENT FOR PARTIAL DISCHARGE FAULT LOCATING FIXTURE

4.1 Structure design scheme of the fixed device of a nine yuan square ultrasonic sensor

Cylindrical holes on the assembly of a nine yuan square ultrasonic sensor are circular in shape, as shown in Figure 5. The left and right sides of the assembly are, respectively, equipped with a bar magnet slot.

The body size of the nine element square ultrasonic sensor array assembly is as follows: length 43 mm; width 40 mm; assembly cylindrical hole diameter 5.2 mm; cylindrical hole depth 5 mm; adjacent to the center distance of the cylindrical hole 8 mm.

4.2 Real figure of the assembly of a nine yuan square ultrasonic sensor

5 CONCLUSION

Partial discharge is an important factor in the degradation of electrical equipment insulation. In addition, it is likely to cause failures in the power system. Pinpointing the location of partial discharge can be timely and accurate to judge the location of the discharge source. It is advantageous to examine and repair, so it is very meaningful to the power system. This article develops he fixed device of a nine yuan round ultrasonic sensor. It is 40 mm in length and 50 mm in width. The diameter of the cylindrical hole on the assembly is 5.2 mm and the depth of the cylindrical hole is 5 mm. Adjacent to the spacing of the center of the cylindrical hole is 8 mm. The ring is circled by the center of the cylindrical hole, whose radius is 11.5 mm. We develop the fixed device of a nine yuan square ultrasonic sensor. It is 43 mm in length and 40 mm in width. The diameter of the cylindrical hole on the assembly is 5.2 mm and the depth of the cylindrical hole is 5 mm. Adjacent to the spacing of the center of the cylindrical hole is 8 mm. We choose aviation aluminum to make the assembly of the ultrasonic sensor, which is a kind of aluminum alloy. Aviation aluminum is a suitable material for the assembly of the ultrasonic sensor. Two kinds of the fixed device of the ultrasonic sensor array and sensors can be combined flexibly, which has a very good fixation effect.

REFERENCES

Li Yan-qing. Study on ultrasonic-based method to detect partial discharge in transformer [D].Baoding: North China Electric Power University, 2004.

Li Ji-sheng, Li Jun-hao, Luo Yong-fen, Li Yan-ming. Study of Phased-Ultrasonic Receiving-Planar Array Transducer for Partial Discharge Location in Power Transformer [J]. Journal of Xi'an Jiaotong University, 2011, 45(4):93–99.

Luo Yong-fen, Li Yan-ming. Phased Ultrasonic Receiving Array Sensor for Partial Discharge Location in Oil [J]. Journal of Xi'an Jiaotong University, 2005, 39(4):402–406.

Hu Ping, Lin Jie-dong, Ma Qing-zeng. The Present Situation and Development of Detection of Partial Discharges in Power Transformers Using Acoustic Emission Technology [J].NDT, 2004, 26(10):502–505.

Xie Qing, Li Yan-qing, Lv Fang-cheng, etc. Method for PD Location in Oil Combining Ultrasonic Phased Array with Wideband Array Signal Processing [J]. Proceedings of the CSEE, 2009, 29(28).

Energy Science and Applied Technology – Fang (Ed.)
© 2016 Taylor & Francis Group, London, ISBN 978-1-138-02833-3

The research based on the design of the electronic control unit of uc/os II engine

G.P. Qi

Chongqing University of Posts and Telecommunication, Chongqing, China

ABSTRACT: Due to the increasingly stringent and various engine requirements of the energy conservation, emissions, the use of the traditional mechanical fuel injection system and ignition system has been unable to meet the requirements, and the use of electronic engine control technology can break through the traditional mechanical system control weaknesses, implement the parametric model and make the engine run at an optimum condition, so as to improve the efficiency of the engine, the power performance and reduce emission pollution. The traditional control system uses proscenium and background systems that have defects in real-time processing, and scalability aspects of the system. It slows down the operation of the entire system, and wasted CPU time. This article considers the uc/os II operating system, which has strong portability, curing and tailoring, stability, high reliability and low cost advantages. It adopts preemptive real-time kernel and manages 64 tasks to support semaphores, event flags, message tank, multi-task management, storage management and timing block management. This article uses the automotive electronics mainstream chip MPC5554 as an example to verify the practicality and efficiency of the technology.

Keywords: Proscenium and background systems; uc/osII operating system; MPC5554

1   INTRODUCTION

Under the premise of learning the working principle of a mono-fuel (CNG) engine, this paper chooses a high-performance 32-bit microcontroller—MPC5554 as the core device; based on an embedded real-time operating system uc/osII, we conducted a control strategy and modular design of real-time multi-tasking, and developed a multi-point sequential gas injection and high-energy direct electronic ignition control system. The engine control system requires a high standard of real-time performance. According to the frequency, function and real-time performance, all the engine control tasks are divided into relatively independent subroutines and are performed based on the time-sharing system. It effectively improves the dynamic characteristics and the control accuracy of the system.

2   TASK ASSIGNMENT

Engine control software is based on the design methodology of real-time multi-tasking to build the engine control system. The concrete implementation of multi-tasking control mechanisms is performed by the scheduler. According to the frequency, function and real-time performance, all the engine control tasks are divided into relatively independent subroutines and are performed based on the time-sharing system. The control system EFI program is divided into five tasks and an interrupt. Task 1 is responsible for starting the injection and ignition control; Task 2 is responsible for completing the converting of an analog signal; Task 3 is responsible for the discrimination of the engine operating conditions and calculation of the next jet and the time delay of ignition control; Task 4 is responsible for the control of engine speed and idle bypass valve; Task 5 is the communication processing tasks, responsible for the completion of uploading data such as engine monitoring, and for the input capture interrupt program to judge the cylinder, speed calculation and jet ignition timing control. Task priority level are arranged in the following order: Capture Interrupt> Task 1> Task 2> Task 3> Task 4> Task 5.

Based on the different functions, the electronic control system can be divided into seven modules: the system initialization module, injection and ignition control tasks, signal acquisition task, pulse width control computing tasks, monitoring and protection tasks, communication processing tasks and an interrupt module. The description of the seven modules is detailed below.

2.1 System initialization module

The module completes the clock initialization of the system, the I/O initialization, analog-to-digital conversion module initialization, e TPU hardware initialization (Vance, J.B, 2011), e TPU channel input capture function initialization and some global variable values initialization.

System clock initialization. This system uses ECU MPC5554 with an 8 MHZ crystal vibration frequency. In order to meet the rapidity of the electric control system, the system clock is configured to 64M through the design of the phase-locked loop FMPLL.

I/O initialization. Because most of the MPC5554's I/O pins are multiplexed, before the system is running, the input and output ports are initialized.

Digital conversion module initialization. Before the signal acquisition task launches e QADC and converts the analog signal, the e QADC must be initialized and be active. After the initialization of e QADC, it should wait for the start of the signal acquisition task cycle.

E TPU hardware initialization. The system uses 13 e TPU channels in a total of three functions: input capture, output comparison and PWM signal generation.

E TPU channel input capture function initialization. In order to capture the rotate speed signal of the crankshaft, the A0 of the e TPU channel is configured as an input capture function. The channel will need to run at the same time when the system is powered, because the rotate speed signal of the crankshaft is the time reference of the entire electronic control system.

2.2 Injection and ignition control tasks

The task is responsible for the startup of injection and the control of ignition delay time. It will start by capturing the interrupt of the program semaphore. In addition to the interrupt, it is the highest priority task.

2.3 Signal acquisition task

The main task is the calculation of digital control, including the inlet pressure signal, the cooling water temperature signal, the throttle position signal, the battery voltage signal, the inlet air temperature and air–fuel ratio signal. Since the task directly affects the next step, i.e. the control amount calculation, it is the highest priority task.

2.4 Pulse width control computing tasks

The main function of pulse width control computing tasks is to calculate the injection time and pulse width; and the ignitions coil moment and duration based on the operating conditions of the engine. It is the core module of the software system, including some subroutines, such as working condition judgment and pulse spectrum checking. This task is activated by input the capturing semaphore of interrupt subroutine.

2.5 Monitoring and protection tasks

The main function of the task is to check the working state of the engine and offer the timely control of the abnormal condition. When the engine speed is higher than 2300r/m, the gas shutoff valves need to be closed; when the rotation speed is under 2200r/m, the gas shutoff valves can be reopened. When the detected throttle opening degree is greater than a certain value, it means that the engine is no longer at an idle state. At this time, the idle by-pass valve of the eTPU_B0 channel needs to be turned off until the engine is in the idle condition to restart the PWM function of this channel.

2.6 Communications processing tasks

This task is mainly responsible for the upload of some monitoring parameters. Upon receiving the flags sending by the host computer, it will continuously upload a certain number of engine parameters. In the data communication module, it designs a virtual Flex CAN communication mechanism for the MAP data calibration experiment to ensure the stability of the data transfer.

2.7 Input the capture of interrupt subroutines

The input capture function is initialized during the initialization process of the system. Every time the falling edge of (the check point of cylinder compression) crankshaft speed signal comes, it will trigger the input of capture interrupt subroutines. This feature includes the following: offering the time reference for injection and ignition control, identifying sub-cylinder signals and recording the current cylinder number, computing the speed, controlling the start of the corresponding injection valve and ignition coil, notifying the judgment of engine working condition, and calculating the corresponding injection pulse width and ignition advance angle.

3 THE CONTROL STRATEGY

The system developed a generic intelligent PID control procedure: close_ loop_ctr (close_ loop_ stuck * p, vuint16_t back value). At the idle conditions and air–fuel ratio closed loop control, the only thing needs to be done is to call this function.

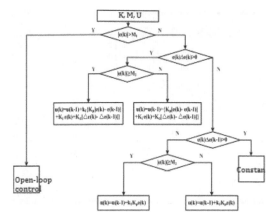

Figure 1. Algorithm flow chart.

The difference is that we need to pass the return value of different sensors and initialize different targets before calling.

At the idle condition, since the controlled quantity is a target rotation speed, we need to pass the speed value (engine), that we calculated during the input of capturing interrupt, to the parameter (back value). Under the air–fuel ratio closed loop control process, the controlled quantity is the air–fuel ratio. We need to pass the detected value of the oxygen sensor—O2_value to the parameter, back value. The variable is the likelihood ratio in the air in the closed-loop control process, the amount of the air–fuel ratio control, and the detected O2_value oxygen sensor is passed to the back value variable parameter. For the idle condition and air–fuel ratio closed loop control, it defines two initial structures: idle and air–fuel ratio. The two structures are the same, in which the target control value disvalue variables, other variables are PID control parameters. The idle condition will target the disvalue target idling speed to 800 or 900. The ideal target is set at 16.7 air–fuel ratio likelihood ratio closed-loop control. The return value of this function is to adjust the value of the correction amount, i.e. the calculation results together with the original function of the current correction amount obtained correction amount, since the beginning of the program to initialize the return value is zero.

4 THE KEY TO THE IMPLEMENTATION OF PART OF THE SYSTEMS

Electronic control systems and e DMA collaborative e QADC completed the acquisition measurement of engine parameters for each condition. After signal acquisition task trigger e QADC, e

DMA automatically configured channel switching commands from a user-defined area in RAM remove the transfer to the on-chip ADC. After the completion of one cycle engine ADC signal acquisition, e DMA will automatically convert the results of e QADC removed from the results of the queue, and delivered to a user-defined RAM area, as a control parameter for the engine electronic control system applications.

There are three different functions used in e TPU of this system: enter a capture, output compare and PWM signals. Therefore, the ECU is powered on, before e TPU function runs, the main CPU must correctly initialize e TPU various peripheral devices. E TPU common initialization process is completed by the main CPU and e TPU, respectively, to complete their responsibilities: the main CPU is responsible for e TPU module initialization, initialization e TPU channels provide initialization parameters, triggers the execution of e TPU function. E TPU is responsible for HSR request corresponding to the initial state and converted.

EFI engine control system calibration data are obtained by repeated calibration experiments, the experiment to deal with large amounts of communication data communication module for the MPC5554 Flex CAN develop a virtual I/O module, the principle of this mechanism is driven by the communication module on adding a ring buffer. When an application wants to send a lot of data and hardware to keep up with the speed of time, these data will be stored in a ring buffer waiting to be sent. In addition, when the hardware receives a lot of data and applications, the data that need not be read are also stored into the buffer.

5 SIMULATION

The main goal of this experiment is to control the precise control of timing and time engine jet and ignition under different conditions, in order to control the effect observation system, set up the injection and the ignition-related MAP diagram, the data as MAP design software. Figure from the top to bottom are the crankshaft speed signal, 1 cylinder jet signal, 2 cylinder, 3 cylinder jet signal, signal.

According to the controlling of the timing of the engine, the 2 cylinder compression TDC signal (jet control time 2 cylinder) should be fourth on cylinder 1 TDC signal after the falling edge of the 3 cylinder crankshaft, TDC signal should be second in 1 cylinder compression TDC signal to a crankshaft after falling edge. As shown in Figure 2, the timing control is completely correct.

Figure 3 and Figure 4 shows the engine acceleration and deceleration waveform under the

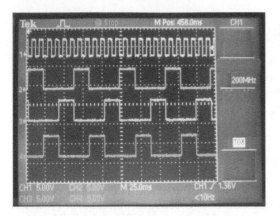

Figure 2. Jet waveform of each cylinder.

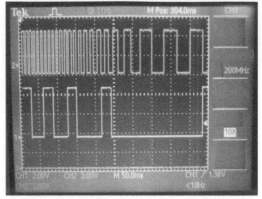

Figure 4. One cylinder jet of the rapid deceleration.

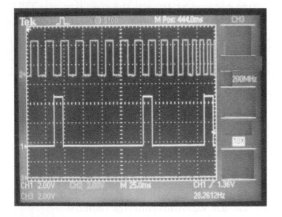

Figure 3. One cylinder jet of rapid acceleration.

condition of acute jet. Jet time reference of each cylinder is respective TDC signals, only the corresponding TDC signal arrival, began to control the cylinder jet time delay.

6 CONCLUSIONS

The hardware system can collect the crankshaft sensor effectively (signal generator simulation) square wave signal; the hardware system can collect the analog signal effectively (power input analog voltage); the system timer can correct implementation of the injection and the ignition delay time of action, the timing and accuracy.

The software system can accurately calculate the speed of crankshaft, can accurately calculate the various analog, can perform 3D MAP interpolation effective and accurate look-up table procedure, can realize the switching control mode and control strategy in various working conditions of the fast, time sequence and time can be controlled according to precise MAP data engine jet and ignition.

On the whole, the overall structure of the system and each submodule meets the requirements for the basic control and ignition engine jet in various working conditions.

REFERENCES

Freescale Semiconductor. 2010 Rev.3.1. MPC5553/MPC5554 Microcontroller Reference Manual[S]. Germany. 1–1170.

Jean J. Labrosse. 2011, MicroC/OS-II the Real-Time kernel Second Edition [M]. CMP Books.

Vance, J.B., Kaul, B.C. Jagannathan, S. Drallmeier, J.A. 2010. Output Feedback Controller for Operation of Spark Ignition Engines at Lean Conditions Using Neural Networks [J]. Proc. of IEEE.16 (2).214–228

Vance, J.B., Singh, A., Kaul, B.C., Jagannathan, S., Drallmeier, J.A. 2011. Neural Network Controller Development and Implementation for Spark Ignition Engines with High EGR Levels [J]. Proc. of IEEE.18 (4).1083–1100.

Zang Huaiquan, Wei Xing. The Application of the Embedded Operating System at Control of Furnace Temperature[C]. Proceedings of the 25th Chinese Control.

Mechanical, manufacturing, process engineering

Energy Science and Applied Technology – Fang (Ed.)
© *2016 Taylor & Francis Group, London, ISBN 978-1-138-02833-3*

CFD simulated engineering model of biodiesel synthesis in a spinning disk reactor

Z. Wen

Faculty of Process and Environmental Engineering, Lodz University of Technology, Lodz, Poland

ABSTRACT: In this paper, a two-disk spinning disk reactor for intensified biodiesel synthesis is described and numerically analyzed, using convection-diffusion reaction species transport model by the CFD software ANSYS©Fluent v. 13.0. The effect of the upper disk's spinning speed is evaluated. The results show that the spinning speed influences the residence time and TG conversion significantly.

Keywords: Spinning disk, biodiesel synthesis, ANSYS©Fluent, numerical simulation, TG conversion

1 INTRODUCTION

Biodiesel production includes the transesterification of oil or fat with alcohol under base conditions in a liquid-liquid environment. Transesterification is a liquid-liquid two-phase reaction.

In liquid-liquid reactions, immiscible reactants must be transferred from one phase to another or contact intimately with each other before reaction occurs, and this mass transfer can become the limiting step. Therefore, the efficiency of mass transfer is of importance for improving production capacity and reducing the process cost and equipment size. It is desired that the mass transfer rate can be increased to a level that allows the reactions to be limited only by intrinsic-reaction kinetics (Krawczyk 1996). To ensure good mass transfer performance, it is important to enhance contact and contact area between the two liquid phases and decrease resistance to mass transfer in the reactor. Conventionally, this can be realized by mechanical stirring or dispersion, e.g. bulk stirring. However, the process is very slow. These contacts can be improved by establishing small-scale liquid structures or eddies within which mass transfer is enhanced (Green et al. 1999). Creating these small-scale liquid structures (micro mixing) is the associated role of mixing and mass transfer equipment and reactors.

The spinning disk reactor (SDR) is one of the process intensification technologies using high gravity fields caused by centrifugation. When a liquid is introduced onto the disk surface at or adjacent to the spin axis, the liquid flows radially outward under the centrifugal force formed by the rotational disk, in the form of a thin film. SDRs are used in many fields. Investigations including mass and heat transfer, and kinetic studies using SDRs have indicated that main residence times, reactant inventories and impurity levels can decrease by up to 99% (Brechtelsbauer et al. 2001, Cafiero et al. 2002).

In the present research, a special version of the SDR designed to explore the possibility of enhancing the efficiency of biodiesel synthesis is described and used, as shown in Fig. 1. ANSYS©Fluent 13.0 was chosen in the present research to simulate the synthesis of biodiesel in the SDR, using the convection-diffusion reaction species transport model. The effect of the upper disk's spinning speed was investigated and the quantitative comparison is also satisfactory.

2 EXPERIMENTAL SETUP AND MECHANISM OF ALKALI-CATALYSED TRANSESTERIFICATION REACTION

The experimental setup was as developed previously by our collaborating partners Zheyan Qiu and Laurence Weatherley (Qiu et al. 2012). The SDR consists of a stable disk, which is coaxially spaced close to a rotating parallel disk separated by 0.2 mm. TG and methanol are pumped from the centers of the stable disk and the spinning disk, with the speed of 0.007 ms-1 and 0.006 ms-1, respectively. The experimental setup is shown in Figure 1.

As mentioned above, biodiesel is mainly produced by transesterification of vegetable oils and animal fats with alcohol in the presence of catalysts. Vegetable oils and animal fats typically consist of TG, which are esters of free fatty acids with the trihydric alcohol, glycerol.

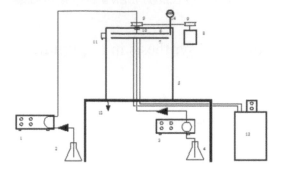

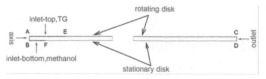

Figure 2. Simplified computational domain of the SDR.

Figure 1. Experimental setup of the intensive SDR for biodiesel synthesis. (1) Peristaltic pump; (2) canola oil vessel; (3) digital piston pump; (4) sodium methoxide vessel; (5) cylinder; (6) rotating disk; (7) stationary disk; (8) variable-speed DC motor; (9) pulley; (10) bearing; (11) sampling point; (12) product drainage; (13) heating circulator; (14) thermometer.

For oil feedstock, rapeseed and soybean oils are most commonly used today, and canola oil was chosen in the present research. In addition, because of its low cost and high product efficiency, sodium hydroxide was chosen as the base catalyst. Basically, methanol is the cheapest alcohol available in the market, hence it was chosen.

Mechanism of alkali-catalysed transesterification reaction. During the transesterification of TG, there are three stepwise and reversible reactions with intermediate formation of diglycerides (DG) and monoglycerides (MG), resulting in the production of methyl esters (RCOOCH3, biodiesel) and glycerol (GL) shown in chemical equations I-III and the overall transesterification reaction shown in chemical equation IV (Freedman et al. 1986, Noureddini & Zhu 1997).

$$TG + CH_3OH \overset{k_1}{\underset{k_2}{\rightleftharpoons}} DG + RCOOCH_3 \qquad (I)$$

$$DG + CH_3OH \overset{k_3}{\underset{k_4}{\rightleftharpoons}} MG + RCOOCH_3 \qquad (II)$$

$$MG + CH_3OH \overset{k_5}{\underset{k_6}{\rightleftharpoons}} GL + RCOOCH_3 \qquad (III)$$

$$TG + 3CH_3OH \overset{catalyst}{\rightleftharpoons} 3RCOOCH_3 + GL \qquad (IV)$$

3 SIMULATION

3.1 Simulation methodology

ANSYS©Fluent v.13.0 was chosen to simulate the biodiesel synthesis process in the SDR. It is important to appropriately mesh the computational

domain. To obtain a satisfactory mesh, it is necessary to simplify the real construction of the SDR. As it can be seen from Figure 1, the region of interest is the space between the spinning disk and the stationary disk. Besides, this construction is axisymmetric, hence only half of it was simulated. The computational domain is shown in Figure 2.

The dimensions of the domain based on the SDR construction are as follows: domain radius AC, 50 mm; gap size AB, 0.2 mm; top inlet radius AE, 1.75 mm; and bottom inlet radius BF, 0.77 mm. As mentioned before, TG was pumped into the reactor through the orifice AE with a velocity of 0.006 ms-1 and methanol along the orifice BF with a velocity 0.007 ms-1, respectively. The corresponding volumetric flow rates of TG and methanol were 5.77×10^{-8} m3 s-1 and 1.3×10^{-8} m3 s-1, respectively.

ANSYS©Fluent predicts the local mass fractions of each species through the solution of convection-diffusion reaction equation for each species. The general form of this equation is given by

$$\frac{\partial}{\partial t}(\rho Y_i) + \nabla \cdot (\rho v Y_i) = -\nabla . J_i + R_i + S_i \qquad (1)$$

The diffusion flux of species i, J_i, which arises due to concentration gradients, can be written as

$$J_i = -\rho D_{i,m} \nabla Y_i \qquad (2)$$

The N_R reactions in which the species participate is given by

$$R_i = M_{w,i} \cdot \sum_{r=1}^{N_R} R'_{i,r} \qquad (3)$$

Concerning the last term in the general Eq. (1), S_i equals to zero because in the present research, there is no additional source for species.

3.2 Modeling assumptions

For the purpose of modeling and in order to simplify the calculation, initial assumptions were made as follows:

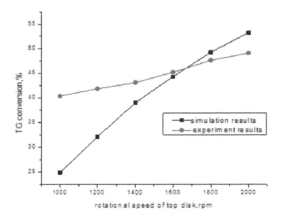

Figure 3. TG conversions obtained by simulation compared with those obtained by the experiment.

1. Feed stocks from the top inlet and the bottom inlet are TG and methanol, respectively.
2. Problem is nearly isothermal because the thermal boundary conditions assume perfect thermal control, thus all physico-chemical properties are nearly constant.
3. At the beginning, the SDR gap is filled with TG at rest and from this point, the process starts by injecting TG and methanol.
4. Reaction rates are determined by Arrhenius expressions, and the effect of turbulent fluctuations is ignored.

3.3 Boundary conditions

In the present research, six cases were calculated and analyzed. The difference among these cases is that rotational speeds of spinning disk are different. The available experimental data formulated the following modeling conditions:

1. Due to the fact that canola oil, as a mixture of fatty acids, consists mainly of oleic acid, triolein ($C_{57}H_{104}O_6$) was chosen as a representation of TG, accordingly diolein ($C_{39}H_{72}O_5$) was chosen as a representation of DG, monoolein ($C_{21}H_{40}O_4$) was chosen as a representation of MG, and methyl oleate ($C_{19}H_{36}O_2$) was chosen as a representation of the resulting methyl ester (biodiesel). All physicochemical data for each substance needed in this research were taken from the Aspen Plus software database.
2. The inlet velocities for TG and methanol are 0.006 ms-1 and 0.007 ms-1, respectively;
3. The boundary temperature is 25 °C.

3.4 Simulation results

TG conversion is a good parameter to verify the accuracy of ANSYS©Fluent numerical simulation.

The comparison between TG conversions obtained by simulation and those obtained by the experiment is shown in Figure 3.

Figure 3 shows the comparison of TG conversions between the simulation results and the experimental data. Undoubtedly we can observe a qualitative agreement between the two values: TG conversions increase with a growing rotational speed of the spinning disk. We can say that the quantitative agreement is also satisfactory, taking into account the complexity of the reaction and uncertainty about reaction kinetic parameters. The discrepancy between the simulation and experiment results probably from idealization of the flow characteristic inside the gap where simple liquid flow was assumed. As observed from the real experimental analysis by a direct inspection of the disk surfaces after disassembling the reactor, there were zones of distinct emulsions present in the gap. This implies that intensification of the process could not be as vigorous as in the case of actual liquid, e.g. the reverse flow in reality was not as clear and regular as in theory.

4 CONCLUSION

The CFD software ANSYS©Fluent v.13.0 was successfully used for the simulation of biodiesel synthesis from canola oil and methanol in the presence of the sodium hydroxide catalyst in the two-disk spinning disk reactor. To accomplish this task, the adequate model of reaction with accompanying mass transfer was formulated and expressed in terms of facilities available within ANSYS©Fluent. The simulation results for several rotational speeds of the spinning disk were compared with the corresponding experimental data obtained from a setup developed by our collaborating partner from Kansas University.

The TG conversion attained in the reactor was significantly affected by the rotational speed of the upper disk. The faster the rotational speed of the upper disk is, the higher the TG conversion will be. Based on the experiment data from Zheyan Qiu and Laurence Weatherley, we can see that the TG conversion obtained by simulation was acceptable.

REFERENCES

Brechtelsbauer C., Lewis N., Oxley P., Ricard F., Ramshaw C., 2001. Evaluation of a spinning disk reactor for continuous processing. Organic Process Research & Development, 5, 65–68.
Cafiero L. M., Baffi G., Chianese A., Jachuck R. J., 2002. Process intensification: precipitation of barium sulfate using a spinning disk reactor. Industrial & Engineering Chemistry Research, 41, 5240–5246.

Freedman B., Butterfield R.O., Pryde E. H., 1986. Trans-esterification kinetics of soybean oil. Journal of the American Oil Chemists' Society, 63, 1375–1380.

Green A., Johnson B., John A., 1999. Process intensification magnifies profits. Chemical Engineering, 106, 66–73. INIST-CNRS, Cote INIST: 5027, 35400008133713.0040.

Jachuck R., 2002. Process intensification for responsive processing. Trans IChemE, 80, 233–238.

Krawczyk T., 1996. Biodiesel-alternative fuel makes inroads but hurdles remain. INFORM, 7, 801–82.

Noureddini H., Zhu D., 1997. Kinetics of transesterification of soybean oil. Journal of the American Oil Chemists' Society, 74, 1457–1463.

Qiu Z., Petera J., Weatherley L.R., 2012. Biodiesel synthesis in an intensified spinning disk reactor. Chemical Engineering Journal, 210, 597–609.

Stankiewicz A. I., Moulijn J.A., 2002. Process intensification: transforming chemical engineering. Chemical Engineering Progress, 96, 22–34.

Energy Science and Applied Technology – Fang (Ed.)
© 2016 Taylor & Francis Group, London, ISBN 978-1-138-02833-3

Numerical simulation on the stability in cutting nickel-based super alloy GH4169

J.J. Liu & G.W. Wang
Institute of Applied Mechanics of Biomedical Engineering, Taiyuan University of Technology,
Taiyuan, Shanxi Province, China
Key Laboratory of Strength of Materials and Structural Impact of Shanxi Province,
Taiyuan University of Technology, Taiyuan, Shanxi Province, China

ABSTRACT: The FEM simulation was developed to investigate the effect of speed on the stability in cutting nickel-based super alloy GH4169. The simulation results indicate that, at a low-speed stage, the chip is continuous because the work hardening prevails over the thermal softening and inertial effect. The transition from continuous to the saw-tooth chip formation occurs at the medium-speed stage since the thermal softening dominates among the three main factors. However, at the high—and ultra-high-speed stage, the inertial effect begins to dominate the deformation process. In addition, the transition of the saw-tooth to continuous chip formation occurs. Consequently, the deformation remains homogeneous. However, the critical speed for the three stages is still an open question.

Keywords: Instability; High-speed cutting; FEM; GH4169

1 INTRODUCTION

In cutting metals, the waste comprises chips or sward and excess metal. Although the chip is a waste compared with the finished part, the shape can offer us important information to predict the lifetime of the cutting tool and the quality of the finished part. When the chip is continuous at a low speed, the roughness of the finished part and the lifetime of the cutting tool are good, but the processing time and charge are too hard to be accepted. At this stage, the work hardening is the leading effect. However, at the medium-speed stage, the time and cost decrease, the roughness and lifetime of the cutting tool deteriorate because of vibration with the formation of a serrated chip. At this stage, the thermal softening dominates, which means the chip loses stability because of the formation of shear bands. Whether we can get the two merits at the same time remains a question. This paper provides us the answer to this question through the numerical simulation.

A super alloy, or high-performance alloy, is an alloy that exhibits an excellent mechanical strength and resistance to creep (tendency for solids to slowly move or deform under stress) at high temperatures, good surface stability, and corrosion and oxidation resistance. Super alloys are commonly used in the parts of gas turbine engines that are subject to high temperatures and require high strength, excellent high-temperature creep resistance, fatigue life, phase stability, and oxidation

and corrosion resistance. Despite these features, the utilization of the nickel-based super alloy is limited due to the poor machinability. In fact, the low thermal conductivity also leads to the difference between the nickel-based super alloys and other tractable materials (Barry, 2001). Therefore, investigating the cutting stability of nickel-based super alloys is of great significant importance, which will make a difference to both the tool life and the quality of the finished part.

2 THE FEM MODEL

In this paper, the Finite Element Model (FEM) was used to simulate the orthogonal cutting process. The FEM is a fully coupled thermal-mechanical model, which was developed using commercial software ABAQUS/explicit and standard. The workpiece is discretized by using the 4-node bilinear element with reduced integration. At the same time, the cutting tool is regarded as an analytical rigid body for simplicity (Zhang, 2008), as shown in Figure 1.

The constitutive relation employed in the numerical simulation is the J-C relation, which is widely used to model the thermal-viscoelastic material.

The material parameters of GH4169 are listed in Zhang (2008). In addition, we choose the extended coulomb friction model to model the tool-chip interaction, and the friction coefficient is set to 0.3 as a constant (Leech, 1985). The separation of the

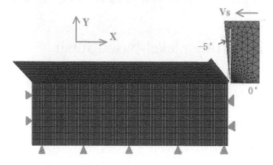

Figure 1. Numerical model employed in this paper (Zhang, 2008).

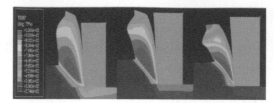

Figure 2. Chip morphology for various speeds: 0.5 m/s, 1.5 m/s, 2.5 m/s.

chip from the workpiece is modeled with the help of a failure zone. Also, the separation of the chip and the workpiece will occur if the critical effective strain is up to 2 (Barry, 2001).

3 THE STABILITY IN CUTTING AT DIFFERENT CUTTING SPEED STAGES

It is well known that the stability in cutting is determined by three leading factors, namely strain hardening (work hardening), thermal softening and inertial effect. However, how the three factors compete and how the shear bands form which lead to the instability in cutting at different speed stage remains to be answered. The context below will answer these questions through the numerical simulation.

3.1 *Stability at the low cutting speed stage*

In order to investigate the effect of the cutting speed on the stability, we set the rake angle and the uncut chip thickness to 95.and $2e^{-4}$ m, respectively. At this speed stage, the strain hardening prevails over the thermal softening, and the inertial effect can usually be ignored. At this time, the chip is continuous, which means the cutting is stable. As the result of the continuous chip, the cutting life and the quality of the finished part is acceptable.

The numerical simulation results for different cutting speeds show that, at the low-speed stage, the chip would not lose stability, as shown in Figure 2.

3.2 *Stability at the medium/high cutting speed stage*

At the medium/high-speed stage, we still keep the rake angle and uncut chip thickness the same as at the low-speed stage. At this speed stage, when the

cutting speed arrives to 3 m/s, the leading factor is not strain hardening but the thermal softening, which means shear localization will occur in the interface between the chip and the rake face of the tool. Although the width of the shear band is small whose magnitude ranges from 10μm to 100μm (Leech, 1985), the propagation speed of the shear band can also cross the whole chip to reach the free surface of the chip before the perturbation go through the shear band in the thickness direction. This is because the inertial effect cannot exert enough influence to the propagation of the shear band. In addition, the shear band still has enough time to traverse the chip. In other words, at this cutting speed, the chip morphology will transfer from continuous to serration-like, as shown in Figure 3.

As the cutting speed increases, the thermal softening becomes more superior and the shear localization becomes more evident, as shown in Figure 4. Also, the inertial effect cannot influence the competition between the thermal softening and the strain hardening.

When the cutting speed continues to increase and arrives to 150 m/s, thermal softening is still the leading factor, but the inertial effect plays a part in the competition between the thermal softening and the strain hardening. In fact, this is because the shear bands do not have enough time to form as before, as the influence of the inertial effect keep reinforcing, as shown in Figure 5.

3.3 *Stability at the ultra-high cutting speed stage*

At the ultrahigh-speed stage, the competition between thermal softening and strain hardening still exists, and thermal softening also prevails over strain hardening. However, the inertial effect will play an important part to dominate this competition by reducing the time that the shear band traverses the whole chip to arrive at the chip's free surface. When the cutting speed reaches to 250 m/s, the shear bands cannot reach the free surface, as shown in Figure 6.

From Figure 6, we can still see the shear bands clearly, but the length is shorter than the bands

Figure 3. Chip morphology for various speeds: 4 m/s, 10 m/s, 30 m/s.

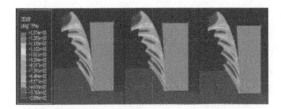

Figure 4. Chip morphology for various speeds: 60 m/s, 70 m/s, 80 m/s.

Figure 5. Chip morphology for various speeds: 150 m/s, 170 m/s, 190 m/s.

Figure 6. Chip morphology for various speeds: 250 m/s, 320 m/s, 360 m/s.

formed at the high-speed stage. This is because the inertial effect reduces the time that shear bands traverse the chip. When the cutting speed is 250 m/s, few shear bands can reach the chip's free surface. However, as the cutting speed arrives at 360 m/s, almost no shear bands can traverse the chip. The stronger the inertial effect is, the shorter the shear band is. A thin temperature distribution layer will form if the cutting speed is large enough. As for this condition, the chip morphology transfer from serration-like to continuous, and we can regard the chip as continuous and the cutting as stable.

High-speed machining has been widely used for its excellent advantages. The concept of high-speed machining was introduced by Salomon. In his view, the cutting temperature will decrease after it attains a peak as the cutting increases. As shown in Figure 6, the chip-tool interface temperature increases with the increasing speed. The results of the numerical simulation demonstrate that Salomon's assumption is not valid for the tool-chip interface temperature.

4 CONCLUSION

In this paper, the numerical simulation model developed by Zhang (2008) was employed to simulate the orthogonal cutting of the nickel-based super alloy GH4169. The numerical simulation model was validated using experimental results, and a good agreement was obtained. It is found that, at the low cutting speed, strain hardening prevails over thermal softening and inertial effect, and the cutting is stable. At the medium/high cutting speed, thermal softening dominates and shear bands have enough time to traverse the chip. Chip morphology transfers from continuous to serration-like. At the ultra-high cutting speed, thermal softening prevails over strain hardening and shear bands can also form. However, the inertial effect reduces the time needed for shear bands to traverse the chip. The stronger the inertial effect is, the shorter the shear band is. When the cutting speed reaches a certain value, a thin temperature distribution layer will form. Thus, we can regard the chip as continuous and the cutting as stable.

ACKNOWLEDGMENTS

This work was supported by the Shanxi Scholarship Council of China (No.2011–032); the Fund Program for the Scientific Activities of Selected Returned Overseas Professionals in Shanxi Province (No.2011–762); the Science and Technology Innovation foundation for postgraduate of TYUT (No. S2014080).

REFERENCES

Bai, Y. L. 1982. Thermo-plastic instability in simple shear. J. Mech. Phys. Solids. pp. 195–207.

Barry, J. & Byrne, G. & Lennon, D. 2001. Observations on chip formation and acoustic emission in machining Ti–6 Al–4V alloy. Int. J. Mach. Tools Manuf. Vol. 41, p. 1055.

Hortig, C. Svendsen, B. 2007. Simulation of chip formation during high-speed cutting. Journal of Materials Processing Technology. 186, 66–76.

Leech, P.W. 1985. Observations of adiabatic shear band formation in 7039 aluminum alloy. Metall. Trans. 16 A, 1900–1903.

Ma, W. Li, X.W. Dai, L.H. Ling, Z. 2012. Instability criterion of materials in combined stress states and its application to orthogonal cutting process. International Journal of plasticity. pp. 18–40.

Molinari, A. & Soldani, X. Miguelez, M.H. 2013. Adiabatic shear banding and scaling laws in chip formation with application to cutting of Ti-6 Al-4V. J. Mech. Phys. Solids.

Salomon, C.J. 1931. German Patent, 523594, 1931–04.

Zhang, D.J. & Wang, C. Liu, G. & Chen, M. 2008. A FEM and experimental study on high speed machining of nickel-based superalloy GH4169. Key Engineering Materials. Vols. 375–376 pp 82–86.

Energy Science and Applied Technology – Fang (Ed.)
© *2016 Taylor & Francis Group, London, ISBN 978-1-138-02833-3*

Transient dynamics analysis for structure of the airborne evaporation cycle system

Yinsai Guo, Yi Zhang & Mingke Cheng
Department of Graduates, Xijing University, Xi'an, China

ABSTRACT: Airborne evaporation cycle system often suffers from shock and vibration under time-varying loads, causing the degradation and failure of mechanical and electrical performance of parts. Transient dynamics analysis is performed by using ANSYS software. The maximum value of deformation and stress, time and region of the maximum value can be obtained by the method of transient dynamics analysis. It verifies that the design structure meets shock and vibration requirements, which provides theoretical basis for further optimization.

Keywords: Airborne evaporation cycle system; transient dynamics analysis; Impulse response; ANSYS

1 INTRODUCTION

Aviation equipment often suffers from shock and vibration under time-varying loads, causing the degradation and failure of mechanical and electrical performance of parts. Therefore, the impulse response of the structure is analyzed. The impulse response of the structure can be analyzed by method of transient dynamics (J.Pitarresi, 2002) (Pereira JT, 2004). Transient dynamics analysis is known as time-process analysis. It is the method for determining the dynamic response to bear any structural loads changing over time. When the structure is bearing under the static load, the transient load, the harmonic load and the action of random combination load, the strain, stress and deformation changing over time are obtained by the method of time-process analysis (Liu Junhua, 2012).

2 BASIC THEORY OF TRANSIENT DYNAMICS

Transient analysis is a technology determining structural response under the action of time—process load. Deformation, speed, acceleration, stress and strain varying with time can be output-ted (Bendsoe M, 1988). A load of time function needs to be inputted in analysis. Load curve on the relationship of the time is divided into the appropriate load step. Every "corner" of the load—time relation curve should be regarded as a load step. Load-time curve is shown in figure 1.

The transient dynamic equilibrium equation of linear structure:

$$[M]\{\ddot{x}\}+[C]\{\dot{x}\}+[K]\{x\}=\{F^a\} \quad (1)$$

$\{x\}$-Displacement vector, $\{\dot{x}\}$-Velocity vector, $\{\ddot{x}\}$-Acceleration vector, $\{F^a\}$-Load vector.

The formula (1) is solved by the method of new mark.

$$\{\dot{x}_{n+1}\}=\{\dot{x}_n\}+\left[(1-\delta)\{\ddot{x}_n\}+\delta\{\ddot{x}_{n+1}\}\right]\Delta t \quad (2)$$

$$\{x_{n+1}\}=\{x_n\}+\{\dot{x}_n\}\Delta t+\left[(\frac{1}{2}-\alpha)\{\ddot{x}_n\}+\alpha\{\ddot{x}_{n+1}\}\right]\Delta t^2 \quad (3)$$

In the formula (2) and (3), α, δ- Integral parameters

Calculate the next time displacement x_{n+1}, and at the time t_{n+1} the control equation:

$$[M]\{\ddot{x}_{n+1}\}+[C]\{\dot{x}_{n+1}\}+[K]\{x_{n+1}\}=\{F^a\} \quad (4)$$

The formula (2) and (3) reordering:

$$\{\ddot{x}_{n+1}\}=a_0(\{x_{n+1}\}-\{x_n\})-a_2\{\dot{x}_n\}-a_3\{\ddot{x}_n\}[K]$$

$$=\{F^a\} \quad (5)$$

$$\{\dot{x}_{n+1}\}=\{\dot{x}_n\}+a_6\{\ddot{x}_n\}+a_7\{\ddot{x}_{n+1}\} \quad (6)$$

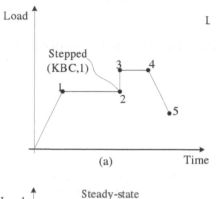

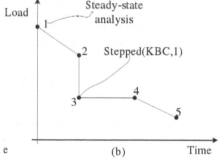

Figure 1. Load-time curve.

In the formula (5) and (6):

$$a_0 = \frac{1}{\alpha \Delta t^2}; a_1 = \frac{\delta}{\alpha \Delta t}; a_2 = \frac{1}{\alpha \Delta t}; a_3 = \frac{1}{2\alpha} - 1; a_4 = \frac{\delta}{\alpha} - 1;$$

$$a_5 = \frac{\Delta t}{2}\left(\frac{\delta}{\alpha} - 2\right); a_6 = \alpha \Delta(1 - \delta); a_7 = \Delta t \delta$$

On the basis of (4), (5) and (6):

$$(a_0[M]) + a_1[C] + [K]\{x_{n+1}\}$$
$$= \{F^a\} + [M](a_0\{x_n\} + a_2\{\dot{x}_n\} + a_3\{\ddot{x}_n\}$$
$$+ [C](a_1\{x_n\} + a_4\{\dot{x}_n\} + a_5\{\ddot{x}_n\}) \qquad (7)$$

On the basis of the formula (4), $\{x_{n+1}\}$ is obtained.

On the basis of the formula (5) and (6), velocity and acceleration are obtained.

On the basis of the formula (3), the initial on node speed or speed is obtained.

3 FINITE ELEMENT MODEL

Structure material selection is Al7075; density is 2810 kg/m³; the yield strength is 455 Mpa; safety factor is 1.5; allowable stress is 303 Mpa; the

(a)

(b)

Figure 2. (a). Boundary condition; (b) Mesh chart.

maximum deformation amount is not more than 1 mm; elastic modulus is 71700 Mpa; poisson's ratio is 0.33. According to the selected components, the weight of the condenser is 5.6 kg; the weight of the compressor is 8 kg; the weight of evaporator is 3.6 kg. Bearing under distributed load of condenser supporting part is 2×10^3 Mpa; bearing under distributed load of compressor supporting part is 1.4×10^3 Mpa; bearing under distributed load of evaporator supporting part is 1.2 x10³Mpa; maximum load of air flow is 1.658×10^3 Mpa. Boundary condition is shown in figure 1.

Mesh is automatically divided which is shown in figure 2. Node number is 13790; element number is 3984.

4 TRANSIENT DYNAMICS ANALYSIS FOR STRUCTURE

According to the National Military Standard GJBl50.18–86 "environmental test methods for military equipment" requirement, the saw tooth wave pulse after the peak is served as the incentive (Wang Shangli, 2012). The saw tooth wave pulse after the peak is a kind of acceleration pulse, A = 20 g, D = 11 ms. Pulse function is inputted in the ANSYS, which is shown in figure 2. Each corner of the curve should be regarded as a load step, so it will be divided into four load steps. The time of the first step load is from 0 ms – 5.5 ms. The count of initial step is 10. The time of the second step load is from 5.5 ms – 16.5 ms. The count of initial step is 15. The time of the third step load is from 16.5 ms – 16.6 ms. The count of initial step is 15. The time of the fourth step load is from 16.6 ms – 33 ms. The count of initial step is 10. Bending deformation is most serious, along the z direction in modal analysis, so the pulse is applied in z direction, which is shown in figure 3.

The result of response can be obtained after transient dynamics analysis. The equivalent stress extreme value—time curve is shown in figure 4. The displacement extreme value—time is shown in figure 5. Green curve shows Maximum; red curve shows Minimum. Maximum of equivalent stress is 5.516 MPa occurring at 16.56 ms. Maximum of displacement is 5.516 MPa occurring at16.56 ms.

The equivalent stress contour and deformation contour are shown in figure 6 and 7. Bending deformation occurs on the support structure of the compressor and the condenser, direction along the Z direction. Under the effect of repeated impact, it is easy to produce fatigue damage.

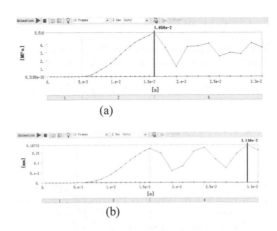

(a)

(b)

Figure 5. (a) The equivalent stress extreme value—time curve; (b) The displacement extreme value—time curve.

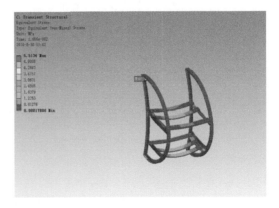

Figure 6. Stress.

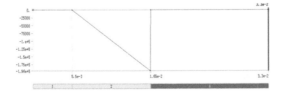

Figure 3. Sawtooth wave after the peak pulse.

	Steps	Time [s]	☑ Acceleration [mm/s²]
1	1	0.	= 0.
2	1	5.5e-003	0.
3	2	1.65e-002	-1.96e+005
4	3	1.66e-002	0.
5	4	3.3e-002	= 0.
*			

Figure 4. The acceleration pulse in Z direction.

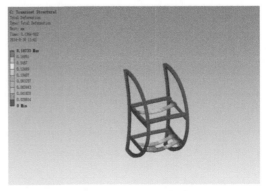

Figure 7. Deformation.

5 CONCLUSIONS

Transient dynamics analysis is performed by using ANSYS software, which proves that the design scheme meets the design requirements. It provides theoretical basis for further optimization.

ACKNOWLEDGEMENTS

This work was supported by the Natural Science Foundation of Shaanxi Province (Program No. 2013 JM8040), the Scientific Research Program Funded by Shaanxi Provincial Education Department (Program No. 2013 JK1204), and Graduate Student Innovation Fund of Xijing University.

REFERENCES

Bendsoe M, Kikuchi N. Generation optimal topologies in structural design using a homogenization method. Computer Methods in Applied Mechanics and Engineering, 1988, vol 71:232–238.

Liu Junhua. Structure optimization and stamping forming simulation of air-conditioned panel. 2012."In Chinese"

Pitarresi J, Phil Geng, Willem Beltman, Yun. Dynamic modeling and measurement of personal computer motherboards. 2002 Electronic Components and Technology Conference, 597–603.

Pereira JT, Fancello EA, Barcelles CS. Topology optimization of continuum structures with material failure constrains. Structural and Multidisciplinary Optimization, 2004, 28:87–98.

Wang Shangli. Research on technologies about the dynamic performance simulation of an air borne Chassis. 2012.

Energy Science and Applied Technology – Fang (Ed.)
© 2016 Taylor & Francis Group, London, ISBN 978-1-138-02833-3

Transverse stress analyses of columns constrained by rebar under local compression

Y. Wang
College of Engineering and Technology, Jilin Agricultural University, Changchun, Jilin, China

ABSTRACT: Simulations of local compression on the centers of plain concrete square columns with rebar meshes and plain concrete square columns were conducted by means of the finite element analysis software ANSYS. The results show the distributions of transverse stress on the cross section xoz is basically consistent between the plain concrete column and the plain concrete column equipped with rebar meshes. At the same time, the local compression area ratio affects the distribution of transverse stress; additionally, the increased compression area ratio significantly reduces the bearing capacity of the specimen with local compression.

Keywords: Local compression; plain concrete columns; rebar meshes; transverse stress; finite element analysis

1 INTRODUCTION

In engineering applications, spiral stirrups or welded wire meshes are usually installed at the ends of concrete structures with local compression in order to increase the concrete's local compression strength as well as to inhibit cracks resulting from stress concentration. However, there are still accidents that occur due to the occurrence of longitudinal cracks in the local compression area or the loss of bearing capacity.

Therefore, this paper by using nonlinear finite element analysis on 3 models with local compression on the center of the plain concrete square columns with rebar meshes to study the failure mechanism of local compression of the concrete.

2 METHODOLOGY

2.1 Material simulation

Using 8-node 3D solid element SOLID65 to simulate concrete element and using 8-node 3D solid element SOLID45 to simulate loading plates (Barbosa, A. F. & Ribeiro, G. O.1998). As the simulated components are plain concrete without strengthening materials with no need to define the constant of strengthening materials. Choose isotropic as the material property of the square concrete columns, elastoplastic constitutive relations is defined as follows: Isotropic Hardening Plasticity, Von Mises and Multilinear Isotropic Hardening is chosen as the plastic property. Isotropic is chosen

as the elastic property of the loading plates(Cook, R. D. 2002; Douglas, M.R. & Lunniss, R.C.1970; Hrabok, M.M.1981; Isenberg, J.1991; Logan, D. L.2002).The elastic modulus of square concrete column take 21641 MPa and poisson ratio take 0.2. Both the steel loading plates and rebar meshes have an elastic modulus of 210000Mpa and a Poisson's ratio of 0.3. Area of reinforcement is 28.26.

2.2 Establishment and solution of finite element model

The simulation specimens are concrete square columns of the size $520 \times 300 \times 300$. The loading plates' sizes are, respectively, $165 \times 165 \times 20$, $80 \times 80 \times 20$, and $40 \times 40 \times 20$.

In the simulations, because all of the columns were symmetric structures locally compressed at their centers by symmetric loads, only ¼ of each model was simulated and analyzed. Assume that the axis of symmetry is the z-axis, the intersection of the z-axis and the top surface of the column is the origin o, and the top surface of the column is the *xoy* plane where x is horizontal and y is vertical.

A rebar mesh consists of ten Φ6 rebars, in which five are laid in the ox direction, and the other five are in the *oy* direction. Five rebar meshes were installed in each column. The top rebar mesh is 30 mm from the top surface of the concrete column, and the distance between neighboring meshes is 50 mm, as shown in Fig. 1- in which Fig. 1a, b, and c are respectively the finite element models for the plain concrete columns embedded with rebar

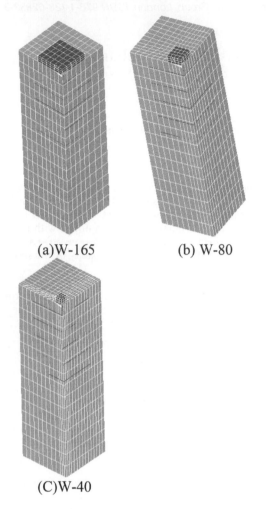

(a)W-165   (b) W-80

(C)W-40

Figure 1. Finite element models.

meshes with local compression areas 165×165 (numbered as W-165), 80×80 (numbered as W-80), and 40×40 (numbered as W-400).

In each finite element model, constraints of symmetry are applied to both the concrete cutting surface and the loading plate cutting surface, and translational degrees of freedom in all three directions at the nodes on the bottom surface are fixed. In the process of nonlinearly solving each model, the applied load is 500 N/mm2, and the step is divided into 100 increments.

3 RESULT

3.1 Transverse stress distribution

All of the finite element models are symmetric in both geometry and load. The analysis is conducted

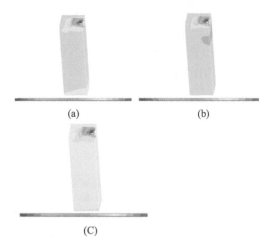

(a)   (b)

(C)

Figure 2. W-165 Transverse stress distribution.

by focusing on the transverse stress component, σ_x, on the longitudinal xoz plane.

Figures 2 through 4 demonstrate the results of the finite element simulations of the plain concrete columns with rebar meshes W-165, W-80, and W-40 under local compression where Fig. a, b, and c illustrate the distributions of transverse stress in the initial, intermediate, and final phases of loading.

3.2 Distributions of transverse stress σx (σy) on the xoz (yoz) plane in plain concrete columns with rebar meshes

Fig. 2 shows the distribution of transverse stress on the longitudinal cross section xoz of the specimen W-165 during the initial (Fig. a), intermediate (Fig. b), and final (Fig. c) phases of loading. Due to the diffusion of the local compressive load, the region right below the local compression undertakes transverse compression. As the depth is increased from the local compression to a certain extent, transverse tension occurs. After it reaches far from local compression, the stress states transitions back to transverse compression. The region of stress concentration around the local compression area increases in size with increased load. Meanwhile, it can be clearly observed from Fig. a, b, and c that a distinct oblique belt of compressive stress exists near the local compression area, below which concrete is in tension. The rebar meshes apply constraints to concrete columns and enhance the anti-cracking strength of specimens under local compression. The local compression bearing capacity of W-165 is 1210kN.

Fig. 3 shows the distribution of transverse stress on the longitudinal cross section xoz of the specimen W-80 under the same local compressive load

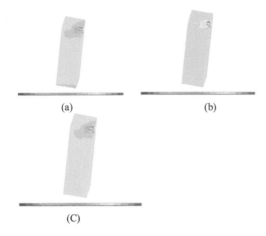

(a)   (b)

(C)

Figure 3.   W-80 Transverse stress distribution.

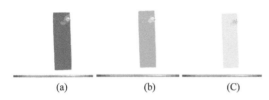

(a)   (b)   (C)

Figure 4.   W-40 Transverse stress distribution.

as that on W-165 during the initial (Fig. a), intermediate (Fig. b), and final (Fig. c) phases of loading. As shown in the figures, the distribution of transverse stress in W-80 is approximately the same as that in W-165. However, the peak transverse tensile stress in W-80 is less than that in W-165, and the location where the peak transverse tensile stress occurs in W-80 is lower than that in W-165. Furthermore, compared to W-165, the transverse stress distribution shape in W-80 changes greatly, and the local compression bearing capacity is significantly reduced. Meanwhile, it can still be clearly observed from Fig. a, b, and c that a distinct oblique belt of compressive stress exists near the local compression area, below which concrete is in tension. The local compression bearing capacity of W-80 is 556kN.

Fig. 4 shows the distribution of transverse stress on the longitudinal cross section xoz of the specimen W-40 under the same local compressive load as that on both W-165 and W-80 during the initial (Fig. a), intermediate (Fig. b), and final (Fig. c) phases of loading. As shown in the figures, the distribution of transverse stress in W-40 is approximately the same as those in W-165 and W-80. However, the peak transverse tensile stress in W-40 is slightly

less than that in W-80, and the location where the peak transverse tensile stress occurs in W-40 is lower than that in W-80. Furthermore, compared to W-80, the transverse stress distribution shape in W-40 changes greatly, and the local compression bearing capacity significantly decreases. Meanwhile, it can still be clearly observed from Fig. a, b, and c that a distinct oblique belt of compressive stress exists near the local compression area, below which concrete is in tension. The local compression bearing capacity of W-40 is 275kN.

Figures 2 through 4 indicate that a loading state of a tension/compression bar still exists in concrete. In addition, the local compression area ratio significantly influences the transverse stress distribution. The more the local compression area ratio increases, the more the peak transverse tensile stress decreases, and the location of the peak transverse tensile stress lowers. The increased local compression area ratio reduces the column's local compression bearing capacity.

4 CONCLUSIONS

The distributions of transverse stress on the cross section xoz are basically consistent between the plain concrete column and the plain concrete column equipped with rebar meshes. At the same time, the local compression area ratio affects the distribution of transverse stress; additionally, the increased compression area ratio significantly reduces the bearing capacity of the specimen with local compression.

REFERENCES

Barbosa, A. F., and Ribeiro, G. O., Analysis of Reinforced Concrete Structures using ANSYS Nonlinear Concrete Model, Computational Mechanics, CIMNE, Barcelona, Spain, 1998, pp. 1–7.

Cook, R. D. e. a., CONCEPTS AND APPLICATIONS OF FINITE ELEMENT ANALYSIS. New York: John Wiley & Sons, Inc., fourth ed., 2002.

Douglas, M. R. and Lunniss, R. C., "An Application of the Finite Element Technique in Bridge Design," Concrete, vol. 4, pp. 197–200, May 1970.

Hrabok, M. M., Stiffened Plate Analysis by the Hybrid Stress Finite Element Method. PhD thesis, University of Alberta, Edmonton, Alberta, Canada, 1981.

Isenberg, J., ed., FINITE ELEMENT ANALYSIS OF REINFORCED CONCRETE STRUCTURES II, (New York, New York), Committee on Finite Element Analysis of Reinforced Concrete Structures, American Society of Civil Engineers, June 1991.

Logan, D. L., A FIRST COURSE IN THE FINITE ELEMENT METHOD. Pacific Grove, CA: Brooks/Cole, third ed., 2002.

Energy Science and Applied Technology – Fang (Ed.)
© 2016 Taylor & Francis Group, London, ISBN 978-1-138-02833-3

Numerical simulation of grinding ball of kainite ductile iron in oil quenching and isothermal tempering

W.R. Wang, B. Qu & D.Q. Wei

College of Mechanical and Electrical Engineering, Guilin University of Electronic Technology, Guilin, Guangxi, China

ABSTRACT: Based on the actual condition of a ductile iron grinding ball in oil quenching and iso-thermal tempering, a three-dimensional model is established by COMSOL and a microstructure transformation program written in MATLAB. The results show an intuition expression on the change in the temperature and the microstructure of the grinding ball during the heat treatment process. The change in the temperature field and the microstructure field during the heat treatment process is observed. This paper provides a theoretical foundation for the selection and optimization of processing parameters during the heat treatment process for the grinding ball of ductile iron. It is recommended to acquire a high-performance grinding ball of ductile iron.

Keywords: oil quenching and isothermal tempering; temperature field; microstructure field; ductile iron; grinding ball

1 INTRODUCTION

Ball mill is the key crushing equipment widely used in building materials, mines, metallurgy, electricity power and other industries. Grinding ball is the most commonly used grinding media in the ball mill and the largest consumption of worn parts in the grinding industry. Based on the working condition of the grinding ball, it has a good impact resistance, abrasion resistance and impact fatigue spelling resistance, which requires a reasonable match of strength, hardness and hardness of grinding ball materials (Wei. 2005).

At present, most of the domestic grinding ball materials are made of high-chromium cast iron or medium-manganese ductile iron (Fu. et al. 2004). However, medium-manganese ductile iron has a low impact toughness and high-chromium cast iron has a higher production cost. Bainite ductile iron has good mechanical properties. Acicular ferrite is the microstructure in the matrix of high carbon austenite and nodular microstructure. The hardness of bainite ductile iron depends mainly on the solid solution content of carbon in bainite. Under the austenitic temperature, graphite carbon of ductile iron diffused into the austenite causes the formation of high carbon austenite. The higher content of silicon in ductile iron can hinder the precipitation of carbides; therefore, the microstructure of the materials has higher hardness. At the same time, retained austenite can buffer the stress concentration, inhibit crack propagation

and improve wear resistance and service life. So, bainite ductile iron has been regarded as one of the major achievements in the metallurgical field since the 1970s. It is a highly competitive material in this twenty-first century. The grinding ball made of bainite ductile iron has a wide market development prospect.

The heat treatment production techniques of the bainite ductile iron grinding ball include isothermal quenching, continuous cooling quenching and alloying as-cast (Chen. et al. 1989). In recent years, many scholars have conducted many studies on the heat treatment production techniques of the bainite ductile iron grinding ball, but only the experimental research has been addressed. It is difficult for the experimental research to fully reflect each segment of the heat treatment process. With the development of the calculator technique, the numerical simulation technology has overcome this shortcoming. The technology can show the change in the information of all kinds of fields during the heat treatment process. At present, the numerical simulation technology in the heat treatment process is mostly used in steel, but the simulation of heat treatment process for ductile iron is used to a less extent.

Based on the actual condition during the heat treatment process, a three-dimensional finite element model of the ductile iron grinding ball is established by COMSOL. The change in the temperature field during the heat treatment process for ductile iron is simulated. The models comprehensively

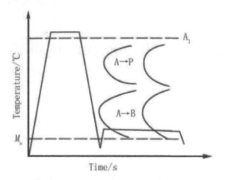

Figure 1. Process of heat treatment.

consider related physical properties dependent on the temperature. Based on the temperature change curve, the change in microstructure transformation is simulated by MATLAB.

2 MATERIALS AND THE HEAT TREATMENT PROCESS

The alloy composition of ductile iron used in the experiment (in wt.%) is as follows: 3.5–3.7C; 2.7–2.9Si; ≤0.5Mn; 0.3–0.5Mo; 0.5–0.8Cu; 0.2–0.25Cr; ≤0.03S; ≤0.07P; 0.03–0.05 Mg; 0.02–0.04Re.

Based on the previous results of the related experiment, the heat treatment process of the grinding ball is described, as shown in Figure 1. The heat treatment process of the grinding ball involved two stages: oil quenching and isothermal tempering. In the oil quenching stage, the grinding ball is placed in an isothermal furnace at 920°C for austenitizing, and then placed in 20# machine oil for rapid cooling so as to avoid pearlite transformation. In the isothermal tempering stage, when the surface temperature of the grinding ball is reduced to 185°C, below the Ms point, the grinding ball is placed in an isothermal furnace at 300°C for kainite transformation. After two hours, the grinding ball is removed from the furnace and cooled down to room temperature (Wei. et al. 2006).

3 ESTABLISHMENT OF THE MATHEMATIC MODEL

3.1 Establishment of the temperature field model

The following assumptions are adopted during the development of the 3-D FEM model:

1. Material properties are homogeneous and isotropic;
2. Transformation latent heat is negligible;

3. Thermal physical parameters are a function of the temperature.

Because of the symmetry of the grinding ball, only 1/8 of the grinding ball is selected to build the model.

The 3-D heat conduction differential equation in the rectangular coordinate system is given in Eq. 1: where ρ = density . ($kg \cdot m^{-3}$); Cp = specific heat ($J \cdot kg^{-1} \cdot °C^{-1}$); T = temperature (°C); t = time(s); x, y, z = coordinate (m); and k = thermal conductivity ($W/(m \cdot °C)$).

The heat treatment process has two stages.

Different stages have different initial conditions and boundary conditions.

The initial condition of the oil quenching stage is as follows: $T = T0 = 920°C$;

The boundary condition is convective heat transfer, which is given in Eq. 2:

$$-\lambda \frac{\partial T}{\partial \mathrm{n}} \Big|_w = h(T_s - T_q) \qquad (2)$$

where $\partial T/\partial \mathrm{n}$ = thermal gradient; w = convective heat transfer boundary; h = surface heat transfer coefficient ($W \cdot m^{-2} \cdot °C^{-1}$); Ts = surface temperature (°C); and T_q = quenching medium temperature (°C).

The initial condition of the isothermal tempering stage is the distribution of the temperature field of the last time in the first cooling stage.

The boundary condition is a constant temperature state, $T_q = T1 = 300°C$.

In this paper, the thermal physical parameters, which are shown in Figure 2 (a), (b), (c), are obtained by using the material simulation software JmatPro. In this paper, the surface heat transfer coefficient is obtained from the literature (Song et al 006).

3.2 Establishment of the microstructure field model

Based on the temperature of the cooling curve and the TTT curve, the amount of the microstructure transformation of the grinding ball at any given time can be simulated. In the process of heat treatment, quenching is a continuous transition process. However, isothermal transformation and continuous transformation can be linked by applying the Scheil superposition principle.

For the diffusion microstructure transformation (e.g. pearlite transformation and bainite transformation), the Avrami equation can be used to calculate the amount of microstructure transformation (M. et al. 1939), which is given in Eq. 3:

$$f = 1 - \exp(-b(T)t^{n(T)}) \qquad (3)$$

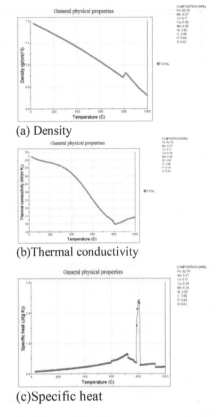

(a) Density

(b)Thermal conductivity

(c)Specific heat

Figure 2.   Thermal physical parameters of ductile iron.

where f = transformation volume; t = isothermal time (s); and b (T), n (T) = new nucleation factor.

Martensitic transformation is a non-diffusion phase transformation. For the non-proliferation phase transition, the Koistinen–Marburger formula can be used to calculate the amount of martensite (R. Sh. et al. 2011), which is given in Eq. 4:

$$f_m(T) = 1 - \exp[-k(Ms - T)^m] \ (T < = M_S) \qquad (4)$$

where $f_m \ (T)$ = the volume fraction of martensite; Ms = martensitic transformation start temperature, Ms = 194.7°C; and k is a constant, $k = 0.011$.

4   NUMERICAL SIMULATION RESULTS AND DISCUSSION

4.1   *Temperature fields*

The cooling curves of six points, which are distributed from the center to the surface along the radius, are shown in Figure 3. The first 71 s is the oil quenching stage and the remaining 129 s is part of the isothermal tempering stage. In the oil quenching

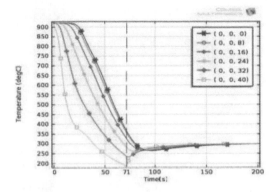

Figure 3.   Cooling curves of six points along the radius in the first 200 seconds of heat treatment.

stage, as shown in Figure 2, the temperature of the center is less than that of the surface, and the cooling curve of the center is more intensive than that of the surface. It means that the temperature gradient of the center is smaller than that of the surface, and the cooling rate of the center is less than that of the surface. The reasons for these differences observed are as follows. The reduction in the temperature near the surface is mainly dependent on the convective heat transfer between the quenching medium and the grinding ball, so the cooling speed is fast. However, the cooling way inside the grinding ball is related to the heat conduction of the materials. So, the cooling speed is slow.

At 71 s, the grinding ball is placed into the isothermal box and enters the second stage. From Figure 3, we can see that the surface temperature of the grinding ball increases gradually to 300°C, and the center temperature of the grinding ball decreases at first and then increases. This is due to the temperature gradient. There is a big gap between the surface temperature of the grinding ball and the external environment temperature, so the surface temperature of the grinding ball increases directly, but the center temperature of the grinding ball is greater than the temperature near the center. Under the action of heat conduction, the center temperature of the grinding ball decreases, while the temperature of the nearby area increases. At the same time, there is a temperature difference between the grinding ball and the external environment. This makes the temperature of the whole grinding ball to increase, until reaching to 300°C, and then remain in the isothermal state.

4.2   *Microstructure fields*

The distribution of the temperature field inside the grinding ball in the rapid cooling stage at 71 s is

163

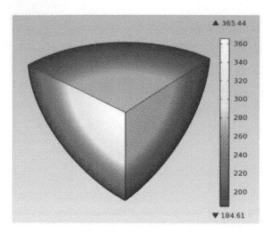

Figure 4. Distribution of the temperature field inside the grinding ball at 71 seconds.

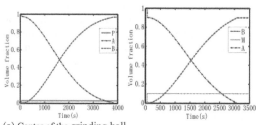

(a) Center of the grinding ball

(b) Surface of the grinding ball

Figure 5. Details of the microstructure transition of the grinding ball during the heat treatment process.

shown in Figure 4. From Figure 3, we can see that the surface temperature is 184.61°C, but the center temperature is still as high as 365.44°C.

Figure 5(a) shows the details of the microstructure transition of the grinding ball center during the heat treatment process. From this figure, we can see that the pearlite transformation occurs first. The time of transformation is about 14.75 s, which is relatively short. The amount of transformation is 0.0264. The kainite transformation occurs in the isothermal tempering stage. With the transformation of the two microstructures, the amount of austenite is decreased. At the end of the transformation, the amount of kainite is 0.9532.

Figure 5(b) shows the details of the microstructure transition of the grinding ball surface during the heat treatment process. It is different from the microstructure transition of the grinding ball center. The grinding ball surface cools too fast to avoid the pearlite transformation. When the

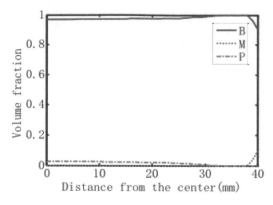

Figure 6. Microstructure distribution of the grinding ball on the radial direction after the heat treatment.

surface temperature of the grinding ball is quickly cooled to below 194.7°C, the martensitic transformation occurs. The time of the microstructure transformation is about 3.8 s, which is extremely short. The amount of the microstructure transformation is 0.0981. The generated martensite content remains unchanged and transforms into tempered martensite after the isothermal tempering stage. The kainite transformation occurs mainly in the isothermal tempering stage, and the transformation ends in the same stage. The amount of kainite is 0.9019.

Figure 6 shows the microstructure distribution of the grinding ball on the radial direction after the heat treatment process. There are three microstructures in the grinding ball after the heat treatment: pearlite, kainite and martensite. Most of them are kainite, about 98%. From the center of the grinding ball to the surface, pearlite content decreases gradually, changes little near the grinding ball center and reduces significantly near the surface. This is due to the characteristics of the different cooling speeds of the surface and the center. Martensite occurs only in the surface region, and its content decreases gradually from the outside to the inside. This is because only the near-surface temperature cools to below the Ms point, and the lower the surface temperature is, the greater the amount of martensite will be.

5   CONCLUSION

1. In the oil quenching stage, due to the common action of heat conduction and heat convection, the cooling rate declines gradually from the surface to the center of the grinding ball. In this stage, very tiny amounts of austenite transform into pearlite, kainite and martensite.

2. At the end of the oil quenching stage, the surface temperature of the grinding ball is 184.61°C, which is 180.83°C lower than that of the center.
3. In the isothermal tempering stage, the surface temperature rises directly to 300°C, while the core temperature decreases at first and then increases, until reaching to 300°C. Only the kainite transformation occurs in this stage, whose concentration is about 98%.

ACKNOWLEDGMENTS

This work was financially supported by the Guang Xi Natural Science Funded Project (2012GXN SFDA053026).

REFERENCES

Avrami, M, 1939, Kinetics of phase change, General Theory, J. Chem. phys, Vol.7, p I103.

Chen. D.L, Ye H.M, liu. G.J, 1998 et al. The research and application of bainite ductile cast irons [M]. Foundry Technology, 6:32–35.

Fu. TS, Chen. S. Y, Liang. J. 2004, ea al. Heat treatment for whet-ball made of bainite ductile iron [J]. Heat treatment of metals, 29(6):59–62.

Sh. Razavi, R., Gordani, G. R., Tabatabaee, S., 2011. Mathematical Modeling of Heat Transfer in Laser Surface Hardening of AISI 1050 Steel [J]. Defect and Diffusion Forum, 312–315: 381–386.

Song G.W, Meng. Q.H, Wang. H.Y, 2008. Heat Transfer coefficient measurement and calculation on the composite surface of parts quenching [M]. Journal of Shenyang Institute of Aeronautical Engineering, 25(1):29–32.

Wei. D.Q, 2005. Development and production of bainite ductile iron grinding ball materials [J]. Journal of Xiangtan Normal University, 27(05):94–97.

Wei. D.Q, Xue X, Yun. Z.D. 2006. Determination of heat treatment pores parameters of the milling ball in low alloyed bainite ductile iron [J], 30(7):59–62.

Energy Science and Applied Technology – Fang (Ed.)
© 2016 Taylor & Francis Group, London, ISBN 978-1-138-02833-3

Shape design and motion analysis of underwater glider

Qian Zhao & Tianhong Yan
College of Mechanical and Electrical Engineering, China Jiliang University, Hangzhou, China

Bo He
Department of Electronic Engineering, Ocean University of China, Qingdao, China

ABSTRACT: Resistance plays a critical role in affecting the performance of underwater glider. This paper proposes a kind of low-resistance underwater glider. First, an optimal shape is obtained by comparison. Then, the AUG's motion equations for steady glide are set up, and the simulation results of hydrodynamic coefficients based on the FLUENT are obtained. Finally, the relationship between motion parameters and controllable variables is simulated using MATLAB software, which provides guidance for the next design.

Keywords: Underwater Glider; Shape Optimal Design; Motion Analysis

1 INTRODUCTION

AUG (Autonomous Underwater Glider) is a type of long-range unmanned vehicle that uses buoyancy control to travel in a saw-tooth trajectory through the water column. With several advantages of low cost, long duration and energy saving, it is widely used in both navy investigation and civilian field.

The idea of developing a glider was first inspired by Stommel in 1989 (Stommel, 1989). Until 2001, three autonomous underwater gliders have been developed. Doug Webb and his colleague via Webb Research Corporation developed Slocum (Webb, 2001), a University of Washington group developed Seaglider (Eriksen, 2001) and Scripps Institution of Oceanography developed Spray (Sherman, 2001). There are two types of Slocum glider. One is electrically powered and the other is thermally powered.

Underwater gliders have no propeller, relying on the changes in net buoyancy to produce power, adjusting the displacement of movable mass to change the position of the gravity center, so as to adjust the pitching and rolling attitude. By using the rudder, it can maintain steady gliding or change its direction flexibly.

CFD (Computational Fluid Dynamics) is a convenient method in researching the hydrodynamic performance of AUG, because simulation experiment condition is easy to set and of low cost, and the experimental period is short.

This paper is organized as follows. In Section 2, we calculate and compare the water resistance and lift-to-drag ratio of 4 different shapes of glider by the CFD method, and choose an optimal shape. Section 3 describes the motion analysis, building the motion equations of steady state, dealing with the fluid simulation experimental data in MATLAB. Finally, Section 4 analyzes the relationship between the motion parameters and the controllable parameters.

2 SHAPE OPTIMAL DESIGN

The underwater glider's streamline shape hull includes the main body and its appendage-wings and rudders. When sailing at a low speed, the external flow field of the glider is laminar flow field (Zeng, 2010), a good laminar shape can achieve both high lift and low drag, which is conducive to reducing energy consumption for underwater gliders.

2.1 *Main body shape*

Keeping the whole length as 2 m and the drainage volume as roughly 50 L, there are four common body shapes, which are as follows (Meng, 2014):

Body 1: The head and the tail are a semicircle sphere, parallel cylinder in the middle. This shape has the most inner space.

Body 2: The head is a semicircle sphere, parallel cylinder in the middle, and the tail is a semi ellipsoid with its eccentricity being 0.97. This shape has a reduction in its space utilization.

Body 3: The head and the tail are both semi-ellipsoid, the eccentricity for the head curve is 0.87, the tail 0.97, and the middle part is a parallel cylinder.

(a) Body1   (b) Body2

(c) Body3   (d) Body4

Figure 1.   Different body types.

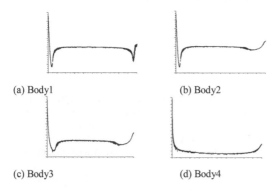

(a) Body1   (b) Body2

(c) Body3   (d) Body4

Figure 2.   Longitudinal section pressure distribution curve of the four body shapes.

Table 1.   Main parameters of the four body types.

Body type	Body1	Body 2	Body 3	Body 4
Length L (m)	2.0	2.0	2.0	2.0
Diameter D (m)	0.2	0.2	0.2	0.24
Length-diameter ratio	10	10	10	8.33
Drag (N)	0.3689	0.2988	0.2799	0.2683
Windward Area (m²)	0.0314	0.0314	0.0314	0.0452
Drag Coefficient Cd	0.3766	0.3051	0.2857	0.1903

This shape promotes the hydrodynamic performance, but has a little difficulty in processing.

Body4: The main body is a revolving body with an elliptical shape, which is 2 m long and the maximum diameter is 0.24 m. Figure 1 shows the 4 different body types.

Using CFD method and FLUENT, this paper analyzes the hydrodynamic characteristics of the four body types, at an attack angle $\alpha = 0°$ and $V = 0.25$ m/s. Figure 2 shows the longitudinal section pressure distribution curve of the four bodies.

Figure 2 shows the comparison of body 1, 2 and 3: the pressure distribution curve of body 4 is relatively flat, and the pressure gradient is smaller, which means body 4 has a better flow field characteristic.

Table 1 shows the comparison of the first three kinds of body shape: the total resistance of body 4 is reduced by 27.27%, 10.21% and 4.14%, and the drag coefficient is reduced by 49.47%, 37.63% and 33.39%.

From the above analysis, we choose body 4 as the main body shape.

Table 2.   Wing parameters.

Name	Wing Span B(m)	Sweep Back angle (°)	Tip Chord length Ct(m)	Chord Length C(m)	Cr/Ct ratio	B/C ratio	Air foil Type
Wings	1.2	25	0.1	0.16	2.2	7.5	naca0005
Tails	0.2	15	0.1	0.13	1.6	2	naca0012

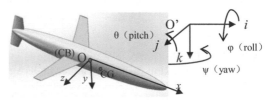

Figure 3.   Overall prototype of the AUG.

2.2   *Wings and tails*

Meng, 2014 verified that the curved surface airfoil has a higher lift-to-drag ratio than the flat profile, with a constant inlet flow and an equal attack angle. We choose the NACA profile as the wings' and tails' profile because of better properties. According to the orthogonal experimental result in that study, Table 2 summarizes the specific parameters of the wings and tails.

Based on the aircraft's design and manufacturing experience, in order to reduce resistance, the biggest cross-section of wings should locate in the 60% length of the main body (Hu, 2005). So, the whole shape of the underwater glider is shown in Figure 3. The model has a total length of 2 m, and the maximum diameter is 0.24 m. The underwater glider's buoyant center is at 1.009 m, total gravity weighs about 60 kg, and total buoyancy is about 62 kg. The center of the gravity (CG) locates 5 mm below the center of the buoyancy (CB). As shown in Figure 3.

3   DYNAMIC MODELLING AND MOTION ANALYSIS

3.1   *Equations of motion in the vertical plane when steady gliding*

In this paper, the external ballast changes the drainage volume of the AUG, making the power to ascend or dive. The moving mass changes the displacement of the CG, generates a pitch moment, and changes the glider's pitch attitude. The hydrodynamic forces (lift, drag and viscous moment) worked on the fixed horizontal wings or body. When they work together, the glider will have a saw-tooth trajectory in the vertical plane. When the external forces and moment worked, the underwater glider is balanced, so that the underwater glider can glide steadily at a constant velocity (K. Fujii, 2005). We specialize the model to the vertical plane, the i-k plane in inertial coordinates and the x-z plane in body coordinates, as shown in Figure 4. We denote α as the angle of attack, θ as the pitch angle, D as the drag, L as the lift, M_{DL} as the viscous moment, G as the gravity, B as the buoyancy, $G = mg$, and the net buoyancy as $m_0 g$, $m_0 g = |G\text{-}B|$.

During a steady gliding, the external forces and moment equilibrium equations are given by

$$\begin{cases} \sum Fx_{ext} = m_0 g \sin\theta + L\sin\alpha - D\cos\alpha = 0 \\ \sum Fz_{ext} = m_0 g \cos\theta - L\cos\alpha - D\sin\alpha = 0 \\ \sum My = M_{DL} - mg\sin\theta z_G + mg\cos\theta x_G = 0 \end{cases} \tag{1}$$

Here, x_G and z_G denote the x axial and z axial displacement of CG, respectively. x_G is a variable with the movement of the moving mass, and $z_G = 5$ mm.

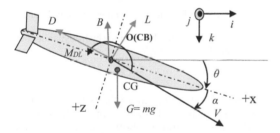

Figure 4. External force on the glider confined to the vertical plane when diving.

3.2 Hydrodynamic force and moment calculation

The hydrodynamic forces and moment coefficients are generally dependent on the angle of attack α and the velocity magnitude V. They are the functions of the attack angle α (Zhang, 2012), so we get the following formula:

$$\begin{cases} D = 1/2\rho V^2 S C_D = (K_{D0} + K_{D\alpha}\alpha^2)V^2 \\ L = 1/2\rho V^2 S C_L = (K_{L0} + K_{L\alpha}\alpha)V^2 \\ M_{DL} = 1/2\rho V^2 S L C_M = (K_{M0} + K_{M\alpha}\alpha)V^2 \end{cases} \tag{2}$$

Here, ρ is the density of water and S is the characteristic area of the glider. C_D, C_L, and C_M are the drag, lift, and pitch moment coefficients, respectively. K_{D0}, K_{L0} and K_{M0} represent the characters of the zero attack angle, and $K_{D\alpha}$, $K_{L\alpha}$ and $K_{M\alpha}$ represent the characters of the non-zero attack angle. Based on Eqs (1) and (2), we can obtain the following relations:

$$\begin{cases} m[(D\cos\alpha - L\sin\alpha)z_G - (D\sin\alpha - L\cos\alpha)x_G]/m_0 = M_{DL} \\ \tan\theta = \tan((D\cos\alpha - L\sin\alpha)/(D\sin\alpha + L\cos\alpha)) \\ V = \sqrt{mg(x_G\cos\theta - z_G\sin\theta)/(K_{M0} + K_{M\alpha}\alpha)} \end{cases} \tag{3}$$

To obtain the hydrodynamic coefficients, the vicinity flow field of AUG is simulated in FLUENT software, with the incoming flow velocity of 0.25 m/s and 0.5 m/s, respectively, and the angle of attack of 0°, 2°, 4°, 6°, 8°, 10° and 12°, respectively. Table 3 summarizes the convergent hydrodynamic coefficients obtained from CFD simulation. Figure 5 and Figure 6 show, respectively, the typical

Table 3. Hydrodynamic coefficients of the glider.

Attack Angle $\alpha(°)$	Gliding Velocity V(m/s)	Lift L(N)	Drag D(N)	Pitch moment M(Nm)	Lift Coefficient: Cl	Drag Coefficient: Cd	Lift to Drag ratio:
0	0.25	0.0127	0.2853	−0.0025	0.0090	0.2022	0.0446
0	0.5	0.0500	1.0571	−0.0092	0.0089	0.1873	0.0473
2	0.25	0.2859	0.2991	0.0308	0.2026	0.2119	0.9560
2	0.5	1.1608	1.1093	0.1229	0.2056	0.1965	1.0464
4	0.25	0.5955	0.3271	0.0631	0.4220	0.2318	1.8207
4	0.5	2.3992	1.2210	0.2486	0.4250	0.2163	1.9650
6	0.25	0.9024	0.3723	0.0951	0.6394	0.2638	2.4237
6	0.5	3.6606	1.4061	0.3775	0.6485	0.2491	2.6034
8	0.25	1.2157	0.4361	0.1274	0.8615	0.3090	2.7876
8	0.5	4.9468	1.6672	0.5046	0.8763	0.2953	2.9672
10	0.25	1.5231	0.5179	0.1587	1.0793	0.3670	2.9412
10	0.5	6.2139	2.0029	0.6241	1.1008	0.3548	3.1024
12	0.25	1.8190	0.6177	0.1886	1.2890	0.4377	2.9448
12	0.5	7.3951	2.4063	0.7408	1.3101	0.4263	3.0733

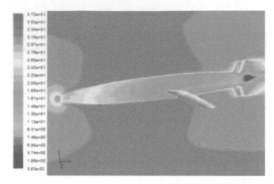

Figure 5.   Pressure distribution.

Figure 6.   Velocity distribution.

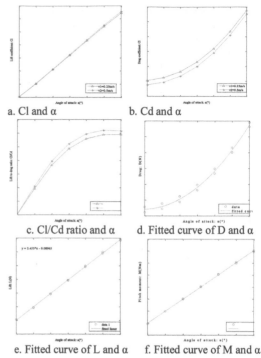

a. Cl and α   b. Cd and α

c. Cl/Cd ratio and α   d. Fitted curve of D and α

e. Fitted curve of L and α   f. Fitted curve of M and α

Figure 7.   Relationship between the hydrodynamic coefficients and the attack angle.

pressure distribution and the velocity vector distribution on the AUG, when the attack angle is zero and the incoming flow velocity is 0.25 m/s.

Figure 5 shows that the pressure distribution on the AUG is smooth, with a small potential difference. In addition, the pressure difference between the upper and lower wing surfaces can produce a big lift up.

Figure 6 shows the vortex separation point near the tail, which helps to reduce the pressure drag, thereby reducing the overall resistance. At the same time, the wing surface flow separation is less, which increases the lift coefficient and the lift-drag ratio. This AUG shape has a good hydrodynamic performance.

The least square method and data fitting method are used to process these data, and then obtain the drag, lift and pitch moment, as a function of the angle of attack α. Figure 7-a,b,c shows the change in hydrodynamic coefficients along with the angle of attack. Figure 7-d,e,f shows these fitted curves of the functions.

Figure 7-a, b and c shows that when the attack angle ranges from 0° to 12°, the hydrodynamic coefficients Cd and Cl both increase with the increase in the angle of attack α. The lift to drag ratio increases first and then drops, reaching a maximum value when the attack angle is 10°.

According to the fitted results of Figure 7-d, e and f, the constants in Eq. (2) can be estimated to be

$$K_{D0} = 4.411, K_{D\alpha} = 0.03556$$
$$K_{L0} = -0.08965, K_{L\alpha} = 2.455$$
$$K_{M0} = -0.01662, K_{M\alpha} = 0.253$$

As the attack angle α is small, $\sin\alpha$ α, $\cos\alpha$ 1, so we can get the relationship between the motion parameters (attack angle α, pitch angle θ, gliding velocity V) and the control variables (x axial displacement x_G and net buoyancy mass m_0). As shown in Figure8~Figure10.

Figure 8 shows that the underwater glider's attack angle α is greatly influenced by the displacement of gravity center x_G, and α decreases with the increase of x_G, and the net buoyancy mass m_0 has a little influence on the angle of attack α.

170

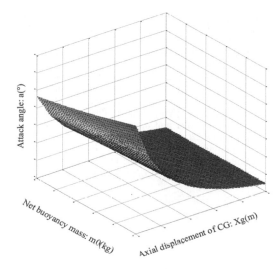

Figure 8. Relationship between α and xG, m0.

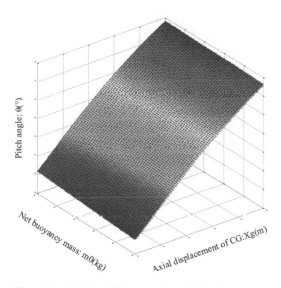

Figure 9. Relationship between θ and xG, m0.

Figure 9 shows that the underwater glider's pitching angle θ is greatly influenced by the displacement of gravity center x_G, and θ increases with the increase of x_G, and the net buoyancy mass m_0 has a little influence on θ. The pitching angle θ has a certain increasing trend with the increase of m_0.

Figure 10 shows that the net buoyancy mass m_0 and the displacement of gravity center x_G both have a great influence on the gliding velocity V, and V increases monotonously with the increase of x_G, m_0.

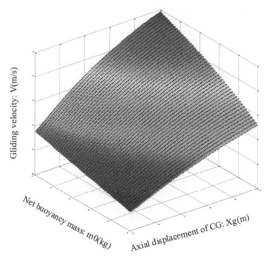

Figure 10. Relationship between V and xG, m0.

4 CONCLUSIONS

This paper proposes a low-resistance underwater glider, having a streamlined shape. We use the CFD method to compute the hydrodynamic coefficients and compare the hydrodynamic performance of four typical underwater glider body shapes. Finally, we obtain the optimal glider shape.

We set up the steady motion equations of underwater gliders. Combining with the computation results from the FLUENT simulation, the relationship between the gliders' motion state parameters α, θ, V, with its design parameters m_0 and x_G, is analyzed. We know that α and θ are both greatly influenced by x_G, but little influenced by m_0, and with the increasing of x_G, α decreases but θ increases. The gliding velocity V is influenced by both m_0 and x_G, and V increases monotonously with x_G, m_0. The result provides the guidance for the next design of gliders.

In future work, we will design the control system first and then manufacture the glider sample, and conduct actual experiments to study the movement properties of the AUG.

ACKNOWLEDGMENTS

This work was partially supported by the Natural Science Foundation of China (51379198, 51075377, 41176076, 31202036), the Zhejiang Natural Science Foundation (R1100015), and the Science and Technology Development Program of Shandong Province (2008GG1055011, BS2009HZ006)

REFERENCES

Eriksen C C, Osse T J, Light R D,et al, 2001. Seaglider: A long range autonomous underwater vehicle for oceanographic research. *IEEE Journal of Oceanic Engineering*, 26(4): 424–436.

Fujii, K, 2005. Progress and future prospects of CFD in aerospace wind tunnel and beyond, *Prog Aerospace Sci.*,41:455–470.

Hu Ke, Yu Jian-cheng, 2005. Design and Optimization of Underwater Glider Shape, *Robot*, 27(2):108–117.

Meng Fan-hao & Yan Tian-hong, 2014. Shape design for autonomous underwater glider based on hydraulic characteristics, *The Ocean Engineering*, 32(2):61–71.

Sherman, R., Davis, E., Owens, W.B., Valdes, J., 2001. The autonomous underwater glider spray. *IEEE J. Oceanic Eng.* 26 (4), 437–446.

Stommel, H., 1989. The slocum mission. *Oceanography* 2(1): 22–25.

Webb, D.C., Simonetti, P.J., Jones, C.P., 2001. Slocum: an underwater glider propelled by environmental energy. *IEEE J. Oceanic Eng.* 26 (4), 447–452.

Zeng Qing-li, Zhang Yu-wen, 2010. The overall design of underwater glider and movement analysis, *Computer Simulation*, 27(1):1–6.

Zhang Feitian, J. Thon, C. Thon, and X. Tan, 2012. Miniature underwater glider: Design, modeling, and experimental results, *IEEE Int. Conf. Robot. Autom.*, 4904–4910.

Energy Science and Applied Technology – Fang (Ed.)
© 2016 Taylor & Francis Group, London, ISBN 978-1-138-02833-3

Analysis of internal flow field of OTS 125–500A low flow rate double-suction centrifugal pump

Wen Yuan
Institute of Petroleum Engineering, Northeast Petroleum University, Daqing, Hei Longjiang, China

Changbin Wang & Kai Zhang
Institute of Metrology and Measurement Engineering, China Jiliang University, Hangzhou, Zhe Jiang, China

Xi Chen
Institute of Petroleum Engineering, Northeast Petroleum University, Daqing, Hei Longjiang, China

ABSTRACT: The objective of this paper is to analyze the OTS 125–500A low flow rate double-suction centrifugal pump. FLUENT code is utilized to simulate the characteristics of the flow field of the centrifugal pump. Five turbulent models commonly used in the numerical simulation are applied. The numerical simulation results are compared to find out the flow charactcristics of the low flow rate centrifugal pump. From the design viewpoint, the results show that the calculation results of five turbulence models can in fact reflect the flow characteristics: at low flow rates, the results of the realizable k-ε model are much closer; at high flow rates, the results of the SST k-ω model are better. Compared with the experimental results, simulation results agree well with it. Therefore, the characteristics of the internal flow field obtained by numerical simulation are high reliable and able to provide a theoretical basis for the optimization design of the double-suction pump.

Keywords: Double-suction centrifugal pump; Turbulence model; internal flow field; Numerical simulation

1 INTRODUCTION

Pump is a widely used general machine. Double-suction pump is one of the most popular pumps with a large delivery head and a large volume flow rate (Qian Zhongdong, 2012, Zhang Jiahui, 2005, Zhang Zheng, 2008). Numerical simulation of internal flow field in centrifugal pumps is significant to improve the hydraulic performance for the purpose of energy saving and the optimization of design realization (Zhang Hui, 2009). The OTS double-suction centrifugal pump is a new generation product. With high efficiency, good anti-cavitation performance and low noise, it is widely used in water conservation, fire protection system, oil refining, chemical industry and other fields.

The objective of this paper is to analyze the internal flow field of the OTS 125–500A double-suction centrifugal pump. Since the internal three-dimensional viscous flow of the turbo-machinery is complex, there is no universal turbulence model at present. The influence of turbulence model selection on CFD numerical simulation results is an important subject. This paper discusses the adaptability

of five turbulence models on the centrifugal pump, drawing the characteristic curves, and then we can find out the most suitable turbulence model for the pump and obtain the pressure, velocity and turbulence energy rules of the internal flow field.

2 NUMERICAL MODEL

2.1 Geometric model

The simulation includes an entire centrifugal pump, besides the suction and discharge pipe and seals to ensure the accuracy. Table 1 summarizes the specific parameters of the basic structure.

2.2 Grid resolution

The commercial code GAMBIT is used to generate the meshes. Because the geometry model is complex, the combination of the structured and unstructured grid methods is chosen. The grid number is selected by experimental data via trail and error. The grid number increases from 1.53 million to 3.5 million, while the head change

Table 1. Structure of the parameters of the OTS 125–500A double-suction centrifugal pump.

Structure of the parameter	Parameter value
Flow rate $Q/(m^3/h)$	328
Head H/m	76
Rotational Speed $n/(r/min)$	1450
Vane Number Z	6
Impeller Outlet Diameter D_2/mm	475
Impeller Outlet Width b_2/mm	45
Impeller Inlet Diameter D_1/mm	68
Cutwater Diameter D_3/mm	484.2

(a) Impeller mesh

is less than 3%. Therefore, taking the calculation complexity into consideration, the number of grids is determined to be 1.53 million.

2.3 Boundary conditions and numerical method

In the numerical simulation, fluid velocity at the inlet and fully developed boundary condition are used, and the walls use no-slip boundary conditions. The multiplex coordinate reference system (MRF) and the Reynolds averaged N-S equation are applied. The SIMPLE algorithm and the first-order upwind discretization are selected, and the residual of iteration is set to less than 10^{-6}.

3 NUMERICAL RESULTS AND ANALYSIS

3.1 External performance prediction

Five turbulent models, which are usually used in numerical simulation, are applied to calculate the delivery head and efficiency performances of the pump. By comparing the numerical results and the experimental ones, the most appropriate turbulence model for the centrifugal pump can be obtained.

The head calculation equation (Wang Fujun, 2004, Zhang Hui, 2009) is as follows:

$$H = \frac{P_{out} - P_{in}}{\rho g} + \Delta Z \qquad (1)$$

where P_{in} = inlet pressure; P_{out} = outlet pressure; ρ = density; and ΔZ = vertical distance between the volute outlet and the impeller inlet.

The efficiency calculation equation is as follows:

$$\eta = \frac{\rho g Q H}{P} \qquad (2)$$

where Q = flow rate and P = shaft power.

(b) Volute mesh

Figure 1. Geometry of the centrifugal pump and grid.

Figures 2 and 3 show the comparison results between the calculated and experimental values. A good consistency between the numerical and experimental results can be found. This indicates that the results obtained by numerical simulation are reliable.

As can be seen from Figure 2, the heads calculated by the k-ε turbulence model are always larger than the experimental ones, while the results calculated by the k-ω turbulence model are opposite. From the design viewpoint, all errors are minor. It can reflect the work of the centrifugal pump well. At low flow rates, the results obtained by the realizable k-ε model are much closer to the experimental ones. At high flow rates, the results calculated by the SST k-ω model are better.

As shown in Figure 3, the trends of efficiency comparison between the simulation and experimental curves are consistent, but all curves show a greater efficiency than those of the actual pump. This can be attributed to the fact that the efficiency

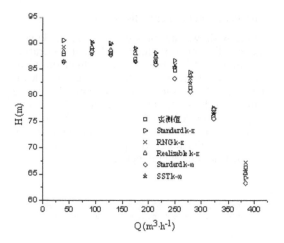

Figure 2. Head comparison between the experimental and numerical curves.

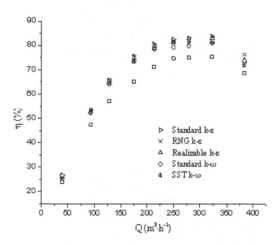

Figure 3. Efficiency comparison between the experimental and numerical curves.

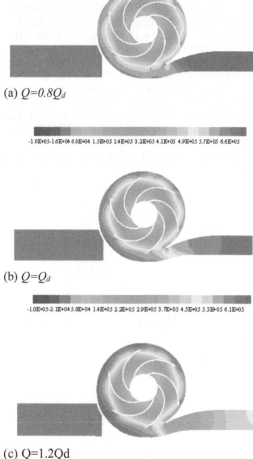

(a) $Q=0.8Q_d$

(b) $Q=Q_d$

(c) $Q=1.2Q_d$

Figure 4. Midsection static pressure distribution under different operating conditions.

obtained by numerical simulation is the hydraulic efficiency, which does not take the volume loss into account. From the design viewpoint, all the efficiency values calculated are close to the experimental ones. At low flow rates, the realizable k-ε model values are closer; at high flow rates, the SST k-ω model shows excellent performance.

3.2 Internal flow field analysis

Static pressure distribution.
Figure 4 shows the midsection static pressure distribution under different operating conditions. In the impeller channel, the work transferred in the impeller to the liquid is converted to pressure energy. The static pressure increases all the way to the impeller discharge. In the blade passage, the working surface pressure is higher than that of the suction surface at the same radius. In the volute, the kinetic energy transforms into static pressure energy. The static pressure increases continuously. Because there is some fluid flowing back into the volute at the volute tongue, where the pressure gradient is large, it causes higher energy losses. With the flow rate increasing, the energy per unit mass gain reduces. The pressure on the blades becomes lower and the pressure difference between the impeller inlet and the volute outlet decreases as well.

Relative velocity distribution

Figure 5 shows the midsection relative velocity vector distribution under different operating conditions. In the blade passage, because the high-pressure region is located near the working surface, the suction surface relative velocity is faster than the working surface relative velocity at the same radius. It induces the "jet-wake" effect next to the suction surface to the impeller discharge. Because of the interference of the volute tongue, the back-flow forms in the volute tongue to the volute outlet direction. The relative velocity in the volute outlet at low flow rate operating conditions is faster than that at high flow rate conditions. With the flow rate increasing, the recirculation zone near the volute tongue gradually expends and much energy is lost.

Turbulent kinetic energy distribution

Figure 6 shows the midsection turbulent kinetic energy distribution under different operating conditions. At low flow rates, the high turbulent kinetic energy region is at the impeller inlet, leading edge of the suction surface and the volute tongue tip. This is because at the end of the volute tongue, the backflow takes place, which results in intense exchange of momentum and large energy losses; from the design viewpoint, the difference in turbulent kinetic energy between the working sur-

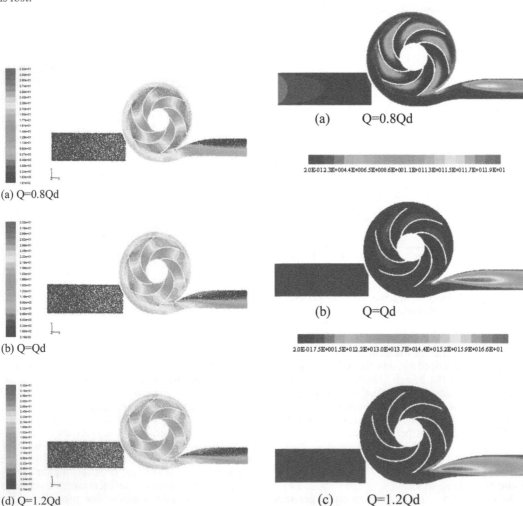

(a) Q=0.8Qd

(b) Q=Qd

(d) Q=1.2Qd

Figure 5.   Midsection relative velocity vector distribution under different operating conditions.

(a)   Q=0.8Qd

(b)   Q=Qd

(c)   Q=1.2Qd

Figure 6.   Midsection turbulent kinetic energy distribution under different operating conditions.

face and the suction surface decreases. However, the value of the suction surface is still higher than that of the working surface. The turbulent kinetic energy increases with the flow rate at the volute tongue and the volute outlet.

4 CONCLUSION

Through the analysis and the study on the external characteristics and internal flow field of the OTS 125–500A low flow rate double-suction centrifugal pump, the following conclusions can be drawn:

1. From the design viewpoint, due to the minor error values, results of the five turbulent models are ideal. Considering the head, efficiency and other performance parameters, at low flow rates, the realizable k-ε turbulence model is better. At high flow rates, the SST k-ω turbulence model is excellent.
2. Through analyzing the static pressure field, velocity field and the turbulent kinetic energy field of the centrifugal pump, it can be found that the pressure is lower at the inlet and higher at the outlet. The lowest pressure region is near the suction surface, where it is easy to cavitate, which is in line with the law of centrifugal flow. Relative velocity is more evenly distributed. There exists "jet-wake" in the impeller, which has an influence on the total efficiency.
3. FLUENT code can well simulate the main characteristics and internal flow field of the OTS 125–500A low flow rate double-suction centrifugal pump. It provides an important theoretical basis for the optimization design of the double-suction centrifugal pump.

REFERENCES

Qian Zhongdong et al. 2012. Optimization of impeller back vanes in double-suction centrifugal pump by CFD technique. *Journal of Drainage and Irrigation Machinery Engineering* 9(30):503–507.
Wang Fujun. 2004. *Computational fluid dynamics analysis*. Beijing. Tsinghua University Press.
Yuan Shouqi et al. 2011. Numerical calculation of internal flow-induced noise in centrifugal pump volute. *Journal of Drainage and Irrigation Machinery Engineering* 3(29):93–98.
Zhang Hui et al. 2009. Numerical Performance Simulation for Internal Flow of Centrifugal Pump. *Water Resources and Power* 27(4):69, 181–183.
Zhang Jiahui. 2005. Numerical calculation and analysis of the internal flow field of low specific speed double-suction centrifugal pump. Lanzhou University of Technology.
Zhang Zheng. 2008. Numerical Simulation of The Inner Flow of The Centrifugal Pump in Fluent. *Science & Technology Yinfomation* 26:93 94.

Energy Science and Applied Technology – Fang (Ed.)
© 2016 Taylor & Francis Group, London, ISBN 978-1-138-02833-3

Performance analysis of the ISD suspension based on a whole vehicle model involving the steering condition

Xiaofeng Yang, Yujie Shen, Yanling Liu & Jun Yang
School of Automotive and Traffic Engineering, Jiangsu University, Zhenjiang, Jiangsu, China

ABSTRACT: This paper discusses the performance improvements of the ISD (Inerter-Spring-Damper) suspension in the steering condition. A whole vehicle dynamic model involving the steering condition was built, and simulations were carried out to test the performance of the ISD suspension. The results show that the RMS of the vertical acceleration decreased by 10.9%, that of the pitch angular acceleration decreased by 15.6%, that of the yaw velocity decreased by 6.8%, and that of the body roll angle decreased by 4.4%. The overall performance of the ISD suspension is better than the passive suspension in the steering condition.

Keywords: vehicle; suspension; inerter; steering

1 INTRODUCTION

As an effective replacement of the mass element, an inerter (Smith, M.C., 2002) is a two-terminal mechanical element and can be realized in the mechanical form and hydraulic form (Papageorgiou, C. *et al*, 2009; Swift, S.J. *et al*, 2013). Owing to the contribution of the inerter, the mechanical elements of the inerter, the spring, and the damper can be strictly corresponded to the electrical elements of the capacitor, the inductor and the resistor. The mechanical vibration-isolated system using the inerter has been widely used in train suspension (Wang, F.C. *et al*, 2006; Wang, F.C. *et al*, 2010), building isolation (Wang, F.C. *et al*, 2010) and other vibration engineering.

Vehicle suspension is an important assembly of bearing body weight and transferring force between the wheel and the body. It plays a decisive role in ride comfort and handling stability of the vehicle. In recent years, scholars have carried out extensive research (Zhang, X.J. *et al*, 2012; Smith, M.C. *et al*, 2004) in the passive vehicle suspension that consists of the three mechanical components: inerter, spring and damper (ISD suspension). However, the existing research findings are all confined to the isolation performance of ISD suspension based on a quarter-car model (Chen, L., *et al*, 2014; Yang, X.F. *et al*, 2014), a half-car model (Chen, L. *et al*, 2012) and a full vehicle model (Zhang, X.L. *et al*, 2013). The enhancement of the ISD suspension performance in the steering condition has not been reported.

This paper discusses the performance improvement of the ISD suspension in the steering condition. The paper is arranged as follows. In section 2, the newly proposed element inerter is briefly introduced. In section 3, the vehicle dynamic model involving the steering condition is built. In section 4, simulations are carried out. Finally, in section 5, some conclusions are drawn.

2 BRIEF INTRODUCTION OF THE INERTER

The inerter is a mechanical element with two terminals, namely the spring and the damper. The force applied to the two terminals of the spring element is proportional to the relative displacement. The force applied to the two terminals of the damper element is proportional to the relative velocity. The inerter is defined as a device where the force applied to the two terminals is proportional to the relative acceleration of the two terminals.

The dynamic equation of the inerter is given in Eq. 1.

$$F = b(\dot{v}_2 - \dot{v}_1) \tag{1}$$

where F is the force applied to the two terminals; b is the inheritance (in kg); and v_1, $v2$ are the velocity of the two terminals.

3 VEHICLE MODEL INVOLVING THE STEERING CONDITION

Ignoring the guiding mechanism of the suspension system, air resistance, rolling resistance and assuming the centroid of the vehicle coincides with the car moving coordinate system origin. We define the forward direction of the vehicle as the X-axis, and the centroid point upwards as the Z-axis, and the left of the driver as the Y-axis. Both the suspension springs and dampers are considered to be linear. The vehicle model involving the steering condition is shown in Figure 1 and Figure 2.

The body vertical displacement equation is given by

$$m_a \ddot{Z}_a = F_{10} + F_{20} + F_{30} + F_{40} \tag{2}$$

The yaw motion equation is given by

$$I_z \dot{\omega}_z = l_f(S_1 + S_2) - l_r(S_3 + S_4) \tag{3}$$

The lateral motion equation is given by

$$mv(\dot{\beta} + \omega_z) = (S_1 + S_2) + S_3 + S_4 + m_a h \ddot{\theta} \tag{4}$$

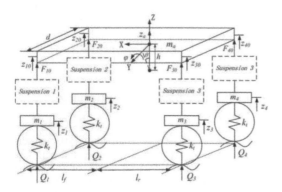

Figure 1.   Vehicle model.

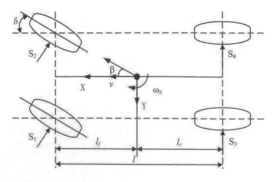

Figure 2.   Steering model.

The body pitching motion equation is given by

$$I_y \ddot{\varphi} = l_r(F_{30} + F_{40}) - l_f(F_{10} + F_{20}) \tag{5}$$

The body roll motion equation is given by

$$I_x \ddot{\theta} = m_a v(\dot{\beta} + \omega_z)h + m_a gh\theta + (F_{20} + F_{40} - F_{10} - F_{30})\frac{d}{2} \tag{6}$$

where m is the vehicle mass; m_a is the sprung mass; I_x is the inertia of the body roll moment; I_y is the inertia of the body pitch moment; I_z is the inertia of the body yaw moment; v is the driving speed; β is the sideslip angle; ω_z is the vehicle yaw rate; δ is the front wheel angle; θ is the vehicle body roll angle; φ is the body pitch angle; S_i is the tire cornering force; F_{i0} is the force of the suspension; d is the tread of the left and right wheels; l_f is the distance between the front wheel and the center of mass; l_r is the distance between the rear wheel and the center of mass; and h is the distance between the sprung mass and the centroid of the roll center.

The equations of the vertical displacement of four corners of the sprung mass are as follows:

$$\begin{cases} Z_{10} = Z_a - l_f\varphi - d/2\,\theta \\ Z_{20} = Z_a - l_f\varphi + d/2\,\theta \\ Z_{30} = Z_a + l_r\varphi - d/2\,\theta \\ Z_{40} = Z_a + l_r\varphi + d/2\,\theta \end{cases} \tag{7}$$

where Z_a is the vertical displacement of the sprung mass and Z_{i0} is the vertical displacement of four corners of the sprung mass.

The dynamic equations of the unsprung mass are as follows:

$$\begin{cases} m_1 \ddot{Z}_1 = k_t(Q_1 - Z_1) - F_{10} \\ m_2 \ddot{Z}_2 = k_t(Q_2 - Z_2) - F_{20} \\ m_3 \ddot{Z}_3 = k_t(Q_3 - Z_3) - F_{30} \\ m_4 \ddot{Z}_4 = k_t(Q_4 - Z_4) - F_{40} \end{cases} \tag{8}$$

where m_i is the unsprung mass; k_t is the tire equivalent stiffness; Q_i is the road input; and Z_i is the vertical displacement of the unsprung mass.

To simplify the calculations, the tire model can be considered as a linear model when the steering angle is small. The equations of the tire considering the impact of the roll tire cornering force are as follows:

$$\begin{cases} S_1 = S_2 = -K_f \alpha_f \\ S_3 = S_4 = -K_r \alpha_r \\ \alpha_f = \beta + \dfrac{l_f}{v} w_z - E_f \theta - \delta \\ \alpha_r = \beta - \dfrac{l_r}{v} w_z - E_r \theta \end{cases} \qquad (9)$$

where K_f is the front tire cornering stiffness; K_r is the rear tire cornering stiffness; α_f is the front tire slip angle; α_r is the rear tire slip angle; and E_f is the steering wheel roll coefficient.

The suspension structure is shown in Figure 1. (For example, consider the left front wheel.)

It can be seen that the ISD suspension consists of the spring k_b, the inerter b and the damper c_b. The inerter b is in series with the damper c and then in parallel with the spring k_b.

Taking the role of the stabilizer bar into consideration, the equations of the force of the ISD suspension are as follows:

$$\begin{cases} F_{10} = k_b(Z_1 - Z_{10}) + b(\ddot{Z}_{b1} - \ddot{Z}_{10}) - \dfrac{K_{af}}{2d}(\theta - \dfrac{Z_2 - Z_1}{2d}) \\ \quad b(\ddot{Z}_{b1} - \ddot{Z}_{10}) = c_b(\dot{Z}_1 - \dot{Z}_{b1}) \\ F_{20} = k_b(Z_2 - Z_{20}) + b(\ddot{Z}_{b2} - \ddot{Z}_{20}) + \dfrac{K_{af}}{2d}(\theta - \dfrac{Z_2 - Z_1}{2d}) \\ \quad b(\ddot{Z}_{b2} - \ddot{Z}_{20}) = c_b(\dot{Z}_2 - \dot{Z}_{b2}) \\ F_{30} = k_b(Z_3 - Z_{30}) + b(\ddot{Z}_{b3} - \ddot{Z}_{30}) + \dfrac{K_{ar}}{2d}(\theta - \dfrac{Z_3 - Z_4}{2d}) \\ \quad b(\ddot{Z}_{b3} - \ddot{Z}_{30}) = c_b(\dot{Z}_3 - \dot{Z}_{b3}) \\ F_{40} = k_b(Z_4 - Z_{40}) + b(\ddot{Z}_{b4} - \ddot{Z}_{40}) - \dfrac{K_{ar}}{2d}(\theta - \dfrac{Z_3 - Z_4}{2d}) \\ \quad b(\ddot{Z}_{b4} - \ddot{Z}_{40}) = c_b(\dot{Z}_4 - \dot{Z}_{b4}) \end{cases}$$
$$(10)$$

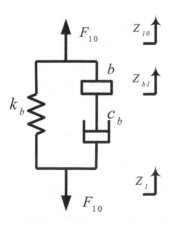

Figure 3.   ISD suspension structure diagram.

where Z_{bi} is the displacement of the inerter; k_1 is the spring stiffness; b is the inertance coefficient; c_b is the damping coefficient; and K_{af} and K_{ar} are the angular stiffness for the stabilizer bar.

4   SIMULATION ANALYSIS

The random road input model is given in Eq. 11:

$$\dot{z}_g(t) = -2\pi f_0 z_g(t) + 2\pi \sqrt{G_0 u} w(t) \qquad (11)$$

where u is the speed; $z_g(t)$ is the displacement of the random road input; G_0 is the road roughness coefficient, which is set as 5×10^{-6} m³cycle⁻¹; and f_0 is the cut-off frequency, which is set as 0.01 Hz.

The parameters of the vehicle model are listed in Table 1.

To consider the performance of the ISD suspension in the steering condition, we assume a step angle of 0.5 rad as the input to the front wheel. The time graphs of the vertical acceleration, the pitch angular acceleration, the yaw velocity, the body roll angle, the suspension deflection and the tire runout are shown in Figure 4 to Figure 9.

In order to get the detailed information of the performance of the ISD suspension, the performance indices are listed in Table 2.

Table 1.   Model parameters.

Name	Parameter
Vehicle mass m/kg	1330
Sprung mass ma/kg	770
Front unsprung mass m1,m2/kg	37
Rear unsprung mass m3,m4/kg	33
Tread of the left and right wheels d/m	1.36
Distance between the front axle to the centroid lf/m	0.955
Distance between the rear axle to the centroid lr/m	1.380
Front wheel side leaning towards coefficient Ef	−0.114
Rear wheel side leaning towards coefficient Er	0
Distance between the centroid and the roll center h/m	0.5
Inertia of body roll moment Ix/kg·m²	293
Inertia of body pitch moment Iy/kg·m²	1074
Inertia of body yaw moment Iz/kg·m²	1591
Tire cornering stiffness Kf, Kr/kN·m⁻¹	35
Angular stiffness of the stabilizer bar Kaf,Kar/kN·m⁻¹	6.695
Spring stiffness of the ISD suspension kb/kN·m⁻¹	22
Damper coefficient of the ISD suspension cb/N·s·m⁻¹	1543
Inertance of the ISD suspension b/kg	659
Tire stiffness kt/kN·m⁻¹	192

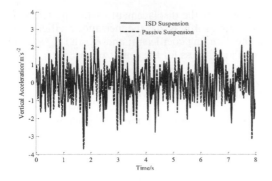

Figure 4.　Time graph of the vertical acceleration.

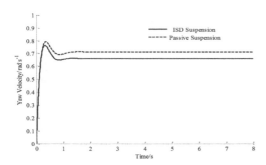

Figure 5. Time graph of the pitch angular acceleration.

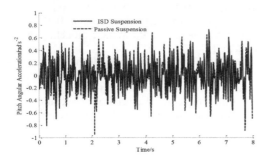

Figure 6.　Time graph of the yaw velocity.

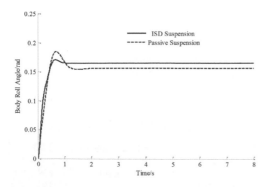

Figure 7.　Time graph of the body roll angle.

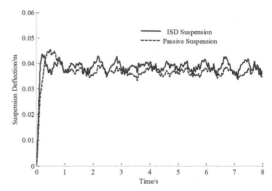

Figure 8.　Time graph of the suspension deflection.

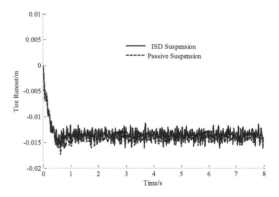

Figure 9.　Time graph of the tire runout.

Table 2.　Performance index.

Name	Index	Passive	ISD	Improvement
Vertical	RMS	0.9876	0.8798	10.9%
acceleration	PEAK	3.6547	3.3543	8.2%
Pitch angular	RMS	0.2872	0.2425	15.6%
acceleration	PEAK	0.9245	0.8314	10.0%
Yaw	RMS	0.6595	0.6147	6.8%
velocity	PEAK	0.7924	0.7625	3.8%
Body roll	RMS	0.1511	0.1444	4.4%
angle	PEAK	0.1758	0.1819	-3.4%
Suspension	RMS	0.0355	0.0339	4.5%
deflection	PEAK	0.0455	0.0453	0.4%
Tire runout	RMS	0.0131	0.0123	6.1%
	PEAK	0.0173	0.0160	7.5%

We considered both the RMS and the peak of the evaluated parameters. For the vertical acceleration, the RMS decreased by 10.9% and the peak decreased by 8.2%. For the pitch angular acceleration, the RMS decreased by 15.6%, the peak decreased by 10.0%. For the yaw velocity, the RMS decreased by 6.8% and the peak decreased by 3.8%. For the body roll angle, the RMS decreased by 4.4%, but the peak increased by 3.4%. For the suspension deflection, the RMS decreased by 4.5% and the peak decreased by 0.4%. For the tire runout, the RMS decreased by 6.1% and the peak decreased by 7.5%.

5 CONCLUSION

This paper has built the whole vehicle model of the ISD suspension involving the steering condition and discussed the improvement of the ISD suspension. From the simulation results, it is shown that all the performances of the suspension are improved except for the peak of the body roll angle. With respect to the RMS, the vertical acceleration decreased by 10.9%, the pitch angular acceleration decreased by 15.6%, the yaw velocity decreased by 6.8%, the body roll angle decreased by 4.4%, the suspension deflection decreased by 4.5%, the tire runout decreased by 6.1%, and the PEAK decreased by 7.5%. It suggests that the overall suspension performance is improved by the ISD suspension.

ACKNOWLEDGEMENTS

This work was supported by the China Postdoctoral Foundation under Grant No. 2014M561591, the 'Six Talents Peak' project of Jiangsu Provincial under Grant No. 2014-JNHB-023, the Postdoctoral Foundation of Jiangsu Province No.1402098C, and the Jiangsu University senior talent project No. 14 JDG153.

REFERENCES

Chen, L., Zhang, X.L., Nie, J.M. Performance analysis of two-stage series-connected inerter-spring-damper suspension based on half-car model[J]. Journal of Mechanical Engineering, 2012, 48(6): 102–108.

Chen, L., Shen, Y.J., Yang, X.F. Design and experiment of vehicle suspension based inerter-spring structure [J]. Journal of Vibration and Shock, 2014, 33(22): 83–87.

Papageorgiou, C., Houghton, N.E., Smith, M.C. Experimental testing and analysis of inerter devices [J]. Journal of Dynamic Systems, Measurement, and Control, 2009, 131: 235–241.

Smith, M.C. Synthesis of mechanical networks: the inerter [J]. IEEE Transaction on Auto Control, 2002, 47(10):1648–1662.

Swift, S.J., Smith, M.C., Glover, A.R., et al. Design and Modelling of a Fluid Inerter [J]. International Journal of Control, 2013, 86(11): 2035 2051.

Smith, M.C., Wang, F.C. Performance benefits in passive vehicle suspensions employing inverters [J]. Vehicle System Dynamics, 2004, 42(4): 235–257.

Wang, F.C., Yu, C.H., Chang, M.L., et al. The performance improvements of train suspension systems with inverters[C]//Proceedings of the 45th IEEE Conference on Design and Control. 2006: 1472–1477.

Wang, F.C., Liao, M.K. The lateral stability of train suspension systems employing inverters [J]. Vehicle System Dynamics, 2010, 48(5): 619–643.

Wang, F.C., Hong, M.F., Chen, C.W. Building suspension with inerters [C]//Proceedings of the Institution of Mechanical Engineers, Part C: Journal of Mechanical Engineering Science, 2010, 224(8): 1605–1616.

Yang, X.F., Shen, Y.J., Chen, L., et al. Design and performance analysis of vehicle ISD suspension based on dynamic vibration absorber theory[J]. Automotive Engineering, 2014, 36(10): 1311 1315.

Zhang, X.L., Nie, J.M., Wang, R.C. Passive skyhook-dampering suspension system based on inerter-spring-damper structural system [J]. Transactions of the Chinese Society for Agricultural Machinery, 2013, 44(10): 10–14.

Zhang, X.J., Ahmadian, M., Guo, K.H. On the benefits of semi-active suspensions with inerters [J]. Shock and Vibration, 2012, 19(3): 257–272.

Energy Science and Applied Technology – Fang (Ed.)
© 2016 Taylor & Francis Group, London, ISBN 978-1-138-02833-3

Study on the quantitative measurement of the tempering martensite for a carbon steel by thermal expansion

Yin Xue, Kejia Liu, Kun Chen & Huifen Chen
School of Materials Science and Engineering, Shanghai Institute of Technology, Shanghai, China

ABSTRACT: A new method is proposed to measure the content of different phases in metallic materials. The content of the metastable phase can be determined by measuring two Thermal Expansion (TE) curve lines, by joining the lines at a high temperature, at which point all the metastable phases degenerate to the annealing phase, i.e. the equilibrium state. Three different heat treatment samples for a kind of low carbon steel were measured by this method; the samples include quenching, annealing and tempering states. The results are checked quantitatively by the diffusion theory. It is shown that TE measurement can be applied to determine the content of different phases of the material to a relatively accurate degree. This method provides a quantitatively analysis of the phase composition of metallic materials.

Keywords: Thermal expansion; Martensite; Tempering; Diffusion Thermal expansion

1 INTRODUCTION

Phase transformation (*PT*) of materials can be used to change the properties of materials, which is conducted by using the heat treatment. Due to the *PT* of materials, not only the mechanical properties, but also the length or the volume is changed. It is inferred that by accurate size measurements and quantitative comparisons of materials under different conditions, one can get the information about the micro-state of the material.

There are two kinds of method for measuring the dimensions of materials: one is the microscopic measurement, which can be measured by X-ray diffraction with the lattice constant measurement; another one is the TE measurement. The TE method mainly focuses on the relative change in the rate of the material size during the change in the temperature.

Many authors have conducted a quantitative measurement on martensite PT for a kind of steel material. For instance, Tang et al. established the numerical model of the tempering process, and verified the numerical simulation results by TE experiments. Bohemen carried out the basic research concerning the volume fraction of tempering M by quantitative analysis of the heating dilation data.

TE curves provide the information about the phase transition of materials, which quantitatively describes the process of the material transition from the non-equilibrium state to the equilibrium one. In this work, we report the study of the quantitative content of *PT* obtained by *TE* curves, for the tempering process of *M* of a steel material, described in the next section.

It is possible to compare the experimental results with the calculation results of *PT*, which is calculated quantitatively for carbon precipitation during the tempering of carbon steel, by diffusion law. The experimental and calculated results are consistent, described in the third section.

2 EXPERIMENTAL MATERIALS AND METHODS

The experimental materials are a low carbon and a low alloy steel. The chemical composition is presented in Table 1.

The samples were fabricated to three cylinder samples of the size $\Phi 5 \times 10$ mm with the following heat treatments: quenching, annealing and quenching with tempering for 2 hours at 200°C, which were labeled as A, B and C, respectively. Sample C is prepared for the observation of the microstructure by grinding, polishing and etching by Intel. *TE* experiments are carried out by the push rod horizontal thermal expansion instrument of brand DIL402C, with a heating rate of 5°C/min from room temperature to 750°C.

Table 1. Chemical composition of a low carbon steel (wt%).

C	Si	Mn	P	S	others
0.2	0.1–0.5	1.0–2.0	<0.030	<0.010	0.1

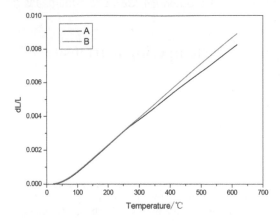

Figure 1. Thermal expansion curves of annealed and quenched samples: A, quenched; B, annealed.

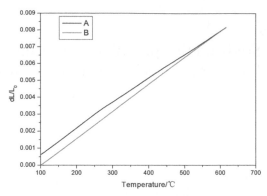

Figure 2. Modified thermal expansion curves of the annealed and quenched steel: A, quenched; B, annealed.

3 QUANTITATIVE ANALYSIS FOR THE THERMAL EXPANSION RESULTS

The principle of TE is that the length or volume of the material increases as the heating temperature increases. In general, the steel with different phases have different relative sizes. The order of the relative size per unit mass is as follows: $L_{martensite} > L_{ferrite} > L_{austenite} > L_{carbides}$. Therefore, when PT occurs, the dimensions of materials will also change. During heating, the PT of steels will lead to an obvious size or volume change. The expansion curves contain size changes of PT, which must deviate from the linear relationship shown in the temperature elongation figure. Therefore, the amount of PT can be determined by the nonlinear changes of the line, which is an effective method for quantitatively researching on the PT of a material.

Figure 1 shows two TE curves of the steel with different states, i.e. the annealed (red) and quenched (black) curves were put together for the convenience of comparison and discussion. The longitudinal coordinate in the figure is dL/L, where L denotes the length of the material in different conditions. The annealing TE line is almost linear. The reason is that the annealed material is in the equilibrium phase with ferrite and cementite (*Cem*) as its phases, and there is no PT during the TE measurement. In contrast, although the quenching curve is linear below 200°C and above 500°C, it deviates from a straight line in the range of 200°C to 500°C. The inflection of the nonlinear increase is caused by PT, i.e. the precipitate PT of the carbon from the martensite to *Cem* during the heating process, i.e. PT occurs in the process. However, as the temperature increases to about 700°C,

the martensite is decomposed into the equilibrium state, with the relative size being the same as the annealed one.

Figure 1 shows the two curves, and the quenched line is not in the equilibrium state, so the line is difficult to be compared with the annealed one. In order to obtain the geometric data of PT quantitatively, one may set L_0 as the equilibrium length, and then $dL_0 = L_t - L_0$ is associated with the equilibrium state. Therefore, the two TE lines in the figure should be joined at a high temperature (e.g. at 727°C), at which point the samples are at the same state, i.e. the equilibrium state, for the martensite has completely transformed to the equilibrium state. In order to compare the dimension of the two samples, one can join the two lines at this point, as shown in Figure 2, rather than at the room temperature (RT), and the difference in the relative dimension of martensite can be obtain by the difference at the RT. This is the key point to obtain the quantitative data of PT.

We can quantitatively compare the size of the material at RT after merging the two TE curves at a high temperature. This method can be applied in the comparison of the length of materials in different heat treatments, which leads to the quantitative results of the microstructures of materials of non-equilibrium states.

However, in practice, we consider the joining point at 620°C for the TE lines as the equilibrium length of the material. This is because that the austenite transition occurs around 727°C, i.e. it may affect the dimension of the sample, which may deviate from the straight line. The two lines have a good linear relationship at the joining point of 620°C, as shown in Figure 2. Compared with the annealed one, the relative volume expansion of the quenched state has an extra size, i.e. $dL/L_0 = 0.062\%$.

186

The lines below 100°C were ignored because TE has a large abnormal deviation around the RT.

4 APPLICATION OF QUANTITATIVE ANALYSIS BY THERMAL EXPANSION

In this section, we study the *PT* of the tempering *M* as an application. The experimental results of the relative length of tempering at 200°C for 2h are shown in Figure 3. Three experimental lines are shown, which represent the quenching, annealing and tempering lines, respectively. The experimental result of the relative length increase (dL/L_0) for the tempering is 0.018%, which accounts for 30% of the relative length increase of the quenching one. This result provides the quantitative degree of the tempering treatment as well as that of martensite transition of the experiments.

The above quantitative experimental results can be calculated by the diffusion theory in order to validate the relationship between the experimental and theoretical results.

We now calculate the carbon diffusion for the tempering. It is well known that the low carbon *M* is the lath martensite (*LM*) and carbides precipitate at the grain boundary after tempering. We may assume that the thickness of the *LM* is *l* and the carbon content at the interface between *LM* and *Cem* is zero. The diffusion can be approximated to one-dimensional diffusion of the constant source within the lath thickness *l*, for the reason that *l* is far less than the lath length. The carbon content $C(x, t)$ in a lath at the coordinate $\times$ $(0 < x < l)$ and at the time t can be expressed by:

$$C(x,t) = \frac{4C_0}{\pi} \sum_{n=0}^{\infty} \frac{1}{2n+1} \sin\frac{(2n+1)\pi x}{l}$$
$$\exp[-(2n+1)^2 \pi^2 Dt / l^2] \qquad (1)$$

where C_0 is the initial carbon content and D is the diffusion coefficient, which can be calculated by the Arrhenius equation, with $D0 = 0.20 \times 10^{-5}$ m2/s, the diffusion constant of M; R = 8.314 J/ (mol·K), the gas constant; Q = 8.40 × 104 J/ mol, the activation energy per mole; and T, the absolute temperature. We therefore obtain the diffusion coefficient of the tempering at 200°C as $D200 = 1.058 \times 10^{-15}$ m2/s.

The total carbon content m in the LM with the thickness l and at time t can be calculated by integrating Eq. (1), where t = 7200s and $C_0 = 0.2\%$. One may assume that the grain thickness l of the LM is of the order of 10 μm, by setting l = 13 μm, the surplus carbon content is m' = 0.060%. This result is consistent with the experimentally measured one (0.062%) as mentioned above.

There are some differences between the assumed thickness l of 13 μm for calculation and the measured thickness l of 3–5 μm by the microscopic measurement, as shown in Figure 4. In this case, the actual diffusion length should be the former. The reason may be that the stable carbide grains are not always formed on the boundary of each lath martensite, but may form among several LM. It is known that crystal nucleation has a critical radius and the crystal growth is stable only when the radius is larger than the critical radius during the nucleation growth process of carbide precipitation in tempering. Therefore, the diffusion length l for the above calculation should be the equivalent diffusion length with the formation of stable nuclei

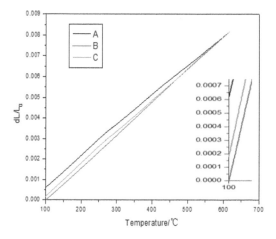

Figure 3. Thermal expansion curves of samples with different heat treatments: A, quenched; B, annealed, C, tempered at 200°C.

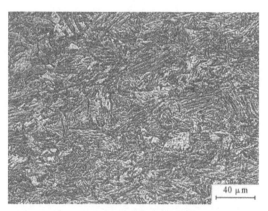

Figure 4. Morphology of the sample tempered at 200°C.

instead of the thickness of single LM. Therefore, the diffusion length should be larger than the thickness of single LM.

It is shown that the diffusion length l is about 5~10 times the thickness of the tempering lath M, which requires additional experimental results and theory for further illustration. However, as this paper mainly focuses on the novel method for the quantitative measurement of tempered M by TE, the nucleation and growth processes of the carbides of the tempered martensite are no longer discussed in detail in this paper.

5 CONCLUSIONS

This paper proposes an experimental method of measurement of the carbon content of the tempered martensite by the *TE* quantitatively. By measuring the *TE* lines that indicate the relationship between the temperature and the length in the metastable and annealed states of materials, we join the lines at high temperatures at which point all phases of steel are in the equilibrium state, so that we obtain the relative size difference by extrapolation, i.e. by measuring the difference of the lines, at the *RT*, quantitatively, thus the length and the content of the metastable phase can be obtained quantitatively. We therefore call this new method the thermal expansion comparison method, and we also gave an example in

detail. This method is demonstrated by the theoretical calculation.

REFERENCES

Crank J. The Mathematics of Diffusion. London: Oxford University Press, 1956.

Lan L.Y., Qiu C.L. & Zhao D.W. Effect of reheat temperature on continuous cooling bainitetrans formation behaviour in low carbon micro alloyed steel. Japer Sci, 48, p. 4356–4364, 2013.

Martelli V., Bianchini G. & Ventura G... Measurement of the thermal expansion coefficient of Aisi 420 stainless steel between 20 and 293 K. Cryogenics, 62, p. 94–96, 2014.

Natter H., Schmeltzer M. &Loffler M.S... Grain-Growth kinetics of Nanocrystal line Iron Studied in Situ by synchrotron Real-Time X-ray Diffraction. J. Phys. Chem. B., 104(11), p. 2467–2476, 2000.

Ozturk B., Fearing V.L. &Ruth J.A. The diffusion coefficient of carbon in cementite, Fe3C, at 450°C. Solid State Ionics, 12, p. 145–151, 1984.

Schneider A., IndenG. Carbon diffusion in cementite (Fe₃C) and Hagg carbide (Fe₅C₂).Clapham-Computer Coupling of Phase Diagrams and Thermochemistry, 31, p. 141–147, 2007.

Tang B.T., Bruschi S. & GhiottiA. Numerical modelling of the tailored tempering process applied to 22MnB5 sheets. Finite Elements in Analysis and Design, 81, p. 69–81, 2014.

Van BohemenS.M.C. Austenite in multiphase microstructures quantified by analysis of thermal expansion. Scripta Materialia, 75, p. 22, 2014.

Energy Science and Applied Technology – Fang (Ed.)
© 2016 Taylor & Francis Group, London, ISBN 978-1-138-02833-3

NVH optimization study based on concept car body

Yu Zhang & Youqiang Cao
State Key Laboratory of Vehicle NVH and Safety Technology, China

Junyi Hu
Department of Mechatronics and Aviation, Zhejiang Institute of Communications, Yuhang, China

ABSTRACT: This study on the low-order frequency optimization for a car Body-In-White (BIW) is based on the concept characteristic model. First, we construct the high-precision BIW finite element concept model containing joint characteristic structures by adopting suitable plate-beam elements. Then, its dynamic characteristic calculation error is controlled within 10%. Second, the clear and concise sensitive structural spatial layout form is acquired by calculating the strain energy distribution of the BIW concept model under the torsion vibration condition. Then, the optimization measures for its dynamic characteristics are proposed. Finally, the body torsional modal frequency is increased by about 2 Hz after verifying the feasibility of these measures based on the detailed structure model. These studies verify that the concept model can assess NVH properties rapidly and is an effective method for BIW structural optimization, and it also provide an accurate guidance for detailed structural optimization problems.

Keywords: BIW, joint branch length, concept model, NVH properties

1   INTRODUCTION

The modern car body structure design is mainly divided into three phases: the conceptual design phase, detailed design and analysis verification phase. The concept body model can simplify the car body's geometric features by using the beam and shell elements. So, it can reduce the size of the elements and raise the efficiency of calculation. In the car body analysis and validation phase, a lot of computations and optimization iterations need to be done in the structure body, thus this process will cost a lot of manpower and computing resources. So, the conceptual model can provide references for detailed development of the model, and reduce the whole development cycle due to its higher computing efficiency and less computing time.

A large number of published articles have mainly aimed at optimization design in the automobile body concept design phase. They have optimized the shape and thickness of the beam sections in the concept design phase. In addition they have optimized the conceptual design with the sensitivity analysis method.

We optimize the BIW model by combining the BIW concept design phase with the detailed car body design phase. We analyze the torsion modal strain energy and find out the weak areas related to structural mechanic characteristics in the BIW concept design phase. Then, low-order modal optimization improvement measures are suggested. Finally, improved the dynamic characteristic by applying these measures into the detailed car BIW phase.

2   JOINT CONCEPT MODEL ACCURACY CONTROL

The car body joint is the important structural part. Its elasticity has a significant impact on the whole car structure body. So, accurate simulation for the car body joint is critical to building a high-precision car body model. The literature takes the T-type joint as the research object, and studies the influence of its branch length for joint stiffness. This model lacks the circular transition areas compared with the real T-type joint, but these transition parts have a significant influence on the stiffness and strength. So, we establish the simplified T-type joint model with circular transition parts, and then analyze the suitable length of the joint branch (see Figure 1).

A previous research shows that the joint structures have a noticeable influence on torsion stiffness of the car body. The torsion condition analysis for the T-type joint can be achieved by adding the torque on top of each branch. After calculation, we find that the maximum twist angle for each branch is larger under the Z-direction torque than the X-or Y-direction torque. In addition, we consider its symmetry and set up two computing conditions

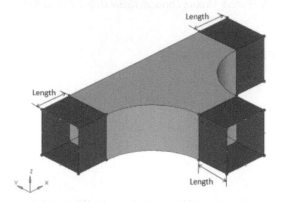

Figure 1.   Simplified T-type joint model.

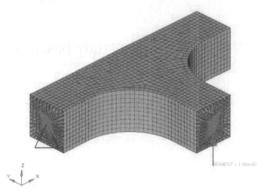

Figure 3.   Condition 2 sketch.

Table 1.   Calculated results according to the branch length (angle unit: 1e-8 rad).

	Length=0	Length=25	Length=50	Length=75	Length=100	Length=125	Length=150
Condition1	0.967	1.092	1.241	1.311	1.343	1.352	1.354
Condition2	1.003	1.200	1.319	1.379	1.388	1.391	1.391

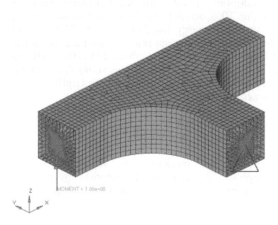

Figure 2.   Condition 1 sketch.

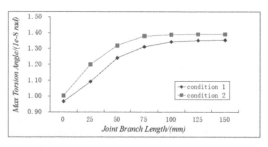

Figure 4.   Relationship between the branch length and the torsion stiffness.

as follows. In Condition 1, we add the Z-direction torque on top of the left horizontal branch and add the full displacement constraints on top of the other two branches (see Figure 2). In Condition 2, we add the Z-direction torque on top of the vertical branch and add the full displacement constraints on top of the other two branches (see Figure 3). Then, the torsion stiffness variation trends along with the branch length are detailed in Table 1 and Figure 4.

From Figure 4, it can be concluded that the torsion stiffness will be stabilized when the branch length is more than 100 mm. So, in order to control the computational accuracy, we set each joint branch length more than 100 mm in the car concept BIW model.

3   CAR CONCEPT MODEL CONSTRUCT AND PRECISION TEST

In order to evaluate the BIW mechanical properties rapidly and present the precise modal, we optimize computational parameters, to establish the BIW concept model from the corresponding detailed model. The detailed model has about 0.24 million elements (see Figure 5) and its weight is 1.77e2 kg. While establishing the BIW concept model, we set the branch length of 4 T-type joints (B-up-joint, B-down-joint, A-up-joint and A-down-joint), all being more than 100 mm. The corresponding concept model has about 0.075 million elements and the weight is 1.82e2 kg. We find out that the weight of the concept model is quite consistent with the detailed model, while the number of elements is reduced significantly. So, the concept model is effective.

Low-order modal is the most important and fundamental parameter for the car body NVH property. The contradictions of freedom modal analysis results between the car concept body and the detailed model are listed in Table 2. The

Figure 5.　Car BIW detailed model.

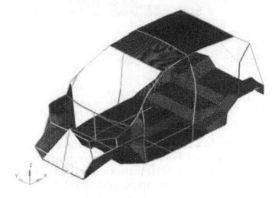

Figure 6.　Car BIW concept model.

Table 2.　Freedom modal analysis results for the concept model and the detailed model.

Modal grade	Concept model frequency	Detail model frequency	Relative error	Vibration mode descriptions
1	29.57Hz	30.74Hz	3.81%	Whole car body torsion
2	33.79Hz	36.39Hz	7.14%	Front end torsion
3	44.32Hz	47.13Hz	6.00%	Front end lateral vibration
4	49.00Hz	48.69Hz	0.63%	Center floor local vibration
5	63.33Hz	58.96Hz	7.41%	Car body vertical vibration

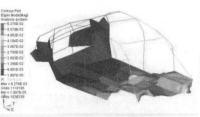

(1) Whole car body torsion of the concept model

(2) Whole car body torsion of the detailed model

Figure 7.　First-order torsional mode shape of vibration.

first-order torsional mode shape of vibration is shown in Figure 7.

From Table 2, it can be concluded that the computational errors between the concept model and the detailed model are within 10%, and the mode shapes of vibration between the two models are consistent. So, the concept model has good computing accuracy.

4　BIW MODAL OPTIMIZATION

4.1　*Modal strain energy analysis for the BIW concept model*

Modal strain energy analysis for BIW can reflect the weak structure areas under the corresponding modal. According to the results of this analysis, the weak areas can be improved and the dynamic mechanic property of BIW can be enhanced.

In the concept BIW model, liner beam elements are used to simulate the open and closed beam structures, which comprised plates and shells in a real car body. At the same time, the simple shell elements are used to simulate the big sheet metal parts. The whole concept model considers the beam elements as principal load-bearing parts. Figure 8 shows that the area with obvious modal strain energy is mainly on the front part of the C-beam. So, the torsion modal of the concept model will be optimized as long as these structures are improved.

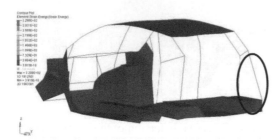

Figure 8. BIW concept model torsion modal strain energy.

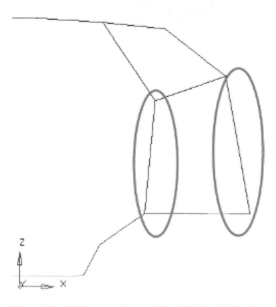

Figure 9. BIW concept model critical optimization areas

4.2 Thickness optimization of the critical beam structures in the BIW concept model

According to the results of the torsion modal strain energy analysis in Figure 8, in order to implement the optimization conveniently and also considering the structural integrity of the C-beam, the thickness of the beams that are under the triangular window of the C-beam (see Figure 9) is increased equivalently. The calculated results are listed in Table 3. The results show that the car body's torsion frequency is increased greatly after changing the plates' thickness from 0.8 mm to 1.5 mm. Then, this improvement can be applied into the optimization procedure for the detailed car body model.

Table 3. BIW concept model optimization calculated results.

	Initial structure	Optimization plan 1	Optimization plan 2
Plate thickness of critical beams in C-beam(mm)	0.8	1.2	1.5
BIW car body torsion modal frequency(Hz)	29.57	29.84	32.96

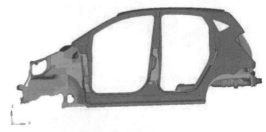

Figure 10. C-beam's outer plate sketch.

4.3 Structure optimization analysis for the BIW detailed model

The C-beam's outer plate is part of the whole car body cover (see Figure 10) with a very large size. As the connection parts between it and other body structures are very complex, adding the thickness of the C-beam's outer plate is difficult to implement in the real engineering procedure. So, the thickness of the inner plate on the side wall of the rear car and the thickness of the inner plate on the back door's circumjacent parts are added. Thus, the following two improvements for the detailed car body are proposed according to the calculated results from Table 3.

1. As the thickness of the whole outer plate cannot be added, the following optimization measure is proposed: the thickness of the plate located in the lower area of the C-beam's triangular window is changed from 0.8 mm to 1.8 mm.
2. The thickness of the inner plate located on the back door's circumjacent parts is changed from 1 mm into 1.5 mm.

After applying the implementations and calculations, the weight of the car body is increased from 1.77e2 kg to 1.88e2 kg, which is increased by 6.21%. In addition, the torsion modal frequency is increased from 30.74 to 32.65, which is increased by about 2 Hz. At the same time, the bend modal frequency is increased from 58.96 Hz to 60.00 Hz. This implement improves the frequency but controls the whole car body's weight, so the goal of optimization is obtained.

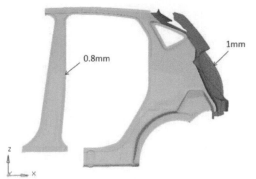

(1)Before optimization

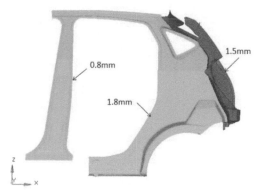

(2)After optimization

Figure 11. Optimization sketch for the detailed car model.

5 CONCLUSION

1. After calculating and analyzing the T-type joint with the circular transition connections, the results show that the computation precision can be assured when the branch length is larger than 100 mm.

2. This research applies the optimization measures into the detailed car body after summarizing the improvements for the torsion modal strain energy weak areas in the car concept body. Finally, the torsion modal frequency is increased by about 2 Hz, while the car's weight is controlled.

3. The car body concept model considers the beam elements as principal load-bearing parts. By well utilizing this characteristic, the optimization implements for the BIW NVH property in the analysis verification phase can be proposed quickly and effectively. This method also has a good engineering application value.

REFERENCES

Cai Shimin.2007.Concept Auto Body Intellectualized CAE Analysis Research of 1D Beam Section's Optimization [D]. Changchun: Jilin University.

Chen Changming.2008.Analysis of Factors Affect Joint Computational Stiffness [J].Vehicle & Power Technology, 4:45–49.

Kong Min. 2010.Cross-Section Optimization for Thin-Walled Beam of Vehicle BIW in Concept Design Phase [J].Machine Design and Research, 26 (6):120–123.

T.M. Cameron. 1997. Sensitivity of Structural Joint Stiffnesses with Respect to Beam Properties: A Hybrid Approach [J]. Computers & Structures, 63(6):1037–1041.

Tang Rensong. 2009. Simulation of BIW Stiffness in Conceptual Design Phase [D].Shanghai: Shanghai Jiao Tong University.

Xu Jianquan.2010. Effects of adding arc bracket joint on the rigidity and strength of bus L-joint structure [J]. Journal of Fujian Agriculture and Forestry University (Natural Science Edition), 39(6): 658–663.

Yasuaki Tsurumi. 2004. First—order analysis for automotive body structure design ~ part 2: Joint analysis considering nonlinear behavior [J]. SAE 01–1659.

Zhang Hongzhe.2010. Research on Intelligent CAE Key Technique in Vehicle Body Concept Design [D]. Changchun: Changchun University of science and technology.

Energy Science and Applied Technology – Fang (Ed.)
© 2016 Taylor & Francis Group, London, ISBN 978-1-138-02833-3

Influence of charge weight on ablation ability of combustion agent

Qigong Zhang, Jinming Li, Zhentao An & Xiumei Cui
Ordnance Engineering College, Shijiazhuang, China

ABSTRACT: The research destruction theory and development trends of destruction technology of metal flow were introduced. The combustion agent was optimized and the ablation device was prepared. Their impact on the efficiency of the working time and the destruction of the shell was analyzed by changing the charge weight of the combustion agent, in order to study its effect on the ablation ability of the destroying device. The results for this destroying device show that the Q235 steel plate used as the shell material can be melted by a charge weight of 60 g, the melt-through effect is better, can make a smooth release of the internal combustion product, maintain the stability of operation, and improve the efficiency of destruction.

Keywords: Destroying Device; Combustion Agent; Ablation Ability; Destruction Method

1 INTRODUCTION

Metal melt flow disposal technology is a new kind of destroyed whole ignition technology, which relies mainly on the ablation ability of the combustion agent, the melt-through metal shell, the ignition internal material, completing the internal destruction of dangerous goods. The technology used in the destruction of dangerous goods packaging protection can improve the efficiency of destroying and destroying security, to maximize the internal control of material harm to the surrounding environment.

The ablation ability of the combustion agent will affect the destruction of equipment in the area of the hole on the package, the internal material combustion product release rate, the internal pressure control package, and prevent the burning detonation phenomenon. In this paper, on the basis of optimizing the combustion agent formula, trends are analyzed, and the capability of the optimal charge weight. Metal melt flow, for the future, destroyed work provides certain reference.

2 COMBUSTION AGENT SELECTION AND OPTIMIZATION

The combustion agent is composed of thermite and a variety of modifiers. By changing the thermite and adding different modifiers, the combustion agent formula can be improved and certain features of thermite reaction can be increased, so as to improve its working efficiency. Thermite by metal combustible agent and the reaction of metal oxidant, which is a mixture of metal combustible agent, guarantees that it is an important factor for a smooth combustion reaction and very important for the selection. The metal combustible agent in unfavorable chooses raw material scarce of metals (picks, bismuth, beryllium), corrosion resistance of metals (calcium) and difficult combustion of metal (manganese). A comparison of commonly used metal combustible agents is made in Table 1. Al gives high heat quantity, high density, combustion products of low melting point, higher boiling point, which is a suitable thermite. Metal oxidant reductive into low melting point and high boiling point of the metal is the main component of the melt metal. By comparing the reaction with Al metal oxidants, as given in Table 2, Pb_3O_4 is large because of its less oxygen content and density, which is used to configure high fever agent, content of combustible agent, and lower the thermite reaction. CuO in the reaction is used to release oxygen, violent reactions. These are not easy to do the main body of thermite reagents. Iron and aluminum heat agent when combustion heat release 3.90 kJ·g − 1, and produces a high temperature melt at about 2400°C, small mechanical sensitivity, not easy to be lit, once lit to go out, can continuous combustion underwater.

In Table 2 the top three groups are CuO + Al, Fe_2O_3+ Al, Fe_3O_4 + Al experiment. The experiment found that these three groups of thermite have the following features:

1. CuO + Al reaction heat yield is the largest, but extremely fast, high pressure, fluid flow splash is serious, need to slow down.

Table 1. Performance of the metal combustibles.

Metal combustible agent	density /g·cm-3	The heating effect of the term/kJ·g-1	The boiling point/°C	Generated oxide	Generated oxide melting point/°C	The boiling point of generated oxide/°C
Al	2.7	3.90	2400	Al2O3	2050	2980
Mg	1.7	4.41	1100	MgO	~2800	~3077
Ca	1.5	3.90	1487	CaO	2572	2850
Ti	4.5	2.39	3000	TiO2	1935	~2227
Si	2.3	2.43	2400	SiO2	1713	2230
B	2.3	2.47	2550	B2O3	800–1100	–

Table 2. Performance of the metal oxides.

Metal oxidant	density /g·cm-3	Oxygen levels (×100)	Term thermal effect /kJ·g-1	The ratio (×100) oxidant	Al	Q/kJ (1kg The term heat)
Fe₂O₃	5.1	30	3.90	74.7	25.3	3887
Fe₃O₄	5.2	28	3.57	76.9	23.1	3493
CuO	6.4	20	3.82	81.6	18.4	4115
MnO₂	5.2	37	4.20	70.7	29.3	4778
Pb₃O₄	9.1	9	1.97	80.5	9.5	1912

2. Fe_2O_3 + Al reaction heat is moderate, the speed and stability.
3. Fe_3O_4 than Fe_2O_3, the relative molecular mass of the thermal effect is relatively small, the thin plate is more suitable for the effective operation.

Considering factors such as heat, injection time and sparks, we select the Fe3O4 + Al and Fe2O3 + Al mixed ratio for the body of the burners.

Considering factors such as heat, injection time and sparks, we select the Fe3O4 + Al and Fe2O3 + Al mixed ratio for the body of the burners. By changing the term, adding the different modifier, improve combustion agent formula, make certain features of thermite reaction increases, thus improve its working efficiency, and optimize the combustion agent formula, at the same time to join the small modification agent to optimize the thermal effect, so the combustion agent is mainly composed of four parts:

Thermite: it accounts for about 70% to 85% of the total weight burned to provide heat for hot melt, to generate the product of the burning rate, ash content, such as splash effect is particularly important. Fe_3O_4 + Al (50%–88%) and Fe_2O_3 + Al (50%–10%) of the thermite mixture is a key part of the burners.

Alloy agent: the agent in the process of injection increases fluid density and momentum. The addition of the alloy agent (content 3%–5%) can effectively reduce the molten alloy after the freezing point and improve the organization performance of the surface metal deposition layer.

Gasification agent: to improve the effect of the cutter for cutting, the pressure in the cutter needs to be increased, to facilitate cutting agent fully reaction and enhance the speed of the jet flow and outlet pressure. Adding suitable gasification agent (content 7%–7%) increases the volume, increases cutting play within the play and outlet pressure can effectively increase the penetration of artifacts, narrow hole angle, and effectively improve the cutting quality. We select the design selection, KNO_3, as the gasification agent.

Thinner agent: in order to adjust the temperature and speed, the reaction is easy to control, also need to add a certain proportion in the middle of the cutting agent not to participate in the combustion synthesis reaction diluent, content from 1% to 3%, can use some metal oxide as the diluent, usually Al_2O_3 is chosen in this design.

After experimental analysis, we obtain the formula ingredients for burners in the ratio of 45:3:5:1.

3 ABLATION ABILITY OF COMBUSTION AGENT

Destruction of equipment is mainly dependent on the burning agent, the heating effect of the formation of metal melt through the metal, dangerous goods. In a given combustion agent composition, the structure of the burning torch and working environment temperature conditions are assumed as followed:

1. Burners form all of the metal melt with the same hot melt and reach the same temperature;
2. Not to consider the surrounding environment of the loss of heat conduction and thermal radiation energy;
3. Melting punch shape approximation for the frustum of a cone body.

The effective utilization of the combustion η is given by

$$\eta = \frac{molten \quad slag}{100} \tag{1}$$

Having ideal effective utilization of the combustion, the combustion agent melt through a certain thickness of the steel plate can be calculated the combustion agent. The thickness of the steel plate for h, the melt whole radius R, and the volume of a steel plate Vt are as follows:

$$V_t = \pi h(R^2 + Rr + r^2)/3 \tag{2}$$

The quality of molten steel Wt is as follows:

$$W_t = V_t \rho_t \tag{3}$$

where Vt = volume of the steel plate and ρt = density of the steel plate.

The required heat of ablation ability is as follows:

$$Q_t = \frac{W_t}{M_t} C_{\rho_t} \cdot T \tag{4}$$

where Mt = molar mass of iron; T = melting point of 1535°C; C_{ρ_t} = heat capacity of iron, with $C_{\rho_t} = 14.1 + 29.7 \times 10^{-3} T$.

The burner's combustion heat release is as follows:

$$Q_{slag} = W_{slag} \cdot q \tag{5}$$

where the combustion agent unit mass combustion heat value, Fe_3O_4 + Al and Fe_2O_3 + Al thermite combustion theory of thermal effects are 3.90 kJ/g and 3.57 kJ/g, for burners, q 3.735 kJ/g.

The assumptions can be listed as follows:

$$Q_{slag} \eta = Q_t \tag{6}$$

$$W_{slag} \cdot q = \frac{W_t}{M_t} C_{\rho_t} \cdot T \tag{7}$$

By substituting Eqs (2) and (3) into Eq. (7), we obtain

$$W_{slag} = \frac{\pi h (R^2 + Rr + r^2) \rho_t}{3 M_t \cdot q \eta} C_{\rho_t} \cdot T \tag{8}$$

where η = effective utilization of the combustion. According to the formula, 100 g burners generated slag quality, which is equal to 71.25 g, and substituting into Eq. (1), we obtain η = 0.7125.

By a type (8) the burning agent the relationship between the quality and capacity of the hot melt. Composition of combustion agent and charge structure, combustion agent ability of hot melt under the control of the charge, the charge, the more vexed hot combustion agent, the more burners melt strongly; therefore, charge is an important factor of consideration to destroy equipment working hours.

4 EXPERIMENT OF ABLATION ABILITY

Destroying device by shell, combustion agent and ignition device and nozzle. Combustion agent by the composite heat agent mixed with a variety of modified agents of extrusion forming, encapsulated in the casing, which has the burners for hollow cylinder, center hole for the pilot hole, side close to the bottom of the firing device, side close to the nozzle, the nozzle composed of high temperature resistance of graphite materials repression,

as shown in Figure 1. Its working principle is that when the flash at the bottom of the device is fired, the flash portion burning, burners, burning agent release large amounts of heat to maintain combustion agent to continue burning, burning agent molten metal iron and burned gas and molten material under the impetus of the air flow from the nozzle jet. This kind of high temperature molten material (temperature range 2300°C to 3500°C) with melt through the metal, ignite flammable material properties, can melt through the metal shell.

The purpose of the experiment is under the condition of burning torch structure parameters unchanged, the destruction of control and the relationship between the thermal capacities. Experiment to destroy play with pressure molding burn medicine column, column diameter 25 mm, inner whole 6 mm. Under the condition of the above-mentioned parameters being the same, to test burners hot melt ability, the stopwatch, using 10 mm thick of Q235 steel as the target board, destruction of nozzle 10 mm steel target distance, experimental record destroyed equipment operation time, opening characteristic parameters. The results are summarized in Table 1.

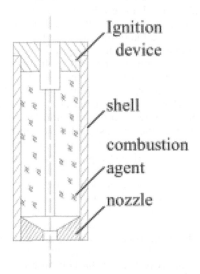

Figure 1. Sketch of the destroying device.

Table 3. Comparison the cutting aperture with different charge weights.

charge weight(g)	working time(s)	The hole diameter(mm)	Inner hole diameter(mm)	Effect of Ablation	the volume of Ablation(mm3)	The quality of molten steel(g)
40	5.26	12.52	2.27	7.25mm	361.24	2.82
50	5.64	13.32	8.91	Melt through	983.03	7.73
60	6.59	14.25	9.15	Melt through	1092.15	8.52
80	6.02	14.69	7.82	Melt through	1025.79	8.00
100	5.48	14.97	6.34	Melt through	940.39	7.34

5 ANALYSIS OF THE EXPERIMENTAL RESULTS

By the previous analysis, the main factors influencing the hot melt ability to control Wslag, control, directly or indirectly affect the pressure inside the bomb and melt flow rate, and thus affect the melt through time, in the control of Wslag = 60 g about to melt through the peak point of time, this time through the longest, the destruction. Table 2 presents the charge within 60 g, with the increase of charge, play inside the combustion agent content, burning time, directly increase the destruction equipment working hours; However, as the control continues to increase, more than 60 g after injection time gradually shortened. This is because the internal pressure increase the heat transfer coefficient, combustion speed, and destruction of equipment shortens the working hours, and found in the experiment, the shell has a loss. By opening characteristic parameters, as given in Table 2, with the increase in transfer quantity outside surface pore size is on the rise, but the inner hole diameter is in charge there was a turning point for 60 g, began to decline. Its reasons mainly include the following: first, the charge weight of 60 g increases gradually, play long gradually increase, the total quantity of heat, gas quantity, injection products splash also gradually increase, so the destruction of equipment injection time gradually lengthen, after the work piece surface of molten pool melting down time increased, at the same time due to scour through artifacts appear flow not reflected on the surface diameter increases. Artifacts, on the other hand, only 10 mm thick and less area has the characteristics of fast heating, cooling slowly. Blow force increases, molten pool residence time shortens, the work piece preheating of the lower metal, can be achieved and not the work piece edge rapidly melting, after so long, obviously increase the penetration, and the increase in the diameter of the holes on the surface is small. After the control continues to increase the burner's combustion speed, blowing force increase, injection time, acting on the work piece a shorter time, diameter of bore decline; overheat, surface burning, carbonization, a shell after 100 g of nozzle erosion is serious, and difficult to guarantee the nozzle with a close cooperation between the projectile and appearance of product with spraying nozzle, splash radius increased. Mainly because the charge is more than 60 g, play long after more than 90 mm, and the convective heat transfer coefficient along the length of the charge increase gradually played the main role, and the charge for the influence of combustion time is relatively weak. Compared with several schemes, control of 60 g, playing for 90 mm long, cutting time, and quality are the best.

6 CONCLUSIONS

Control design affects the combustion mode within the play and, in turn, influences the melt-through effect, and explosive load is too small to melt through the work, when the charge is too large, it leads to the erosion effect that occurs in the play, the melt-through effect is poor, is more likely to lead to explosion and other security-hidden danger. The value of different agents charge is not the same. The experimental analysis shows that for the purposes of this formula, charge control at around 60 g is advisable, and the melt through time, through the effects are better.

ACKNOWLEDGMENTS

ZHANG Qi-gong (1990–), Shenyang, Liaoning. Graduate student reading. Main research field for the protection and security technology.

REFERENCES

Wang Peng, Zhang Jing. 2011. Thermo dynamic analysis, composition design and experimental study on metal-cutting pyrotechnic composition [J].Chinese Journal of Energetic Materials, 19(4): 459–463.

Wang Peng, Zhang Jing., 2010. Review on pyrotechnic cutting technology [J]. Chinese Journal of Energetic Materials.18 (4): 476–480.

Wu Yiying, Wang Ruilin, Xin Wentong etal. 2014. Research on grain design of cutting ammunition [J]. Hot Working Technology, 4: 166–169.

Wu Yiying, Wang Ruilin, Xin Wentong etal. 2014. Combustion and cutting efficiency of cutting ammunition influencing by charge diameter [J]. Electric Welding Machine, 3:26–30.

Yi Jiankun, He Wuyi, Wu Tengfang etal. 2004. Preliminary investigation into application of thermite in destruction of ammunition [J]. Engineering Blasting, 10(4):21–25.

Yi Jiankun,Wu Tengfang,He Wuyi. 2005. Experimental Study on the composition of thermite applied in the ammunition disposal [J]. Blasting, 22(1): 107–111.

Energy Science and Applied Technology – Fang (Ed.)
© 2016 Taylor & Francis Group, London, ISBN 978-1-138-02833-3

Simulation analysis of vibration-impact invalidation of low voltage apparatus part

X.D. Miao & Bing Zhang
Shanghai University of Engineering Science, Shanghai, China

Y. Huang
Shanghai Second Polytechnic University, Shanghai, China

ABSTRACT: In order to solve the problem of the low voltage apparatus part invalidation mechanism under the working conditions of shock and vibration, this paper presents the simulation analysis of the low voltage apparatus part invalidation mechanism under vibration and impact based on the finite element method. According to the consistency between the simulation and testing results, the feasibility of the method for the invalidation mechanism of vibration and impact is validated. According to the simulation analysis, structure design and material is optimized to avoid failure resulting from the impact and vibration.

Keywords: vibration; impact; invalidation; simulation; low voltage apparatus part

1 INTRODUCTION

A low-voltage apparatus is a main important part of the electrical appliance industry. In the manufacturing process, it works as a switch, control, protection, detection, indication and alarm, which directly reflects a country's advanced manufacturing industry and the level of automation equipment. In recent years, with the rapid development of the national grid, the use of low-voltage electrical appliances for the safe operation of the power grid is very important, such as circuit breakers, contactors, relays and switches. The security of the power system and its development strategy must be put to a new height, to ensure the safety and stable operation of the national economy, and the power system of long-term, rapid, steady growth, in order to achieve more safety and more intelligent protection for low-voltage electrical appliances. It is particularly important for the function of the low-voltage electrical apparatus. The authority and justice to detection and the analysis of vibration for low-voltage electrical accessories not only plays a role in evaluation, control, prevention and information feedback, but also becomes indispensable basis to trade exchange at the international level. Therefore, even in the modern quality management, it is an important factor in the quality inspection and analysis for vibration and impact of low-voltage electrical appliances and accessories.

In the application of the low-voltage apparatus, the efficiency will be affected by all kinds of vibration and impact from the external environment. Due to the specificity of the shape of the low-voltage electrical apparatus, some parts will produce stress concentration, as a result of which the damage will occur. At present, the projects with the low-voltage apparatus test are mainly focused on the general examination, the drop test, the temperature rise test, the insulation resistance and withstand voltage test, the rated and decomposition ability test, the short-circuit making and breaking capacity test, the short-time withstand current test, the action characteristic test, the life test and the sensitivity test, but the overall analysis of vibration and impact is relatively less. Taking into account the integrity of the low-voltage appliance shell on its functions, important influence on the security and stability of its work, the vibration-impact analysis for the low-voltage apparatus part is very important.

2 THE METHODS OF SIMULATION

The purpose of this paper is to provide a method for the analysis of the low-voltage apparatus part under vibration-impact conditions, by using Ansys software to analyze the stress concentration. According to the simulation results, the analysis of the stress concentration part of the apparatus will

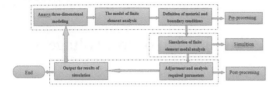

Figure 1. Ansys simulation analysis process of low-voltage electric appliance fitting vibration.

be reflected. In addition, we strengthen the design for the relatively weak parts in the structure, reduce the stress concentration by a series of structure optimization, and improve the service life and the stability of the low-voltage apparatus part.

Simulation analysis research methods of the low-voltage apparatus part mainly consists of the following parts: pre-processing, simulation and post-processing, as shown in Figure 1. In the pre-processing stage, the 3D model of Ansys is established, and then the element properties are defined, followed by the finite element mesh generation and material and boundary condition parameters. In the simulation stage, the vibration-impact load and the solving type is analyzed, and then the solver is used to solve the problem and to export the result. In the post-processing stage, the results must satisfy the required results for the output.

3 THE ANALYSIS OF SIMULATION

3.1 Three-dimension modeling of Ansys

According the design of low-voltage apparatus parts, the 3D model is established in ANSYS using from top to down. The modeling method of top-down is used in the cylinder. This modeling method needs only the defining of the most advanced graphical model. In the bottom-up modeling method, the key points are first defined, and then the line is formed, as well as the surface and the body. In the process of modeling, in order not to affect the results of the vibration and impact analysis as the premise, we simplify the relevant parts of the UG 3D model to improve the efficiency of modeling.

The 3D model of the low-voltage apparatus is created directly by the Ansys software, effectively avoiding further into the part geometry in the ANSYS pigment loss problem through the UG software modeling. In addition, the Boolean operation of its powerful tool, Ansys, can be achieved between plus or minus, classification, segmentation, lap bonding and complex operation, greatly improving the efficiency of the complicated 3D model. The UG 3D model is shown in Figure 2, and the Ansys 3D model simplification diagram is shown in Figure 3.

Figure 2. Structure of the low-voltage apparatus part.

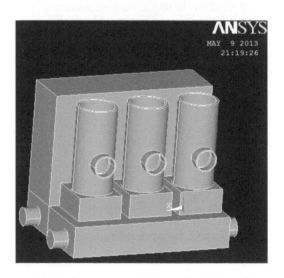

Figure 3. Simple 3D model of the low-voltage apparatus part.

3.2 Defining the element attribute

The element attribute is characteristic of the model, which must be specified before the ANSYS mesh. It includes 3 parts: the element type, material properties and real constants. The design of the accurate definition element attribute model makes this kind of approximation model to approximate the entity.

3.3 Determination of the model of the unit type

In the simple model of the low-voltage appliance part structure, the 3D model can be easily classified as a hexahedron. For determining the model of the unit type in more than 150 different ANSYS database unit types, the solid SOLID185 is chosen. To determine the cell type, and the corresponding shape function is determined. In addition, the simulation mainly focuses on the analysis of stress concentration; therefore, in order to obtain the results with high accuracy, this design uses two units, so that the change in stress concentration tends to be linear.

3.4 Defining the material property unit

Most element types need material properties. Depending on the application, the material can be linear or nonlinear, isotropic or orthotropic, constant or temperature dependent. In the research of simulation analysis of the low-voltage electric appliance fitting vibration and impact, we need to specify the material density DENS, the elastic modulus EX and Poisson's ratio of PRXY. According to the properties of different parts of the low-voltage apparatus, two units of material properties are defined in the ANSYS model, corresponding to the density of DENS in the cylindrical structure model of the material designated as 1850, the elastic modulus of EX designated as 10200, Poisson's ratio PRXY designated as 0.4, and the density of DENS in other parts of the material in the model designated as 7840, the modulus of elasticity EX designated as 2E+011, and Poisson designated as 0.3 PRXY. The different parts of the material model corresponding to the entity are defined as different characteristics, so that the material properties in the model unit is more close to the real, which lays the foundation for the accurate analysis of the finite element mesh division and stress.

3.5 Mesh and solve

The model is meshed by using the free meshing method. Then, the continuous acceleration load is exerted on the bottom of the model in three steps, which represents the vibration and impact. The model is finally solved. The stress simulation results after post-processing is shown in Figure 4, which indicates the different stress densities of the structure of the low-voltage apparatus part. According to the results, the maximum stress density is 256Mpa and 80Mpa, which is much greater

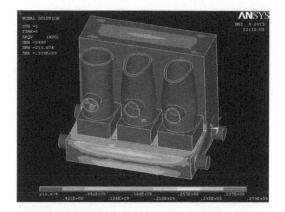

Figure 4.   Von stress results of the 3D model of the low-voltage apparatus part.

than the allowance stress density of the material Q235 and plastic, respectively. It means that the structure will be damaged.

4   RESULTS AND DISCUSSION

Based on the above analysis, the following suggestions are made: for the test case, the ultimate stress exceeded the limit of the stress concentration of the material; therefore, we recommend the use of higher-strength steel material or process under the allowed conditions, thickened guide thickness of epoxy resin for electrical insulating materials, due to a certain angle, axial force of the column results not only under external loading conditions, but also by the large bending moment and torque, therefore leading to quick denaturing and fracture.

REFERENCES

Hongwu Liu, ET al, 2010. Intelligent contactor the phenomenon of false action in the condition of vibration simulation analysis. Low voltage electrical apparatus, 11.

Jianping Chen, Zhijie Huang. 2009, Analysis the phenomenon of burning and improvement about the type of VD4 vacuum circuit breaker closing latching electromagnet. Electrical technology (10).

Energy Science and Applied Technology – Fang (Ed.)
© 2016 Taylor & Francis Group, London, ISBN 978-1-138-02833-3

Optimization of beam structure on the baler

Zhen Hong, Yuancheng He & Zhiming Meng
Luzhou Vocational and Technical College, Luzhou, China

ABSTRACT: This article describes the Pro/E MECHANICA module structure of the beam 150t mechanical automatic hydraulic baler statics analysis, combined with stress analysis, deformation analysis and manufacturing process of the technical requirements. The organization is optimized to obtain the appropriate size while meeting the requirements of the machine to work, save material, and easier manufacturing, so as to achieve the purpose of reducing costs.

Keywords: Baler, Stress Analysis, Deformation analysis, optimal design

1 INTRODUCTION

Structural optimization technology for further promoting the use of engineering design still has a practical value to solve huge optimization design of the finite element model, structural optimization, cross-cutting issues and more scientific design (Li Quanyong, 2000). Currently, there are all kinds of computer-aided design software, by combining computer-aided design engineers, a final modification is better able to solve such problems of optimal design methods. To meet the more-aided design of mechanical products, the application of Pro/E in the MECHANICA module can analyze the stress distribution and deformation condition of structural components, basically consistent with the actual situation, to optimize the mechanical structure, providing a great convenience to people (Ruan Jing, 2012).

A 150t automatic hydraulic baler, for example, withstands a certain load on the beam, and the structure is relatively large, which is more important, with the necessity of performing FEM analysis and structural optimization. The material used on the beam, including 16Mn, Q345 physical parameters and performance parameters, is substantially the same. It also belongs to the plastic material; in the case of gradually increasing stress, deformation failure will occur and must be plastically deformed.

2 THEORETICAL ANALYSIS OF THE ORIGINAL STRUCTURE

Taking into account the reliability of the factor structure, the introduction of the safety factor K, combined with the structure, suffered a cyclic alternating stress, with the work running fast enough, the frequency of the actual work not being high, and the value of K taken as 1.3. So, the members suffering considerable stress (Eq. on the left hand side is the actual maximum stress) must be less than or equal to the allowable stress $[\sigma_s]$, namely:

$$\sqrt{\frac{1}{2}[(\sigma_1 + \sigma_2)^2 + (\sigma_2 - \sigma_3)^2 + (\sigma_3 - \sigma_1)^2]} \leq [\sigma_s]$$

where $[\sigma_s] = \dfrac{\sigma}{K}$

The physical parameters of the generation materials calculated the allowable stress as $[\sigma_s] =$ 265.4 MPa.

In other words, when the crossbar after optimization by the maximum load, any one of the beams must be less than the entire stress 265.4 MPa to ensure that materials are not due to plastic deformation failure.

3 THE ORIGINAL STRUCTURE MODELING AND ANALYSIS

Pro/E has strong modeling capabilities, based on stretching, scanning, mixing, rotation feature modeling, full-related data, parametric features such as three-dimensional modeling technology advantage favored by designers of the current project (Hu Zhiliang).

The original structure has a peripheral length of 2050 mm and a width of 1350 mm, an intermediate flange sockets designed with a primary pressure cylinder, and the remaining portion with a 50 mm steel plate consisting of stiffener composition. The model is shown in Figure 1.

Figure 1.　Three-dimensional model of the beam.

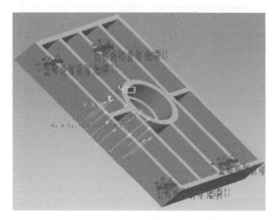

Figure 2.　Beam static structural constraint graph.

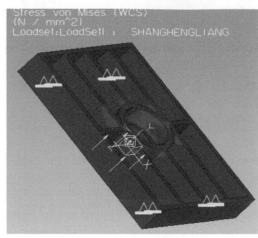

Figure 3.　Front beam for optimizing the stress analysis chart.

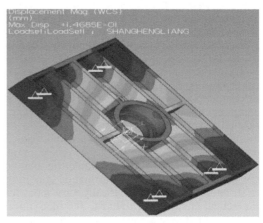

Figure 4.　Optimized rear beam deformation (displacement) analysis chart.

PRO/E in accordance with the size of a three-dimensional model is set up, with setting constraints as follows: the four corners of the hole on the beam displacement are set to 0, and we define constraints of gravity of 9.8 m/s^2, the applied load, the load stress calculated by the following formula:

$$P = F/s$$
$$S = \prod (R^2 - r^2)$$

Substituting the values, we obtain the following: the applied stress load size is 3.35×106 Pa, with its constraints and load shown in Figure 2, the overall stress analysis cloud shown in Figure 3 and deformation shown in Figure 4.

From the above-mentioned analysis results, optimized beams suffered also around the flange sockets with a maximum stress of 42 MPa, far less than the allowable stress, and strain generated in the vicinity of the most serious of the center hole, a maximum of 0.19 mm, almost negligible, the model proved to be fully optimized to meet the functional requirements. However, from the data, there are many optimization space, in which case, after the process without changing its optimized efficiency also slightly higher than the original, mainly reflected in the material cost and time-consuming workers, after this calculation only advantages in optimizing the material, comparing before and after optimization are given in Table 1.

By comparing the data on each crosspiece less than the original 0.38 tons, saving material costs of 1280 yuan, reducing the waste of resources; at the same time, in accordance with the data, this optimization is not the best, there is a part of the optimization of space, way can be used to change the structure or thinning of the plate thickness to achieve.

Table 1. Comparison of the economic benefits before and after optimization.

Property status	Volume (m^3)	Quality (t)	Required material fee (yuan)
Before optimization	0.402	3.15	18900
Optimized	0.353	2.77	16620

4 OPTIMAL DESIGN CONCLUSION

1. Three-dimensional design software Pro/E was used to structure the stress deformation analysis, combined with the actual situation of the workshop production process can be easily realized modeling, modification and other stress situations a series of simulation and optimization model design process. We can also check repeatedly until a suitable optimization.
2. After the optimal design of the model, we can also optimize the design to make some adjustments based on market prices, to re-evaluate its economic benefits, and to obtain various types of the optimal design to meet various user requirements.

ACKNOWLEDGMENTS

Fund Project: Sichuan Provincial Key Laboratory project (GK201310) About the author: Hong Zhen (1981–), male, master, Luzhou, Sichuan people, mainly engaged in the research hydraulic integration.

REFERENCES

Hu Zhiliang, Qiu Xing Kai, Xu Junliang. Based on Pro / E Mechanica mechanical structure optimization. Computer application technology, (12): 41–43.

Li Quanyong mechanical structure design optimization Retrospect and Prospect. Guilin Institute of Electronic Science Daily, 2000, (12): 114–119.

Ruan Jing can optimize the design, Heyuan Cheng, Xu Jun, 150-ton hydraulic ram baler. Modern machinery, 2012, 2.

Energy Science and Applied Technology – Fang (Ed.)
© *2016 Taylor & Francis Group, London, ISBN 978-1-138-02833-3*

Design of a long straight elbow pipe welding machine

Chaocheng Gong & Deyang Luo
College of Mechanical Engineering, Southwest Jiaotong University, Sichuan, China

ABSTRACT: This paper presents a design of an elbow girth welding machine. The design uses a kind of special gas pipeline as its main welding object. With respect to the mechanical design, it consists of a positioner, mobile trolley, adjustable bracket, welded baffles, welder mount, with the design being more innovative. In its control system, it uses the Programmable Logic Controller (PLC) as its control unit and the touch screen as its interactive interface. The numerous field welding experiments show that the elbow girth welding machine runs smoothly with a high degree of automation and a good welding effect. It meets the requirements of production quality and efficiency, and is an ideal professional equipment for pipeline welding.

Keywords: elbow girth welding machine, gas pipeline, PLC, touch screen

1 INTRODUCTION

With the continuous development of the pipeline industry and large-scale application of pipelines, the demand for pipe welding is increasing. Pressure pipes, commonly used as transmission media of oil, gas and other high pressure gas, play an important role in the national economy. Considering its safety and reliability, the welding requirements must be guaranteed. At present, various generic pipeline welding equipment has emerged at home and abroad; however, for the welding of some special pipes, their main welding method is manual welding in many cases and their welding quality and efficiency cannot be guaranteed. So, the development of an appropriate special welding machine has become an important issue.

This paper studies and designs a kind of elbow girth welding machine, using submerged arc welding as its welding mode to realize the welding of the long straight pipe and the elbow. The main advantage of the welding is that it is reliable to clamp the long work piece, and its holding structure and mechanical transmission have adopted a relatively new design idea. In its control system, it uses the Programmable Logic Controller (PLC) as its control unit and the touch screen as its interactive interface. It runs with a high degree of automation, and the welding efficiency and quality are guaranteed.

2 THE WELDING REQUIREMENTS OF THE WORKPIECE

This paper aims to achieve the welding of large quantities of long straight pipes. The welding

Figure 1. Long straight elbow pipe.

Table 1. Work piece size table.

Pipe diameter（mm）		Pipe wall thickness	Pipe length	Welding form
I Series	II Sreies	（mm）	（mm）	
Φ51	Φ48	≤δ9	≤9000	SAW
Φ60	Φ57	≤δ11	≤9000	SAW
Φ73	Φ76	≤δ13	≤13000	SAW
Φ89	Φ88.9	≤δ13	≤16000	SAW
Φ114	Φ108	≤δ16	≤16000	SAW
Φ140	Φ133	≤δ20	≤12000	SAW
Φ168	Φ159	≤δ28	≤9000	SAW
Φ219.1	Φ219	≤δ36	≤2000	SAW

elbow's maximum diameter and minimum diameter are 219 mm and 48 mm, respectively, and its minimum length of the work piece is 1500 mm. The long straight pipe elbow schematic is shown in Figure 1, and the specific diameter and length are given in Table 1. The welding requirements of the work piece is to weld the long straight pipe and the elbow section. The work pieces are heavier and longer, and the welding pass rate is over 98%. In order to meet the requirements of the passing rate, we use a submerged arc as its welding method, and use multi-lane welding to fill the seam. However, for this type of work piece how to clamp and weld becomes a matter of concern.

3 DESIGN AND WORKING PRINCIPLE OF THE WELDER

In the mechanical structure, the welding machine is mainly composed of a positioner, mobile trolley, adjustable stand, welded baffles and welder mount. In addition, it is equipped with a flux recovery machine and a torch cooling device for automatic recycling of the flux and cooling of the torch. The overall program of the equipment is shown in Figure 2.

3.1 Positioner

In order to achieve the displacement of the work piece, a lot of welding equipment have a variable-bit machine structure, while the structures of the variable-bit machine are different. In this design, the positioner uses the motor reducer as its power source to drive the timing belt gear system, and then achieves the rotation of the work piece. The speed and angle of the work piece rotation adopt the frequency control and the encoder control. The variable-bit machine includes the work piece clamping device, the rotating structure and the support structure. The design of the work piece clamping device uses two manual three-arrested chucks to achieve clamping. The three-arrested chuck uses the screw with opposite rotating to drive, so when the screw is rotating, the chuck can open and close at the same time to enable the work piece to clamp and relax, and the two chucks separated by 500 mm can ensure the concentricity of the work piece and provide the reliable holding force. The rotating body and the large gears are installed to conduct gear meshing transmission, and the installation of the rotating body and the large gear must ensure the concentricity: six limit wheels on the circumference can limit. The friction between the limit wheel and the barrel rolls, so that the limit wheels can rotate. The support structure is used to support the entire rotating part of the positioner. The positioner structure diagram is shown in Figure 3.

3.2 Adjustable bracket and welded baffles seat

The drive way of the adjustable bracket is consistent with the principle of the clamping device of the variable-bit machine, which uses the lead screw with the opposite thread. The difference is that the contact place between the upper part of the adjustable bracket and the work piece is the roller. The adjustable bracket is mainly used for positioning before the work piece is clamped. When the work piece is placed on the adjustable bracket, we use the manual hand wheel to adjust the position of the work piece in the vertical direction. When adjusted to the proper position, the clamping device of the positioner begins to clamp the work piece.

The welded baffles seat is critical during the welding of the work piece because the surface of the tubular work piece is not easy to stack the flux, especially the pipes of small diameter. The design must meet the requirement that there is no occurrence of interference during one rotation, enabling to stack the flux. For work pieces of different diameters, the flux baffle can be adjusted up and down to accommodate different diameter sizes of work pieces in order to ensure the welding quality of the submerged arc. It should be noted that the welded baffles are not simply made in ordinary steel. They must be heat protected; otherwise, they will melt because of the high temperature during welding. For protection, the design uses the asbestos material with a better insulation effect on the baffle surface.

3.3 The welder mount

The welder mount mainly comprises a base plate, bracket 1 and bracket 2. The bottom plate is used to mount welder bracket 1, the adjustable bracket, the positioner, the flux receiving box and other structures. Bracket 1 is welded by a rectangular square tube, with cylinder 1 and the linear guide. Similarly, bracket 2 is also made of a hollow rectangular square tube with the wire feeders, the welding funnel, the cross board, the torch and other structures. The welding torch can move forward and backward in a wide range through cylinder 1 and the linear guide. The welding torch is mounted on the cross board, achieving fine-tuning

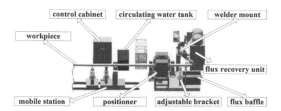

Figure 2. Overall program of the equipment.

Figure 3. Positioner structure.

by the cross sub-carriage. The opening and closing of the welding flux in the flux funnel is controlled by cylinder 2. During welding, the torch is launched by the cylinder, and adjusted to the suitable location above the work piece. After welding, the cylinder will return, and the flux receiving box will receive the welding flux. While the recycling inlet of the flux recovery unit is connected to the flux receiving box, the discharge port is connected to the flux funnel, thus the flux can be recycled, and the degree of automation is improved.

3.4 Mobile trolley

The mobile trolley is divided into the power trolley and the non-power trolley, and both of them will realize the long-distance transport of the work piece. The power trolley uses the motor as its power source, and its movement is driven by the sprocket chain and its speed is adjusted by the frequency converter. For the long work pieces, the other end of the work piece needs to be supported during welding. Thus, the trolley is mounted on a support structure, and its basic principle is consistent with the adjustable bracket. In order to prevent the slipping of the work piece during conveying, the trolley is fitted with a pneumatic clamping mechanism for clamping the work pieces. The difference between the clamping device of the positioner and the work piece is that its clamping mechanism involves the use of the cylinder rather than manual clamping to provide a clamping force. The clamping mechanism consists of two biaxial cylinders installed backwards, so the inward movement of the chuck can achieve clamping; simultaneously, the outward movement can realize the release. The whole clamping mechanism is mounted on the trolley base, moving with the trolley. The reason why

we use the cylinder as its clamping force is that the manual clamping may not be sufficient to provide a greater force to clamp the work piece, which will lead to loosening or even shedding during the movement of the work piece on the tracks. Thus, in order to ensure safety and reliability, the pneumatic clamping way is adopted. The structure of the non-powered trolley is relatively simple compared with the power trolley, and it is just the connections between the support structure and the base, mainly to support the work piece. Figure 6 shows a power trolley.

4 CONTROL SYSTEM DESIGN

The electrical control system of the long straight elbow pipe welding machine consists of a programmable controller PLC (Siemens S7–200), a human-machine interface touch screen SMART-700 IE, with PLC as a system control unit and the operation panel as a command initiate unit, and the control panel includes a touch screen and all switch buttons. When the system starts, PLC will conduct logic processing according to the signal received from the outside according to the program that have been set, and then it outputs the command to control the operation of the cylinder, the motor rotation, the recovery operation and the welding oscillator action. The device operates through the off button on the SMART-700IE touch screen and

Figure 6.   Power trolley.

Figure 4.   Adjustable bracket and flux baffles.

Figure 5.   Welder mount.

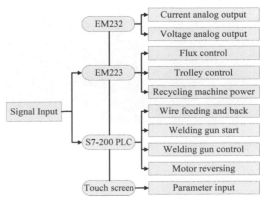

Figure 7.   PLC control principle.

each switch button on the panel. The touch screen communicates with the PLC through the PPI cable. The parameters of different diameters of the work piece, the torch oscillation and the welding speed can be set via the touch screen. This will not only make up for the defect that the PLC human-machine dialogue is more single, but also simplify the PLC control program. The PLC electric control block diagram is shown in Figure 7.

5   SPOT WELDING TEST

Spot welding test is done for the pipe work piece whose outer diameter is 90 mm. The welding operation sequences are as follows: we lift the work piece to the machine, clamp the work piece with pneumatic head feed, manually adjust the torch to align the welding, open the flux switch to release the flux, and the welding power ignite the arc automatically after pressing the start button; at this time, the rotary motor will whirl, and start welding swing by setting the process according to the welding procedures. The welding power will automatically extinguish when it reaches to the setting welding angle. The rotary motor will stop rotating, and we close the flux switch and return head pneumatically. We clear the slag, feed head pneumatically again and adjust the height of the torch. We then open the flux switch and press the start button to the next welding, until all welding is accomplished. Then, we return the head pneumatically and remove the work piece. The field test welding shows that the welding is good and the quality is high. Figure 8 depicts the welding effect.

6   CONCLUSIONS

The numerous field welding experiments show that the elbow girth welding machine runs smoothly with a high degree of automation and a good welding effect. It meets the requirements of production quality and efficiency and is an ideal professional equipment for pipeline welding. It has the following advantages:

1. It has a wider applicability, and is effective to clamp and weld the pipes with a length of more than 1.5 m or in, and the diameter is less than 220 mm.
2. The design is relatively new, and can be used as a reference for the design of other pipeline welding equipment.
3. It used the programmable logic controller (PLC) as its control unit and the touch screen as its interactive interface with a more convenient operation and a high degree of automation.

REFERENCES

Chen Yichuan, 2006, the special welding design theory of modern automation, Modern welding, No. 48, p. 6–9.
Guang Yang, 2004, Status and Prospects of welding automation technology. Maschinen Markt, No. 11: p. 36–37.
Luo Wen, Huan Xi, 2008, Electrical control and PLC technology.

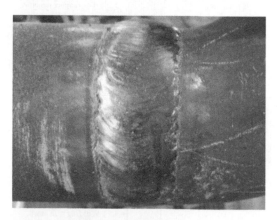

Figure 8.   Welding effect diagram.

Control and automation

Energy Science and Applied Technology – Fang (Ed.)
© 2016 Taylor & Francis Group, London, ISBN 978-1-138-02833-3

Cooperative control for multiple coordinated mobile robots

N. Zhao, J.L. Liu & J. Yuan
Institute of Oceanographic Instrumentation of Shandong Academy of Sciences, Qingdao, China
Shandong Provincial Key Laboratory of Ocean Environment Monitoring Technology, National Engineering and Technological Research Center of Marine Monitoring Equipment, Qingdao, China

ABSTRACT: The distributed Lyapunov control method is investigated for multiple coordinated mobile robots. A back stepping control method based on Lyapunov is proposed for coordinated multiple mobile robots. The motion trajectories are predefined and programmed into the controller hardware's of each robot. The distributed cooperative control law is designed for each robot to carry out the trajectory-tracking of predefined trajectory using the Lyapunov stability method. Finally, numerical simulations show the effectiveness of the proposed cooperative control method.

Keywords: Multiple mobile robots, cooperative control, back stepping

1 INTRODUCTION

Cooperative control of multiple mobile robots is a typical problem for multi-robot coordination and cooperation. It can complete the complex task, significantly improving the efficiency of the completed task. The virtual structure and following-leader approach is mainly used in the current research. Leonard, 2001 used a distributed virtual structure formation framework to achieve a cooperative control of mobile robots. Ren, 2003studied the cooperative control problem under a limited overland communication, (Do, 2008) designed and distributed controller estimation for non-holonomic robots. Jiang, 1995studied the cooperative control problem of non-holonomic robots. However, the virtual structure requires the position and attitude information of the corresponding virtual points on the rigid body as their target tracking points. There is a strong coupling between the robots, which cannot achieve a flexible formation transformation. In addition, the following-leader approach can control robots with a certain angle and tracking distance, which can achieve formation deformation with good flexibility, but requires the real-time communication between the robots. So, the sensor performance of the angle and distance measured is high, and the communication traffic increases rapidly as the distance increases. In this paper, we present a fully distributed multi-robot cooperative control method. Depending on the work tasks, we predefine the trajectories of robots and pre-program the trajectories into each robot controller. Then, we design control laws for each robot to achieve the predefined trajectory tracking. In this method, there is no communication coupling between the robots, which simplifies the communication links between the robots. Finally, the simulation results show the effectiveness of the control method.

2 PROBLEM DESCRIPTION

In order to realize the coordinated operation of multiple robots, the desired motion trajectories of each robot as well as the desired velocities should be defined first as the desired trajectory and the desired speed. The control law of each robot is designed, respectively, to track the predefined tracking trajectory, which can achieve the desired motion of each robot. Figure 1 shows the position and attitude error.

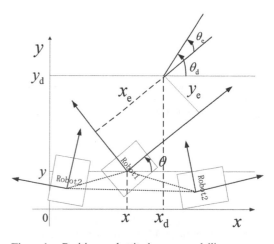

Figure 1.   Position and attitude error modelling.

First, the inertial coordinate framework and the robot-fixed coordinate framework are established, respectively. We can control the mobile robots with different speeds by two rear wheels to control the robot speed and heading, and the robot's state is represented by the midpoint of the two driving wheel positions and heading in the inertial coordinate's angle.

Let $p = (x, y, \theta)^T$, $q = (v, \omega)^T$, where (x, y) is the coordinates in the inertial coordinate framework of the robot; θ is the heading of the robot, which is the angle between the forward direction and the positive direction of the x axis; and v, ω is the displacement speed and the angular speed, respectively, and they are the control input in the kinematic model of the robot. So, the kinematic equation is given by

$$\dot{p} = \begin{bmatrix} \dot{x} \\ \dot{y} \\ \dot{\theta} \end{bmatrix} = \begin{bmatrix} \cos\theta & 0 \\ \sin\theta & 0 \\ 0 & 1 \end{bmatrix} q \quad (1)$$

Then, the position and attitude error is given by

$$\dot{p}_e = \begin{bmatrix} \dot{x}_e \\ \dot{y}_e \\ \dot{\theta}_e \end{bmatrix} = \begin{bmatrix} \cos\theta & \sin\theta & 0 \\ -\sin\theta & \cos\theta & 0 \\ 0 & 0 & 1 \end{bmatrix} \begin{bmatrix} x_d - x \\ y_d - y \\ \theta_d - \theta \end{bmatrix} \quad (2)$$

Furthermore, we obtain

$$\dot{x}_e = \omega y_e - v + v_d \cos\theta_e$$
$$\dot{y}_e = -\omega x_e + v_d \sin\theta_e \quad (3)$$
$$\dot{\theta}_e = \omega_d - \omega$$

The trajectory tracking problem of the mobile robot with the kinematic model controls the input

$$q = (v, \omega)^T$$

where $v = v(x_e, y_e, \theta_e, v_d, \omega_d, \dot{v}_d, \dot{\omega}_d)$ and $\omega = \omega(x_e, y_e, \theta_e, v_d, \omega_d, \dot{v}_d, \dot{\omega}_d)$, makes the closed-loop trajectory of Eq. (3) and control input v, ω uniformly bounded and $\lim_{t\to\infty} \|(x_e, y_e, \theta_e)^T\| = 0$ for any initial error of $(x_e(0), y_e(0), \theta_e(0))$.

3 BACK STEPPING CONTROL LAW DESIGN

Assuming that v and ω are uniformly continuous and bounded, we can design the control law for v and ω, which makes Eq. (3) uniformly bounded. If v and ω do not converge to zero at the same time,

then the closed-loop system (3) still converges to zero, that is $\lim_{t\to\infty} \|(x_e, y_e, \theta_e)^T\| = 0$, where

$$v = c_x x_e + v_d \cos\theta_e, \quad c_x > 0,$$
$$\omega = \omega_d + y_e v_d \int_0^1 \cos(s\theta_e)\,ds + c_\theta\theta_e, \quad c_\theta > 0$$

In which c_x and c_θ are adjustable parameters for the designed controller.

Proof, Firstly, let $x_e = c_3 \omega y_e$, $\theta_e = 0$, then the subsystem of equation (3) about y_e is stabilized.

So we introduce a new variable $\bar{x}_e = x_e - c_1 \omega y_e$ where $c_1 \in R_+$.

We obtain the derivation of $\bar{x}_e$ along the time that is

$$\dot{\bar{x}}_e = \dot{x}_e - c_3\dot{\omega}y_e - c_3\omega\dot{y}_e$$
$$= \omega y_e - v + v_d\cos\theta_e - c_1\dot{\omega}y_e -$$
$$c_1\omega(-\omega x_e + v_d\sin\theta_e)$$

We choose a Lyapunov function, which is given by

$$\dot{V}(t, x_e, y_e, \theta_e) = -c_1\omega^2 y_e^2 + \bar{x}_e(-y_e\omega + \alpha - v) + \frac{1}{\gamma}\theta_e[\gamma y_e v_d \int_0^1 \cos(s\theta_e)\,ds + \omega_d - \omega]$$

where $\gamma > 0$.

We obtain the derivation of V along the time, that is

$$\dot{V}(t, x_e, y_e, \theta_e) = -c_1\omega^2 y_e^2 + \bar{x}_e(-y_e\omega + \alpha - v) + \frac{1}{\gamma}\theta_e[\gamma y_e v_d \int_0^1 \cos(s\theta_e)\,ds + \omega_d - \omega]$$

where $\alpha = \omega y_e + v_d\cos\theta_e - c_1\dot{\omega}y_e - c_1\omega(-\omega x_e + v_d\sin\theta_e)$.

We choose the tracking control:

Law $v = \alpha - y_e\omega + c_2\bar{x}_e$ and

$$\omega = \omega_d + \gamma y_e v_d \int_0^1 \cos(s\theta_e)\,ds + c_3\gamma\theta_e$$

where $c_2, c_3 > 0$, which makes

$$\dot{V}(t, x_e, y_e, \theta_e) = -c_1\omega^2 y_e^2 - c_2\bar{x}_e^2 - c_3\theta_e^2 < 0.$$

4 NUMERICAL SIMULATIONS

According to the operating task characteristics of the coordinated multiple mobile robots, we choose the horizontal formation trajectories as the desired formation trajectories, and carry out the formation simulations using three mobile robots. The initial positions of the three robots are, respectively,

$[x_{10}, y_{10}, \theta_{10}]^{\mathrm{T}} = [0.4\mathrm{m}, 0, \frac{\pi}{2}\,\mathrm{rad})]^{\mathrm{T}},$

$[x_{20}, y_{20}, \theta_{20}]^{\mathrm{T}} = [0.6\mathrm{m}, 0, \frac{\pi}{2}\,\mathrm{rad}]^{\mathrm{T}},$

$[x_{30}, y_{30}, \theta_{30}]^{\mathrm{T}} = [0.8\mathrm{m}, 0, \frac{\pi}{2}\,\mathrm{rad}]^{\mathrm{T}}.$

The initial positions of the predefined trajectory-tracking of Robot1 is $[x_{10}^{d}, x_{10}^{d}, \theta_{10}^{d}]^{\mathrm{T}} = [1\mathrm{m}, 0.4\mathrm{m}, \frac{\pi}{4}\,\mathrm{rad}]^{\mathrm{T}}$. The initial positions of the predefined trajectory-tracking of Robot2 is $[x_{20}^{d}, x_{20}^{d}, \theta_{20}^{d}]^{\mathrm{T}} = [1\mathrm{m}, 0.6\mathrm{m}, \frac{\pi}{4}\,\mathrm{rad}]^{\mathrm{T}}$. The initial positions of the predefined trajectory-tracking of Robot3 is $[x_{30}^{d}, x_{30}^{d}, \theta_{30}^{d}]^{\mathrm{T}} = [1\mathrm{m}, 0.8\mathrm{m}, \frac{\pi}{4}\,\mathrm{rad}]^{\mathrm{T}}$, the distance being $0.2m$ among the three robots. We choose $\omega_{\mathrm{d}} = 0$, $v_{\mathrm{d}} = 0.2\mathrm{m/s}$, $c_x = 3, c_\theta = 10$ and the sample time as $\tau = 8\mathrm{ms}$. We carry out three different simulations according to the different defined trajectories, the angle between the trajectory-tracking and the positive direction of the x axis is, respectively, $\frac{\pi}{4}\,rad, 0\,rad, -\frac{\pi}{4}\,\mathrm{rad}$, as shown in Figure 2. In this figure, we mark the positions of the three robots at the time periods 200s, 500s and 700s, showing that the formation is carried out using the proposed control method.

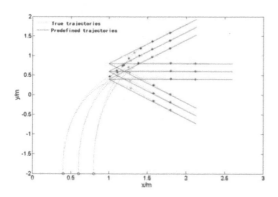

Figure 2. Parallel trajectory-tracking of multiple coordinated robots.

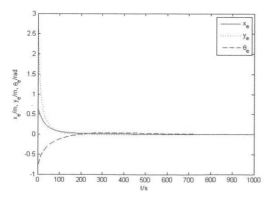

Figure 3. Position and attitude error of the robots.

5 CONCLUSIONS

The cooperative control of multiple mobile robots is investigated. We design the back stepping control law through selecting the Lyapunov function in order to implement a parallel formation of multiple robots. The simulation results show that by using the control method, the multiple robots can form the desired formation in the free obstacle case and achieve the desired trajectories.

REFERENCES

Do, K.D. 2008. Formation tracking control of unicycle-type mobile robots with limited sensing ranges. IEEE Transactions on Control Systems Technology, 16(3): 527 538.

Jiang, Z. P. Pomet J. B. 1995. Backstopping-based adaptive controller for uncertain no holonomic systems. In Proceeding of 34th IEEE Conference on Decision and Control, New Orleans, LA, 1573–1578

Lawton, J. Beard, R.W. Young, B. J. 2003. A decentralized approach to formation maneuvers. IEEE Trans on Robotics and Automation, 19(6):933–941

Leonard, N.E. Fiorelli, E. 2001. Virtual leader, artificial potentials and coordinated control of groups. IEEE Conference on Decision and Control, 2968–2973

Ren, W. Beard, R. W.A.2003. Decentralized scheme for spacecraft formation flying via the virtual structure approach. AIAA Journal of Guidance, Control and Dynamics

Wang, P.C. 1991. Navigation strategies for multiple autonomous mobile robots moving in formation. Journal of Robotic Systems, 8(2):177–195.

Energy Science and Applied Technology – Fang (Ed.)
© 2016 Taylor & Francis Group, London, ISBN 978-1-138-02833-3

Design of smart home wireless terminal control system based on ARM and ZigBee

F.L. Zhang & H. He
Tianjin University of Technology, Tianjin, China

Z.H. Zhang
Transmission and Transmitter Department, Tianjin Radio and TV Station, Tianjin, China

ABSTRACT: Based on ARM and ZigBee technologies, this paper aims to build a smart home wireless control system, outside of the system remote control of the home, via the Internet of thing and internal of that set up home wireless LAN via ZigBee wireless communication technology, to realize the intelligent management of information appliances and security appliances. The second part is the design and development of software. On the basis of the hardware platform, this paper completed the design of a home wireless terminal node and a coordinator node.

Keywords: ARM; ZigBee; smart home; LAN; coordinator

1 INTRODUCTION

With the continuous development of computer network and modern communication technology, smart home has attracted increasing attention. Smart home is a high-tech product to bring people into the information age, and it is the state's high quality of life that people pursue an inevitable requirement. Smart home will not only be an indispensable necessity of life in the future, but also has a broader market prospect, and the smart home will vigorously promote the development of the Internet of Things.

This design is based on the ARM and ZigBee smart home technology wireless terminal control system, which can remote control home appliances. For the development of the Internet of Things, this design will provide an opportunity. The results of final commissioning of the system can meet the basic needs of the users. The reasonable design can achieve the desired objectives.

2 DESIGNING AN OVERALL PROGRAM

The entire system design combined with the entire functionality of the system is shown in Figure 1.

The whole system design is completed by taking various factors into account. It not only needs to achieve the function of the system, but also to improve the scalability and reduce the costs. In this design, the household control terminal adopts

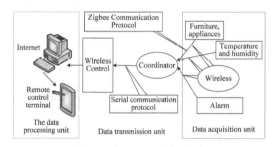

Figure 1.   Schematic design of the whole system.

embedded processors, and the wireless Local Area Network (LAN) in the family adopts ZigBee technology as the wireless networking technology. Meanwhile, embedded devices not only has low power consumption, small size, low cost advantages, but also has good stability and practicality, which can be easily connected to the Internet and be independent of the computers. ZigBee technology also has the low cost, high capacity network self-organization and more unified standard for each node equipment. The design of the whole system meets the demand analysis and functional requirements.

In the smart home control system, according to the various modules of the system, the entire program is designed. Overall, the entire program can be divided into three units: the data acquisition unit, the data transmission unit and the data processing unit.

2.1 The data acquisition unit

This section uses the latest wireless communication technology called ZigBee, and each wireless node simulates furniture and electrical appliances within the family. The whole home wireless network is a wireless local area network by a number of Zig-Bee nodes. In other words, users remote control the terminal sending control information which is sent to the coordinator node in the wireless local area network after processing control information, and the coordinator node forwarded control information to the appropriate wireless nodes based on address information to control and monitor home appliances.

2.2 The data transmission unit

The entire data transfer part is the core of the whole system. On the one hand, it connects to the coordinator in the wireless Local Area Network (LAN) through the serial port (UART), and receives transmitting control information by the coordinator. On the other hand, it connects to the Internet through a network port, so that the user remote controls home within the family via the Internet and wireless control terminal.

2.3 The data processing unit

The data processing unit is a unit that the user controls the family environment through a variety of intelligent terminal sending control commands, such as smart phones and PDAs.

3 SYSTEM HARDWARE DESIGN

3.1 The main control board circuit design

In this design, the main control panel comprises SAMSUNG's 16/32 S3C2440 A microprocessor, which uses the ARM920T core, clocked at 400 MHz. Its hardware structure block diagram is shown in Figure 2.

In this design, UART0 connects to the ZigBee coordinator module, and UART1 connects to GPRS modules, and the external internet network is introduced through the DM9000 card. According to the various external data received, S3C2440 processor carries out the analysis and processing work, and the processing result is sent to the corresponding sensor device node or the user via the communication interface.

3.2 Wireless module circuit design

In the system, the wireless module is mainly responsible for the establishment of the wireless

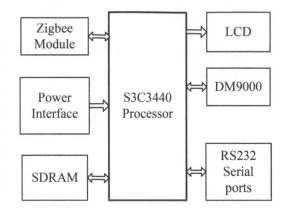

Figure 2.   Control board hardware block diagram.

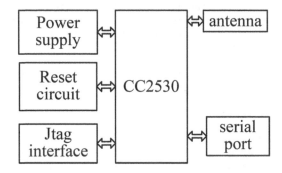

Figure 3.   CC2530 application circuit block diagram.

network within the family, transmitting users' sending control commands and uploading the collected data of node devices. Currently, in the market, more extensive use of the wireless module is the CC2530 chip produced by TI that integrates an enhanced 8051 processor and RF transceiver chip, complied with the IEEE802.15.4/ZigBee Union standard, IEEE 802.15.4, ZigBee2007/Pro/ RF4CD and other protocol standards. Meanwhile, CC2530 can work in multiple modes, in line with the system's low-power requirements. The CC2530 application circuit block diagram is shown in Figure 3.

4 THE WIRELESS NODE SOFTWARE DEVELOPMENT

4.1 The wireless coordinator node software design

The wireless coordinator node is the core of the ZigBee wireless communication network. There is one network coordinator in the ZigBee network, and its main function is to establish the network, set parameters, management and maintenance information. The wireless coordinator software design flow chart is shown in Figure 4.

After the completion of the system power up and initialization of coordination hardware and z-stack, the coordinator will detect channel energy and scan the channel, and then choose a large idle energy channel as the strongest network channel, thereby establishing a network and generating network number PANID and configuring network parameters. After the successful establishment of the wireless network, the coordinator enters the listening state. If it gets the network application of the child node, the coordinator will allow the addition and assign the network address, establishing a binding document.

4.2 *The terminal node software design*

The main task of the wireless terminal node is to collect information that controls and queries information, and upload sensor data sent by the user. The wireless terminal node software design flow chart is shown in Figure 5.

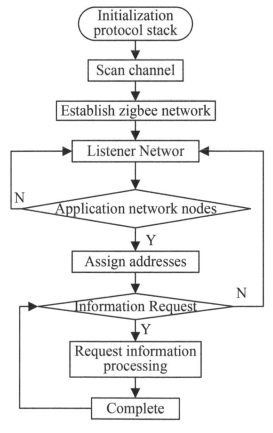

Figure 4. Wireless coordinator software design flow.

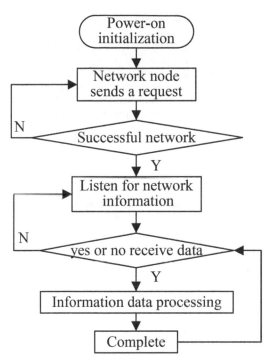

Figure 5. Wireless terminal node software design flow chart.

The next step after module power-up and initialization is the terminal node channel scanning to search the channel that can be added to the wireless network, and then it sends a network request to the coordinator node, and then it checks whether it connects to the network or not. If the network is successful, the coordinator node will assign a network address; if it is not successful, it continues to send network request information. After gaining success to join the wireless network, the terminal node will listen for network information, query whether to receive the data or not, if it queries data successfully, it further determines the data type, when it receives the user query information, it obtains the corresponding sensor node data, and sends the data to the coordinator node through the wireless network. When it receives the user control command, the corresponding terminal called sensor nodes perform the control operations.

5 CONCLUSIONS

The design of the smart home system uses the ARM and ZigBee technology to build a wireless network within the family. The hardware platform is based on S3C2440 as the central part and the transplant embedded Linux operating system to complete the design of intelligent gateway to achieve a real-time monitoring and remote control of household equipment status through the Internet. With a safe and fast transmission, low equipment cost, simple and flexible networking, the system is stable, functional and has low power consumption, thus reaching intelligent management of home devices. The system has broad application prospects in real-world applications.

REFERENCES

Guo, W.T. & He, Y. G. 2011. Research and design of smart home remote monitoring system, Computer Measurement and Control, (09): 2109–2112.

He, H. &Zheng, Y. &Li, J.W. 2013. Design of Remote Vibration Monitoring System Based on Wireless Technology, Applied Mechanics and Materials, 380:884–888.

Liu, L.J. & Zhang, G. M. 2011. Design of smart home management system based on ZigBee wireless technology, Computer Technology and Development, 21(12):250–253.

Li, J.B. & Hu, Y.Z. 2011.Design of ZigBee network based on CC2530, Electronic Design Engineering, 19(16):108–111.

Wu, W.Z. & Li, W. L. 2011. The smart home system based on ARM and Zigbee, Computer engineering and design, 3(5):30–34.

Zeng, S.W & Zhang, Y. & Qiu, W. Q. 2011. Smart home control system design based on the Internet of Things, Modern Electronics Technique, (09): 168–171.

Energy Science and Applied Technology – Fang (Ed.)
© 2016 Taylor & Francis Group, London, ISBN 978-1-138-02833-3

Lubricating oil automatic filter based on high gradient magnetic field

Fengren Wang, Yangwei Li & Qin Guo
School of Logistics Engineering, Wuhan University of Technology, Wuhan, China

ABSTRACT: By optimizing the structure of an oil filter, this paper develops a lubricating oil filter that can adsorb ferromagnetic impurities in oil based on a high gradient magnetic field. Meanwhile, it can clean the sewage automatically.

Keywords: Automatic lubricating oil filter; High gradient magnetic field

1 INTRODUCTION

Inevitably, in the process of mechanical operation, friction and wear are caused by relative motion, leading to increased energy consumption, reduced economic benefits, reduced machine precision, poor performance or even scrapping. The lubricating oil has lubrication, cooling, anti-rusting, cleaning, sealing and buffering functions, and its quality is directly related to the mechanical work quality and efficiency. However, lubricating oil easily suffers from mechanical friction wear particle pollution, reducing its quality.

Existing lubricating oil filters mostly rely on a single paper filter hole to adsorb harmful impurities in oil, the smaller the pores of the filter paper is, the better the filtering effect will be. However, the oil's passing ability is bad. Second, those greater than or less than the paper filter pore impurities do not have a permanent filter effect, but only those that are equivalent to paper filter microporous and embedded into the paper filter pore impurities have a permanent filter effect, and the filtration efficiency is extremely low. In order to ensure the safe use of machine parts inside, higher equipment is needed to implement the filtering effect.

In view of the above shortcomings, we designed a kind of magnetic filter by using a high gradient magnetic field. By the reasonable design, it can effectively clean and automatically exclude ferromagnetic impurities that exist in the lubricating oil.

2 HIGH GRADIENT MAGNETIC FIELD

2.1 *The selection of magnetic field*

This filter uses the electromagnetic field as a magnetic field, although it is convenient to make permanent magnet, but electromagnetic field can get satisfying magnetic field strength by changing the current, at the same time, the magnetic impurity can be cleaned easily, making the work more efficient, therefore the project selected the electromagnetic field.

2.2 *Arrangement of high gradient magnetic field*

The oil filter device is arranged in a poly magnetic porous medium in the magnetic field to increase the gradient of the magnetic field. In the magnetic field, the magnetic particles of attraction are proportional to the external magnetic field intensity H as well as its gradient gradH. So, on the basis of the unchanged magnetic field intensity, we can strengthen the gradient to increase the adsorption capacity to improve its adsorption effect.

3 POSITION ARRANGEMENT OF INLET AND OUTLET OIL HOLES

Different oil inlet layouts have a large effect on the adsorption of ferromagnetic materials. The general layouts of the form are "upper into lower out type", "lower into left out type" and "same side into and out type".

The following are the trajectory of iron powder particles of three types.

Comparing the above iron powder particle trajectories of oil hole's position arrangement, it can be found that the iron powder movement route whose oil hole's position is "same side into and out type" is the longest, which is the most conducive to the adsorption of ferromagnetic materials and achieves a better adsorption effect.

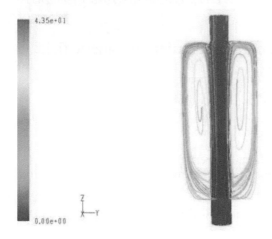

Figure 1.   Upper into lower out type.

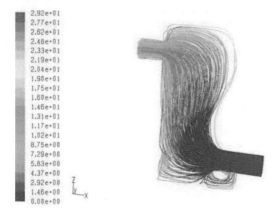

Figure 2.   Lower into left out type.

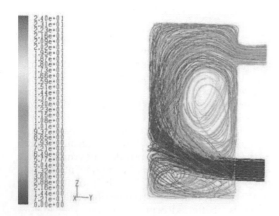

Figure 3.   Same side into and out type.

4   STRUCTURE

4.1   *The structure of components*

This filter is mainly composed of 1-water inlet, 2-discharge fan, 3-electric magnetic core, 4-baffle, 5-oil inlet, 6-oil outlet, 7-discharge and extra SCM.

4.2   *The main structural features*

1. In the case of electricity, the electric magnetic core can produce a larger magnetic field. It has the advantage of adsorbing ferromagnetic materials in lubricating oil.
2. Dampers can move up and down along the electric magnetic core and separated dampers space up and down. Its movement can remove their adsorbed ferromagnetic materials on the electric magnetic core. With the rotation of the removal fan and the inlet water flushing, it can achieve the function of removing iron filings automatically.
3. All components are controlled by the SCM, which can be more effective in realizing a collaborative work, better adsorption and achieving chip removal operation.

5   CALCULATION

5.1   *Calculation of electric magnetic field strength multilayer solenoid axis*

$$B_{z0} = \frac{1}{2}\mu_0 n_1 I[\frac{\frac{1}{2}l+z}{\sqrt{r^2+(\frac{1}{2}l+z)^2}} + \frac{\frac{1}{2}l-z}{\sqrt{r^2+(\frac{1}{2}l-z)^2}}]$$

(1)

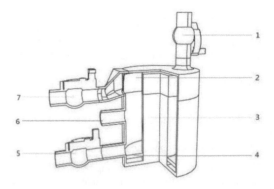

Figure 4.   Structure.

222

$$B_r(r,z) = -\frac{B'_{z0}}{2}r + \frac{B''_{z0}}{2^2 \cdot 4}r^3 - \frac{B^{(5)}_{z0}}{2^2 \cdot 4^2 \cdot 6}r^5 + .. \qquad (2)$$

Formula (1) and (2):

I——field current

l——length of the solenoid

n_1——number of turns per unit length of the solenoid;

r——solenoid radius;

z——distance of a point on the axis center O to the solenoid;

B_{z0}——axial magnetic field strength;

$B_r(r,z)$——magnetic field intensity of the external axis.

By using Eqs (1) and (2), we can calculate the solenoid axis magnetic field strength.

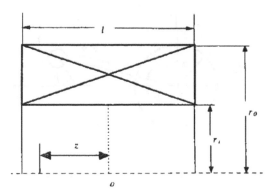

Figure 6. Calculation of the various parameters in the formula of the actual coil magnetic field.

5.2 Magnetic field gradient magnetic core

Suppose there is a radius of ferromagnetic medium wire a, centered on the coordinate axis, the external magnetic field strength H_0 can be uniformly magnetized medium wire to obtain magnetization M, with the speed V_0 of the fluid medium flow wire, the fluid viscosity index η, the density Q_f, the magnetic susceptibility K_f, the fluid with particle density Q_p, the magnetic susceptibility K_p, and the radius b.

When $H_0 < H_s$,

$$H_{2r} = \left[1 + \frac{a^2}{r^2}\right]H_0\cos\theta$$

$$H_{2\theta} = -\left[1 - \frac{a^2}{r^2}\right]H_0\sin\theta$$

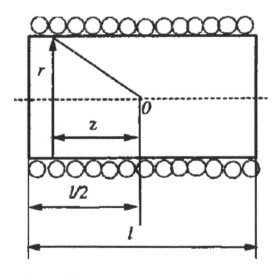

Figure 5. Single layer solenoid.

When $H_0 > H_s$,

$$H_{2r} = \left[1 + k\frac{a^2}{r^2}\right]H_0\cos\theta$$

$$H_{2\theta} = -\left[1 - k\frac{a^2}{r^2}\right]H_0\sin\theta$$

H_{2r}, $H_{2\theta}$ ures ——field component in the polar coordinates;

H_s——saturated magnetic field strength of the medium wire;

K——ratio of the saturation magnetization and the external magnetic field strength;

M_s——saturation magnetization.

By the medium wire field expression for the field, the medium wire in the background field strength is related to the volume magnetic susceptibility and the size of the external magnetic field intensity, and the medium wire. Joining the magnetic medium in the magnetic field after field, the great change in the surrounding medium wire, the intensity of the magnetic field and the medium wire r are inversely proportional to the square of the distance, and changes with H. Therefore, the medium wire surrounding the magnetic field by the uniform magnetic field is not the uniform magnetic field.

When $H_0 < H_s$, the magnetic field gradient $\mathrm{grad}H_2 = -2H_0a^3/r^3$ increases the external magnetic field intensity and decreases the particles, with the medium wire distance increasing the values of the magnetic field gradient. This is due to the non-uniform magnetic field gradient, which causes the particles to be removed from the oil.

5.3 The attractive force of ferromagnetic material in the magnetic field

$$F_m = -V_p(X_p - X_m)H\mathrm{grad}H$$

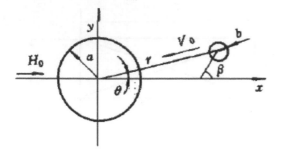

Figure 7.   Ferromagnetic medium wire force diagram.

V_p—volume of the ferromagnetic impurities (m^3);
X_p—permeability of ferromagnetic impurities (H/m);
X_m—permeability of the oil (H/m);
H—external magnetic field intensity (A/m);
$gradH$—gradient magnetic field strength of the point (A/m^2).

From the above equation, it can be seen that the ferromagnetic impurities by attraction and the external magnetic field intensity is proportional to H and directly proportional to the gradient of the magnetic field strength gradH. The traditional magnetic filter improves the magnetic attraction by improving the external magnetic field strength H, so as to improve the filter ability; the high gradient magnetic filter by means of the magnetic medium mainly improves the external magnetic field gradient gradH to increase the attraction, possibly achieving more efficient filtering effect of pollutants.

6   CONCLUSION

The lubricating oil of the high gradient magnetic automatic filter design changes the traditional design ideas of the filter by adopting the filter magnetic adsorption method and by changing the magnetic field gradient in the form of high gradient magnetic field increased the adsorption effect based on the magnetic field, and magnetic field on the basis of adsorption improves the structure of electromechanical combination to achieve automatic cleaning chip function. Compared with the traditional filter, the filtering effect is better and more convenient to use.

REFERENCES

Liao Huajun. 2012, Analysis and structure optimization of permanent magnet type high gradient magnetic filter field [D]. University of Science and Technology Liaoning.
Wang Huajun. 1999, Calculation of magnetic field of solenoid [J]. Journal of Sichuan University of Science & Engineering, 04:23–25.
Xu Shuhui, Ge Zhengyu. 2006, Mechanism Analysis and Experimental Study on Highgradient Magnetic Filter [J]. Meikuang jixie coal mine machinery, 01:32–34.
Xu Shuhui, Zhou Minglian. 2006, Experimental study on an Eight Gradient Cooler-magnetic Filter for Hydraulic System [J]. Chinese hygraulic & pneumatics, 04:58–61.

Energy Science and Applied Technology – Fang (Ed.)
© 2016 Taylor & Francis Group, London, ISBN 978-1-138-02833-3

A preliminary study on the soft landing trajectory design and control strategy of Chang'e 3

Linan Zhang

Department of Electrical and Electronic Engineering, North China Electric Power University, Beijing, China

ABSTRACT: Chang'e 3 has successfully launched at 13:30 on December 2nd, 2013, becoming the focus of the whole world. China has become the third nation in implementing the moon's soft landing after America and the former Soviet Union. The article establishes a corresponding model based on the genetic algorithm, analyzing to determine the pericynthion and the apocynthion of the Chang'e 3's landing-preparing orbit, the choice of the 15*100 km elliptical orbit and the optimal control strategy of landing trajectory and its different stages. In addition, the error analysis of this model has been discussed.

Keywords: trajectory optimization, genetic algorithm, parameter normalization, contour map

1 INTRODUCTION

When Chang'e 3 is at a high speed, in order to achieve a soft landing within the predetermined region on the moon accurately, the key is the design of landing trajectory and control strategy. To meet the states of all the key points of each stage, it is of great importance to try to reduce the fuel consumption of the soft landing process.

The models established in this paper are based on the following three hypotheses:

1. It is assumed that there is no influence of other celestial bodies in space on the moon and Chang'e 3;
2. When Chang'e 3 is on the landing-preparing orbit, there is no precession;
3. The moon is supposed to be a sphere without considering the impact of the rotation of the moon.

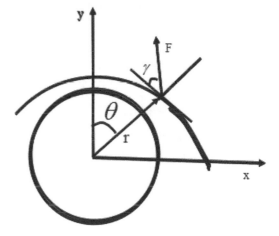

Figure 1. Model diagram.

2 THE DETERMINATION OF THE POSITION OF THE PERICYNTHION AND THE APOCYNTHION AND THE CORRESPONDING VELOCITY

2.1 *The establishment of a genetic algorithm model*

The model of the system. The two-body model is used to describe the motion of the system. The established landing coordinate system is shown in Figure 1.

Assuming that the landing trajectory is in the longitudinal plane, the selenocentric coordinate is set to be the origin of the coordinate O. O_y refers to the starting breaking points of the power down section and O_x points to the starting direction of the movement of Chang'e 3. The r, θ, w and m are, respectively, set as the selenocentric distance, the polar angle, the angular velocity and the quality of Chang'e 3, with v as the velocity of the lander along the direction of r, F as the thrust of the retroengine (a fixed constant or 0), I_{sp} as the specific impulse, μ as the moon's gravitational constant, and μ as the thrust direction angle between the engine thrust and the local horizontal line.

The initial conditions of the power down section are determined by the pericynthion of the elliptical orbit, while the terminal conditions are to achieve soft landing of Chang'e 3 on the lunar surface. Setting the initial time $t_0 = 0$ and the terminal time t_f indefinitely, the corresponding initial conditions are $r0 = r_f + h_0$, v_0, and the terminal constraints are $r_f = r_l$, $v_f = 0$ and $w_f = 0$, where r_l is the moon radius; h_0 is the initial orbit height; and w_0 is the orbital angular velocity. The optimal orbit design is in the premise of satisfying the above conditions to adjust the thrust magnitude and direction, which requires the following performance indices to reach the maximum: $\bar{J} = \int_0^{t_f} \dot{m}\, dt$.

In other words, it requires to transform it into a local minimum problem of this function under the multi-constrained conditions mentioned above.

Normalization. During e trajectory optimization process, because the orders of the state variables' magnitude are of great difference, optimizing the computation process will lead to the loss of the effective digits. This shortcoming can be overcome by normalization to improve the accuracy of the calculation. Several parameters are defined as follows:

$$r_{rRf} = r_0, m_{rBf} = m_0.$$

So

$$\bar{r} = \frac{r}{r_{ref}}, \bar{v} = \frac{v}{v_{ref}}, v_{ref} = \sqrt{\frac{\mu}{r_{ref}}},$$

$$\overline{I}_{sp} = \sqrt{\frac{r_{ref}}{\mu}}, \overline{F} = \frac{F}{F_{ref}},$$

$$F_{ref} = m_{ref} * \frac{v_{ref}^2}{r_{ref}},$$

$$\bar{m} = \frac{m}{m_{ref}}, \bar{w} = w * \sqrt{\frac{r_{ref}^3}{\mu}},$$

$$\bar{t} = \frac{t}{t_{ref}}, t_{ref} = \frac{r_{ref}}{v_{ref}}, \bar{\theta} = \theta.$$

Then, the kinetic equations of Chang'e 3 can be described as follows:

$$\dot{\bar{r}} = v$$

$$\dot{\bar{v}} = \left(\frac{\overline{F}}{\overline{m}}\right) * \sin\phi - 1/\bar{r}^2 + \bar{r} * \bar{w}^2$$

$$\dot{\bar{\theta}} = \bar{w}$$

$$\dot{\bar{w}} = -\left[\left(\frac{\overline{F}}{\overline{m}}\right) * \cos\phi + 2 * \bar{v} * \bar{w}\right]/\bar{r}$$

$$\dot{\bar{m}} = -\overline{F}\sqrt{I_{sp}}.$$

The corresponding initial conditions and the terminal constraints are given by:

$$\bar{r}_0 = 1, \bar{v}_0 = 0, \bar{w}_0 = w_0 * \sqrt{r^3/\mu};$$

$$\bar{r}_0 = \frac{r_l}{r_0}, v_f = 0, w_f = 0.$$

The performance indices can be rewritten as:.

$$\bar{J} = \int_0^{\bar{t}_f} \dot{\bar{m}}\, d\bar{t}$$

Parameterization method. Being discretized, the soft landing trajectory can be averagely divided into n sections and each node of each section is set as the thrust direction Φ. Then, the thrust direction angle Φ of the n+1 node and the terminal moment t_f can be set as the parameters to be optimized. Using the adaptive genetic algorithm to predict the optimal Φ value of the next time point according to the Φ value of the initial conditions at the initial time, optimal solutions of the Φ values of the same time interval can then be calculated iteratively. The moment of each node can be obtained by the following formula:

$$t_i = t_0 + i * \frac{t_f - t_0}{n}, (i = 0, 1, \ldots, n).$$

Thus, the direction angle of the thrust of each node has a corresponding node time. If assuming that the thrust direction angle can be expressed as a polynomial: $\phi(t) = \lambda_0 + \lambda_1 * t + \lambda_2 * t^2 + \lambda_3 * t^3$, then the corresponding node moments and the thrust direction angles can be used to fit this polynomial to obtain the coefficients of the polynomial $\lambda_i = 0, 1, \ldots, n$), and then to gain the thrust direction angles of the entire landing trajectory.

Genetic algorithm (GA) is a kind of adaptive global optimization and stochastic search algorithm, which does not depend on the model of the problem and initial conditions, and does not have any differentiable or continuous requirements of the domain of solution. So, it is applicable to the study of solving the optimal path problem of the lander. However, how the terminal constraints reflect to the fitness function of GA must be dealt as well. Considering the accuracy of trajectory optimization, using the dynamic penalty function method to process the terminal constraints, the fitness function can be expressed as follows:

$$f(x) = -\sigma_1[(v(t_f) - v_f)^2 + (w(t_f)r(t_f) - \omega_f r_f)^2]^{1/2} - \sigma_2 |r(t_f) - r_f| + m_0 + J$$

where X is the individuals of the population; $v(t_f)$, $\omega(t_f)$ and $r(t_f)$ are the state values of the terminal moment; and σ_1, σ_2 are the penalty factors. With optimization, they become larger, which gradually strengthens the constraints.

Using binary codes for this search, the crossover operation adopts the single-point crossover and the mutation operation employs the binary variation.

After finding the optimal path by the searching algorithm, the horizontal distance between the pericynthion and the predetermined landing point can be calculated. Then, the lander orbit plane of Chang'e 3 can be determined. The pericynthion's distance of 15 km and the apocynthion's distance of 100 km are known. 'μ' is set as the standard gravitational parameter of the moon, and 'a' is set as the orbital semi-major axis of the elliptical orbit of Chang'e 3. According to the velocity formula derived from the Vis Viva equation:

$$v = \sqrt{\mu\left(\frac{2}{r} - \frac{1}{a}\right)}$$

The velocity size v_n at the pericynthion and the velocity size v_n at the apocynthion can be obtained.

2.2 The solution of the model

As shown in Figure 1, the direction pointing toward the pericynthion is the Y-axis, the direction pointing toward the velocity of the lander at the pericynthion is the X-axis. According to the documentation and the analysis, it can be deduced that

$$F = \begin{cases} \varphi = \arctan\left(\dfrac{\lambda_v}{\lambda_\omega}\right) \\ F_{\max}, L(t) > 0 \\ 0, L(t) < 0 \text{ and } L(t) = \dfrac{\lambda_\omega \sin\gamma}{m} - \dfrac{\lambda_\omega \cos\gamma}{m} - \dfrac{\lambda_m}{I_{sp}} - \dfrac{1}{I_{sp}}, \\ indefinite, L(t) = 0 \end{cases}$$

This indicates that the lander's thrust reverser works with either the maximum thrust or the minimum thrust. Establishing the differential equations based on this using the genetic algorithm, the optimal trajectory can be determined.

2.3 Results

The results show that the total time of the main deceleration phase and the attitude adjustment stage is 454.8677s, and the quantity of fuel is 1.4862t. The time-dependent forced direction variation curve is shown in Figure 2.

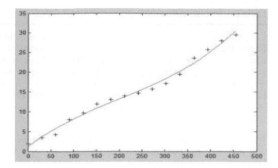

Figure 2. Time-dependent forced direction variation curve.

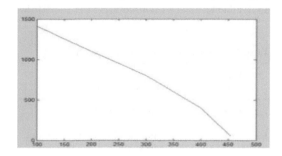

Figure 3. Time-dependent speed variation curve.

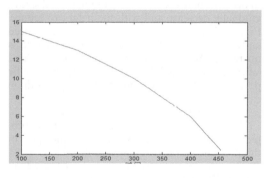

Figure 4. Time-dependent height variation curve.

The function of the force direction γ about the time is as follows:

$$\gamma(t) = 0.00001t^3 - 0.0002t^2 + 0.0845t + 1.3080$$

The jet direction of the lander may be under control according to this equation, making the lander to land along the optimal trajectory. From the figure, it can be seen that the angle between the visible jet direction and the Y-axis is gradually increased from 0° to about 32°, causing the lander's velocity direction and magnitude to change, the angle between the speed direction and the Y-axis

Table 1. Orbital elements.

The orbital semi-major axis a	The orbital eccentricity e	The orbital inclination i	Longitude of the ascending node Ω (reckoned from the 0° longitude of the moon)	Argument of pericenter ω	Mean anomaly making the epoch M
1794.53 km	0.0237	90°	340.49°	29.01°	29.12°

Table 2. Latitudes and longitudes of the vertical projection points of the pericynthion and the apocynthion on the moon.

	Longitude	Latitude
pericynthion	19.51 W	29.01 N
apocynthion	170.49E	29.01S

Table 3. Velocity size and direction of Chang'e 3.

	Velocity Size (km/s)	Velocity direction
v_n	1.692	Along 19.51 W meridian to the north
V_d	1.613	Along the 170.49E meridian to the south

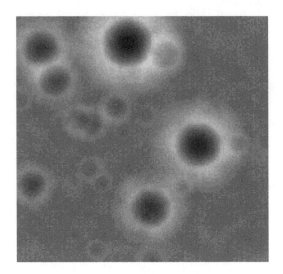

Figure 5. Digital elevation map at a distance of 2400 m.

to become increasingly larger, and the speed size to gradually decrease and with the height to reduce. Figures 3 and 4 show the time-dependent speed and height variation curves when the device is operating. According to this and the longitude and latitude of the scheduled landing point, the specific position of the pericynthion and the apocynthion can be calculated.

Orbital elements are computed in Table 1.

The latitudes and longitudes of the vertical projection points of the pericynthion and the apocynthion on the moon are computed in Table 2.

The velocity size and direction of Chang'e 3 are given in Table 3.

3 THE OPTIMAL CONTROL STRATEGY OF THE LANDING TRAJECTORY AND THE STAGES OF THE LANDING PROCESS OF THE CHANG'E 3

3.1 *The coarse obstacle avoidance phase*

The distance between Chang'e 3 and the moon of the coarse obstacle avoidance section ranges from 2.4 km to 100 m. The main requirement is to avoid large craters, realize hovering at a distance of 100 m above the designed landing point, and determine the initial landing place.

We use the Matlab software to determine the contour map of the elevation map, as shown in Figure 5, and grid it to easily gain the fit landing grid. To obtain the results more accurately, the variance of each grid should be obtained, from which the grid with a smaller variance and a shorter distance away from the original location of the lander is selected.

The result is shown in Figure 6.

Based on the above analysis, the best landing area should be selected in the area of the horizontal axis of 800~1200 m and the vertical axis of 1000~1400 m. According to the analysis of variance, the variance is 0.0086, i.e. the $1000 \leq x \leq 1200$-and-$1200 \leq y \leq 1400$ area should be found, so that the lander controls the moving direction according to its own analysis results, moving at a distance of 100 m along the Y-axis.

3.2 *The fine obstacle avoidance*

The fine obstacle avoidance segment ranges from 100 m to 30 m away from the lunar surface, requiring Chang'e 3 to hover at a distance of 100 m from the lunar surface, take pictures of

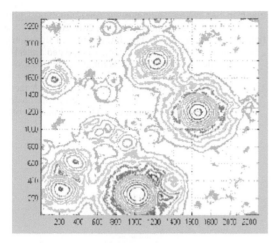

Figure 6.　Transformation of the elevation map shown in Figure 5 into a contour map.

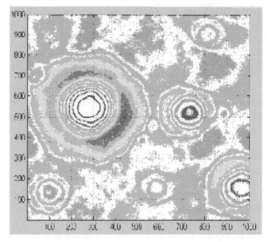

Figure 8.　Transformation of the elevation map shown in Figure 7 into a contour map.

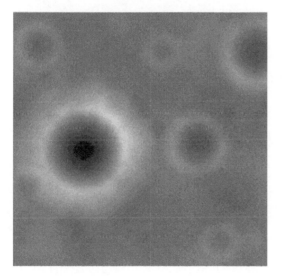

Figure 7.　Digital elevation map at a distance of 100 m.

the 100 m-range region near the landing site, and get a three-dimensional digital elevation map, as shown in Figure 7.

By analyzing the three-dimensional digital elevation map, it can be realized that avoiding large craters to determine the best landing site and then to achieve the horizontal speed of 0 m/s at the distance of 30 m above the landing point. To avoid craters and other rugged areas, the meshing roughness (the variance) δ and the horizontal distance R from the original location of the lander can be used for analysis and determination in order to select the best landing site.

Agreeing with the ideas of the coarse obstacle avoidance, taking into account the lander's vol-ume of 1 to 2 cubic meters and the error caused by the movement of the lander, the final choice of the scheduled landing point is the 4 m*4 m area. In other words, the figure is about to be split into several 4 m*4 m sub-blocks, the variance of each block can be calculated. Considering the distance from the grid to the original position of the lander, it is relatively easy to select the optimal landing area. According to the results, it can be seen that the optimal landing zone lies in the $50 \leq x \leq 54$-and-$28 \leq y \leq 32$ area, so the moving direction and distance of the lander is to move 2 m along the X-axis and 20 m along the Y-axis.

Considering the above factors, the time of the last three stages can be found as about 306 s. Simplifying the force of the lander at this stage, the equation can be written as follows:

$$v_e m(t) = (m_0 - \int m(t)dt)g'.$$

Here, the quality of consumption is 0.4131t, so the total fuel consumption of the whole process is 1.8993t.

4　CONCLUSIONS

In this article, the main consideration is the most influential factor of the terrestrial gravitational perturbation, the lunar non-spherical gravitational perturbation and the errors caused by the difference between the actual position and the estimated position. By setting $\varepsilon_1(J1)$ as the relative error caused by the terrestrial gravitational perturbation and $\varepsilon_2(J2)$ as the relative error caused by the oblateness of the moon, then we get

$$\varepsilon_1(J1) = O(10^{-4})$$
$$\varepsilon_2(J2) = O(10^{-5})$$

The difference between the actual position and the estimated position ranges from a few meters to several hundred meters, with the relative error ranging from $O(10^{-4})$ to $O(10^{-5})$.

So, this model can be extended to solve some of the problems with multiple variables and multiple constraints.

REFERENCES

Da-yi Wang, Tie-shou Li. The Numerical Solution of Boundary Value Problems of the optimal lunar soft landing by Xing-rui Ma. Aerospace Control, Section 3, 2000.

Lin Liu, San Wang. Summary of lunar satellite orbit mechanics. Progress in Astronomy, 2(4), 2003.12.

Jian-feng Zhu, Shi-jie Xu. The optimization of the lunar soft landing orbit based on adaptive simulated annealing genetic algorithm. Acta Aeronautica ET Astronautica Sinica, 28(4), 2007.7.

Zheng-Ji Song, Dian-Wu Geng, Xing-Wei Jiang. A recognition algorithm of the danger zone on the basis of the elevation map. Journal of Harbin Institute of Technology, 38(6), 2006.6.

Energy Science and Applied Technology – Fang (Ed.)
© *2016 Taylor & Francis Group, London, ISBN 978-1-138-02833-3*

Research on low power control technology of EM-MWD downhole instrument

Cheng Zhang & Haobin Dong
School of Automation, China University of Geosciences, Wuhan, China

ABSTRACT: Due to the large power consumption of the Electromagnetic Measurement While drilling (EM-MWD) downhole system and the limit of battery capacity, the drilling efficiency will be reduced while replacing the battery frequently, so the low power control technology has to be an effective way to extend the downhole battery's working hours of the EM-MWD system. This paper analyzes the influence factors of the downhole instrument's consumption theoretically, and adopts low power hardware technology and software dynamic power management technology, respectively, from the quiescent dissipation and dynamic power consumption. The results show that the battery operating time of the downhole instrument prolonged to 55% under the circumstances in which the EM-MWD downhole instrument collects and launches each data every 5 minutes.

Keywords: Electromagnetic Measurement While Drilling, downhole system, battery-powered, low power design

1 INTRODUCTION

In the drilling process of special drilling, the key to ensure drilling quality is how to control the trajectory well. The technology in the application of measurement while drilling of transmitting the real-time data from the downhole system can effectively reduce downtime and improve work efficiency (W.K. Han, 2013). The Wireless Measurement While Drilling (MWD) system can withstand the test of harsh conditions, and ensure data transmission. It can also reduce the drilling time and energy consumption, and improve the production efficiency (W.Y. Chen, 2011).

The Electromagnetic Measurement While Drilling (EM-MWD) has three main functions: Mud Pulse Telemetry (MPT), electromagnetic transmission and sound transmission. The acoustic method is mainly used in horizontal wells because of its fast attenuation and large interference. MPT has a problem of less information, so it is difficult to meet the data demands of intelligent drilling. In addition, it cannot use when the gas content in the mud exceeds 20%. The EM-MWD has an advantage of ignoring the influence of properties of drilling fluid (Y.T. Shao, 2007, X. Li, 2010). At present, the wireless EM-MWD has become a hot research topic because of its high signal transmission rate, short measurement time, low cost and transfer data without drilling fluid.

MWD technology has been developed rapidly from the 1980s in the 20th century. Since the 1990s, some companies such as Schlumberger, Halliburton

Sperry Sun and Weatherford have launched a series of commercialization of EM-MWD products. Some foreign EM-MWD products are listed in Table 1.

At present, the research of EM-MWD technology in the domestic sector has made a breakthrough progress, but it also needs to improve continuously. Some domestic EM-MWD products are listed in Table 2. We need to make an important progress in the following aspects: (1) increasing the transmission distance of the distance of the electromagnetic signal through the optimization of operating frequency and efficiency of the downhole signal transmitter, deepening the research of weak signal detection and processing technology; (2) increasing the depth measurement in the EM-MWD system by extending the antenna or using the signal repeater mode; (3) developing the high-capacity and high-temperature batteries suitable for gas drilling to improve the signal transmission power and the continuous working time of the EM-MWD system; (4) realizing a two-way communication.

The EM-MWD equipment usually uses batteries or generators as the power supply method. Battery power supply always uses a high energy battery of high temperature resistance. It has the advantages of simple and low electromagnetic interferences. Generator power supply is usually driven by the turbine through mud pressure. The utility model has the advantage of providing continuous long-term supply, but has a serious electromagnetic interference.

Owing to the limit of the actual work environment and requirements of equipment size, the EM-MWD downhole instrument is mostly powered by

Table 1. Some foreign EM-MWD products.

Company	Transmit frequency (Hz)	Maximum data transfer rate (b/s)	Maximum operating temperature (°C)	Pressure (MPa)	Working hours (h)
Schlumberger	0.1875–12	12	125	85	200
Halliburton	2.0–20	10	150	124	200
Weatherford	2.0–20	12	150	137	120
Geo Link	2.0–20	4	150	137.9	250
Blackstar	2.0–20	6	150	137	80–130
ZTS	1.2–10	10	120	105	Turbine power supply

Table 2. Some domestic EM-MWD products.

Prototype models	Working frequency (Hz)	Transmission rate (bps)	Maximum operating temperature (°C)	Working hours (h)	Maximum vertical depth (m)
SEMWD-2000	2–12.5	0.5–6.25	125	200	3600
CEM-1	10–20	-	125	216	3081
DREMWD	3.5–11	12	-	-	904
CQEMWD	1.25–5	2.5–5	125	150	3841

batteries. The downhole equipment's power consumption, battery capacity, and the working way of the instrument will determine the instrument's working time. In addition, the time to replace the batteries will be a big waste of manpower and material resources. In order to extend the battery's operating time and improve the drilling efficiency, the technical problem of how to control the power consumption of the system must be solved in the development of the downhole instrument.

2 THE ANALYSIS OF THE SYSTEM POWER CONSUMPTION

The power consumption of the system is divided into static power and dynamic power. Static power consumption is standby power, which is decided by the power of the circuit structure and the device. It can be reduced by selecting low power devices or the design and optimization of the circuit. Dynamic power is the system transmitting power, which is mainly decided by the frequency of the transmitting signal.

To increase the working hours and improve the drilling distance, it will be better to reduce the average current of the system when the power supply is limited. We need to configure the system dynamically to make each function module in the system at the lowest power state needed to meet the performance requirements. We can also control the power of each component independently in order to reduce power consumption (Y.L. Zhang, 2014).

3 STATIC POWER ALGORITHM

The EM-MWD downhole instrument is mainly composed of a power supply management circuit,

sensors, microprocessor and signal transmitting circuit, as shown in Figure 1.

Low power MCU. The processor adopts a low power 8-bit single chip microcomputer named STM8S208, which has a power supply voltage working range of 2.95~5.5 V and a low working current. It can work on an internal low power clock frequency such as 128k and has a variety of low power mode. Its lowest current consumption is 4.5 μA. The STM8 series single chip is integrated with a plurality of integrated module that can run independently. It can also save the space and cost of the circuit design and reduce the power consumption of the system.

Each module of the MCU can be run independently, and can also work independently in a dormant state such as timer, input/output port, A/D conversion and watchdog. Any module can wake up the CPU through an interrupt if we need the CPU to start working in order to make the system run at the lowest power consumption. Usually, the CPU is set to a dormant state by the software. When needed, we can use the interrupt to wake up the CPU from dormancy. After completion of the work, it can enter the sleep state correspondingly.

We need to minimize the power as follows: (1) close the peripheral that is not required to use and use the PCG function to close the corresponding clock; (2) all unused pin must be set to a low level output to avoid any unnecessary loss; (3) select the appropriate waking-up mode and power supply voltage; and (4) choose the appropriate clock frequency.

Sensor module. The downhole instrument uses a pressure sensor and a nine-axis sensor to collect the data. If the sensors are always in the working state, it will consume most of the power, and it

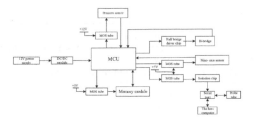

Figure 1. Block diagram of the downhole instrument.

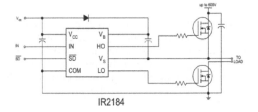

Figure 2. Half of the H-bridge.

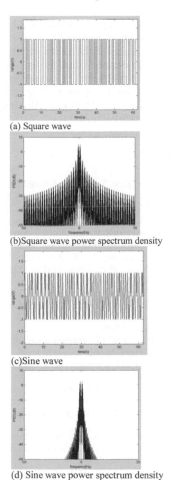

(a) Square wave

(b)Square wave power spectrum density

(c)Sine wave

(d) Sine wave power spectrum density

Figure 3. Manchester coding waveform and its power spectrum density.

does not need to work continuously to collect the data in actual use. So, we can select modular sensors to achieve the purpose of saving electricity by controlling the switching of the sensor's power to make the sensors work intermittently. The system uses the MOS switch circuit to control the sensor supply, which will only open the sensors in time of need.

Signal transmitting circuit. The instrument transmits the signal through an H-bridges consisted of two half bridge chips named IR2184 and 4 pieces of MOS tubes. The H-bridge has a high emission efficiency and easily controls the launching state. The MOS tubes have advantages of high input impedance, low drive current, fast switching speed, high frequency characteristics, and excellent thermal stability (J. Huang, 2014). The half of the H-bridge is shown in Figure 3.

4 DYNAMIC POWER MANAGEMENT

Hardware control of transmitting power. There are two ways of launching: one is the square wave produced by the H-bridge; the other is a sine wave caused by the amplifier. Usually, three kinds of high efficiency power amplifier are used: class D, class E and class F. Class D is always used in the low frequency band, which is suitable for the 5–40 Hz low frequency used in the EM-MWD instrument. At low output power, the efficiency is less than 85% of the class D amplifier (within 8 W) (S.Y. Huang, 2008), and the transmit power is less than 6 W in the EM-MWD.

We generate a waveform with a frequency of 1 Hz and a length of 1000s through MATLAB, and analyze its power spectral density, as shown in Figure 3. As in the range of f ± fc (0–2 Hz), the square wave's power is 85.66% of the total power, and the sine wave is 98.94%. Assuming the efficiency of the class D power amplifier to be 85%, we can find the efficiency of the H-bridge to be slightly larger than the sine wave. The H-bridge is also simple to be built and easy to control, so the system uses the H-bridge for launching.

Software dynamic power control. The core of power management is realized by the software. The key is to the allocation of system resources according to the system working condition (Y.H. Wang, 2010). In the software design, we set the chip into a low-power state when it does not work, and awake it when it needs to work. The serial only works when a data transmission occurs between the host computer and the exploring tube. The H-bridge is closed when we collect the data through the sensors, and then the MCU should be set into a low-power mode after the completion of data processing and transmitting.

In the software design, the MCU will be set into a low-power state when it does not work. When

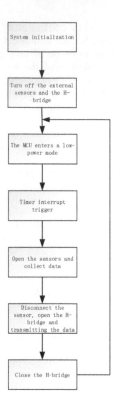

Figure 4.   Software flow chart.

necessary, we will wake up the MCU to start working through the timer interrupt. The software flow chart is shown in Figure 4. The working flow is as follows. First, the system is powered on and initialized. Then, we shut off the power supply of the external sensor through MOS tubes and set the MCU into a low-power mode. When arriving at the set time, the MCU is awakened and the sensor supply is opened to collect the data and transmit.

5   EXPERIMENT AND TEST

According to the above research on low power consumption, we conducted a test on the EM-MWD downhole instrument. We use a 12V lithium battery as the power supply and a series of 1 ohms resistance as the sampling resistance to measure the current. The transmitted signal is the square wave, its peak voltage is 24V and frequency is 10 Hz, and the collecting and transmitting time is 30 s. The average current is about 30 mA without the external load. We set the MCU into a low-power mode and turn off the external sensors. The measured average current is about 5 mA. The experimental results are outlined in Table 3.

Table 3.   Experimental results.

Working time interval (min)	Using time after reducing power consumption (h)	Using time before reducing power consumption (h)
5	140	91
10	242	123
20	396	151
30	508	164

As shown in Table 3, we can find that the using time of the battery is mainly affected by the system static working current when the working time interval is large. The use of the low-power design in this paper has reduced the quiescent current of the system and prolonged the working time.

6   CONCLUSION

The EM-MWD downhole instrument has taken a series of measures to reduce the power consumption from the aspects of the hardware circuit design and software flow control, and the feasibility is validated by the experiment. The experiments show that the use of the low-power design can significantly increase the using time of the battery and improve its working efficiency.

REFERENCES

Chen. W.Y. The Key Technology Research on Signal Measurement of Measurement While Drilling System. (MS., Chongqing University, China 2011).
Han W.K. L.S. Kumar, Y.L. Guan, etal. Design of coded digital telemetry system for acoustic downhole channel with drilling noise. Information, Communications and Signal Processing, IEEE, 2013:1–5.
Huang. S.Y, S.Y. Shi. The analysis of class D power amplifier efficiency parameter. Journal of Sichuan Vocational and Technical Collefe, 2008, 18(2): 115–116.
Huang. J, Q. He. The research of constant current source circuit based on VMOS. Journal of Electric Power, 2014, 29(01):43–47+51.
Li. X, A.G Yao, Y.S. Li. Transmission characteristics of new electromagnetic—measurement while drilling system. Coal Geology & Exploration, 2010, 38(2):22–25+94.
Shao Y.T, A.G. Yao, G.M. Zhang. Application and development of electro-magnetic telemetry in drilling operation. Coal Geology& Exploration, 2007, 35(3):77–80.
Wang. Y.H. Design of dynamic power management for embedded system. (MS., South-Central University for Nationalities, China 2010).
Zhang.Y.L, Y. Gao, J.L. Wang. Low power design of field measurement system based on solar power. Instrument Technique and Sensor, 2014, 51 (01): 107–110.

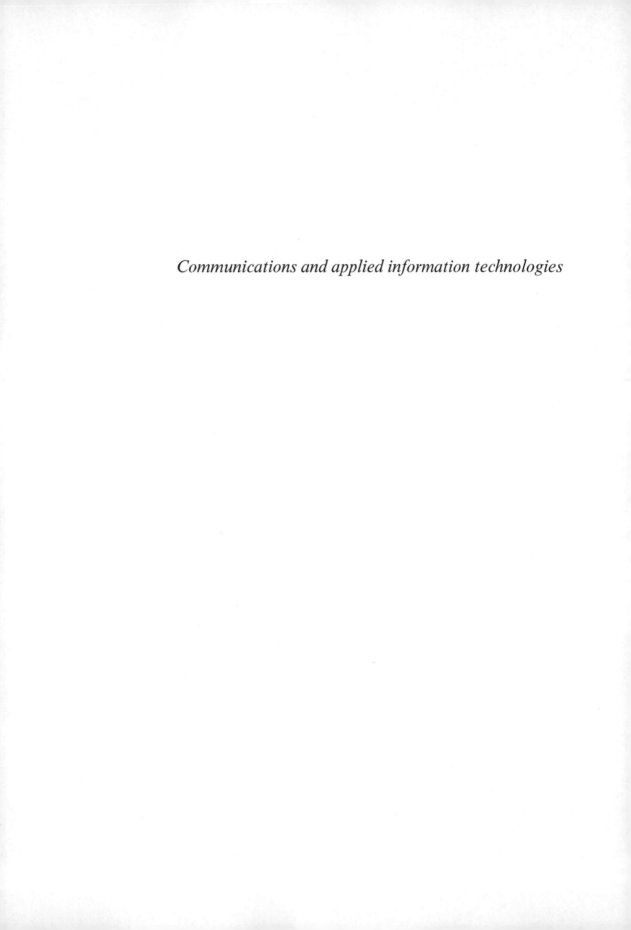

Communications and applied information technologies

Energy Science and Applied Technology – Fang (Ed.)
© 2016 Taylor & Francis Group, London, ISBN 978-1-138-02833-3

Performance study of efficient modulation technique to satellite broadband communication system

Z.K. Li

College of Communication Engineering, PLA University of Science and Technology, Beijing, China

Y.H. Pan

Nanjing, Telecommunications Technologies Institute, Nanjing, China

B. Shang

College of Communication Engineering, PLA University of Science and Technology, Beijing, China

ABSTRACT: A high-order modulation technique has been widely used in satellite broadband communication and digital transmission over nonlinear satellite channels to meet the requirements of power and spectral efficiency. This paper investigates the performance of the Mary Amplitude-Phase Shift Keying (APSK) digital modulation in Additive White Gaussian Noise, and presents the basic modulation principle. In addition, the theoretical upper bound of SER (Symbol Error Rate) of high-order APSK is derived, while the performance of 64 APSK is also discussed. The theoretical analysis and simulation results indicate that the new technology can achieve the targets of providing higher data rates for usage at the existing bandwidth.

Keywords: Highly efficient modulation scheme; satellite broadband communication; APSK; performance analysis

1 INTRODUCTION

There are some outstanding issues regarding the limited available spectrum bandwidth resources, the larger bandwidth cost, increased difficulty in obtaining the spectrum efficiency and insufficient quantity of channel with the increasing digital broadband satellite businesses as well as the expansion of satellite system capacity. High-speed data transmission has become the urgent demand of satellite communication to improve bandwidth efficiency and achieve the real-time satellite communication in transmitting a large amount of information. Therefore, the use of a new type of high-order modulation scheme, especially the amplitude and phase shift keying modulation, becomes an important development direction.

APSK and QAM are both the amplitude phase shift keying modulation, which can ensure the anti-interference performance and make full use of the signal constellation diagram surface. QAM has a highly frequency spectrum utilization, which is widely used in large and medium digital microwave communication systems and satellite digital communication, especially for the situation of limited spectrum and bandwidth resources. The constellation shape of QAM is usually rectangle or cross, but it is better to take the star shape whose boundary is more close to circular.

This paper investigates the 64 APSK modulation scheme that designs for efficient transmission over satellite channels owing to its intrinsic robustness against nonlinear amplifier distortions. The traditional DVB-S2 standard adopted QPSK, 8PSK, 16 APSK and 32 APSK modulation schemes. The technique based on the 64 APSK modulation scheme has shown increased performance in contrast to DVB-S2, and it is mainly targeted at professional applications due to the higher requirements in terms of the available Signal to Noise Ratio (SNR). This paper first presents the basic modulation principle. Second, it derives an expression of 32 APSK and 64 APSK to approximate the error probability after a detailed derivation of 16 APSK error probability. Finally, it compares with the 16 APSK and 64QAM schemes, and analyzes the performance of 64 APSK modulation transmitted in the Additive White Gaussian Noise (AWGN) channel through detailed computer simulations.

2 PERFORMANCE ANALYSIS

2.1 *Modulation principle*

The so-called APSK, a variation of both M-PSK and QAM, is an amplitude and phase modulation method different from traditional rectangular QAM, which was also called stellate QAM. Compared with QAM of the same order, APSK uses fewer amplitude levels as well as fewer phase shifts. These amplitude levels are less susceptible to noise and reduce the symbol error rate. MAPSK (M = 16, 32, 64) arranges the symbols into more than two concentric rings with a constant phase offset. 16 APSK is a double-ring PSK version with 16 different phase positions, which has four symbols in the center ring and twelve symbols in the outer ring. Symbols of each level are uniformly distributed. Likewise, 64 APSK has four multiple amplitudes in four concentric rings and four constant phase offsets. Four amplitude levels allow the new technique to operate at a higher power level and achieve the higher data rates than DVB-S2. The universal constellation of 16 APSK and 64 APSK is shown in Figure 1.

2.2 *Derivation of APSK symbol error rate expressions*

The same modulation scheme has the same derivation process of symbol error rate expressions. This paper presents a detailed derivation of 16 APSK symbol error rate, for example.

According to reference, the union bound for the symbol error rate of M-ary APSK is given in general terms as:

$$P_e = \frac{1}{M}\sum_{i=1}^{M}P(E|S_i) \le \frac{1}{M}\sum_{i=1}^{M}\sum_{j=1,j\ne i}^{M}P(S_i \to S_j) \quad (1)$$

where E is the error event; $P(E|S_i)$ is the probability of the transmitted symbol S_i that is erroneously detected; and $P(S_i \to S_j)$ is the probability of the transmitted symbol S_i that is erroneously

detected. $P(S_i \to S_j)$ can be computed using the Q-function:

$$P(S_i \to S_j) = Q\left(d_{ij}/\sqrt{2N_0}\right) \quad (2)$$

where d_{ij} is the Euclidean distance between S_i and S_j. The single-sided noise spectral density is N_0. The investigated constellation of 16 APSK is shown in Figure 2.

The sector regions of Figure 2 indicate the correct judgment domains of the symbols. From the figure, it can be seen that each symbol in one ring has the same probability of being erroneously detected. 16 APSK has two constellation rings with radii R_1 and R_2, respectively. The radius ratios are denoted as $\gamma_1 = R_2 / R_1$. Expression (1) can be written as:

$$P_e = \frac{1}{16}\sum_{i=1}^{16}P(E|S_i) = \frac{1}{16}[4P(E|S_1)+12P(E|S_5)] \quad (3)$$

where

$$P(E|S_1) = P(S_1 \to S_2)+P(S_1 \to S_3)+P(S_1 \to S_4)$$
$$+P(S_1 \to S_5)+P(S_1 \to S_6)$$
$$P(E|S_5) = P(S_5 \to S_1)+P(S_5 \to S_4)+P(S_5 \to S_6)$$

The Euclidean distance d_{ij} of (2) is given by:

$$d_{1,2} = \sqrt{2}R_1, \quad d_{1,4} = \sqrt{R_1{}^2+R_2{}^2 - 2R_1R_2\cos\left(\frac{\pi}{6}\right)}$$

$$d_{1,5} = R_2 - R_1, \quad d_{4,5} = \left(2\sin\frac{\pi}{12}\right)R_2 \quad (4)$$

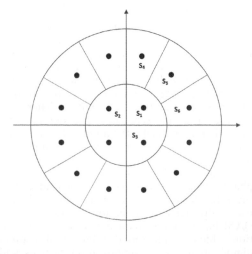

Figure 2. Investigated constellation diagram of 16 APSK.

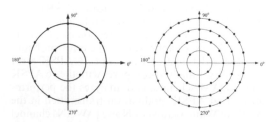

Figure 1. Constellation of 16 APSK and 64 APSK.

Substituting (2) and (4) into (3) and making use of the distance relations $d_{1,2}=d_{1,3}, d_{1,4}=d_{1,6}$, we obtain the approximate symbol error rate expression for 16 APSK:

$$P_e' \le \frac{1}{2}Q\left(\sqrt{\frac{\left(1+\gamma_1^2-2\gamma_1\cos\frac{\pi}{6}\right)}{2\alpha'}\frac{E_s'}{N_0}}\right) + \frac{1}{2}Q\left(\sqrt{\frac{1}{\alpha'}\frac{E_s'}{N_0}}\right)$$

$$+Q\left(\sqrt{\frac{(1-\gamma_1)^2}{2\alpha'}\frac{E_s'}{N_0}}\right) + \frac{3}{2}Q\left(\sqrt{\frac{2\gamma_1^2\sin^2\frac{\pi}{12}}{\alpha'}\frac{E_s'}{N_0}}\right)$$

$$(5)$$

where

$$E_s' = \left(4+12\gamma_1^2\right)R_1^2/16 = \alpha' R_1^2, \ \alpha' = \left(1+3\gamma_1^2\right)/4$$

As the same derivation process, the expression of 32 APSK can be expressed as:

$$P_e'' \le \frac{1}{4}Q\left(\sqrt{\frac{\left(1+\gamma_1^2-2\gamma_1\cos\frac{\pi}{6}\right)}{2\alpha''}\frac{E_s''}{N_0}}\right)$$

$$+\frac{1}{2}Q\left(\sqrt{\frac{(1-\gamma_1)^2}{2\alpha''}\frac{E_s''}{N_0}}\right) + \frac{3}{4}Q\left(\sqrt{\frac{2\gamma_1^2\sin^2\frac{\pi}{12}}{\alpha''}\frac{E_s''}{N_0}}\right)$$

$$+\frac{1}{4}Q\left(\sqrt{\frac{1}{\alpha''}\frac{E_s''}{N_0}}\right) + Q\left(\sqrt{\frac{2\gamma_2^2\sin^2\frac{\pi}{16}}{\alpha''}\frac{E_s''}{N_0}}\right)$$

$$+\frac{5}{4}Q\left(\sqrt{\frac{\left(\gamma_2^2+\gamma_1^2-2\gamma_1\gamma_2\cos\frac{\pi}{12}\right)}{2\alpha''}\frac{E_s''}{N_0}}\right)$$

$$+\frac{1}{2}Q\left(\sqrt{\frac{\left(\gamma_2^2+\gamma_1^2-2\gamma_1\gamma_2\cos\frac{5\pi}{48}\right)}{2\alpha''}\frac{E_s''}{N_0}}\right)$$

$$(6)$$

where

$$E_s'' = \left(4+12\gamma_1^2+16\gamma_2^2\right)R_1^2/32 = \alpha'' R_1^2,$$
$$\alpha'' = \left(1+3\gamma_1^2+4\gamma_2^2\right)/8$$

The 32 APSK is characterized by three constellation rings having 4, 12 and 16 uniformly spaced APSK points with radii R_1, R_2 and R_3, respectively. The radius ratios are denoted as $\gamma_2 = R_3/R_1$. Our analysis focuses on a 4+12+20+28-APSK signal constellation. We obtain the approximate SER expression for 64-APSK, which is given by:

$$P_e \le \frac{1}{8}Q\left(\sqrt{\frac{\left(1+\gamma_1^2-2\gamma_1\cos\frac{\pi}{6}\right)}{2\alpha}\frac{E_s}{N_0}}\right)$$

$$+\frac{1}{4}Q\left(\sqrt{\frac{(1-\gamma_1)^2}{2\alpha}\frac{E_s}{N_0}}\right) + \frac{3}{8}Q\left(\sqrt{\frac{2\gamma_1^2\sin^2\frac{\pi}{12}}{\alpha}\frac{E_s}{N_0}}\right)$$

$$+\frac{1}{8}Q\left(\sqrt{\frac{1}{\alpha}\frac{E_s}{N_0}}\right) + \frac{3}{8}Q\left(\sqrt{\frac{\left(\gamma_2^2+\gamma_1^2-2\gamma_1\gamma_2\cos\frac{\pi}{10}\right)}{2\alpha}\frac{E_s}{N_0}}\right)$$

$$+\frac{1}{2}Q\left(\sqrt{\frac{(\gamma_2-\gamma_1)^2}{2\alpha}\frac{E_s}{N_0}}\right) + \frac{5}{8}Q\left(\sqrt{\frac{2\gamma_2^2\sin^2\frac{\pi}{20}}{\alpha}\frac{E_s}{N_0}}\right)$$

$$+\frac{5}{8}Q\left(\sqrt{\frac{\left(\gamma_2^2+\gamma_3^2-2\gamma_3\gamma_2\cos\frac{\pi}{14}\right)}{2\alpha}\frac{E_s}{N_0}}\right)$$

$$+\frac{3}{4}Q\left(\sqrt{\frac{(\gamma_3-\gamma_2)^2}{2\alpha}\frac{E_s}{N_0}}\right) + \frac{7}{8}Q\left(\sqrt{\frac{2\gamma_3^2\sin^2\frac{\pi}{28}}{\alpha}\frac{E_s}{N_0}}\right)$$

$$(7)$$

where

$$E_s = \left(4+12\gamma_1^2+20\gamma_2^2+28\gamma_3^2\right)R_1^2/64 = \alpha R_1^2,$$
$$\alpha = \left(1+3\gamma_1^2+5\gamma_2^2+7\gamma_3^2\right)/16, \gamma_3 = R_4/R_1$$

In this expression, we assume $R_1 = 1$, $\gamma_1 = 2.75$, $\gamma_2 = 4.52$ and $\gamma_3 = 6.3$. Furthermore, we can obtain the theoretical relationship diagram of the bit error rate and the signal to noise ratio of 64 APSK (solid line in Figure 5) after the conversion of the SER to the BER.

3 PERFORMANCE SIMULATION

3.1 Simulation model

First, we determine the simulation block diagram of the 64 APSK system, which is shown in Figure 3.

Information symbol sequence of sixty-four possibilities is generated by using the uniform random number generator. Figure 4 shows the mapping of these symbols of 64 APSK into the corresponding points of the signal. Noise random number generators make the noise component $[n_c, n_s]$, and the decision device selects the signal point which is nearest to the vector. Finally, the counter records the erroneous symbol numbers. Figure 5 shows the SER simulative diagram of different signal to noise ratios transmitting 1000000symbols. From the figure, we can see that the simulative SER is less than the union bound of the theoretical SER.

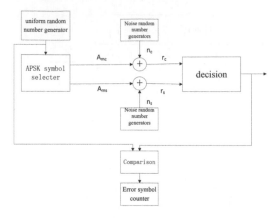

Figure 3. Simulation block diagram of the APSK system.

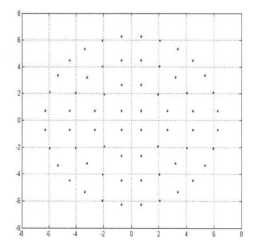

Figure 4. Mapping constellation of 64 APSK.

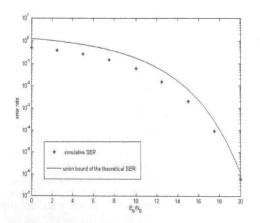

Figure 5. Relationship diagram of the error rate and Eb/N0 of 64 APSK.

3.2 *Performance comparison*

Compared with 16 APSK, we discuss the performance of the 64 APSK modulation scheme in AWGN. Figure 6 shows that the SNR deviation of 16 APSK and 64 APSK is about 7dB when the SER is less than 10^{-5}. Besides, we can draw the conclusion that as the order increases, E_b/N_0 also increases as well as the transmission bandwidth decreases, when maintaining the constant SER and data transmission rate. In other words, we increase the power to save the bandwidth.

We then compare the performance of 64QAM and 64 APSK, as shown in Figure 7. From the figure, we can see that the SER of 64 APSK is less than that of 64QAM. In other words, the performance of 64 APSK is better. Furthermore, 64 APSK easily compensates the nonlinear repeater, which can adapt to the satellite transmission channel of the worst linear characteristic.

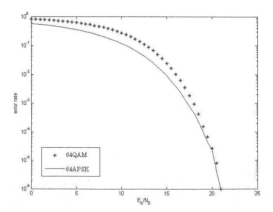

Figure 6. Contrast diagram of the error rate between 16 APSK and 64 APSK.

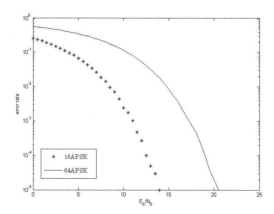

Figure 7. Contrast diagram of the error rate between 64QAM and 64 APSK.

4 CONCLUSION

The technique based on 64 APSK modulation scheme will have a bright future in a satellite broadband high data transmission system due to its advantages of making a full use of spectrum resources. This paper mainly studied the performance of 64 APSK. The research indicated that the new modulation scheme has an obvious improvement in performance in contrast to the 64QAM modulation scheme. The proposed technique has an increasingly higher data transmission rate in lower SNR as the constellation order increases, which will make further efforts to make great contributions to saving the satellite bandwidth and enhancing the capacity of the satellite.

REFERENCES

Afelumo, O. & Awoseyila, A. B. 2012. Simplified evaluation of APSK error performance. Electronics Letters. 48(14):886–888.

De, Gaudenzi. R. & Martinez, A. 2006. Performance analysis of Turbo-Coded APSK modulations over nonlinear satellite channels. IEEE Trans. 5(9):2396–2407.

ETSI EN 302 307 V1.1.1 (2005–03). Digital Video Broadcasting (DVB): Second generation framing structure, channel coding and modulation systems for broadcasting interactive services, news Gathering and other broadband satellite applications.

Fan, C. X. & Cao, L. N. 2008. Principles of Communications. Beijing: National Defense Industry Press.

Gaudenzi, R. D. & Martinez, A. 2006. Turbo-coded APSK modulations design for satellite broadband communications. International Journal of Satellite Communications and Networking. 24(4):261–281.

Sung, W. & Kang, S. 2009. Performance analysis of APSK modulation for DVB-S2 transmission over nonlinear channels. Int. J Satell Commun Netw. 27(6):295–311.

Wang, L. N. & Le, G. X. 2000. MATLAB and Communication Simulation. Beijing: The People's Posts and Telecommunications Press.

Zhuang, N. & Yu, H. Y. 2011. Symbol error ratio computation method for cross QAM signals. Computer Engineering. 37(13):279–281.

Energy Science and Applied Technology – Fang (Ed.)
© 2016 Taylor & Francis Group, London, ISBN 978-1-138-02833-3

Performance investigation of blind equalization and frequency offset estimation scheme in optical coherent MSK system

Lihong Yan, Jing He, Ming Chen, Jin Tang & Lin Chen
Key Laboratory for Micro/Nano Optoelectronic Devices, Ministry of Education,
College of Computer Science and Electronic Engineering, Hunan University, Hunan, China

ABSTRACT: The performance of blind equalization and frequency offset estimation scheme in coherent optical Minimum-Shift Keying (MSK) communication is investigated. At the receiver, blind equalization scheme is used to compensate fiber dispersion and nonlinearity in the electrical field. Meanwhile, a frequency offset scheme based on Fast Fourier Transform (FFT) and Chirp Z-Transform (CZT) can be used to frequency offset estimation and recovery. The simulation results show that the 10-Gb/s MSK signals can be recovered successfully using the blind equalization and frequency offset estimation scheme.

Keywords: Blind equalization, Minimum-Shift Keying (MSK), Fast Fourier Transform (FFT), Chirp Z-Transform (CZT)

1 INTRODUCTION

Recently, to achieve high spectrum efficiency, advanced optical modulation formats have been studied widely in the optical transmission system (G. W. Lu 2009, D. Xie 2014, J. Y. Mo 2005). More attention has been paid to digital modulation technology including the constant envelope modulation technique with minimum frequency spectrum occupancy. Among various advanced optical modulation formats, the spectrum efficiency of minimum-shift keying (MSK) is higher than that of the on-off keying (OOK) and Phase Shift Keying (PSK) at the same bit rate (J. Y. Mo 2006). It has the properties of continuous phase, high spectrum efficiency and low side-lobes (J. Y. Mo 2005, J. Y. Mo 2006, Y. Y. Zhan 2013). To achieve high spectrum efficiency and compensate transmission impairments, Digital Signal Processing (DSP) has been applied to the optical coherent MSK system. In this paper, the blind equalization (H. J. Wang 2014) and frequency offset estimation scheme (H. J. Leng 2012) for coherent optical Minimum-Shift Keying (MSK) communication is investigated. The simulation results show that the blind equalization and frequency offset estimation scheme at the receiver can compensate transmission impairments and frequency offset effectively.

2 PRINCIPLE OF BLIND EQUALIZATION AND FREQUENCY OFFSET SCHEME

For each $nT_s \leq t < (n+1)T_s$, by using the Laurent decomposition (P. Laurent, 1986, W. Shieh 2008),

the transmitted signal of the MSK signal can be expressed as

$$s(t) = \sqrt{\frac{2E_s}{T_s}} \sum_{n \in Z} x(n) c_0 (t - kT_s) \qquad (1)$$

where T_s is the symbol duration and $c_0(t)$ is the nonzero shaping pulse in the interval $[0, 2T_s]$.

Based on a coherent optical MSK system, Fig. 1 shows a block diagram of the principle of the blind equalization and Frequency Offset Estimator (FOE). First, the generated MSK signals at the transmitter are transmitted into the Standard Single-Mode Fiber (SSMF). Then, the blind equalization is used to compensate fiber dispersion and nonlinearity of the received MSK signals. Here, we use spatial-diversity equalizers (FSE) with 9 T/2 taps. In addition, $g(n)$ is the tap coefficient column vector of the equalizer, $y(n)$ is the column vector of the equalizer input, and $z(n)$ is the equalizer output. Subsequently, a frequency offset esti-

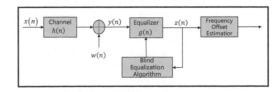

Figure 1. Block diagram of the blind equalization and frequency offset estimator.

mator is applied to estimate the phase rotation caused by Frequency Offset (FO). After the blind equalization and frequency offset recovery, a decoder block is utilized to demodulate the received MSK signals.

To implement the blind equalization for the received MSK signals, the output $z(n)$ of the equalizer can be obtained by the sequence $x(n)$:

$$z(n) = \sum_{k \in Z} f_k x(n-k) + v(n) \qquad (2)$$

where $f_k = \sum_l g_l h_{k-l}$, $v(n) = \sum_{k \in Z} g_k w(n-k)$, for each k and $g(z) = \sum_k g_k z^{-k}$ is the transfer function of the equalizer. For the blind equalization, the cost function can be expressed as

$$J(g) = E\left[\left(|z(n)|^2 - 1\right)^2\right]$$
$$+ E\left[\left(\rho(h)|z(n) + z(n-1)|^2 - 1\right)^2\right] \qquad (3)$$

with

$$\rho(h) = \left[\sin(\pi h / 2) / \sin(\pi h)\right]^2 \qquad (4)$$

In the MSK modulation format, $\rho(h) = 1/2$. $J(g)$ is the minimum when the output of the equalizer coincides with a delayed and rotated version. After the blind equalization, the frequency offset estimator is used to compensate the frequency offset. Based on the maximization of the periodogram of $z^4(n)$, the frequency offset estimator can be described as

$$\Delta f = \frac{1}{4} \arg \max_{\phi \in [-R_S/2, R_S/2]} f(\phi) \qquad (5)$$

with

$$f(\phi) = \left| \frac{1}{N} \sum_{n=0}^{N-1} z^4(n) e^{-2j\pi n\phi} \right|^2 \qquad (6)$$

where R_S is the sampling is rate and N is the number of samples. It is similar to the FFT, so we can use the FFT to obtain a peak frequency. It is four times the corresponding frequency offset. The frequency offset estimation scheme consists of two steps. First, a coarse step is introduced to calculate the range of the frequency offset with a small number of samples. Second, a fine step employs Chirp Z-Transform (CZT) to search the maximum value around the roughly estimated frequency offset (H. J. Leng 2012).

3 SIMULATION RESULTS AND DISCUSSION

To investigate the performance of the blind equalization and frequency offset estimation scheme, numerical simulations are performed in a 10-Gb/s coherent optical MSK system. The laser operates at a wavelength of 1550 nm without the laser linewidth, and the dispersion coefficient and Differential Group Delay (DGD) of SSMF are 16.75 ps/nm/km and 0.2 ps/km, respectively. The noise figure of the EDFA1 and EDFA2 is 4 dB.

The relationship between the BER and the Optical Signal Noise Ratio (OSNR) is shown in Figure 2. Assuming that the frequency offset is ignored, taking into account the blind equalization scheme, the Bit Error Rate (BER) performance of the Back to Back (BTB) and transmission over 100 km SSMF are measured. From Figure 2, it can be seen that the two measured BER curves have

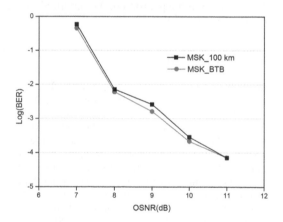

Figure 2. BER curves of the BTB and 100 km SSMF transmission using the blind equalization scheme.

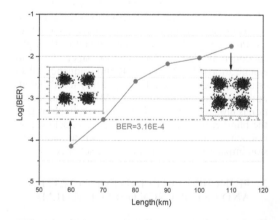

Figure 3. BER versus length of SSMF at the OSNR of 9 dB.

similar characteristics. The simulation results indicate that the blind equalization scheme can eliminate the transmission impairments and recover the received MSK signals effectively.

The impacts of the length of SSMF on the BER are shown in Figure 3. Considering the incorporation of the CD and PMD in the simulation, the distortion compensation capability of the blind equalization scheme can be investigated. Considering that the SSMF length is 60 km and 110 km, respectively, Fig. 3 shows the corresponding constellations of the received MSK signals. It is shown that the BER increases as the length of SSMF varied from 60 km to 110 km. After 70 km SSMF transmission, at the OSNR of 9 dB, the coherent optical MSK system using the blind equalization scheme can achieve a BER of 3.16×10^{-4}.

After the back-to-back (BTB) and 100 km SSMF transmission, considering that the frequency offset estimation scheme is used, Fig. 4 shows the measured BER for the coherent optical MSK system. The line-width of the laser is ignored, and the frequency offset is set as 0.6 GHz. Compared with the BTB case, at a BER of 1×10^{-3}, there is only 0.3 dB OSNR penalty after 100 km SSMF transmissions. The simulation results show that the frequency offset estimation scheme is insensitive to fiber dispersion and ASE noise. In addition, it can compensate the FO successfully.

The Normalized Frequency Variance (NFV) of the frequency offset estimation scheme versus OSNR is shown in Figure 5. In addition, the NFV is defined as $E[|\Delta \hat{f} T_s - \Delta f T_s|^2]$, where $\Delta \hat{f}$ is the estimated frequency. Here, the Δf is set as 0.2 GHz and 1.25 GHz, respectively. The frequency offset Δf can be randomly chosen from -1.25 GHz to 1.25 GHz. It covers most of the frequency offset

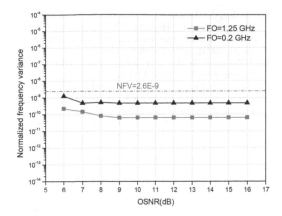

Figure 5. Normalized frequency variance of the frequency offset estimation algorithm versus OSNR.

ranges that the scheme can estimate. As can be seen from Figure 5, the NFV of the FOE scheme is lower than 2.6×10^{-9}. It is about 6.4×10^{-11} at a Frequency Offset (FO) of 0.2 GHz and 4.8×10^{-10} at a FO of 1.25 GHz, respectively. The simulation results show that the frequency offset estimator can reach a high accuracy at a low OSNR.

4 CONCLUSION

In this paper, the performance of blind equalization and frequency offset estimation scheme for coherent optical Minimum-Shift Keying (MSK) communication is investigated and analyzed. The simulation results show that the transmission impairments can be eliminated by the blind equalization. Meanwhile, the frequency offset estimation scheme can compensate the frequency offset ranging from -1.25 GHz to 1.25 GHz effectively in a 10-Gb/s coherent optical MSK communication system.

ACKNOWLEDGMENT

This work was supported by the National Natural Science Foundation of China (61307087, 61377079), the National "863" High Tech Research and Development Program of China (2011 AA010203), and the Fundamental Research Funds for the Central Universities and Young Teachers Program of Hunan University.

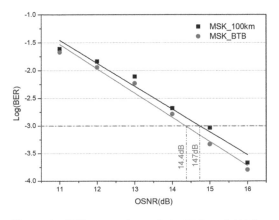

Figure 4. BER comparison of the BTB and 100 km SSMF transmission using the frequency offset estimation scheme.

REFERENCES

D. Xie, J. He, L. Chen, J. Tang, and M. Chen, 2014: Chinese Optics Letters, Vol. 12 (2014) No. 4, 040604.

G. W. Lu, T. Sakamoto, A. Chiba, T. Kawanishi, K. Miyazaki, and K. Higuma, 2009: European Conference on Optical Communication (Vienna, September 20–24, 2009), 5287111(2009), p.1.

H. J. Wang, J. He, L. Chen, and J. Tang, 2014: Optical Engineering, Vol. 53 (2014) No. 4, 046107.

H. J. Leng, S. Yu, X. Li, M. Y. Lan, P. Liao, T. Y. Wang, and W. Y. Gu, 2012: IEEE Photonics Technol. Lett, Vol. 24 (2012) No. 9, p. 787.

J. Y. Mo, Y. Dong, Y. J. Wen, S. Takahashi, Y. X. Wang, and C. Lu, 2005: European Conference on Optical Communication (Glasgow, Scotland. September 25–29, 2005), (2005), p.781.

J. Y. Mo, Y. J. Wen, Y. Dong, Y. X. Wang, and C. Lu, 2006: Optical Fiber Communication Conference. (California, American, March 5–10, 2006), 1636824 (2006).

P. Laurent, 1986: Communications, IEEE Transactions on, Vol. 34 (1986) No. 2, p. 150.

W. Shieh, H. Bao, and Y. Tang, 2008: Opt. Express, Vol. 16 (2008) No.2, p. 841.

Y. Y. Zhan, M. Zhang, M. T. Liu, and X. Chen, 2013: Chinese Optics Letters, Vol. 11 (2013) No. 3, 030604.

Energy Science and Applied Technology – Fang (Ed.)
© 2016 Taylor & Francis Group, London, ISBN 978-1-138-02833-3

Software-defined vehicular networks: Opportunities and challenges

M. Zhu, Z.P. Cai & M. Xu
College of Computer, National University of Defense Technology, Changsha, Hunan, China

J.N. Cao
Department of Computing, The Hong Kong Polytechnic University, Hong Kong, China

ABSTRACT: Vehicular Ad Hoc Networks (VANET) are an important network environment to improve travailing safety, comfort and efficiency. However, due to the highly dynamic network topology and large network scale, efficient network management in VANET remains a big challenge. To simplify network management, Software-Defined Networking (SDN) has rapidly emerged as a widely accepted networking paradigm for wired networks. In this paper, we propose Software-Defined Vehicular Networks (SDVN) that integrates the SDN paradigm with VANET. We show possible opportunities that SDVN may bring to VANET, such as a simpler network management, a better network performance and a faster network innovation. We also point out some new research challenges caused by the inherent characteristic of VANET in SDVN.

Keywords: Vehicular Ad Hoc Networks; Software-Defined Networking; Open Flow

1 INTRODUCTION

Vehicular Ad Hoc Networks (VANET) are regarded as an important component in the intelligent transportation system (ITS) (Faezipour et al. 2012). They promise to enhance traveling safety, comfort and efficiency via communications between vehicles. Due to the high mobility of vehicles, VANET have some unique characteristics such as frequent topology changes, intermittent network connection and dynamic network density. These inherent characteristics make it difficult to manage VANET efficiently.

To simplify network management, Software-Defined Networking (SDN) has been proposed recently, which has rapidly become a widely accepted networking paradigm (Kirkpatrick 2013). In SDN, network control functionality is decoupled from forwarding functionality and becomes directly programmable in a logically centralized controller. This networking programmability can not only simplify network management, but also accelerate network innovation. In SDN-based networks, network operators can add new networking functions by just updating software in the controller instead of updating all software in each switch. Besides, SDN can support fine-grained flow control, so that new protocols can be tested in the real-word environment together with a normal traffic.

Currently, SDN-based solutions have been widely adopted in various kinds of networks, such as data center networks (Curtis et al. 2011), wide area networks (Jain et al. 2013), wireless access networks (Dely et al. 2011) and wireless sensor networks (Luo et al. 2012). However, no previous research has considered the introduction of the SDN paradigm to VANET.

In this paper, we present Software-Defined Vehicular Networks (SDVN), which integrate the SDN with VANET to simplify network management and accelerate network innovation. We show opportunities brought by SDVN, such as easier network management, better network performance and faster network innovation. Besides, we also point out new research challenges that should be addressed in SDVN. To our best knowledge, this is the first work that tries to integrate SDN with VANET.

The rest of the paper is organized as follows. We introduce some background of SDN and VANET in section II. In section III, we present the networking architecture of SDVN. We show possible benefits that SDN may bring to VANET in section IV and new challenges that need to be addressed in section V. Section VI concludes this work.

2 BACKGROUND

2.1 *Networking issues in VANET*

VANET is an instantiation of Mobile Ad Hoc Networks (MANET). MANET have no fixed infrastructure and instead rely on ordinary nodes to perform routing of messages and network management functions. However, VANET faces unique challenges in transmitting messages for various applications. Here, we list three key issues in VANET that make network management difficult.

a. Rapid network topology change. Since communication links in VANET have a very short duration time, a connection from the source to the destination may quickly break down. Thus, traditional routing protocols for MANET (e.g. AODV and OLSR) have poor performance in VANET. Besides, a rapid network topology change may cause a huge routing overhead in maintaining network topology.

b. Frequent network fragmentation. In VANET, traffic lights always separate vehicles into groups. Link quality among vehicles in different groups is too poor and causes network fragmentation. To reduce the affection of network fragmentation, delay tolerant routing protocols such as GPSR (Karp et al. 2000) have been proposed. The GPSR uses geographic positioning information instead of link information to select next forwarding hops. Without maintaining network topology, routing in GPSR has a lower overhead. However, since all forwarding decision is made according to local information, the GPSR may cause a long end-to-end packet delay.

c. Dynamic network density. The network density of VANET varies greatly in different scenarios. In the traffic jam situation, VANET is regarded as a dense network. While in suburban traffics or in night traffics, there are a few vehicles on the road and the network density is low. This dynamic network density requires adaptive and flexible network management that can automatically switch the network configuration between the high-dense network model and the sparse network model.

2.2 Basic principles of SDN

SDN is a novel networking paradigm that was first introduced to simplify network management in campus networks (McKeown & Anderson et al. 2008). The main principle of SDN is to separate the software control plane from the data forwarding plane using a centralized controller to provide network control functions. In SDN-based networks, the architecture of network devices is relatively simple, which can reduce the cost and improve the performance.

OpenFlow is a typical solution for SDN (Limoncelli 2012). It uses flow-entry to define which action a switch should take for an incoming packet. When a packet comes in, the switch first checks whether there is a flow-entry in its flow table that matches the packet. If there is one, the switch will take the action according to the flow-entry. Otherwise, the switch will send the packet to the controller and query a flow-entry so that the following packets in the same flow can be forwarded according to the flow-entry.

In SDN-based networks, the controller maintains a global network topology and utilizes this global view to generate global optimized routing paths. Switches do not need to exchange routing information with each other, thus the routing overhead between switches is low. Besides, it is easy for controller to dynamically switch the routing strategy according to network situations via inserting and deleting flow entries in switches. All these features may bring benefits for VANET to achieve an easier network management and a better network performance.

3 SOFTWARE-DEFINED VEHICULAR NETWORKS

Vehicular Ad Hoc Networks (VANET) utilize vehicle-to-vehicle communication to improve traveling safety, comfort and efficiency. Due to the frequent topology changes and intermittent network connectivity, the management of the network efficiently is a challenge. To improve the efficiency of network management, we propose Software-Defined Vehicular Networks (SDVN), a SDN-based networking architecture for VANET. With the help of SDN, networking functions become directly programmable to network operators. The network programmability can improve the efficiency of network management in VANET.

The overview architecture of SDVN is shown in Figure 1. SDVN consists of two types of nodes:

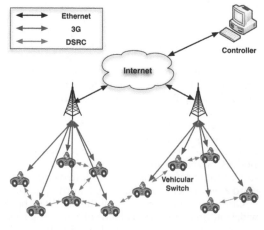

Figure 1. Architecture of SDVN.

vehicular switches and the controller. Different from a conventional vehicular router, there is no routing decision logic in the vehicular switch. Thus, the only function that vehicular switches support is just forwarding packets according to forwarding rules. Here, the forwarding rules are generated by the controller and sent to vehicular switches. The controller provides all networking services including routing, quality-of-service and access control. Compared with the traditional VANET, network management in SDVN is much easier. Network operators can configure the whole network via the controller instead of individually configuring each vehicular router. Besides, the networking functions can be directly reprogrammed and tested in SDVN. Thus, new networking functions can be supported and applied in the real-world network with a shorter period of time.

The communication devices on vehicular switches can be classified into two categories. One kind is the short-range communication device, such as the dedicated short-range communication (DSRC) device and the WiFi device. The other kind is the long-range communication device, such as the 3G device or the 4G Long Term Evolution (LTE) device. The short-range communication device can support the vehicle-to-vehicle high-bandwidth communication within a short communication range. Thus, they are used in data transmitting between vehicles and roadside units. The long-rang communication device can support the vehicle-to-Internet limited-bandwidth communication within a much longer communication range. Due to the high price of using cellular networks, the long-rang communication device is just used for Internet access from vehicles.

The centralized controller is connected to the Internet via Ethernet. Vehicular switches can consult the controller across the Internet via cellular networks. This is the control channel between vehicular switches and the controller. Using the control channel, vehicular switches can send request for packet-forwarding rules to the controller and receive the reply as well. In addition to request-and-reply messages, states of vehicles (e.g., vehicle's geographical coordination, velocity and one-hop communication neighbors) can also be sent to the controller through the control channel. Due to the large coverage of present-day cellular networks and seamless handoff between base-stations, the cellular network can provide stable controller access for vehicular switches in most of the urban environments.

Similar to the conventional VAENT, the application range of SDVN covers both safety-related applications such as vehicle collision avoidance and non-safety-related applications such as trip planning and media content sharing. Most safe-ty-related applications require real-time packet processing, while comfort applications pay more attention on network throughput. To meet different requirements of various applications, SDVN uses fine-grained flow management to support the quality of service.

4 OPPORTUNITIES

By exploring the use of the SDN paradigm in VANET, SDVN brings some opportunities for VANET, such as a simpler network management, a better network performance and a faster network innovation.

4.1 Simpler network management

SDN provides an efficient network management mechanism. Instead of individually configuring the network component on each switch, all network management in SDN can be done on the controller. In addition, switches in SDN have a simple architecture and can be controlled through standard interfaces. Thus, there are no control differences among various vendors' switches. This not only allows remote network management, but also greatly reduces the workload.

VANET is a network that consists of vehicles within a community. The number of vehicles, which depends on the size of the community, ranges from hundreds to thousands, or even more. Vehicles have a relatively high mobility, thus vehicles will frequently join in or leave the network. Network management in VANET, such as adding or removing new vehicles and network protocol upgrading, involves a huge cost using traditional mechanisms that configure parameters of each vehicular switch individually.

In SDVN, the controller controls all vehicular switches using standard interfaces. Network operators can remotely configure all vehicular switches using just a laptop. In addition, the network function update can be accomplished by just updating the software in the controller. With the help of SDN, network management of VANETs will be more efficient.

4.2 Better network performance

SDN can provide a global view of the whole network in the logically centralized controller. The controller can get networking statistics from switches, including packet count and flow duration time. These networking statistics can help networking applications, such as routing, load balancing and quality-of-service, make global optimizations and finally improve whole network's performance.

In traditional VANET, all vehicles are organized in a distributed way. Due to the highly dynamic network topology, packet routing performances such as throughput and end-to-end packet delay are poor. Many routing algorithms have been proposed to improve the routing performances, such as AODV, OLSR, GPSR and road-based forwarding (Saleet et al. 2011). However, since there is no central controller and each vehicular switch just chooses the routing path according to its local information, communication congestion and unbalanced message propagation are a serious matter of concern.

In SDVN, the central controller will collect states of all vehicles so that the global network topology can be easily obtained. With a digital map, the controller can also provide real-time traffic information on each road. All these data will be used to make global optimization in load balancing and quality-of-service, and finally improve the whole network's performance.

4.3 Faster network innovation

SDN is a promising network management paradigm to facilitate network innovation through network programmability. Because SDN supports fine-grained flow management, packets of different protocols can be treated using different forwarding rules. In SDN, new network protocols such as new routing protocol can be tested in the real-world network with a normal traffic. This feature can accelerate the development of new network applications.

In traditional VANET, a new network protocol must first be tested in simulation environments and small-scale test beds, and then be deployed in real-world networks. Due to the long-term period of developing, testing and deploying, innovation of VANET is very slow. Although a large number of protocols have been proposed for VANET, only a few have been tested in the real-world environment.

In SDVN, new protocols can be tested in the real-world environment with a low cost. All we need to do is just to label the testing flow and process it using the proposed protocol. Other application flows are still processed using normal protocols. With the help of SDN, VANET may encounter a fast development period.

5 CHALLENGES

Due to inherent characteristics of VANET such as high mobility of vehicles and a large network scale, there are some research challenges in SDVN. Safety-related applications in VANET require the real-time routing process for messages. Therefore, the controller has to shorten the flow setup time to meet the real-time requirement. Dynamic network topology caused by the high mobility of vehicles may lead to a short link duration. Vehicular switches need to consult the controller frequently, which brings a higher overhead. For some scenarios in which VANET contains a large number of vehicles, one controller may be not enough to provide network services, thus the cooperation between multiple controllers must be considered.

5.1 Low latency requirement

Since the vehicular switch has no routing computation logic, when the first packet of a flow arrives, the vehicular switch will request the controller for the flow-entry. When the controller receives the packet, it will generate a flow-entry for it and send the flow-entry back to the switch so that all the following packets of this flow will be forwarded automatically according to the flow-entry. Communications between the vehicular switch and the controller may introduce an additional latency that is called the flow-entry setup time.

This flow-entry setup time may bring a challenging issue for safety applications in VANET. In VANET, safety applications such as collision avoidance break warning or traffic light warning need to transmit messages to neighbor vehicles in the real time (less than 200 ms) (Sichitiu et al. 2008). However, the flow-entry setup time may be longer than 350 ms using the wireless channel. Thus, new application-specific mechanisms must be proposed in SDVN to meet this low latency requirement.

5.2 Short link duration

As mentioned previously, the first packet of a flow will lead to an installation of a flow-entry in the vehicular switch. The following packets of the flow will be processed automatically according to the flow-entry. This flow-entry will be stored in the flow table for a constant period, called the time-to-live. If no packets of that flow come for a time-to-live period, the flow-entry will be discarded. Otherwise, the flow-entry will stay in the flow table for another time-to-live period.

A flow-entry's duration time is the duration from a flow-entry's setup to its expiration. The average duration time of the flow-entry reflects the networking efficiency. If most of the flow-entries have a long duration time, the consulting frequency of the control plane is low, which means the overhead is low. Otherwise, if most of the flow-entries have a short duration time, the switch then needs to consult the control plane frequently, causing a huge overhead.

Unfortunately, the link duration time in VANETs is always short due to the high mobility of vehicles. It is shown that multi-hop paths that only use vehicles moving in the same direction on a highway have a lifetime comparable to the time needed to discover the path (Blum et al. 2004). This may lead to a short flow duration time and finally causes a high communication overhead. So, new solutions must be found to address this issue.

5.3 Large network scale

The network scale and node density of VANET varies depending on different scenarios. In the city centers or highways around big cities, the network scale could be quite large. As we have mentioned earlier, SDVN adopts a logically centralized controller. For a small network scale VANET, one controller may be enough to provide networking services for all vehicular switches. However, for scenarios in which the network scale is too large, we need to consider using multiple controllers to share the networking load.

In SDVN that has multiple controllers, new challenges may rise, such as consistency problem, load balancing problem and handover problem. Due to the inherent characteristic of VANET, these problems in SDVN may be more challenging than those in other SDN-based networks, such as SDN-based wide area networks. For example, the high mobility of vehicles leads to a fast node distribution change, so that efficient load balancing algorithms are required.

6 CONCLUSION

VANET is a special type of MANET that uses vehicle-to-vehicle communications to improve traveling safety, comfort and efficiency. Most research has been done to improve network performance of VANET, but due to the long-term developing, testing and implementing period, only a small number of these new protocols or algorithms have been tested and implemented in the real-world environment. SDN is a promising network management paradigm to facilitate network innovation through network programmability. Over the past few years, SDN has been widely accepted and adopted in various network environments such as data center networks and wide area networks. However, no efforts have been made for integrating the SDN paradigm with VANET.

In this paper, we propose Software-Defined Vehicular Networks (SDVN), which is a new networking architecture of SDN-based VANET. To our best knowledge, this is the first work trying to integrate the SDN paradigm with VANET. We show possible opportunities that SDVN may bring to VANET, such as a simpler network management, a better network performance and a faster network innovation. We also point out some new research issues in SDVN for further research in the future.

ACKNOWLEDGMENTS

The authors would like to thank the National Natural Science Foundation of China under Grant Nos. 61379144, 61379145 and 61272485 for their support.

REFERENCES

Blum J., Eskandarian A. & Hoffman L. 2004. Challenges of Intervehicle Ad Hoc Networks. IEEE Transactions on Intelligent Transportation Systems, 5(4): 347–351.

Curtis A.R., Kim W. & Yalagandula P. 2011. Mahout: Low-overhead datacenter traffic management using end-host-based elephant detection, in Proceedings of IEEE INFOCOM'11, 1629–1637.

Dely P., Kassler A. & Bayer N. 2011. OpenFlow for Wireless Mesh Networks, in Proceedings of ICCCN'11, 1–6.

Faezipour, M., Nourani, M., Saeed, A. & Addepalli, S. 2012. Progress and challenges in intelligent vehicle area networks, Communications of the ACM, 55(2): 90–98.

Jain S., Zhu M., Zolla J., Hölzle U., Stuart S., Vahdat A., Kumar A., Mandal S., Ong J., Poutievski L., Singh A., Venkata S. & Wanderer J. & Zhou J. 2013. B4: Experience with a Globally-Deployed Software DefinedWA, in Proceedings of ACM SIGCOMM'13, 3–14.

Kirkpatrick K. 2013. Software-defined networking, Communications of the ACM, 56(9): 16–19.

Karp B. & Kung H.T. 2000. GPSR:Greedy perimeter stateless routing for wireless networks, in Proceedings of MobiCom'00, 243–254.

Luo T., Tan H.P. & Quek T.Q.S. 2012. Sensor OpenFlow: Enabling Software-Defined Wireless Sensor Networks," IEEE Communications Letters, 16(11): 1896–1899.

Limoncelli T.A. 2012. OpenFlow: A Radical New Idea in Networking, ACM Communication, 10(6): 1–7.

McKeown N. & Anderson T. 2008. OpenFlow: enabling innovation in campus networks, ACM SIGCOMM Computer Communication Review, 38(2): 69–74.

Saleet H., Langar R., Naik K., Boutaba R., Nayak A. & Goel N. 2011. Intersection-Based Geographical Routing Protocol for VANETs: A Proposal and Analysis, IEEE Transactions on Vehicular Technology, 60(9): 4560–4574.

Sichitiu M.L. & Kihl M. 2008. Inter-Vehicle Communication Systems a Survey, IEEE Communications Surveys & Tutorials, 10(2): 88–105.

Energy Science and Applied Technology – Fang (Ed.)
© *2016 Taylor & Francis Group, London, ISBN 978-1-138-02833-3*

Research on management information system: A novel approach

Yichao Li
Huaxin College, Shijiazhuang University of Economics, Hebei, China

ABSTRACT: Over the past years, Management Information System (MIS)-related research has become a hot topic. In this paper, we design and implement a novel neural network and multi-agent-based intelligent MIS with a detailed discussion and efficient analysis. We propose a framework of a intelligent agent-based neural network classification model to solve the problem of gap between two applicable flows of intelligent multi-agent technology. We first discuss the crucial concepts of agent technology and agent-based system. Later, by analyzing the intelligent MIS, we propose the methodology and detailed process of the novel intelligent system. Finally, with the experimental results and established analysis by Dynamic Enterprise Modeling (DEM), we point out the real-world application of our proposed approach.

Keywords: Multi-Agent Analysis, Management Information System (MIS), Neural Network

1 STRUCTURE OF THE MULTI-AGENT INTELLIGENT MANGEMENT INFORMATION SYSTEM

1.1 The basic systematic architecture

From the perspective of information processing, the overall intelligent system can be considered as a distributed system composed of multi-layered, diverse, intelligent business units that cooperate with one another to accomplish a complex function. The business unit has the information processing and communicating ability to overcome single or multiple tasks. Moreover, it is layered and can be nested. A coordinated, multi-level, multi-agent system is built up with the knowledge, initiative and collaboration agent that has a greater processing efficiency. The architecture is shown in Figure 1. The entire system's structure is divided into three layers: interface layer, business process layer and system resource layer. The intelligent layer is a crucial part of the proposed MIS for the human-machine interface. The business process layer as a multi-agent system is the core of the entire system. The agent of different sizes, different levels and functions cooperates with each other to build up a nested community to treat the business. The system can be divided into the following types: (1) atomic agent, with a single tasking; (2) composite agent, composed of atomic agents; (3) sub-agent, constructed with numbers of load agent and atomic agent, its internal agent being managed and controlled. All the agents in the system coordinate and communicate with each other.

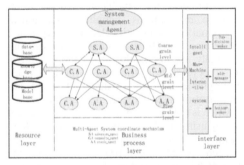

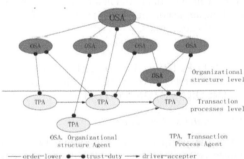

Figure 1. Architecture and dynamic business model.

The coordination mechanism for multi-agent systems is the key to determining the strength of the whole business process layer's functionalities. The system resource layer is the databases, knowledge base and model base used to store data, knowledge, models and other information resources.

1.2 The Approach to achieve system

According to the architecture model of the intelligent management information system, with the goal-driven strategy, the entire system development procedures are as follows (Mitra, S, 2002): (1) Requirements Analysis. The problem reduction method is adopted to decompose system requirement tasks. The decomposed relationship is represented by And OR Tree, and the allocation of resources, resource dependence, data flow and control flow path, time sequence and interaction time are labeled. (2) System Design. First, we get the And OR relationship according to granularity of problem solving, solution strategy and needs analysis. We design the atomic agent from the tasks represented by the nodes of And OR Tree's, and induce the logical form of the composite agent and the sub-system agent bottom-up. The structure association and sematic association are described by the hierarchy An OR sematic network, and used to record the location and condition of the sematic trigger. Next, we use 6-tuple to describe the behavior between agents. Then, we design communication mechanisms including communication, original communication language types and the determine formats of the communication content. Finally, we design the coordination control and conflict resolution strategies of the system. (3) Agent implementation. According to the hierarchical organizational structure in the And OR images type, which is induced from step 2, we get the agent system by using the Mode-Agent mapping system. (4) Management information system based on the multi-agent system is constructed to develop the agent, which is spread out in all corresponding network nodes according to the network structure of the application units.

2 ENTERPRISE-SUPPORTED MULTI-AGENT DYNAMIC MANAGEMENT INFORMATION SYSTEM

2.1 Analysis of structure

From the perspective of the system theory, the enterprise is a system composed of several departments, and the departments itself are also systems including a number of sub-sections. In the long term, each sector runs for a common purpose, but the short-term goal of each department is not the same or can even conflict. Corporate sectors can be regarded as a collection to complete certain business functions. As a result, departments could be abstracted as a black box with input and output interfaces and specific function. In addition, the successful completion of department functions requires not only departmental inside efforts but also the coordination between departments. Thus, the key factors to achieve the coordination between the departments are the intelligence, autonomy, responsible and coordination of human, the main body of departments. In general, the abstract of the department is extremely similar to the agent characteristic, thus we can use an abstract representation of one agent to the various departments of enterprises. In summary, the department and the agent characteristic feature abstract results are very similar, thus we can represent the various departments of enterprises by the agent in the abstract.

2.2 Analysis of business processes (Roya, A, 2009)

Business process could be subdivided into a series of business processing elements. The business processing is a coarse-gained business process, with certain stability and atomicity, such as product demand analysis, product design, product manufacturing and product sales are part of business processing elements, and atomicity is an important feature of the business process. Analysis and access to the business process element should be independent of the company's existing departments. Business process is certain stability. The adjustment of business processes can be understood as the recombination of business process, similar to the adjustment of the departmental functions.

2.3 Multi-Agent System-Based Dynamic Model

Combining the above analysis of the enterprise's organization structure and business process, the specific business sector of enterprise must be separated from its own business processing function in order to support the dynamic enterprise model, and the multi-agent system architecture supporting the dynamic business model is shown in Figure 1 (right). The whole system is divided into two layers: organizational structure layer and business process layer. The organizational structure layer has a novel structure. The specific business department is described with the Organizational Structure Agent (OSA), corresponding to the organizational structure agent. That the hierarchy relationship between organizational structure agents is consistent with the company's existing relationships between the departments ensures that the organizational structure layer is consistent with the current organizational structure. In the business process layer, the business processing unit is represented by the agent of the business processing element (TAP), and the entire business processing is represented by the combination of TAP orderly (Stuart J. R, 2003).

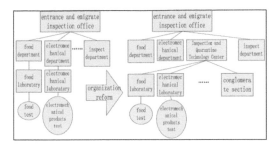

Figure 2. Example for Organization and Business Restructuring.

3 EXPERIMENT AND EXAMPLE DESIGN

All kinds of testing laboratories of a provincial Entry-Exit Inspection and Quarantine Bureau (CIQ) were managed by their corresponding administrative law enforcement offices instituted before the institutional reform, such as food laboratories belonged to the food department and mechanical and electrical laboratories belong to the EMD. Besides, their main tasks were to detect the entry-exit commodity assigned by the corresponding administrative law enforcement offices. After China joined the WTO, it has been trying to adapt to the novel international atmosphere. Therefore, a new independent business unit, inspection and quarantine technology center, was set up to manage all the testing laboratories separated from administrative offices. It takes all legal entry-exit commodity inspection tasks commissioned by the administrative law enforcement offices and detects services commissioned by other units or individuals in society, as shown in Figure 2. The entire inspection and quarantine intelligent management information system docs not nccd to rc-dcsign and develop, but only OSA and TPA in the system were treated to achieve the organization and business restructuring, as shown in Figure 2.

4 SUMMARY AND CONCLUSION

The present research introduces agent technology to the intelligent management information system.

It aims at researching the theory and application of the multi-agent intelligent management information system to design a framework model of the multi-agent intelligent management information system and provide its implementation strategy. The system has the virtues of sharing resources, easy to expand, high reliability and great flexibility; furthermore, the agent can coordinate with each other. In the future, we plan to use the mathematical optimization (Chen, Guang-Sheng, 2013, Perera, Kanishka, 2006, Cao, Yuping, 2014) to modify our system.

REFERENCES

Bobek, S. and Perko, I., 2006. Intelligent Agent Based Business Intelligence. University of Maribor, Faculty of Economics and Business, Razlagova ulica14, SI-2000 Maribor, Slovenia.

Chen, Guang-Sheng, et al. "Existence of three solutions for a nonlocal elliptic system of-Kirchhoff type." Boundary Value Problems 2013.1 (2013): 1–9.

Cao, Yuping, and Chuanzhi Bai. "Multiplicity of Nontrivial Solutions for a Class of Nonlocal Elliptic Operators Systems of Kirchhoff Type." Abstract and Applied Analysis. Vol. 2014. Hindawi Publishing Corporation, 2014.

Mitra, S. Pal, S.K. and Mitra, P. 2002. Data Mining in Soft Computing Framework: A Survey. IEEE Trans. Neural Networks, 13(1):3–14.

Perera, Kanishka, and Zhitao Zhang. "Nontrivial solutions of Kirchhoff-type problems via the Yang index." Journal of Differential Equations 221.1 (2006): 246–255.

Roya, A., Norwati, M., Nasir, S. (2009). Training Process Reduction Based On Potential Weights Linear Analysis To Accelerate Back Propagation Network, Accepted by International Journal of Computer Science and Information Security (IJCSIS), Vol. 3, No. 1, and July.

Stuart J. R. and Peter N., 2003. Artificial Intelligence: A Modern Approach. Second edition, Upper Saddle River, NJ: Prentice Hall, ISBN 0–13–790395–2.

Energy Science and Applied Technology – Fang (Ed.)
© 2016 Taylor & Francis Group, London, ISBN 978-1-138-02833-3

Deceptive jamming suppression method for LFM fuze based on STFRFT

Penglei Nian, Guolin Li & Fei Li
Naval Aeronautical and Astronautical University, Yantai, China

ABSTRACT: The chirp rate of SVLFM (Slope Varying Linear Frequency Modulation) fuze cannot vary at random, which must be within a certain percentage of the LFM chirp signal rate, so the fuze can still be affected by the false target deceptive jamming. In response to above-mentioned problem, a new method based on STFRFT (Short-Time Fractional Fourier Transform) is proposed in this paper. The chirp rate of SVLFM fuze varies at each PRI (Pulse Repetition Interval), so the target echo and the deceptive signal can be separated in the fractional Fourier domain. Because of the characteristic that the maximum amplitude of STFRFT is linear with the window width when the rotation angle is matching with the chip rate, but the maximum amplitude stays unchanged if the chirp rates is not matching with the chip rate, then the target echo can be distinguished from the false target signal effectively on a low SJR. The simulation results show that the principle is right and the method has a good ability to suppress deceptive jamming.

Keywords: linear frequency modulation; chirp rate; STFRFT; deceptive jamming suppression

1 INTRODUCTION

The LFM signal has been used in a high resolution radar and fuze because of high distance resolution and detection range. Various types of jamming signals have been used to confuse the fuze receiver, especially a destructive jamming signal known as a digital radio frequency memory repeat jammer (Dun Peng 2013, GE Wang Qinglin 2012, Hu Min 2010), which could be composed of replicas of the transmitted fuze signal. The jamming equipment can construct a replica of the transmitted fuze signal or transmit the jamming signal at any time through an appropriate time delay unit.

There are some methods used for suppressing the deceptive jamming through the analysis of the time-frequency characteristic. A method for eliminating smart jamming (Lu Gang 2011), based on the transmitting verification signal, was proposed, but the eliminating equipment is too big to use in fuze. The method (Mehrdad Soumekh, 2006) can separate the echo signal from single false target jamming, but is out of function when the jamming signal consists of several components. For the SVLFM (M. Greco 2008) fuze, the chirp signal rate varies at each PRI, and the method makes the best use of the autocorrelation and cross-correlation of the echo signal and the jamming signal. However when the chirp rate of the deceptive jamming signal is close to that of the echo signal, and the SJR is lower, the output of the jamming can still jam the fuze. The FRFT

(Tao Ran 2010, Tang Pengfei 2013, Wang Cunwei 2010, Wen Jingyang 2012) can be useful for separating the LFM signal component, but when the chirp rate of each component is close to each other, it will enhance the computational complexity

In this paper, we propose a method to distinguish echo signal from the jamming signal based on the STFRFT. First, the target echo signal and the deceptive signal can be separated in the fractional Fourier domain, and then we can get the output of each component in the shot-time fractional domain, which can be compared with the output in the fractional domain. As a result, the echo signal can be distinguished from the jamming signal.

2 THE INFLUENCE OF THE FALSE TARGET DECEPTIVE JAMMING

The mth pulse period signal of the SVLFM fuze can be written as follows:

$$s_m(t) = rect(\frac{t}{T_m})\exp\left[j\pi(\mu + \xi_m)t^2 + j\varphi \right] \tag{1}$$

where $rect(\frac{t}{T}) = \begin{cases} 1, |t| < \dfrac{T}{2} \\ 0, others \end{cases}$ and μ is the chirp rate.

The chirp rate perturbation parameter ξ_m is chosen via a random number generator that is known to the user. B is the bandwidth, and the pulse width is given as: $T_m = \frac{B}{\mu + \xi_m}$, which always changes with ξ_m.

The jamming equipment can get the fuze signal and transmit the deceptive signal, which is highly related to the fuze signal. So, the frequency modulation rate should be changed at each PRI. However, ξ_m can only change in a certain percentage of the chirp rate μ. In this case, the chirp rate of the deceptive signal can be close to that of the echo signal.

Figure 1 shows the pulse compression of different ξ_m, and the difference in ξ_m can reduce the relevance between the deceptive signal and the echo signal. However, when the chirp rate of the deceptive jamming signal is close to that of the echo signal and the SJR is lower, the output of the jamming can still jam the fuze. So, we must try to suppress the jamming.

3 ANTI-DECEPTIVE JAMMING BASED ON THE STFRFT

3.1 Signal separation based on the FRFT

The αth order FRFT of a signal $s(t)$ is defined as

$$
S_\alpha(u) = \int_{-\infty}^{+\infty} K(u,t)s(t)dt
$$
$$
= \begin{cases} \sqrt{1 - j\cot\alpha} \int_{-\infty}^{+\infty} s(t)e^{j\pi(u^2\cot\alpha - 2ut\csc\alpha + t^2\cot\alpha)}dt, & \alpha \neq n\pi \\ s(u), & \alpha = 2n\pi \\ s(-u), & \alpha = (2n+1)\pi \end{cases}
$$
$$(2)$$

Where, α is the rotation angle, n is the integer. α Is the parameter determining the chirp rate employed by the transform? If the chirp rate is μ and $\alpha = -arc\cot\mu$, the chirp signal will be a δ function in fractional Fourier domain. If α and μ is mismatching, there will no δ function. When the signal contains several components, we can get every component step by step based on each

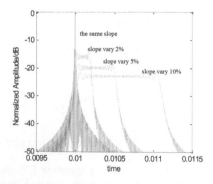

Figure 1. Pulse compression of different signals.

signal component power. So through FRFT, echo signal and deceptive jamming component can be separated from the signal received by the fuze.

3.2 Analysis of STFRFT principle

At some PRI, the chirp rate perturbation parameter $\xi_m = 0$, the signal at some PRI is written as follows:

$$
s_t(t) = rect(\frac{t}{T})A\exp[j\pi\mu t^2 + j\varphi] \tag{3}
$$

Where, we can define $\varphi = 0$, through fractional Fourier transform, $s_t(t)$ will be transformed into

$$
S_a(u) = \int_{-\infty}^{+\infty} K(u,t)s(t)dt
$$
$$
= A\sqrt{1 - j\cot\alpha}\, e^{j\pi u^2\cot\alpha}\int_{-\frac{T}{2}}^{\frac{T}{2}} e^{j\pi t^2(\mu + \cot\alpha) - j2\pi ut\csc\alpha}dt \tag{4}
$$

Because the optimal rotation angle is $\alpha_m = -arc\cot\mu$ and $\alpha \in (0,\pi)$, so the FRFT can be defined as

$$
S_{\alpha_m}(u) = A\sqrt{1 - j\cot\alpha}\, e^{j\pi u^2\cot\alpha_m}
$$
$$
\times \int_{-\frac{T}{2}}^{\frac{T}{2}} e^{j\pi t^2(\mu + \cot\alpha_m) - j2\pi ut\csc\alpha}dt
$$
$$
= \frac{A}{\sqrt{\sin\alpha_m}}e^{j\frac{\alpha_m}{2} + j\frac{3\pi}{4} - j\pi\mu u^2}\int_{-\frac{T}{2}}^{\frac{T}{2}} e^{-j2\pi ut\csc\alpha}dt
$$
$$
= \frac{AT}{\sqrt{\sin\alpha_m}}e^{j\frac{\alpha_m}{2} + j\frac{3\pi}{4} - j\pi\mu u^2}\sin c(uT\csc\alpha) \tag{5}
$$

Where, $\sin c(x) = \frac{\sin\pi x}{\pi x}$. From formula (5), it is known that if $u = 0$, the value of $S_{\alpha_m}(u)$ modulus will be the maximum.

$$
M(|S_{\alpha_m}(u)|) - \frac{AT}{\sqrt{\sin\alpha_m}} \tag{6}
$$

If the value of A and α is fixed, $M(|S_{\alpha_m}(u)|)$ is linear with the pulse width T. The window function is defined as $u(t) = rect(\frac{t}{T_w}), T_w = \frac{T}{2}$. By multiplying the signal with window before taking FRFT, the STFRFT is obtained

$$
S_{w\alpha_m}(u) = \int_{-\infty}^{+\infty} K(u,t)u(t)s(t)dt
$$
$$
= A\sqrt{1 - j\cot\alpha}\, e^{j\pi u^2\cot\alpha_m}
$$
$$
\times \int_{-\frac{T}{2}}^{\frac{T}{2}} rect(\frac{t}{T_w})e^{j\pi t^2(\mu + \cot\alpha_m) - j2\pi ut\csc\alpha}dt
$$
$$
= \frac{AT_w}{\sqrt{\sin\alpha_m}}e^{j\frac{\alpha_m}{2} + j\frac{3\pi}{4} - j\pi\mu u^2}\sin c(uT_w\csc\alpha) \tag{7}
$$

From formula (7), it is known that $M(|S_{w\alpha_m}(u)|) = \frac{AT_w}{\sqrt{\sin\alpha_m}}$ is linear with the window width T_w. If the value of α is fixed, but the chirp rate is $\mu + \xi_m(\xi_m \neq 0)$, the output of the STFRFT can be expressed as

$$S_{w\alpha_m}(u) = A\sqrt{1 - j\cot\alpha}e^{j\pi u^2\cot\alpha_m}$$
$$\times \int_{-\frac{T}{2}}^{\frac{T}{2}} rect(\frac{t}{T_w})e^{j\pi t^2(\mu+\xi_m+\cot\alpha_m)-j2\pi ut\csc\alpha}dt$$
$$= \frac{A}{\sqrt{\sin\alpha_m}}e^{j\frac{\alpha_m}{2}+j\frac{3\pi}{4}-j\pi\mu u^2}\int_{-\frac{T_w}{2}}^{\frac{T_w}{2}}e^{j\pi\xi_m t^2 - j2\pi ut\csc\alpha}dt$$

(8)

When $\xi_m \neq 0$, $S_{w\alpha_m}(u)$ is a Fresnel integral. The envelope of $S_{w\alpha_m}(u)$ can be defined as

$$E[S_{w\alpha_m}(u)] = |C(P) - C(Q) + jS(P) - jS(Q)| \quad (9)$$

where $P = -\frac{u\csc\alpha}{\sqrt{\xi_m}} + \frac{T_w\sqrt{\xi_m}}{2}$, $Q = -\frac{u\csc\alpha}{\sqrt{\xi_m}} - \frac{T_w\sqrt{\xi_m}}{2}$, $C(\cdot)$ and $S(\cdot)$ are the Fresnel integral function.

We cannot obtain the accurate value from the Fresnel integral, but based on computer simulation, we can know that the maximum value of the signal after the FRFT remains unchanged and the output width of the main lobe is linear with the window width T_w. Figure 2 shows the output of the fractional Fourier domain (FRFD) and the short-time fractional Fourier domain (STFRFD).

Figure 2(a), (b) shows the $\xi_m = 0$, and the output amplitude in the STFRFD is 50 percent of that in the FRFD. Figure 2(c) (d) shows the output of the FRFD and the STFRFD when $\xi_m \neq 0$. Figure 2(e), (f) shows the local amplification of Figure 2(c), (d). From these, we can know that the maximum amplitude remains unchanged, but the main lobe width reduced by 50 percent.

3.3 Deceptive jamming suppression

For the SVLFM fuze, only the chirp rate of the echo signal is the same as that of the transmit signal, and the false target deceptive jamming signal is different with the echo signal, so the optimal rotation angle of is FRFT is obtained. Each component of the received signal can be separated step by step first in the FRFD. The maximum amplitude of each signal after the FRFT can be compared with that after the STFRFT. It can be judged that the signal component is the echo signal if the amplitude value changed significantly, and the signal component is the jamming signal if the amplitude value remains unchanged. The principle of anti-deceptive jamming for the SVLFM fuze is shown in Figure 3.

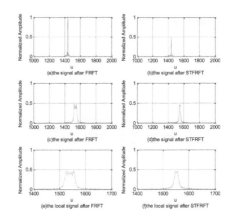

Figure 2. Different signals after the FRFT and the STFRFT.

Figure 3. Jamming suppression principle.

In order to make it a significant change between the amplitude in the FRFD and that in the STFRFD, the window function can be defined as

$$u(t-\tau_i) = rect(\frac{t-\tau_i}{T_w}) \quad (10)$$

where the window width is half of the minimum of each pulse width $T_w = \frac{1}{2}\min_m\{T_m\}$ and τ_i is the delay of the ith component after signal separation.

4 COMPUTATIONAL EXAMPLES AND ANALYSIS

We assume that the echo pulse width is $1\,\mu s$, the frequency width is 200 MHz, so the chirp rate is 2×10^{14}Hz/s and the deceptive jamming signal consists of two components: one is a 5 percent difference from the echo signal rate and the other is a 10 percent difference. The signal to jamming ratio (SJR) is -10dB and the signal to noise ratio (SNR) is -5dB. The sampling rate is 1GHz. Figure 4 shows the output of the echo signal and the deceptive jamming signal in the FRFD.

Through the FRFT, each signal component can be separated step by step according to the signal power. Figure 5 shows each signal component, with 5(a), (b) indicating the jamming components and 5(c) indicating the echo signal.

Figure 6(a) shows the signal indicated in the FRFD in Figure 5(a), and Figure 6(b) shows that

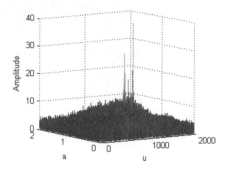

Figure 4. Output of the signal FRFT.

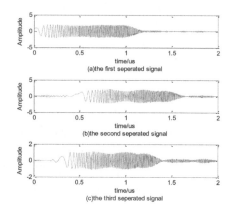

Figure 5. Separated signal.

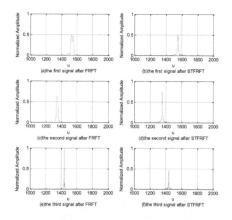

Figure 6. Contract of the signal FRFT and STFRFT.

in the STFRFD. Comparing with each other, it is known that the maximum amplitude remains unchanged, but the main lobe width narrows down to half. The signal in Figure 5(b) is the same as that in Figure 5(a). However, the maximum amplitude of the main lobe in Figure 6(e) is only half that in Figure 6(f), so it can be made sure that the signal in Figure 5(c) is the echo signal.

5 CONCLUSIONS

The SVLFM fuze has an anti-deceptive jamming performance to some extent, but when the chirp rate of the jamming signal is close to that of the echo signal, the fuze can be disturbed, leading to loss of performance. To solve the problem, this paper first separates each signal component by the FRFT, and then distinguishes the echo signal from the jamming signal based on the characteristic that the maximum amplitude of the STFRFT is linear with the window width for the same LFM signal, but the maximum amplitude does not change with the window width if two frequency modulation ratios are different. The simulation results show the feasibility of the method.

ACKNOWLEDGMENTS

This work was supported by the National Natural Science Foundation of China (NO.61102165).

REFERENCES

Dun Peng, Zhu Xiaosong, Xue Wancheng, 2013. De-Correlation Separation Method of LFM Signal Based on RFRFT [J]. Electronic Information Warfare Technology, 28(1): 17–20.

GE Wang Qinglin, Ying Ying, Li Jing, 2012. A Kind of Convolution Modulation Method that Generate Radar Multi-False Target [J]. Modern Defence Technology, 40(1): 137–139.

Hu Min, Li Guolin, Zhang Ying, 2010. Study on Digital Multidelay Smart Jamming Signal [J]. Telecommunication Engineering. 50(6): 21–27.

Lu Gang, Tang Bin, Luo Shuangcai, 2011. Adaptive cancellation of DRFM false targets for LFM radar [J]. Systems Engineering and Electronics, 33 (8):1760–1764.

Mehrdad Soumekh, 2006. SAR—ECCM using phase-perturbed LFM chirp signals and DRFM repeat jammer penalization [J]. IEEE Transactions on Aerospace and Electronic Systems, 42(1): 191–204.

Greco, M., Gini, F., Farina A. 2008. Radar detection and classification of jamming signals belonging to a cone class [J]. IEEE Transactions on Signal Processing, 56(5):1984–1993.

Tao Ran, Li Yanlei, Wang Yue, 2010. The short-time fractional Fourier transform and its applications [J]. IEEE Trans on Signal Process, 56(11): 2568–2580.

Tang Pengfei, Yuan Bin, Bao Qinglong, 2013. Design and simulation of digital channelized receivers in fractional Fourier domain [J]. Journal of Systems Engineering and Electronics, 24(1): 36–43.

Wang Cunwei, Wang Yong-liang, Li Rong-feng, 2010. An Effective Technique for Smart Jamming Elimination [J]. Journal of Air Force Radar Academy, 24(4):244–246.

Wen Jingyang, Zhang Huanyu, Wang Yue, 2012. Parameters Estimation Algorithm of LFM Pulse Compression Radar Signal [J]. Transaction of Beijing Institute of Technology, 32(7): 746–750.

Energy Science and Applied Technology – Fang (Ed.)
© *2016 Taylor & Francis Group, London, ISBN 978-1-138-02833-3*

Research of an envisioned smart doorbell

Yue Zhang, Mochen Xia & Boyang Yu
Beijing University of Post and Telecommunication, Beijing, China

ABSTRACT: The doorbell is a ubiquitous appliance in people's home. Present-day doorbells can notify homeowners about a visitor's appearance, but with a limitation that homeowners can only get notifications when they are at home. To fill this gap, we envisioned a smart doorbell that can notify homeowners when a visitor appears, and further build a real-time communication between the visitor and the homeowners without a location constraint. The system consists of an outdoor interactive facility, homeowners' smart phones and networks that interconnect them. It can detect homeowners, identify visitors, make proper policies, establish communications and record visits. The working process of the system is presented in this paper, and some security issues are discussed. Further research should devote more attention to the implementation of the smart doorbell system.

Keywords: smart doorbell; smart device

1 INTRODUCTION

Nowadays, the doorbell is a common device in residences. Original doorbells provide visitors a key to press in order to notify the hosts, while electronic doorbells offer additional functions such as voice communication or video communication between the outdoor and indoor devices, which is ubiquitous in the gates of apartments. A major limitation of current doorbells is that it can reach its goal, of informing homeowners of visitors' arrivals, only when at least one host is at home. If nobody is at home, the existing doorbell cannot notify anyone, nor can it record any information such as time and the visitor's identity. To remedy this limitation, Tarik Taleb proposed a 3GPP enhanced eDoorbell application prototype based on the IMS to provide a video-based communication between visitors and homeowners (who possibly have a remote access available). In his paper, the accessibility of context information, such as current location of homeowners, their schedule, and their personal preferences, enables a sound basis for the definition and enactment of customizable management policies that decide the best appropriate home resident to whom the notification of a visit should be forwarded (Taleb et al. 2010). However, in this solution, the methods of getting house residents' location are based on the IMS and Femto, which are not deployed in many residences. Besides, a method for identifying visitors is also required in order to support its policy management. In addition to Taleb's research, Yeon-Joo Oh implemented a video door phone system to enable user mobil-

ity, by which homeowners can monitor visitors at their front door using communication-capable devices such as PCs, PDAs and home servers with Internet access, regardless of physical constraints. The architecture is based on the Session Initiation Protocol and a home gateway system connected to a conventional intercom or a video door phone. His work demonstrated the feasibility of a SIP-based video door phone system (Oh et al. 2006). Hyunjeong Lee's paper discussed and simulated an answering doorbell service that can analyze the appropriate homeowners in the house to whom a notification of the visit is sent (Lee et al. 2006). In his design, the policy-making mechanism relies on a sensor platform that performs location sensing, which might be costly and not be accurate enough to detect the homeowners' indoor position.

In this paper, we envisioned a smart doorbell system, which not only took advantages of previous research findings, but also eliminated some existing problems. Based on Yeon-Joo Oh's paper, we designed a communication sub-function in the system that relies on the SIP (Oh et al. 2006). To alleviate some constraints in Tarik Taleb's prototype (Taleb et al. 2010), the location detection in our design is based on prevailing WLAN networks instead of Femto and location sensors. Besides, we also proposed a method to identify visitors using the smart doorbell system.

The system includes an outdoor micro-computer, homeowners' smart phones with the installation of related applications, and a WLAN network with Internet access. The system is composed of five sub-functions: the first aims to detect whether any homeowner is at home, which is achieved by querying in the local network. The second function is to identify visitors through their identifiers based on Bluetooth. The third function is to make different policies for different visitors by context information and a pre-set preference. The fourth function is used to establish and manage a real-time communication supported by the SIP. The last function is used to record the visits. With all the above functions achieved, the smart doorbell system can recognize visitors, provide notification to homeowners, and set up a long-distance real-time session regardless of homeowners' location, in a user-friendly way. The specific details are organized as follows. Section 2 introduces the hardware facilities and network of the system. Section 3 discusses the settings and functionalities of the system. Section 4 summarizes the operation process of the system. Section 5 presents some discussion. Finally, Section 6 concludes this paper.

2 NETWORK AND HARDWARES

The smart doorbell system is an envisioned IoT system. Its basic function is to provide immediate notification to house residents whenever a visitor appears, regardless of house residents' physical location. In this case, we need three components in the system: an outdoor facility that interacts with visitors, smart devices carried by homeowners, and networks that interconnect them.

2.1 Network

Nowadays, WLANs (Wireless Local Area Network) are widely deployed in residents due to its ease of achieving by a wireless router. The WLAN is based on various versions of IEEE 802.11. It is a wireless network connecting two or more devices, and can provide a connection to the wider Internet. With its development, the newest IEEE 802.11ac standard provides a very high transfer speed (Kelly et al. 2014), which is needed for the functioning of the smart doorbell system.

As a requisite in the whole system, the WLAN provides a local interconnection for the outdoor facility and homeowners' terminals, which are currently at home and within the WLAN's range. By installing the WLAN at home, the outdoor facility can connect to the network through a wireless router instead of through a cable, which is convenient for its deployment. Besides, due to the characteristic that homeowners' smart terminals will automatically connect to the WLAN when they are within its range, we can easily detect which homeowners are at home by querying which terminals are connecting to the WLAN.

The Internet is used to connect the outdoor facility with homeowners' terminals when they are away from home. The outdoor terminal can connect to the Internet through the Internet access point provided by the WLAN at home. In the meantime, a homeowners' smart terminal can access the Internet through the cellular network or WLANs when they are away from home. Thus, the connection between visitors and homeowners can be conveniently set up without the constraints of homeowners' location.

Considering the system's requirements and all the above advantages, the WLAN and the Internet are two indispensable components in the system that support the interconnection between the outdoor facility and homeowners' terminals.

2.2 Hardwares

The hardwares in the system include an outdoor facility and homeowners' smart phones. The outdoor terminal is a micro-computer with a microphone, a webcam, a WLAN, Bluetooth modules, and a touch-screen monitor. It connects to the homeowners' smart terminals directly through the WLAN when they are at home, and via the Internet when they are away from home. Homeowners' smart terminals, with the application being installed and the cellular network being supported, are also indispensable to ensure the functioning of the system.

3 SETTINGS AND FUNCTIONALITIES

3.1 Settings

The operation of the whole system is based on proper configurations. Homeowners should set their own identifiers, visitors' identifiers, and some other settings in order to enable the system to make suitable polices.

Homeowners' identifiers are used to detect whether anyone of them is at home. Due to the widespread of smart phones, most people carry them all the time. Thus, we use the locations of smart phones to conjecture the locations of homeowners. In order to make the system detect whether any smart phone is at home, house residents must pre-set their smart phones' identifiers in the system, which are the MAC addresses. Then, the system can continuously keep informed about which homeowner is at home by querying in the local network, which will be discussed later.

Besides their own identifiers, homeowners should also set the identifiers of their friends and relatives in order to make the system identify some visitors. Unlike homeowners' own identifiers, we use Bluetooth MAC addresses to identify visitors. This is because not all visitors are able to access the WLAN in the house, thus not all visitors' identifiers can be obtained through querying in the local network. By pairing with a visitor's smart phone using Bluetooth, the system can obtain the phone's Bluetooth MAC address, and further compare it with pre-set Bluetooth MAC addresses to detect whether the visitor is a friend, a relative or a stranger.

Permission settings are also necessary. When recording visitors' identifiers, homeowners should also set visitors' permission levels based on personalization, which not only limits every visitor's ways of long-distance communication with homeowners but also defines which homeowners are within a visitor's reach. For example, a relative may be set to have authority to start a video call with any homeowners, while a neighbor can only communicate with one specific homeowner by messages. Besides, homeowners should also set permissions for strangers, to specify valid actions for them. A blacklist, which contains undesirable persons who have no rights forwarding any kind of communication, can also be added due to homeowners' preferences.

Security settings are an indispensable part in the system. Before the system can function, every homeowner has to sign up for an account with a password, and an administrator must be designated to be responsible for managing homeowners' accounts. All settings in the system can be set via applications on a homeowner's smart phones, as long as he/she has logged in with his/her account name and password. To further ensure the security of the system, homeowners' accounts can be bound with their identifiers; therefore, even if a hacker has got an account and its password, he/she cannot hack into the system without the homeowner phone's MAC address.

3.2 Functionalities

The system is composed of five sub-functions: the first aims to detect whether any homeowner is at home, which is achieved by querying in the local network. The second function is to identify visitors through their identifiers based on Bluetooth. The third function is to make different policies for different visitors by context information and a pre-set preference. The fourth function is used to establish and manage a real-time communication supported by the SIP. The last function is used to record the visit.

Homeowner detector. When a visitor appears and presses the doorbell, the system will first detect whether any resident is at home. We detect homeowners' smart phone in order to speculate their location. Due to the characteristic that homeowners' smart phones always connect the home WLAN automatically, we can know whether a homeowner's smart phone is at home by querying in the WLAN. The IPv4 protocol and the ARP protocol allow the system to scan each device in the home WLAN and compare their MAC addresses with pre-set ones, which are homeowners' identifiers, to obtain information about which homeowner is at home.

Visitor identification. The second function is to identify visitors. We regard to visitors' smart phones', Bluetooth MAC addresses are set as their identifiers. The system will ask the visitor to pair his/her smart phone with the outdoor microcomputer using Bluetooth, in order to get the visitor's Bluetooth MAC address. Then, the system will query the address in the database, to get the visitor's identification, or mark the visitor as a stranger.

Policy management. Policy management is another important sub-function. It is responsible for making suitable decisions, such as forwarding notifications to proper homeowners, asking visitors for identifications or declining a visit. The policy management is based on the above-discussed necessary preference settings. The steps are summarized in Section 4.

Communication establishment. There are three kinds of homeowners' and visitors' communications: messages, voice call and video call. After the system sets up a communication, the real-time voice and video captured by the outdoor terminal and the user's smart phone, as well as message information, will be first encoded to a suitable form, and then transferred within the home WLAN or over the Internet. The real-time data transmission is supported by the RTP/RTCP, and the management of virtual communication channels is supported by the SIP. The RTP provides end-to-end network transport functions suitable for applications transmitting real-time data, such as audio, video or simulation data, over multicast or unicast network services. The RTP does not address resource reservation nor does it guarantee quality-of-service for real-time services. The data transport is augmented by a control protocol (RTCP) to allow monitoring of the data delivery in a manner scalable to large multicast networks, and to provide minimal control and identification functionality. The RTP and RTCP are designed to be independent of the underlying transport and network layers. The protocol supports the use of RTP-level translators and mixers (Frederick et al. 2003). The Session Initiation Protocol (SIP) is an application layer control. (signaling) protocol for creating, modifying and terminating sessions with one or more participants. These sessions include

Internet telephone calls, multimedia distribution and multimedia conferences (Rosenberg et al. 2003). With the above protocols, the system can establish and manage a real-time communication between homeowners and visitors, in the form of voice call, video call or messages. Based on these protocols, Yeon-Joo Oh implemented a real-time visitor communication.

Communication system (Oh et al. 2006). In his work, the feasibility of a SIP-based video door phone system was demonstrated.

Recording. In the end, the record of the visit will be stored in the system, containing time, visitor's identity and visitor's actions. If the visitor is a stranger or a person whose identification was not recorded in the system, the system will advise the visitor to type in his/her profile, in order to save it in the system after homeowners' verification.

4 OPERATION PROCESS

The operation process is summarized in Figure 1. When a visitor presses the doorbell, the system will first detect the homeowners' locations. In the condition of one or more homeowners at home, the system will only notify these residents by ringing their phones and showing the identity and real-time image of the visitor on the phone's screen. In normal cases, residents will send a short voice message to the outdoor facility by the application and go to open the door. In special cases, such as the visitor is a mischievous or an unwanted person, residents can choose to mute and ignore the ringing. Conversely, if nobody is at home, the system will ask the visitor to pair his/her phone with the outdoor facility through Bluetooth to get the visitor's identification. Then, the system will make policies for the visitor based on homeowners' settings and the visitor's identity. The visitor may only have a permission to leave a message or may have a further option to forward a real-time communication with homeowners. Visitors who have no authority to call homeowners and those who have the permis-

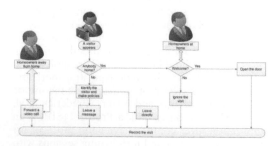

Figure 1. Operation process.

sion but choose not to forward a call can leave a message for homeowners, or leave directly. If the visitor choose to call homeowners, the system will send a communication request to all homeowners who are within the visitor's reach, decided by the visitor's permissions set by the homeowners. Homeowners who receive the request can accept it and turn to a real-time communication, or can also turn down the request. The first one who accepts the request will first join a session with the visitor, and the system will notify others that the visitor has been taken care of. The latter ones who see the notification can also join the session. If nobody accepts the request, the visitor can only leave a message. In the end, the system will record the visit with every detail such as time, visitor's identity and visitor's actions.

5 DISCUSSION

This paper presents a smart doorbell system with new features that notify to the homeowners about visitor's appearance when they are away from home. Previous papers proposed some kind of long-distance real-time communication system applied to doorbells. We take advantages of Yeon-Joo Oh's paper in terms of a SIP-based real-time communication design. Meanwhile, in his paper, it was stated that the design needs to be extended to include a mechanism for automatically detecting the residents' current position and available communication devices in the home (Oh et al. 2006). Besides, in Tarik Taleb's paper, the methods to get house residents' location are based on the IMS and Femto (Taleb et al. 2010), which are not deployed in many residents, and a method for identifying visitors is also required in order to support its policy management. To remedy this common limitation, we developed a pattern that can detect homeowners' location and verify visitors' identity in a way that is effective and easy to be achieved, and this pattern could also be applied to other platforms.

It can never be denied that there are some security issues in the smart doorbell system. When detecting homeowners and identifying visitors, the system uses their MAC addresses and Bluetooth MAC addresses as the identifiers, which are vulnerable because both can be easily forged. The smart doorbell may falsely recognize a fake MAC address as a real host, or regard a fake Bluetooth MAC address as a real relative, which are risky. To make up this weakness, we can introduce BlueID technology, a practical system that identifies Bluetooth devices by fingerprinting their clocks, to make sure the authentication can hardly be spoofed (Huang et al. 2014). However, the employing of BlueID still needs consideration because the cost of adopting BlueID may be high, and the risk of fake MAC addresses and Bluetooth addresses may not be so high because hackers must get the MAC addresses of real homeowners and visitors to make the forgery accepted by the system, and thus this risk is unlikely to lead to property damage. However, in the future, if a function that could remotely open the door for visitors is added to the system, the risk of fake identifiers could be eliminated to ensure the personal and property safety of homeowners.

6 CONCLUSION

This paper introduced a smart doorbell that overcomes the traditional doorbell's limitation that homeowners could only be notified when they are at home. In our design, the system can detect which homeowners are at home, verify visitors' identify, make suitable policies, set up a real-time communication between homeowners and visitors, and finally record the visit.

In this paper, we first envisioned a physical layer that contains hardware and networks for the system. Later, we discussed some necessary settings supporting the system's operation. Then, a pattern for detecting homeowners and identifying visitors is proposed, and a method for setting up a real-time long-distance communication is applied to the system. Thereafter, a recording mechanism is discussed. Then, the whole process dealing a visit is performed. Finally, we discussed the relationship between this paper and other research findings, and also some security issues.

This paper presents no experimental results. The main objective of this paper is to provide envisioned design and ideas for relative companies and manufacturers. The next step of this research is to manufacture prototypes and solve implementation issues.

REFERENCES

Frederick R, Jacobson V, Design P. RTP: A transport protocol for real-time applications [J]. IETF RFC3550, 2003.

Huang J, Albazrqaoe W, Xing G. BlueID: A practical system for Bluetooth device identification[C]// INFOCOM, 2014 Proceedings IEEE. IEEE, 2014: 2849–2857.

Kelly V. New IEEE 802.11 ac™ Specification Driven by Evolving Market Need for Higher, Multi-User Throughput in Wireless LANs [J]. IEEE Standards Association, 2014.

Lee H, Kim J, Huh J. Context-aware based mobile service for ubiquitous home[C]//Advanced Communication Technology, 2006. ICACT 2006. The 8th International Conference. IEEE, 2006, 3: 4 pp.-1854.

Oh Y J, Paik E H, Park K R. Design of a SIP-based real-time visitor communication and door control architecture using a home gateway [J]. Consumer Electronics, IEEE Transactions on, 2006, 52(4): 1256–1260.

Rosenberg J, Schulzrinne H, and Camarillo G, et al. RFC 3261: SIP: session initiation protocol [J]. 2003.

Taleb T, Kunz A, Schmid S, et al. Call-Handling by an IMS-HNB Based Interactive doorbell[C]//Wireless Communications and Networking Conference (WCNC), 2010 IEEE. IEEE, 2010: 1–6.

Energy Science and Applied Technology – Fang (Ed.)
© 2016 Taylor & Francis Group, London, ISBN 978-1-138-02833-3

Intelligent Compressive Sensing with adaptive observations

J.B. Xu & S.W. Zhou
Hunan University, Changsha, China

ABSTRACT: Compressive Sensing (CS) is a new signal acquisition framework, which allows for signal recovery from far fewer than what is achieved through traditional sampling methods. In this paper, a sequential adaptive compressed sensing procedure for signal support recovery is proposed and analyzed. The procedure makes use of sparse sensing matrices to perform sketching observations that are able to quickly identify irrelevant signal components. The results show that adaptive compressed sensing enables recovery of weaker sparse signals than those that can be recovered using traditional non-adaptive compressed sensing approaches.

Keywords: Sequential compressed sensing, signal sampling, Euclidean distance, stopping rule

1 INTRODUCTION

The rapid development of information technology makes a sharp increase in demand for information, signal bandwidth, sampling rate and processing speed simultaneously. However, The traditional methods of signal samples are based on the Nyquist sampling theorem. In acquiring the signal, the sampling frequency must be greater than twice the highest frequency signals to accurately reconstruct the original signal. However, the actual sampling data obtained, many of which are not important, for example, in the signal or image processing, retaining only some of the important data, the reconstructed signal or image does not cause visual differences, which process first high-speed sampling and then recompression wastes a lot of resources (Yaakov Tsaig, 2006).

Compressive sensing is a new theory. In the case of a sparse or compressible signal, signal sampling and signal compression occurs simultaneously. This makes it prominent in the field of signal processing with advantages and broad prospects. In the data acquisition and signal processing, the compressive sensing theory breaks through the traditional limitations of the Shannon theorem, by changing the mode of data collection and achieving further development and innovation of the traditional theory. It uses the correlation between the data, greatly reduces data transmission and storage in the network, and saves a lot of network resources.

In 2004, Candes and Donoho proposed the CS (compressive sensing) theory (Donoho D L, 2006), and proved that as long as the signal is sparse in an orthogonal space, we can sample the signal at a low frequency and reconstruct it with a high probability. Since the theory was put forth, a lot of institutions and scholars have been engaged in the research, thousands of related papers in this area have been written and a great progress has been currently made. The CS theory brought a great change in the field of signal sampling (Jin Wang, 2012). In addition, it is promising in compressed imaging, converting analog information and bio-sensing. Rice University developed a "single-pixel camera" based on the CS theory applied successfully to optical imaging. The DISP team of Duke University developed a new multi-spectral imager. Another research group of Xi'an University of Technology proposed a method using ultra-low rate sampling to detect ultra-wideband echo signals.

In this paper, we discuss how a sequential compressed sensing processing signal problem in the network environment proposes a new "stopping rule" method, and verifies the method's correctness by experiments. We apply the sequential compressed sensing model by adopting the new stopping rule to process sample images, in order to show that this new model is indeed feasible.

In Section 2, we describe the basic principles of compressive sensing and emphasize on the sequential compressed sensing signal processing procedure. Then, we propose a new stopping rule method and verify its rationality in theory. In Section 3, through a series of experiments, we prove the correctness of the method and test the practicality of the new CS model. In Section 4, we conclude this paper with a summary.

2 COMPRESSIVE SENSING

2.1 *Compressive Sensing theory*

Compressive sensing has become popular in recent years. It is a new research direction between mathematics and information technology, put forward by Candes, and is a new sample technology different from the Nyquist sampling theorem. According to the Nyquist theorem, only when the sampling rate reached more than twice the signal bandwidth, the original signal can be reconstructed from the sampling signal with a high probability.

With the increase in the bandwidth, the popular requirements for the signal sampling rate, the transmission speed, and the storage space also grow higher. For relieving the pressure on the signal transmission speed and the storage space, the common solution now is signal compression (Kim S J, 2007). However, the traditional signal transmission creates severe waste, because a large number of the sampling data are abandoned during compression, which are far from being insignificant or generate redundant information for the signal. The traditional process of data sampling and compression is shown in Figure 1.

According to the compression sensing, a measuring matrix, which is incoherent with the transformation matrix, may be used to project the lineation of the variation factor as a low-dimensional observation vector, which reserves the required information for signal reconstruction. By a further study on sparse optimization, an original high-dimensional signal can be rebuilt precisely and efficiently. The signal processing of compressed sensing is shown in Figure 2.

First, if the signal $X \in R^N$ is compressible in a certain orthogonal basis or the tight frame ψ, the sparse transformation $\theta = \psi^T X$ can be obtained, where θ is the equivalent or approximate sparse representation of ψ. Second, we design a smooth, irrelevant to the transform base ψ, $M \times N$ dimen-

sional measurement matrix Φ, and use Φ to observe θ to get observation collection $Y = \Phi \theta = \Phi\psi^T X$. This process can also be expressed as the observation matrix A^{CS} that observes X non-self-adaptively: $Y = A^{CS}X$, where $A^{CS} = \Phi \psi^T$, with A^{CS} being called the information operator of CS. Lastly, we solve the optimization problem in the sense of 0-norm to get the exact or approximate solution $\hat{X}$ of X; the solution vector $\hat{X}$ obtained has the most sparse representation on the base ψ, which is given by

$$\min \|\psi^T X\|_0 \text{ s.t. } A^{CS}X \approx \Phi\psi^T X = Y \qquad (1)$$

The compressive sensing theory mainly involves three aspects. (1) For the signal $X \in R^N$, we find an orthogonal basis or a tight frame base to make the representation of X on the base ϖ sparsest, which is called the signal sparse representation problem. (2) We design a smooth and transformation matrix ψ irrelevant $M \times N$ dimensional measurement matrix Φ, to guarantee the sparse vector θ without losing important information when dropping from the N-dimensional to M-dimensional, which is called the low-rate signal sampling problem. (3) We design a fast reconstruction algorithm, to recover the signal from the linear observation $Y = A^{CS}X$, which is called the signal reconstruction problem.

2.2 *Sequential Compressed Sensing*

Previously, we have shown that for a sparse signal of the sparse degree k, when using the compressed sensing approach to treatment, in general, the signal can be well restored with 4k sample values. However, different signals will have different characterisrics and structures. Some signals may well be recovered to the original signal with a sampling value less than 4k, but some signals may require multiple 4k sampling values in order to recover. Thus, for different signals of the sparse degree k, when sampling with a fixed 4k sample values, devices and the network load will be increased with too much sampling, and a good sampling results cannot be achieved with a few samples.

Sequential Compressive Sensing Procedure

The sequential compressive sensing (SCS) method can well solve the above problems. For different signals, the SCS method can select adaptively an appropriate sample value to reduce it as far as possible under the condition of ensuring the signal recovery effect. As shown in Formula 2, x is the sparse signal, Φ is the observation matrix, consisting of Φ_0, $\Phi_1 \ldots \Phi_n$, and y is the observed signal, consisting of $y_0, y_1 \ldots y_n$ correspondingly.

The basic principle of compressed sensing sequence is sampling the original signal gradually

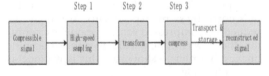

Figure 1. Traditional process procedure.

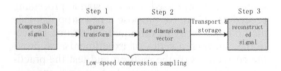

Figure 2. Compressed sensing procedure.

and sequentially, and then reconstruct it correspondingly for signal recovery, when the recovered signal meet the requirements of the "stopping rule", ending the sampling process. The sampling procedure can be described as follows: the compression side sends the sampling signal y_0 obtained through the measurement matrix Φ_0 on X to the decompression side.

$$y = \begin{pmatrix} y_0 \\ y_1 \\ . \\ . \\ . \\ y_n \end{pmatrix} = \begin{pmatrix} \phi_0 \\ \phi_1 \\ . \\ . \\ . \\ \phi_n \end{pmatrix} x = \phi x \qquad (2)$$

The decompression side recovers the signal y_0 through the corresponding reconstruction algorithm to get s_0; if the signal s_0 meets the corresponding requirements, it ends the sampling process; otherwise, it continues to increase the number of samples and enters the next iteration, i.e. the compression side measures x with the measurement matrix Φ_1, getting the sampling signal y_1 and then sends y_1 to the decompression side. The decompression side recovers (y_0+y_1), getting the reconstructed signal s_1. Similarly, if the signal s_1 meets the corresponding requirements, it ends the sampling process; otherwise, it continues to increase the number of samples and enters the next iteration. The Nth step of the sampling process is as follows: the compression side measures x with the measurement matrix Φ_{N-1}, getting the sampling signal y_{N-1} and then sends y_{N-1} to the decompression side. The decompression side recovers $(y_0+y_1+...+y_{N-1})$, getting the reconstructed signal s_{n-1} and then judges whether the signal s_{n-1} meets the corresponding requirements; if yes, then it ends the sampling process; otherwise, it enters the next iteration.

Especially, if the dimensions of all measurement matrices $\Phi_1,\Phi_2,...\Phi_n$ are the same, it means the step length of the sequential compressed sensing is fixed. In other words, every increasing number of samples is the same. Otherwise, the step length of the sequential compressed sensing is non-fixed.
Stopping Rule

The effective setting of the stopping rule is a very important aspect of sequential compressed sensing; when stopping rule requirements are met, the sampling process is ended. Here, we introduce the assumed stopping rule of this paper, and prove its effectiveness through an experiment in Section 3.

To a digital signal represented as a two-dimensional matrix, we take the mean value of each column of the matrix as a component of a vector. Obviously, this vector reflects some features of the matrix. We label it as the feature vector of the

signal. According to the principle of compressed sensing, the more the number of samples, the closer the recovered signal to the original signal. At the same time, the feature vector of the recovered signal gets closer to that of the original signal. The comparison between the recovered signal and the original signal is complex and difficult at the decompression side of sequential compressed sensing. This is because only a part of the original signal is transmitted to the decompression side. However, we can compare the respective feature vector easily. In Figure 3, vector a represents the feature vector of the original signal and vector b represents the feature vector of the recovered signal, and vector c represents the difference between a and b.

When the length (norm) of vector c reaches close to 0, it means the length of vector b will reach closely equal to vector a. In other words, the length (norm) of vector c is exactly equal to the Euclidean distance between a and b. The implementation code in Matlab is represented by the norm (a-b).

After the setting of the new stopping rule, the procedure of sequential compressed sensing can be described concretely as follows.

Step 1: the compression side transmits the sampling signal y_0 obtained through the measurement matrix Φ_0 on X and the additional feature vector z to the decompression side. The decompression side recovers the signal y_0 through the corresponding reconstruction algorithm to get s_0, and calculates its feature vector z_0 and the value of the Euclidean distance between z_0 and z. If the value meets the corresponding requirements, then it ends the sampling process. Otherwise, it continues to increase the number of samples and goes to the next step.

Step 2: the compression side transmits the sampling signal y_1 obtained through the measurement matrix Φ_1 on X, and the decompression side recovers the signal (y_0+y_1) through the corresponding reconstruction algorithm to get s_1, and calculates its feature vector z_1 and the value of the Euclidean distance between z_1 and z. If the value meets the corresponding requirements, then it ends the sam-

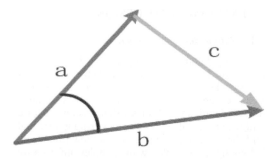

Figure 3.   Vector subtraction.

269

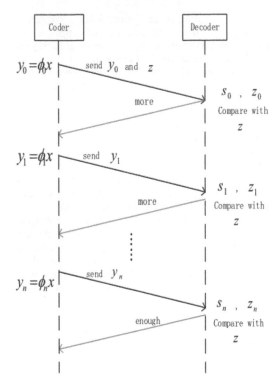

Figure 4. Sequential compressed sensing process.

pling process. Otherwise, it continues to increase the number of samples and goes to the next step.

Step N: the compression side transmits the sampling signal y_{N-1} obtained through the measurement matrixΦ_{N-1} on X, and the decompression side recovers the signal $(y_0+y_0+...+y_{N-1})$ through the corresponding reconstruction algorithm to get s_{N-1}, and calculates its feature vector z_{N-1} and the value of the Euclidean distance between z_{N-1} and Z. If the value meets the corresponding requirements, then it ends the sampling process. Otherwise, it continues to increase the number of samples and goes to the next step.

Figure 4 shows the sampling procedure.

3 EXPERIMENTAL DESIGN

3.1 *Sample image introduction*

The sample images of this paper used in the experiments are Lena, Camera-man, Shepp-logan, Monderian, all obtained with a 256×256 matrix. Of these, Lena represents a smooth image, with a moderate mix of detail, smooth areas, shadows and textures, and the energy of each frequency band is very high, and there are high frequencies and low frequencies, belonging to samples with complex information. Camera-man represents a hopping image, including

characters and vision, which forms a clear hopping zone where characters and vision touch, and the other area shows a smooth throughout. Shepp-logan represents medical images, which are relatively simple without continuity over color, but there are many graphics, and has a smooth curve, with moderate complexity and being favorable for the test sample. Monderian represents a simple image, which is automatically generated by computer and composed of a number of rectangles, color is distanced by a color block, the structure is clear and the data is simple.

3.2 *MSE And PSNR*

MSE (Mean Squared Error) is an equivalent precision measurement, which is measured under the same conditions. Besides, there is a better method representing the error, which is called the standard deviation, defined as a measurement of the square of the average value of the sum square.

If there are n measured values with error $\varepsilon 1$, $\varepsilon 2$ ......εn, respectively, then the mean squared error σ of the set of measurement values is given by

$$\delta = \sqrt{\frac{\xi_1^2 + \xi_2^2 + ...\xi_n^2}{n}} = \sqrt{\frac{\sum \xi_i^2}{n}} \tag{3}$$

Its Matlab implementation statement is σ = sum (sum(abs(X1-X2)^2)), where X1 and X2 are two measurement values obtained under the equivalent precision measurement.

PSNR (Peak Signal to Noise Ratio) is the peak signal to noise ratio, which means the ratio of the signal's maximum possible power and the destructive noise power influencing the precision of the image, and is an objective evaluation criterion. It is generally used for the measurement between the maximum value signal and the background noise. Usually, after the image is compressed, the output image will be different from the original image to some degree. In order to measure the image quality after processing, we usually refer to the PSNR to measure whether a handler can be satisfying. It is the log value of the value, which represents the mean square error between the original image and the processed image relative to $(2^n-1)^2$ (the square of the maximum signal, where n is the number of bits per sample), and expressed as dB. Its mathematical expression is given in Formula 4:

$$PSNR = 10 \times \log_{10}\left(\frac{(2^N-1)^2}{MSE}\right) \tag{4}$$

Its Matlab implementation statement is given as follows: PSNR = 10*log10((2^n-1) ^2/MSE). The larger the PSNR value the lower the distortion.

3.3 Verifying the correctness of the stopping rule

The experimental tests mentioned above make the Euclidean distance (i.e. the mode of two vector subtraction) between the feature vector of the recovered signals and the feature vector of the original signal as a valid stopping criterion. The results are shown in Figures 5 and 6.

In Figure 5, the horizontal axis refers to the sample number, while the vertical axis expresses the Euclidean distance of two vectors. The initial sample number is 100. As shown in the figure, the sample number increases with the reduced Euclidean distance; when the sample number is small, the Euclidean distance rises relatively fast; and when the sample number reaches a certain degree, the Euclidean distance rises relatively low and tends to be steady.

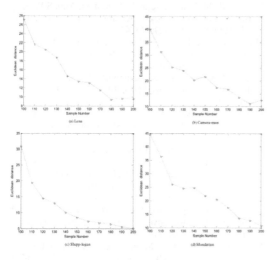

Figure 5. Euclidean distance and sample number.

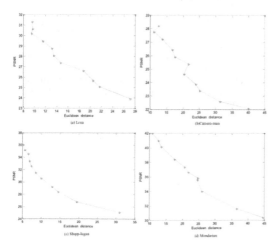

Figure 6. Euclidean distance and PSNR.

In Figure 6, the horizontal axis refers to the Euclidean distance, while the vertical axis expresses the signal-to-noise ratio. From the figure, it can be seen that the signal-to-noise ratio increases with the reduced Euclidean distance. When the Euclidean distance is large, the signal-to-noise ratio rises relatively slow; however, the signal-to-noise ratio rises faster when the Euclidean distance becomes smaller (indicating the growing closeness between the two proper vectors). With a large Euclidean distance, the sampling frequency can be increased by a larger ratio appropriately, while with the smaller one, the sampling frequency may be increased by a smaller ratio until the sampling ends at a certain degree of the Euclidean distance.

3.4 Verifying the practicality of the new model

We have proved the correctness of the stopping rule. Here, we apply the stopping rule to the sequential compressive sensing model, and then test the excellence of the model by an experiment, by performance comparison of the new model (non-fixed step model) with the fixed-step model. The experiment was divided into two parts: (1) For the same number of times of measurement, we compared the PSNR value; (2) For the PSNR value roughly equal, we compared the number of times of measurement.

In order to enhance comparability, we set the dimension of the initial observation matrix of the non-fixed step model and the fixed model step as 100, and the step length of the fixed-step model as 10, i.e. the increased sample value of each time is 10 stationary. However, it depends on the situation of the stopping rule in the new model. In the experiment, when the value of the Euclidean distance is relatively large, we set the step length equal to 15. When the value of the Euclidean distance is relatively small, we set the step length equal to 10.

(1) For the same number of times of measurement, comparing the PSNR value (number = 5):

As shown in Table 1, in the case of the same number of times of measurement, when the non-fixed step model is compared with the fixed-step model for different pictures, the PSNR values are found to be larger. In the fixed-step model, after the fifth sampling, the sample numbers for all pictures are 150, while in the non-fixed step model, the sample numbers are 165, 160, 160 and 165, respectively. This is because the new model can choose the step length flexibly based on the size of the Euclidean distance. The contrast of the image recovered is shown in Figure 7. Pictures a(1), b(1), c(1), d(1) were recovered from the fixed-step model, while pictures a(2), b(2), c(2), d(2) were recovered from the non-fixed step model.

Table 1. Experimental results.

Picture	Fixed-step model PSNR	New model PSNR	Fixed-step sample number	New sample number
Lena	27.43528	34.61805	150	165
Camera	25.76081	31.55806	150	160
ShepRes	31.59293	40.51256	150	160
RectRes	35.94016	48.03102	150	165

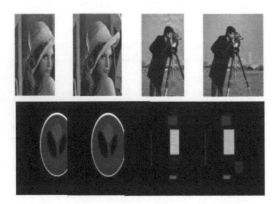

Figure 7. Image comparison.

Table 2. Number of times comparison.

Picture	Fixed-step model PSNR	New model PSNR	Fixed-step times	New model times
Lena	27.435286	27.3681	5	4
Camera	25.76081	25.6425	5	4
Shepp	31.59293	31.5339	5	4
RectR	35.940167	35.7114	5	4

(2) In the case of the PSNR value being roughly equal, comparing the number of times of measurement

The experimental results are given in Table 2.

For different images, the respective PSNR values are set as the second column of Table 2, whose corresponding number of sample times is five .While in the new model, to achieve the same PSNR level, the corresponding number of sample times is only four. By comparison between the new model and the fixed-step model, to achieve same required SNR level, the experiment data show that the new model can more quickly meet the requirements.

4 CONCLUSIONS

In this paper, we exploited a new stopping rule for the sequential compressive sensing process, and the rule showed better performance than the non-adaptive compressive sensing case. The results of our simulation have demonstrated its advancement and intelligence. Therefore, this system can greatly improve the efficiency and time. Future work will focus on the improvement of the stopping rule.

REFERENCES

Baron D, Wakin M B, Duarte M F, et al. Distributed compressed sensing [J]. Preprint, 2006.

Donoho D L. Compressed sensing [J]. Information Theory, IEEE Transactions on, 2006, 52(4): 1289–1306.

Dmitry M. Malioutov, Sujay R. Sanghavi, Alan S.Willsky. Sequential Compressed Sensing. IEEE Journals & Magazines 2010. 4(2):435–444.

Hans Scholten, Pascal Bakker. Opportunistic Sensing in Wireless Sensor Networks. ICN 224–229, 2011

Haupt J, Baraniuk R, Castro R, et al. sequentially designed compressed sensing[C]//Statistical Signal Processing Workshop (SSP), 2012 IEEE. IEEE, 2012: 401–404.

Haupt J, Bajwa W U, Rabbat M, et al. Compressed sensing for networked data [J]. Signal Processing Magazine, IEEE, 2008, 25(2): 92–101.

Jinping Hao, Filippo Tosato, Robert J. Piechocki. Sequential Compressive Sensing in Wireless Sensor Networks. Vehicular Technology Conference (VTC Spring), 2012 IEEE 75th. IEEE, 2012.

Jin Wang, Shaojie Tang, Baocai Yin, Xiang-Yang Li. Data gathering in wireless sensor networks through intelligent compressive sensing. Infocom, Proceedings IEEE, 2012. Page(s):603–611.

Kim S J, Koh K, Lustig M, et al. An efficient method for compressed sensing[C]//Image Processing, 2007. ICIP 2007. IEEE International Conference on. IEEE, 2007, 3: III-117-III–120.

Malioutov D M, Sanghavi S, Willsky A S. Compressed sensing with sequential observations[C]//Acoustics, Speech and Signal Processing, 2008. ICASSP 2008. IEEE International Conference on. IEEE, 2008: 3357–3360.

Yaakov Tsaig, David L. Donoho. Extensions of compressed sensing. Signal Processing. Volume 86, Issue 3, March 2006, Pages 549–571.

Energy Science and Applied Technology – Fang (Ed.)
© 2016 Taylor & Francis Group, London, ISBN 978-1-138-02833-3

Enhancing reliable time-triggered data transmission for safety critical tasks in TTEthernet

W.J. Zha, F. Hu & H.P. Chen
School of Shanghai Jiao Tong University, Shanghai, China

ABSTRACT: In this paper, we study the Time-Triggered Ethernet (TTEthernet) systems where periodic communication tasks are transmitted. We address the problem of the improvement and optimization of the communication level schedule. In this context, most of the recent studies have focused on how to guarantee the determinacy for transmission delays. However, the reliability of data transmission should also be taken into consideration. In this work, we model the current TTEthernet system and find the flaws lurking in the TTEthernet. Then, we modify the configuration of the TTEthernet switch to enhance the reliability of data transmission for the whole system. Furthermore, we classify the periodic communication tasks into soft real-time tasks and hard real-time tasks to improve the reliability of the specific tasks. The experimental results show that our approaches are effective for the reliability of data transmission with simple modification, which is adaptive to the industrial size cases.

Keywords: TTEthernet, Scheduling, Real-Time, Reliability

1 INTRODUCTION

In the past few decades, Ethernet is used in all walks of life. The end systems in Ethernet communicate with package switching, which is based on the IEEE 802.3 standard. The packages are sent by end systems and forwarded link by link through switches to the destination end system. Generally speaking, when packages are forwarded through the same switch simultaneously, they will be stored in a queue, waiting to be transmitted through relative ports by a certain mechanism. Due to the queuing mechanism, the time that packages spend waiting to be transmitted is uncertain. On this condition, the end-to-end delays in Ethernet are undetermined. It is concluded that the traditional Ethernet is not suitable for applications with strict timing restriction because of the non-deterministic timing behavior. Many researchers want to find an approach to extend Ethernet to satisfy the strict temporal constraint. The strict temporal constraint means that the end-to-end delays for packages are bounded.

To solve the non-deterministic timing behavior problem and satisfy the strict temporal constraint, Ethernet adopts the Time-Triggered Ethernet Switch (TTEthernet Switch) (K. Steinhammer, 2006) that is based on the Time-Triggered Protocol (TTP) (H. Kopetz, 1994) to receive and forward packages in the network. Therefore, we refer to the Ethernet with the TTEthernet Switch as the Time-

Triggered Ethernet (TTEthernet). TTEthernet is able to offer the deterministic timing behavior for package transmission with the help of the TTEthernet Switch and it is also compatible with standard Ethernet to offer transmissions for packages without temporal requirements, as shown in Figure 1.

The TTEthernet Switch receives packages sent by end systems with TTEthernet controllers or standard Ethernet controllers. The two kinds of packages are distinguished by a two-byte Type Field in the package header as follows:

Event-Triggered Package is sent by a standard Ethernet controller, which does not need temporal requirements.

Time-Triggered Package is sent by a TTEthernet controller periodically. The end-to-end delays for Time-Triggered packages must be guaranteed to be bounded.

To satisfy a strict deterministic timing behavior, all Time-Triggered tasks are sent and forwarded

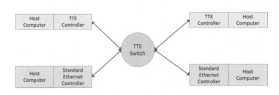

Figure 1. Time-Triggered Ethernet System.

at a predefined time. Thus, there must be a static schedule table to specify the time at which Time-Triggered tasks are forwarded through switches. At every end system, the TTEthernet controller sends tasks at a predefined activation time. In addition, at every TTEthernet Switch, tasks are checked to make sure that they arrive at predefined receive windows. A receive window is a predefined time interval in a period between which the task should arrive at the switch. If a task arrives at the switch during its relative receive window, the switch keeps the task and forwards it at the time as static schedule tables set. In contrast, if a task does not arrive during the receive window, the switch drops the task. With this mechanism, the transmission of Time-Triggered tasks is kept timing deterministic. TTEthernet does not guarantee the end-to-end delays for Event-Triggered tasks, and Event-Triggered tasks can only be transmitted when there are no Time-Triggered tasks. Thus, TTEthernet combines the Time—and Event-Triggered tasks together without influencing the deterministic timing behaviors of Time-Triggered ones.

However, the reliability of data transmission for Time-Triggered tasks is not taken into consideration. This is because whenever a task arrives beyond the receive window, the task will be dropped. It is well known that the end system activates a task at a predefined time. However, the end system will not activate the task on time once the computer controller is disturbed by exceptions or interrupts, such as external input, emergency instruction or application exception. On this condition, it is hard for a Time-Triggered task to arrive at the switch during the receive window when the task is late at the sending end. Because TTEthernet obeys the static schedule tables strictly, the Time-Triggered tasks that are delayed by interrupts and exceptions are directly dropped by the switches, and the data integrity and reliability are seriously affected. In this work, we aim to modify the TTEthernet Switch to reduce the number of lost Time-Triggered tasks when interrupts and exceptions occur, in order to enhance the reliability of data transmission.

Contributions. We analyze the current TTEthernet and find the flaws lurking in it. We propose this imperfection and implement a simulation model to prove it, and the evaluation result illustrates our viewpoint. Then, we modify the scheduler of the TTEthernet Switches to try to cover the shortage. With the simulation model, we can improve the reliability of data transmission for at least 30%. At present, we cannot guarantee the data integrity for all tasks. So, we add an attribute to the tasks and classify the tasks into two parts. We present an approach to make sure that the hard real-time tasks will not be dropped, and our evaluation result verifies our idea. The approaches presented in this paper modify TTEthernet and achieve the high data transmission reliability.

Outline. This paper is organized as follows. Section 3 shows the overview of the TTEthernet system and gives a motivation example. Section 4 presents the modifications on TTEthernet to improve the performance. Section 5 discusses the TTEthernet for further enhancement from the aspect of the real-time tasks. Finally, Section 6 concludes the paper.

2 APPLICATION MODEL AND MOTIVATION

In this section, we build a TTEthernet system simulation to observe and study how TTEthernet guarantees the deterministic timing behavior. From the simulation, we discover the potential problems in the system.

Problem Formulation. A TTEthernet system consists of a set of End Systems (ESes) interconnected by full duplex physical links and Network Switches (NSes). We model a TTEthernet system as an undirected graph $G(V, E)$, where $V = ES \cup NS$ is the set of end systems and network switches, and E is the set of physical links. In our simulation, ES are divided into two parts: source nodes (SRCs, $SRC = \{src_1, src_2, \ldots, src_m,\}$), which are used to send and dispatch tasks, and receive nodes (RCVs, $RCV = \{rcv_1, rcv_2, \ldots, rcv_m,\}$), which are used to accept tasks and submit the data to the application level. Actually, there is no difference between the SRCs and the RCVs with respect to implementation because the physical links in TTEthernet are full duplex. We distinguish them to accurately elaborate their functionality. In addition, the Network Switches $NS = \{ns_1, ns_2, \ldots, ns_m,\}$ forward all accepted tasks to the next nodes according to the static schedule table. It is easier to infer from the subscripts of the equations that the number of SRCs, RCVs and NSes are m, n and l, respectively.

We refer to the process where a package is sent from the source node, forwarded by switches and received by the receive node as a task represented by τ, and distinguished by subscript τ_n. Because the SRCs send packages periodically, the tasks are periodic too. In order to comprehend and implement the model, every source node is responsible for sending only one package in order to facilitate the system. This means there is a one-to-one correspondence between a source node and a task. We define the task set as $Tasks = \{\tau_1, \tau_2, \ldots, \tau_m\}$. The number of tasks is the same as the number of SRCs.

Thus, we define a task τ_n as the following tuple, and explain the meaning for every argument to elaborate how a task is transmitted:

$$\tau_n \in Task, \quad \tau_n(N_{src}, N_{dest}, len, path, t_a, t_s, t_r, d, P) \quad (1)$$

where

> N_{src} represents the end system that activates and sends τ_n. It is the source node for τ_n.
>
> N_{dst} represents the end system that receives τ_n. It is the destination node for τ_n. A destination node accepts more than one task, but a task owns its sole destination.
>
> len represents the length of the package that τ_n is responsible for. The length of a package is concerned with the transmission time in switches.
>
> path is a list that consists of NSes through which τ_n must be transmitted, such as path $= \{ns_{k1},ns_{k2},\ldots,ns_{kn}\}$. The source node N_{src} sends τ_n to sn_{k1} which is the first switch in path. In addition, τ_n is transmitted by switches in the path in turn, and finally it is forwarded by sn_{kn}, which is the last switch in the path to the receive node N_{dst}.
>
> t_a is a list that consists of arrival time for τ_n at each switch in the path, respectively, such as $t_a = \{t_{a1},t_{a2},\ldots t_{an}\}$. Arrival time is the time at which τ_n arrives at the switch. t_{a1} is the time τ_n that arrives at ns_{kn}, t_{an} is the time τ_n that arrives at ns_{kn}. t_{a1}^n is used to represent the arrival time of τ_n at ns_k.
>
> t_a is a list that consists of start time for τ_n at each switch in the path, respectively, such as $t_s = \{t_{s1},t_{s2},\ldots,t_{sn}\}$. Start time is the time at which τ_n begins to be transmitted at the switch. As the same meaning as t_s, the start time in t_s is the time τ_n that starts to be transmitted at switches in path. It is worthy to note that the start time is predefined to guarantee the deterministic timing behavior. When τ_n arrives before the start time, which means $t_{an} \leq t_{sn}$, the switch ns_x forwards τ_n at the time t_{sn}. Otherwise, ns_x will not accept it. In addition, t_{s1}^n is used to represent the start time of τ_n at ns_k.
>
> t_f is a list that consists of finish time for τ_n at each switch in the path, respectively, such as $t_f = \{t_{f1},t_{f2},\ldots,t_{fn}\}$.

Finish time is the time at which τ_n finishes transmitting through the switch port. A task is transmitted through the switch port byte by byte, so it takes some time to finish the transmission:

$$c_k^n = len(\tau_n)/sr_k, \qquad (2)$$

where $len(\tau_n)$ is the length of τ_n, and sr_k denotes the service rate at the switch port. We define c_k^n as the transmission time at the switch port ns_k for τ_n. So, we can get the following equation:

$$t_{f1}^n = t_{s1}^n + c_k^n, \qquad (3)$$

where t_{f1}^n represents the finish time of τ_n at ns_k; P is the period of τ_n; and d is the deadline of τ_n. The deadline is the same as the period, which means τ_n must finish in one period.

With the tuple for a task, we can infer how the task is transmitted in the TTEthernet system. Actually, the transmission for tasks is the duty of switches. Whenever a task comes, the switch looks up the static schedule table to find the corresponding disposition of the task. Here, we give the tuple for an item in the static schedule table to explain how a switch deals with a task:

$$item_k (\tau_n,rw,t_s,port), \qquad (4)$$

where τ_n indicates the task this switch will handle. If τ_n arrives, the other data in this item will be used to decide how to forward. rw is the receive window for τ_n at this switch. The receive window is a time interval in one period, such as (t_s,t_1). τ_n will be accepted only if it arrives between the receive window, otherwise, it will be dropped. We use rw_k^n to represent the receive window for τ_n at sn_k. t_s represents the start time for τ_n. The switch starts to forward τ_n at time t_s. port represents the output port for τ_n. When the switch accepts τ_n, it forwards τ_n through the port.

3 MOTIVATION EXAMPLE

Here we provide an example that is modeled by OPNET (4) to demonstrate a TTEthernet system and find the flaws in it.

Figure 2 shows the topology of the model in this example, and Table 1 complements tasks and their attributes in the network. From column Task, we can see that there are 9 Time-Triggered tasks, from τ_1 to τ_9. For simplicity, every source node activates and sends only one task. So, there are 9 source nodes in column SRC, from src_1 to src_9. With

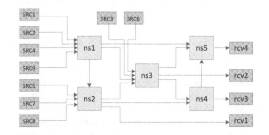

Figure 2. Topology of the TTEthernet system in the simulation.

Table 1. Topology and tasks.

	src_1	src_2	src_3	src_4	src_5	src_6	src_7	src_8	src_9
Task	t_1	t_2	t_3	t_4	t_5	t_6	t_7	t_8	t_9
len(B)	100	600	100	225	225	100	350	225	475
path	ns1,	ns1,	ns3,	ns1,	ns2,	ns3,	ns2,	ns2,	ns1,
	ns2,	ns5	ns4	ns2	ns4,	ns4,	ns3	ns3,	ns3,
	ns3				ns5	ns5		ns4	ns5
dest	rcv2	rcv4	rcv3	rcv1	rcv4	rcv4	rcv2	rcv3	rcv4
P(ms)	1	2	2	1	2	1	5	5	10

Figure 3. Optimal scheduling result of 9 Time-Triggered tasks.

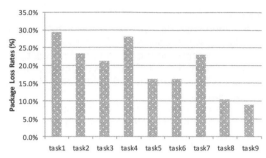

Figure 4. Original miss ratio of tasks: baseline of the experimental results.

the scheduling algorithm (L. Zhang), an optimal schedule is shown in Figure 3.

If tasks are sent and forwarded as shown in Figure 3, none of the tasks will wait in queues or miss their deadlines. Furthermore, the end-to-end delays for all the tasks are deterministic. However, this perfect situation is not always available. The source nodes are computers, which are shown in Figure 1. This means the source nodes are not only responsible for activating and sending tasks, but also deal with interrupts or computation, as described in Section 1. Because one computer is responsible for multi-jobs, sending tasks that we focus on in this research cannot be guaranteed on schedule. Thus, the task probably does not arrive at the switch between the receive window as expected. In addition, the task will be dropped or will miss the deadline. On this condition, the reliability of transmission and the integrity of data are influenced. We evaluate the transmission reliability and data integrity by the package loss rate defined as follows:

$$ml_n = 1 - N_r^n / N_a^n, \tag{5}$$

where ml_n represents the package loss rate for task τ_n, and N_r^n and N_a^n represent the number of tasks accepted by the receive node and the total number of tasks that are sent by the source node in a long time, such as 1000 seconds. Suppose that the total number of packages sent from src_9 is 100,000, and only 89,342 tasks are received at rcv_4. So, the package loss rate for τ_9 is 10.685%. If the package loss rate is higher, the transmission reliability and data integrity are lower. Therefore, what we try to reduce the package loss rates for tasks and the system to improve the reliability of transmission and integrity of data.

We suppose that all tasks are sent at most time, as shown in Figure 3, but the probability that a task is sent at random time in a period is 30%. If a task is sent at random time, it will be probably dropped by the switches and will not be accepted by the receive node. Under this circumstance, we run the model for 1000 seconds on a simulator name OPNET, and the results are shown in Figure 4. We can see that the package loss rates for all tasks are high, and the average rate is 22%. It is inferred that the data integrity is affected seriously. We discuss the solution in Section 4 and Section 5.

4 SOLUTION FOR THE LOSS OF PACKAGE

We have demonstrated an example to show the high package loss rate in TTEthernet when tasks are sent at random time. We have demonstrated an example to show the high package loss rate in TTEthernet when tasks are sent at random time. This is because current TTEthernet strictly keeps the tasks being sent and forwarded as the static schedule tables set. This mechanism guarantees constant end-to-end delays for tasks and deterministic timing behaviors; meanwhile, it leads to package loss that affects the transmission reliability and data integrity of the system. A task that arrives at the switch a little later and misses the receive window is dropped as given in following equation:

$$\forall ns_k \in path_n, t_{ak}^n \notin rw_k^n, N_m^n = N_m^n + 1. \tag{6}$$

From the equation, we can see that the receive window is the sole judgment criteria to determine whether a task should be dropped or not. If we find another baseline to decide to drop a task or not, we may decrease the rate of the package loss. In this section, we discuss 2 algorithms to attempt to reduce the package loss rate whenever the tasks are sent.

NoRW Algorithm. NoRW Algorithm is short for No Receive Window Algorithm, which means switches forward tasks without considering the receive window if the tasks miss the receive window. If a task that misses the receive window arrives at the switch, it will not be dropped as long as it has not missed its deadline. In addition, it will be accepted and put into the switch queue waiting to be forwarded. Besides, if the task does not miss the receive window, it will be put into the queue and forwarded at a predefined start time. Under this circumstance, a task is probably accepted and forwarded to the receive node even if its source node controller is disturbed by interruptions.

276

However, there is still a problem: when a task should be dropped and when a task should be forwarded further. Here we offer two solutions to handle this problem.

NoRW Solution 1. Every time the switch accepts a task that has missed the receive window, it puts the task in the queue for further forwarding if the task has not missed the deadline at the very moment. No matter how little time remains, the switch just transmits the task. Otherwise, the switch drops the task and the number of package loss increases:

$$N_m^n = \begin{cases} N_m^n, ||| & t_{a_k}^n < d_n \\ N_m^n + 1, & \text{otherwise} \end{cases}, \quad (7)$$

where t_{a1}^n represents the arrival time of τ_n at switch ns_k, and d_n is the deadline of τ_n.

NoRW Solution 2. The switch is required to forecast. When the switch accepts a task that has missed the receive window, it computes the remaining time to the deadline for that task. If the remaining time is enough for the task to be transmitted through the rest of the switches, the task will be put into the queue and forwarded to the next node. Otherwise, the switch drops the task and the number of package loss increases by 1:

$$N_m^n = \begin{cases} N_m^n, ||| & t_{a_k}^n + c_k^n * N_{ns} \le d_n \\ N_m^n + 1, & \text{otherwise} \end{cases}, \quad (8)$$

where N_{ns} represents the number of rest switches through which τ_n is transmitted, and c_k^n is the transmission time for τ_n at switch ns_k. We set the same service rate for every switch, so $c_k^n * N_{ns}$ denotes the total time that τ_n spends for transmission through the rest of the switches.

In the case of NoRW Solution 2, the number of packages that are dropped is less than the number of package loss in NoRW Solution 1. This is because the forecast in NoRW Solution 2 will drop the tasks that are destined to miss the deadlines in advance. Then, these tasks will not affect and block other tasks. In the case of NoRW Solution 1, tasks that are destined to miss the deadlines have not missed but disorganize the transmission of other tasks. Furthermore, tasks that should have transmitted as static schedule tables set probably miss the deadlines.

The comparison of the package loss rate for original TTEthernet scheduling and 2 NoRW scheduling is shown in Figure 5. The horizontal ordinate demonstrates the 9 Time-Triggered tasks from τ_1 to τ_9. In addition, the 3 kinds of package loss rates, original TTEthernet, NoRW Solution 1 and NoRW Solution 2, respectively, are classi-

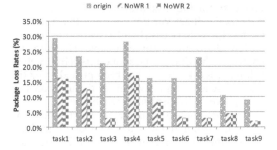

Figure 5. Package loss rates for the original TTEthernet scheduling and 2 NoRW Scheduling.

fied by 3 figures. The vertical ordinate indicates the package loss rates. The precise package loss rates for tasks and the system are given in Table 2. The last column improvement indicates the improvements of the transmission reliability and data integrity with NoRW Solution 2, compared with the original TTEthernet scheduling. Finally, the last row sys indicates the average package loss rates for the 3 kinds of scheduling. We can see that NoRW Solution 1 and NoRW Solution 2 both decrease the package loss rates remarkably, and NoRW Solution 2 has achieved a 55.6% improvement of the data integrity than the original TTEthernet scheduling.

We can see that the package loss rates are decreased by NoRW solutions effectively. Besides, NoRW Solution 2 can decrease the rate of package loss further than NoRW Solution 1. This is because the forecast in NoRW Solution 2 drops the tasks that are destined to miss the deadlines in advance. Under this circumstance, other tasks will not be influenced.

It should be noted that the package loss rates for some tasks will increase instead of decrease. Because of the NoRW transmission for some tasks, these tasks might block other tasks that are transmitted as the static schedule tables set, which leads them to miss the deadlines. However, from the aspect of the whole system, the average package loss rate is decreased remarkably, and the side effect of this NoRW solution is tolerable.

EDF Algorithm. After we analyzed the log in NoRW Solution 2, we suggest that the rate of package loss can be decreased further. As for the algorithm in NoRW Solution 2, when there are at least two tasks waiting in the queue in the switch, the task with a later deadline queueing before the task with an earlier deadline will be forwarded earlier. However, on this condition, the task with an earlier deadline must wait for a long time before the switch is in a position to forward it. It is more urgent than the task with a later deadline, but it waits longer time than the latter one. Thus, it is very easy for the task with an earlier deadline to miss its deadline. We can conclude that the First Come First Service (FCFS) algorithm in the NoRW algorithm is not reasonable. Therefore, we try to use the Earliest Deadline First (EDF) algorithm (L. Kruk), which has been proved to be the optimal scheduling algorithm in the uniprocessor to improve the transmission reliability.

Here we use an example to explain how the EDF algorithm reduces the number of package loss on the basis of NoRW Solution 2. There are 3 tasks, τ_m, τ_n and τ_k, in the system. The deadline of τ_m is 2 and the deadline of τ_n is 1. This means τ_n is more urgent that τ_m. However, if the switch forwards them with the FCFS in NoRW Solution 2, τ_n will miss the deadline, as shown in Figure 6. This is because τ_m arrives at switch3 before τ_n and τ_n must wait a long until the finish of both τ_m and τ_k. However, the EDF algorithm is effective on this condition, as shown in Figure 6. Even though τ_m arrives at switch3 before τ_n, the switch3 still forward τ_n whose deadline is earliest in the queue. Besides, we can see that with the EDF algorithm, all the 3 tasks are finished before their deadlines, and the package loss rates are decreased.

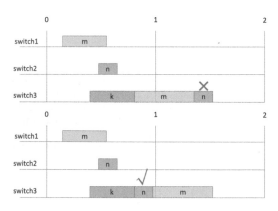

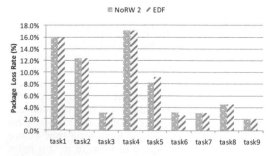

Figure 6.   Example of EDF Scheduling.

Figure 7.   EDF Algorithm reduces the package loss rates for some tasks.

So, we add the EDF algorithm to our TTEthernet system on the basis of NoRW Solution 2. The model runs for 1000 second and the comparison of package loss rates of NoRW Solution 2 and the EDF algorithm is shown in Figure 7. The two kinds of bars represent the package loss rates of 9 tasks with NoRW Solution 2 and the package loss rates with the EDF algorithm. We can see that the package loss rates of τ_1 and τ_6 decline, but the package loss rate of τ_5 increases remarkably. From this, we can see that the average package loss rate decreases a little with the EDF algorithm compared with NoRW Solution 2.

5   REVISED MODEL TO IMPROVE REAL TIME

From the evaluation results in Section. 4, the NoRW algorithm can improve the data transmission reliability by reducing the package loss rates remarkably. However, we try to reduce the package loss rate further by the EDF algorithm in this section. It seems that the EDF algorithm is not as effective as we expected. We analyze the system run-time logs, and figure it out that it is a little awkward for the EDF algorithm to play a role in the TTEthernet system. There two main causes as follows:

(1) There must be at least 2 tasks waiting in the queue. We still take Figure 6 as an example. If there is no τ_k at switch3, τ_m will be forwarded as soon as it arrives at switch3, and τ_n must wait until the finish transmission for τ_m. Because τ_n cannot preempt τ_m, which is being transmitted, it can only be inserted into the queue before some tasks that are waiting to be transmitted and less urgent than itself. However, τ_k is not always available as shown in the example. Most tasks are sent and forwarded as the static schedule tables set and the example in Figure 6 is an extreme case.

(2) When the package loss rate for one task is decreased, the package loss rates for other tasks could not be increased. We change the deadline for τ_m to 1.5 instead of 2 and run the example in Figure 6. We find that τ_n is accepted by the receive node, meanwhile τ_m misses its deadline. On this condition, the EDF algorithm makes no sense from the aspect of the system. Furthermore, we cannot ensure which task will be accepted and which task will be dropped. The non-deterministic timing behavior is not advocated in TTEthernet.

Priority-Based Algorithm. From the analysis, we can infer that the EDF algorithm cannot improve the transmission reliability and data integrity for the tasks and the system. In addition, the EDF algorithm is non-deterministic, because it is possible to increase the package loss rate for some tasks that are crucial for the system. The Time-Triggered tasks, or real-time tasks, are classified into two types: hard real-time tasks and soft real-time tasks. We call the crucial tasks hard real-time tasks.

The hard real-time tasks must finish before their deadlines, or they will provoke a great loss to the system. Besides, the soft real-time tasks are not that influential. Even though they are restricted by deadlines, they will not affect the system seriously when they miss the deadline. For the flight control instruction in an unmanned helicopter, it is a hard real-time task. If the flight control instruction misses the deadline, the helicopter in the air is out of control and is in danger of falling down. The video capture instruction is a soft real-time task. Even though it misses its deadline in this period, the only influence is the high quality of the video. Now that the hard real-time tasks are more vital than soft real-time tasks, it is natural to guarantee that the hard real-time tasks does not miss the deadline. When the two types of tasks conflict, the soft real-time tasks are sacrificed. On this condition, we refer to the hard real-time tasks as priority tasks. The priority tasks will never miss the deadline when they conflict with soft real-time tasks.

We modify the model to add a new attribute called priority to the Time-Triggered tasks. The value of the attribute priority is used to distinguish hard real-time tasks and soft real-time tasks. The hard real-time task's value for priority is **1** and the soft real-time task's value is **0**. When the switch accepts a task, it checks the value of the attribute priority. If the value is 1, the switch forwards it right now. Even though there is a soft real-time task being transmitted, the switch interrupts the transmission and forwards the hard real-time task. When the transmission for the hard real-time task finishes, the switch re-transmits the interrupted task. However, the hard real-time task can only preempt other soft real-time tasks. If a hard real-time task arrives at a switch while another hard real-time task is transmitted, the former one must wait in the queue. We call this the schedule priority-based algorithm.

We add the attribute priority to τ_4 to the system on the basis of the EDF algorithm. We can see that the improvement of τ_4 is the least, so we choose τ_4 as the hard real-time task, and other tasks are all soft real-time tasks. The model runs for 1000 seconds, and the results are shown in Figure 8. The priority (τ_4) bars represent the package loss rates for the 9 tasks when τ_4 is a hard real-time task, and the EDF bars are the package loss rates by the EDF algorithm for a comparison. Precise data are given in Table 4. We can see that the package loss rate for τ_4 is zero. This means the priority-based algorithm is useful for the hard real-time task.

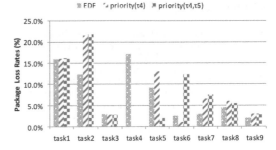

Figure 8. Package loss rates for no priority tasks, one priority task and two priority tasks.

It is concluded that the package loss rate for a task can be thoroughly eliminated when the task is the only hard real-time task in the system. Then, we find the package loss rates when there is more than one hard real-time task in the system. Then, we define τ_9 as another hard real-time task and run the model for 1000 seconds. The results are shown in Figure 8, represented by the priority (τ_4, τ_5) bars. We find that the package loss rate for τ_4 is still zero but the package loss rate for τ_5 is not zero, which is also lower than other soft real-time tasks. This result is reasonable, because one hard real-time task cannot preempt another hard real-time task. When the two tasks conflict in a switch, one of them must wait, and probably misses its deadline. When a task is only the hard real-time task in the system, it never misses the deadline. When there are more hard real-time tasks, the decrease in the package loss rate is more ineffective. When all Time-Triggered tasks are all hard real-time tasks, they will be scheduled with the FSFC algorithm, as shown by the result in Section 4.

6 CONCLUSION

In this work, we model a TTEthernet system and find that the performance for transmission reliability and data integrity is not guaranteed when interruptions or exceptions disturb the system. Therefore, we discuss 2 kinds of algorithms to attempt to reduce the package loss rates for tasks. The evaluation results demonstrate that the NoRW Algorithm is remarkably effective in improving the performance in the TTEthernet system, which can reduce 50% of the package loss rate on average. Even though the EDF algorithm is not useful in our example, it is shown to decrease the package loss rates for tasks. Then, we find a priority-based algorithm to eliminate the package loss rate and ensure data integrity for the hard real-time task. The evaluation results show that it is indeed significant for the system.

REFERENCES

H. Kopetz and G. Grunsteidl. TTP-A protocol for Fault-Tolerant Real-Time Systems. IEEE Computer, 1(27):14–23, August 1994.

Information on http://www.opnet.com.

IEEE Standard 802.3, 2000 Edition. Carrier Sense Multiple Access with Collision Detect on (CSMA/CD) Access Method and Physical Layer Specifications.

L. Kruk, J. Lehoczky, S. Shreve, and S.-N. Yeung. The Annals of Applied Probability.

L. Zhang, D. Goswami, R. Schneider, and S. Chakraborty. Task-and Network-level Schedule Co-Synthesis of Ethernet-based Time-triggered Systems. In Proc. 19th Asia and South Pacific Design Automation Conference (ASP-DAC), pages 119–124, January.

K. Steinhammer, P. Grillinger, A. Ademaj, and H. Kopetz. A Time-Triggered Ethernet (TTE) Switch. in Proc. the conference on Design, automation and test in Europe, pages 794–799, 2006.

Energy Science and Applied Technology – Fang (Ed.)
© 2016 Taylor & Francis Group, London, ISBN 978-1-138-02833-3

The design and application of data disaster recovery based on IPSec VPN

Bo Liu, Wei Wang & Wang Li
Shandong Provincial Key Laboratory of Computer Networks, Shandong Computer Science Center,
National Supercomputer Center, Jinan, China

ABSTRACT: Taking into account that some business systems do not have special network and the requirements of disaster recovery are low based on the analysis of business systems, we propose a design of disaster recovery based on IPSec VPN using the public network. It is proved that the design is available, maintainable and safe after one year of testing.

Keywords: Disaster; Data, Recovery; IPSec; VPN

1 BACKGROUND OF DISASTER RECOVERY

With the rapid development of information technology, enterprises have paid more attention to information data and information systems. For this reason, the integrity of information data and sustainability of the system operation become increasingly important. The huge growth of information data and high concentration of business systems not only make the information data increasingly important, but also increase the probability of all kinds of risk. With the openness of the network, the risk of data loss is growing.

The best way to reduce the damage caused by a disaster is the data disaster recovery. A set of high cost performance solutions of the disaster recovery is summed up by studying a typical and successful case of the Shandong province disaster service center in China.

2 INTRODUCTION OF DISASTER RECOVERY

2.1 *Construction of the disaster recovery system*

Disaster recovery is a process of making a backup copied from data, data processing system, network system, infrastructure, technology support ability and the ability of operation in order to recover the data before the disaster. During the normal operation of the system, part of the data set is copied from one hard disk to the other hard disk, so that we can recover the data and make the system run normally after the disaster.

A complete disaster backup system is mainly composed of the disaster backup infrastructure environment, the communication system, the data backup system, the strategy of disaster backup and the disaster recovery plan.

It is important to make up the data recovery strategy after choosing the storage backup software and hardware. In addition, different users need to set their own recovery strategy according to different requirements.

2.2 *Full backup*

The way of full backup is to back up all files, including system files, user files and application files. It is a simple backup form, but needs more time.

2.3 *Incremental backup*

The way of incremental backup is to back up the updated data. It takes a short time, but the backup form is relatively complex.

2.4 *Synchronous replication*

The synchronous replication means that the data in the backup center is consistent with that in the data center. So, when any node in the replication environment updates the data, the change will be immediately reflected in other nodes. The real-time capability of the synchronous replication is very strong, and the backup data can be consistent with the production data with no data loss. On the one hand, the synchronous replication can guarantee the consistency and integrity of the data; on the other hand, it increases the burden of the network and application system. Because it needs to wait for the remote site in the synchronous replication, the operation time is too long to affect the performance of the application.

2.5 *Asynchronous replication*

The asynchronous replication means that the local production data is copied to the other place in a background process after the host system writes data on the local disk, so all data in the nodes of the replication environment are not consistent in a certain period. When any node in the replication environment updates the data, the change will be reflected in other nodes after some time. On the one hand, it is difficult to guarantee the consistency and integrity of data; on the other hand, it reduces the load of network and application. Because the data cannot be synchronized all the time, a small amount of data is always lost after the disaster.

3 THE DATA DISASTER RECOVERY CENTER IN SHANDONG PROVINCE

The data disaster recovery center in Shandong province provides services for user business and e-government network business. In accordance with the "two districts three center" pattern design, the disaster recovery center promises to speed the recovery rate of user business and reduce the data loss caused by regional disaster.

The data disaster backup center also helps the users to choose the most suitable solution for disaster according to the different IT business systems.

According to the balance of risk cost and the local/long-distance disaster technology design principle, the center designed a set of data disaster recovery method based on IPSec VPN after a preliminary test. Finally, the method verified the feasibility and reliability through a concrete practical application.

4 THE DISASTER RECOVERY METHOD BASED ON IPSEC VPN

4.1 *IPSec VPN*

VPN (Virtual Private Network) is a Virtual Private Network, which can take advantage of cheap ways to access public Network (mainly on the Internet, but also include the transmission medium such as frame relay and TAM Network), and it can also use a variety of technology of upper-layer protocol attachments to provide customers with Network services similar to a private network performance.

Compared with the traditional way of line connected with cost advantages, it is adopted by many companies and telecom operators. Besides, through expanding the scope of private network VPN technology, it improves the flexibility of user network operation and management greatly.

IPSec Security (IP) is a framework structure of open standards, and it can ensure Security and confidentiality of communications by using encryption Security services on the network. The IPSec Internet network provides a completely transparent, powerful, flexible, security services to end users based on the current core network protocol in the TCP/IP protocol. In addition, the IPSec protocol mainly takes the identification, integrity checks, encryption and other security measures to ensure the transmission security of data.

The IPSec protects the security of OSI upper protocol data by inserting a predefined content in the packet head way, and the encapsulation process of packet made by IPSec VPN in the tunnel mode, as shown in Figure 1.

The upper part of the figure indicates a common IP packet and the lower part of the figure indicates a special packet encrypted by the IPSec. The IPSec inserts an IPSec head between the original IP header and the IP load, so that it can not only encrypt the original IP load, but also realize verification between the IPSec head and the original IP load to ensure the integrity of the data.

4.2 *The typical applications of disaster recovery based on IPSec VPN*

The city of Rongcheng is located in the east of Shandong Province in China, about 550 km from the provincial capital Jinan. The electronic teaching institution is fully responsible for the education administrative management and business management; besides, it also provides electronic teaching resource services and technical services in order to promote the extensive use of modern education technology in the field of education. In addition, the disaster recovery center in Shandong Province organizes and implements the project of disas-

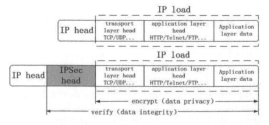

Figure 1. Encapsulation process of packet made by IPSec VPN in the tunnel mode.

ter recovery based on IPSec VPN for Rongcheng municipal commission of education in May 2013.

The business servers of Rongcheng municipal commission of education are deployed on HP C7000, which made the ESX cluster with 3 blades, including the WEB, video conference, and other business systems running on it. The business systems mainly contain win2003, win2000, win2008 and Linux system, and the databases mainly contain SQL2005, SQL2000 and mysql. Besides, the HP C7000 connects with EVA4400 through the optical fiber, and there are about 2.7 T bare spaces, with an actual usage of about 300 G, so that the daily incremental data is not big.

Rongcheng municipal commission of education copies data to the disaster recovery center storage using the Acronis Backup&Recover 11 software by means of IPSec VPN. After the disaster, the data will be recovered from the disaster recovery center to ensure the normal operation of the system. The data disaster recovery center in Shandong Province sets up the VPN link and the concurrent tunnels through the firewall equipment, and then maps the disaster recovery storage space to the backup servers of Rongcheng municipal commission of education. The topology structure of the scheme is shown in Figure 2.

This project was completed in April 2013 and tested until April 2014. In this project, the data was transmitted through VPN, and the channel peak was found to be at 3.9 MB/s using the IPSec protocol, and finally the data was copied onto other remote storage through the backup software. After the test, we found that some files, such as teaching video, that take up a larger space and have a better continuity can transmit at the rate of 3.3 MB/s, and the files, such as teaching software installation files, that take up less space and have a low continuity can transmit at the rate of 2.4 MB/s. We can learn that the transmission rate based on IPSec VPN is not significantly affected by different net-

works relative to the SAN from the table; in addition, there is no damage to the system performance with a less economic input. The detailed test data is shown in Table 1.

Table 1. Detailed test data in disaster recovery based on IPSec VPN.

No	File type	File size	Peak rate	Cost time	Average speed
1	Teaching video (3 folder,15 subfolders)	4.28GB	3.9MB/s	22 min,8 s	3.3MB/s
2	Teaching software installation files (486 folders, 5780 subfolders)	4.31GB	2.7MB/s	30 min,39 s	2.4MB/s

5 CONCLUSIONS

Because the advantage of IPSec VPN is that we can get an independent transport channel as long as the network is available, the transmission rate is not affected by the public network; therefore, it is very suitable for remote data disaster. Through the long time running and testing, we can fully confirm that the data disaster recovery based on IPSec VPN has reliable performance for data disaster protection. We found that this method is less dependent on the software and hardware and has a low capital investment requirement and simple scheme, so this kind of disaster recovery method has a higher cost performance, and it can also provide an important reference function for the large-scale disaster recovery in the future.

REFERENCES

Cheng, hua. 2008. Application in seismic data transmission based on IPsec VPN: 117–120.
Hao, jianming. 2009. The research and practice of information system disaster recovery: 83–86.
Qing,ke. 2012. Practical study of Cisco IPSec VPN: 13–16.
Wu, ruiqiang. 2012. Some considerations of constructing data and system disaster recovery center: 55–58.
Yang, yixian. 2010. The information system disaster technical review: 21–25.
Zhang, wei. 2011. The information system disaster recovery with data backup: 32–34.
Zou, hengming. 2009. Information system of disaster response: 16–27.

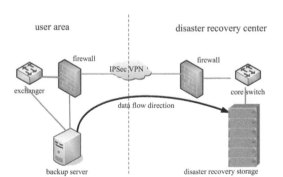

Figure 2. Topology structure of disaster recovery based on IPSec VPN.

Applied and computational mathematics

Energy Science and Applied Technology – Fang (Ed.)

Analysis of the numerical properties of LOD-FDTD method with arbitrary-order in space

Min Su & Peiguo Liu

Department of Electronic Science and Engineering, National University of Defense Technology, Changsha, China

ABSTRACT: In this paper, a 3D second-order-in-time, arbitrary-order-in-space Locally One-Dimensional Finite-Difference Time-Domain (LOD-FDTD) method is studied. The formulation of the arbitrary-order LOD-FDTD method is mainly analyzed. By analyzing the elaborate factors, the numerical properties are derived. The results show that the arbitrary-order LOD-FDTD is stable and has a smaller error relative to the conventional LOD-FDTD method and other FDTD methods.

Keywords: LOD-FDTD method; arbitrary-order; ADI-FDTD method; unconditional stable; numerical dispersion

1   INTRODUCTION

Since the traditional finite-difference time-domain (FDTD) method was proposed by Yee in 1966, due to its simplicity and flexibility, it has been widely used to solve a wide range of electromagnetic problems, such as electromagnetic compatibility and electromagnetic dispersion. However, the traditional FDTD method is not an implicit method, and the time step size is restricted by the Courant-Friedrich-Levy (CFL) condition, which affects its computational efficiency for fine mesh structures. In recent years, some unconditional-stable FDTD methods have been developed to overcome the CFL condition, such as CN-FDTD method, the ADI-FDTD method and the LOD-FDTD method. To minimize the numerical dispersion errors, a arbitrary-high-order LOD-FDTD method is developed. It is a cautious choice to achieve the best results between the computational efficiency and accuracy.

In this article, we present a comparison of the LOD-FDTD method with the arbitrary-order accuracy-in-time and conventional LOD-FDTD method and other FDTD methods. In Section 2, the formulae of the arbitrary-order LOD-FDTD method is proved. The method is divided into three-time steps. The numerical stability analysis for the LOD-FDTD method is presented in the next section, and then the normalized numerical phase velocity errors of all angles of propagation and different even orders such as 2nd order, 4th order, 6th order, 8th order and 10th order is studied, CFLN and CPW are studied as well. The results show that the normalized numerical dispersion phase velocity errors of the arbitrary-order

LOD-FDTD method are evidently decreased compared with the conventional LOD-FDTD method. Section 4 concludes this paper.

2   FORMULATIONS

The formulae of the arbitrary-order LOD-FDTD method are derived in this section. For limpidity and simplicity, the 3D wave propagation is considered in a linear, lossless isotropic medium and isotropic medium with permittivity ε and permeability μ.

The time-dependent Maxwell's curl equations are written in the following form:

$$\frac{\partial U}{\partial t} = [A]U + [B]U + [C]U \tag{1}$$

where $U = \left[E_x, E_y, E_z, H_x, H_y, H_z \right]^T$ and

$$A = \begin{pmatrix} 0 & A_E \\ A_H & 0 \end{pmatrix} \quad B = \begin{pmatrix} 0 & B_E \\ B_H & 0 \end{pmatrix}$$

According to the X, Y and Z directions, the application of the second-order-in-time and Crank-Nicolson formulation (1) is approximated with the following formula:

$$
\begin{aligned}
U^{n+1} &= \frac{([I]+\frac{\Delta t}{2}[A])([I]+\frac{\Delta t}{2}[B])([I]+\frac{\Delta t}{2}[C])}{([I]-\frac{\Delta t}{2}[A])([I]-\frac{\Delta t}{2}[B])([I]-\frac{\Delta t}{2}[C])} \phi^n \\
&= [\Lambda] U^n
\end{aligned}
\tag{2}
$$

Here, U^{n+1} represents the field values at $t = n\Delta t$, while Δt is the time step, $[\Lambda]$ is a matrix of the whole time step, and

$$[\Lambda] = \begin{bmatrix} 0 & 0 & 0 & 0 & -\dfrac{1}{\varepsilon}\dfrac{\partial}{\partial z} & \dfrac{1}{\varepsilon}\dfrac{\partial}{\partial y} \\[2mm] 0 & 0 & 0 & \dfrac{1}{\varepsilon}\dfrac{\partial}{\partial z} & 0 & -\dfrac{1}{\varepsilon}\dfrac{\partial}{\partial x} \\[2mm] 0 & 0 & 0 & -\dfrac{1}{\varepsilon}\dfrac{\partial}{\partial y} & \dfrac{1}{\varepsilon}\dfrac{\partial}{\partial x} & 0 \\[2mm] 0 & \dfrac{1}{\mu}\dfrac{\partial}{\partial z} & -\dfrac{1}{\mu}\dfrac{\partial}{\partial y} & 0 & 0 & 0 \\[2mm] -\dfrac{1}{\mu}\dfrac{\partial}{\partial z} & 0 & \dfrac{1}{\mu}\dfrac{\partial}{\partial x} & 0 & 0 & 0 \\[2mm] \dfrac{1}{\mu}\dfrac{\partial}{\partial y} & -\dfrac{1}{\mu}\dfrac{\partial}{\partial x} & 0 & 0 & 0 & 0 \end{bmatrix}$$

By uresplitting and computing Eq. (2) into three steps, we can get the following formulae:
Stepures 1: $n \to n+1/3$

$$E_y^{n+1/3} = E_y^n - \frac{\Delta t}{2\varepsilon}\left(\frac{\partial H_z^{n+1/3}}{\partial x} + \frac{\partial H_z^n}{\partial x}\right)$$

$$E_z^{n+1/3} = E_z^n + \frac{\Delta t}{2\varepsilon}\left(\frac{\partial H_y^{n+1/3}}{\partial x} + \frac{\partial H_y^n}{\partial x}\right)$$

$$H_y^{n+1/3} = H_y^n + \frac{\Delta t}{2\mu}\left(\frac{\partial E_z^{n+1/3}}{\partial x} + \frac{\partial E_z^n}{\partial x}\right) \qquad (3a)$$

$$H_z^{n+1/3} = H_z^n - \frac{\Delta t}{2\mu}\left(\frac{\partial E_y^{n+1/3}}{\partial x} + \frac{\partial E_y^n}{\partial x}\right)$$

Stepures 2: $n+1/3 \to n+2/3$

$$E_x^{n+2/3} = E_x^{n+1/3} + \frac{\Delta t}{2\varepsilon}\left(\frac{\partial H_z^{n+2/3}}{\partial y} + \frac{\partial H_z^{n+1/3}}{\partial y}\right)$$

$$E_z^{n+2/3} = E_z^{n+1/3} - \frac{\Delta t}{2\varepsilon}\left(\frac{\partial H_y^{n+2/3}}{\partial y} + \frac{\partial H_y^{n+1/3}}{\partial y}\right)$$

$$H_x^{n+2/3} = H_x^{n+1/3} - \frac{\Delta t}{2\mu}\left(\frac{\partial E_z^{n+2/3}}{\partial y} + \frac{\partial E_z^{n+1/3}}{\partial y}\right) \qquad (3b)$$

$$H_z^{n+2/3} = H_z^{n+1/3} - \frac{\Delta t}{2\mu}\left(\frac{\partial E_x^{n+2/3}}{\partial y} + \frac{\partial E_x^{n+1/3}}{\partial y}\right)$$

Stepures 3: $n+2/3 \to n+1$

$$E_x^{n+1} = E_x^{n+2/3} - \frac{\Delta t}{2\varepsilon}\left(\frac{\partial H_y^{n+1}}{\partial z} + \frac{\partial H_y^{n+2/3}}{\partial z}\right)$$

$$E_y^{n+1} = E_y^{n+2/3} + \frac{\Delta t}{2\varepsilon}\left(\frac{\partial H_x^{n+1}}{\partial z} + \frac{\partial H_x^{n+2/3}}{\partial z}\right)$$

$$H_x^{n+1} = H_x^{n+2/3} + \frac{\Delta t}{2\mu}\left(\frac{\partial E_y^{n+1}}{\partial z} + \frac{\partial E_y^{n+2/3}}{\partial z}\right) \qquad (3c)$$

$$H_y^{n+1} = H_y^{n+2/3} - \frac{\Delta t}{2\mu}\left(\frac{\partial E_x^{n+2/3}}{\partial z} + \frac{\partial E_x^{n+2/3}}{\partial z}\right)$$

3 STABILITY AND NUMERICAL DISPERSION ANALYSIS

3.1 Verification of Unconditional Stability

Using the Matlab software, we can get six eigenvalues of the matrix $[\Lambda]$ as follows:

$$\lambda_1 = \lambda_2 = 1, \lambda_3 = \xi_1 + j\sqrt{1-\xi_1^2}, \lambda_4 = \xi_1 - j\sqrt{1-\xi_1^2},$$
$$\lambda_5 = \xi_2 + j\sqrt{1-\xi_2^2}, \lambda_6 = \xi_2 - j\sqrt{1-\xi_2^2} \qquad (4)$$

Here,

$$\xi_1 = \frac{\begin{bmatrix} 1 + b^3 d^3 P_x^2 P_y^2 P_z^2 - dbP_x^2 - dbP_y^2 - dbP_z^2 \\ -b^2 d^2 P_x^2 P_y^2 - b^2 d^2 P_x^2 P_z^2 - b^2 d^2 P_y^2 P_z^2 + 4db\sqrt{db}P_x P_y P_z \end{bmatrix}}{A_x A_y A_z}$$

$$\xi_2 = \frac{\begin{bmatrix} 1 + b^3 d^3 P_x^2 P_y^2 P_z^2 - dbP_x^2 - dbP_y^2 - dbP_z^2 \\ -b^2 d^2 P_x^2 P_y^2 - b^2 d^2 P_x^2 P_z^2 - b^2 d^2 P_y^2 P_z^2 - 4db\sqrt{db}P_x P_y P_z \end{bmatrix}}{A_x A_y A_z}$$

Also,

$$A_\alpha = 1 + bdP_\alpha^2, B_\alpha = 1 - bdP_\alpha^2, d = \Delta t / (2\mu),$$

$$b = \Delta t / (2\varepsilon), P_a = \frac{-2\sin(\dfrac{k_\alpha \Delta \alpha}{2})}{\Delta \alpha},$$

$$\frac{\partial}{\partial \alpha} = jP_\alpha = -2j\frac{1}{\Delta \alpha}\sum_{l=1}^{L} d_{(2l-1)/2}\sin\left(\frac{2l-1}{2}k_\alpha \Delta \alpha\right),$$

$$\alpha = x, y, z, k_x = k\sin\theta\cos\varphi, k_y = k\sin\theta\sin\varphi,$$

$$k_z = k\cos\theta.$$

It is not difficult to show that $|\lambda_1| = |\lambda_2| = |\lambda_3| = |\lambda_4| = |\lambda_5| = |\lambda_6| = 1$. The result shows that the method is unconditionally stable.

Caseures 1, $\lambda = \lambda_3$ or $\lambda = \lambda_4$: the numerical dispersion value is given as

$$\tan^2(\frac{\omega\Delta t}{2})$$
$$= \frac{\begin{bmatrix} bdP_x^2 + bdP_y^2 + bdP_z^2 + b^2 d^2 P_x^2 P_y^2 \\ + b^2 d^2 P_x^2 P_z^2 + b^2 d^2 P_y^2 P_z^2 - 2bd\sqrt{bd}P_x P_y P_z \end{bmatrix}}{1 + b^3 d^3 P_x^2 P_y^2 P_z^2 + 2bd\sqrt{bd}P_x P_y P_z}$$

Caseures 2, $\lambda = \lambda_5$ or $\lambda = \lambda_6$: the numerical dispersion value is given as

$$\tan^2(\frac{\omega\Delta t}{2})$$
$$= \frac{\begin{bmatrix} bdP_x^2 + bdP_y^2 + bdP_z^2 + b^2 d^2 P_x^2 P_y^2 \\ + b^2 d^2 P_x^2 P_z^2 + b^2 d^2 P_y^2 P_z^2 + 2bd\sqrt{bd}P_x P_y P_z \end{bmatrix}}{1 + b^3 d^3 P_x^2 P_y^2 P_z^2 - 2bd\sqrt{bd}P_x P_y P_z}$$

288

Table 1.

Order D	$D_{1/2}$	$D_{3/2}$	$D_{5/2}$	$D_{7/2}$	$D_{9/2}$
2	1				
4	9/8	-1/24			
6	225/192	-25/384	3/640		
8	1.196289	-0.079753	0.009570	-0.000698	
10	1.211243	-0.089722	0.013843	-0.001766	0.000119

Caseures 3, $\lambda = \lambda_1$ or $\lambda = \lambda_2$: the numerical dispersion value is given as

$$\tan^2(\frac{\omega \Delta t}{2}) = \frac{\left[bdP_x^2 + bdP_y^2 + bdP_z^2 + b^2 d^2 P_x^2 P_y^2 + b^2 d^2 P_x^2 P_z^2 + b^2 d^2 P_y^2 P_z^2 \right]}{1 + b^3 d^3 P_x^2 P_y^2 P_z^2}$$

Forure investigating the numerical dispersion of the high-order LOD-FDTD method, we present the P_a for the 2nd order, 4th order, 6th order, 8th order and 10th order with the first-order space partial differential operator, and then $\partial U / \partial \alpha$ is approximated as follows:

$$\frac{\partial (U \mid_{i,j,k}^n)}{\partial \alpha} = \frac{1}{\Delta \alpha} \sum_{l=-M+1}^{M} D_{(2l-1)/2} U \mid_{i+(2l-1)/2,j,k}^n$$

Here,

$$D_{(2l-1)/2} = \frac{(-1)^{l-1}(2M-1)!!^2}{2^{2L-2}(M+l-1)!(M-l)!(2l-1)^2}$$

where M = 1, 2, 3.... The coefficients D are given in Table 1.

3.2 Numerical property results

In order to simplify the explanation, let us set case 3 as an example in the following discussion. In this section, the cell sizes of the three directions are taken as $\Delta x = \Delta y = \Delta z = \lambda / CPW$. We suppose that the normalized numerical phase velocity value is 1. We then set $\theta = 90^0, \phi = 0^0$ in the X direction; $\theta = 90^0, \phi = 90^0$ in the Y direction; and $\theta = 0^0$ in the Z direction. Given the cells per wavelength (CPW) and the CFL number (CFLN), the numerical property accuracy of the arbitrary high-order method is analyzed with four factors: 1. all angles of propagation; 2. different order; 3. different CFLN; and 4. different CPW.

Figure 1 proposes the two methods (LOD-FDTD and ADI-FDTD) of the normalized phase velocity errors versus all propagation angles with the second order at CFLN = 5 and CPW = 20. It clearly shows that the minimum phase velocity error is obtained at $\theta = 45^0$ and $\phi = 45^0$, and it

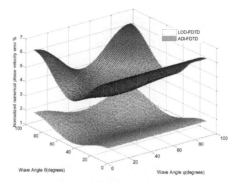

Figure 1. Normalized phase velocity error versus propagation angle at CFLN = 5 and CPW = 20.

increases at other angles from the point. Furthermore, the proposed LOD-FDTD method significantly reduces the normalized phase velocity error than the ADI-FDTD method.

The time step of the conventional FDTD method is employed in Figure 2 (namely CFLN = 1). For an easy illustration, here we consider the two-dimensional case. From Figure 2, it can be seen that the numerical dispersion error of the high-order LOD-FDTD method is between the conventional LOD-FDTD method and the conventional FDTD method. It is clear that the numerical dispersion error of the high-order LOD-FDTD method is 48% larger than that of the conventional FDTD method, but it is 12% smaller than that of the conventional LOD-FDTD method.

The normalized numerical phase velocity error of the proposed LOD-FDTD method with a different propagation angle φ (in degrees) is shown in Figure 3 The inset is the local magnification of Figure 3 with the propagation angle φ ranging from 40^0 to 50^0. The inset clearly shows the normalized numerical phase velocity error of the arbitrary-order method. We can find that the phase velocity error is decreased when the arbitrary-order schemes are employed. The 4th order scheme decreases the dispersion error from 1.75% to 1.55%. However, for other order schemes (e.g. 6th, 8th and 10th), the dispersion error is not better than that of the 4th order scheme.

In order to investigate the dispersion error for other time step sizes, synthetic studies with different CFLN are undertaken, where we set $\theta = 45^0$, $\phi = 45^0$ and CPW = 50. The results are shown in Figure 4, from which we can find that the CFLN increases as the normalized phase velocity error increases. Besides, it is clearly shown that for the 2nd, 4th and 8th order schemes, the dispersion error is equal to 1.

Figure 5 shows the normalized phase velocity errors with different CPW. Here, we set $\theta = 45^0$, $\phi = 45^0$ and CFLN = 3. It is clearly shown that the phase velocity

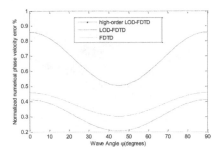

Figure 2. Normalized phase velocity error versus propagation angle at CFLN = 1 and CPW = 20.

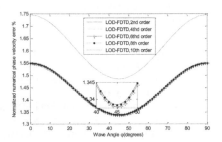

Figure 3. Normalized phase velocity error versus propagation angle at CPW = 20 and CFLN = 3.

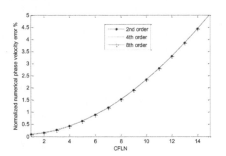

Figure 4. Normalized phase velocity error versus CFLN at CPW = 50.

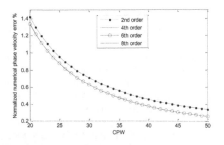

Figure 5. Normalized phase velocity error versus CPW at CFLN = 3.

error of the 2nd order scheme is the topmost from the beginning to the end. The figure also shows that the CPW increases as the error decreases. Besides, the CPW is larger and the curve trend becomes slow.

4 CONCLUSIONS

This paper studied the numerical properties of the LOD-FDTD method with arbitrary-order accuracy. The unconditional stability of the arbitrary-order method is mainly proved. Then, the normalized phase velocity error accuracy of the arbitrary method is studied from all angles of propagation, different orders, CFLN and CPW. From the above results, it is found that the dispersion error of the arbitrary-order LOD-FDTD method is smaller and computational efficiency than that of the conventional LOD-FDTD method and other methods.

ACKNOWLEDGMENTS

This work was supported by the National Natural Science Foundation of China under Grant No. 61372027

REFERENCES

A. Taflove. 2005. Computational Electrodynamics, "The Finite-Difference Time-Domain Method". Norwood, MA: Artech House.

Erping Li. 2007.Numerical Dispersion Analysis with an Improved LOD-FDTD Method. IEEE Microw. Wireless Compon. Lett.VOL. 17, NO. 5. pp.319–321.

F. Zheng & Z. Chen. 2000. toward the development of a three dimensional unconditionally stable finite-different time-domain method., IEEE Trans. Microwave Theo. Tech. pp. 1550–1558.

G. Eason & B. Noble. 1955. On certain integrals of Lipschitz-Hankel type involving products of Bessel functions. Phil. Trans. Roy. Soc. London, vol. A247, pp. 529–551.

I. Ahmed & eng-Kee. 2007. Development of the three-dimensional unconditionally stable LOD-FDTD method IEEE transactions on antennas and propagation, VOL. 56, NO. 11. November 2008,pp 3596–3600.

M. Li & J. Nuebel. 2000. EMI from airflow aperture arrays in shielding enclosures- experiments, FDTD, and MoM modeling. IEEE Trans. Electromagn. Compat. 42(3):265–275.

R Dehkhoda & A. Tavakoli. 2008. An efficient and reliable shielding effectiveness evaluation of a rectangular enclosure with numerous apertures. IEEE Trans. Electromagn. Compat. 50(1): 208–212.

Stoer J. & Bulirsch R. 1993. Introduction to numerical analysis [M]. 2nd Ed. New York Springer-Verlag.

Weiming Fu. 2005. Stability and Dispersion Analysis for Higher Order 3-D ADI-FDTD Method. IEEE transactions on antennas and propagation, Vol. 53, No. 11. pp 3691–3696.

Yong-Dan Kong. 2011. High-Order Split-Step Unconditionally-Stable FDTD Methods and Numerical Analysis. IEEE transactions on antennas and propagation, Vol. 59, No. 9. pp 3280–3289.

Energy Science and Applied Technology – Fang (Ed.)
© 2016 Taylor & Francis Group, London, ISBN 978-1-138-02833-3

Influence of nonlinear phase on the linear propagation of partial-coherent Airy beam

Y. Meng & X.Q. Fu
Key laboratory for Micro/Nano Optoelectronic Devices of Ministry of Education, College of Computer Science and Electronic Engineering, Hunan University, Changsha, China

ABSTRACT: This work presents the analysis on the spatial coherence of Airy beam with a nonlinear phase by using the theoretical model of the mutual intensity and generalized Van Cittert-Zernike theorem. It is shown that the enlargement of B-integral (higher nonlinear phase) weakens the spatial coherence of Airy beam gradually. In addition, it reveals that the truncation coefficient and linear propagation are relevant elements to the spatial coherence of Airy beam. The coherence enhances with the decreasing propagation distance and increasing truncation coefficient.

Keywords: Airy beam; spatial coherence; mutual intensity

1   INTRODUCTION

Due to the unique properties of non-diffracting, Airy beam (Siviloglou, G. A. et al. 2008) has been the key project to the researchers all the time, with most of them focusing on the following characteristics of Airy beam: diffraction-free (Altucci, C. et al. 2002), self-accelerating (Dolev, I. et al. 2012, Zhang, Y. et al. 2014), self-healing (Broky, J., et al, 2008) and self-focusing (Mei, Z. R. 2014). Meanwhile, the generation and transmission characteristics of the Airy beam are also the priority of researchers (Chamorro-Posada, P., et al, 2014, Jiang, Y. et al. 2012). Besides, its application is another hotspot for investigation (Abdollahpour, D. et al. 2010, Baumgartl, J. et al. 2008). It has been a common method for the experiment to produce Airy beam by using the high-power Gaussian beam. However, this way will inevitably impose a nonlinear phase on the Airy beam, which is bound to bring a certain influence on the Airy beam transmission (Yin, L. H. & Agrawal, G. P. 2007). Conversely, most studies have focused on the full-incoherent Airy beam, while a few have investigated on the partial-coherent Airy beam (Zi, Y. et al. 2013). However, the advantages of the partial-coherent beam, such as resistance to speckle (Guo, L. et al. 2010), have attracted the wide attention of the researchers. So, it is very necessary to study the partial-coherent Airy beam with a nonlinear phase.

This paper theoretically studies the transmission properties of the partial-coherent Airy beam with a nonlinear phase, mainly focusing on the influence of the nonlinear phase on the coherent characteristics the transmission characteristics of the Airy beam.

2   COHERENCE EVOLUTION

Researches have shown that a new nonlinear phase would be produced when an intense Gaussian beam propagates from the laser to the Fourier lens to generate the Airy beam (Yin, L. H. & Agrawal, G. P. 2007), and the generated Airy beam can be expressed as (Wu, Y. F. et al. 2013)

$$E(x) = \exp\left[ax - 8a^3B^2 + i\left(B + 2aBx + 2a^3B - \frac{16}{3}a^3B^3\right)\right]$$
$$\times \mathrm{Ai}(x - 4a^2B^2 + i4a^2B)$$

(1)

where B = $(2\pi/\lambda)$, the so-called B-integral, represents the value of the nonlinear phase; λ is the vacuum wavelength of the Gaussian beam; I is the intensity of the Gaussian beam; and γ is the nonlinear index. Research has shown that the B-integral needs to be less than about 3 to avoid beam degradation (Hunt, J. T. et al. 1993). In addition, without any constraint, the intensity of this beam can be obtained as

$$I(x) = E \times E^* = \exp\left(2ax - 16a^3B^3\right)$$
$$\times \mathrm{Ai}(x - 4a^2B^2 + i4a^2B)$$
$$\times \mathrm{Ai}(x - 4a^2B^2 - i4a^2B)$$

(2)

We assume that the degree of spatial coherence function of the light source of the Airy beam meets the classical shell model, and the expression can be written as (Mandel, Leonard & Wolf, Emil 1995)

$$\mu(x_1 - x_2, 0) = \exp\left[-\frac{(x_1 - x_2)^2}{2\delta_0}\right] \tag{3}$$

where μ is the degree of coherence of x_1, x_2 on the plane of $z = 0$ and δ_0 is the initial coherence length. The value of δ_0 is the half width (at $1 = $ e-intensity point) of $|\mu|$. The initial coherence of a beam is related to the different values of x1, x2, instead of x1 or x2.

The generalized Van Cittert–Zernike theorem has been widely used in the statistical optics to calculate the intensity of beam of two points on the observing plane, which can be written as (Goodman, J. W., Wiley-Interscience)

$$J(\varepsilon_1, \varepsilon_2, z) = \frac{\kappa(\overline{\varepsilon})}{(\lambda z)^2} \times \exp\left[-i\frac{2\pi}{\lambda z}(\Delta \varepsilon \overline{\varepsilon})\right]$$

$$\times \int_{\infty}^{\infty} I(\overline{x}) \exp\left[i\frac{2\pi}{\lambda z}(\Delta \varepsilon \overline{x})\right] d\overline{x} \tag{4}$$

$$\kappa(\overline{\varepsilon}) = \int_{\infty}^{\infty} \mu(\Delta x) \exp\left[i\frac{2\pi}{\lambda z}(\overline{\varepsilon} \Delta x)\right] d\Delta \overline{x} \tag{5}$$

where $x = (x_1 + x_2)/2$, $\Delta x = x_1 - x_2$, $\varepsilon = (\varepsilon_1 + \varepsilon_2)/2$, $\Delta \varepsilon = (\varepsilon_1 - \varepsilon_2)$.

By using Eq. (1)-(5), we can obtain the mutual intensity of two points on the observing plane, which can be, respectively, written as

$$J_{Ai} = (\overline{\varepsilon}, \Delta \varepsilon, Z, B) = \sqrt{\frac{Z^4 \delta_0^2}{8\pi^2 (2a + iZ\delta\varepsilon)}}$$

$$\times \exp[-16a^3 B^2] \times \exp\left[-2\delta_0^2 (Z\overline{\varepsilon})^2\right] \times \exp\left[-i(Z\overline{\varepsilon}\Delta\varepsilon)\right]$$

$$\times \exp\left[\frac{a^5\left(\frac{8}{3} + 64B^2\right) - \left(\frac{4}{3}a^3 + 8a^3 B^2\right)(Z\Delta\varepsilon)^2 - \frac{a}{2}(Z\Delta\varepsilon)^4}{4a^2 + (Z\Delta\varepsilon)^2}\right]$$

$$\times \exp\left[i \times \frac{4a^2(Z\Delta\varepsilon) + \left(\frac{2}{3}a^2 + 4a^2 B^2\right)(Z\Delta\varepsilon)^3 - \frac{1}{12}(Z\Delta\varepsilon)^5}{4a^2 + (Z\Delta\varepsilon)^2}\right] \tag{6}$$

where $Z = 2\pi/\lambda z$.

For the convenience of calculation, we specially selected two points meeting the condition of distribution centered on the optical axis, so $\varepsilon = 0$, $\Delta\varepsilon = 2\varepsilon_1 = -2\varepsilon_2 = X$.

Furthermore, with the definition of the function of the degree of coherence, the measurement to evaluate whether the coherence is good or not can be written as (Mandel, Leonard & Wolf, Emil, in Cambridge University)

$$\mu(\varepsilon_1, \varepsilon_2) = \frac{J(\varepsilon_1, \varepsilon_2)}{\sqrt{J(\varepsilon_1, \varepsilon_1)} \cdot \sqrt{J(\varepsilon_2, \varepsilon_2)}} \tag{7}$$

We can obtain the function of the degree of coherence of the Airy beam with the nonlinear phase, which can be written as

$$\mu_{Ai} = (X, Z, B) = \sqrt{\frac{2a}{2a + iZ \cdot X}} \times \exp\left[-\frac{2}{3}a^3 - 16a^3 B^2\right]$$

$$\times \exp\left[\frac{a^5\left(\frac{8}{3} + 64B^2\right) - \left(\frac{4}{3}a^3 + 8a^3 B^2\right)(ZX)^2 - \frac{a}{2}(ZX)^4}{4a^2 + (ZX)^2}\right]$$

$$\times \exp\left[i \times \frac{4a^2(ZX) + \left(\frac{2}{3}a^2 + 4a^2 B^2\right)(ZX)^3 - \frac{1}{12}(ZX)^5}{4a^2 + (ZX)^2}\right] \tag{8}$$

From this expression, it can be seen that the coherence is closely related to the truncation coefficient, linear propagation and the nonlinear phase. In order to more clearly understand how these factors influence the spatial coherence of the Airy beam, we select different parameters to describe the clear image of $|\mu_{Ai}| \sim X$ under the following chart for specific analysis.

From Figure 1(a), it can be seen that the width of $|\mu_{Ai}|$ gets gradually narrower with the truncation coefficient increasing from 0.1 to 0.5 when $Z = 1$ and $B = 1$. In other words, with the increasing truncation coefficient, the coherence length of the Airy beam declines. Figure 1(b) shows $|\mu_{Ai}|$ with several Z when $a = 0.1$ and $B = 1$. It shows that the width of $|\mu_{Ai}|$ broadens gradually when the transmission distance increases. From this, it is easy to conclude that the increasing transmission distance can enhance the spatial coherence of the beam. Figure 1(c) shows the relationship between the spatial coherence of the Airy beam and the nonlinear phase with $a = 0.1$ and $Z = 1$. It should be noted that the greater value of B means the higher nonlinear phase imposed on the Airy beam generated by the high-power Gaussian beam. From this image, it can be seen that the width of $|\mu_{Ai}|$ gets increasingly narrower when the value of B-integral increases. It means that the nonlinear phase reduces the spatial coherence of the Airy beam, which urges the researchers to find a more effective way to generate the Airy beam for the sake of better coherence. Figure 1 (a)(b)(c) shows that any tiny change in the three factors can bring a remarkable influence to the coherence. We can change the three factors to get a more appropriate Airy beam with a suitable coherence.

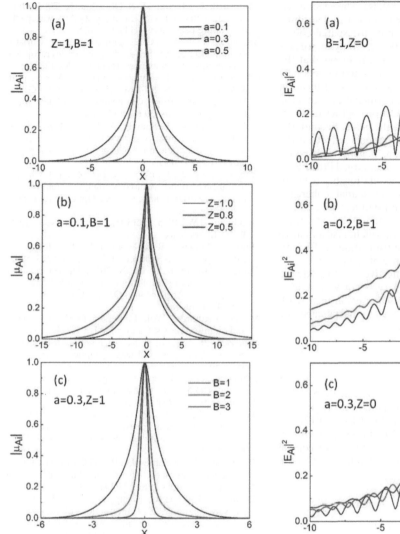

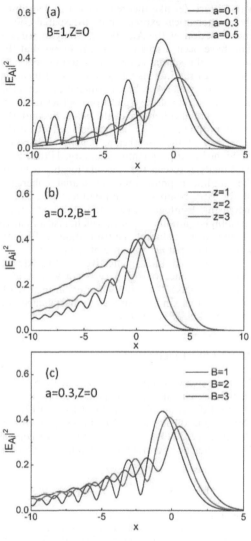

Figure 1. The degree of spatial coherence of the Airy beam as a function of the horizontal axis for several different parameters with (a): Z = 1, B = 1; (b): a = 0.1, B = 1; (c): a = 0.3, Z = 1.

Figure 2. Intensity distribution |EAi|2 of the Airy beam with different parameters with (a): B = 1, Z = 0; (b): a = 0.2, B = 1; (c): a = 0.3, Z = 0.

The change in the parameters is just several ways that essentially changes the intensity distribution of the Airy beam, as shown in Figure 2. These parameters can have an impact on the spatial coherence of the Airy beam accordingly. The decreasing transmission distance Z, the increasing truncation coefficient a and the increasing B-integral are doomed to trigger the Airy beam to change, which together become the important factors that weaken the spatial coherence of the Airy beam. It is extremely important for us as researchers who devote themselves to studying the Airy beam and its coherence to notice that the nonlinear phase is an incontrovertible factor to the spatial coherence of the partial-coherent Airy beam.

3 CONCLUSION

In conclusion, based on the fact that the nonlinear phase would be imposed when the high-power Gaussian beam is used to generate the Airy beam, we investigated the spatial coherence of this special beam with a high nonlinear phase. The generalized Van Cittert-Zernike theorem is used to determine the mathematical expression of the spatial coherence function of the Airy beam. Many important data have been discussed. First, it shows that the width of the spatial coherence function gets narrower with the increasing value of the truncation coefficient of the Airy beam, which means that the increasing value of the truncation coefficient weakens the spatial coherence of the beam. Second, the linear propagation has the opposite effect on the coherence of the Airy beam. It gets much better while the linear propagation is enlarged little by little. Finally, it reveals the vital relationship between the nonlinear phase and the spatial coherence of the Airy beam with the nonlinear phase. The increasing B-integral means that the Airy beam has a higher nonlinear phase, which, just similar to the increasing value of the truncation coefficient, weakens the spatial coherence of the beam.

This research finds an effective way to study the spatial coherence of the Airy beam with a high nonlinear phase. It helps researchers in their work related to this kind of Airy beam, particularly the characteristic of coherence.

ACKNOWLEDGMENTS

This research was supported in part by the Specialized Research Fund for the Doctoral Program of Higher Education of China (20110161110012), the Department of Science and Technology of Hunan Province (2013TP4026), and Hunan Provincial Natural Science Foundation of China (12JJ7005).

REFERENCES

Abdollahpour, D., Suntsov, S., Papazoglou, D.G. & Tzortzakis, S. 2010. Spatiotemporal Airy light bullets in the linear and nonlinear regimes, Physical review letters 105(25): 253901.

Altucci, C., Bruzzese, R., De Lisio, C., Porzio, A., Solimeno, S. & Tosa, V. 2002. Diffractionless beams and their use for harmonic generation, Optics and lasers in engineering, 37(5): 565–575.

Baumgartl, J., Mazilu, M. & Dholakia, K. 2008. Optically mediated particle clearing using Airy wavepackets, Nature photonics, 2 (11): 675–678.

Broky, J., Siviloglou, G.A., Dogariu, A. & Christodoulides, D.N. 2008. Self-healing properties of optical Airy beams, Optics express, 16(17): 12880–12891.

Chamorro-Posada, P., Sánchez-Curto, J., Aceves, A.B. & McDonald, G.S. 2014. Widely varying giant Goos–Hänchen shifts from Airy beams at nonlinear interfaces, Optics letters, 39 (6): 1378–1381.

Dolev, I., Kaminer, I., Shapira, A., Segev, M. & Arie, A. 2012. Experimental observation of self-accelerating beams in quadratic nonlinear media, Physical review letters, 108(11): 113903.

Guo, L., Tang, Z., Liang, C., & Tan, Z. 2010. Characterization of tightly focused partially coherent radially polarized vortex beams, Chinese Optics Letters, 8(5): 520–523.

Goodman, J.W. 1985. Statistical Optics, New York, Wiley-Interscience.

Hunt, J.T., Manes, K.R. & Renard, P.A. 1993. Hot images from obscurations, Applied optics 32(30): 5973–5982.

Jiang, Y., Huang, K. & Lu, X. 2012. The optical Airy transform and its application in generating and controlling the Airy beam, Optics Communications, 285(24): 4840–4843.

Mandel, Leonard & Wolf, Emil 1995. Optical coherence and quantum optics. Cambridge University press.

Mei, Z.R. 2014. Light sources generating self-focusing beams of variable focal length, Optics letters, 39(2): 347–350.

Siviloglou, G.A., Broky, J., Dogariu, A. & Christodoulides, D.N. 2008. Ballistic dynamics of Airy beams, Optics letters, 33(3): 207–209.

Wu, Y., Fu, X., Wang, W., Yu, W., & Wu, H. 2013. The effects of nonlinear phase for the generation and propagation of finite energy Airy beams, J. Opt., 15:105203.

Yin, Lianghong & Agrawal,G.P. 2007. Impact of two-photon absorption on self-phase modulation in silicon waveguides, Optics letters 32(14): 2031–2033.

Zhang, Y., Belić, M.R., Zheng, H., Chen, H., Li, C., Li, Y. & Zhang, Y. 2014. Interactions of Airy beams, nonlinear accelerating beams, and induced solitons in Kerr and saturable nonlinear media, Optics express, 22(6): 7160–7171.

Zi, Y., C.S.W.C., Ji-Xiong, H.K. & L.P. 2013. Investigation on partially coherent Airy beams and their propagation, Acta Phys. Sin: 094205.

Energy Science and Applied Technology – Fang (Ed.)
© 2016 Taylor & Francis Group, London, ISBN 978-1-138-02833-3

Simulation of fountain based on particle system

L.J. Dai & L.F. Miao
College of Mathematics, Physics and Information Engineering, Zhejiang Normal University, Zhejiang, China

ABSTRACT: The simulation of the natural landscape in computer has been one of the important research contents in virtual reality and computer graphics. This thesis adopts the particle system to simulate fountain by analyzing the physical movement of the fountain particle. In the simulation of fountain particle motion, this thesis adds the effects of the wind and gravity field, and simulates the spray and mist generation, according to the collision detection of fountain particles and the surrounding environment. The experimental results show that the simulation method has effectively enhanced the realism and immersion of fountain simulation.

Keywords: Particle System; Fountain Simulation; Spray and Mist; Virtual Reality

1 INTRODUCTION

The natural landscapes such as slow floating clouds, spectacular fountain, complex fog and others have a lot of details and changing shape. The simulation of the natural landscape in computer has been one of the most important research fields of virtual reality and computer graphics.

The simulation of the natural landscape can be divided into two ways: one is based on the physical model and the other is based on the particle system. In the physical modeling method, it is difficult to show the abundant details and dynamic properties of the natural landscape by using analytic surface for simulating the natural landscape. The particle system uses a large number of particles that have their own attributes and lifetime as the basic units to build a complex landscape, and updates the properties of particles for dynamic simulation.

Particle system is the typical algorithm based on the dynamic stochastic growth principle. Particle system, proposed by Reeves in 1983, is mainly used to solve the display of large objects in computer, which are composed of many micro objects with regular activity. In the simulation of an irregular natural scene, particle system has a simple data structure and its algorithms are easy to implement.

In the simulation of the natural scene, the effect of water simulation is essential to the scene reality. He Xiao, Ming yun He, Zhong jian Bai added uniformly accelerated motion to simplify the motion of particles, and used texture color fusion to draw the particles to realize the fountain simulation. Huai xin Wang, Hua geng Wan, Shu xing Xiao, based on the particle system, successfully realized the real-time simulation of a large-scale music fountain, by using music data to build the particle motion of music fountain and improving fountain particles' drawing method. Jing mi Zhao, Hui Zhang, Guo qin Zheng proposed a new method of fountain simulation, based on the particle system API. They adopted a linear particle instead of a point particle and the texture mapping algorithm to achieve a variety of fountain simulations. By analyzing the physical movement of fountain particles, this thesis realizes the fountain simulation based on the particle system. In the simulation of fountain particle motion, this thesis adds the effects of the wind and gravity field, and simulates the spray and mist generation, according to the collision detection of fountain particles and the surrounding environment. The experimental results show that the simulation method has effectively enhanced the realism and immersion of fountain simulation.

2 ANALYSIS OF FOUNTAIN PHYSICAL PROCESS

Fountain is not only an important part of the modern landscape, but also a waterscape art. A spectacular fountain sculpture and a mellow fountain flow fully demonstrate the dynamic and static beauty. There are many famous fountain landscapes in the world, such as the Fountain of the Four Rivers in Rome, the Geneva Lake Fountain, and the Versailles Fountain.

The physical process of the fountain includes the following steps:

1. Water is blown out from the fountain under the pressure.
2. Water from the fountain flows in the projectile track in the air.
3. Water falls into the pool, and enters into the next cycle.

When the fountain flows in the air, it will be influenced by gravity, wind and air resistance. In the process of movement, fountain flows may turn into spray and mist by collision with the surrounding objects.

3 FOUNTAIN PARTICLE SYSTEM

3.1 *The basic process of fountain particle system*

The principle of the particle system uses a large number of particles with the same properties, as the particle swarm simulates an irregular fuzzy object. The particle system makes the particle movement to meet certain rules of dynamics by controlling the particle properties. In this thesis, the process of the fountain particle system is shown in Figure 1.

3.2 *Fountain particle*

Fountain particles have many properties such as position, size, speed, life, lifetime, texture, transparency and so on. The fountain particle is the smallest unit of the fountain flow, which ban be regarded as an N-dimensional vector:

$$R^n = \{Position, Speed, Size, LifeTime, \\ Texture, Transparency \cdots \}$$

Fountain flow is composed of a large number of water particles. Therefore, the complexity of fountain particles seriously affects the rate of fountain simulation. There are many types of fountain particles, such as spherical, point and rectangle. The spherical particle is similar to the droplet in reality. However, the spherical particle is complex to draw and will reduce the rate of fountain simulation. The point particle can achieve real-time simulation, but cannot show the texture properties of fountain particles. The rectangular particle can not only ensure the simulation speed, but also show the texture details of the droplet.

In reality, droplets have different sizes and transparency. Sometimes, many droplets will merge and form a small column. In view of the above-mentioned phenomenon, this thesis combines the sequence diagram with the random function to realize the diversity of droplet particles. When generating droplet particles, the random function will select a diagram of the sequence diagram as the texture of droplet particles. Some sequence diagrams are shown in Figure 2.

The properties of spray particles and mist particles are basically the same as those of droplet particles. The differences between them are as follows:

1. Particle size. In this thesis, the size of spray particles and mist particles is one-quarter of the droplet particle size.
2. Transparency. The transparency of spray particles and mist particles is higher than that of droplet particles, and does not change with time.
3. Lifetime. The lifetime of spray particles and mist particles is fixed.
4. Force. The influence of wind on spray particles and mist particles is far larger than the gravity.

The textures of spray particles and mist particles are shown in Figure 3.

3.3 *Simulation of wind field*

Wind is caused by air flow, which is formed by solar radiation. In this thesis, according to the

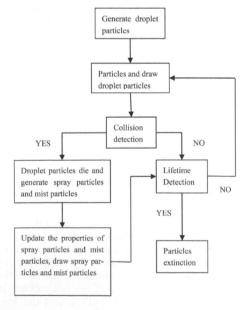

Figure 1. Process of fountain particle system.

(a) (b) (c)

Figure 2. Sequence diagram of droplet particles.

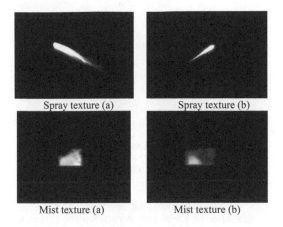

| Spray texture (a) | Spray texture (b) |
| Mist texture (a) | Mist texture (b) |

Figure 3.   Spray texture and mist texture.

change in wind speed, wind can be divided into stable wind and gradient wind. The wind that the speed and direction of the wind do not significantly change with time is called the steady wind. In the 3D coordinate system, the steady wind can be expressed as:

$$V_{wx} = V_{w0x} + rand() \qquad (1)$$

$$V_{wy} = V_{w0y} + rand() \qquad (2)$$

$$V_{wz} = V_{w0z} + rand() \qquad (3)$$

The wind that the wind speed gradually increases or decreases with time is called the gradient wind. In the 3D coordinate system, the gradient wind can be expressed as:

$$V_{wx} = V_{w0x} + \int_{0}^{\Delta t} a_x dt \qquad (4)$$

$$V_{wy} = V_{w0y} + \int_{0}^{\Delta t} a_y dt \qquad (5)$$

$$V_{wz} = V_{w0z} + \int_{0}^{\Delta t} a_z dt \qquad (6)$$

where V_{wx}, V_{wy}, V_{wz} are the wind speed on the X axis, Y-axis and Z-axis, respectively; V_{w0x}, V_{w0y}, V_{w0z}, are the initial velocity of the wind in the axis direction; a_x, a_y, a_z are the acceleration of the wind speed; and Δt is a unit time.

3.4   *Speed of the particle*

Particles in the air are influenced by gravity, wind and air resistance. Air resistance is relevant to the particle's speed, direction and volume. The calculation of air resistance is complex. Besides, air resistance has a small effect on the particle motion. Thus, in this thesis, air resistance in the particle motion is ignored. The speed of the fountain particle can be expressed as:

$$V_x = V_{x0} + V_{wx} + rand() \qquad (7)$$

$$V_y = V_{y0} + V_{wy} + \int_{0}^{\Delta t} Gdt + rand() \qquad (8)$$

$$V_z = V_{z0} + V_{wz} + rand() \qquad (9)$$

where V_x, V_y, V_z are the speed of the particle in the 3D coordinate axis direction; V_{x0}, V_{y0}, V_{z0} are the initial velocity of the particle; V_{wx}, V_{wy}, V_{wz} are the wind speed; and G is the gravity.

3.5   *Position of the particle*

A large number of fountain particles constitute the fountain flow. The position of the particle is relevant to the fountain particle's speed and time. The position of the fountain particle can be expressed as:

$$P_x = P_{x-1} + \int_{0}^{\Delta t} V_{x-1} dt \qquad (10)$$

$$P_y = P_{y-1} + \int_{0}^{\Delta t} V_{y-1} dt \qquad (11)$$

$$P_z = P_{z-1} + \int_{0}^{\Delta t} V_{z-1} dt \qquad (12)$$

where P_x, P_y, P_z are the next position of the particle; P_{x-1}, P_{y-1}, P_{z-1} are the current position of the particle; V_{x-1}, V_{y-1}, V_{z-1}, are the current speed of the particle; and Δt is a unit time.

4   EXTINCTION OF THE PARTICLE

The extinction of the particle means that the fountain particle is removed from the particle system. The fountain particle's extinction situations can be divided into 4 cases: (1) the life of the particle is over its lifetime; (2) the particle position is out of the set range; (3) the transparency of the particle is 0; and (4) particle collision is with the surrounding objects.

5(a) Under the effect of gravity

5(b) Under the effect of gravity and steady wind

5(c) Under the effect of gravity and gradient wind

5(d) With spray and mist

Figure 4.   Different effects of the fountain simulation.

5   EXPERIMENTAL RESULTS

This thesis realizes the above method in personal computer (1.9GHZ AMD CPU, HD7500 Graphic card, 4G memory) based on the Maya particle system. The different effects of the fountain simulation are presented in Figure 4.

6   CONCLUSIONS

This thesis presents a real-time fountain simulation method based on the particle system. The method uses a sequence diagram as the fountain particle texture to achieve the diversity of fountain particles and realizes the simulation of spray and mist by collision detection. The method proposed in this thesis improves the realism and immersion of the fountain simulation.

There are several aspects to be improved: study on (1) the transmission of the droplet in different light conditions and (2) the integration of different droplets into a new droplet.

ACKNOWLEDGMENT

This work was supported by the National Natural Science Foundation of China (No. 61170315).

REFERENCES

Fang, J.W. 2006. Real-time Simulation of Fountain Based on Hardware-Accelerated and Particle System. Computer Engineering and Application 16:117–120.
Jiang, H.H. & Tang, B.P. & Zhang, G.W. 2009. 3D Fountain Simulation Based on Particle System and OpenGL. Computer Measurement & Control 17(9):1717–1723.
Luo, J. & Wang, L. 2006. Real-time Simulation of Fountain Based on Large-scale Particle System. Computer & Information Technology 1:25–27.
Reeves, W. T. 1983. Particle Systems-A Technique for Modeling a Class of Fuzzy Objects. Computer Graphics 17(3):359–376.
Ren, Q.D. 2010. Fountain Simulation Based on GPU. Science Technology and Engineering 10(36):9110–9114.
Wang, H.X. 2010. Real-time Simulation of Large-scale Music Fountain. Journal of Image and Graphics 15(3):524–529.
Wang, H.X. 2010. Music Fountain Simulation Based on Pen Interaction. Hang Zhou: Zhejiang University.
Wei, K.P. 2007. Real-time simulation of three-dimension fountain based on texture mapping and particle system. Computer Engineering and Design 28(11):2586–2588.
Wei, L.F. & Li, Y.J. 2010. Simulation and realization of fountain based on particle system. ELECTRONIC TEST 2:23–26.
Xiao, H. 2007. Fountain Simulation Based on Particle System in OpenGL. Computer Simulation 24(12):201–204.
Zhang, L.L. 2009. Real-time Simulation of Fountain Based on Particle System. Xi an: Xi a University of Science and Technology.
Zhao, J.M. 2006. Fountain Simulation Based on Particle System. Application Research of Computers 1:244–247.

Energy Science and Applied Technology – Fang (Ed.)
© 2016 Taylor & Francis Group, London, ISBN 978-1-138-02833-3

Studying the impact of interference on HF radar target SNR

Yining Mao, Wu Xie & Hong Xie

Harbin Engineering University, Heilongjiang, Harbin, China

ABSTRACT: The nearby lightning interference is a typical sudden shock interference, which exhibits significant local features in the time domain. It generally occupies a relatively short duration of action, but contains a lot of energy, thus significantly raising the noise floor of the Doppler spectral analysis and hindering the detection of the target signal. Therefore, how to effectively eliminate a high-frequency shock interference has become a important topic for researchers. A multiple Doppler signal recovery method is proposed in this paper based on the singularity features of impulse interference in the Doppler domain and the wavelet analysis method in the Doppler domain impulse interference detection. Different recovery methods are compared, and we successfully suppress the impact that the impulse interference has on a high-frequency ground wave radar.

Keywords: interference, radar, SNR

1 INTRODUCTION

This paper mainly studies the principle of the HF surface wave radar (HFSWR) shock jamming and method, and solves the impact of interference detection and location problem in theory.

We propose the implementation scheme of HFSWR interference shock[3]. First, this paper simulates the FMICW signal and lightning interference signal, and lays a foundation for the subsequent recognition of the impact of the interference, positioning and data recovery section. Second, this paper analyzes and expounds the HFSWR to detect the impact problems of interference, and points out that the radar system bandwidth limits the impact interference detection in time domain signal and puts forward the principle of wavelet transform to detect the impact jamming. Finally, this paper give the implementation scheme of HFSWR interference shock. The main content of this article includes the Doppler signal recovery program and the impact interference suppression method of performance analysis. This paper gives two kinds of algorithm simulation results.

2 HFSWR SIGNAL PROCESSING

We determine the truncation FMICW signal, which is represented as

$$s(t) = \sum_{n=0}^{p-1} rect\left(\frac{t - nT - mT_s}{T_0}\right)$$
$$\times \cos\left[2\pi\left(f_c t - \frac{\alpha(t - mT_s)^2}{2}\right)\right] \quad (1)$$

$mT_s < t < (m+1)T_s, m \in [0, N-1]$, $pT < T_s$ where T_s is the frequency modulation cycle; N is the frequency modulation cycle number; p is the number of truncation pulse frequency modulation cycle; $rect(\frac{1}{T_0})$ is the rectangular pulse by the duration of T_0; T is truncation of the pulse repetition period; f_c is the carrier frequency; $\alpha = \frac{B}{T_s}$ is the frequency modulation slope; and B is the FM signal bandwidth.

We set R as the radar and the point target radial motion as the velocity v. The target echo signal can be approximated as

$$s(t) = \sum_{n=0}^{p-1} rect\left(\frac{t - nT - mT_s - \tau_0}{T_0}\right)$$
$$\times \cos\left[2\pi\left(f_c t + f_d t - \frac{\alpha(t - mT_s - \tau_0)^2}{2}\right)\right],$$

$mT_s < t < (m+1)T_s \quad m \in [0, N-1]$ \hfill (2)

where $\tau_0 = \frac{2R}{c}$, $f_d = \frac{2v}{c}f_c$ and c is the speed of light.

3 IDENTIFICATION AND POSITIONING OF THE SHOCK INTERFERENCE

Wavelet transform has a wide range of applications in signal detection and feature extraction. It is made of gens points or linear combination of the wavelet function and the signal. The transform and inverse transform of wavelet function are given by

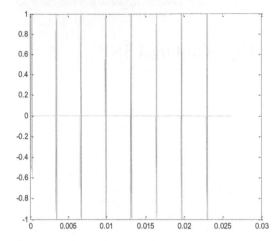

Figure 1. The simulation of a frequency sweep cycle of a normal echo signal.

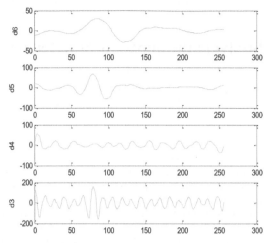

Figure 2. Impact of the jamming signal wavelet detection results (1).

$$W_\psi f(a,b) = |a|^{\frac{1}{2}} \int_{-\infty}^{+\infty} f(t)\psi(\frac{t-b}{a})dt \qquad (3)$$

$$f(t) = \frac{1}{C} \int_{-\infty}^{+\infty} \int_{0}^{+\infty} W_\psi f(a,b)\psi_{a,b}(t)dadb \qquad (4)$$

The formula $\psi(t)$ is the wavelet function, which meets the conditions $\int_{-\infty}^{+\infty} \psi(t)dt = 0$, where $\psi_{a,b}(t)$ is the wavelet functions. By the wavelet function $\psi(t)$ through a series of translation, scaling meets the conditions $\psi_{a,b}(t) = |a|^{\frac{1}{2}} \psi(\frac{t-b}{a})$, where a is the telescopic scale factor and b is the translation factor.

The wavelet transform the signal, showing the signal of high-frequency components that are relatively concentrated on the time domain and the frequency composition that is relatively concentrated on the frequency domain. Clearly, the characteristics of the wavelet transform of the singular signal detection are particularly advantageous.

The wavelet basis function is used for a 6 layer signal decomposition result diagram.

From Figure 3, the d1 layer detail coefficients can be loaded into the radar echo signal in the position and orientation of lightning interference.

4 HFSWR PRINCIPLE AND IMPLEMENTATION METHOD OF IMPACT INTERFERENCE SUPPRESSION

The impact of interference suppression should follow the following principles [2]: (1) try to weaken the interference signal on the shock disturbance period; and (2) to ensure that the normal echo signal is not affected to a great extent.

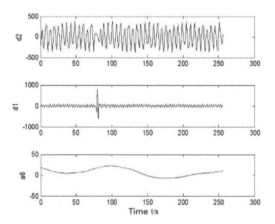

Figure 3. Impact of the jamming signal wavelet detection results (2).

4.1 Interpolation method of recovery

The impact of the Doppler signal interference can be abandoned in the process of sampling, and then through the interpolation approximate normal value in the original position. This is the interpolation method to restore the principle of Doppler signals.

4.2 Interpolation zero of recovery

Because the location by the lightning interference signal amplitude increased significantly, we consider the impact directly interfering with the location of zero [5], and the results are simulated.

From the figure, it can be seen that the two methods have a good recovery effect and a signal-

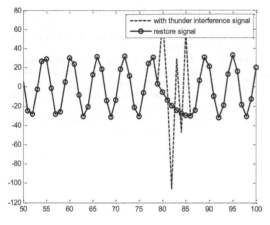

Figure 4.   Data recovery in the time domain.

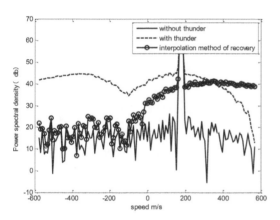

Figure 5.   Data recovery in the Doppler domain.

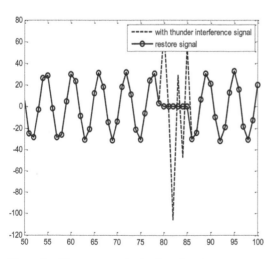

Figure 6.   Data recovery in the time domain.

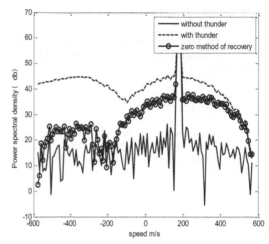

Figure 7.   Data recovery in the Doppler domain.

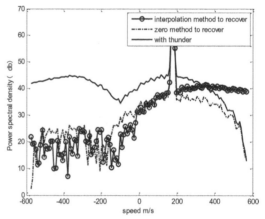

Figure 8.   Comparison of the effect of different recovery methods.

to-noise ratio, and are significantly improved; however, the interpolation method of recovery is better than the other method.

5   SUMMARY

The impact interference signal is in effect at the time of the interference signals and the normal echo signal super position. First, we detect the impact disturbance signal and determine its position; second we get rid of the position signal; finally, we return to the normal signal. In this paper, we adopt two methods to suppress the interference shock. By the simulation of lightning interference and suppression, it is concluded that by inhibiting the impact interference, the signal-to-noise ratio is improved significantly.

ACKNOWLEDGMENT

This work was supported by the Heilongjiang Province Natural Science Foundation Projects (F201339).

REFERENCES

Barnum J R, Simpson E E. Over-the-horizon radar sensitivity enhancement by impulsive noise excision[C]. Radar Conference, 1997. IEEE National. IEEE, 1997: 252–256.

Information Theory, IEEE Transactions on, 1992, 38(2): 617–643.

Mallat S, Hwang W L. Singularity detection and processing with wavelets [J].

Marple L. A new autoregressive spectrum analysis algorithm [J]. Acoustics, Speech and Signal Processing, IEEE Transactions on, 1980, 28(4): 441–454.

O. Rioul, M. Vetterli. Wavelets and Signal Processing [J]. IEEE Signal Processing Magazine. 1991, 8(4):14–38.

Remote Sensing Symposium, 2002. IGARSS'02. 2002 IEEE International. IEEE, 2002, 1: 515–517.

Methods and algorithms optimization

Energy Science and Applied Technology – Fang (Ed.)
© 2016 Taylor & Francis Group, London, ISBN 978-1-138-02833-3

Comparison of CUDA and OpenCL performance on FDTD simulation

Mingyu Zhang, Shujuan Geng & Junpeng Zhang
School of Information and Electrical Engineering, Shandong Jianzhu University, Jinan, China

ABSTRACT: It usually takes much time to produce large-scale electromagetic simulations with the Finite Difference Time Domain (FDTD) method, but now General Purpose Graphics Units (GPU) has provided a solution to solve this problem. It was found, by analyzing its algorithm characteristics, that FDTD is an inherent data parallel algorithm. Compute Unified Device Architecture (CUDA) and Open Computing Language (OpenCL) offer two different ways for simulating FDTD on GPUs. OpenCL is a new open standard that can be used to program CPUs, GPUs and other platforms while CUDA is special for NVIDIA. In this paper, we compared the representation of CUDA and OpenCL on the FDTD algorithm. Compared with the traditional calculation, when the Yee cell number reaches millions, the acceleration of GPU based on OpenCL can accelerate up to 20 times, while the acceleration of GPU based on CUDA can reach 30 times. The result proved that these two methods are very fit to simulate electrical-large object, furthermore CUDA is about 1.5 times faster than OpenCL.

Keywords: Comparison, cuda, opencl, performance

1 INTRODUCTION

Ever since the finite difference time domain (FDTD) was put forward by K.S. Yee in 1966, it has been widely used in the field of electromagnetic science and engineering, which mainly uses FDTD algorithm characteristics. In the space and time component, the FDTD method makes use of the alternating sampling disperse method on electric field E (or magnetic field H). As a result, each E (or H) component will be surrounded by four H (or E) components, by which Maxwell equations are transformed into one-dimensional finite difference equations to solve issues related to an electromagnetic field, and thus the FDTD method is easy to use. In theory, solely by assigning the parameter to every Yee cell in the space, one can obtain the target of any shape, or any material structure desired. With such strong commonality, FDTD can be used to deal with those problems that are concerned with complex shape, structure and heterogeneous medium dispersion. However, FDTD is an inherent data parallel algorithm. In order to meet the needs of accuracy, lots of grids and a large number of iterative calculations must be set up. More importantly, in order to ensure the accuracy of the FDTD algorithm, the grid size must be very strictly controlled. Generally, the space step is smaller than 1/12 of the wavelength. When FDTD is applied to complicated objects, the time step must also meet the relevant conditions. All this will lead to huge memory and runtime requirements when the object is much bigger than wavelength, which, to some extent, limits the development of FDTD. Above all, it is necessary to find some appropriate solution in order to improve the electromagnetic simulation efficiency, which, if there is any, will bring profound influence to the FDTD popularly used in electromagnetic simulation computation.

The FDTD algorithm determines that its electromagnetic field calculation is only related to this point and the surrounding four points, thus FDTD has the inherent advantage of parallel computing. In recent years, parallel computing technology used in the FDTD algorithm has developed more and more fast. Previous parallel computing technology mainly included network parallel, GPGPU, and DirectX. Since the appearance of CUDA, and OpenCL technology, FDTD parallel computation has been greatly improved. With one computer and a graphics card, usable parallel programs can be developed. The CUDA architecture contains a unified shadier assembly line, which enables computing program to arrange each mathematical logic unit on the GPU. Also, parallel computing can be performed using the unified instruction. In the process of computing, not only can the memory be randomly read from and written to, but also the cache controlled by the program can be accessed, so that efficient general-purpose computing can be implemented. Since the release of GeForce GTX 8800, NVIDIA has produced a compiler to compile CUDA, the result of which is CUDA C has become the first specially designed language of GPU to use a general computing program on GPU. CUDA has broad application prospects in medical imaging, computational fluid dynamics, and environmental science. While OpenCL is a new open industry standard, which can be used to develop various general programs running on CPU, GPU and other platforms. CPU and GPU can be linked together to reduce system energy consumption and hardware cost. At the same time OpenCL can be used in a variety of host languages such as C, C++, and Delphi. In other words, OpenCL can be easily mastered by people with various development backgrounds. Thus developers can concentrate on the accelerating algorithm with no need to master another new program. OpenCL was proposed in June 2008 at the WWDC conference by Apple, later it was handed over to the Khronos Group, who announced building a general open standards working group based on Apple's proposal to create OpenCL industry norms. By November 18, 2008, technical details of the OpenCL 1.0 specification were completed, closely followed by the OpenCL1.1, 1.2 version. The fast development proved that OpenCL has wide development prospects.

This paper expounds the FDTD algorithm principle, CUDA and OpenCL architecture. Furthermore, it realizes the simulation of two-dimensional TM wave on CPU, and on GPU based on OpenCL and CUDA. When the number of Yee cells reaches millions, the GPU based on OpenCL can achieve a stable speed of 800–900 Megacells/s, and an acceleration of about 20 times compared with that of CPU. While the GPU based CUDA can reach 1300–1400 Megacells/s, an acceleration of about 30 times. This will have a profound influence upon electromagnetic simulation computation.

2 FDTD ALGORITHM

2.1 Two-dimensional FDTD basic algorithm

Maxwell's equations on electromagnetic field coupling in time and space are:

$$\frac{\partial D}{\partial t} = \frac{1}{\sqrt{\varepsilon_0 \mu_0}} \nabla \times H \tag{1a}$$

$$\frac{\partial H}{\partial t} = -\frac{1}{\sqrt{\varepsilon_0 \mu_0}} \nabla \times E \tag{1b}$$

Taking two-dimensional TM wave as an example, Maxwell's equations can be changed are as follows:

$$\frac{\partial D_z}{\partial t} = \frac{1}{\sqrt{\varepsilon_0 \mu_0}} (\frac{\partial H_y}{\partial x} - \frac{\partial H_z}{\partial y}) \tag{2a}$$

$$\frac{\partial H_x}{\partial t} = -\frac{1}{\sqrt{\varepsilon_0 \mu_0}} \frac{\partial E_z}{\partial y} \tag{2b}$$

$$\frac{\partial H_y}{\partial t} = \frac{1}{\sqrt{\varepsilon_0 \mu_0}} \frac{\partial E_z}{\partial x} \tag{2c}$$

Through adopting samples to Yee cells, discrete Maxwell's equations changed into difference equations are as follows:

$$\frac{D_z^{n+1/2}(i,j) - D_z^{n-1/2}(i,j)}{\Delta t} =$$
$$\frac{1}{\sqrt{\varepsilon_0 \mu_0}} (\frac{H_y^n(i+1/2,j) - H_y^n(i-1/2,j)}{\Delta x})$$
$$- \frac{1}{\sqrt{\varepsilon_0 \mu_0}} (\frac{H_x^n(i,j+1/2) - H_x^n(i,j-1/2)}{\Delta x}) \tag{3a}$$

$$\frac{H_x^{n+1}(i,j+1/2) - H_x^n(i,j+1/2)}{\Delta x}$$
$$= -\frac{1}{\sqrt{\varepsilon_0 \mu_0}} \frac{E_z^{n+1/2}(i,j+1) - E_z^{n+1/2}(i,j)}{\Delta x} \tag{3b}$$

$$\frac{H_y^{n+1}(i+1/2,j) - H_y^n(i+1/2,j)}{\Delta x}$$
$$= \frac{1}{\sqrt{\varepsilon_0 \mu_0}} \frac{E_z^{n+1/2}(i+1,j) - E_z^{n+1/2}(i,j)}{\Delta x} \tag{3c}$$

Fig1 and formula 3 show that the FDTD algorithm has the inherent advantage of parallel computing, since its electric or (magnetic) iterative calculation is only related to the surrounded magnetic or (electric) iterative calculation and the value of the last step, does not need to consider the whole electromagnetic field distribution. When parallel computing is executed, each calculating node requires very little memory swapping, and parallel propulsion can be promoted without influencing each other, and thus high efficiency of parallel computing is ensured.

2.2 Numerical stability and dispersion conditions

FDTD uses finite difference equations to take the place of Maxwell's equations. Therefore it must be ensured that the differential equations which are

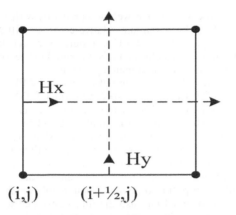

$$(i,j) \qquad (i+\tfrac{1}{2},j)$$

Figure 1. Two-dimensional TM Yee cells structure.

obtained are convergent and stable so as to meet the design intention and obtain the correct results.

According to the Courant stability condition, the two-dimensional space discrete interval Δt, and space intervals ∂x, ∂y must meet the following conditions.

$$c\Delta t \le \cfrac{1}{\sqrt{\cfrac{1}{(\partial x)^2} + \cfrac{1}{(\partial y)^2}}} \qquad (4)$$

In order to reduce the corresponding numerical dispersion, interval Δt must also agree with formula 5.

$$\Delta t \le \frac{T}{12} \qquad (5)$$

3 HARDWARE FOUNDATION AND EXECUTION ENVIRONMENT

3.1 CUDA hardware foundation and environment

CUDA (Compute Unified Device Architecture), is a kind of parallel programming model. Unlike previous parallel environments, to develop parallel programs, CUDA need only a graphics card whose configuration is higher than G8x, making full use of the GPU nuclear superiority to realize parallel data processing. The GPU developed by NVIDIA which is suitable for parallel computing has numerous Scalar Processors (SP), which constitutes the hardware foundation of CUDA. In the CUDA model, the defined kernel function is similar to C language. It was developed by NVIDIA as merely an extension of C language so as to be easily mastered and used by developers. CUDA includes such storages as the global memory, the shared storage, the local storage memory, the read-only memory, and the texture memory.

3.2 OpenCL hardware foundation and environment

OpenCL (Open Computing Language) is the first general programming standard oriented to heterogeneous systems. Similar to CUDA, in order to develop a parallel program it just needs a graphics card whose configuration is higher than G8x and one computer. OpenCL has four kinds of models, the platform model, the execution model, the memory model, and the programming model. The platform model is used to describe coordinated implementation of a single processor or more processors which can execute OpenCL code. The execution model defines how to configure the OpenCL environment on a host computer and how to implement the kernel on the equipment, assigning work items to single calculating cells by defining the core implementation in order to maintain processing components concurrent executing. The memory model includes global memory, constant memory, local memory and private memory. The programming model supports data level parallel and task level parallel, making full use of the computing power of the devices in a heterogeneous platform to achieve parallel computing.

4 TWO-DIMENSIONAL FDTD SIMULATION ON CUDA AND OpenCL

4.1 Algorithm and implement design

In this paper we used Tesla C2070 GPU processors and Intel core i7–2600 - k (TM) CPU to simulate a two-dimensional TM wave. Tesla C2070 has 448 core processors, with 6 GB GDDR5 memory whose single precision floating point performance reaches 1.03 Tflops and whose double precision floating point performance can reach 515 Gflops. Such high configuration shows that it is very suitable for parallel computing. In addition, Intel core i7–2600 - k (TM) CPU has a basic frequency of 3.4 GHz and an 8 core processor. In two-dimensional TM wave simulation calculation, there are three field components Ez, Hx, Hy, with a Dipole Source as the excitation source, wavelength 1.31 um, center frequency 229 Thz, boundary absorption conditions PML, spatial step length 0.05 um, and time step 0.000117851130 ps, in order to meet the Courant stability condition.

GPU and CPU must cooperate with each other very well. Just as figure 2 shows, the program is divided into HOST parts and equipment (DEVICE). The part surrounded by a dotted line is DEVICE, the rest is HOST. HOST is in charge of memory allocation and it initializes the field value, while DEVICE is responsible for the propulsion of the electric field to the magnetic field and of the magnetic field to electric propulsion. If GPU is possessed for a long period of time by the cell, it may be mistaken for a hardware problem and even crash. For this reason, CPU is in charge of advancing the calculations in the time domain.

Taking Ez calculating as an example, in the time domain propulsion direction, GPU reads the Ez value of the previous step from the global memory. Hx and Hy values can be obtained in the same manner. The Ez value can be calculated using formula 3 and then saved for the next step field calculation. Both CUDA and OpenCL lack a synchronization mechanism. This coupled with FDTD algorithm characteristics, determines that the FDTD iterative calculation in the time domain can only be carried out by using the electromagnetic field value of the previous step to push that of the next step. Thus DEVICE is classified into three kernels. Ez—kernel takes charge of Hx&Hy - > Ez, Hx—kernel computes Ez - > Hy, Hy—kernel is in charge of Ez - > Hy computing.

4.2 Simulation results and analysis

In this paper, we used Tesla C2070 GPU processors and Intel core i7–2600 - k (TM) CPU to simulate a two-dimensional TM wave so as to analyze computing speed and accuracy based on CUDA and OpenCL. When we compared this with that of CPU, the maximum relative error was 10E-5, which proved that GPU computes very reliable. When the calculation data is presented in the form of images, their data match perfectly.

Table 1 shows the computing speed of CPU and that of GPU based respectively on CUDA and OpenCL with different number of Yee cells, the unit is MegaCells/s. In Formula 6, I and J are two variables for the two-dimensional TM wave, T is time iteration steps, and t is calculating time. CPU computing time only includes the time consumed when the electromagnetic field values recur, while GPU computing time includes not only recursive time consumed but also the time consumed when creating graphics card storage and copying data between the CPU and GPU.

$$Speed = \frac{I*J*T}{10^6 * t} \qquad (6)$$

Table 1 shows the calculation results with different numbers of Yee cells. It can be seen that when the Yee cell number is small, the advantages of GPU are not apparent. This is because GPU time is mainly spent on the kernel invoke and the superposition of the time step. But with the Yee cells number increasing the advantage of GPU becomes obvious. When the Yee cell number reaches a certain limit, the GPU multiprocessor SP occupancy rate almost reaches the limit, and thus the computing speed becomes stable. The computing speed of GPU based on OpenCL is stable at 880–1000 MegaCells/s, while that of GPU based on CUDA is stable at 1300–1400 MegaCells/s. In the actual simulation computation, the grid number is usually very large, up to millions or even more, which shows that this method is very suitable for actual

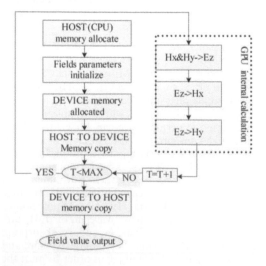

Figure 2. Flowchart of two-dimension TM wave.

Table 1. CUDA, OpenCL, CPU computing speed.

Yee cells	CPU	OpenCL	CUDA
20000	23.28	13.2	19.86
80000	28.65	45.36	50.32
180000	35.65	98.36	135.36
320000	40.65	500.45	800.36
720000	44.65	992.5	1314.49
2000000	48.78	1007.03	1307.18
3380000	45.05	983.75	1360.08
5120000	44.44	971.58	1409.07
8000000	43.45	990.72	1415.02
11520000	43.27	980.68	1395.92
15680000	43.55	986.49	1406.47
20480000	43.65	963.67	1398.51
25920000	42.68	899.21	1406.51
32000000	43.28	881.65	1396.87

electromagnetic simulation. Figure 3 is drawn by Origin8 Pro SR4 from the data of table 1.

Table 2 shows the speedup ratio of the GPU based on CUDA and OpenCL when compared with CPU with different number of Yee cells. When the Yee cells number is very small, the speedup is not very obvious. When there are 20,000 Yee cells, the GPU computing speed is even below that of CPU, mainly because the Yee cell number is too small to form enough threads for multiprocessor parallel computing. But when the Yee cell number reaches a certain number, the speedup of OpenCL can be stable at 20 times while the speedup of CUDA can be stable at 30 times. In our practical electromagnetic simulation computation, the grid number is enormous, and thus suitable for solving practical electromagnetic problems. Figure 4 is drawn by Origin8 Pro SR4 from data of table 2. From this figure we can easily make the conclusion that CUDA is obviously better in terms of acceleration.

Table 2. CUDA, OpenCL speedup.

Yee cells	OpenCL speedup	CUDA speedup
20000	0.570446735	0.853092784
80000	1.583246073	1.756369983
180000	2.759046283	3.796914446
320000	12.31119311	19.68905289
720000	22.22844345	29.43986562
1280000	22.12323633	29.75308642
2000000	20.64432144	26.79745797
3380000	21.83684795	30.19045505
5120000	21.86273627	31.70724572
8000000	22.8013809	32.56662831
11520000	22.66420153	32.2606887
15680000	22.65189437	32.29552239
20480000	22.07720504	32.03917526
25920000	21.06865042	32.95477976
32000000	20.37084104	32.27518484

5 CONCLUSION

This paper analyzes FDTD algorithm parallelism and numerical dispersion conditions, and it realizes two-dimensional TM wave simulation. PML is used as the absorbing boundary conditions. The simulation results of CPU and GPU based on CUDA and OpenCL matches perfectly, which shows that this algorithm is suitable for simulation of practical electromagnetic problems. The GPU based on OpenCL can reach an acceleration of 20 times, while the GPU based on CUDA can achieve an acceleration of 30 times, which proves that CUDA is more applicable for FDTD algorithm with PML as the boundary condition.

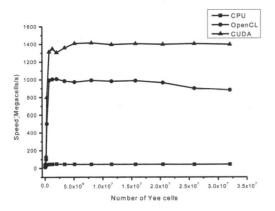

Figure 3. CUDA, OpenCL, CPU computing speed with different number Yee cells.

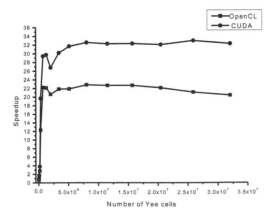

Figure 4. OpenCL, CUDA speedup with different number Yee cells.

REFERENCES

A. M. Shreim and M. F. Hadi, "Integral PML absorbing boundary conditions for the high-order M24 FDTD algorithm," Progress In Electromagnetics Research, vol. 76, pp. 141–152, 2007.

A. Vaccari, A. Cala'Lesina, L. Cristoforetti, and R. Pontalti, "Parallel implementation of a 3D subgridding FDTD algorithm for large simulations," Progress In Electromagnetics Research, vol. 120, pp. 263–292, 2011.

B. Fornberg, J. Zuev, and J. Lee, "Stability and accuracy of time-extrapolated ADI-FDTD methods for solving wave equations," Journal of computational and applied mathematics, vol. 200, pp. 178–192, 2007.

C.-H. Huang, C.-C. Chiu, C.-L. Li, and Y.-H. Li, "Image reconstruction of the buried metallic cylinder using FDTD method and SSGA," Progress In Electromagnetics Research, vol. 85, pp. 195–210, 2008.

D. Liuge, L. Kang, and K. Fanmin, "Parallel 3D finite difference time domain simulations on graphics processors with CUDA," in Computational Intelligence and Software Engineering, 2009. CiSE 2009. International Conference on, 2009, pp. 1–4.

H. Scherl, B. Keck, M. Kowarschik, and J. Hornegger, "Fast GPU-based CT reconstruction using the common unified device architecture (CUDA)," in Nuclear Science Symposium Conference Record, 2007. NSS'07. IEEE, 2007, pp. 4464–4466.

J. A. Silva-Macêdo, M. A. Romero, and B.-H. V. Borges, "An extended FDTD method for the analysis of electromagnetic field rotators and cloaking devices," Progress In Electromagnetics Research, vol. 87, pp. 183–196, 2008.

J.-Z. Lei, C.-H. Liang, W. Ding, and Y. Zhang, "EMC analysis of antennas mounted on electrically large platforms with parallel FDTD method," Progress In Electromagnetics Research, vol. 84, pp. 205–220, 2008.

J. Li, L.-X. Guo, and H. Zeng, "FDTD investigation on bistatic scattering from a target above two-layered rough surfaces using UPML absorbing condition," Progress In Electromagnetics Research, vol. 88, pp. 197–211, 2008.

J. C. Phillips, J. E. Stone, and K. Schulten, "Adapting a message-driven parallel application to GPU-accelerated clusters," in High Performance Computing, Networking, Storage and Analysis, 2008. SC 2008. International Conference for, 2008, pp. 1–9.

J. E. Stone, D. Gohara, and G. Shi, "OpenCL: A parallel programming standard for heterogeneous computing systems," Computing in science & engineering, vol. 12, p. 66, 2010.

L.-X. Guo, A.-Q. Wang, and J. Ma, "Study on EM scattering from 2-D target above 1-D large scale rough surface with low grazing incidence by parallel MoM based on PC clusters," Progress In Electromagnetics Research, vol. 89, pp. 149–166, 2009.

K. Karimi, N. G. Dickson, and F. Hamze, "A performance comparison of CUDA and OpenCL," arXiv preprint arXiv: 1005. 2581, 2010.

M. Unno, Y. Inoue, and H. Asar, "GPGPU-FDTD method for 2-dimensional electromagnetic field simulation and its estimation," in Electrical Performance of Electronic Packaging and Systems, 2009. EPEPS'09. IEEE 18th Conference on, 2009, pp. 239–242.

M. Izadi, A. Kadir, M. Z. Abidin, C. Gomes, and W. F. W. Ahmad, "An analytical second-FDTD method for evaluation of electric and magnetic fields at intermediate distances from lightning channel," Progress In Electromagnetics Research, vol. 110, pp. 329–352, 2010.

N. Funabiki and Y. Takefuji, "A neural network parallel algorithm for channel assignment problems in cellular radio networks," Vehicular Technology, IEEE Transactions on, vol. 41, pp. 430–437, 1992.

Parberry, M. B. Kazemzadeh, and T. Roden, "The art and science of game programming," in ACM SIGCSE Bulletin, 2006, pp. 510–514.

P. Harish and P. Narayanan, "Accelerating large graph algorithms on the GPU using CUDA," in High performance computing–HiPC 2007, ed: Springer, 2007, pp. 197–208.

P. Jaaskelainen, C. S. de La Lama, P. Huerta, and J. H. Takala, "OpenCL-based design methodology for application-specific processors," in Embedded Computer Systems (SAMOS), 2010 International Conference on, 2010, pp. 223–230.

P. Pospíchal, J. Jaros, and J. Schwarz, "Parallel genetic algorithm on the cuda architecture," in Applications of Evolutionary Computation, ed: Springer, 2010, pp. 442–451.

S. A. Manavski and G. Valle, "CUDA compatible GPU cards as efficient hardware accelerators for Smith-Waterman sequence alignment," BMC bioinformatics, vol. 9, p. S10, 2008.

S.-Q. Xiao, Z. Shao, and B.-Z. Wang, "Application of the improved matrix type FDTD method for active antenna analysis," Progress In Electromagnetics Research, vol. 100, pp. 245–263, 2010.

S. Yang, Y. Chen, and Z.-P. Nie, "Simulation of time modulated linear antenna arrays using the FDTD method," Progress In Electromagnetics Research, vol. 98, pp. 175–190, 2009.

T. Namiki, "A new FDTD algorithm based on alternating-direction implicit method," Microwave Theory and Techniques, IEEE Transactions on, vol. 47, pp. 2003–2007, 1999.

T. W.-H. Sheu, R. Y. Chung, and J.-H. Li, "Development of a Symplectic Scheme with Optimized Numerical Dispersion-Relation Equation to Solve Maxwell's Equations in Dispersive Media," Progress In Electromagnetics Research, vol. 132, pp. 517–549, 2012.

W.-Q. Jiang, M. Zhang, H. Chen, and Y.-G. Lu, "CUDA implementation in the EM scattering of a three-layer canopy," Progress In Electromagnetics Research, vol. 116, pp. 457–473, 2011.

Y. H. Liu, Q. H. Liu, and Z.-P. Nie, "A new efficient FDTD time-to-frequency-domain conversion algorithm," Progress in Electromagnetics Research, vol. 92, pp. 33–46, 2009.

Y. J. Zhang, A. Bauer, and E. P. Li, "A novel coupled T-matrix and microwave network approach for multiple scattering from parallel semicircular channels with eccentric cylindrical inclusions," Progress In Electromagnetics Research, vol. 53, pp. 109–133, 2005.

310

Energy Science and Applied Technology – Fang (Ed.)
© 2016 Taylor & Francis Group, London, ISBN 978-1-138-02833-3

Estimation of the entropy of network traffic using space-saving algorithm

F. Fu, H.Q. Li, W.P. Li & Q. Xiao
Key Laboratory of Spacecraft in-orbit Fault Diagnosis and Maintenance, Xi'an, China

ABSTRACT: Research on detecting network traffic anomalies via entropy theory has attracted tremendous interest, but it is difficult to calculate entropy in a large-scale network. This paper presents a new algorithm for estimating the entropy of network traffic, and offers a new way to determine the number of counters in the space-saving algorithm. This algorithm first obtained heavy hitters of network traffic using the space-saving algorithm, then estimated the counts of rest elements by the distribution characteristic of heavy hitters, and finally calculated the estimated entropy of network traffic. The experimental results show that this method can significantly reduce storage memory and time while calculating the entropy of network traffic with small estimation errors.

Keywords: Network traffic; Entropy estimation; Space-Saving algorithm; Heavy hitters

1 INTRODUCTION

Entropy-based anomaly detection has attracted substantial attention of researchers in recent years (Eiland 2006 & Lakhina 2005 & Lawniczak 2009). The attraction of entropy metrics is their capability of condensing an entire feature distribution into a single number, and at the same time, retaining important information about the overall state of the distribution (Ziviani 2007 & Tellenbach 2011). However, some difficulties appeared while calculating the entropy of a large-scale network environment with the rapid increase of network users and bandwidth. For example, the MAWI dataset (Borgnat 2009), collecting traces over trans-Pacific backbone links, had over 30 million packets and 50,000 IP addresses in 15 minutes. The time and space costs are huge for accurately calculating entropy of a large-scale network traffic, so estimated entropy is a better choice to ensure real-time capability of anomaly detection.

Wagner 2005 presented an entropy-based worm detection method, which estimated entropy by using data compression algorithms, but it needed a large storage space for all IP addresses to calculate Kolmogorov complexity. Compressed counting was proposed by Li 2009 to estimate Shannon entropy by certain functions of the αth moments. It needed the α parameter of data within limits, but the attack data did not always met the criteria. Hierarchical sampling over sketches (HSS) was used to estimate a class of metrics over data streams, which was proposed by (Bhuvanagiri

2006), by the heavy-tailed distribution caused a significant error.

In this paper, we present an algorithm called the HHEE (Heavy Hitters Entropy Estimation), based on the space-saving algorithm in data mining, to estimate entropy in a large-scale network with less storage memory and calculating time. The algorithm first obtained heavy hitters of network traffic using space-saving algorithm, then estimated the counts of rest elements by the distribution characteristic of heavy hitters, and finally calculated the estimated entropy of network traffic.

We rigorously evaluated the algorithm with the real traffic from the WIDE backbone networks. It could effectively reduce the space and time required for the entropy calculation in the large-scale network environment, even when containing a large number of abnormal and attack traffic.

2 IMPROVEMENT OF SPACE-SAVING ALGORITHM

2.1 Space-Saving algorithm

The frequent items problem (also known as the heavy hitters problem) is one of the most vigorously studied problems in data streams. Several algorithms have been proposed to handle the top-k, the frequent elements problems, and their variations. These techniques can be classified into counter-based and sketch-based techniques (Metwally 2005). Counter-based algorithms decide for each new arrival whether to store the item or not, and if so, what counts to associate with it, such as LossyCounting and SpaceSaving. The sketch algorithms use hash functions to define a linear projection of the input, such as CountSketch and CountMin. Cormode 2009 gave an experimental comparison of these algorithms, and drew a clear conclusion that the SpaceSaving algorithm has surprisingly clear benefits over others.

Formally, given an alphabet, A, a frequent element, E_i, is an element whose frequency, or number of hits, F_i, in a stream S whose current size is N, exceeds a user-specified support ΦN, where $0 \leq \Phi \leq 1$, whereas the top-k elements are the k elements with highest frequencies. The underlying idea of the space-saving algorithm is to maintain partial information of interest, i.e. only m elements are monitored. The counters are updated in a way that accurately estimates the frequencies of the significant elements, and a lightweight data structure is utilized to keep the elements sorted by their estimated frequencies. The counter at the ith position in the data structure is denoted as $count_i$. The counter $count_i$ estimates the frequency, F_i, of some elements, E_i.

2.2 Determining the number of counters

It is crucial to determine the number of counters, as the value of m, because it directly affects memory requirements and approximate accuracy. In the original space-saving algorithm, the value of m is specified by the user according to the required error rate, ε, which is obtained from the prior statistic distribution knowledge of the dataset. In this paper, we determined the value of m depending on the support ratio, Φ, and proved its validity. Any element, E_i, with frequency $F_i > \Phi N$ is guaranteed to be reported.

First, we will introduce three lemmas of the space-saving algorithm.

Lemma 1 Among all counters, the minimum counter value, min, is no greater than N/m.

Lemma 2 An element E_i with $F_i >$ min must be reported.

Lemma 3 Space-saving uses a number of counters of min(|A|, 1/ε) to find all frequent elements with error ε. Any element, E_i, with frequency $F_i > \Phi N$ is guaranteed to be reported.

From the above lemma, we known that the number of counters equals the minimum value between |A| and 1/ε. When |A| is smaller than 1/ε, the number of counters, m, equals |A|, which means each element has a counter, the result is precise and the space-saving algorithm does not reduce any storage memory or time. When |A| is greater than

or equal to 1/ε, m equals 1/ε. In that case, Lemma 3 did not consider the relationship between ε and Φ. Each heavy hitter in the result will be overestimated, if ε is greater than Φ. All heavy hitters will be found, but their frequency counts are not reliable, thus they are not appropriate for entropy calculation. Hence, we consider to determine the value of m depending on the support ratio, Φ. From the above lemma, we deduce the following.

Lemma 4 Assuming no specific data distribution, space-saving uses a number of counters of min(|A|, 1/Φ) to find all frequent elements. Any element, E_i, with frequency $F_i > \Phi N$ is guaranteed to be reported.

Proof. There are two possibilities for the relationship between |A| and 1/Φ.

When |A| is smaller than 1/Φ, m equals |A|. From Lemma 3, any element E_i with frequency $F_i > \Phi N$ is guaranteed to be reported. Thus, Lemma 4 is valid.

When |A| is greater than or equals to 1/Φ, m equals 1/Φ. From Lemma 1, $min \leq N/m = N\Phi$, any heavy hitter E_i with frequency $F_i > N\Phi$, since $F_i > min$. From Lemma 2, any element E_i with $F_i > min$ must be reported. Thus, Lemma 4 is valid.

From Lemma 4, we had a more intuitive approach to determine the appropriate value of m. In a stream with a large amount of elements, it will reduce the value of m compared with the ε approach, as well as less storage memory and calculating time.

3 HHEE ALGORITHM

3.1 Algorithmic framework

The HHEE algorithm can calculate the entropy of any attribute in a large-scale network, so we chose the IP address attribute for an example in the following analysis, which can also be used for other attributes such as port. Entropy estimation is needed in a large-scale network environment because there are massive amounts of IP addresses in it, and the space consumption is huge if saving all of the IP addresses. The framework of the HHEE algorithm is shown in Figure 1, which can be divided into three main steps. Step 1: It finds n_1 heavy hitters whose frequency, F_i, is greater than ΦN, in the stream S by the improved space-saving algorithm, and also obtains the number of all IP addresses, n, in the stream S. Step 2: It obtains the distribution parameter values of heavy hitters, to estimate the frequency of the remaining n_2 IP addresses, $n_2 = n - n_1$. Step 3: It calculates the entropy of the IP address in the stream S.

As shown in Figure 1, the input of the algorithm is IP addresses of network packets and the output of the algorithm is the estimation entropy value, *EH*. The HHEE algorithm loops in Step 1 until the packet input is complete, meanwhile it updates the stream-summary data structure of the space-saving algorithm, and adds the count value of packets, *N*, and the count value of IP addresses, *n*. When all packets in the stream *S* had been inputted, we can get all the heavy hitters with their frequency values. This is because the space-saving algorithm ensures all heavy hitters in the correct order. It passes the frequency value of each heavy hitter to Step 2 as input. Step 2 fits the distribution parameter values of heavy hitters, and gets the frequency value of each IP address through estimating the frequency of non-heavy hitters. Step 3 combines the precise part of entropy, which calculates from the heavy hitter IP addresses, and the estimate part of entropy, which calculates from the non-heavy hitter IP addresses.

3.2 *Entropy values of the estimated part*

The distribution characteristic of IP addresses or port attributes in the network traffic can be described as Zipf's law (Adamic & Westphal 2002). We obtain the distribution parameters a and b, and then estimate the frequency of the remaining IP addresses, as shown in Figure 2.

We draw the data on a log-log graph, with the axes being log (rank order) and log (frequency) and the plot is linear. We estimate the frequency of the remaining elements through the parameters a and b of the approximate line, which can be obtained from the frequency value of heavy hitters. In the HHEE algorithm, only the heavy hitter IP addresses are stored, but the remaining IP addresses are not, so as to achieve the purpose of reducing the required storage space in the process of entropy calculation. The effect of the improve-

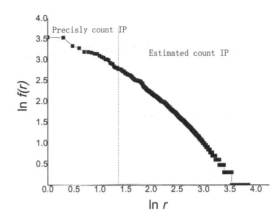

Figure 2. Estimate frequency of the remaining IP addresses.

ment in the large-scale network environment is significant.

4 COMPARISION AND ANALYSIS

We implement our algorithms on the MAWI dataset over trans-Pacific backbone links, which contains numerous attacks with many unknown types. The purpose of the HHEE Algorithm is to obtain the distribution feature of network traffic with less storage cost, for abnormal analysis and testing. So, the estimated entropy value must be close to the actual entropy value, even if there are a large number of abnormal in the network traffic. We divide the MAWI dataset by months, and compare the estimated and actual entropy values, as shown in Figure 3.

In Figure 3, we selected 100,000 packets as a stream node, so the monthly data had 233 nodes. We set the support ratio as $\Phi = 0.001$ and the number of counters as $m = 1000$. There were 7498 IP addresses in the traffic data, and 172 heavy hitter IP addresses. Normally, calculation of entropy takes 7498storage space, but the HHEE algorithm estimated that the entropy took only 1000storage space, with average errors of 0.11.

We kept the value of Φ constant, and adjusted the packets count, *N*, of each stream, to analyze its impact on estimation error. In Figure 4, we compared the estimation error of different N values equal to 10^4, 10^5, 10^6. When *N* was equal to 10^4, on average, there was 1707 IP addresses and 180 heavy hitters in each stream, with an error of 0.01. Compared with the precise calculation, the HHEE algorithm needed to save 1000 IP addresses, with some improvement of memory consumption. When *N* was equal to 10^5, on average, there was 7523 IP addresses and 173 heavy hitters in each

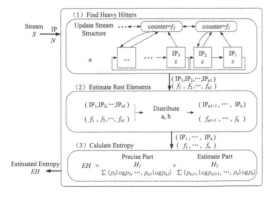

Figure 1. Framework of the HHEE algorithm.

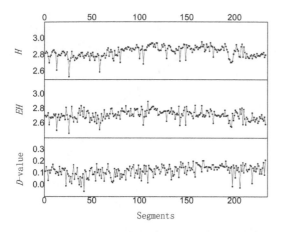

Figure 3. Comparison of the estimated and actual entropy values.

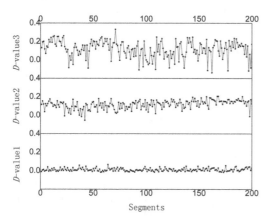

Figure 4. Comparison of estimation error with different N values.

stream, with an error of 0.11. The HHEE algorithm needed to save 1000 IP addresses, with great improvement of memory consumption. When N was equal to 10^6, on average, there was 31996 IP addresses and 148 heavy hitters in each stream, with an error of 0.11. The HHEE algorithm still needed to save 1000 IP addresses, with enormous improvement of memory cost. However, as shown in Figure 4, the error curve was extremely volatility at that time. As the number of packets in each stream increased, the estimation error value was larger, but the memory consumption decreased rapidly. According to the number of packets, we chose the appropriate support value of Φ, in order to achieve the balance between the accuracy and the memory consumption.

5 CONCLUSION

Estimating the value of entropy in a large-scale network can be of great use for network attack and anomaly detection. This paper extended the previous space-saving algorithm, and we proposed an algorithmic framework for entropy estimation with less storage memory and calculating time, and analyzed the influence of estimated error between the support ratio and packet number. Our approach is mainly focused on the heavy hitters in the network traffic to reduce memory consumption. We expect that there might be more heuristic algorithms in the future, which is one of our future research studies.

REFERENCES

Adamic, L.A. & Huberman, B.A. 2002. Zipf's law and the Internet. *Glottometrics* 3(1): 143–150.
Bhuvanagiri, L. & Ganguly. S. 2006. Estimating entropy over data streams. *Annual European Symposium; Proc., 2006.* London: UK.
Borgnat, P. & Dewaele,G. 2009. Seven Years and One Day: Sketching the Evolution of Internet Traffic. *IEEE Communications Society; Proc., 2009.* Rio de Janeiro: BRAZIL.
Cormode, G. & Hadjieleftheriou, M. 2009. Finding the frequent items in streams of data. *Communications of the ACM* 52(10): 97–105.
Eiland, E.E. & Liebrock, L. M. 2006. An application of information theory to intrusion detection. *IEEE International Workshop on Information Assurance; Proc., 2006.* Egham: UK.
Lakhina, A. & Crovella, M. 2005. Diot C. Mining anomalies using traffic feature distributions. *Computer Communication Review* 35 (4): 217–228.
Lawniczak, A.T. & Wu,H. & Stefano, B.N. 2009. Detection of Packet Traffic Anomalous Behaviour via Information Entropy. *Complex Networks* 207: 197–208.
Li, P. 2009. Compressed counting. ACM-SIAM Symposium on Discrete Algorithms;Proc., 2009. New York: NY.
Metwally, A. & Agrawal, D. & Abbadi, A. 2005. Efficient computation of frequent and top-k elements in data streams. *Database Theory* 3363(2005): 398–412.
Tellenbach, B. & Burkhart, M. 2011. Accurate network anomaly classification with generalized entropy metrics. *Computer Networks* 55(15): 3485–3502.
Wagner, A. & Plattner, B. 2005. Entropy based worm and anomaly detection in fast IP networks. *IEEE WET ICE / STCA security workshop; Proc., 2005.* Washington: DC.
Westphal, C. 2002. A user-based frequency-dependent IP header compression architecture. *Global Telecommunications; Proc., 2002.* Taipei: Taiwan.
Ziviani, A. & Monsores, M.L. 2007. Network anomaly detection using nonextensive entropy. *Communications Letters* 11(12): 1034–103.

Energy Science and Applied Technology – Fang (Ed.)
© 2016 Taylor & Francis Group, London, ISBN 978-1-138-02833-3

Controlling the work-in-process in a tandem manufacturing system by hedging point policy

Minru Zhu

College of Computer, Nanjing University Posts and Telecommunications, Nanjing, China

ABSTRACT: The hedging point policy is often used for the optimal control of a manufacturing system. Till now, this policy usually focuses on the production surplus of the manufacturing system. However, the production surplus, can-not describe the quantity of the work-in-process. For example, the number of the work-in-process in a buffer between two consecutive machines may be zero, even if the local production surplus is a quite large positive or negative number. So in this paper, the control policy focusing on the work-in-process in the tandem manufacturing system is researched into, which is helpful to control the work-in-process directly and conveniently.

Keywords: Tandem Manufacturing System, Optimal Control, Hedging Point Policy

1 INTRODUCTION

If the material flow in a manufacturing system is considered as a continuous fluid, the dynamics of the system can be described as linear differential equations (J. Kimemia 1983). Usually, the state variables in the manufacturing system are composed of two parts: the production surplus and the machine's state (i.e. operational or failed). The control variable is the production rate of the machine. The dynamical equation of the system says that the system surplus' derivative with respect to the time equals the difference between the production rate and the demand rate. The objective function is usually the mean cost due to the production surplus in a sufficiently long period of time.

To obtain the optimal control of the manufacturing systems, the dynamical programming method or the maximum principle method are used in (S. B. Gershwin 1994, O. Maimon 1998, Liu Xiaoming 2009). In the dynamical programming method, the Hamilton-Jacob-Bellman (HJB) equation can be obtained according to the optimal principle, and the hedging point control policy is got from the HJB equation.

The problem of optimizing the hedging point for the single-machine and single-part-type system has been solved by Gershwin, Bielecki and Kumar (T. Bielecki 1988, R. Akella 1986), Khmelnitsky (E. Khmelnitsky 2011). In such system, the positive production surplus is the work-in-process and the negative production surplus is the backlog. But in a tandem manufacturing system, the production surplus corresponding to each machine, i.e.

the local surplus, is usually not equal to the work-in-process, except for the last machine. Therefore, the control is not directly on the work-in-process if the local surpluses are chosen as the first part of the system state.

In this paper, the optimal control directly on the work-in-process in a tandem manufacturing system is researched into. Therefore, the local surplus corresponding to the last machine and the work-in-process corresponding to other machines are chosen as the first part and the machine's states are chosen as the second part of the system state. The objective is to minimize the costs caused by the work-in-process and the backlog. At first, the hedging point policy for the control on the work-in-process is proposed for a simple two-machine tandem system, and then the behavior of the system under this control is analyzed in Section 2. In Section 3, the results are extended to the general case, i.e. the tandem manufacturing system with more than two machines. Section 4 is the conclusion part.

2 DISCUSSED PROBLEMS

A tandem two-machine system is composed of two machines and produces only one type of part (Fig.1). There are two buffers in the system. The first one exists between the two machines and the second one is after the second machine. Suppose that the capacities of both buffers are infinite.

Let $\mu_i(t)$ denote the production rate of machine i ($i = 1,2$) at time t. and the maximum production rate of

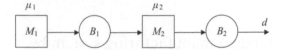

Figure 1. The tandem two-machine system M1 and M2 denote the two machines. B1 and B2 denote the two buffers.

the machine is μ_i. $\alpha_i(t)$ Denote the state of machine i at the time t, where 1 denotes that machine is operational and 0 denotes the machine is failed. $x_1(t)$ Denote the work-in-process level in the buffer 1 at time t. Let $x_2(t)$ the production surplus of machine 2. Let d denote the demand rate for the product. $c(x_1(t), x_2(t))$ Be the cost function. So the optimal control problem of a tandem two-machine manufacturing system can be formulated as follows:
Min

$$E\left\{\int_{t_0}^{T} c(x_1(s), x_2(s))ds\, x_1(t_0), x_2(t_0), \alpha_1(t_0), \alpha_2(t_0)\right\} \quad (1)$$

S. t.

$$dx_1(t)/dt = u_1(t) - u_2(t) \quad (2)$$
$$dx_2(t)/dt = u_2(t) - d \quad (3)$$
$$u_i(t) \le \alpha_i(t)\mu_i \quad i = 1,2 \quad (4)$$

Suppose the cost-to-go function is:

$$J(x_1, x_2, \alpha, t) = (T - t)J^* + W(x_1, x_2, \alpha) \quad (5)$$

Then the HJB equation is as follows:

$$J^* = \min_{m \in \Omega(\alpha)}\left\{c(x_1, x_2) + \sum_j W(x_1, x_2, j, t)\lambda_{ja}\right.$$
$$\left. + \frac{\partial w}{\partial x_1}(x_1, x_2, \alpha)(u_1 - u_2) + \frac{\partial w}{\partial x_2}(x_1, x_2, \alpha)(u_2 - d)\right\} \quad (6)$$

Where λ_{ja} denote the probability that the machine state change from α to j.

Suppose that $W(x_1, x_2, \alpha)$ is a quadratic function with respect to (x_1, x_2), namely

$$W(x_1, x_2, \alpha) = \frac{1}{2}(x_1 + x_2)^2 - Z_1(\alpha)x_1 - Z_2(\alpha)x_2 + D(\alpha) \quad (7)$$

Then the mathematical programming problem has the following form of linear programming:
Min

$$(x_1 - Z1)(u_1 - u2) + (x_2 - Z_2)(u^2 - d) \quad (8)$$

S. t.

$$u_i(t) \le \alpha_i(t)\mu_i, \quad i = 1,2 \quad (9)$$

For $\alpha_1\alpha_2 = (1,1)$, this linear programming problem normalized as
Min

$$(x_1 - Z_1)u_1 + [(x_2 - x_1) - (Z_2 - Z_1)]u_2 + 0 \cdot v_1 + 0 \cdot v_2 \quad (10)$$

S. t.

$$u_1 + v_1 = \mu_1, \quad u_2 + v_2 - \mu_2 \quad (11)$$

Where v_1 and v_2 are the slack variables.

Let, $u_R = (u_1, u_2)^T$, $u_N = v_1, v_2)^T$ denote the basic part and non-basic part of the decision vector; $C_B = (x_1 - Z_1, (x_2 - x_1) - (Z_2 - Z_1))^T$, $c_N = (0,0)^T$, the cost coefficients vector; $B = diag\{1,1\}$ and, $N = diag\{1,1\}$, the basic part and non-basic part of the coefficient matrix in the constraints. According to the simplex method, the necessary and sufficient condition for $u_N = 0$ to be part of the optimal solution of Equations (10)–(11)

$$C_N^T(x_1, x_2) - C_B^T(x_1, x_2)B^{-1}N$$
$$= [-(x_1 - Z_1) - (x_2 - x_1) + Z_2 - Z_1)] > 0 \quad (12)$$

Namely $x_1 > Z_1$ and $x_2 < x_1 + (Z_2 - Z_1)$, which means the domain 1 in Fig.2. In this domain, the control is $u = (u_1, u_2)^T = (\mu_1, \mu_2)^T$, which is the basic part (i.e. u_B) of the optimal solution of Equations (10)–(11).

Similarly, we can get the controls for the states in other three domains in the (x_1, x_2) plane (Fig. 2). The optimal control in the four domains can be summarized as follows.

$$u = (u_1, u_2)^T$$
$$= \begin{cases} (0,0)^T, & \text{if } x_1 - Z_1 > 0 \text{ and } x_2 > x_1 + Z_2 - Z_1, \\ (0, \mu_2)^T & \text{if } x_1 - Z_1 > 0 \text{ and } x_2 < x_1 + Z_2 - Z_1, \\ (\mu_1, \mu_2)^T & \text{if } x_1 - Z_1 < 0 \text{ and } x_2 < x_1 + Z_2 - Z_1, \\ (\mu_1, 0)^T & \text{if } x_1 - Z_1 < 0 \text{ and } x_2 > x_1 + Z_2 - Z_1. \end{cases} \quad (13)$$

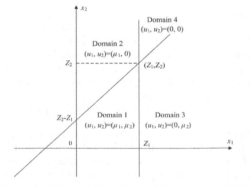

Figure 2. The state space and the hedging point control on the work-in-process in a tandem two-machine system when the two machines are both operational.

If (x_1, x_2) is on the line $x_1 = 0$ or $x_1 = Z_1$, the control is

$$u = (u_1, u_2)^T = \begin{cases} (0,0)^T, & \text{if } x_2 > Z_2, \\ (\min(\mu_1,\mu_2), \min(\mu_1,\mu_2))^T, & \text{if } x_2 < Z_2. \end{cases}$$
(14)

If (x_1, x_2) is on the line $x_2 = x_1 + Z_2 - Z_1$, the control is

$$u = (u_1, u_2)^T$$
$$= \begin{cases} (\mu_1, (\mu_1 + d)/2)^T, & \text{if } x_1 < Z_1 \text{ and } x_2 = x_1 + Z_2 - Z_1, \\ (0,0)^T, & \text{if } x_1 > Z_1 \text{ and } x_2 = x_1 + Z_2 - Z_1. \end{cases}$$
(15)

If, $(x_1, x_2) = (Z_1, Z_2)$, the control is $u = (u_1, u_2)^T = (d, d)^T$.

For, $(\alpha_1, \alpha_2) = (0,1)$, $u_1 = 0$ holds. The control is

$$u = (u_1, u_2)^T = \begin{cases} (0,0)^T & \text{if } x_2 > x_1 + Z_2 - Z_1, \\ (0, \mu_2)^T & \text{if } x_2 < x_1 + Z_2 - Z_1, \\ (0, d/2)^T & \text{if } x_2 = x_1 + Z_2 - Z_1. \end{cases}$$
(16)

For $(\alpha_1, \alpha_2) = (0,1)$, $u_2 = 0$ holds. The control is

$$u = (u_1, u_2)^T = \begin{cases} (0,0)^T & \text{if } x_1 > Z_1, \\ (\mu_1, 0)^T & \text{if } x_1 < Z_1, \\ (0,0)^T & \text{if } x_1 = Z_1. \end{cases}$$
(17)

For $(\alpha_1, \alpha_2) = (0, 0)$, obviously the control is $u = (u_1, u_2)^T = (0, 0)^T$.

Now we can analyze the behaviors of the system under the above-mentioned control.

For, $(\alpha_1, \alpha_2) = (1, 1)$, suppose that the original state $(x_1(t_0), x_2(t_0))$ satisfying, $x_1(t_0) < Z_1$, $x_2(t_0) < x_1(t_0) + (Z_2 - Z_1)$, So the control is $(u_1, u_2)^T = (\mu_1, \mu_2)^T$, the states at the time $t > t_0$ $x_1(t) = x_1(t_0) + (\mu_1 - \mu_2)(t - t_0)$, $x_2(t) = x_2(t_0) + (\mu_2 - d)(t - t_0)$. The buffer state will move along $x_2(t) = x_2(t_0) + [x_1 - x_1(t_0)](\mu_2 - d)/(\mu_1 - \mu_2)$. At time $t = t_0 + [x_1(t_0) - x_2(t_0) + (Z_2 - Z_1)]/(\mu_2 - d - \mu_1)$, if $(\mu_2 - d)/(\mu_1 - \mu_2) > (Z_2 - x_2(t))/(Z_1 - x_1(t))$. It will reach $x_2 = x_1 + Z_2 - Z_1$. Then move along it till reaching the hedging point (Z_1, Z_2). If $(\mu_2 - d)/(\mu_1 - \mu_2) < (Z_2 - x_2(t))/(Z_1 - x_1(t))$, at the time $t = t_0 + (Z_1 - x_1(t_0))/\mu_1 - \mu_2)$, it will reach the line $x_1 = Z_1$. Then move along the line till reaching the hedging point (Z_1, Z_2). If $(\mu_2 - d)/(\mu_1 - \mu_2) = (Z_2 - x_2(t))/(Z_1 - x_1(t))$, it will reach the hedging point (Z_1, Z_2). at $t = t_0 + (Z_1 - x_1(t_0))/(\mu_1 - \mu_2)$. If the original state satisfying $x_2(t_0) > x_1(t_0) + (Z_2 - Z_1)$ and $x_1(t_0) < Z_1$, the control is $(u_1, u_2)^T = (\mu_1, 0)^T$, the states at the time $t > t_0$ are $x(t) = x_1(t_0) + \mu_1(t - t_0)$ $x_2(t) = x_2(t_0) - d(t - t_0)$. The buffer state will move long $x_2(t) = x_2(t_0) - [x_1 - x_1(t_0)]d/\mu_1$ and reach the line $x_2 = x_1 + Z_2 - Z_1$ at the time $t = t_0 -$

$[x_1(t_0) - x_2(t_0) + Z_2 - Z_1]/(d + \mu_1)$. Then move along the line till reaching the hedging point (Z_1, Z_2). If the original state satisfying $x_1(t_0) > Z_1$ and $x_2(t_0) < x_1(t_0) + (Z_2 - Z_1)$, the control is $(u_1, u_2)^T = (0, \mu_2)^T$ and the states are $x_1(t) = x_1(t_0) - \mu_2(t - t_0)$, $x_2(t) = x_1(t_0) + (\mu_2 - d)(t - t_0)$ at the time $t > t_0$. The buffer state will move along the line $x_2 = x_2(t_0) - [x_1 - x_1(t_0)](\mu_2 - d)/\mu_2$ and at the time $t = t_0 + (Z_1 - x_1(t_0))/\mu_2$ reach the line $x_1 = Z_1$. Then it will move along $x_1 = Z_1$ till reaching the hedging point (Z_1, Z_2). If the original state satisfying $x_1(t_0) > Z_1$, $x_2(t_0) > x_1(t_0) + (Z_2 - Z_1)$ the control is $(u_1, u_2)^T = (0, 0)^T$ and the states at the time $t > t_0$ are $x_1(t) = x_1(t_0)$ and $x_2(t) = x_2(t_0) - d(t - t_0)$. The buffer state will move along the line $x_1 = x_1(t_0)$ and at the time $t = t_0 - (x_1(t_0) - x_2(t_0) + Z_2 - Z_1)/d$ reach $x_2 = x_1 + Z_2 - Z_1$. Then move along the line till reaching the hedging point (Z_1, Z_2). Once the buffer state reaches the hedging point (Z_1, Z_2), the control is $(u_1, u_2)^T = (d, d)_T$. If $(\alpha_1, \alpha_2) = (1,1)$, which tends to keep the state staying on the hedging point. However, if anyone of the machines fails, the buffer state will leave the hedging point. If the machine state is $(\alpha_1, \alpha_2) = (0,1)$, the control is $(\mu_1, \mu_2) = (0, d/2)$, which drives the buffer state move along the line $x_2 = x_1 + Z_2 - Z_1$ in the negative direction of the x_1 axis and x_2 axis till reaching the x_2 axis and keep on moving in the negative direction of the x_2 axis, or the change of the machine state. If the machine state is $(\alpha_1, \alpha_2) = (1,0)$, the control is $(u_1, u_2) = (0,0)$, which drives the buffer state move along the line $x_1 = Z_1$ in the negative direction of the x_1 axis till the change of the machine state.

Summarizing the above analysis, it can be seen that in the steady state, the buffer state must belong to one of the following four situations: at the hedging point (Z_1, Z_2), in the line $x_1 = Z_1$, in the line $x_2 = x_1 + Z_2 - Z_1$, and in the domain $0 \le x_1 < Z_1$ and $x_2 < x_1 + Z_2 - Z_1$.

3 HEDGING POINT CONTROL ON THE WIP IN A GENERAL TANDEM SYSTEM

In this section, the hedging point control is extended to a general tandem manufacturing system, in which there are n tandem machines producing only one type of product (Fig.3). Then the mathematical formulation of this problem is as follows.

Min

$$\min E\{\int_{t_0}^{T} c(x_1(s), L, x_n(s))ds \,|\, x_1(t_0), \cdots, x_n(t_0), \alpha_1(t_0), \cdots, \alpha_n(t_0)\}$$
(18)

Figure 3. The general tandem manufacturing system with n machines.

S. t.

$$dx_i(t)/dt = u_i(t) - u_{i+1}(t), \quad i = 1, \cdots, n-1 \qquad (19)$$

$$dx_n(t)/dt = u_n(t) - d \qquad (20)$$

$$u_i(t) \le \alpha_i(t)\mu_i, \quad i = 1, \cdots, n \qquad (21)$$

Similar to the situation of the two-machine system, the optimal control problem for the general tandem manufacturing system can be transformed to the following linear programming problem:

Min

$$\sum_{i=1}^{n-1}(x_i - Z_i)(u_i - u_{i+1}) + (x_n - Z_n)(u_n - d) \qquad (22)$$

S. t.

$$u_i(t) \le \alpha_i(t)\mu_i, \quad i = 1, \cdots, n. \qquad (23)$$

For $(\alpha_1, \cdots, \alpha_n) = (1, \cdots, 1)$, let $v_1, \cdots, v_n$ are the slack variables, this linear programming problem is

Min

$$(x_1 - Z_1)u_1 + \sum_{i=2}^{n}[(x_i - x_{i-1}) - (Z_i - Z_{i-1})]u_i + \sum_{i=1}^{n} 0 \; v_i \qquad (24)$$

S. t.

$$u_i + v_i = \mu_i, i = 1, \cdots, n \qquad (25)$$

If $u_R = (u_1, \cdots, u_n)^T \; u_N = (v_1, \cdots, v_2^T) \; C_B = [x_1 - Z_1, (x_2 - x_1) - (Z_2 - Z_1), \cdots, (x_n - x_{n-1}) - (Z_n - Z_{n-1})]^T$, $c_N = (0, \cdots, 0)^T \; B = diag\{1, \cdots, 1\} \; N = diag\{1, \cdots, 1\}$ the necessary and sufficient condition for $u_N = (v_1, \cdots, v_2)^T = (0, \cdots, 0)^T$ the optimal solution of Equations (24)–(25) is

$$C_N^T(x_1, x_2) - C_B^T(x_1, x_2)B^{-1}N$$
$$= [-(x_1 - Z_1), -(x_2 - x_1) + (Z_2 - Z_1),$$
$$\cdots, -(x_n - x_{n-1}) + (Z_n - Z_{n-1})] > 0 \qquad (26)$$

i.e. $x_1 - Z_1 < 0 \; x_i < x_{i-1} + (Z_i - Z_{i-1}) \; i = 2, \cdots, n$. In this domain, the control is

$$u = (u_1, \cdots, u_n)^T = (\mu_1, \cdots, \mu_n)^T \qquad (27)$$

Which is the basic part of the optimal solution.

Similarly, we can get the controls for the states in other domains in the n-dimension space as follows

$$u = (u_1, u_2, \cdots, u_i, \cdots, u_n)^T \qquad (28)$$

$$u_1 = \begin{cases} 0, & \text{if } x_1 > Z_1 \\ \mu_1, & \text{if } x_1 < Z_1 \end{cases}, \qquad (29)$$

$$u_i = \begin{cases} 0, & \text{if } x_i > x_{i-1} + Z_i - Z_{i-1} \\ \mu_i, & \text{if } x_i < x_{i-1} + Z_i - Z_{i-1} \end{cases}, \quad i = 2, \cdots, n \qquad (30)$$

The control in the open domains can drive the buffer state to the boundaries.

On the boundaries separating these open domains, the control tends to keep the state on the boundaries and drive it to the hedging point. Suppose that there are some consecutive $x_i, x_{i+1}, \Lambda, x_{i+k-1}$ satisfying

$$x_i = \begin{cases} Z_1, & \text{if } i = 1 \\ x_{i-1} + Z_i - Z_{i-1}, & \text{if } i > 1 \end{cases} \qquad (31)$$

$$x_{i+j} = x_{i+j-1} + Z_{i+j} - Z_{i+j-1}, \quad j = 1, \Lambda, k-1 \qquad (32)$$

If $i = 1$, the control is

$$u_i = u_{i+1} = \cdots u_{i+k-1} = u_{i+k} \qquad (33)$$

If $i > 1$, the control is

$$u_{i+j} = u_{i-1} - (j+1)\Delta u = u_{i-1} - (j+1)(u_{i-1} - u_{i+k})/(k+1) \qquad (34)$$

If $(x_1, \cdots, x_n) = (Z_1, \cdots, Z_n)$, the control is

$$u = (u_1, \cdots, u_n)^T = (d, \cdots, d)^T \qquad (35)$$

Which tends to keep the buffer state on the hedging point $(Z_1, \cdots, Z_n)$. Obviously, it is a special case of Equation (33).

If some machines are down, i.e. there exist $\alpha_i = 0$ holds for $i = i_1, \cdots, i_k$ and $\alpha_i = 1$ holds for $i \ne i_1, \cdots, i_k$. So the linear programming problem is

Min

$$(x_1 - Z_1)u_1 + \sum_{i \ne i_1, \cdots, i_k}^{n} [(x_i - x_{i-1}) - (Z_i - Z_{i-1})]u_i \qquad (36)$$

S. t.

$$u_i(t) = 0, \quad i = i_1, \cdots, i_k \qquad (37)$$

$$u_i(t) \le \mu_i, \quad i \ne i_1, \cdots, i_k \qquad (38)$$

Obviously this linear programming problem is a small-scale version of Equations (22)–(23). So the controls are the same as Equations (29), (30), (33) and (34) for $i \neq i_1, \cdots, i_k$ and $u_i(t) = 0$ for $i = i_1, \cdots, i_k$.

4 CONCLUSION

In this paper, the optimal control problem directly on the work-in-process in a tandem manufacturing system is researched into. At first, a tandem two-machine system is considered. In this simpler situation, a hedging point policy is proposed for the control on the work-in-process and its objective is to minimize the mean costs due to the work-in-process inventory and the penalty for the backlog. This control tends to drive the buffer state of the system to the boundaries separating the state space, keep it on the boundary, and drive it to the hedging point. The behaviors of the tandem two-machine system under this control are analyzed in detail. And then the hedging point control policy is extended to the general tandem manufacturing system with multiple machines. Under this control, the optimal value of the objective function is completely determined by the value of the hedging point.

REFERENCES

E. Khmelnitsky and Ernst Presman, 2011. "Optimal production control of a failure prone machine", Operations Research, Vo.182, p 67–86.

J. Kimemia and S. B. Gershwin, 1983, "An algorithm for the computer control of a flexible manufacturing system", IIE Transactions, vol.15, no.4, p.353–362.

Liu Xiaoming, Zhaotong Lian, 2009, "Cost-effective inventory control in a value-added manufacturing system", European Journal of Operational Research, vol.196, p.534–543.

O. Maimon, E. Khmelnitsky and K. Kogan, 1998. "Optimal Flow Control in Manufacturing Systems: Production Planning and Scheduling", Kluwer Academic Publishers.

R. Akella and P. R. Kumar, 1986. "Optimal control of production rate in a failure prone manufacturing system", IEEE Transactions on Automatic Control, vol. AC-31, no.2, p.116–126

S. B. Gershwin, 1994. Manufacturing System Engineering, Prentice Hall Inc.

S. B. Gershwin, 1997. "Design and operation of manufacturing system --- control—and system—theoretical models and issues". Proceedings of the American Control Conference, Albuquerque, New Mexico, p. 1909–1913.

S. B. Gershwin, 2000, "Design and operation of manufacturing systems", IIE Transactions, vol.32, p.891–906.

T. Bielecki and P. R. Kumar, 1988. "Optimality of zero-inventory policies for unreliable manufacturing systems", Operations Research, vol.36, no.4, p.532–541.

Energy Science and Applied Technology – Fang (Ed.)
© 2016 Taylor & Francis Group, London, ISBN 978-1-138-02833-3

Evolution of low power laser affected by the plasma with Gaussian profile

Yanglin Tang & Xiquan Fu
College of Computer Science and Electronics Engineering, Hunan University, Changsha, China

ABSTRACT: The equation describing the propagation of probe beam in plasma channel has been established and the approximate solution of the envelope equation is obtained by a modified WKBJ method. The theoretical solutions show the probe beams can be transformed into ring-shaped beams by the effect of plasmas having Gaussian density profile. The influence of electron density, plasma channel width and plasma channel length on the evolution of low power probe beams in Gaussian plasmas has been investigated clearly.

Keywords: ring-shaped beam; plasma channel; plasma defocusing; WKBJ method

1 INTRODUCTION

The research of the propagation of intense laser pulses in plasma channels has obtained lots of attentions, due to various applications such as lightning channeling in air (Couairon & Mysyrowicz 2007), the fast ignition scheme (Wang et al. 2013), harmonics generation (Ding et al. 2013) and the laser Wakefield acceleration of electrons (Goers et al. 2014; Schroeder et al. 2013). For many of these applications, it is desirable to study the influence of the plasma defocusing on the evolutions of the probe beams in plasmas. There are some studies investigated the effect of plasma defocusing on the evolution of probe beam in plasmas. A method to generate ring-shaped beams by adjusting the plasma electron density in experiment has been recently reported (Tan et al. 2013, 2014). As we know the mechanism for the filaments remain stable over long distance is attributed to the plasma defocusing (Sprangle et al. 2002). So the plasma defocusing is a very important effect in the research of laser-plasma interactions.

As the propagation of probe beams in plasmas will be affected by nonlinear phenomena such as preformed plasma channel, channel-coupling non-linearity, Wakefield effect, relativistic nonlinearity and ponder motive nonlinearity (Sharma et al. 2003). Various methods has been used to theoretically analyze the propagation of laser in plasma such as the variation method (Upadhyay et al. 2008), the source dependent expansion technique (Esarey et al. 1997) and the moment method (Lam et al. 1977). However the fundamental influence of preformed plasma channel is still not researched clearly and systematically. A simple theoretical model and experiment are urgent need to study the impact of the profile of plasma electron density, the number of electron density and the plasma width on laser plasma interactions. The nonlinear effects of laser plasma interactions will be investigated more clearly and systematically, until the role of the preformed plasma channel is studied clearly. In the paper, the influence of the plasma defocusing on the propagation of a low power probe beam has been quantitatively analyzed by using a modified WKBJ method.

2 THEORETICAL ANALYSIS

The laser beam is consider to be linearly polarized and the radiation field is written as

$$\vec{E}(\vec{r},z,t) = \vec{e}_x \frac{E_0(\vec{r},z,t)}{2} \exp(ikz - i\omega_0 t) + c.c. \quad (1)$$

Where $\vec{E}_0(\vec{r},z,t)$ is the complex amplitude, k is the laser wave number, ω_0 is the laser frequency and $\vec{e}_x$ is the unit vector along the axis. As the power of probe beam is very low, the relativistic nonlinearity and ponder motive nonlinearity are negligible. Thus, the vector equation describing the propagation of the low power probe beam in plasma is given by

$$\left[\nabla^2 - \frac{1}{c^2} \frac{\partial^2}{\partial t^2} \right] \vec{E}(\vec{r},z,t) = \frac{\omega_p^2(\vec{r})}{c^2} \vec{E}(\vec{r},z,t) \quad (2)$$

Transforming independent variables $z \to z$, $\zeta \to \zeta - \beta_g ct$, where $\beta_g = v_g/c$ is the normalized group velocity. Substituting Eq. (1) into Eq. (2) yields

$$\left\{ \nabla_\perp^2 + 2ik\frac{\partial}{\partial z} + \left(1 - \beta_g^2\right)\frac{\partial^2}{\partial \zeta^2} - \frac{\omega_p^2}{c^2} \right\} \psi(\vec{r}, z, \zeta) = 0 \quad (3)$$

where $\psi(\vec{r}, z, \zeta) = eE_0(\vec{r}, z, t)/m_0 c\omega_0$ is the normalized electric field amplitude, e is the electron charge, m_0 is the rest mass of electron, $\omega_0 = (4\pi n_e(\vec{r})/e^2/m_0)^{1/2}$ is the plasma frequency, c is the light velocity in vacuum, $n_e(\vec{r})$ is the distribution of electron density of plasma, respectively. In deriving Eq. (3), the slowly varying envelope and paraxial approximations, i.e., $|\partial \psi/\partial z| \ll |k\psi|$ and $|\partial \psi/\partial \zeta| \ll |\omega_0 \psi|$, have been used. Assuming the laser beam to be continuous beam, the influence of the GVD can be neglected. Thus Eq. (3) reduces to

$$\left(\frac{\partial^2}{\partial x^2} + \frac{\partial^2}{\partial y^2} + 2ik\frac{\partial}{\partial z} - \frac{\omega_p^2(x,y,z)}{c^2} \right) \psi(x,y,z) = 0 \quad (4)$$

Substituting critical plasma density $n_c = m_0 \omega_0^2/4\pi e^2$ into Eq. (4) yields

$$\left[\frac{\partial^2}{\partial x^2} + \frac{\partial^2}{\partial y^2} + 2ik\frac{\partial}{\partial z} - k^2\frac{n_e(x,y,z)}{n_c} \right] \psi(x,y,z) = 0 \quad (5)$$

Assume

$$\psi(x,y,z) = A(x,y,z)\exp(i\varphi(x,y,z)) \quad (6)$$

Substituting Eq. (6) into Eq. (5) yields

$$\left(2k\frac{\partial \varphi}{\partial z} + \left(\frac{\partial \varphi}{\partial x}\right)^2 + \left(\frac{\partial \varphi}{\partial y}\right)^2 - \frac{\partial^2}{\partial x^2} - \frac{\partial^2}{\partial y^2} + \frac{n_e}{n_c}k^2 \right) A = 0 \quad (7)$$

$$2k\frac{\partial A}{\partial z} + 2\frac{\partial A}{\partial x}\frac{\partial \varphi}{\partial x} + 2\frac{\partial A}{\partial y}\frac{\partial \varphi}{\partial y} + A\frac{\partial^2 \varphi}{\partial x^2} + A\frac{\partial^2 \varphi}{\partial y^2} = 0 \quad (8)$$

As $n_e \ll n_c$, Wentzel-Kramers-Brillouin-Jeffreys (WKBJ) approximation (Fu & Guo 2002) is used to derive the WKBJ solution.

$$\varphi_w(x,y,z) = \varphi(x,y,0) - \frac{k}{2n_c}\int_0^z n_e(x,y,z')dz' \quad (9)$$

$$A_w(x,y,z) = A(x,y,0) \quad (10)$$

By submitting WKBJ solutions into Eq. (8), it is easy to know the first item of Eq. (8) is much more

than the other items, and last four items can be work as perturbation. The modified approximate solution of amplitude Eq. (11) can be derived from Eq. (8) by the amplitude of the second and third items of Eq. (8) replaced by the WKBJ solution of amplitude and the phase of the last four items replaced by the WKBJ solution of phase.

$$A(x,y,z) = A(x,y,0)$$
$$\times \exp\left[-\frac{1}{k}\left(\int_0^z \frac{\partial A(x,y,0)}{\partial x}\frac{\partial \varphi(x,y,z')}{\partial x} \right.\right.$$
$$+ \frac{\partial A(x,y,0)}{\partial y}\frac{\partial \varphi(x,y,z')}{\partial y} \bigg)dz'$$
$$- \frac{1}{2k}\int_0^z \left(\frac{\partial^2 \varphi(x,y,z')}{\partial x^2} \right.$$
$$\left.\left. + \frac{\partial^2 \varphi(x,y,z')}{\partial y^2} \right)dz' \right] \quad (11)$$

The modified WKBJ solution Eq. (11) is accurate for $n_e \ll n_c$, and the error is smaller than the WKBJ solution because of the consideration of electron density gradient. The distribution of probe beam and electron density profile make no difference to the validity of the modified WKBJ solution of amplitude, so Eq. (11) can be used to analyze the evolutions of probe beams in plasmas.

Consider a Gaussian beam $A(x,y,0) = A_0\exp[-(x^2+y^2)/w^2]$ transmits though a plasma channel with Guassian density profile $n_e(x,y,z) = n_0\exp[-(x^2+y^2)/\sigma^2]$, where A_0 w n_0 and σ are the initial axial amplitude, beam width, the initial axial electron density and plasma channel width, respectively. The initial phase of probe beam is considered to be constant $\varphi(x,y,0) = c$. Eq. (11) can be reduced to

$$A(x,y,z) = A_0\exp\left[-\frac{x^2+y^2}{w^2} + \frac{n_0}{n_c}\frac{(x^2+y^2)z^2}{w^2\sigma^2}\exp \right.$$
$$\times \left(-\frac{x^2+y^2}{\sigma^2} - \frac{x^2+y^2}{w^2} \right)$$
$$- \frac{z^2 n_0}{8n_c}\left(\frac{2}{\sigma^2} - \frac{4(x^2+y^2)}{\sigma^4} \right)$$
$$\left.\times \exp\left(-\frac{(x^2+y^2)}{\sigma^2} \right) \right] \quad (12)$$

Where z is the propagation distance in plasma. The axial intensity of the probe beam I_c can be deduced from Eq. (12).

$$I_c = I_0\exp\left(-\frac{z^2 n_0}{2\sigma^2 n_c} \right) \quad (13)$$

322

I_0 is the initial axial intensity of the probe beam.

The modified approximate solution of amplitude Eq. (12) is analyzed for w = 1mm with different electron densities, plasma widths and the propagation distance, respectively.

Figure 1 shows the spatial distribution of the laser intensity with different electron densities after propagating through 5mm plasma channel. The rule of the probe beam modulated by the preformed plasma channel with Gaussian density profile has been researched. Figure 1(a)–(d) show that the intensity of the central zone decreased and the peak intensity increased, as n_0 increased from 0 to 0.2 n_c. The modulation of probe beam by the preformed plasma channel is distinct due to a big electron density n_0. When $n_0 = 0.2 n_c$ a ring-shaped beams is generated. As is well known, plasma defocusing is the major physical mechanism of laser filamentation, the plasma channel with Gaussian profile will also deflect the probe beam. The refractive index of plasma channel is relevant to

electron density $N^2 = 1 - n_e/n_c$, so the Gaussian electron density profile results in a nonuniform refractive index with minimum at center and maximum at the periphery. According to the Fermat principle, the periphery of the plasma is focusing, while the plasma center is defocusing, which results in generating a ring-shaped beam.

When n_0 increased, the refractive index gradient increased and a more obvious ring-shaped beam performed, because the degrees of the defocusing of the plasma center and the focusing of the plasma periphery are proportional to the refractive index gradient. Figure 1(e) shows that the axial intensity of laser beam can be adjusted by changing n_0. So Eq. (13) can be used to detect the number of electron density of Gaussian plasma.

Figure 2 shows the spatial distribution of the beam intensity with different plasma channel widths after propagating in 5mm plasma channel. Figure 2(a)-(d) show that the intensity of the central zone decreased and the peak intensity increased, when plasma width σ decreased from 1.4mm to 0.8mm. The modulation

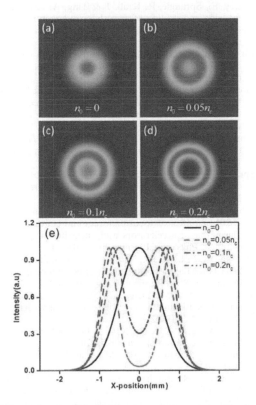

Figure 1. Spatial distribution of laser intensity for w = 1 mm, σ = 1 mm z = 5 mm and electron density (a) $n_0 = 0$, (b) $n_0 = 0.05 n_c$, (c)$n_0 = 0.1n_c$ and (d) $n_0 = 0.2 n_c$, respectively. (e) The corresponding one dimensional graph.

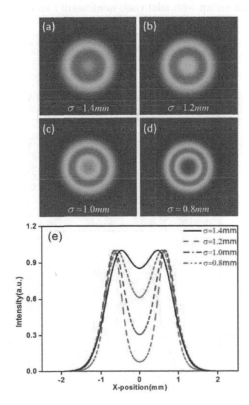

Figure 2. Spatial distribution of laser intensity for w = 1 mm, $n_0 = 0.1 n_c$, z = 5 mm and plasma width (a) σ = 3 mm, (b) σ = 4 mm, (c) σ = 5 mm and (d) σ = 6 mm, respectively. (e) The corresponding one dimensional graph.

of probe beam by the plasma channel is prominent owe to the small channel width σ. Because a smaller plasma width results in a bigger electron density gradient and a bigger refractive index gradient. The defocusing of the plasma center and the focusing of the plasma periphery become more evident due to a smaller plasma width σ.

Figure 3 displays the distribution of the laser intensity with different propagation distances. Figure 3(a)-(d) show that the generated RSBs become more evident when the probe beam propagate further in plasmas. As the laser beam propagates further, the effect of the plasma defocusing becomes more violent. In Eq. (12) we can see that z^2 has the same influence on the solution with n_0. So the dark spot size of the generated RSB can be controlled handily by regulating propagation distance.

3 CONCLUSIONS

So as to study the impact of plasma channel on the evolution of probe beam, an envelope equation has been set up with relativistic nonlinearity and pon-

dermotive nonlinearity neglected. The approximate solution has been obtained to study the propagation of the probe beam, it finds that the evolution of the probe laser intensity can be controlled by adjusting the electron density, plasma channel width and plasma channel length. We only theoretically analyzed the Gaussian beam propagates in plasma channel with Gaussian profile, but our method has general character and other non-Gaussian conditions can also be analyzed by our method. The rule of probe beam modulated by plasma channel with Gaussian profile is quite clear.

REFERENCES

Couairon, A. & Mysyrowicz, A. 2007 Femtosecond filamentation in transparent media. Physics reports 441 (2), 47–189.

Ding, P. J., Liu, Z. Y., Shi, Y. C. h., Sun, S. H., Liu, X. L., Wang, X. S., Guo, Z. Q., Liu, Q. C., Li, Y. H. & Hu, B. T. 2013 Spectral characterization of third-order harmonic generation assisted by a two-dimensional plasma grating in air. Phys. Rev. A 87 (4), 043828.

Esarey, E., Sprangle, P., Krall, J. & Ting, A. 1997 Self-focusing and guiding of short laser pulses in ionizing gases and plasmas. IEEE J. Quantum Electron. 33 (11), 1879–1914.

Fu, X. Q. & Guo, H. 2002 Laser-plasma electron-density measurement using x-ray interferometry. Phys. Rev. E 65 (6), 067401.

Goers, A. J., Yoon, S. J., Elle, J. A., Hine, G. A. & Milchberg, H. M. 2014 Laser wakefield acceleration of electrons with ionization injection in a pure n5+ plasma waveguide. Applied Physics Letters 104 (21), 214105.

Lam, J. F., Lippmann, B. & Tappert, F. 1977 Self-trapped laser beams in plasma. Phys. Fluids 20 (7), 1176–1179.

Sprangle, P. Peñano, J. R. & Hafizi, B. 2002 Propagation of intense short laser pulses in the atmosphere. Phys. Rev. E 66, 046418

Schroeder, C. B., Esarey, E., Benedetti, C. & Leemans, W. P. 2013 Control of focusing forces and emittances in plasma-based accelerators using near-hollow plasma channels. Physics of Plasmas 20 (8), 080701.

Sharma, A., Prakash, G., Verma, M. P. & Sodha, M. S. 2003 Three regimes of intense laser beam propagation in plasmas. Phys. Plasmas 10 (10), 4079–4084.

Tan, C., Fu, X. Q. & Deng, Y. B. 2013 Generation of ring-shaped beams by a graded-index plasma lens. J. Opt. 15 (12), 125202.

Tan, C., Wang, Q. K. & Fu, X. Q. 2014 Topological insulator sb 2 te 3 as an optical media for the generation of ring-shaped beams. Optical Materials Express 4 (10), 2016–2025.

Upadhyay, A. K., Singh, R. G., Singh, V. & Jha, P. 2008 Effect of transverse ponderomotive nonlinearity on the propagation of ultrashort laser pulses in a plasma channel. Phys. Plasmas 15 (12), 124503.

Wang, W. W., Cai, H. B., Jia, Q. & Zhu, S. Q. 2013 Collisional effects on the generation of fast electrons in fast ignition scheme. Physics of Plasmas 20 (1), 012703.

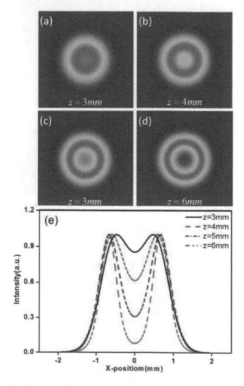

Figure 3. Spatial distribution of laser intensity for $w = 1$ mm, $n_0 = 0.1$ n_c, $\sigma = 5$ mm and the propagation distance (a) $z = 3$ mm, (b) $z = 4$ mm, (c) $z = 5$ mm and (d) $z = 6$ mm, respectively. (e) The corresponding one dimensional graph.

Energy Science and Applied Technology – Fang (Ed.)
© 2016 Taylor & Francis Group, London, ISBN 978-1-138-02833-3

A robust MDS algorithm for TDOA-based passive source localization by using a single calibration emitter

Wei Wu, Hongyi Yu & Li Zhang
Zhengzhou Information Science and Technology Institute, Zhengzhou, China

ABSTRACT: Sensor position error is known to degrade the passive source localization accuracy significantly. To deal with this problem, first, the noisy sensor position is improved by a single calibration emitter. Then, the improved sensor positions are used to estimate the target position by the algebraic MDS method. The simulation results demonstrate that the proposed solution is able to attain the CRLB when the sensor position error is small, and has a more robust performance than the two-step WLS method with large sensor errors.

Keywords: passive location; TDOA; sensor position error; calibration

1   INTRODUCTION

The problem of passive localization of an emitting target has been an important research area in many applications such as navigation, microphone arrays, sensor networks and speaker tracking. We usually use the locating parameters, such as the Angle Of Arrival (AOA), the Time Of Arrival (TOA) and the Time Difference Of Arrival (TDOA), to construct a set of nonlinear equations to locate the target. In this paper, we focus on TDOA localization.

In modern localization applications, the sensors are put on the satellites of unmanned aerial vehicles, which means we may not able to get the true sensor positions in practice. A large number of literature is available for alleviating the effect of sensor position errors on localization accuracy (Wei Hewen 2008, Sun Ming 2012, Wei Hewen 2010, HO K C 2008, Ho K C, Kovavisaruch L, Parikh H, 2004, Yang Le, Ho K C, 2010).

This paper first proposes a novel method of utilizing a single calibration source to improve the accuracy of sensor positions. Then, in the second step, the MDS method is used to solve the locating problem by considering sensor position errors.

This paper is divided into six sections. Section 2 presents the formulation of the location problem. Section 3 presents the derivation of the proposed solution. Section 4 evaluates the CRLB of the estimation. Section 5 confirms the theory by simulations. Section 6 concludes this paper.

2   PROBLEM FORMULATION

The localization scenario is shown in Figure 1. The unknown target position is $\mathbf{u} = [x, y, z]^T$, from

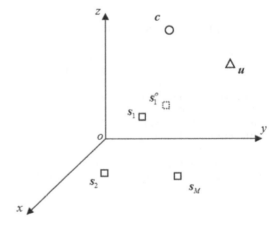

Figure 1.   Source localization scenario.

which the signal is received by M sensors with the known position $\mathbf{s}_i = [x_i, y_i, z_i]^T = \mathbf{s}_i^o + \varphi_i$, where $i = 1, 2, \ldots, M$ whose true positions are unavailable as $\mathbf{s}_i^o = [x_i^o, y_i^o, z_i^o]^T$. The position error is φ_i, which is assumed to be a zero mean random variable with a covariance matrix $\mathbf{Q}_\beta$. We define

$$\mathbf{s} = \left[\mathbf{s}_1^T, \mathbf{s}_2^T, \ldots, \mathbf{s}_M^T\right]^T = \mathbf{s}^o + \varphi, \quad \mathbf{s}^o = \left[\mathbf{s}_1^{oT}, \mathbf{s}_2^{oT}, \ldots, \mathbf{s}_M^{oT}\right]^T,$$

$$\varphi = \left[\varphi_1^T, \varphi_2^T, \ldots, \varphi_M^T\right]^T.$$

Let the TDOA of a signal received by the sensor i and sensor 1 be t_{i1}. Then, the RDOA measurement can be expressed as

$$r_{i1} = v \cdot t_{i1} = d_{i1} + n_{i1} \tag{1}$$

where v is the signal propagation speed; d_{i1} is the true value of the RDOA measurement; and n_{i1} is the RDOA measurement error. The d_{i1} is defined as $d_{i1} = d_i - d_1 = \| \mathbf{u} - \mathbf{s}_i^o \| - \| \mathbf{u} - \mathbf{s}_1^o \|$. We define $\mathbf{r} = [r_{21}, r_{31}, \ldots, r_{M1}]^T = \mathbf{d} + \mathbf{n}$, $\mathbf{d} = [d_{21}, d_{31}, \ldots, d_{M1}]^T$, $\mathbf{n} = [n_{21}, n_{31}, \ldots, n_{M1}]^T$, and assume that $\mathbf{n}$ is a zero mean random variable with a covariance matrix $\mathbf{Q}_\alpha$.

In order to reduce the impact of sensor position errors, the RDOAs of a calibration source with the known position $\mathbf{c} = [x_c, y_c, z_c]^T$ can be measured as

$$\mathbf{r}^c = \mathbf{d}^c + \mathbf{n}^c \tag{2}$$

where

$$\mathbf{r}^c = [r_{21}^c, r_{31}^c, \ldots, r_{M1}^c]^T = \mathbf{d}^c + \mathbf{n}^c, \; \mathbf{d}^c$$
$$= [d_{21}^c, d_{31}^c, \ldots, d_{M1}^c]^T, \; \mathbf{n}^c = [n_{21}^c, n_{31}^c, \ldots, n_{M1}^c]^T$$

We assume that $\mathbf{n}^c$ is a zero mean random variable with a covariance matrix $\mathbf{Q}_c$.

3 THE PROPOSED SOLUTION

The proposed solution has two stages. The first stage estimates the sensor position errors and gets the improved sensor positions using TDOA measurements from the calibration. The second stage uses the multidimensional scaling method to establish a linear solution equation to solve the position of the unknown target.

First stage: from the definition of d_{i1}^c, we have $d_{i1}^c + d_1^c = d_i^c$. Squaring both sides, substituting $d_1^c = \| \mathbf{c} - \mathbf{s}_1^o \|$ and $d_i^c = \| \mathbf{c} - \mathbf{s}_i^o \|$, and simplifying yields

$$d_{i1}^{c2} + 2 d_{i1}^c d_1^c = \mathbf{s}_i^{oT} \mathbf{s}_i^o - \mathbf{s}_1^{oT} \mathbf{s}_1^o - 2 \left(\mathbf{s}_i^o - \mathbf{s}_1^o \right)^T \mathbf{c} \tag{3}$$

The true RDOA d_{i1}^c and the true sensor position $\mathbf{s}_i^o$ are not available, so we use the noisy ones. By setting $d_{i1}^c = r_{i1}^c - n_{i1}^c$, $\mathbf{s}_i^o = \mathbf{s}_i - \varphi$ and ignoring the second-order error terms of sensor position errors and RDOA noise, we can rewrite (3) as

$$2 \left(r_{i1}^c + d_1^c \right) n_{i1}^c = r_{i1}^{c2} + 2 r_{i1}^c d_1^c - \mathbf{s}_i^T \mathbf{s}_i + \mathbf{s}_1^T \mathbf{s}_1 + 2 \left(\mathbf{s}_i - \mathbf{s}_1 \right)^T \mathbf{c} + 2 \left(\mathbf{s}_i - \mathbf{c} \right)^T \varphi_i - 2 \left(\mathbf{s}_1 - \mathbf{c} \right)^T \varphi_1 \tag{4}$$

It should be noted that the true distance d_1^c depends on the true sensor position as well. Here, we expand d_1^c at the noisy position $\mathbf{s}_1$ and retain the linear terms as

$$d_1^c = \| \mathbf{c} - \mathbf{s}_1^o \| = \| \mathbf{c} - \mathbf{s}_1 + \varphi_1 \| \approx r_1^c + \rho_{\mathbf{c}, \mathbf{s}_1}^T \varphi_1 \tag{5}$$

where $r_1^c = \| \mathbf{c} - \mathbf{s}_1 \|$ and $\rho_{\mathbf{c}, \mathbf{s}_1} = (\mathbf{c} - \mathbf{s}_1)/\| \mathbf{c} - \mathbf{s}_1 \|$. By substituting (5) into (4) and ignoring the second-order error terms, we can get

$$2 \left(r_{i1}^c + r_1^c \right) n_{i1}^c = h_i - 2 \left(\mathbf{c} - \mathbf{s}_i \right)^T \varphi_i - 2 \left(\mathbf{s}_1 - \mathbf{c} - r_{i1}^c \rho_{\mathbf{c}, \mathbf{s}_1} \right)^T \varphi_1 \tag{6}$$

where $h_i = r_{i1}^{c2} - \mathbf{s}_i^T \mathbf{s}_i + \mathbf{s}_1^T \mathbf{s}_1 + 2 \left(\mathbf{s}_i - \mathbf{s}_1 \right)^T \mathbf{c} + 2 r_{i1}^c r_1^c$. By transforming (6) into the matrix equation, we get

$$\mathbf{h}_c = \mathbf{G}_c \boldsymbol{\varphi} + \mathbf{D} \mathbf{n}_c \tag{7}$$

where $\mathbf{h}_c$ is an $(M-1) \times 1$ vector with its $(i-1)$th element equal to h_i; $\mathbf{G}_c$ is an $(M-1) \times 3M$ matrix and its $(i-1)$th row is $\mathbf{G}_c[i-1,:] = 2[(\mathbf{s}_1 - \mathbf{c} - r_{i1}^c \rho_{\mathbf{c},\mathbf{s}_1})^T, \mathbf{0}_{3(i-2) \times 1}^T, -(\mathbf{s}_i - \mathbf{c})^T, \mathbf{0}_{3(M-i) \times 1}^T] \mathbf{D} = 2 \cdot \mathrm{diag}([r_{21}^c + r_1^c, r_{31}^c + r_1^c, \cdots, r_{M1}^c + r_1^c])\cdot$

We know that $\mathbf{n}_c$ has a covariance matrix $\mathbf{Q}_c$ and $\boldsymbol{\varphi}$ is the zero-mean Gaussian distributed random vector with the covariance matrix $\mathbf{Q}_\beta$. From the Bayesian Gauss–Markov theorem (S. M. Kay, 1993), the linear minimum MSE (LMMSE) estimator of $\boldsymbol{\varphi}$, denoting it as $\hat{\boldsymbol{\varphi}}$, is given by

$$\hat{\boldsymbol{\varphi}} = \left(\mathbf{Q}_\beta^{-1} + \mathbf{G}_c^T \left(\mathbf{D Q}_c \mathbf{D}^T \right)^{-1} \mathbf{G}_c \right)^{-1} \mathbf{G}_c^T \left(\mathbf{D Q}_c \mathbf{D}^T \right)^{-1} \mathbf{h}_c \tag{8}$$

In addition, the covariance matrix of $\hat{\boldsymbol{\varphi}}$ is given by

$$\mathrm{cov}\left(\boldsymbol{\varphi} - \hat{\boldsymbol{\varphi}} \right) = \left(\mathbf{Q}_\beta^{-1} + \mathbf{G}_c^T \left(\mathbf{D Q}_c \mathbf{D}^T \right)^{-1} \mathbf{G}_c \right)^{-1} \tag{9}$$

Then, we can obtain the improved sensor positions as

$$\hat{\mathbf{s}} = \mathbf{s} - \hat{\boldsymbol{\varphi}} = \mathbf{s}^o + \boldsymbol{\varphi} - \hat{\boldsymbol{\varphi}} \tag{10}$$

From (10), we can find that $\hat{\mathbf{s}}$ has a same covariance matrix with $\hat{\boldsymbol{\varphi}}$, which means $\mathrm{cov}(\mathbf{s}) - \mathrm{cov}(\hat{\mathbf{s}})$ is a positive semi-definite and guarantees that $\hat{\boldsymbol{\varphi}}$ is better than the original ones.

Second stage: in order to generate a MDS equation, we first give the form of a position coordinates matrix $\mathbf{Z}$ as

$$\mathbf{Z} = \begin{bmatrix} x_1^o - x & y_1^o - y & z_1^o - z & id_{11} - id_{01} \\ x_2^o - x & y_2^o - y & z_2^o - z & id_{21} - id_{01} \\ \vdots & \vdots & \vdots & \vdots \\ x_M^o - x & y_M^o - y & z_M^o - z & id_{M1} - id_{01} \end{bmatrix} \tag{11}$$

where $d_{01} = -d_1$, and the scalar product matrix $\mathbf{B} = \mathbf{Z Z}^T$, whose (m, n) entry, is given by

$$[\mathbf{B}]_{mn} = 0.5[(d_{m1} - d_{n1})^2 - (\mathbf{s}_m^o - \mathbf{s}_n^o)^T (\mathbf{s}_m^o - \mathbf{s}_n^o)] \quad (12)$$

Then, a new method (S. M. Kay, 1993) without the use of eigenvalue factorization is applied to generate a linear equation with respect to the unknown parameter $\mathbf{z} = [x, y, z, d_{01}]^T$, which is given by

$$\mathbf{BA}\begin{bmatrix} 1 \\ \mathbf{z} \end{bmatrix} = \mathbf{0}_M \quad (13)$$

where $\mathbf{A} = \mathbf{P}$ and the ith column of $\mathbf{P}$ is $\mathbf{P}[:,i] = [1 \ \ x_i^o \ \ y_i^o \ \ z_i^o \ \ d_{i1}]^T$. After changing the form of (13), we get

$$\mathbf{Hz} = \mathbf{h} \quad (14)$$

where $\mathbf{H} = \mathbf{BA}_2$, $\mathbf{h} = -\mathbf{BA}_1$ and $\mathbf{A} = [\mathbf{A}_1 \vdots \mathbf{A}_2]$.
Practically, the TDOAs and sensor positions are measured with additive noises and after putting the noisy measurement r_{i1} and $\mathbf{s}_i$ into (14) yields the error vectors:

$$\varepsilon = \hat{\mathbf{H}}\mathbf{z} - \hat{\mathbf{h}} = \hat{\mathbf{B}}\hat{\mathbf{A}}\begin{bmatrix} 1 \\ \mathbf{z} \end{bmatrix} \quad (15)$$

From (23)-(25), we obtain the error vector with respect to the noise vector $\mathbf{n}$ and φ as

$$\varepsilon = \mathbf{G}_1\mathbf{n} + \mathbf{G}_2\varphi \quad (16)$$

A weighted LS solution to the target is given by

$$\hat{\mathbf{z}} = (\hat{\mathbf{H}}^T\mathbf{W}\hat{\mathbf{H}})^{-1}\hat{\mathbf{H}}^T\mathbf{W}\hat{\mathbf{h}} \quad (17)$$

where the weighting matrix $\mathbf{W}$ is given by

$$\mathbf{W} = \mathrm{E}[\varepsilon\varepsilon^T]^{-1} = [\mathbf{G}_1\mathbf{Q}_\alpha\mathbf{G}_1^T + \mathbf{G}_2\mathbf{Q}_\beta\mathbf{G}_2^T]^{-1} \quad (18)$$

The covariance matrix of $\hat{\mathbf{z}}$ is given by

$$\mathrm{cov}(\hat{\mathbf{z}}) = (\mathbf{H}^T\mathbf{W}\mathbf{H})^{-1} \quad (19)$$

4 THE CRLB

The CRLB gives the lowest possible variance for any unbiased estimate of a parameter vector as

$$\mathrm{CRLB} = \begin{bmatrix} \mathbf{X} & \mathbf{Y} \\ \mathbf{Y}^T & \mathbf{Z} \end{bmatrix}^{-1} \quad (20)$$

where

$$\mathbf{X} = \left(\frac{\partial \mathbf{d}}{\partial \mathbf{u}}\right)^T \mathbf{Q}_\alpha^{-1} \left(\frac{\partial \mathbf{d}}{\partial \mathbf{u}}\right), \ \mathbf{Y} = \left(\frac{\partial \mathbf{d}}{\partial \mathbf{u}}\right)^T \mathbf{Q}_\alpha^{-1} \left(\frac{\partial \mathbf{d}}{\partial \mathbf{s}^o}\right),$$

$$\mathbf{Z} = \left(\frac{\partial \mathbf{d}}{\partial \mathbf{s}^o}\right)^T \mathbf{Q}_\alpha^{-1} \left(\frac{\partial \mathbf{d}}{\partial \mathbf{s}^o}\right) + \mathbf{Q}_\beta^{-1}$$

and

$$\frac{\partial \mathbf{d}}{\partial \mathbf{u}} = \left[(\mathbf{u} - \mathbf{s}_2^o)/d_2 - (\mathbf{u} - \mathbf{s}_1^o)/d_1 \cdots (\mathbf{u} - \mathbf{s}_M^o)/d_M - (\mathbf{u} - \mathbf{s}_1^o)/d_1\right]$$

$$(21)$$

$$\frac{\partial \mathbf{d}}{\partial \mathbf{s}^o} = \begin{bmatrix} (\mathbf{u} - \mathbf{s}_1^o)/d_1 & \cdots & (\mathbf{u} - \mathbf{s}_1^o)/d_1 \\ -(\mathbf{u} - \mathbf{s}_2^o)/d_2 & & \mathbf{0}_{3\times 1} \\ & \ddots & \\ \mathbf{0}_{3\times 1} & & -(\mathbf{u} - \mathbf{s}_M^o)/d_M \end{bmatrix} \quad (22)$$

We can get the minimum localization variance by stacking up the first three diagonal elements in (20).

5 SIMULATIONS

This section provides the performance of the proposed method by comparing it with the two-step WLS method. The TDOA measurements are generated by adding the zero mean Gaussian noise as well as the noisy sensor positions to the true values. The covariance matrix of TDOA is $\mathbf{Q}_\alpha = \sigma_r^2 \mathbf{J}_{(M-1)\times(M-1)}$, where $\sigma_r^2 = 0.001$ and $\mathbf{J}$ is a matrix whose diagonal elements are set to 1 and 0.5 otherwise. The covariance matrix of $\mathbf{Q}_\beta$ is $\sigma_s^2 \mathbf{I}_{3M\times 3M}$ and $\mathbf{I}$ is an identity matrix. The location accuracy is $\mathrm{RMSE} = \sqrt{\mathrm{E}[(\hat{\mathbf{u}} - \mathbf{u})^T (\hat{\mathbf{u}} - \mathbf{u})]}$ with the number of total runs $N = 10000$. The true sensor positions are listed in Table 1.

In the first simulation, we examine the estimation of the RMSE as a function of the sensor position noise power for a target at the position $[2000, 2500, 3500]^T$ m, with a calibration emitter at $[1000, 1500, 2000]^T$ m.

In Figure 2, the RMSE results clearly show a better performance of the proposed method than

Table 1. True positions (in meters) of receivers.

Sensor number	x_i^o	y_i^o	z_i^o
1	350	−100	−350
2	400	150	−100
3	−300	500	200
4	350	200	100
5	−100	−100	−100

327

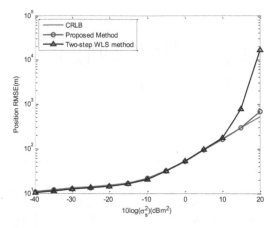

Figure 2. The estimation accuracy of the proposed method in comparison with the two-step WLS and CRLB.

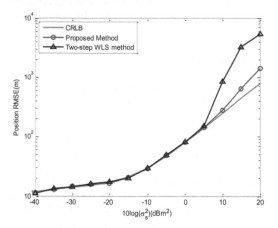

Figure 3. The estimation accuracy comparison with unequal position noise powers.

the two-step WLS method. The proposed method is able to attain the CRLB when the sensor position noise is not large. Besides, the two-step WLS method deviates from the CRLB at a sensor position noise power of about 10dB, while the proposed method deviates slightly from the CRLB at about 15dB. The threshold effect by the nonlinear problem occurs at about 5dB later compared with the two-step WLS method.

In order to make the simulation more general, we let the sensor position have different noise powers $Q_\beta = \sigma_s^2 \text{diag}([1,1,1,2,2,2,5,5,5,10,10,10,3,3,3])$. The estimation results are shown in Figure 3.

The results shown in Figure 3 demonstrate a similar conclusion of the first simulation. Although the RMSE of the proposed method almost deviates from the CRLB at the same noise power as the

two-step WLS method, it still gives a clear improvement when the noise power is comparatively large.

6 CONCLUSIONS

This paper investigates the use of a single calibration emitter to alleviate the effect of localization accuracy by sensor position errors. Then, an algebraic closed-form MDS solution is given to get the target position estimation. The simulation results show that the proposed method has a good localization accuracy, which is able to attain the CRLB when the sensor position is not large. By comparison, the proposed method deviates from the CRLB later than the two-step WLS method, which means that it exhibits more robust performance.

7 APPENDIX

The derivation of G_1 and G_2 in (16) is given by

$$G_1 = T_1 - BAT_2 + T_3 \ , \ G_2 = K_1 - BAK_2 + K_3 \qquad (23)$$

where $T_1 - T_3$ is as defined previously (Wei Hewen 2010) and

$$K_1 = - \begin{bmatrix} \sum_{m=1}^{M} a_m \left(s_1^o - s_m^o\right)^T & -a_2 \left(s_1^o - s_2^o\right)^T & \cdots & -a_M \left(s_1^o - s_M^o\right)^T \\ -a_1 \left(s_2^o - s_1^o\right)^T & \sum_{m=1}^{M} a_m \left(s_2^o - s_m^o\right)^T & \cdots & -a_M \left(s_2^o - s_M^o\right)^T \\ \vdots & & \vdots & \vdots \\ -a_1 \left(s_M^o - s_1^o\right)^T & -a_2 \left(s_M^o - s_2^o\right)^T & \cdots & \sum_{m=1}^{M} a_m \left(s_M^o - s_m^o\right)^T \end{bmatrix} \qquad (24)$$

$$K_2 = \begin{bmatrix} 0_{3M}^T \\ a_1 I_3 & \cdots & a_M I_3 \\ 0_{3M}^T \end{bmatrix} , \ K_3 [i,:] = [C_{i1} q \ , \cdots , \ C_{iM} q] \qquad (25)$$

where $anA \begin{bmatrix} 1 \\ z \end{bmatrix} n [a_1, a_2, ..., a_M]^T$, $v = (PP^T)^{-1} \begin{bmatrix} 1 \\ z \end{bmatrix}$, v_i is the ith element of v, $q = [v_2, v_3, v_4]$, $C = B(I - AP)$, $C(i:j)$ is a matrix constructed by the ith to the jth columns of C, and C_{ij} is the (i,j) entry of C.

ACKNOWLEDGMENT

This work was supported by the National Natural Science Foundation of China under Grant No. 61201381.

REFERENCES

Ho K C, Yang LE, 2008. On the use of a calibration emitter for source localization in the presence of sensor position uncertainty [J]. IEEE Trans on Signal Process, 56(12), p.5758–5772.

Ho K C, Kovavisaruch L, Parikh H, 2004. Source localization using TDOA with erroneous receiver positions[C] // IEEE ISCAS, Vancouver Canada, 3: II/453–II/456.

Kay, S M. 1993, Fundamentals of Statistical Signal Process Estimation Theory [M]. Englewood Cliffs, NJ: Prentice-Hall.

Sun Ming, Yang Le, Ho K C, 2012. Efficient joint source and sensor localization in closed-form [J]. IEEE Signal Processing Letters, 19(7), p.399–402.

Tufts D W, 1990. The effects of perturbations on matrix-based signal processing [C] // ASSP Workshop on Spectrum Estimation and Modeling, p.159–162.

Wei Hewen, Wan Quan, 2008, et al. A novel weighted multidimensional scaling analysis for time-of-arrival-based mobile location [J]. IEEE Transactions on Signal Processing, 56(7), p.3018–3022.

Wei Hewen, Peng Rong, 2010, et al. Multidimensional Scaling Analysis for Passive Moving Target Localization with TDOA and FDOA Measurements [J]. IEEE Trans on Signal Process, 58(3), p.1677–1688.

Yang Le, Ho K C, 2010. Alleviating Sensor Position Error in Source Localization Using Calibration Emitters at Inaccurate Locations [J]. IEEE Transactions on Signal Processing, 58(1), p.67–83.

Energy Science and Applied Technology – Fang (Ed.)
© *2016 Taylor & Francis Group, London, ISBN 978-1-138-02833-3*

Personalized E-commerce recommendation algorithm based on customer features and AGA

Jun Zhang & Ping Wu
Computing Center of East China Normal University, Shanghai, China

ABSTRACT: In this paper, a novel personalized recommendation algorithm is proposed, which has a higher precision and more humanization. It can make use of more information such as customer features, properties of products, online behavior log and real-time browsing log fort recommendation. It analyzed customer's behavior log combined with the database of product properties to get customer features. Then, it obtained the recommendation value combined with the customer features and generated recommended population of entity similarity matrix. In the end, it recommended products combined with the purchase intention and the recommendation groups calculated by the AGA. The experiment based on the Movie-Lens data set shows that the algorithm has a better performance.

Keywords: personalized recommendation, customer features, AGA, E-commerce

1 INTRODUCTION

The collaborative filtering algorithm (EuiHong, 2005), the item-based algorithm (Sarwar B, 2001) and the algorithm based on product properties (Middleton S E, 2004) are widely used in the personalized recommendation field. However, as the E-commerce sites is gradually refined and the requirement for recommendation is increased, the deficiencies of the aforementioned algorithm emerge apparently, such as the low accuracy, efficiency and humanization. How to recommend satisfied products to customers has become more important than ever in the recommended algorithm field.

To solve the above problems, this paper proposed a new algorithm based on the customer features and the AGA. It obtains information regarding the customer's interest in products and considers the behavioral features, such as promotional sensitivity and loyalty to make the recommendation more personalized and precise.

2 THE ALGORITHM MODEL

2.1 Description of the algorithm

The first step to analyze the customer's online browsing history log is to get the purchase intention and the second step is to quantify the customer features by analyzing the customer's online behavior log and the database of product properties. The third step is to combine the customer fea-

tures and the property similarity degree of different products, generating the initial populations of recommendation. The fourth step is to calculate the recommendation groups by using the AGA. The last step is to integrate with the purchase intention and the recommendation groups, products are recommended to the customer.

2.2 The customer features

Customer features are objective and quantitative descriptions of the customer analyzed by the customer's online behavior logs including the purchased logs and the browsing history logs. One of

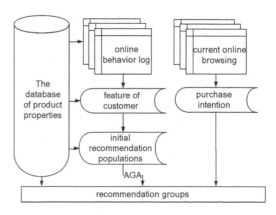

Figure 1. The personalized recommendation algorithm model.

the B2C companies defines the customer features as follows.

In our method, the customer features consisted of two parts: the customer's preference for basic properties of the products and the customer behavior features. They are introduced as follows:

(i) The preference for basic properties can be defined as an appropriate value based on analyzing the customer's preference degree for the products' basic properties according to the purchased logs in a certain period of time.

(ii) Customer behavior features embodied as the customer's individual characteristics connected with the customer's shopping behaviors, such as the promotional sensitivity, publicity sensitivity and shopping loyalty. These factors play important roles during the customer's online consumption. Behavior features reflect the preference for the artificial attributes of products.

The customer features (CF) reflected the preference for the basic properties of the product and the behavior of the customer. CF_i is the customer's preferences for product attributes i that include the basic properties and the artificial properties (Li Feng 2007):

$$CF_i = \frac{S_i}{\sum_{j=1}^{n} S_j} \qquad (1)$$

Table 1. The example of customer features by a B2C website.

Dimensions	Property
Basic properties	Gender, age, height, children
Purchasing power	Income, car, house
Behavior features	Promotional sensitivity, shopping loyalty
Social networks	Networks, affiliated groups
Psychological Characteristics	Hesitant, impulsive, targeted
Interests	Fashion, digital, food, animation

Table 2. The customer features.

	i_1	i_2	...	i_{n-1}	i_n
u_1	CF_{11}	CF_{12}	...	$CF_{(n-1)}$	CF_{11}
u_2	CF_{12}	CF_{22}	...	$CF_{2(n-1)}$	CF_{11}
...	...	...	CF_{ij}	...	...
u_{m-1}	$CF_{(m-1)1}$	$CF_{(m-1)2}$	...	$CF_{(m-1)(n-1)}$	$CF_{(m-1)n}$
u_m	CF_{m1}	CF_{m2}	...	$CF_{m(n-1)}$	CF_{mn}

The larger the value is, the more favorite the product property is. Here, n is the total number of product properties; and S_i is the number of property i appeared in the product purchased in a certain period of time.

The customer features can be reflected as $C = \{CF_1, CF_2, ... CF_i ...\}$, where i represents the product property that appeared in the product purchased.

2.3 Product property

The product attributes presented in this paper contain properties of products, such as color and size, and some artificial properties including discounts and degree of publicity. The product properties selected in this algorithm are built on the objective and identifiable properties of the products. These attributes match the database of the product properties. In order to describe the product properties of the model, the product properties database can be represented by vectors as follows:

where N is the number of a physical product n's type; parameter i is the number of a product's feature; and f_{ij} is expressed as the jth eigenvalues of i. k means that each eigenvalue of a feature is not fixed. Each feature value represents as a binary; for example, if the product has the features, its value is 1; otherwise it is 0 (Wang Pei, 2004).

2.4 The purchasing intention of customers

The products that are clicked frequently by the customer can be considered as the potential purchased products for the customer. As the type of the product is one of the properties of this product, we consider the propensity to buy as an entity. For the entity set that contains the type feature $K = \{k_1, k_2, ... k_n\}$, the purchase intention of customer K_p is defined as $K_p = k_i$. The k_i represents the entity that the customer clicks most frequently (WANG Shuxi, 2013).

2.5 The similarity matrix of the same product properties

The similarity between entities about a certain property can be calculated by the evaluation of the customer through the similarity measure methods such as the similarity of cosine. This similarity of the same property between the different products can be calculated by the method of cosine of the angle between the two vectors. We consider the vectors i, j to represent the property k of the product M and N (Zhang Yu). The similarity of the property k is $s(M_k, N_k)$:

$$s(M_k, N_k) = s(i, j) = \cos(i, j) = \frac{i \cdot j}{\|i\| * \|j\|} \qquad (2)$$

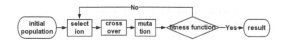

Figure 2.   The process of GA.

2.6   The recommendation value matrix of the product

The recommendation value (RV) of a product can be calculated by summing the value that multiplied the value of the customer preference and the value of the properties of the product. Here, n is the main properties of the product:

$$RV_p = \sum_{i=1}^{n} P_{pi} * CF_{pi} \qquad (3)$$

3   THE PROCESS OF OUR RECOMMENDATION ALGORITHM

3.1   Obtaining the purchase intention and the customer features

We choose the customer's current purchasing intention as the product, which has the highest click rate among all the product types that customer had clicked. For example, for the type property set T = {T-shirt, PC, camera...}, according to the statistics of the customer's current browsing log, we get the customer's purchase intention as $K_p = k_i = Camera$. By analyzing the customer's online behavior log, we calculate the features of the customer by formula (1). The features of the customer are $\{CF_1, CF_2, ...CF_i ...\}$, and then we sort the features of the customer according to the preference rate. By considering the precision and efficiency, we just take a few of n that rank the top on one property for recommendation in the actual use.

3.2   Obtaining the recommendation group by the AGA method

After obtaining the preferred features of the customer, we sort laterally according to the preference and sort each property longitudinally according to the number of the customer's choice by the further analysis of the customer's purchasing log. We select the first n rows data as the reference group of the matrix.

Then, for each property in the reference group, we rank it by the property similarity and select the first i values to add to the recommendation group (Burke R, 2002). So, we get a $n(i+1)$ rows, and j columns of the recommendation matrix as the initial matrix of the Genetic Algorithm (GA). Then, we calculate the RV for each product by using formula (3) through the GA and get the optimal products.

GA is a searching method for global optimization, which is based on biological evolutionary mechanisms such as natural selection, heredity and mutation. This heuristic is routinely used to generate useful solutions to optimize and search problems (Mitchell, Melanie, 1996). The GA belongs to the larger class of evolutionary algorithms, which generate solutions to optimize problems using techniques inspired by natural evolution, such as inheritance, mutation, selection and crossover.

We choose the Adaptive Genetic Algorithm (AGA) as our method in this paper. The GA with adaptive parameters is a significant and promising variant of GAs. The probabilities of pc and pm greatly determine the solution accuracy and the convergence speed of this method. Instead of using fixed values of pc and pm, the AGA utilizes the population information in each generation and adaptively adjusts the pc and pm in order to maintain the population diversity as well as to sustain the convergence capacity. In the AGA, the adjustment of pc and pm depends on the fitness values of the solutions (Srinivas. M, 1994). The formula of pc and pm is as follows:

$$P_c = \begin{cases} \dfrac{k_1(f_{max} - f)}{f_{max} - f_{avg}}, & f \geq f_{avg} \\ k_2, & f \leq f_{avg} \end{cases},$$

$$P_m = \begin{cases} \dfrac{k_3(f_{max} - f')}{f_{max} - f_{avg}}, & f' \geq f_{avg} \\ k_4, & f' \leq f_{avg} \end{cases} \qquad (4)$$

3.3   Recommending products according to the recommendation groups

We can get multiple best individuals using the GA. Then, sorted by the RV, we take the first best N individuals as the recommendation basis group. Combining the purchase intention and the basis group, we get N kinds of products. We then match the products with the database of product property. If the product exist, we recommend the product. If it does not exist, we then add a new product to the basis group from the remaining recommendation group in ranking order. After filling the basis group, we continue the process until the last product in the recommendation group.

4 COMPUTATIONAL EXAMPLES AND ANALYSIS

4.1 *Experimental methods*

We use the MovieLens data set as our experimental data provided by the US GroupLens project team. The data set contains 100000 movie remark scores on 1682 movies by 943 users. The score ranges from 1 to 5, and each user remarks at least 20 films present in this data set. We first import the movies that have a labeled property value (0 or 1) into the database, and make the genres of the movie as the property simulation to the basic properties of the product and the user's information simulation to the customer's behavior features. We choose 1000 items of the data set as our experimental data set, with 80% as the training set and 20% as the testing set.

In order to verify the accuracy of the algorithm, we use the Mean Absolute Error (MAE) to measure the result (HERLOCKER L J, 2004). We measure the accuracy by calculating the deviation of the predicted score and the actual score. The smaller the MAE value is, the higher the prediction accuracy is and the better the recommend quality is (KARYPIS G, 2001). If the prediction score of the algorithm is $p_1, p_2, \ldots p_n$ and the actual score is $q_1, q_2, \ldots q_n$, then the MAE is defined as follows:

$$MAE = \frac{\sum_{i=1}^{N} |p_i - q_i|}{N} \tag{5}$$

4.2 *Experimental results*

First, we consider the impact of the AGA on the algorithm performance used for calculating the RV. We compare the AGA with the GA in same method as follow.

From Figure 3, it can be seen that the AGA method used to calculate the optimal recommend values in this paper is significantly better than the traditional GA method in relation to both convergence speed and accuracy. It not only ensures the completeness and accuracy of the search, but also avoids local search to obtain the global optimal value. Therefore, the algorithm based on the AGA is more reasonable and effective.

For the whole recommendation algorithm, we compare the algorithm based on product properties and the algorithm based on customer features proposed in this paper. The result is shown in Figure 4.

With the number of customer features increasing, the MAE value continues to decline and the rate of decline is slowed down. The reason is that

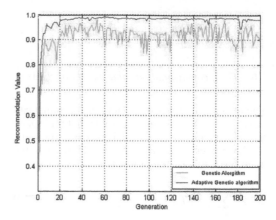

Figure 3. The comparison between the GA and the AGA.

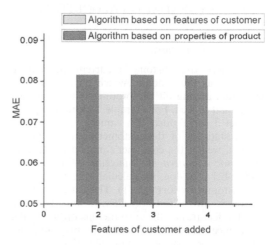

Figure 4. The comparison of the MAE between the recommendation algorithms based on product properties and customer features.

as the customer features increases, the filter conditions increases as well, the number of recommendation group declines, leading to the decline in the AGA initial population number. Overall, the accuracy of the proposed algorithm is better than that of the traditional personalized recommendation algorithm.

5 CONCLUSIONS

In this paper, we proposed an improved method for the recommendation algorithm that can be more robust and efficient during the recommendation. With the development of E-commerce, how to recommend products efficiently and accurately will become increasingly important for the

E-commerce. Our recommendation method takes the customer features as a starting point, analyzes the customer's online log and recommends in line with the customer's own interests of products and behavior features. The algorithm can reduce the recommended range, increase the recommended efficiency and make the recommendation more personalized. As a result, it can improve the product sales and satisfy the customer's shopping need. However, the algorithm needs to analyze the customer's online log, so the recommendation is not real time while improving the efficiency. In the future, we will continue to improve this algorithm to make it more efficient while making the response time faster.

REFERENCES

Burke R. Hybrid recommender systems: survey and experiments. User Modeling and User-Adapted Interaction, 2002.

Karypis G. Evaluation of item-based top-n recommendation algorithms. Proceedings of the 10th International Conference on Information and Knowledge Management. New York:ACM, pp:247–254, 2001.

Herlocker L J, Konstan A J, Terveen G L, et al. evaluating collaborative filtering recommender systems. ACM Transactions on Information Systems, 22(1), pp: 5–53, 2004.

Li Feng, LI Junhuai, Wang Ruilin, Zhang Jing. Personalized recommendation algorithm based on product features. Computer Engineering and Applications, 43(17), pp: 194–197, 2007.

The R&D system of Jingdong. The decryption of Jingdong technology. Publishing House of Electronics Industry, pp: 234–240, 2014.

Han EuiHong (Sam), Karypis G. Feature-based recommendation system. Proceedings of the 14th ACM International Conference on Information and Knowledge Management, pp: 446–452, 2005.

Middleton S E, Shadbolt N R, Roure D C. Ontological userProfiling in recommender systems. ACM Transactions on Information Systems, pp: 54–58, 2004.

Mitchell, Melanie. An Introduction to Genetic Algorithms. Cambridge, MA: MIT Press, 1996.

Sarwar B, Karypis G, Konstan J. Item-based collaborative filtering recommendation algorithms. Proceedings of the Tenth International Conference on Word Wide Web, ACM Press, pp: 285–295, 2001.

Srinivas. M, Patnaik. L. Adaptive probabilities of crossover and mutation in genetic algorithms. IEEE Transactions on System, Man and Cybernetics, 24(4), pp: 656–667, 1994.

Wang Pei. Recommendation based on personal preference. Computational Web Intelligence. Singapore: World Scientific Publishing Company, pp: 101–115, 2004.

WANG Shuxi, Li Anyu. A Commodity Recommend Scheme Based on Customer's Purchase Intentions. Journal of Integration Technology, 5, pp: 15–18, 2013.

Zhang Yu, Liu Yudong, Ji Zhao. Vector similarity measurement method. Technical Acoustics, 8, pp: 534–536, 2009.

Energy Science and Applied Technology – Fang (Ed.)
© 2016 Taylor & Francis Group, London, ISBN 978-1-138-02833-3

Researching on the improvement and case of grey prediction model

Yuansheng Huang & Guangli Wang
North China Electric Power University, Boading, China

ABSTRACT: Prediction is the important basis and foundation of tasks such as system planning. When predicting for the smooth progress of production work, it needs to select the appropriate model according to different samples. The traditional grey prediction model predicts directly without dealing with the original data. This paper first uses the buffer operator processing method to deal with the raw data. Then, it uses the traditional grey prediction model with the processed data to obtain the prediction value. Finally, it goes back to the initial prediction value. Through the example analysis, this paper proves that the improved grey prediction model can improve prediction precision.

Keywords: Grey prediction model; Buffer operator processing method; Prediction

1 INTRODUCTION

The objective world is the world of information, which includes a large number of known and unknown information. If the system with all known information is called the white system and the system with all unknown information is called the black system, then the grey system is the system with the known and unknown partial information. It is in between the white system and the black system with the features of uncompleted information.

Predicting a system with both known and uncertain information is to completely predict the grey process that changes within a certain range. Grey prediction uses potential rules such as data collection to establish the grey model to predict the grey system. It uses known information to establish the differential equation and calculate the unknown data.

2 THE BASIS THEORY OF THE GREY SYSTEM

2.1 *The basic concept of the grey system*

The grey system theory was first proposed by a China scholar, Professor Deng Julong. It takes a small sample, with poor information and uncertainty, as the research object. It not only overcomes the disadvantages of classical statistical analysis methods, but also makes up the shortcomings of using the mathematical statistics method for analysis. Through the generation and development of known information, it extracts valuable

information and grasps the running behavior and evolution.

2.2 *Grey generating operation*

1. Accumulated Generating Operation
   As the basis of grey modeling, accumulated Generating Operation (AGO) successively adds the same sequence of data to generate new data. The series before accumulation is called the original series and those after accumulation is called the generated series.
2. Inverse Accumulated Generating Operation
   Inverse Accumulated Generating Operation means making the adjacent two data minus. Besides, the data obtained is an inverse accumulated generating value. The Inverse Accumulated Generating Operation is an inverse operation of the Accumulated Generating Operation.

3 MODEL RESEARCH

3.1 *Traditional Grey Prediction Model*

After irregular random variables grey generation according to the grey system theory, the original data sequence becomes the exponential law series. Then, the first-order exponential equation can be used to establish the differential equation according to the series after generation, which is called the grey model. The GM (1, 1) model is one of the most commonly used grey models. As a special case of GM (1, n), GM (1, 1) contains a first-order exponential equation that only contains single variables.

The concrete modeling process is as follows:
(1) Establishing the original data sequence $X^{(0)}$ as

$$x^{(0)} = \{x^{(0)}(1), x^{(0)}(2), ..., x^{(0)}(n)\} \tag{1}$$

(2) Establishing the accumulated generating sequence $X^{(1)}$ as

$$x^{(1)} = \{x^{(1)}(1), x^{(1)}(2), ..., x^{(1)}(n)\} \tag{2}$$

$$x^{(1)}(t) = \sum_{i=1}^{t} x^{(0)}(i) \tag{3}$$

(3) Establishing the original form and differential equation of GM (1, 1)
The original form of GM (1, 1) is as follows:

$$x^{(0)}(t) + ax^{(1)}(t) = u \tag{4}$$

where a is the development coefficient, which reflects the development trend of the data sequence; u is the coordination coefficient, which reflects the change in the relationship between data.
Establishing the first-order differential equation as:

$$x^{(1)}(t) = \frac{u}{a} + ce^{at} \tag{5}$$

That is,

$$\frac{dx^{(1)}(t)}{dt} + ax^{(1)}(t) = u \tag{6}$$

$$\frac{dx^{(1)}(t)}{dt} = x^{(0)}(k+1) \tag{7}$$

where x is the mean value of k and $k+1$. The above equation can be written in the matrix form:

$$Y_n = B\hat{A} \tag{8}$$

where

$$Y_n = \begin{bmatrix} x^{(0)}(2) \\ x^{(0)}(3) \\ ... \\ x^{(0)}(n) \end{bmatrix} \tag{9}$$

$$A = \begin{bmatrix} a \\ u \end{bmatrix} \tag{10}$$

$$B = \begin{bmatrix} -\frac{1}{2}[x^{(1)}(1) + x^{(1)}(2)] & 1 \\ -\frac{1}{2}[x^{(1)}(2) + x^{(1)}(3)] & 1 \\ ... & ... \\ -\frac{1}{2}[x^{(1)}(n-1) + x^{(1)}(n)] & 1 \end{bmatrix} \tag{11}$$

(4) Determining the equation parameters
Obtaining parameters according to the principle of the least squares:

$$\hat{A} = (B^T B)^{-1} B^T Y_n = \begin{bmatrix} \hat{a} \\ \hat{u} \end{bmatrix} \tag{12}$$

(5) Establishing the time response function
Through the above calculation, the time response function of the model can be obtained as follows:

$$x^{(1)}(t) = [x^{(1)}(1) - \frac{\hat{u}}{\hat{a}}]e^{-\hat{a}t} + \frac{\hat{u}}{\hat{a}} \tag{13}$$

(6) Determining the original value
Making the inverse accumulated generating on $x^{(1)}(t)$ and reducing the value, we get:

$$\hat{x}^{(0)}(t+1) = \hat{x}^{(1)}(t+1) - \hat{x}^{(1)}(t)$$
$$= (1 - e^{\hat{a}})(x^{(0)}(1) - \frac{\hat{u}}{\hat{a}}) \bullet e^{-\hat{a}t} \tag{14}$$

3.2 Improved Grey Prediction Model

The buffer operator can deal with the original data sequence. The result shows that it can effectively weaken the influence of the abnormal value and the stringent change trend of the original data. In addition, the buffer operator emphasizes the principle of "new information priority". It can be constructed on the basis of information of the existing data sequence. In addition, it can make the new information remain unchanged under the action of the buffer operator.
The process is as follows:

$$y^{(0)}(t) = \frac{1}{n-t+1}(x^{(0)}(t) + x^{(0)}(t+1) + ... + x^{(0)}(n))$$
$$= \frac{1}{n-t+1}\sum_{j=t}^{n} x^{(0)}(j), t = 1, 2, ..., n \tag{15}$$

4 EXAMPLE ANALYSIS

In order to prove the progressiveness of the improved method, this paper takes the total grain output from 2004 to 2012 as the research object. The data is listed in Table 1. This paper, respectively, the uses traditional grey prediction model and the improved grey prediction model to predict the data from 2011 to 2012 according to the data from 2004 to 2010.

Table 1. The total grain output from 2004 to 2012.

2004	2005	2006	2007	2008
46946	48402	49804	50160	52870
2009	2010	2011	2012	
53082	54647.7	57120.8	58958	

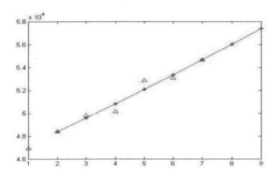

Figure 1. Prediction data based on the traditional model.

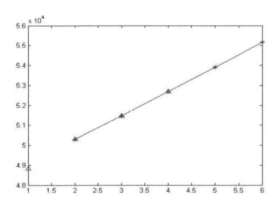

Figure 2. Prediction data based on the improved model.

4.1 Traditional model prediction

Using the traditional model to predict and prediction data can be obtained using the Matlab7.1 software. The concrete result is shown in Figure 1.

$$\hat{x}^{(1)}_{(2011)} = 56011.3 \text{ and } \hat{x}^{(1)}_{(2012)} = 57388.1$$

4.2 Improved model prediction

The data sequence, which is dealt with by the buffer operator process method, is as follows:

$$y^{(0)}_{(1)} = \frac{1}{4}(46946.9 + 48402.2 + 49804.2 + 50160.3) = 48828.4$$

$$y^{(0)}_{(2)} = \frac{1}{4}(48402.2 + 49804.2 + 50160.3 + 52870.9) = 50309.4$$

$$y^{(0)}_{(3)} = \frac{1}{4}(49804.2 + 50160.3 + 52870.9 + 53082.1) = 51479.4$$

$$y^{(0)}_{(4)} = \frac{1}{4}(50160.3 + 52870.9 + 53082.1 + 54647.7) = 52690.3$$

Prediction data can be obtained using the Matlab7.1 software. The concrete result is shown in Figure 2.

$$y^{(0)}_{(5)} = 53918.0 \text{ and } y^{(0)}_{(6)} = 55179.1$$

That is

$$y^{(0)}_{(5)} = \frac{1}{4}(52870.9 + 53082.1 + 54647.7 + x^{(2)}_{(2011)}) = 53918.0$$

$$y^{(0)}_{(6)} = \frac{1}{4}(53082.1 + 54647.7 + x^{(2)}_{(2011)} + x^{(2)}_{(2012)}) - 55179.1$$

$$\hat{x}^{(2)}_{(2011)} = 56075.5, \hat{x}^{(2)}_{(2012)} = 57915.3$$

4.3 Comparison of prediction results

From Table 2, we can see that the error of prediction data through the buffer operator process is less than the traditional model. It proves that the

Table 2. Comparison of prediction results.

	Traditional grey prediction model				Improved grey prediction model		
Original data	Prediction value	Absolute residual	Relative residual (%)		Prediction value	Absolute residual	Relative residual (%)
57120.8	56011.3	1109.5	1.94		56075.5	1045.3	1.83
58958	57388.1	1569.9	2.66		57915.3	1042.7	1.77

improved prediction model can improve prediction precision.

5 CONCLUSION

The grey prediction model predicts the future data only through the past data, but factors that affect the future data are varied. The traditional grey prediction model ignores it. Therefore, its prediction precision is lower. This paper compares the prediction results between the traditional grey prediction model and the improved grey prediction model through an actual example. It proves that the improved grey prediction model can improve prediction precision.

REFERENCES

Deng Julong, 2000, Positional target in grey target theory [J]. The Journal of Grey System, (1): 1–8.

Ren Gonghang, Liu Li, Miao Xinqiang, 2010. Prediction and implementation of the improvement grey model to electrical load [J]. Machinery Design& Manufacture (2):p. 232–234.

Wang Changjiang. 2006. Selection of Smoothing Coefficient via Exponential Smoothing Algorithm [J]. Journal of North University of China (Natural Science Edition), 27(6): p. 58–61.

Energy Science and Applied Technology – Fang (Ed.)
© 2016 Taylor & Francis Group, London, ISBN 978-1-138-02833-3

The NSPC model for predicting the spread of Ebola

Jinjing Shen
North China Electric Power University, Baoding, China

ABSTRACT: With the widespread outbreak of Ebola, it is essential to draw up a medical program for eradicating the epidemic. On the basis of the traditional epidemic model—SIR model, we construct the NSPC model to forecast the future trends of Ebola. We describe the fluctuation of four classified cases by establishing four corresponding difference equations. Then, according to historical data, we can adopt the nonlinear least squares method to obtain the spread equation and the trend chart.

Keywords: NSPC model; the nonlinear least squares method

1 INTRODUCTION

1.1 Background

West Africa's epidemic of Ebola virus disease, whose death toll has hit an all-time high, has attracted international attention. To make reliable infectious disease emergency response plans, the first important thing is to make an accurate forecast for trends in disease development.

1.2 Historical research

The establishment and spread of infectious diseases is a complex phenomenon with many interacting factors. The role of mathematical epidemiology is to model the establishment and spread of pathogens. The most typical ones are the SIR model and the SIS model.

These models are known as compartmental models in epidemiology, and serve as a base for a mathematical framework to understand the complex dynamics of these systems, which hope to model the main characteristics of the system.

On the basis of the SIR model, we put up with the NSPC model, with a new crowd classification by adopting a more concrete definition given by the WHO files.

1.3 The classification of the crowd

According to the related definition given by the WHO, we can divide the crowd in affected areas into four groups as follows:

Non-case: any suspected or probable case with a negative laboratory result, i.e. "Non-case" showing no specific antibodies, RNA or specific detectable antigens.

Suspected case: any person, alive or dead, suffering or having suffered from a sudden onset of high fever and having had contact with:

- a suspected, probable or confirmed Ebola case;
- a dead or sick animal (with Ebola)

Probable case: any suspected case evaluated by a clinician.

Confirmed case: any suspected or probable case with a positive laboratory result.

1.4 Some assumptions

① The drug or vaccine effect on each person is the same, regardless of the difference in efficacy taken by age, physical fitness, district and other factors.

As a result of the uncertainty of these factors' influence, to simplify our model, we can only average the efficiency of the effect of drugs and vaccine.

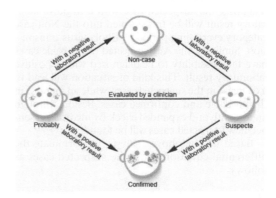

Figure 1. Conversion chart of the four groups.

② Population fluctuations only influenced by the Ebola virus, namely we do not take the natural fertility and mortality rates into account.

In the disease outbreak period, the increase and decrease in the population is mainly affected by Ebola.

③ We make Ebola spread forecasts under the premise that there are no effective medicine and vaccine supply for the ward people.

Only on the basis of this assumption, we can estimate the actual transmission of the Ebola virus spread.

④ We do not take the movement of people between regions into account.

In Ebola-hit countries, both government and medical institution monitor the movement of persons closely.

⑤ In different districts, for the cases who are diagnosed as suspected and probable, they have the same ratio to contact with healthy people

The contact ratio may be different in the different districts as a result of the population density difference. To simply our model, we just consider the average contact ratio as the value of the same contact ratio.

1.5   *The establishment of the NSPC model*

In order to develop a reliable plan of medicine manufacturing and delivery, we first need to learn the spread of the Ebola virus in the future few months. Furthermore, we need to claim that our prediction model is built on the basis of existing data. In other words, we make predictions about the Ebola epidemic in the absence of effective drug and vaccine.

To represent the decrease and increase in the number of the four groups, referring to the SIR model, we aim at these four groups to establish the NSPC model in the form of differential equation, respectively.

We first study the group classified as the Suspected case. Under the condition of professional laboratory diagnosis, the suspected case with a negative laboratory result will be transformed into the Non-case category every time period of Ebola virus transmission. Similarly, the Any suspected or probable cases have the possibility to be diagnosed with a positive laboratory result. This kind of situation will lead to a decrease in the suspected cases, while an increase in the probable and confirmed cases. Besides, through an in-depth and expanded check by medical personnel, new suspected cases will be found.

Based on the above analysis, we formulate the differential equation about the suspected cases as follows:

$$\frac{dS}{dt} = -K_{SN}S - K_{SP}S - K_{SC}S - K_S S + \lambda S + CT \quad (1)$$

- where K_{XY} is the ratio that the group X convert into Y, which is confirmed by the laboratory result. K_{XY} contains $K_{SN}, K_{SP}, K_{SC}, K_{PN}, K_{PC}$.
- where CT denotes the new suspected cases of routine inspection, and λ denotes the contact ratio with the suspected cases.
- where t is the time period in days

In the same way, we can separately establish the differential equation about the Non-case, Probable case and Confirmed case as follows:

$$\frac{dN}{dt} = -K_{SN}S + K_{PN}P \quad (2)$$

$$\frac{dP}{dt} = K_{SP}S - K_{PC}P - K_P P \quad (3)$$

$$\frac{dC}{dt} = K_{SC}S + K_{PC}P - K_C C \quad (4)$$

The formulas (1), (2), (3), (4) make up our NSPC model.

The following should be noted:

- For the suspected cases of people, considering they just have similar symptoms, people could still contact with them.
- For the Probable cases and Confirmed cases, who are evaluated clinically and have a positive laboratory result, they are under medical personnel monitoring. We assume they have no way of contacting with natural persons.

1.6   *Solution to the difference equation*

Step 1: To get the specific function expressed as S, N, P and C on t, respectively, we first simplify the differential equation by parameters Polymerization. Function (1) and (3) convert into the following expressions (5) and (6):

$$\frac{dS}{dt} = K_{SS}S + CT, (K_{SS} = -K_{SN} - K_{SP} - K_{SC} - K_S + \lambda) \quad (5)$$

$$\frac{dP}{dt} = K_{PP}P + K_{SP}S, (K_{PP} = -K_{PC} - K_P) \quad (6)$$

Step 2: Then, we transform the formulas (5), (2), (6) and (4) into the form of difference equations. With the use of the *least squares method*, we can determine the value of parameters:

$$K_{SS} \quad K_{SN} \quad K_{PN} \quad K_{PP} \quad K_{SP} \quad K_{SP} \quad K_{SC} \quad K_{PC} \quad CT$$

Step 3: Finally, we solve the first-order differential equations by the *nonlinear fitting method*. With the help of the MATLAB toolbox *nilinfit* function, we can get the fitting results.

Besides, we can receive the S, N, P and C function expression on t (time) change:

$$S = a_s e^{K_{SS}t} + c_S \tag{7}$$

$$N = a_{n1} e^{K_{SS}t} + a_{n2} e^{K_{PP}t} + c_n \tag{8}$$

$$P = a_{p1} e^{K_{SS}t} + a_{p2} e^{K_{PP}t} + c_p \tag{9}$$

$$C = a_{c1} e^{K_{SS}t} + a_{c2} e^{K_{PP}t} + a_{c3} e^{-K_c t} + C_c \tag{10}$$

2 CASE STUDY

Among those Western African countries with a widespread Ebola transmission, the outbreak in Sierra Leone is the most serious, as is shown in the latest data according to the WHO. So, we choose Sierra Leone as a typical object for research.

2.1 Parameter determination

We counted 14 districts of Sierra Leone's population of infection and death toll separately, with the time range from the August 2014 to February 2015. A larger amount of statistics help us to get a more accurate and realistic trend forecast. These 14 administrative regions are Bo, Bombali, Bonthe, Kailahun, Kambia, Kenema, Koinadugu, Kono, Moyamba, Port Loko, Pujehun, Tonkolili, Western Area urban, Western Area Rural.

As a result of the limitation of space, this paper only lists the fitting results of the BO district as follows:

*The value of the parameter can be calculated by solving the differential equation according to the above-mentioned steps. The results are shown in Table 1.

By observing the number of population changes in the shape of the figures, we can get the following conclusion:

- In the absence of an effective drug and vaccine, the number of people in all four categories has been a surge in the trend.
- Ideal fitting results verify the validity of our method.

*At the same time, through the nonlinear fitting, we can get the specific formula for the number of changes in the Non-case, Suspected case, Probable

Table 1 The value of the parameter.

Parameter		
$K_{SS} = 0.0156$	$K_{SN} = 0.5391$	$K_{PN} = 0.3095$
$K_{PP} = 0.0203$	$K_{SP} = 0.0212$	$K_{SC} = 0.02675$
$K_{PC} = 0.1663$	$KT = 3.6421$	

Trend of changes

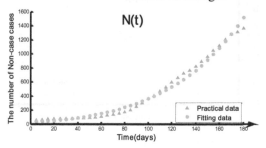

Figure 2. Non-cases.

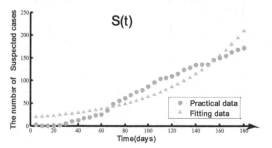

Figure 3. Suspected cases.

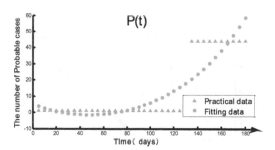

Figure 4. Probable cases.

case and Confirmed case. The prediction equations are as follows:

$$S = 12.2967 e^{K_{SS}t} + 5.6199 \tag{11}$$

$$N = 98.5534 e^{K_{SS}t} + 98.4208 e^{K_{PP}t} - 114.7598 \tag{12}$$

$$P = 4.601 e^{K_{SS}t} + 19.1881 e^{K_{PP}t} - 18.4470 \tag{13}$$

$$C = 15.2926 e^{K_{SS}t} - 211.8422 e^{K_{PP}t} - 160.02 \tag{14}$$

where S, N, P, C, respectively, represent the number of people who are classified into the Suspected case, Non-case, Probable case and Confirmed case.

By expressions we output, we can separately predict the concrete number of the Non-case, Suspected case, Probable case and Confirmed case in

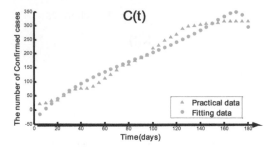

Table 5. Confirmed cases.

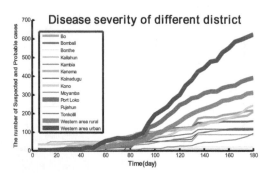

Figure 6. Severity considering the suspected case and the probable case.

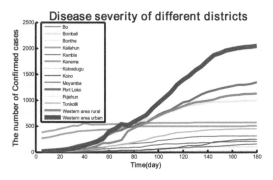

Figure 7. Severity considering the confirmed case.

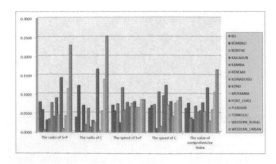

Figure 8. Comprehensive severity by districts.

the BO district of Sierra Leone, in a certain period of time in the next few days. Similarly, we can calculate the prediction equation for the other 13 districts of Sierra Leone.

Subject to the given indicator, we can compute the severity of the disease among different districts in Sierra Leone. We classify the Suspected case and the Probable case as one kind, and the confirmed case as the other kind.

2.2 Disease severity trends

As shown in Figures 5 and 6, when we take the Suspected case and the Probable case into account, the most serious Ebola epidemic area is identified in the districts Western urban, Western rural, Port Loko, BO, Bombali and Kenema.

When we take the Confirmed case into account, the most serious Ebola epidemic area is identified in the districts Western urban, Western rural, Port Loko, BO, Kailahun, Bombali and Kenema.

Then, we compute a total severity index considering all the three groups (relatively Suspected case, Probable case and Confirmed case).

REFERENCES

Chen Ji-rong, Yang Fang-ting, Zhan Shou-yi. Processing on the Parameters and Initial Values of SARS Simulation Model for Beijing. Journal of system simulation, 2003(7):995–998.

http://www.who.int/csr/resources/publications/ebola/ebola-case-definition-contacten.pdf

http://en.wikipedia.org/wiki/Ebola_virus_epidemic_in_West_Africa

http://health.gov.sl/wpcontent/uploads/2015/02/Ebola-Situation-Report_Vol-253.pdf

http://www.historyofvaccines.org/content/articles/different-types-vaccines

http://en.wikipedia.org/wiki/Attenuated vaccine

Paul Fine, Ken Eames, David L Heymann. "Herd Immunity": A Rough Guide. Clinical Infectious Diseases. Oxford Unvi Press Inc, 2011(52): 911–916.

WHO Ebola Response Team. Ebola Virus Disease in West Africa the First 9 Months of the Epidemic and Forward Projections. New England Journal of Medicine. Massachusetts Medical SOC, Waltham Woods Center 2014(371): 1481–1495.

Energy Science and Applied Technology – Fang (Ed.)
© 2016 Taylor & Francis Group, London, ISBN 978-1-138-02833-3

Application of particle swarm optimization in Levy flight for OFDM system resource allocation

Dong Wei & Mincang Fu
Information Engineering Department, Engineering College of CAPF, Xi'an, China

ABSTRACT: Resource allocation is a very important link in Orthogonal Frequency-Division Multiplexing (OFDM) system, while the existing algorithms fail to take account of system throughput and fairness between users simultaneously. The paper proposed to combine Levy flight search and Particle Swarm Optimization on the basis of analysing and comparing the existing algorithms, which has overcome the disadvantage of local optimal easily taken place on traditional Particle Swarm Optimization. The simulation results show that the algorithm introduced in this paper has solved resource allocation issue properly and realized maximization of capacity of the system based on ensuring fairness between users.

Keywords: OFDM, Particle Swarm Optimization, fairness between users

1 INTRODUCTION

Orthogonal Frequency Division Multiplexing (OFDM) has been widely used in high definition digital television broadcasting system, BWA, WLAN (IEEE802.11), WLAN (IEEE802.16), Wimax and so on due to characteristics such as resistance to multipath interference, strong frequency selective fading, high spectral efficiency.

Fading parameters of different users are relatively independent in multi-user OFDM system. The probability of deep fading of a certain subcarrier to all users is extremely low. Thus a deep fading subcarrier of a user can be allocated to other users to realize dynamic resource allocation so as to improve utilization ratio of frequency resource. Studies on OFDM system resource allocation have become hot at home and abroad at present. For single-user OFDM system, water fill algorithm and greedy algorithm can realize optimum allocation (T M Cover, 1991) (Davids, 2004). For multi-user OFDM system, dynamic resource allocation of subcarriers, power and bit according to channel condition becomes relatively more complicated. Self-adaptive resource allocation for multi-user OFDM system is mainly divided into two categories at present: the first one is rate adaptive (RA), which makes the sum of data rate of all users in the system and sub-channels maximum under a constant total power; the other one is margin adaptive (MA), which makes the sum of transmitted power of all users in the system and sub-channels minimum under a constant user data rate. Literature (Wang Jun-Fang, 2006 Lio B, 2009, Lei Ming,

2010) adopts canonical genetic algorithms, which has simplified the computation complexity compared with traditional algorithms. The algorithm can satisfy adaptive allocation issues of different requirements. Literature (Wang Zhao, 2013) adopts improved genetic algorithm, which has improved convergence rate and made significant improvement on system performance. Literature (Ye Lan-Lan, 2012) allocates subcarriers with the best channel capacity of channel capacity matrix firstly at subcarriers allocation. Power distribution adopts water fill algorithm, considerably improved user data transmission rate. Literature (Hu Shan-Feng, 2013) adopts ant colony algorithm to solve subcarrier allocation under proportional rate constraints in the case of equal power allocation. It is a suboptimal allocation. The algorithm realizes downlink link capacity maximization for multi-user under OFDM system and maintains proportional fairness between users. Literature (Chen Yan, 2012) proposes a modified Particle Swarm Optimization POSBA with segmented adjustment strategy, which has optimized convergence rate and system performance.

The paper adopts a modified Particle Swarm Optimization to solve OFDM system resource allocation problems. Particle Swarm Optimization is prone to having local optimum being affected by random oscillation in the later stage of the search. It has a poor robustness as well which will lead to waste of resources. In order to solve the problem, the paper introduced Levy flight strategy to Particle Swarm Optimization. After each particles location updating, individual position will be updated continuously with Levy flight strategy and calculate the objective function so as to avoid local optimum and to improve the solving capability of Particle Swarm Optimization.

2 OPTIMIZED MATHEMATICAL MODEL FOR MULTI-USER OFDM SYSTEM

Assume that there are K users and N subcarriers in the multi-user OFDM system and the system total bandwidth is B. $h_{k,n}$ is the channel gain of user $k(k \in \{1,2,...,K\})$ on subcarrier $n(n \in \{1,2,...,N\})$. N_0 is noise power spectral density, and then carrier-to-noise ratio is $Nh_{k,n}/N_0B$. To define $P_{k,n}$ as the power of user K on subcarrier n, then the rate of OFDM symbol of the kth user can be expressed as R_k

$$R_k = \sum_{n=1}^{N} \frac{1}{N} c_{k,n} \log_2\left(1 + \frac{NP_{k,n}h_{k,n}^2}{\Gamma_k BN_0}\right) \qquad (1)$$

Where as in the formula, $\Gamma_k = -\ln(5BERk)/1.6$ is SNR gap, which indicates the gap between theoretical signal-to-noise ratio and the actual signal-to-noise ratio after reaching to a required rate. $c_{k,n} = \{0,1\}$ is the symbol allocated to user k by subcarrier. $c_{k,n} = 1$ means allocation and $c_{k,n} = 0$ means no allocation. Namely:

$$c_{k,n} = \begin{cases} 0 & \text{if subcarrier n is allocated k} \\ 1 & else \end{cases} \qquad (2)$$

When the optimization objective is maximized system capacity resource allocation, considering proportional fairness between users, the mathematical model of optimization problem can be expressed as:

$$\max_{P_{k,n}c_{k,n}} \sum_{k=1}^{K} \sum_{n=1}^{N} \frac{1}{N} c_{k,n} \log_2\left(1 + \frac{NP_{k,n}h_{k,n}^2}{\Gamma_k BN_0}\right) \qquad (3)$$

$$s.t. R_1 : R_2 : \cdots : R_{k-1} : R_k = \gamma_1 : \gamma_2 : \cdots : \gamma_{k-1} : \gamma_k \qquad (3a)$$

$$c_{k,n} \in \{0,1\} \ \forall \, k,n \qquad (3b)$$

$$\sum_{k=1}^{k} c_{k,n} = 1 \quad \forall \, n \qquad (3c)$$

$$P_{k,n} \geq 0 \quad \forall \, k,n \qquad (3d)$$

$$\sum_{k=1}^{k} \sum_{n=1}^{n} P_{k,n} \leq P_{total} \qquad (3e)$$

Where as in the formula, constraint condition (3a) refers to constraints on proportional fairness among users. $\{\gamma_i\}_{i=1}^{k}$ is a group of randomly preset values. The rate ratio of each user can be consistent

with the preset value. (3b) and (3c) indicate that a subcarrier can only be allocated to a user. (3d) indicates that the power allocated to the subcarrier is non-negative value. (3e) indicates that the sum of each subcarriers should not exceed the total power system.

3 PARTICLE SWARM OPTIMIZATION INTRODUCED TO LEVY FLIGHT

In the process of OFDM system resource allocation, Particle Swarm Optimization is considered as an optimization tool. A large number of scientific test results show that Particle Swarm Optimization based on groups and fitness concept has relatively faster calculation speed and better optimization ability for large-scale mathematical optimization problems. However, it is prone to having local optimum being affected by random oscillation in the later stage of the search. It has a poor robustness as well which will lead to waste of resources. To solve the problem, the paper introduced cuckoo algorithm as complementary, which has reduced the probability of local optimum occurrence.

3.1 PSO algorithm

Particle Swarm Optimization (PSO) is an intelligence algorithm proposed by Russell Eberhart, an electrical engineer and James Kenned, a social psychologist in the United States based on a simple social model in 1995 inspired by the social behaviors of birds and fish in nature.

Supposed that a particle swarm is composed of m particles in the n-dimensional space. $X_i^t = (X_{i1}^t, X_{i2}^t, \cdots X_{in}^t)$ is the coordinate position of particle $i(i \in \{1,2,...,m\})$ at moment t. The velocity of particle i is defined as particle moving distance at each iteration, expressed as $V_i^t = (v_{i1}^t, v_{i2}^t, \cdots v_{in}^t)$. The flight speed and position of particle i at moment t at $j(j \in \{1,2,...,n\})$ dimensional space should be updated according to the formula as below:

$$v_{ij}^t = w v_{ij}^t + c_1 r_1 \left(P_{ij} - x_{ij}^{t-1}\right) + c_2 r_2 (g_j - x_{ij}^{t-1}) \qquad (4)$$

$$v_{ij}^t = \begin{cases} v_{max} & v_{ij}^t > v_{max} \\ -v_{max} & v_{ij}^t > -v_{max} \end{cases} \qquad (5)$$

$$v_{ij}^t = X_{ij}^{t-1} + v_{ij}^t \qquad (6)$$

Where as in the formula, w is a weighting coefficient, with a value range between 0.1–0.9. c_1 and c_2 are known as learning factor. Generally $c_1 = c_2 = 2$. r_1 and r_2 is a random number between (0, 1). P_{ij} is the optimal historical location of the current particle, namely the individual extremum P_{best} g_i is the optimal historical location among the whole particle swarm, namely global extremum g_{best} v_{max} is particle velocity extremum to improve the search results.

Generally weighting coefficient w can be expressed as below:

$$w = w_{max} - \frac{w_{max} - w_{min}}{iter_{max}} \times k \qquad (7)$$

Where as in the formula, w_{max} is the Max. weight, w_{min} is the Min. weight, $iter_{max}$ is the Max. number of iterations and k is the current number of iterations.

Among parameters of Particle Swarm Optimization, weighting coefficient is an important parameter. When w is relatively bigger, it is advantageous to improve the global searching ability of the algorithm but with a poor partial searching ability; and when w is relatively smaller, it can enhance the partial searching ability but with a poor ability in searching for new field. In order to achieve better optimization effects, the paper adopted a self-adaptive adjustment strategy of w based on particle individual adaptive value. Inertia weight coefficient will change automatically (Kandasamy Illanko, 2009) along with particle target value variation.

$$w = \begin{cases} w_{min} + \dfrac{(w_{max} - w_{min})(f_i - f_{min})}{f_{avg} - f_{min}} & f_i \le f_{avg} \\ w_{min} & f_i > f_{avg} \end{cases} \qquad (8)$$

f_i is the current objective function value of the particle. f_{avg} is the average objective value of all particles. f_{min} is the Min. objective value of all particles.

3.2 Levy flight profile

Levy flight is an equation of random walk. The next position is to be determined by current position together with transition probability. The update formula is:

$$X_i^{t+1} = X_i^t + \alpha \cdot L(\varepsilon) \qquad (9)$$

α is a step factor. is point to point multiplication. $L(\varepsilon)$ is random search vector of parameter ε, $L(\varepsilon): \mu = t^{-3}$, $1 < \varepsilon$ 3 This distribution shows the change of location is in line with the power-law distribution of random fluctuation process. Location update strategy of Levy flight can be expressed as:

$$X_i^{t+1} = X_i^t + \alpha \cdot L(\varepsilon): \ 0.01\frac{\mu}{v}(X_i^t - X_b^t) \qquad (10)$$

Where as in the formula, X_b^t is the optimal birds nest location experienced? Both μ and v obey normal distribution:

$$\left(\delta_\mu = \left[\frac{\Gamma(\varepsilon)\sin(0.5\pi(\varepsilon - 1))}{2^{\frac{\varepsilon-2}{2}}\Gamma(0.5\varepsilon)(\varepsilon-1)} \right]^{\frac{1}{\varepsilon-1}} \right), v : N(0,1) \qquad (11)$$

3.3 Steps of Particle Swarm Optimization introduced to Levy flight

The steps of Particle Swarm Optimization introduced to Levy flight in solving OFDM system resource allocation:

Step 1: Set OFDM system parameters;

Step 2: Initialization parameters: population size, initial particle velocity, location, the maximum number of iterations, etc.;

Step 3: Calculate objective function value of each particle and store the optimal historical location;

Step 4: Determine whether meet the termination conditions of settings. If yes, go to Step 8. If no, go to Step 5;

Step 5: Update the birds nest position X_i^t by adopting formula (4), (5) and (6);

Step 6: Update the individual and the global optimal position respectively;

Step 7: Update the birds nest position by adopting formula (10) and update the individual and the global extremum continuously. Go to Step 3;

Step 8: Output the optimal solution. Complete the process.

4 SIMULATION RESULTS

To verify feasibility and superiority of the proposed algorithm, simulation comparison was conducted by multi-user OFDM system self-adaptive resource allocation algorithm based on genetic algorithm (Kandasamy Illanko, 2009) and artificial fish swarm algorithm (Cheng Yong-Ming, 2009) and the algorithm mentioned in the paper. The paper performed simulation by adopting MATLAB. The wireless channel has been simulated to a frequency selective channel with 6 paths. Parameter settings are as shown in the table below (The channel gain matrix is randomly generated between 0 and 1):

Figure 2 gives the system a 16 user, the total capacity of different algorithms in the allocation between users. The algorithm proposed in this paper is normalized rate deviation given the smallest proportion, algorithm given the bigger proportion, algorithm given the biggest proportion. This

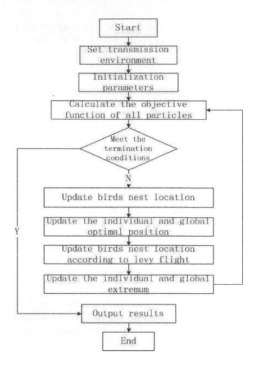

Figure 1. Algorithm flow chart.

Table 1. Parameters setting

Parameter name	Parameter values	Parameter name	Parameter values
System bandwidth	1 MHz	Population size	80
Subcarrier number	64	Maximum number of iterations	200
User number	1–16	w_{max}	0.9
Total power	0.1 W	w_{min}	0.4
BER	10^{-3}	N_0	10^{-8} W/Hz

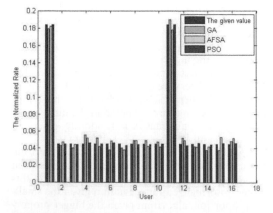

Figure 2. The normalized user capacity with 16 users.

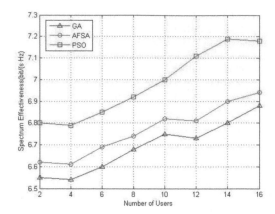

Figure 3. The spectral efficiency under different number of users.

is because algorithm in this paper assumed that $R_1 : R_2 : \cdots R_{k-1} : R_k = \gamma_1 : \gamma_2 : \cdots : \gamma_{k-1} : \gamma_k$ in the sub carrier allocation process, guaranteed the proportional fairness among users.

As we can see from Figure 3, the spectrum efficiency of all of the three algorithms is as number of users increases and continuously improve. The algorithm of spectral efficiency in this paper is better than the other two algorithms.

5 CONCLUSIONS

The paper studied OFDM system resource allocation and combined characteristics of ergodic of Levy flight with Particle Swarm Optimization specifically for the disadvantages existing in traditional Particle Swarm Optimization algorithms during the practical application, which has conquered the disadvantage of Particle Swarm Optimization of being easily trapped in local optimum and greatly improved the ability of optimization algorithm. It has been applied in OFDM system resource allocation. The simulation results show that the algorithm proposed in the paper can solve OFDM system resource allocation properly. It has improved the spectrum efficiency and increased the system capacity while ensuring fairness between users.

REFERENCES

Chen Yan. Resource Allocation of OFDM Systems Based on Enhanced Particle Swarm Optimization Algorithm (J). *Application of Computer System*, 2012, 21(7):114–118.

Cheng Yong-Ming, Jiang Ming-Yang. Adaptive Resource Allocation in Multiuser OFDM System Based on

Improved Artificial Fish Swarm Algorithm (J). *Application Research of Computers*, 2009, 26(6):2092–2094.

Davids, et al. Ordered. Sub carrier Selection Algorithm for OFDM-Speed wlans. IEEE Trans. on Wireless Communications, 2004, 3(9):1452–1458.

Hu Shan-Feng, Li Hong-Lie. Subcarrier Allocation Research of OFDM System Based on Ant Colony Algorithm (J). *Computer Simulation*, 2013, 30(2):343–358.

Kandasamy Illanko, Kaamran Raahemifar and Alagan Anpalagan. Sub-Channel and Power Allocantion for Multiuser OFDM with Rate Constraints using Genetic Algorithm(J).IEEE Pacfic Rim Conference on Communications, 2009: 571–575.

Lio B, Jiang M, Yuan D. Adaptive Resource Allocation in Multiuser OFDM System Based on Genetic Algorithm (J). IEEE Proc, 2009(1):270–273.

Lei Ming, Zhou Li, Xie Yao. Adaptive Bit and Power Allocation Algorithm for OFDM System (J).*Computer Simulation*, 2010, 27(4):197–200.

T M Cover and J Thomas. *Elements of Information Theory* (M). New York: Wiley, 1991.

Wang Jun-Fang, Zhu Guang-Xi, Jing Jiang. Bit Loading Algorithm in Multi-user OFDM System Based on Genetic Algorithm. *Journal of Huazhong University of Technology,* 2006, 34(2):27–29.

Wang Zhao, Li You-Ming, Zhao Cui-Ru. OFDMA Resource Allocation Based on Adaptive Genetic Algorithm (J). *Journal of Ningbo University (NSEE)*, 2013, 26(1):23–27

Ye Lan-Lan, Li Jun, Jing Ning. A Novel Power Allocation Algorithm in Multiuser OFDM Systems (J). *Journal of China University of Metrology*, 2012, 23(4): 379–382.

Energy Science and Applied Technology – Fang (Ed.)
© 2016 Taylor & Francis Group, London, ISBN 978-1-138-02833-3

Motor imagery EEG signal identification based on the improved EMD algorithm

Kewen Zhai & Jianping Liu
Information Engineering Department, Engineering College of CAPF, Xi'an, China

ABSTRACT: EMD algorithm is an effective method in the feature extraction of BCI EEG signals. However, the existence of end point effect affects the promotion of classification accuracy. An end point inhibition method based on similarity distance is developed for the processing of EEG signals. The improved algorithm in this paper is of higher timelines and much lower computation assumption in comparison with mirror image extension method. The feasibility and correctness of the improved algorithm is verified by experiment using the data from international competition.

Keywords: Empirical Mode Decomposition (EMD), end effect, Brain-Computer Interface (BCI)

1 INTRODUCTION

Relinquishing the normal interaction between the human brain and the outside world through the peripheral nerves and muscles, brain-computer interface (BCI) achieves the direct communication between the human brain and the external environment through brain signal analysis, information extraction and external device control (P R Kennedy,2003). However, the signal data obtained by the BCI features a large amount, unsteady signal, low validity and complexity. How to efficiently conduct the feature extraction of the EEG signals becomes a key technology to the EEG signal identification.

In 1998, Huang et al. put forward a new signal processing method—Empirical Mode Decomposition (EMD) algorithm (Rilling. G, 2008). Based on the data's local features, EMD algorithm decomposes signals into several Intrinsic Mode Functions (IMFs) and then conducts the Hilbert conversion of these IMFs to obtain the instant features of signals (Yongjun Deng,2001). However, EMD algorithm has a defect, that is, its significant end effect, which might influence the decomposition effect. Concerning the end effect of the EMD algorithm and combining the fixed features of EEG signals, this paper puts forward an end effect inhibition method based on the minimum similarity distance. The improved EMD algorithm was employed to process the international competition data in 2003. The first order IMF energy was extracted as the characteristic quantity, which was put into SVM classifier for classification. It showed good classification effects.

2 EMD ALGORITHM

The ultimate goal of EMD algorithm is to decompose the original complex signals into the sum of a series of IMFs and make all IMFs meet the following two conditions for the convenience of the Hilbert conversion.

The number of the extreme points must be identical to the number of zero crossing points or the two vary from each other by the difference of one at most in terms of a group of data series;

The average value enveloped by the local maximum value and the local minimum value must be zero in terms of any data point.

The original signal $x(t)$ undergoes the screening process of EMD. The process is described as below:

Confirm all the extreme points of the original signal $x(t)$, obtain upper envelope curve of $x(t)u(t)$, and the lower envelope curve of $x(t),l(t)$ through the function fitting based on cubic spline interpolation. The average value of both is the average envelope curve of the original signal $m_1(t)$ which is:

$$m_1(t) = \frac{u(t) + l(t)}{2} \qquad (1)$$

Subtract the average envelope $m_1(t)$ from $x(t)$, and the new sequence $h_1(t)$ can be obtained, which is:

$$h_2(t) = x(t) - m_1(t) \qquad (2)$$

At the moment, $h_2(t)$ usually does not meet the conditions of IMF. So $h_2(t)$ is regarded as the original signal and the iteration of above process is repeated. When the iterations reach "k," then;

$$h_{1k}(t) = h_{1k-1}(t) - m_{1k}(t) \tag{3}$$

Where, $h_{1(k-1)}(t)$ and $h_{1k}(t)$ stands for the signal obtained after "k-1" and "k" iterations; $m_{1k}(t)$ stands for the average envelope curve of $h_{1(k-1)}(t)$. If $h_{1k}(t)$ meets the conditions of IMF, then $h_{1k}(t)$ is the first order IMF component, which is expressed as $c_1(t)$ and contains the highest frequency component of the original signal. During the practical application, s_d, the standard deviation of two adjacent iteration results, is chosen as the rule to terminate the iteration rule, namely:

$$s_d = \sum_{t=0}^{T} \frac{\left| h_{1(k-1)}(t) - h_{1k}(t) \right|^2}{h_{1k}^2(t)} \tag{4}$$

When s_d is smaller than 0.2 or 0.3, the iteration ends.

Subtract the first IMF component $c_1(t)$ from the original signal $x(t)$, and the new signal $r_1(t)$, can be obtained, namely:

$$r_1(t) = x(t) - c_1(t) \tag{5}$$

Regard $r_1(t)$ as the original signal $x(t)$ and repeat the "a" step. Then a series of IMF components, including $c_2(t)$ and $c_3(t)$ can be obtained. When the remnant item, $r_n(t)$, becomes the monotone function, the whole EMD algorithm ends. Then, the original signal, $x(t)$, is decomposed into a series of IMF components and the sum of a remnant item, namely:

$$x(t) = \sum_{i=1}^{n} c_i(t) + r_n(t) \tag{6}$$

3 EMD END EFFECT

3.1 Cause of the EMD end effect

An important step for the EMD algorithm is to construct the signal's upper and lower envelope curves so as to obtain the signal's instant average envelope. Huang et al. put forward the construction of the upper and lower envelope curves through the fitting of cubic spline curves based on the signal's maximum and minimum value point. Since the signal at the end is usually not the extreme point, the cubic spline curves can easily be diverged at the two ends of the data sequence. In fact, the screening process of EMD also decides that the divergence is not just limited to the small scope of the two ends, but can be spread inward

during the screening process of various IMFs, thus resulting in large distortion of the final result. This is the EMD end effect (Trad D, 2011).

3.2 Improvement of the EMD end effect

Currently, the end effect inhibition method is achieved mainly through the seeking of the proper extreme points at the two ends of the signal. In this way, the envelope curve obtained through the signal fitting can envelop the whole signal. The linchpin to find the proper extreme points is to clarify the change tendency of the original signal waveform at the two ends. Since the EEG changes have strict rules to follow, it is much more proper to predict the overall changes of the signal according to the signal's change tendency at the two ends. If a section of wavelet is obtained within the signal, the change tendency of the waveform is the most similar to that of the waveform around the end. Then the former and latter waveform of the wavelet can be used to define the other waveforms. The similarity degree of the waveforms of the two ends can be measured through the similarity of the magnitude and shape (Fang ji Wu, 2008).

Definition of several concepts: (1) Definition of the distance between two points:

$$D_i = x(t_1) - y(t_2) \tag{7}$$

(2) Definition of the matching distance between two sections of curves:

$$D_{dis} = \sum_{i=1}^{n} \left| x(t_1 + l\Delta t) - y(t_2 + i\Delta t) \right| \tag{8}$$

Where, n stands for the local signal length; $\{ x(t_1 + \Delta t), x(t_1 + 2\Delta t), ..., x(t_1 + n\Delta t) \}$ For the local signal on $x(t)$; $\{ y(t_2 + \Delta t), y(t_2 + 2\Delta t), ..., y(t_2 + n\Delta t) \}$ for the local signal on $y(t)$; and Δt for the minimum time interval.

From the above equation, it can be known that, the smaller the matching distance between the two sections of curves, the higher the similarity of the magnitude of the two.

(3) Definition of the waveform similarity coefficient between the waveforms on the two ends:

$$\rho(x(t), y(t)) = \frac{\text{cov}(x(t), y(t))}{\sqrt{\sigma(x)} \sqrt{\sigma(y)}} \tag{9}$$

Where, cov($\cdot$) stands for the covariance and $\sigma(\cdot)$ for the variance $(0 \leq \rho \leq 1)$.

From the above equation, it can be known that the larger the similarity coefficient ρ of the two sections is, the higher the similarity between the shapes of the two.

The matching distance and the waveform similarity coefficient between the two sections of curves show the difference degree between the two sections of signals in terms of magnitude and shape. The two measuring standards are both superficial and cannot fully measure the similarity degree of the two sections of signals. This paper comprehensively considers the magnitude and shape of signals and defines the similarity coefficient of the two sections of signals (Chenxi Deng, 2007).

(4)Definition of the similarity coefficient of the two sections of signals

$$p = \frac{D_{uni}}{\rho + \varepsilon} \qquad (10)$$

Where, $D_{uni} = \dfrac{D_{dis}}{\max(D_{dis})}$

Stands for the normalized matching distance;

ε For the infinitesimal positive number, which can prevent the occurrence of the situation that the denominator is "0" when "ρ" is "0".

From the equation with the given similarity coefficient, it can be known that, the smaller P is, the closer the two sections of signals are to each other.

Prolongation of the data. Signal prolongation includes the prolongation of the left and the right end. Take the left end as an example to illustrate the method put forward in this paper.

Step1: Assume the original signal as $x(t)$ and label the left end of $x(t)$ as s_0. Label the subscript corresponding to all the maximum points of the signal $x(t)$ from left to right as $\{M_0, M_1, M_2, ... M_n\}$, the subscript corresponding to all the minimum points as $\{N_0, N_1, N_2, ... N_n\}$.; the waveform from s_0 to N_0 as ω_0 (ω_0 is a wave section waiting to be matched); the length as N; the length from s_0 to M_0 as L (it is assumed that the left end first has the maximum value point and then has the minimum value point) and $L < N$ (See Fig. 1).

Step2: Adopt M_0 as the reference point of ω_0 and conduct "n-1" translation matching from the left to the right. M_0 Is overlapped with the maximum value points $\{M_1, M_2, ..., M_n\}$. In total, "n-1" matching wave sections whose length is N

$\{\omega_1, \omega_2, \omega_3, ... \omega_n\}$ are generated, and the matching distance D_{dis} of $\{\omega_1, \omega_2, \omega_3, ..., \omega_n\}$, the waveform similarity coefficient ρ are calculated respectively.

Step3: Adopt the wave section when p_j is the minimum as the optimal matching waveform section in the calculation of p_j. Assume the coordinate of the maximum value of M_0 as M_p. Then the coordinate corresponding to S_0 is assumed to be $S_p = M_p - L$;;

Step4: From the former point of S_p, the practical waveform information is prolonged to S_0 in the left. The information prolongation length depends on the specific situation of the information processed. This paper assumes the length to be $2N$;

Step5: The above is the prolongation process of the left data end. The prolongation of the right data end is the same. After the data prolongation, the EMD algorithm can be adopted to decompose them and conduct follow-up processing.

4 EXPERIMENTAL RESULT

This paper adopts the experimental plan designed by data set III in the international BCI contest in 2013(right and left hand motor imagery).The data were obtained by Graz University of Technology through the above experimental plan.

The experiment chose 140 groups of EEG data as the research objects. The authentic classification of every test group was known for the convenience of dividing the training set and the test set, and the calculation of the classification accuracy. Every test group contained three channel data, including C3, C4 and CZ. The test length of each group was 9 s and the sampling frequency was 128 Hz. The motor imagery in the test started from t = 3. Therefore, the valid time period was from the 3rd s to the 9th s. In order to obtain better classification results, this paper conducts feature extraction and classification of 640 sampling points from the 4th s to the 8th s. According to the ERD/ERS features of the single hand motor imagery EEG signals, EEG energy values of different rhythms were adopted as the feature values as the input of the follow-up feature classifier for identification and classification (Martisius I, 2011).

Conducted the EMD algorithm of a group of test data and calculated the instantaneous frequency of various IMF components. The decomposition results are shown in Fig. 2.

According to the characteristics of the end effect improvement method, the first-order IMF components after the decomposition of EMD were extracted and their energy values were calculated as the input of SVM. Divided 140 groups of experimental data randomly into two groups with 70 groups of data in each group. One group was the

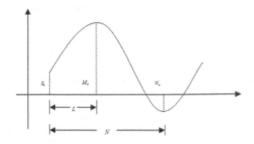

Figure 1. Sketch map of matching waveform.

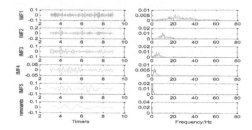

Figure 2.   Breakdown drawing of EMD.

Table 1.   The comparison of three algorithms.

Algorithm name	Time	Classification accuracy
Improved EMD algorithm	205.4 s	74.2%
Mirror image prolongation method	276.3 s	88.5%
Minimum Similarity distance method	98.4 s	87.1%

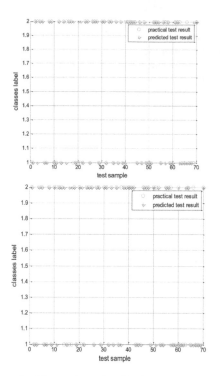

Figure 3.   The result of classification.

test set and the other was the training set. Two classification tests were conducted. The experimental result is shown in Fig. 3. Stands for the authentic classification of the test set, * for the predicted classification of the test set and the overlapped part of the two for the coincidence of the authentic classification and the predicted classification. In other words, the classification is either correct or wrong.

This paper conducted the same experiment of the unimproved EMD algorithm and the EMD algorithm based on the mirror image prolongation method. The results of the three algorithms are shown as below:

Based on the comparison, the motor imagery classification accuracy rate obtained through the improved EMD algorithm is 87.1%. The operation time of the program is one second longer than that of the unimproved EMD algorithm. The complexity of the original algorithm is reduced and the timeliness of the practical applications is improved.

5   CONCLUSIONS

This paper puts forwards an EMD end effect inhibition method based on the minimum similarity distance. Before the decomposition of the EMD algorithm, the signals were prolonged through calculating the similarity coefficient of the signal of the two ends and the signal of the whole wave section. The method is suitable for the processing of signals strictly adhering to specific rules. The experiment proved that the method can improve the processing effect of EEG

signals, increase the classification accuracy rate and reduces the time of the program operation.

REFERENCES

Chenxi Deng, Jian Wang, Jinfeng Fan, et al. A self-adaptive method dealing with the end issue of EMD [J]. Acta Electronica Sinica, 2007, 35(10): 1944–1948.

Fang ji Wu, Liang sheng Qu. An improved method for restraining the end effect in empirical mode decomposition and its applications to the fault diag nosis of larger otatingmachinery [J]. Journal of Sound and Vibration, 2008, 314:586–602.

Martisius I, Damasevicius R, Jusas V, Birvinskas D. Using Higher Order Nonlinear Operators for SVM Classification of EEG Data [J]. ELEKTRONIKA IR ELEKTROTECHNIKA, 2011, 119(3):99 –102.

P R Kennedy, K D Adams. A decision tree for brain-compute err interface devices [J]. IEEE Transaction on Neural Systems and Rehabil itat ion Engineering, 2003, 11(2): 148–150.

Rilling. G, Flandrin. P. One or two frequencies the empirical mode decomposition answers [J]. IEEE Trans. on Signal Processing, 2008, 42:1–11.

Trad D, Al-ani T, Monacelli E, Jemni M. Nonlinear and no stationary framework for feature extraction and classification of motor imagery [J]. IEEE International Conference on Rehabilitation Robotics, 2011, 6(4):1–6.

Yongjun Deng et al. EMD algorithm and boundary issue in Hilbert conversion [J]. Chinese Science Bulletin, 2001, 46 (3): 257–263.

Energy Science and Applied Technology – Fang (Ed.)
© 2016 Taylor & Francis Group, London, ISBN 978-1-138-02833-3

Optimizing analysis of all-terrain vehicle articulated mechanism based on the complex roads

Zhongliang Meng
School of Mechanical Engineering, Zhaozhuang University, Zhaozhuang, China

Gaojian Lv
School of Mechanical Engineering, Guizhou University, Guiyang, China

ABSTRACT: Aiming at the practical conditions of the all-terrain vehicle articulated steering mechanism under complicated road conditions, this paper applies the multi-body dynamics software Recurdyn based on the mathematical modeling and motion of the steering mechanism to analyze the motion and stress of key points and optimize the articulated point position, steering and pitching position of the model structure. After adjusting the structure of two articulated points' optimal position, the results show the maximum rotary speed and minimum stress after optimization.

Keywords: Articulated tracked vehicle, Virtual prototype, steering mechanism, articulated point

1 INTRODUCTION

The all-terrain articulated tracked vehicle is a universal platform for a fast maneuver engineering operation and multi-functions, which includes the front and rear vehicle body. The articulated system connects the front and rear vehicle body when it is running. It needs to realize the motion that features three and five degree-of-freedomdirections under different conditions, and meet the performance requirements of grade ability and across the vertical obstacles. The articulated system plays an important role in meeting the overall vehicles' traffic abilities, maneuver abilities and carrying capacities. The structure design and the analysis optimization of the articulated system are the technical difficulties for the key technology. Based on the established overall vehicle assembly model, the paper carries out the motion simulation through the motion module using the multi-body dynamic software Recurdyn, and optimizes the design of the articulated point position through designing modules. The overall vehicle model is shown in Figure 1. The overall vehicle comprises the front vehicle, the rear vehicle and five free articulated devices. The front vehicle includes the front vehicle body and the front vehicle chassis, and the rear vehicle includes the rear vehicle body and the rear vehicle chassis.

2 ESTABLISHING THE ARTICULATED SYSTEM MODEL (CHEN JIN-TAO, 2007, WANG GUO-QIANG, 1997, TATSURO MURO, 20001, CHEN SHU-YAN, 2008)

The simplified articulated system is shown in Figure 2. The five degree-of-freedom direction

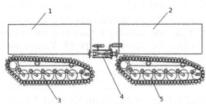

1. Front vehicle, 2. Rear vehicle, 3. Front vehicle chassis, 4. Five degree-of-freedom direction articulated device, 5. Rear vehicle chassis

Figure 1. Diagram for the overall vehicle structure.

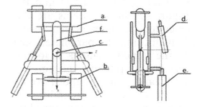

1. Connecting mechanism a, 2. Rotation mechanism b, 3. Articulated point c, 4. Front balance cylinder d, 5. Connecting mechanism f

Figure 2. The main view and the left view of the articulated system assembly.

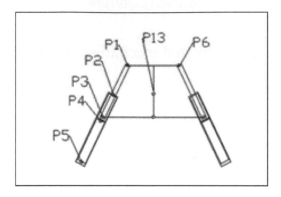

Figure 3. Steering system diagram.

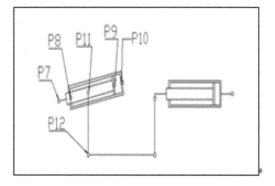

Figure 4. Pitching system diagram.

articulated steering mechanism is mainly composed of the front and rear balance cylinders, left and right steering cylinders and the articulated connection mechanism.

Three important parts in the optimization are design variables, objective functions and constraint conditions. In this optimization, the whole design variables include two parts: one is the distance between two points P1 and P6, as shown in Figure 3, for the steering system, and the other is the distance between P11 and P12, as shown in Figure 4. The two items are their respective independent parameters that need to be determined finally, and the parameters need to be set. Here, the standard rule is as follows: under the steering and pitching systems and within a certain distance range of the above two design variables, there exists the condition of minimum stress and maximum rotary speed (i.e. minimum turning radius and maximum pitch angle). In the optimizing setting, the constraint condition provides the range of a certain objective function value and sets the condition of the objective function.

3 STRUCTURE OPTIMIZATION BASED ON DIFFERENT ROADS

In fact, the all-terrain vehicle needs to adapt to various complicated road conditions. Therefore, this section, based on three different road conditions, considers the stress and turning radius of the hydraulic system as the optimizing condition, simulates the optimizing structure, and selects the optimal value (He Yan, 2008). This is because the all-terrain tracked vehicle needs to adapt to a variety of rough topography. The road conditions will be set in two aspects of this optimization. First, the terrain is a steep slope, which has the advantage of the pitching hydraulic system; second, three different road conditions are set, which includes ordinary roads, sandy roads and snowy roads. The motion on the road for tracked vehicles includes the pitching motion and the steering motion. It runs up and turns around at the same time. Therefore, the motion simulation includes two drives and needs to hold the time point of the pitching motion, namely the time is the hydraulic system of stretching out and drawing back. The period goes through three stages: the front vehicle body begins to run up and the hydraulic rod shrinkages. When it runs on the slope, the front vehicle track is on the road. When it runs up the slope totally, the hydraulic rod restores to its original position. The function is set to be step(time, 3, 0, 5, –100) + step (time, 5, 0, 8, 300) + step (time, 8, 0, 10, –200). The time function for the steering system is set to be step (time, 9, 0, 13, and 150). The simulation motion time is set to be 15 seconds.

4 THE POSITION OPTIMIZATION OF THE ARTICULATED POINT

The objectives of the optimization are to make the articulated mechanism turn towards the articulated point P1 and the pitching articulated point P7 meet the requirements of stress under three road conditions. In other words, the objectives are to make the power provided by the hydraulic system within the allowed conditions, make the steering articulated point P13 and pitching articulated point P11 meet the requirements of the rotary speed, and make the turning radius and the pitching angle within the allowed scope (J. Yamakawa, 2004). The two design variables, the distance between P1 and P6 and the distance between P11 and P12, are the minimum values (both values for stress and the rotary speed are maximum), thus we can receive the maximum stress of the articulated points P1 and P7 and the maximum rotary speed of the articulated points P11 and P13: On the sandy road: FP7 = 162820.276 N, WP7 = 0.214 rad/s,

FP1 = 329440. 329 N and WP13 = 0.239 rad/s. On the ordinary road: FP7 = 192824.160 N, WP7 = 0.214 rad/s, FP1 = 396841.102 N and WP13 = 0.239 rad/s. On the snowy road: FP7 = 213204.477 N, WP7 = 0.214 rad/s, FP1 = 346803.983 N and WP13 = 0.239 rad/s.

Through analyzing the data sets, we can know that the stress of the articulated point P1 in the steering system on the ordinary road is maximum, so the optimization of the steering system shall be carried out on the ordinary road. In the pitching system, the stress of the articulated point P7 on the snowy road is maximum, so the optimization of the pitching system shall be carried out on snowy road. The two articulated points P7 and P13 have the same rotary speed under three road conditions, so the choice of road has no effect on the optimization of the rotary speed and will not be considered.

5 OPTIMIZATION OF THE STEERING SYSTEM AND THE PITCHING SYSTEM

Before optimization, the distance between P1 and P6 is 380 mm, the maximum stress of P1 point is 276657364 N, and the maximum rotary speed of P13 point is 0.199 rad/s. In practice, we hope that the maximum stress value of P1 is less than 360000 N and the rotary speed of P13 is greater than 0.21, so we find the best position through optimizing the distance between P1 and P6. First, we set the design parameter. The variable is the distance between P1 and P6 with the modification range from 280 to 500. We then set the performance index, the constraint and the objective in the above range of the minimum internal stress and the maximum rotary speed, respectively. The best distance obtained through optimizing calculation is 203.32 mm. The calculated results are shown in Figure 5. After 25-time optimizing calculation, the results show that the minimum stress is 322481 N and the maximum rotary speed is 0.238.

By using the same method, we optimize the distance between P11 and P12, and find that the max-

imum stress value of P7 point is less than 160000 N and the rotary speed is greater than 0.175. The distance through optimization is 190 mm.

6 RESULT VERIFICATION

We find the two optimizing distances to be 203 mm and 190 mm, respectively. After setting the distance between P1 and P6 and the distance between P11 and P12 to be the values of simulation research on the ordinary and snowy roads, respectively, the results are found to be consistent with the four objective parameters. We need to choose the roads to verify the results. Through the road simulation calculation, the results are obtained, as shown in Figure 6 to Figure 8, from which we can see that the result satisfies the purpose of the optimization.

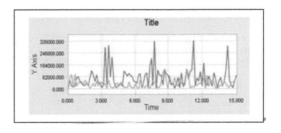

Figure 6.   The stress result figure after the steering system optimization.

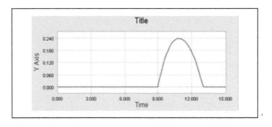

Figure 7.   The rotary speed figure after the steering system optimization.

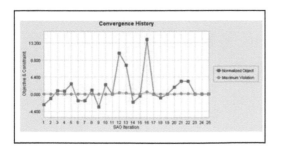

Figure 5.   The optimizing result for the steering system.

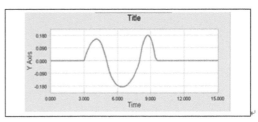

Figure 8.   The rotary speed result figure after the pitching system optimization.

7 CONCLUSION

The paper applies the stress analysis and theory simplification using the multi-body dynamics software Recurdyn to determine the key points of the stress of the articulated steering mechanism of all-terrain vehicles that can run normally under the complicated road conditions. The paper analyses the motion and stress of the key points according to the practical motion, carries out the motion simulation analysis on this basis and optimizes the design of the articulated position of the articulated point. Based on the optimization of the steering system and the pitching system on the ordinary concrete road, snowy road and sandy road, it is found that the rotary speed is maximum and stress is minimum for the articulated point of two systems under three different types of road.

REFERENCES

Chen Jin-tao, Li Li & Wang Jun-jie, Research on Stationary Turning Performance of Articulated Tracked Vehicles. Computer Simulation, 24(12), pp.155–158,2007.

Chen Shu-yan. The crawler walking device configuration and mobility of mobile robot research. Yangzhou: Yangzhou University press, 2008.

He Yan, Zhang Xiao-yan, Dynamics Modeling and Simulation for Tracked Vehicles, Proceedings of the 27th Chinese Control Conference, 2008.

James H. Lever, Daniel Denton & Gary E. Phetteplace, Mobility of a lightweight tracked robot over deep snow, Journal of Terramechanics, 43(6), pp.527–551,2006.

Yamakawa, J. & Watanabe, K. A spatial motion analysis model of tracked vehicles with Torsion bar type suspension, Journal of Terramechanics, 41(2), pp.113–126,2004.

Tatsuro Muro, Soiehiro Kawahara & Takahiro Mitsubayashi, ComParison between Centrifugal and vertical vibro-compaction of high-lifted decomposed weathered Granite sandy soil using a tracked vehicle. Journal of Terramechanics, 38(7), pp.15–45,2001.

Wang Guo-qiang, Chen Yue-sun & Ma Ruo-ding, articulated structure parameters on turning performance of tracked vehicle. Journal of jilin university of technology, 86(02), pp. 7–12,1997.

Energy Science and Applied Technology – Fang (Ed.)
© 2016 Taylor & Francis Group, London, ISBN 978-1-138-02833-3

Research on feature selection method in Chinese text automatic classification

Ying Hong & Zengmin Geng
Computer Information Center, Beijing Institute of Fashion Technology, Beijing, China

ABSTRACT: This paper introduced the importance and workflow of Chinese web page classification. It studied the defects of chi-square statistic algorithm and improved it. At last, it verified the improved chi-square statistic algorithm combined with Bayes algorithm through a series of experiments. The experimental results show that the improved algorithm improved the accuracy of Chinese web page classification.

Keywords: text classification, feature selection, Bayes algorithm, chi-square statistic

1 INTRODUCTION

Chinese web page automatic categorization analyzes characteristics of web page unclassified and divided it into corresponding category (Kousu Ling, 2007). Web page automatic classification has the advantage of rapid classification without human intervention. It has been widely used in user behavior analysis, personalized recommendation service, precision marketing and other fields. It is a very practical technology.

This paper focuses on the study of Chinese web page classification techniques. It introduced the related technologies of Chinese web page classification. It studies on feature selection algorithm deeply and puts forward the corresponding improvement ideas. Finally, it has verified the validity of improvement ideas through experiments.

2 WORKFLOW OF CHINESE TEXT CLASSIFICATION

The Workflow of Chinese web page classification is shown in Fig. 1:

As we can see from Fig. 1, the workflow of Chinese web page classification is divided into two parts: training process and classification process (Jianming Cui, 2013). In the training process, we will select feature in training set and get a collection of feature items after Chinese segmentation. Then, we will create a feature vector space and all instances are represented as vectors. At last, we use classification algorithm to construct classifier and train the classifier using the training set. In the classification process, the web page to be classified

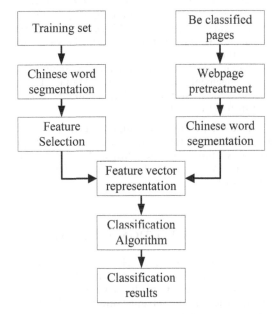

Figure 1. The workflow of Chinese web page classification.

will be expressed in vector form after a Chinese word processing. Then, we use the trained classifier to classify them.

3 CHINESE WORD SEGMENTATION

Chinese word segmentation is a unique concept in Chinese text classification (Kangxin Fan,

2009). There are no obvious segmentation signs between words in Chinese text. The computer needs to find out automatically the dividing line between words using the Chinese word segmentation. Therefore, the Chinese word segmentation is the foundation of Chinese web page classification technology.

The Chinese word segmentation tool used in this paper is Pangu word components. The version number is V2.1.0.0, It is an open source word components. The main features are:

1. Pangu word components can automatically identify some of the unknown word which is not in the dictionary.
2. Pangu word components can solve the problem of segmentation ambiguity according to frequency.
3. Pangu word components can effectively identify Chinese place names and names.
4. Punctuation, conjunctions, auxiliary word and other needs to be filtered out in Chinese word segmentation. Pangu word components offer a file named StopWord.txt. If users want to filter the words, they just need to add the words to the file.
5. Pangu segmentation provides a dictionary management tool named Dict Manag. Users can add, modify and delete dictionary words through this tool.

As used herein, the Chinese stop word list is released by Harbin Institute of Technology Social Computing and Information Retrieval Research Center, which contains 767 Chinese stop words.

4 FEATURE SELECTION ALGORITHM

Feature selection means to select the representative words from all words contained in a document to constitute the set of feature items (Xiaoying Su, 2014). The feature set must meet the following three requirements:

1. Feature items can accurately describe the document content.
2. Feature items can distinguish the target document from other documents.
3. Make the dimension of feature vectors as small as possible.

The feature selection algorithms commonly used are Document Frequency (DF), Information Gain (IG), Mutual Information (MI), chi-square (CHI), etc. The results show that: CHI algorithm is best, followed by the IG algorithm, MI algorithm and DF algorithm is the worst (Peng Di, 2014).

5 THE CHI-SQUARE STATISTIC ALGORITHM (CHI)

The chi-square statistic algorithm illustrates the importance of features by measuring the degree of correlation between feature items t and category c. The prerequisite is assuming that the relationship between c and t meets x^2 distribution. It is shown in equations (1):

$$x^2(t,c) = \frac{N(AD-CB)^2}{(A+C)(B+D)(A+B)(C+D)} \qquad (1)$$

In the formula, A is document frequency included in category c and has the feature item t. B is document frequency which has the feature item t but is not included in category c. C is document frequency included in category c and does not include the feature item t. D is document frequency does not be included in category c and does not has the feature item t. N is the total number of training document.

High frequency characteristics that appear in any of the categories are determined having a contribution for judgment on the category during feature extraction using the chi-square statistic algorithm (Haifeng Liu, 2013). When t and c are independent, the value of x^2 is 0. The larger the value of x^2, the higher the degree of correlation between characteristics t and category c.

6 IMPROVED TEXT FEATURE EXTRACTION ALGORITHM

The correlation of characteristics and categories are two situations: positive and negative (Zhe Zhao, 2013). When $AD-CB > 0$, features and categories are related. When $AD—CB < 0$, features and categories are negative. Positive correlation capability and negative correlation capability of features and categories are treated equally when we calculate value of $x^2(t, c)$ using equations (1). This paper considers selecting feature items which positive correlation with the category can get better classification results. The chi-square statistic algorithm ignores the impact of word frequency. So in this paper, we has been improved the calculation of $x^2(t, c)$. The improved algorithm we called NCHI is shown in equations (2):

$$x^2(t,c) = \frac{N(AD-CB)^2}{(A+C)(B+D)(A+B)(C+D)} \log fq(t,c) \qquad (2)$$

Among them, $fq(t,c)$ is frequency of feature words t in category c. The purpose of multiplied by the frequency in the formula is to improve the chi-square value of high-frequency words in the same category.

7 EXPERIMENTAL ANALYSIS

Training and test sets preparations. In experiments, we use Bayes algorithm and conducted two experiments to verify the advantage of improved text feature extraction algorithm. In this paper, we use news dataset provided by Sogou laboratory as Experimental corpus. We select six categories amount to 4500 corpus including 2756 training examples and 1744 test examples from the data set.

Evaluating index. In this paper, the three indicators are used to evaluate the classifier on a single class of classification performance. They are precision (p), recall (r), and F1 value.

$$p = \frac{a}{a+c} \qquad (3)$$

$$r = \frac{a}{a+b} \qquad (4)$$

Among them, parameter a is the number of relevant records retrieved. Parameter b is number of relevant records not retrieved. Parameter c is the number of irrelevant records retrieved. F1 is weighted average of precision and recall. It is shown in equations (5):

$$F1 = \frac{2 \times r \times p}{r+p} \qquad (5)$$

Experimental results and analysis. Bayes algorithms and the original chi-square statistic algorithm are used in the first set of experiments. Bayes algorithm and the improved chi-square statistic algorithm are used in the second set of experiments. The experimental results are shown in table 2.

As we can see from Table 2, the classification accuracy has been increased obviously when we used the chi-square statistic improved algorithm.

Table 1. The distribution of text category and the number of samples set.

Text category	Number of training sets	Number of test sets
News	345	211
Finance	331	215
Sports	352	231
Education	342	222
IT	340	212
Car	340	209
Military	352	220
Health	354	224

Table 2. The distribution of text category and the number of samples set.

The experimental group	Feature selection algorithm	Average precision (%)	Average recall (%)	F1 value (%)
First	CHI	73.32	75.41	71.83
Second	NCHI	82.26	81.31	82.12

8 CONCLUSION

This paper introduced the Chinese word segmentation process and studied the feature selection algorithm. It improved Chi-square statistic algorithm and designed two set of experiments to demonstrate the effect of the improved algorithm. The results show that the improved algorithm improves the accuracy of classification.

ACKNOWLEDGEMENT

This work was financially supported by the Foundation for Beijing teacher team construction-youth with outstanding ability project (No.YETP1414), the twelfth five-year plan important subject of Beijing education science research (No. AJA11174) and the scientific research program of Beijing institute of fashion technology (No.2012 A-17).

REFERENCES

Haifeng Liu, Zhan Su, Shousheng Liu: Improved CHI text feature selection based on word frequency information. Computer Engineering and Applications, 2013, 49(22) p.110–114(In Chinese).

Jianming Cui, Jianming Liu, Zhouyu Liao: Research of text categorization based on support vector machine. Computer Simulation, 2013, 30(2) p. 299–302(In Chinese).

Kousu Ling: Research on feature selection in Chinese text classification. Computer Simulation, 2007, 24(3) p.289–291(In Chinese).

Kangxin Fan: Design of NB combination text classifier based on various feature selection. Computer Engineering, 2009, 35(24) p.191–193(In Chinese).

Peng Di, Liguo Duan: New Naïve Bayes text classification algorithm. Journal Data Acquisition and Processing, 2014, 29(1) p.71–75 (In Chinese).

Xiaoying Su, Yanpeng Hu, Junhui Yang, Ming Li: A new probabilistic classifier design for text categorization. Computer Technology and Development, 2014, 24(3) p. 46–48(In Chinese).

Zhe Zhao, Yang Xiang, Jisheng Wang: Text classification based on parallel computing. Journal of Computer Applications, 2013, 33(S2) p.60–62, 66(In Chinese).

Energy Science and Applied Technology – Fang (Ed.)
© 2016 Taylor & Francis Group, London, ISBN 978-1-138-02833-3

Numerical simulation for contamination of batch transportation in reducer pipe

Tao Yu, Peng Du, Xi Chen & Wen Yuan
Northeast Petroleum University, Heilongjiang, China

ABSTRACT: This paper simulated and analyzed the cause of the mixed oil segment in a straight pipe. The goal of this paper is to show how the concentration distribution of a contaminated product changes when the pipeline diameter changes by connecting to a reducer. Based on a mathematical model of the contamination in batch transportation, using FLUENT, the numerical simulation was accomplished for batch transportation of diesel and gasoline in a straight pipe, a concentric reducer pipe and an eccentric reducer pipe. The analysis was carried out to determine the concentration distribution of the contaminated product in these pipes. The results show that the length of the mixed oil segment in the concentric reducer pipe is shorter than that in the eccentric reducer pipe. Thus, the concentric reducer is an ideal choice.

Keywords: Numerical simulation; contamination regulation; reducer pipe

1 INTRODUCTION

Two kinds of oil in a transportation batch pipeline blends at the contact surface, and the concentration distribution of the contaminated product is influenced by various factors such as the physical properties of oil and the operating conditions of oil. Consequently, the control of the contaminated oil volume and the cutting of contaminated soil becomes a big problem. So, the study on the effect of different pipes on the regulation of contamination is very important.

This article took diesel and gasoline transportation for an example, combined with the engineering practice, and made a numerical simulation for the process of oil blending through a straight pipe, a concentric reducer pipe and an eccentric reducer pipe. It analyzed the regulation of the contamination in these three pipes when transported in different orders. The results provided a certain theory basis for the choice of different reducer fittings in pipeline construction, which have a certain reference value.

2 MODEL FOUNDATION

2.1 Mathematical models

Continuity equation is given by

$$\frac{\partial \rho}{\partial t} + \text{div}\left(\rho \vec{u}\right) = 0 \qquad (1)$$

Momentum conservation equation is given by

$$\frac{\partial(\rho u_i)}{\partial t} + \text{div}\left(\rho u_i \vec{u}\right) = \text{div}\left(\mu \text{grad} u\right) - \frac{\partial p}{\partial x_i} + S_i \qquad (2)$$

where S_i is the generalized source item in the momentum conservation equation

Energy equation is given by

$$\frac{\partial(\rho T)}{\partial t} + \frac{\partial(\rho u T)}{\partial x} + \frac{\partial(\rho v T)}{\partial y} + \frac{\partial(\rho w T)}{\partial z}$$
$$= \frac{\lambda}{C_p}\left\{\frac{\partial^2 T}{\partial x^2} + \frac{\partial^2 T}{\partial y^2} + \frac{\partial^2 T}{\partial z^2}\right\} \qquad (3)$$

In the process of transportation, in order to reduce the contaminated oil volume, the oil flow should be made turbulent, so the numerical simulation adopted the standard $k - \varepsilon$ model.

K is given by the following equation:

$$\rho\frac{\partial K}{\partial t} + \rho u_j\frac{\partial K}{\partial x_j} = \frac{\partial}{\partial x_j}\left[\left(\mu + \frac{\mu_t}{\sigma_k}\right)\frac{\partial K}{\partial x_j}\right]$$
$$+ \mu_t\frac{\partial u_i}{\partial x_j} - \rho\varepsilon \qquad (4)$$

ε is given by the following e:

$$\rho\frac{\partial\varepsilon}{\partial t}+\rho u_j\frac{\partial\varepsilon}{\partial x_j}=\frac{\partial}{\partial x_j}\left[\left(\mu+\frac{\mu_t}{\sigma_\varepsilon}\right)\frac{\partial\varepsilon}{\partial x_j}\right]$$
$$+\frac{c_1\varepsilon}{K}\mu_t\frac{\partial u_i}{\partial x_k}\left(\frac{\partial u_i}{\partial x_k}+\frac{\partial u_k}{\partial x_i}\right)-\frac{c_2\rho\varepsilon^2}{K} \qquad (5)$$

$$\mu_t=C_\mu\rho\frac{k^2}{\varepsilon} \qquad (6)$$

$$\varepsilon=v\overline{\left(\frac{\partial u_i'}{\partial x_k}\right)\left(\frac{\partial u_i'}{\partial x_k}\right)} \qquad (7)$$

2.2 Geometry models

The size of the three kinds of pipe is shown in the below figure, each of whose length is 10.5 m. The length of the reducer is 0.5 m, whose diameter varied from 500 mm to 350 mm. The grid adopted a structured hexahedral mesh.

2.3 Physical property

The physical properties of gasoline and diesel adopted is shown in the below table.

2.4 Settings

We adopted the multiphase flow (VOF) model for calculation. Briefly, we chose the "velocity inlet" as inlet boundary conditions, the "outflow" as outlet boundary conditions and standard solid non-slip boundary conditions.

In order to make the simulation results close to the actual one, first, we chose the steady model and set 0 to the volume fraction of the later oil. When reaching a convergence, we then chose the unsteady model and set 1 to the volume fraction of the later oil. The simulation for the contamination process of the transportation products was accomplished.

3 NUMERICAL SIMULATION AND ANALYSIS

3.1 Numerical simulation of straight pipe

When gravity is not considered, the mixed oil distribution is symmetric, the later oil inserted into the

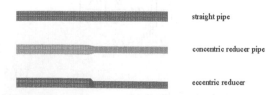

Figure 1. The size of the three kinds of pipes.

forward oil in a wedge shape. The oil head of the wedge gradually lengthens. One reason is that the flow velocity on the cross section of the pipe is uneven and forms convection diffusion. Another reason is that there is turbulent diffusion along the radial and axial directions. Turbulent diffusion is more severe than molecular diffusion. The radial turbulence diffusion makes the liquid particle exchange severe, "destroying the oil wedge head" to make different kinds of oil product blended. Turbulent diffusion makes the concentration on the cross section to be uniform, which helps to suppress concentration diffusion in the axial direction, and to reduce the blend of oil (Steinbrenner J. P. 1990).

When gravity is considered, the mixed oil distribution is not symmetric in the radial direction. When the density of the later oil is greater than that of the forward oil, due to the density difference, the tip of the wedge is close to the lower wall of the pipe (Benek J. A. 1985).

3.2 Numerical simulation and analysis of three kinds of pipes

Gasoline flows ahead of diesel.

When gasoline flows ahead of diesel, at the moment of 6 s, the oil mixed acutely in the region of the reducer, and the concentration of the mixed oil increased. This is because the diameter decreased, the velocity increased, and the two kinds of oil exchange severely (Versteeg H. K. 1995).

The curve shows the contamination concentration distribution in the axial direction. The concentration changed acutely in the region of the

Table 1. Physical properties of gasoline and diesel.

Physical properties	Gasoline	Diesel
Density(kg/m³)	720	830
Viscosity(m²s⁻¹)	$4\times e^{-6}$	$6\times e^{-6}$

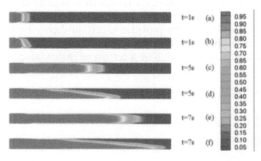

Figure 2. The cloud picture of the concentration distribution in the straight pipe. (a), (c), (e): Took not gravity into account (b), (d), (f): Took gravity into account

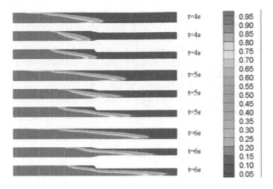

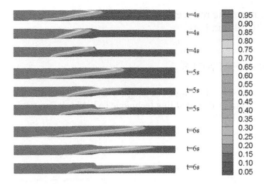

Figure 3. The cloud picture of the concentration distribution on the vertical section when gasoline flows ahead of diesel.

Figure 5. The cloud picture of the concentration distribution on the vertical section when diesel flows ahead of gasoline.

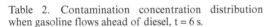

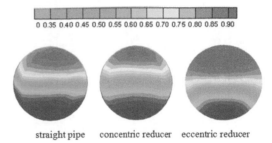

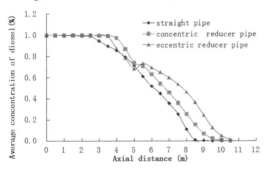

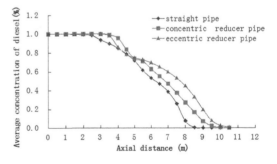

straight pipe   concentric reducer   eccentric reducer

Figure 4. The cloud picture of the concentration distribution on the X = 5.25 m section when gasoline flows ahead of diesel.

Figure 6. The cloud picture of the concentration distribution on the X = 5.25 m section when diesel flows ahead of gasoline.

Table 3. Concentration distribution of contamination when diesel flows ahead of gasoline, t = 6 s

Table 2. Contamination concentration distribution when gasoline flows ahead of diesel, t = 6 s.

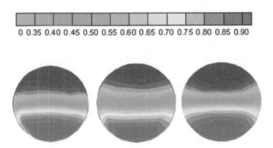

reducer. The length of the mixed oil segment is shorter in the concentric reducer pipe than that in the eccentric reducer pipe.

Diesel flows ahead of gasoline.

When diesel flows ahead of gasoline, the length of the mixed oil segment is shorter in the concentric reducer pipe than that in the eccentric reducer pipe.

4 CONCLUSIONS

Using the FLUENT software, the simulation of the concentration distribution of contamination was accomplished and the results were found to be reasonable.

When the pipe diameter decreased, the length of the mixed oil segment in the concentric reducer is shorter than that in the eccentric reducer. So, the concentric reducer can be an ideal choice to reduce the oil mix loss.

REFERENCES

Benek J.A., Buning P G, Steger J L, 1985. A 3-D Chimera grid embedding technique [A]. AIAA Paper [C], 85–1523.

Influence of variable diameter pipe in cold and hot crude oil transportation [J]. Oil & gas storage and transportation, 2011, 30(3):183–186.

Steinbrenner J.P., Chawner J R, Fouts C L, 1990. Multiple block grid generation in the interactive environment [A]. AIAA Paper [C], 90–1602.

Versteeg H.K., Malalasekera W, 1995. An Introduction to Computational Fluid Dynamics: The Finite Volume Method [M]. London: Longman Group Ltd, 25–37.

Energy Science and Applied Technology – Fang (Ed.)
© 2016 Taylor & Francis Group, London, ISBN 978-1-138-02833-3

An ant colony optimization algorithm for the MANET based on environment analysis

Shiping Fan & Jinchuan Yan
Information and Communication Department, Chongqing University of Posts and Telecommunications, Chongqing, China

ABSTRACT: This paper presents a modified algorithm that makes some improvement in the control of special nodes, the structure of data packet, the node routing table, and the pheromone update mode and path establishment. The algorithm proposes special node concept and makes special control on nodes with a high mobility, low energy and altitude delay. In addition, it changes the traditional way that is based only on the hop count information update. This algorithm contains two kinds of parallel updating mechanism, which are based on path and local information to balance the load effectively. NS-2 simulation is used to prove whether the new algorithm improves the survival time of the network.

Keywords: mobile Ad-hoc network; routing; ant colony optimization

1 INTRODUCTION

The Mobile Ad-hoc Network (MANET) is a self-organizing network with mobility. As a multi-hop, non-center and self-organizing wireless network, the terminal can move freely and enable communication in the network. It plays a great role in the sensor network and the IOT field. Mobility can be in the network process, so that the topology changes frequently and dramatically.

The improved routing protocol based on the ant colony algorithm will consider network congestion degree, information transmission distance, the residual capacity and other information. In the AntNet algorithm, after forward ants reach their destination node, they transform into backward ants that leave pheromone, changing weight selection on the link. The ARA broadcasts the transition of forward ants with a less overhead. If the intermediate node receives repeated forward ants, then they are discarded. The ARAMA algorithm is based on the node information in the request packet in perception, the detection and evaluation of node information in the path, the change in routing information from the intermediate node to the destination node, and the routing resource balance of related nodes. The CAARAMA algorithm proposed by Jhong on the basis of the signal strength is based on the perceived neighbor node communication range whether it has been removed to choose alternative paths in advance to reduce packet loss. The AntHocNet protocol is an on-demand routing of the multi-path hybrid routing protocol. The EAAR protocol is based on the AntHocNet update pheromone to add the destination node energy and the minimum number of references.

2 DESCRIPTION OF ALGORITHM

According to the network with limited energy and higher mobility, the paper presents an ant colony optimization algorithm based on the environment analysis.

This routing protocol takes the distance, residual capacity and delay into consideration. It balances the load of network capacity and reduces the death node capacity to a certain extent.

It controls the special node with a high mobility, low energy and altitude delay.

The multi-path algorithm stores a plurality of routing information, such as timely feedback routing, recovery routing, and reduce the number of packet loss. The control of ant colony algorithm does not need to send out the broadcast control information, mostly to reduce the cost.

This algorithm in the bidirectional path changes the pheromone concentration. When the destination node is normal, it will choose a few good reverse routing paths to establish the reverse routing. It is effective in preventing the two-way channel capacity, which is not balanced in the environment.

2.1 *The pheromone update*

In this algorithm, pheromone concentration plays a very important role in path choice. We use two pheromone tables: routing table and neighbor table. The routing table (Table 1) is used to record the next hop information and pheromone. The pheromone update mainly includes the jump update and the normal update. The jump update is used when a new ant packet arrives, while the normal update changes with intervals.

The neighbor table (Table 2) is used to record the energy and probability of neighbors.

We define the effective energy $E_e^i(t)$, the node congestion rate $S_a^i(t)$ and the rate of change $C_m^i(t)$ as follows:

$$E_e^i(t) = \frac{E_r^i(t) - E_p^i}{E_{max}^i} * \alpha \tag{1}$$

$$S_c^i(t) = \frac{L(t)}{L_{max}} * \beta \tag{2}$$

$$C_m^i(t) = \frac{\left|N^i(t) \bigcup N^i(t-1) - N^i(t) \bigcap N^i(t-1)\right|}{N^i(t)} * \theta \tag{3}$$

We summarize the effective energy $P_e(t)$, the node congestion rate $P_a(t)$, the rate of change $P_m(t)$, and the probability of selection P_m in the path as follows:

$$P_e(t) = \prod_{i=1}^{k} E_i^e(t) \tag{4}$$

$$P_a(t) = \sum_{i=1}^{k} S_a^i(t) \tag{5}$$

$$P_m(t) = \sum_{i=1}^{k} C_m^i(t) \tag{6}$$

$$P_m = E_m * \sum_{i=1}^{m} \tau_{im} \tag{7}$$

Table 1. Routing Structure.

Destination node	Next hop node				
	1	2	3	...	M
1	τ_{11}	τ_{12}	τ_{13}	...	τ_{1m}
2	τ_{21}	τ_{22}	τ_{23}	...	τ_{2m}
3	τ_{31}	τ_{32}	τ_{33}	...	τ_{3m}
...	...	...	...	...	...
N	τ_{n1}	τ_{n2}	τ_{n3}	...	τ_{nm}

Table 2. Local information table.

	Next hop node			
	1	2	...	M
Pheromone	$> \tau_{i1}$	$> \tau_{i2}$	...	$> \tau_{im}$
E	E_1	E_2	...	E_m
P_m	P_1	P_2	...	P_m

The normal update is defined for each time period:

$$t_{di}(t + \Delta t) = (1 - \rho) * \tau_{di} \tag{8}$$

For each period of time Δt, all pheromone paths regularly volatilization ρ is expressed by the volatile coefficient of pheromone. The volatile way makes the path information without the use of long-term hormone gradually. When the path is selected, the update mechanism can be expressed as:

$$t_{di}(t) = (1 - \tau) * \tau_{di} + \tau * \Delta \tau_{di} \tag{9}$$

Upon the arrival of a new packet time according to the routing decision, all the pheromone updates are performed. τ is the weight factor of path information. Larger values of τ indicate that the path information plays a bigger role in the whole process.

τ_{di} represents the incremental in the closest routing:

$$\Delta \tau_{dj} = \left(\supset_1 P_e + \frac{\supset_2}{P_{c+1}} + \frac{\supset_3}{P_{m+1}} + \frac{\supset_4}{D_j} \right) * Q \tag{10}$$

2.2 Route establishment

This algorithm is a hybrid multi-path algorithm. In the establishment of routing, it is accomplished through the active on-demand establishment mode.

Source Node: When the source S needs to send data packets to nodes D, it checks the source S node routing table whether it contains the path that can reach routing D first. If so, then it selects the routing according to the following equation:

$$\begin{cases} \text{next two hops} & q > q_0 \\ P_i = \dfrac{P_i}{\sum P_i} & q \leq q_0 \end{cases} \tag{11}$$

where q_0 is the value and q is the random number.

If the node does not point to a destination node routing, then it sends two forward ants FANT to establish routing by local information.

Receives: when a node receives a forward ant FANT, first, it checks the forward ant routing sequence whether they are the same. Forward ants by path probability select the corresponding routing sequence. When the sequence is the same as the current sequence, it shows that it forms a loop. The path-finding ants become an invalid message, and they are discarded directly. Then, it checks the number of hops of forward ants. If forwarded message are more than a certain number of hops, it indicates that this routing will be forwarded by a lot of nodes, including the utilization ratio of routing, the timeliness of poor quality, and energy consumption. If it is not appropriate to continue the rest of forwarding node energy, the ants will be discarded.

When the forward ants meet the conditions, the routing carries on the inspection of the receiving node. It checks whether the destination node receives the FANT. When the received nodes are the intermediate nodes, it operates at intermediate nodes. When the received nodes are the destination nodes, it is not necessary to carry out further routing operation, the relevant operation directly to destination nodes.

The wireless self-organized network communication node is often a complicated work environment; the communication node is in an abnormal state. Greater influences on the communication mainly include a low energy node, variable nodes and a high congestion rate. As the middle route of some control over the actual need in the process of this special node.

When the remaining energy $E_e^i(t) \leq a$, it may increase the death. When the congestion rate $S_a^i(t) \geq b$, it easily leads to excessive delay. When the mobility rate $C_m^i(t) \geq c$, information is easy to lose. If the node is the destination node, then the reverse routing is established. If the node is the intermediate node, it is directly discarded. The threshold value is defined according to the actual one.

Intermediate node: If the received node is the intermediate node, it checks whether the node is special or not. As an intermediate node, when the rate of change is too large, then the nodes are often mobile, not as a long-term routing. When the residual energy of node is too small, then this node will die, and cannot send too much information. When the node congestion rate is too high, the intermediate node queuing delay is too long, but it is also easy to lose information. If the check is the intermediate node, then it is discarded directly. If it is not the intermediate node, then it checks whether it is related to the routing table containing the routing path reaches D. If so, it will select routing to send in accordance with Eq. (11). If the node does not point to a destination node routing, it sends 2 forward ants looking for routing to its neighbor nodes. It checks whether the own routing table is the first time to receive information from the source node. If this is the first time it receives the source node information, then it establishes the reverse routing. If it is not first received packet to the source node, then the existing routing and comprehensive will be compared. If the condition is satisfied, then this routing is accepted, an alternate path will do, or otherwise it is discarded. Thus, pheromone update is performed.

Destination node: It checks whether it is a special node. When the node is a special node that does not need to wait for the next forward ants, it establishes reverse routing directly, updates the reverse

path information, and rejects all of the following forward ants. When the node is not a special node, it checks whether the first one receives the source node routing information. If it does receive, it puts the information into the routing table. If not, a comprehensive comparison is made to meet the conditions of acceptance of this routing, to perform backup path selection. Otherwise, it is discarded. At a certain period of time, it will accept forward ants in accordance with the formula of comprehensive comparison of path information. After the end of the period, it chooses the best 3 and establishes the reverse routing. It checks whether the node receiving the reverse routing of BANT is the source node. If an intermediate node receives the establishment of reverse routing request message, it checks the pheromone update mechanism. Then, it joins and continues down a jump to send the reverse routing information. If the source node receives the reverse routing, it detects whether the destination node in its route is a special node. If so, then it sends the message directly, and does not wait. Otherwise, at the beginning of time, it waits for the rest of the reverse packets, a comprehensive comparison of the received path within a certain period of time is made, choosing the best path to send the message and for updating.

2.3 Routing maintenance and link failure

Network routing requires regular maintenance to ensure the accessibility of network nodes. When the maximum information of the node of a path element is less than the maximum value of 30%, it sends the route request message. According to the recovery, reversing the ants will modify the pheromone concentration value. The path information only in the process of updating pheromones, no additional by random transmit way of finding new routing. When all link pheromones of the nodes are lower than the initial pheromone value of 30%, it sends the request routing without feedback, judging that the routing has failed. It then deletes all existing routing, sends a broadcast message, and as a newly joining node it starts to seek the road. When all link pheromones of the nodes are lower than the initial pheromone value of 30%, it sends the request routing without feedback, judging that the routing has failed. It then deletes all existing routing, sends a broadcast message, and as a newly joining node it starts to seek the road. It broadcasts to neighbors routing and can search for a new route better, to avoid falling into the local optimal solution.

When the node is found with another node with a link failure, first, it determines whether the next node is only the next path to the destination routing, or whether it is the optimal route. If so, the upward jump node, respectively, sends link failure information. The last road is to the same judge and deal with. If not, then it deletes the node routing information in the routing table.

3 SIMULATION AND ANALYSIS

To test the algorithm proposed in this paper, a further comparative analysis of the relative performance agreement between the SARA, AOMDV and ARAMA in terms of delay, the number of death nodes, packet delivery ratio is performed.

The moving scene is 100 networks, the range of movement is 1500*1500 meters, nodes have a communication range of 250 meters, the simulation time is 100 s, and the network nodes are 10 m/s, 20 m/s, 30 m/s, 40 m/s and 50 m/s, respectively. The maximum pause time is 0 second. The number of CBR sources is 20 and the size of date packet is 512Byte. The MAC layer is 802.11. The transceivers for sending and receiving power consumption are 660 mw and 395 mw. Others in table 3.

Figure 1 shows the transmission delay at different speeds. The SARA and AntHocNet are much better than the AOMDV. The AntHocNet tries to send routing information to a small path delay. The SARA performs better, which is mainly reflected in two aspects: First, the SARA mechanism for a multi-path, the path to the destination node formed to be more than the other two forms. In the path selection on a particular node as a priority to avoid the intermediate node, it prevents the formation of the open circuit in the process. The second is the route maintenance mechanism, in which invalid routing can regularly check by comparing the two methods.

Table 3. Fixed parameters.

α	β	θ	ρ	τ	ꓛ	a	b	c
1.2	1	1	0.05	0.2	0.25	0.2	0.8	0.8

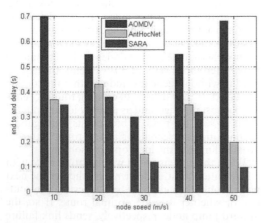

Figure 1. End-to-end delay change with node speed.

Figure 2 shows the delivery rate. The AntHoc-Net is based on the traditional routing algorithm, in which the pheromone will accumulate on a good path, and the selection will be a local busy road. The situation is particularly prominent; efficiency will be less random especially at lower rates. The SARA performed better in the first routing mechanism by way of multi-path, and will exclude the particular node, to ensure the normal communication routing information; followed in computing nodes pheromone when the waiting time is added on each path considered, when stacked one on the way to its pheromone will be reduced, reducing the probability of being selected.

Figure 3 shows the number of death nodes. Energy is below a certain algorithm SARA worth nodes avoidance and protection, in terms of the number of nodes in the performance of the more prominent death. Meanwhile, the local pheromone update highlights that the impact of energy on the adjacent node's local information for neighboring nodes is generally low energy and cannot be

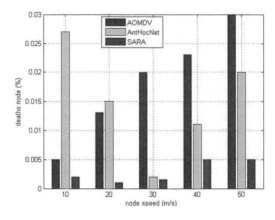

Figure 2. Delivery rate change with node speed.

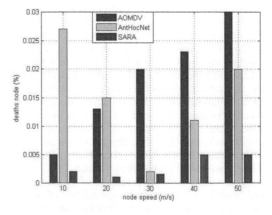

Figure 3. Death node change with node speed.

included in the selection. Besides, the AntHoc-Net in some cases would have excessive accumulation of pheromone with a sharp increase in the death of a node and poor performance under this dimension.

4 CONCLUSION

In this paper, based on the existing ant routing algorithm, the paper presents a multi-path environment based on the ant colony algorithm. It considers the agreement protocol of energy consumption, the rate of change of the node, the amount of cache in the path when in the similar path delay. The algorithm proposes a special node concept. It makes special control on nodes with a high mobility, low energy and altitude delay. The simulation experiment for the algorithm in the NS-2 simulation environment shows that, compared with the traditional protocol, there is a certain degree of improvement in the algorithm with respect to the node mortality, end-to-end delay and the survival time of the network.

REFERENCES

Di Caro G. AntNet: Distributed Stigmergetic Control for Communication Networks. Journal of Artificial Intelligence Research, 317–365, 1998.

Di Caro, Ducatelle F, Gambardella L M. AntHocNet: An Adaptive Nature-Inspired Algorithm for Routing in Mobile Ad Hoc Networks. *European Transactions on Telecommunnications*, 16 (5), 443–455. 14, 2006.

Hussein O, Saadawi T. Ant Routing Algorithm for Mobile Ad-Hoc Networks ARAMA. *Proceedings of the IEEE Performance Computing and Communications Conference (IPCCC)*, 15–17, 2003.

Jhong J D, Cross-layer ant based algorithm routing for MANETs. Proceedings of the 5th International Conference on Mobile Technology, Applications, and Systems (Mobility). Yilan, China, Article No.85, 2008.

Misra, Sanjay K Dhurandher, Mohammad S, et al. An Ant Swarm-Inspired Energy-Aware Routing Protocol for Wireless Ad-Hoc Networks. *The Journal of Systems and Software*, 83 (11), 2188–2199, 2010.

Marina M K. Ad Hoc On-Demand Multipath Distance Vector Routing. *Wireless Communications and Mobile Computing*, 6: 969–988, 2006.

Mesut Gunes, Udo Sorges, Imed Bouazizi. ARA: The Ant-Colony Based Routing Algorithm for MANETs. *Proceedings of the 2002 International Conference on Parallel Processing Workshops*.79–85, Aachen 2002.

Energy Science and Applied Technology – Fang (Ed.)
© 2016 Taylor & Francis Group, London, ISBN 978-1-138-02833-3

Multicarrier phase-coded design based on the genetic algorithm

N.N. Cao & T.D. Pei

Dalian University, Liaoning, Dalian, China

ABSTRACT: The MCPC signal is an orthogonal frequency division multiplexing coded radar signal. This paper analyzes the effects of autocorrelation function and adjacent-channel interference on ambiguity function. With the cross correlation being ignored, the suppression of the autocorrelation side lobe is studied. A method based on the genetic algorithm is presented by constructing cost function. The simulation results and performance analysis illustrate that the algorithm proposed in this paper is efficient and feasible.

Keywords: Multicarrier; Autocorrelation; Genetic Algorithm

1 INTRODUCTION

The multicarrier phase-coded signal (MCPC) proposed by Levanon in 2000 is OFDM technology combined with phase coding. It has a high spectrum efficiency and is orthogonal mutually; meanwhile, its ambiguity diagram is an ideal thumbtack shape. Besides, it can balance between spectrum efficiency, ambiguity function and the peak-to-mean envelope power ratio. The radar is not put into the practical stage as it is still immature (DENG Bin 2011). The MCPC signal has higher design requirements because of numerous parameters and flexible structure.

When designing a signal, the emission signals that have a low cross correlation need to be orthogonal mutually to avoid interference and to get independent information from the echo of various targets. Furthermore, the signal has a low autocorrelation side lobe level to get a high resolution of multiple targets. In other words, the MCPC signal should have the following characteristics: low autocorrelation function side lobe and weak cross-correlation. To deal with this problem, this paper proposes a method using the genetic algorithm to optimize the phase code. The genetic algorithm, which has the characteristics of intelligent global optimization and a fast convergence, is based on the probability model, and it is an effective optimal search algorithm to solve nonlinear optimization. In this paper, the autocorrelation function side lobe is as the objective function under the condition of neglecting cross-correlation, and then the cost function is constructed, which is used to optimize the signal in the genetic algorithm. The simulation results and performance analysis illustrate that the algorithm proposed in this paper is efficient and feasible.

2 REQUIREMENTS OF THE MCPC SIGNAL

The MCPC waveform is proposed as the multicarrier radar waveform. The general form of the MCPC signal can be written as

$$s(t) = \sum_{n=0}^{N-1} u_n(t)\exp(j2\pi n\Delta f t) \tag{1}$$

where $u_n(t) = \sum_{m=0}^{M-1} a_{n,m}r(t - mt_b)$; N is the number of subcarriers, with each subcarrier modulated by M bits phase-coded; $a_{n,m} = e^{j\theta_{n,m}}$ is the element m of the sequence modulating carrier n; $r(t) = \begin{cases} 1, 0 \le t \le t_b \\ 0, \text{else} \end{cases}$ is the envelope of the element signal; t_b is the chip duration; and Δf is the subcarrier spacing. In order to ensure the orthogonality among subcarriers, $\Delta f = \frac{1}{t_b}$. The structure of the MCPC signal is shown in Figure 1.

The ambiguity function (AF) of the MCPC signal can be written as

$$
\begin{aligned}
\chi(\tau, \upsilon) &= \int_{-\infty}^{+\infty} s(t)s^*(t + \tau)\exp(j2\pi\upsilon t)dt \\
&= \sum_{n=1}^{N}\sum_{p=1}^{N}\exp[-j2\pi(p-1)\Delta f \tau]\int_{-\infty}^{+\infty} u_n(t)u_p^*(t + \tau) \\
&\quad \cdot\exp[j2\pi(n-p)\Delta f t + j2\pi\upsilon t]dt \\
&= \chi_{auto}(\tau, \upsilon) + \chi_{cross}(\tau, (n-p)\Delta f + \upsilon)
\end{aligned}
\tag{2}
$$

The AF can be divided into two parts: $\chi_{auto}(\tau, \upsilon) = \sum_{n=1}^{N}\exp[-j2\pi(n-1)\Delta f \tau]\chi_n(\tau, \upsilon)$ is the

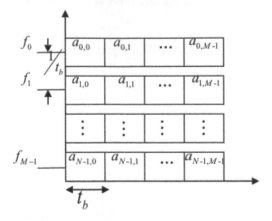

Figure 1. The structure of the MCPC signal.

main part of the AF obtained with the condition of $n = p$, and

$$\chi_{cross}(\tau, \upsilon) = \sum_{n=1}^{N} \sum_{\substack{p=1 \\ p \neq n}}^{P} \exp[-j2\pi(p-1)\Delta f \tau]\chi_{n,p}(\tau, \upsilon) \text{ is}$$

the interference between the subcarriers obtained with the condition of $n \neq p$, which can be ignored as it is very small to the AF when Δf is very little.

$\chi_n(\tau, \upsilon)$ is the auto AF of the nth subcarrier signal, and $\chi_{n,p}(\tau, \upsilon)$ is the cross AF between the nth and the path subcarriers. According to Deng Bin (2011), by making $i = \left\lfloor \dfrac{\tau}{t_b} \right\rfloor$, we therefore have

$$\chi(\tau, \upsilon) \approx \chi_{auto}(\tau, \upsilon)$$
$$= \sum_{n=1}^{N} \exp[-j2\pi(p-1)\Delta f \tau]$$
$$\cdot \sum_{i=(M-1)}^{M-1} \chi_1(\tau - it_b, \upsilon) \sum_{m=1}^{M-|i|} a_{n,m} a_{n,m+|i|}^* \exp(j2\pi \upsilon t_b)$$

$$(3)$$

where

$$\chi_1(\tau, \upsilon) = \exp[j\pi \upsilon(t_b - \tau)]Sa[\pi \upsilon(t_b - |\tau|)]\left(1 - \frac{|\tau|}{t_b}\right)$$

$|\tau| \leq t_b$, otherwise $\chi_1(\tau, \upsilon) = 0$. In addition, it is the AF of one chip.

When $\upsilon = 0$, we get the auto-correlation function (ACF) $\chi(\tau, 0)$ of the signal $s(t)$. Optimizing the objective function is an effective method to construct the optimal phase sequence. In the design of the orthogonal signal, in the case of ignoring the cross-correlation, the objective function can be selected to minimize the autocorrelation side lobe. So, the cost function can be expressed as

$$E = \max\left(\chi(\tau, 0), |\tau| \leq t_b; \tau \neq 0\right)$$

$$(4)$$

3 CODE OPTIMIZATION

The cost function is a nonlinear multivariable problem. The Genetic Algorithm (GA) is an effective algorithm to solve this problem. Through using the GA to optimize formula (4), the method based on the genetic algorithm is proposed.

The genetic algorithm, which is based on a simulated biological group evolution, is a search algorithm to solve optimization in computational mathematics. Through heredity, variation and competition between individuals, it is able to exchange information and further approximates the optimal solution of the problem. Because the multi-phase code sequence is a multivariate pseudo-random sequence, each state can be mapped by using a binary number. In this paper, we consider four phase codes. If $\phi_{n,m} \in (0, \frac{\pi}{3}, \pi, \frac{3\pi}{2})$, it will be expressed as [00 01 10 11]. The flow chart is shown in Figure 2.

1. Randomly generating initial population;
2. Selection: the fitness value of each individual will be calculated by formula (4), Based on the principle of evolution, the excellent individual

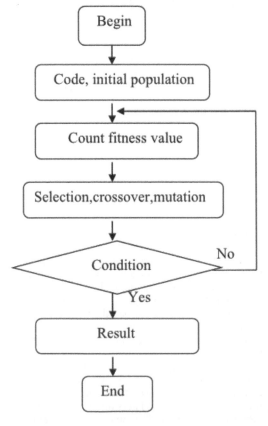

Figure 2. Flow chart of the GA.

will produce genetic to future generations using the roulette selection algorithm;

3. Crossover: the individuals of the population were randomly matched in pairs, and each pair of individuals must be in a certain probability to proceed a single point of intersection;

4. Mutation: changing some genes of population in the probability of 0.05;

5. When the fitness value gets a number that we want, the optimization operation will stop; otherwise, it goes back to step (2). Besides, it will repeat the cycle until the condition is fulfilled.

4 SIMULATION RESULTS

Assuming the number N of subcarriers of the MCPC signal to be 3, each carrier frequency element's length M is 60. From the above GA, we select 80 initial sequences as the initial population and formula (4) as the cost function, crossover probability and mutation probability as 0.5 and 0.05. After the 100 iterations of the GA, the optimization phase sequence is obtained, as shown in

Table 1. The optimization phase sequence.

$\phi_{n,m}$	phase sequence
$\phi_{1,m}$	3,2,1,0,0,1,2,1,2,0,3,3,1,2,1 ,2,3,0,1,1,0,1,1,2,3,1,3,2, 0,0,0,1,3,3,2,0,2,3,2,3,1,2 ,2,0,2,0,2,1,3,1,0,0,3,3,0, 0,0,0,2,0
$\phi_{2,m}$	2,0,2,3,1,2,1,3,2,1,1,0,3,0,2 ,0,1,2,3,1,1,0,3,2,0,2,1,1, 2,1,2,1,1,2,2,0,0,1,0,1,0,1 ,3,3,1,0,2,3,0,1,0,0,1,1,1, 2,0,3,3,3
$\phi_{3,m}$	3,2,2,1,2,3,0,3,1,0,2,1,2,1,1 ,0,0,3,1,3,1,0,1,2,3,1,1,3, 2,3,0,3,3,1,2,3,2,2,2,1,0,2 ,3,2,0,2,2,3,0,1,3,2,1,0,3, 2,0,0,1,0

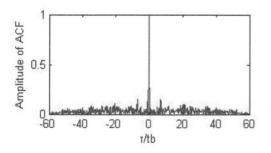

Figure 3. Amplitude of autocorrelation (N = 3, M = 60).

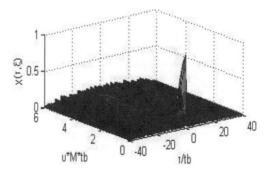

Figure 4. Ambiguity diagram (N = 3, M = 60).

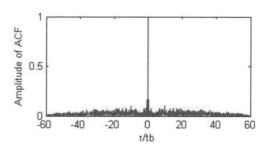

Figure 5. Amplitude of autocorrelation (N = 7, M = 60).

Table 1: 0, 1, 2, 3 in the table, respectively, express $0, \frac{\pi}{2}, \pi, \frac{3\pi}{2}$.

The average of the autocorrelation side lobe peak is about 0.03978, as shown in Figure 3. The ambiguity diagram has an ideal thumbtack shape. Under the condition of keeping other parameters constant and changing M as 40, the average of the autocorrelation side lobe peak becomes 0.0481. The further analysis showed that the size of the autocorrelation side lobe can be changed and formula (4) will impose more restrictions by increasing the number of elements and carriers. The experiment simulation proves that through changing the parameters of the genetic algorithm, such as iterative algebra and crossover/mutation probability, we may obtain better optimization results, but more time is needed to obtain the results.

5 CONCLUSION

This paper presents a new method to design the signal, the multicarrier phase-coded design, based on the genetic algorithm. It uses the nonlinear optimization of the GA to find the code sequence that meets the requirements of cost function. The simulation experiment proves that this method is effective.

REFERENCES

Deng Bin, Sun Bin. 2010, Ambiguity Function Analysis For MCPC Radar Signal. Proc. Of the 2 st Int. Conf. On Industrial Mechatronics and Automation (ICIMA),pp. 650–653.

Deng Bin, Wei Xi-zhang, Li Xiang. 2011, Parameter Designing of Random Shifted Phase-coded MCPC Radar Signal. Journal of College of Electronic Science and Engineering, 33(2), pp. 68–72.

Deng Bin. 2011, Research on the Signal Designing and Processing of Multi-Carrier Phase Coded Radar. National University of Defense Technology, the degree of Doctor

E. Mozeson, N. Levanon. 2003, Multicarrier radar signal with low peak-to-mean envelope power ratio. Proc. Of the 2 st Int. Conf. On Radar, Sonar Navigation, pp. 71–77.

Huo Kai, Jiang Wei-dong. 2011, A New OFDM Phase-coded Stepped-frequency Radar Signal and Its Characteristic. Journal of Electronics & Information Technology, 33(3), pp. 677–683.

Levanon N. 2000, Multifrequency Complementary Phase-Coded Radar Signal. Proc. Of the 6th Int. Conf. On Radar, Sonar Navigation, pp. 276–284.

Li Xiaoming, Luo Ding. 2011, Orthogonal Code Design Based on the Genetic Algorithm for MIMO Radar. Electronic Sc.

Energy Science and Applied Technology – Fang (Ed.)
© 2016 Taylor & Francis Group, London, ISBN 978-1-138-02833-3

Improving the bubble replacement algorithm in multi-core with shared L1 instruction cache

Xin Wan & ManMan Peng

College of Computer Science and Electronics Engineering, Hunan University, Changsha, China

ABSTRACT: With the multi-core processors becoming the mainstream and the development of cache resources, managing cache resources has become a multi-core key issue. Cache replacement policy is an important factor of cache resource management. The most commonly used algorithm is the LRU replacement algorithm, which has some shortcomings .When the Cache capacity is less than the program working set, it is prone to lead to conflict miss. Furthermore, the LRU replacement algorithm does not consider the frequency for data block access. The bubble replacement algorithm used in the multi-core shared caches overcomes this problem to some extent by using the frequency of data block access and information on recently accessed data. On this basis, this paper presents the Improvement of Bubble Replacement Algorithm in Multi-core with Shared L1 Instruction Cache (IBA-WSIC) cache management, adopting the advantage of bubble replacement algorithm and improvement of the algorithm. The experimental results show that this management effectively improves the cache hit rate and the IPC.

Keywords: Cache resource management; Cache partitioning; Cache sharing; replacement algorithm

1 INTRODUCTION

In order to solve the speed gap between the CPU and the memory unparallel growth rate, cache memory has come into use. Cache, an important bridge between the processor and main memory, plays a crucial role in a computer system's performance optimization.

In multi-core processors, cache management is particularly important. If you hit a fetch data in the cache, then you only need one or two processor cycles to get the data; however, if the data is not needed for the cache, then it often requires an order of magnitude of processor cycles to get the data. Cache replacement algorithms are the main factors affecting the performance of the cache. The RAND, FIFO and LRU are most commonly known cache replacement policy. The most widely used policy is the Least Recently Used replacement policy (LRU), which operates by considering recent replacement history and replaces the recently least visited block out of cache. Such methods and procedures better fit to a higher locality principle, which can cause a cache invalidation rate. In terms of hardware implementation, this method is comparatively simple. However, a recent research trend indicates that the LRU's strategy poor performance is mainly due to memory-constrained applications, and some of the reasons are as follows. (1) In the LRU policy, the data will be directly inserted into (upgraded to) the MRU position when deletion (hit), then the data in the recently visited block become invalid data, which often requires to cross the road most of the Cache to become the candidate blocks for replacement. (2) In the LRU policy, when a selection of a replacement block is carried out, consideration is given only to access information, and not data access frequency. (3) The existence of the cyclic load data access mode in Load, if the available space does not fully accommodate the Cache load working set, the LRU policy may cause Cache shake.

To overcome this problem, many solutions have been proposed. These include a pseudo-LRU replacement algorithm, which uses only low-Cache associatively. The other approach is one that considers the impact of light and other factors on the IPC for Cache. The IPC index in the optimal target proposed an optimal shared Cache partitioning method. In one paper, the author proposed a bubble replacement algorithm by considering the most recent information on the frequency of data blocks to be accessed in order to improve the hit rate of multi-core architecture shared Cache. In this paper, we adopt the advantage of the bubble replacement algorithm and improve the algorithm. Section 2 provides a detailed description. A multi-core private L1 data cache and two cores that share the same L1 instruction cache structure is also described.

2 OVERVIEW OF IBA-WSIC CACHE MANAGEMENT

2.1 *Introduction to replacement algorithm*

The main issues that impact the cache performance includes cache capacity, replacement policy and associated group degrees. The LRU replacement algorithm uses a locality principle of inference: If a block has been recently accessed, then it is likely to be accessed again. In this regard, the best choice is to replace the least recently used block. The current replacement policy can be seen by dividing it into three key parts: insertion strategy, eviction policy and promotion strategy.

Insertion strategy is a strategy where the new position into the cache block is placed in such a way that it positions in the current group. the LRU policy always cache block's new entrants on the MRU position. Eviction policy is a policy giving a means for the cache block to be replaced within the group in such a way that the LRU policy always works by inserting the evicted cache block in the LRU cache block location. Promotion strategy refers to the idea of how to change the location of strategies in the cache block within the group when the cache block is hit. In other words, cache block hit by LRU policy is always placed directly into the MRU position. The work in the proposed bubble method to replace the proposed algorithm is used only in the L2 cache, in which the L1 cache uses the LRU algorithm. In this paper, a modified bubble replacement algorithm is presented.

Improved bubble replacement algorithm:

a. Insert a new block strategy
   Cache block into the new position is generally of two kinds, namely LRU and MRU. The bubble replacement algorithm insertion strategy works by moving the data to the bottom of the block when it is accessed only once or when it is accessed few times than all other blocks. Therefore, the bubble queue is considered to be the one that was frequently data block. The LRU insertion strategy did not consider the frequency of data blocks accessed (blocks of data to be transferred in the Cache LRU stack MRU position). The insertion strategy and the LRU insertion strategy showed a significant difference.
b. Hit the block strategy
   The enhanced block hit has two strategies, namely MP and SIP. The strategy to promote and move frequently accessed data blocks bubble to the top of the queue and moving less frequently accessed data blocks to the bottom of the bubble queue, which leads recently accessed data block to gradually advancing to the top of the bubble queue. By doing so, the frequency of data blocks to be accessed and information on recently accessed block is constructively taken into account.
c. The selection strategy of replaced block
   Replacement policy refers to accessed data block that is not in the cache, selecting a block from the cache to be replaced with a new block of data. In order to find out the number of times a block accessed least as replacement block more accurately, for each block adds a counter bit method used here to determine the number of the accessed time of the block. The bubble algorithm to each cache group (set) set a group tag hit H, initialized to 0, when hit, H is set to 1, when does not hit, H is 0. When the value of H is 1, the replaced data block is selected from the bottom of the bubble queue; when the value of H is 0, the replaced data block is selected from the group that is the least number of block (the minimum bit count block), and the remaining data blocks sequentially advance a position to the top of the queue. The width of the counter needs analysis based on the results of a typical load, but the paper makes the following conclusion: in each period, the average life visits Last level cache block is generally relatively small, with a 4-bit counter to save sufficient accuracy can be guaranteed. Therefore, this article uses the same four-width counter. When the counter is the maximum number, keeping the value of the same number, the location of the call rises again until the top.

2.2 *The cache structure*

Between multi-core processors and main memory often contain more than one level cache. L1 Cache tightly coupled with the processor core, most of them are private for the processor, and more often last level Cache (LLC) is large capacity Cache. Studies have shown that the shared Cache structure can effectively use resources, but fully shared the bulk Cache whose available space is very large, vulnerable to influence by the growing of the delay line, resulting in slower access, while full sharing can also cause the thread negative interference. Private organization access is fast, but they lack the flexibility, when the load is unbalanced, there will be a large working set of core Cache shake. The working set has a smaller idle core Cache. "Ineffective use of valuable chip Cache resources, thus limiting the overall throughput".

It is known that the technical of L1 cache is very difficult and also has highest production costs, and improving the capacity of L1 cache increases the technical difficulty and production cost, very low cost, so making full use of the L1 cache is very important for the entire cache management. In a multi-core processor, each core typically has private L1 instruction and data caches. In this paper, two cores share a L1 level instruction cache and private

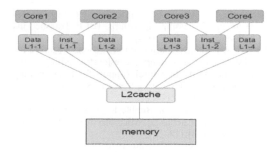

Figure 1.   Cache structure.

Table 1.   Basic configuration of the simulator.

Parameters	Value
Number of cores	4
Core type	Out-of-order
Base frequency	2 GHz
L1 cache (I & D)	2-way set associative, 2-cycle latency
Coherence protocol	MOESI snooping protocol
Memory	200 cycles access time
OS	Linux 2.6.27

L1 data cache to manage L1 cache, L2 cache for quad-core shared cache. Shared instruction cache to a certain extent improve the cache utilization. The Cache structure is shown in Figure 1.

3   EXPERIMENTAL RESULT ANALYSIS

3.1   *Experimental environment*

In this paper, the experiment was performed in the multi-core simulator Multi2 sim. Multi2Sim is a simulation framework for CPU-GPU heterogeneous computing written in C. It includes models for superscalar, multi-threaded, and multi-core CPUs, as well as GPU architectures. The simulation system is a quad-core processor, with each core having a proprietary data Cache, two cores sharing instruction Cache (quad-core into two groups), quad-core share L2 Cache. The x86 instruction set architecture is used. The mainstream business is 64 KB L1 cache, so this experiment L1 cache is 64 KB. Table 1 presents the basic configuration of the simulator.

3.2   *Test load*

The benchmark is selected from SPEC CPU 2006, taking into account that the cache performance is simulated and tested, so we selected some representative benchmarks. SPEC 2006 divided into CINT 2006 (integer application) and CFP 2006 (floating point applications), so we selected three applications from each one, so that the whole experiment is more comprehensive. New management methods are tested for the following two aspects: cache hit rate and IPC.

3.3   *Analysis of experimental results*

Impact on the hit rate. Figure 2 shows the improvement of the hit ratio for IBA-WSIC methods with the bubble replacement algorithm and the LRU algorithm. From the figure, it can be seen that the hit rate does not show a big difference when comparing the original bubble algorithm with the LRU algorithm in

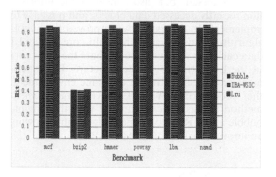

Figure 2.   Hit ratio improvement.

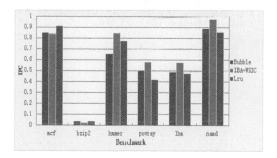

Figure 3.   The improvement of the IPC.

these applications. The new management approach with respect to the original bubble algorithm and LRU algorithm hits has shown some improvement, as the effect of the application of bzip2 new management methods and original method are the same. Furthermore, the new cache management method enhances the overall average of about 2%.

The impact on the IPC. CPU performance is actually affected by the CPU frequency and the IPC in real time. Accurate CPU performance criteria should be as follows: CPU performance equals to IPC (instructions executed per clock cycle each CPU number) multiplied by the frequency (MHz clock speed), IPC importance of performance observed. As can be seen from Figure 3, using the IBA-WSIC method, for most of the test procedure, the IPC is improved, except the applications bzip2 and mcf, whose effects were somewhat inferior to the original algorithm, with the overall improvement on average about 10%, so as to enhance the overall performance.

4 CONCLUSION

With the increase in the number of processor cores, Cache resource management has become increasingly important. The bubble algorithm takes into account the frequency of data blocks to be accessed, but also considers the information recently accessed data blocks, and it is easy to implement. In this paper, we improved the bubble algorithm; and presented the multi-core private L1 data cache, in which two cores share the same core of the L1 instruction cache structure. The experiment results show that in the same environment, the proposed IBA-WSIC cache management method in this paper compared with the original bubble and LRU replacement algorithm, showing an improved cache hit rate and IPC.

REFERENCES

Chang J. Cooperative caching for chip multiprocessors [D]. Madison, Wisconsin, USA: University of Wisconsin at Madison. 2007: 385–396.

Kharbutli M, Solihin Y. Counter-based cache replacement and bypassing algorithms. IEEE Transactions on Computers, 2008, 57(4): 433–447.

Kim C. An adaptive: non—uniform cache structure for wire-dominated on-chip caches[c].Proceeding of the Int Conf on Architectural Support for Programming Languages and Operating Systems. San Jose, California: ACM, 2002: 211–222.

Lin Xiao. Algorithm of Bubble Replacement in Multi-Core Shared Caches [J]. Microelectronics & Computer, 2011, 28(4):118–121.

Liu C. Organizing the last line of defense before hitting the memory wall for craps [C]. Proceedings of the 10th Int Symp on High Performance Computer Architecture Madrid, Spain: IEEE. 2004: 176–185.

Method for implementing a pseudo least recent used (LRU) mechanism in a four-way cache memory within a data processing system [EB/OL] (2001), [2010-04-26]. http://www.patentstorm.us/patents/6240489/de-scription.html.

Subramanian R. Adaptive caches: effective shaping of cache behavior to workloads [C]. Proceedings of the 39th Annual IEEE/ACM Int Symp on Microarchitecture, Orland, FIorida, USA: IEEE, 2006.

Suo Guang, Yang XueJun Dual-core processor performance optimal partition shared Cache [J]. Microelectronics & Computer, 2008, 25(9): 28–30.

Ubal Rafael, Sahuquillo Julio, Petit Salvador, L pez Pedro. Multi2 sim: A simulation framework to evaluate multicore-multithread processors. Proceedings of the 19th International Symposium on Computer Architecture and High Performance Computing. Washington: IEEE Computer Society, 2007: 62–68.

Energy Science and Applied Technology – Fang (Ed.)
© 2016 Taylor & Francis Group, London, ISBN 978-1-138-02833-3

The numerical simulation of variable diameter high temperature phase change thermal energy storage container

H.T. Cui, K.K. Sun, N. Li & H.L. Zhao
School of Mechanical Engineering, Hebei University of Science and Technology, Hebei, China

ABSTRACT: The heat storage process of a high-temperature thermal energy storage container used in solar thermal power generation was numerically simulated. The change curve of the PCM temperature and liquid rate over time in the process were obtained. The high-temperature thermal energy storage container design using the variable diameter heat retainer was present after the analysis of the curve and on the premise of heat storage, the performance of heat storage was simulated. The results indicate that the variable diameter scheme greatly reduced the total heat storage time and also effectively reduced the effect on the performance of heat storage, which is caused by "dead zones" during the same diameter regenerative terminal stage, and it even made the PCM liquid rate distribution more uniform.

Keywords: solar thermal power generation; high temperature thermal energy storage container; phase change material; numerical simulation

1 INTRODUCTION

Solar thermal power generation is a technology that converts the solar energy into the electric energy. The collector gathers the solar radiation, and then generates heat. Through the heat exchange device, the heat will convert into the high temperature and high pressure steam, and then the steam drives the turbine power generation and produces the electric energy (WU, Y.T. et al., 2007; Cui, H.T. et al., 2012). However, because the solar energy is a kind of intermittent energy sources, and the radiation intensity is influenced by the circadian and changes in the weather, solar thermal power stations usually rely on heat storage to ensure the stable operation of the power system in the evening and cloud. The storage system is an indispensable part in the solar thermal power station, which plays a role in buffer capacity, stable load and process regulation. Its working condition directly affects the normal operation of the solar thermal power generation system, so the physical model establishment and the characteristics research of the heat storage system are the premises of improving the solar energy heat utilization efficiency.

At present, many experts and scholars investigating on the regenerator structure mostly focus on the impact of the heat accumulator's arrangement and diameter on the heat storage performance (Cui, H.T. et al., 2010; 2014; 2015; Sharmaa, R.K. et al.,2015), but the joint regenerative performance of different diameters in the same accumulator is less studied. This paper simulated the regenerative process of the same regenerator temperature phase change thermal storage device, discussed its performance and summarized the change rule of the PCM liquid rate. Then, it proposed the design using the same diameter heat accumulator and simulated its regenerative properties. Also, this paper provides a theoretical basis for the optimization design of the high-temperature phase change thermal energy storage container.

2 NUMERICAL SIMULATION

2.1 *Accumulator's structure and simplification*

A solar thermal power generation regenerative heat exchanger consists of a high-temperature regenerator cabinet, high-temperature phase change thermal storage tubes and phase change thermal storage materials. The tubes are ceramic composite steel pipes, which use an encapsulated aluminum-silicon alloy as their heat storage material. A lot of tubes with the same specifications are arranged in a square (Cui, H.T. et al., 2012).

Figure 1 shows a structure diagram of a high-temperature regenerator with the same diameter, whose length, width and height are, respectively, 1500 mm, 1200 mm and 700 mm. The internal regenerator diameters are 180 mm, and their spacings are 50 mm, whose lower left side is the air inlet and the upper right side is the air outlet. Figure 2 shows a structure diagram of a high-temperature regenerator with a varying diameter, whose length, width and height are, respectively,

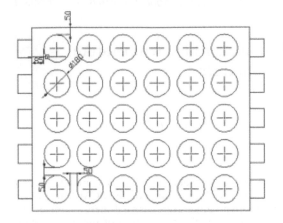

Figure 1. The structure diagram of the container in the same pipe diameter heat storage.

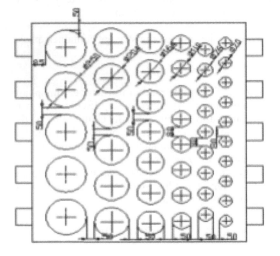

Figure 2. The structure diagram of the container in the variable pipe diameter heat storage.

1300 mm, 1550 mm and 700 mm. The internal regenerator spacings are about 50 mm. The vertical spacings of the first three columns are 50 mm, but the latter three columns are 80 mm. In order to make the outlet fluid develop fully, the length and radius of the pipes used in the air inlet and outlet is, respectively, 100 mm and 60 mm.

To facilitate the use of the FLUENT software to simulate the regenerative process, the physical model made the following assumptions (Cui, H.T. et al., 2015):

1. the phase change material is isotropic;
2. the phase change material's specific heat capacity, thermal conductivity and density is constant, and does not change with temperature;
3. the heat loss of the outer surface is ignored and affected by the wall thickness;

4. the Bousssinesq assumption is met, and the fluid density changes only in the buoyancy items are considered;
5. the phase change material is a melted incompressible Newtonian fluid;
6. the impact of natural convection is considered, and the natural convection is laminar flow.

2.2 The choice of the phase change material

In this paper, we selected Al-12Si as the high-temperature phase change material. It has many advantages in many aspects, such as thermal stability, slow decay, and steady temperature change, little volume change during the heat storage and release process. Its physical parameters are shown in the paper by Liu, J. et al. 2006.

2.3 Numerical simulation of heat accumulator

GAMBIT software establishes the accumulator geometric model, and then meshes it. The software puts the grid file outputted from GAMBIT into the FLUENT software, and checks it. We select the 3D separate unsteady solver and solidification/melting model to simulate the phase transition. In the boundary condition setting, we set the air inlet and the outlet velocity to 10 m/s, use the SIMPLC algorithm to solve and initialize the whole area that the initial temperature is set to 300 K. In order to get the PCM temperature and liquid rate variation with time, we need to arrange the monitor in proper position before iteration, namely to monitor the changes in PCM temperature and liquid rates. In the iteration, the time step should not be set too big or too small, and only ensure that the equation can get the stable convergence in the maximum steps. FLUENT uses the melting/freezing model to simulate the accumulator with 180 mm pipes. The temperature changes can be analyzed from the temperature curve, and the distribution of the liquid phase change can be obtained from the liquid phase distribution nephogram. In addition, under the same operating conditions, the performance of the accumulator with varied diameters can be simulated. Then, we will get the differences through the comparative analysis of the same diameter and variable.

3 THE SIMULATION RESULTS AND ANALYSIS

Figure 3 shows the heat storage condition temperature change in the PCM area over time at the same pipe diameter, whereas Figure 4 shows the temperature change at the variable pipe diameter heat storage. By comparison of the above two figures, it can be seen that the PCM temperature change in the heat storage process is substantially consist-

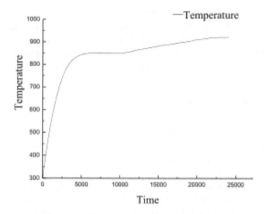

Figure 3. Temperature change in the PCM area in the melting process at the same pipe diameter heat storage.

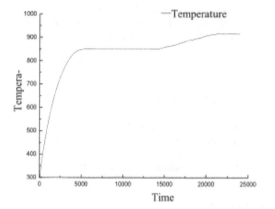

Figure 4. Temperature change in the PCM area in the melting process at the variable pipe diameter heat storage.

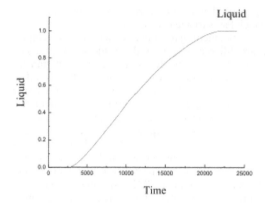

Figure 5. Liquid fraction change in the PCM area in the melting process at the same pipe diameter heat storage irregular triangular arrangement.

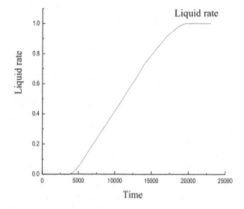

Figure 6. Liquid fraction change in the PCM area in the melting process at the variable pipe diameter heat storage irregular triangular arrangement.

ent in both cases. The region warms very quickly in the early phase change heat storage, i.e. the temperature quickly reaches the phase transition temperature in a relatively short period of time. Because the heat storage starts with temperatures lower than the phase transition temperature of the PCM, the process is in the sensible heat regenerative phase. The temperature difference between the hot air and the heat storage is too large, so the heat conducts quickly and the temperature varies faster. Thereafter, the PCM is in a latent heat storage phase, so that its temperature changes slowly and maintains at the phase transition temperature. However, in the same diameter condition, the heat storages closed to the air inlet have smaller diameters, so that they can accomplish the melting process faster, and then start the sensible host storage stage earlier, which makes the overall temperature of the PCM to get out of the stable development. Along with the accomplishment of the latent heat

process, the PCM temperature begins to rapidly rise, finally tending to flatten.

Figure 5 shows the heat storage condition liquid fraction change in the PCM area over time at the same pipe diameter, whereas Figure 6 shows the liquid fraction change at the variable pipe diameter heat storage. It can be seen from those two figures that the liquid fraction is positively related with time in both cases. However, in the initial stages of the melting, the liquid fraction of the same diameters appears at 2490s, which is significantly earlier than 3400s in the case of the varied diameters. This is because the regenerator diameters are 90 mm near the hot air inlet in the latter, but the diameters are 125 mm on the contrary, which leads to a larger resistance in the varied diameters. Then, it will take more time to reach the phase transition temperature. Along with the melting, the rate significantly accelerates and then slows down at the same diameter. This is because the effect of the thermal conduction becomes weak, but

383

the natural convection is correspondingly enhanced, which leads to a faster melting.

As the melting reaches a higher level, the temperature difference between the inside and the outside also gets much smaller, and then the heat transfer rate becomes slow. However, the situation is different in the varied diameters. The melting velocity of each regenerator is substantially consistent due to the decreasing arrangement. When the melting tends to the end, the liquid phase curve being around 75%, the curve tends to be flat, and the velocity is relatively slow. Considering that the diameter of the regenerator closed to the hot air outlet is larger than the variable diameter, the resistance of the same diameter is larger than that, which leads to the more severe "dead zone" phenomenon. So, it takes more time to accomplish the thermal storage. With the continuation of heat storage, the process is finished at 1960000s, but with the same diameter design at 21700s. It can be obviously seen that the whole melting time in varied diameter design decreases by 15.67% compared with the same diameter design, and the whole storage time decreases by 9.67%.

Figure 5 shows the liquid fraction change in the PCM area over time at the same pipe diameter heat storage condition, whereas Figure 6 shows the liquid fraction change at the variable pipe diameter heat storage. It can be seen from those two figures that the liquid fraction is positively related to time in both cases. However, in the initial stages of the melting, the liquid fraction of the same diameters appears at 2490s, which is significantly earlier than 3400s in the case of the varied diameters. This is because the regenerator diameters are 90 mm near the hot air inlet in the latter, but the diameters are 125 mm on the contrary, which leads to a larger resistance in the varied diameters. Then, it will take more time to reach the phase transition temperature. Along with the melting, the rate significantly accelerates and then slows down in the same diameter design. This is because the effect of the thermal conduction becomes weak, but the natural convection is correspondingly enhanced, which leads to a faster melting. As the melting reaches a higher level, the temperature difference between the inside and the outside also gets much smaller, and then the heat transfer rate becomes slow. However, the situation is different in the varied diameters. The melting velocity of each regenerator is substantially consistent due to the decreasing arrangement. When the melting tends to the end, the liquid phase curve is around 75%, the curve tends to be flat, and the velocity is relatively slow. Considering the diameter of the regenerator closed to the hot air outlet is larger than the variable diameter, the resistance of the same diameter is larger than that, which leads to the more severe "dead zone" phenomenon. So, it takes more time to accomplish the thermal storage. With the continuation of heat storage, the process is

finished at 1960000s, but the same diameter design at 21700s. It can be obviously seen that the whole melting time in the varied diameter design decreases by 15.67% compared with the same diameter design, and the whole storage time decreases by 9.67%.

4 CONCLUSIONS

This article describes the parameter settings in the FLUENT melting/freezing model for solving the problem of phase transitions, and simulates the regenerative process of the high-temperature phase change used in solar thermal power generation. The total melting time and the change rules of temperature and liquid rate over time can be obtained. Then, the fraction of the liquid rate has been analyzed by comparing the distribution nephogram of the same and varied diameter designs at the same liquid rate, which determines the optimization design of the high temperature phase change thermal storage used in solar thermal power generation. It also lays a foundation of optimization design on high-temperature thermal storage.

ACKNOWLEDGMENT

This work was supported by the Natural Science Foundation of Hebei Province (E2014208005), and was the major project of the Hebei Province Department of Education (ZH2012079).

REFERENCES

Cui, H.T. et al. 2010. The enhancement of high porosity metal foam to phase change energy storage. Journal of Hebei University of science and technology 31(2):93–96.

Cui, H.T. et al. 2012. Numerical simulation on melting and solidification process of aluminum-silicon alloy. Journal of Hebei University of Science and Technology 33(5):406–411.

Cui, H.T. et al. 2014.The numerical simulation of the heat storage process about battery incubator in the communication base station. Journal of Hebei university of science and technology 35(1):64–68.

Cui, H.T. et al. 2015. Numerical simulation of solar thermal power generation with high temperature phase change thermal energy storage. Journal of Hebei University of Science and Technology 36(2):150–155.

Liu, J. et al. 2006.Selection of high temperature phase change material Al-Si alloy and its experimental study on compatibility with metal containers. Acta energiae solaris sinica. 27(1):36–39.

Sharmaa, R.K. et al. 2015. Developments in organic solid-liquid phase change materials and their applications in thermal energy storage. Energy Conversion and Management, 95(4):193–228.

WU, Y.T. et al. 2007. High temperature heat storage technology in solar thermal power. Solar Energy 3(3):23–25.

Energy Science and Applied Technology – Fang (Ed.)
© 2016 Taylor & Francis Group, London, ISBN 978-1-138-02833-3

An intelligent medical guidance system based on multi-words TF-IDF algorithm

Y.S. Lin & L. Huang
Cooperative Innovation Center for Internet Healthcare, Zhengzhou, Henan, China
School of Information Engineering, Zhengzhou University, Zhengzhou, Henan, China

Z.M. Wang
Cooperative Innovation Center for Internet Healthcare, Zhengzhou, Henan, China

ABSTRACT: The traditional human-aided medical guide service costs heavy human resources. The online medical guide service is almost human-aided and has an uncertain waiting time for patients. Existing medical guidance systems still have insufficiency for improvement in reliability and correctness. This paper proposes an intelligent medical guide system based on the multi-word TF-IDF algorithm, which is implemented on the Android mobile platform. This system applies the improved TF-IDF algorithm and the cosine similarity algorithm to calculate the possibility of disease for patients, and gives the reasonable user guidance. The experimental results proved that the improved TF-IDF algorithm proposed in this paper increased the correctness and reliability of medical guidance.

Keywords: Internet healthcare; Intelligent medical guidance; TF-IDF algorithm; Cosine similarity algorithm

1 INTRODUCTION

Human-aided medical guide service is adopted by most of the current hospital, which involves a vast number of human resources. In fact, some Internet medical service websites and mobile apps use online human-aided medical guide service, which cannot assure a timely response to the user. Therefore, the use of the computer intelligent medical guide system instead of the human-aided medical guide service will effectively save the running costs of hospital and medical websites.

At present, the intelligent medical guide system has two main ways: (1) intelligent medical guide system, which is based on knowledge (Wu, 2010 & Yang, 2014); (2) intelligent medical guide system, which is based on similarity calculation (Liang, 2014). The intelligent medical guide system constructs the knowledge base by the knowledge and experience of medical experts, and uses fixed diagnosis condition to judge the disease that the user may suffer from and to provide the medical guide service in the form of an expert system. The realization of the knowledge base and diagnosis condition is according to the knowledge and experience of medical experts in different fields and departments. The input and maintenance of the knowledge base mainly rely on the professionals, and the upgrade of the knowledge base some-

times needs to change the original knowledge base and diagnosis condition; therefore, the realization and maintenance of the knowledge base is complicated and inefficient. Currently, the function of the medical guide system, which is based on the knowledge base, is not perfect. For example, the literature (Yang, 2014) deduces the department by using the symptom only, and does not predict the disease. The intelligent medical guide system, which is based on similarity calculation, develops a vector space model by symptoms that the user inputs and symptoms of disease, judges the disease that the user may suffer from and provides the medical guide service using the similarity calculation. Compared with the intelligent medical guide system, which is based on knowledge base, it is more efficient, easy to extend and easy to maintain (see Liang, 2014). However, the similarity calculation method ignores the effect of the reliability of comparability of disease and medical guide results, when several symptoms that the user inputs occur in certain diseases at the same time.

To deal with the problems in the intelligent medical guide system, this paper designs and realizes an intelligent medical guide system based on the improved TF-IDF algorithm. The system develops the vector space model by natural language symptoms that the patient inputs and symptoms of disease, calculates the weight of the symptoms through the improved TF-IDF algorithm, and then calculates the disease that the user may suffer from by the cosine similarity algorithm, in order to provide users the disease forecast before diagnosis, guide the user for the right medical treatment, and improve the efficiency of the patient visit.

2 INTELLIGENT MEDICAL GUIDE SYSTEM

2.1 *Preparing the new file with the correct template*

This paper designed an intelligent medical guide system based on the TF-IDF algorithm. The system framework is shown in Figure 1, which consists of three modules: (1) UI module, which includes user register, login, input symptoms and display medical guide results; (2) NLP (Nature language symptom) module, which is the first segment of the natural language symptom that is input by the user; word segmentation uses the IK Analyzer, which is an open-source and based on java; (3) MGC (Medical guide calculation) module, which is the core module of this system.

2.2 *Analysis and improvement for the algorithm*

This paper designed an intelligent medical guide system based on the TF-IDF algorithm. The system framework is shown in Figure 1, which consists of three modules: (1) UI module, which includes user register, login, input symptoms and display medical guide results; (2) NLP (Nature language symptom) module, which is the first segment of the natural language symptom, which is

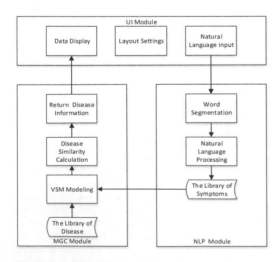

Figure 1. System framework.

input by the user; word segmentation uses the IK Analyzer, which is an open-source and based on java; (3) MGC (Medical guide calculation) module, which is the core module of this system. VSM m and d for VSM vector modeling, respectively. Thus, $\vec{d} = (W_{1,d}, W_{2,d}, ..., W_{t,d}, W_k)$, W_k is the weight of the same symptoms, $W_{i,m}$ is the weight of ith symptoms from m except the same symptoms, $W_{i,d}$ is the weight of ith symptoms from d except the same symptoms, $W_{i,m}$ and $W_{i,d}$ use the TF-IDF algorithm (Liang, 2014) to calculate the weights, and W_k use the modified TF-IDF algorithm to calculate the weight. The cosine of the included angle between two vectors is used to calculate the similarity between user input and matching diseases. The similarity of Sim (m, d) is given by

$$Sim(m, d) = \cos\theta = \frac{\vec{m} \cdot \vec{d}}{\|\vec{m}\| \times \|\vec{d}\|} \tag{1}$$

3 INTELLIGENT MEDICAL GUIDE SYSTEM

The TF-IDF algorithm is a method for calculating the weight, which is a kind of effective method always used in information retrieval and data mining. Huang (2011) proposed the use of the TF-IDF with semantic information to calculate similarity. Quang (2011) proposed the use of TF-IDF to identify the user's handwriting. Yu (2012) proposed the use of TF-IDF to calculate similarity in the radar intelligence analysis.

3.1 *Conventional TF-IDF algorithm*

Term Frequency and Inverse Document Frequency consists of the weight of the keywords of the TF-IDF method. TF is the frequency of a word that appears in the document, and IDF is the frequency of a word that appears in all the documents, and is also a measure of the importance of a word.

The formula for the calculation of weight is given as follows:

$$W = TF \times IDF \tag{2}$$

where, in the Intelligent Guide System, TF means the frequency of a symptom for a given word that appears in the disease, and the formula is given by

$$TF = \frac{n}{m} \tag{3}$$

where n is the number of the symptom that appeared in one of the diseases, and m is the number of symptoms to be included in this disease.

The word's frequency has the same value due to a symptom that just appears once in a disease. The more common the symptom, the more difficult to judge the name of the disease. On the contrary, the less common the symptom, the more important the diagnosis of the disease. Thus, in this paper, we use the Baidu index (Liang, 2014), to calculate the weight of a single symptom word. The formula is given by

$$TF = \frac{\min(u_i)}{u_i} \qquad (4)$$

where u_i is the symptom of the Baidu index and $\min(u_i)$ is the smallest of all symptoms of the disease in the Baidu index.

IDF is the frequency of a symptom that appears in all diseases, and the formula of IDF is given as

$$IDF = \log\left(\frac{A}{A_{in}+1}\right) \qquad (5)$$

where A is the total number of the diseases and A_{in} is the number of the diseases that contain the same symptom at the same time.

Thus,

$$W = \frac{\min(u_i)}{u_i} \times \log\left(\frac{A}{A_{in}+1}\right) \qquad (6)$$

3.2 Analysis and improvement of the algorithm

The following two criteria are commonly used to determine the size of the weights for the symptoms from the perspective of the statistical properties:

1. The smaller the Baidu index of symptoms values, the higher the weight value of the symptoms;
2. The fewer the number of symptoms in all diseases, the higher the weight value of the symptoms.

It should be noted that the determination of the disease cannot be judged by one symptom, but it can be obtained after combining a variety of symptoms and factors. There can be a variety of clinical symptoms for each disease in the disease library; meanwhile, one or more symptoms may be the same for a variety of diseases.

Suppose the user enters two symptoms, say n_1 and n_2, then there are three cases of the diseases in the disease library that contains the symptoms n_1 and n_2: only symptom n_1 is included for the disease; only symptom n_2 is included for the disease; or both the symptoms n_1 and n_2 are included. The weight value of the disease that contains the symptoms n_1 and n_2 cannot be distinguished effectively due to the effect of the first two cases. When the

two symptoms n_1 and n_2 are included at the same time for some other disease, say disease D, then the probability of suffering the disease will be increased for the users. However, when applying the original TF-IFD algorithm, the probability of suffering this disease D is lower, and it is unable to effectively extract the disease that the user may be suffering from. This is because the weights of the symptoms n_1 and n_2 are calculated separately, i.e. the number of occurrences for the symptoms n_1 and n_2 in all diseases are considered separately, and not taking into consideration the case that the symptoms n_1 and n_2 occur at the same time. The disease that the users may be suffering from can be clearly distinguished from the medical perspective when two or more symptoms occur at the same time. However, the TF-IDF algorithm does not take this into consideration. Based on the above analysis, when two or more symptoms are output at the same time, the original algorithm cannot effectively analyze the disease that the user may be suffering from.

Therefore, when two or more symptoms occur simultaneously, it cannot separately consider the calculation of the weights of different symptoms; the combination of the weight values of a variety of symptoms should be considered. Based on the above analysis, we propose a third criterion as follows:

3. Take the symptoms as a combination symptom when two or more symptoms occur simultaneously, and the IDF values are the number of co-occurrences of the symptoms.

Based on the above three criteria, we propose an improved weight calculation method: the TF-IDF algorithm is based on the multi-word. The formula for calculating the weight values of symptoms using the improved TF-IDF algorithm is given by

$$W = TF \times IDF = \frac{\sum_{i=1}^{k} t_i}{N} \times \log\left(\frac{A}{A_{in}}\right) \qquad (7)$$

where $K(K > 1)$ is the number of co-occurrence of the symptoms; N is the number of symptoms to be included in disease D; t_i is the number of symptom i that occurs in disease D; A is the total number of the diseases; and A_{in} is the number of the diseases that contain k symptoms at the same time.

4 THE SIMULATION

We implemented our designated intelligent medical guide system on the Android mobile platform, and used the MySQL as the back-end database.

4.1 The purpose of the experiment

We implemented our intelligent medical guide system, respectively, under the traditional TF-IDF algorithm and the modified TF-IDF algorithm. Then, the reliability of our system and of the modified algorithm could be verified through comparing the experimental results. During the experiment, a correct medical guidance means that the disease with the highest possibility in the possible disease list, which is responded by the system to the user, is exactly the disease when the patient is diagnosed by a doctor. Correctness means the percentage of the system diagnosing the disease correctly over the whole sample set. Reliability implies the value of the highest possibility of the disease when the intelligent medical guide system is used. A high reliability demonstrates that the disease that is diagnosed with the highest possibility by our intelligent medical guide system should be the exact disease that a patient suffers from, and the opposite of the above is also true.

4.2 The data of the experiment

The source of the data of our experiment comes from two parts. Some of the data are derived from the medical guidance data, which is collected by the diagnosis platform of the people's hospital of Henan province. This data set contains the diseases and the symptoms that the patient consulted and the diagnosis result that the nurse gave to them, and we call this data set as H. The other part of the data is collected from consultations from the patient and the corresponding exact correct answers from the doctor in websites such as Haodaifu and Youwenbida, and we call this data set as T.

The data sets H and T consist of our whole experimental data. The data set H contains 300 items, while the data set T contains 100 randomly chosen items. In our experiment, we value the answers from the nurses and doctors when they provide correct medical guidance, and then we can verify the correctness of the intelligent medical guide system using the data set H and the improving ratio of the reliability of the system using the data set T.

4.3 Analysis of the experiment

After testing the data set H using two systems, we implement under the traditional TF-IDF algorithm and the modified TF-IDF algorithm, respectively. We computed, compared and analyzed the result of the medical guidance. Figure 1 shows the result of the correctness under the two algorithms. Our result proves that the modified algorithm can achieve a higher correctness of 86%, compared with the traditional ones of 79.7%. The average

Table 1. Margin settings for A4 size paper and letter size paper.

Algorithm	Correctness
Traditional TF-IDF	79.7%
Modified TF-IDF	86%

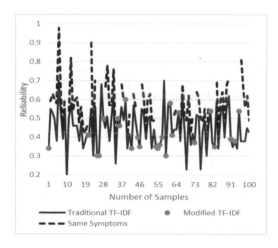

Figure 2. Caption of a typical figure. Photographs will be scanned by the printer.

improved percentage of the correctness is 6.3%. Through the analysis, we know that a higher correctness cannot be achieved due to the insufficiency of synonymity in the natural language process.

Later, we apply the data set T to test the result of the medical guidance using the two systems mentioned above, and also the comparison and analysis of the reliability of the result is listed below.

We divide the user inputs into two categories: user inputs containing a single symptom and user inputs containing multiple symptoms. The experimental results are shown in Figure 2. We can find that when user inputs contain a single symptom, the reliability of each algorithm shows no difference when compared with another one; however, when the user inputs contain multiple symptoms, the reliability of the modified algorithm shows obvious improvement compared with the traditional ones, with the peak point of the percentage of the improvement being 35% and the average percentage of the improvement being 11.8%, which means that the modified algorithm shows an excellent advantage in the reliability of the medical guidance over the traditional TF-IDF algorithm.

5 CONCLUSION

As the traditional TF-IDF algorithm cannot tackle the problem of processing multiple symptoms in the intelligent medical guidance system efficiently, this paper proposes an intelligent medical guidance system based on the multi-words TF-IDF algorithm. The experimental results show that the modified TF-IDF algorithm can achieve higher correctness than the traditional TF-IDF algorithm. According to reliability, on the condition of multi-works inputs, the modified algorithm still shows obvious benefits. At the same time, our system cannot process the symptoms sufficiently in the natural language process, and this could be the focus of our future study.

REFERENCES

Huang C.H. et al. 2011. A text similarity measurement combining word semantic information with TF-IDF method, Chinese Journal of Computers, 2011, 34(5): 98–106.

Liang, L. 2014. Research on the intelligent medical guide system basing on the VSM weight improved algorithm. M.E. Zhengzhou University.

Quang, A.B. et al. 2011. Writer identification using TF-IDF for cursive handwritten word recognition: Kaizhu Huang (ed.), International Conference on Document Analysis and Recognition. Beijing, 844–848.

Song, R. 2013. The research and implementation of method for domain Chinese word segmentation. D.E. Beijing University of Technology.

Su, Y. et al. 2011. The improvement of VSM model based on semantics. Computer Applications and Software, 28(8):158–161. (VSM)

Wu, J.C. 2010. Research of intelligent guide medical system based on MAS. M.E. Jinan University.

Yang, B. 2014. Design and implementation of a comprehensive outpatient medical guide system based on patient's symptoms. M. E. Southwest Jiaotong University.

Yu, M et al. 2012. Research on intelligence distribution based on TF-IDF classifier, Computer Engineering and Design, 33(5):1822–1826.

Energy Science and Applied Technology – Fang (Ed.)
© 2016 Taylor & Francis Group, London, ISBN 978-1-138-02833-3

Numerical simulation of three different vortex shedders in power law

Peng Du, Xi Chen, Wen Yuan, Xiaohan Feng & Tao Yu
Northeast Petroleum University, Heilongjiang, China

ABSTRACT: Using the Gambit modeling software, this paper established a trapezoidal, T-shaped and circular cylinder with a slut in three different shapes of vortex shedders by grid. The numerical simulation of the non-Newtonian fluid was carried out on Fluent. In the power law fluid, the T-shaped vortex shedder cannot form a stable vortex, while the others are better and the trapezoidal vortex shedder has more accurate measurement results. The comparison between two different fluids under the various parameters of vortex shedders shows that, of the three different shapes, trapezoidal has the best performance and is an ideal vortex shedder.

Keywords: Vortex shedder; Numerical simulation, Power Law

1   THE PRINCIPLE OF VORTEX FLOWMETER

In a fluid field setting, a vortex shedder can produce regular vortex alternately, and this phenomenon is called the Karman vortex street. In 1878, Strouhal studied the wind pendulum and found that the Strouhal number had some relationship with vortex shedding frequency:

$$f = S_t \frac{v}{d} \qquad (1)$$

where f = vortex shedding frequency; S_t = Strouhal number associated with the vortex shedder's structure size; v = average velocity of flow; and d = diameter of vortex shedder melting flow:

$$q_v = \frac{f}{K} \qquad (2)$$

where q_v = volume flow and K = condition index of the vortex shedder.

In other words, for the cortex shedder in a certain case, the measure of flow would only relate to the vortex shedding frequency. The Strouhal number is a constant related to the vortex shedder. Therefore, choosing a superior size of the vortex shedder is very important.

2   NUMERICAL SIMULATION

2.1   *Model and mesh*

The model is established by using the Gambit software, whose pipe diameter is 50 mm, and the three different vortex shedders' structure size is shown in Figure 1. The trapezoidal vortex shedder has the following parameters: d = 14 mm, a = 2 mm, L = 20 mm, b = 2 mm. The T-shaped vortex shedder has the following parameters: d = 14 mm, d = 14 mm, a = 3 mm b = 10 mm, L/d = 1.75. The circular cylinder with a slut vortex shedder has the following parameters: d = 14 mm, s = 1.4 mm. We maintain the same size of the model as far as possible to reduce the accidental error.

The meshing grid cell is Quad and the grid type is submap. Because the cylinder with a slut vortex shedder flow field is more complex, the mesh field is divided into nine zones with a dense grid cell for better vortex. At the time of dividing grid, the other models are divided into several different areas with a dense grid cell, to improve the calculation accuracy. The three different vortex shedder's mesh is shown in Figures 2 and 3.

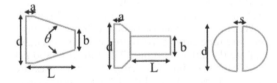

Figure 1.   Three different vortex shedder's structure size.

Figure 2.   Trapezoidal vortex shedder mesh.

Figure 3. T-shaped and circular cylinder with a slut vortex shedder mesh.

2.2 The option of boundary conditions

The material is power law. We ensure that the Re number ranges from 2.0×10^5 to 8.5×10^5. The calculation method of the Re number in power law is given by

$$\mathrm{Re} = \frac{\rho D^n v^{2-n}}{\dfrac{k}{8}\left(\dfrac{6n+2}{n}\right)^n} \qquad (3)$$

where v = apparent viscosity; ρ = fluid density; k = consistency coefficient; and n = power law index.

All of these parameters can be defined in Fluent. For the model viscous, we choose RNGk–ε. The time step changed with Re number ranging from 0.000125 to 0.0005.

2.3 Trapezoidal vortex shedder

First, the trapezoidal vortex shedder is numerically simulated. We set a total of 15 taps, to find the best location for the taps. After the shedder of 0.5D, we set the first position. In addition, we set two taps below and above it. Then, after the three taps every 0.5D, we set up a tap. We define the upper left for tap 1, from top to bottom of tap point 2 to 15, respectively. Finally, in the upstream and downstream of the flow field, respectively, we set a tap and measure the static pressure of them. So, we can get the pressure loss. The trapezoidal vortex shedder pressure flow field and the location of the pressure tap in Re = 5×10^5 are shown in Figure 4.

We read the numerical data and draw every tap's pressure time curve carried on the origin. Then, we analyze the curve with the FFT transform and get the vortex shedding frequency. The time–pressure curve of taps 1 to 5 and the FFT transform curve are shown in Figures 5, 6, 7, 8 and 9.

The time–pressure curve of the tap should be a sine curve. However, with respect to the position of different taps, the changes in pressure in the flow field are also different. This is because some taps' time–pressure curve is just a curve similar to the sine curve but not similar to the sine curve of taps 3, 4 and 5. It is shown that the position of taps

Figure 4. Trapezoidal vortex shedder pressure flow field and location of the pressure tap.

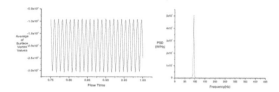

Figure 5. The time–pressure and FFT transform curves of tap 1.

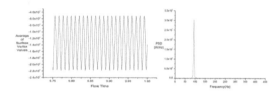

Figure 6. The time–pressure and the FFT transform curves of tap 2.

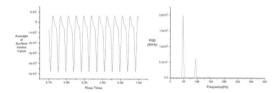

Figure 7. The time–pressure and FFT transform curves of tap 3.

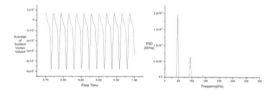

Figure 8. The time–pressure and FFT transform curves of tap 4.

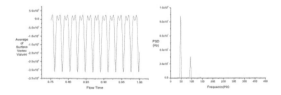

Figure 9. The time–pressure and FFT transform curves of tap 5.

3, 4 and 5 in the flow field is not stable, and that the measurement results will have a deviation. So, the position of taps 1 and 2 is ideal.

Then, we get the vortex shedding frequency by the FFT transform of PSD. However, if the power spectra have two or more peaks, the energy of the main peak is not accounted for the main components. The measured vortex shedding frequency will also have a deviation. This can also mean that the position of taps 1 and 2 is ideal. The rest of the 15 taps, taps 5, 6, 7 and 8 have just one peak in the FFT transform curve. The vortex shedding frequencies of the 15 taps are shown in Figures 10 and 11.

The stability analysis of the taps. The stability of the vortex radial pressure is given by

$$f_p = \frac{\sqrt{\frac{1}{n}(\sum_{i=0}^{n}((p_{ci} - \bar{p}_c)^2 + (p_{ti} - \bar{p}_t)^2)}}{p} \times 100\% \qquad (4)$$

where f_p = stability of the vortex radial pressure; $\bar{p}_c$ = average of the tap's peak value of the pressure signal; $\bar{p}_t$ = average of the trough; p_{ci} = peak of the pressure value; p_{ti} = trough of the pressure value; p = inlet pressure; and n = number of cycle. The stability of the 15 taps is shown in Figure 12.

It is shown that the stability of taps 5, 9 and 15 is more stable. This is because the vortex is more stable with the flow of the flow field. However, the stability of all taps is good. The stability of the vortex shedding cycle in the different taps is given by

$$f_t = \frac{\sqrt{\frac{1}{n}\sum_{i=1}^{n}(T_i - \bar{T})^2}}{\bar{T}} \times 100\% \qquad (5)$$

where f_t = stability of the vortex shedding cycle; T_i = vortex shedding cycle; $\bar{T}$ = average of the vortex shedding; and n = cycle number. The stability of the 15 taps is shown in Figure 13.

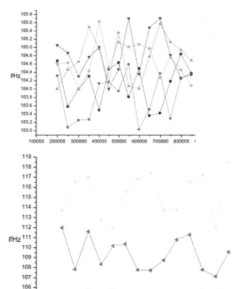

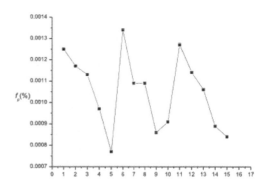

Figure 11. The frequency and Re number of taps 6 to 15.

Figure 12. The stability of the 15 taps.

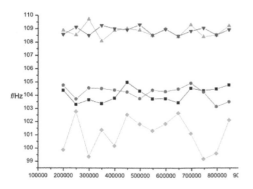

Figure 10. The frequency and Re number of taps 1 to 5.

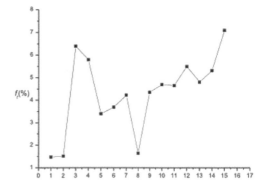

Figure 13. The stability of the 15 taps.

Figure 14. The position of the taps and cylinder with a slut vortex shedder pressure flow field.

Figure 15. The position of the taps and the T-shaped vortex shedder pressure flow field.

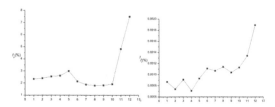

Figure 16. The stability of the vortex radial pressure and the vortex shedding cycle in the cylinder with a slut vortex shedder flow field.

It is found that the stability of taps 1, 2 and 8 is better. We calculate the Strouhal number by the data of vortex shedding frequency. The results are compared with the experimental data, which is 0.1995. We know that the position of taps 1, 2 and 8 is the best. The error rate of the trapezoidal vortex shedder is found to be ranging from 8.98% to 17.7% in different Re numbers. So, we can get a preliminary conclusion that the location of the pressure taps, which is set in full development and vortex stability, is better.

2.4 T-shaped and cylinder with a slut vortex shedder

According to the preliminary conclusion, the position of the taps and the shedder pressure flow field in $Re = 5 \times 10^5$ is shown in Figures 14 and 15.

The time–pressure curve of all taps can be a sine curve, except taps 11 and 12. This is because the vortex is more unstable under the action of resistance. The stability of the cylinder with a slut vortex shedder is shown in Figure 16.

We can see that the position of taps 7, 8, 9 and 10 in the cylinder with a slut vortex shedder flow field is best. Similarly, the position of taps 5, 6, 7 and 8 in the T-shaped vortex shedder flow field is the best. Our conclusion is verified. The error rate of the cylinder with a slut vortex shedder ranges from 9.75% to 18.9%, and that of the T-shaped vortex shedder ranges from 21.3% to 27.6%.

3 CONCLUSIONS

The results show that the location of the pressure taps is not fixed. The location of the pressure taps, which is set in full development and vortex stability, is better.

Among the three, trapezoidal has the best performance, and has been proved to be the ideal vortex shedder in this paper.

REFERENCES

Japanese industrial standard on vertex flowmeter calibration, 2002.
Xu, X. 2004(1). Vortex flowmeter and its application.
Yun, F. Jia. 2009(45). Dynamic stress field distribution characteristics of vortex flow sensor.

Network technology and application

Energy Science and Applied Technology – Fang (Ed.)
© 2016 Taylor & Francis Group, London, ISBN 978-1-138-02833-3

A multi-layer group mobility model for mobile ad hoc networks

Ming Hu, Mingjiao Lan, Xingquan Wang & Jie Zhou
Communication Training Base, General Staff Department of the PLA, Zhangjiakou, China

ABSTRACT: Tactical Internet is an important application of mobile ad hoc networks. Because of the task of battle and the restriction of organization, the Tactical Internet nodes' movement has the characteristics of multi-layer group mobility. In this paper, we propose a group mobility model, namely the Multi-layer Group Mobility (MGM) model, which can respond to the Tactical Internet environment. Moreover, the MGM model is compared with the Reference Point Group Mobility model and the Random Waypoint Mobility model in the physical link and routing protocol aspects. The simulation results show that the MGM model can reflect the topology change and suit the simulation research of mobile ad hoc networks better in the Tactical Internet environment.

Keywords: mobile ad hoc network; Tactical Internet; multi-layer group mobility model

1 INTRODUCTION

Mobile ad hoc networks (MANET) are a collection of randomly moving wireless devices (also called nodes). These networks are particularly suitable for emergency situations such as battle communication, disaster area rescue, wild research and other situations in which infrastructure networks are impossible to operate. Military communication is a primary application field of the MANET, and the most important application is Tactical Internet, which is constituted by radio stations, routers, computer hardware and software. In Tactical Internet, the movement of mobile nodes has the characteristics of layer and distribution, which means that the movement of lower nodes is influenced by the movement of upper nodes, and all nodes have the characteristics of group mobility.

During the protocol simulation of the MANET, the movement of nodes changes the network topology, and the mobility model has a profound effect on the performance of the protocol. So, the mobility model reflecting the characteristics of nodes movement is needed for protocol performance evaluation in Tactical Internet. A group mobility model in the universal environment, namely the Reference Point Group Mobility (RPGM), is proposed, but as a single-layer group mobility model, the RPGM cannot be used in Tactical Internet because its nodes have multi-layer group mobility characteristics. In this paper, we propose a Multi-layer Group Mobility (MGM) model, which can respond to the Tactical Internet environment, and comparison with the Random Waypoint Mobility (RWM) model and the RPGM model is made in both physical link and routing protocol aspects. The simulation results show that the MGM model can really reflect the movement characteristics of nodes and preferably evaluate the performance of the protocol in Tactic Internet.

2 MOBILITY MODEL

The mobility model must really reflect the movement characteristic of mobile nodes. As can be seen from the analysis of MANET mobility models, according to the nodes' movement relativity, there are two types of mobility models: entity mobility model and group mobility model.

In the entity mobility model, a single node's movement and state is independent of others'. Now, there are many representative entity mobility models such as the Random Walk Mobility model, the Random Waypoint Mobility model and the Gauss-Markov Mobility model. Other entity mobility models are improved on these models. Among these models, the RWP is the most widely used entity mobility model, based on which many MANET routing protocols are researched.

In the RWP model, first, each node stays at a random period time Tp in the current position, $T_p \in [T_{min}, T_{max}]$, then it randomly selects a target position in the simulation region, and moves to the target position with a random speed V, $V \in [V_{min}, V_{max}]$. After reaching the target position, the node stays at a random period time Tp, then randomly selects a target position again, and the process is repeated until the end of the simulation.

The major shortcoming of the entity mobility model is that each node moves independently without considering the links among the nodes, so the entity mobility model cannot preferably simulate the group movement of MANET nodes.

In the group mobility model, a single node's movement and state depend on other nodes' within the group, so the movement shows group mobility characteristics, which adapts particularly to the military application. In order to simulate the characteristics of the nodes' group mobility, many kinds of group mobility models are proposed such as the RPGM, the Column Mobility model, the Confined Group Mobility model, and the Double Rank Group Mobility model. Among these group mobility models, the basic model is the proposed RPGM.

In the RPGM model, each group has a logic center called the Reference Point (RP), whose movement and state characteristics affect all the nodes in the group. Each node in the group adds a random offset as own speed and angle based on the speed and angle of the RP. Figure 1 shows an example of the RPGM model.

The speed and angle are expressed as follows:

$$|\vec{V}_{member}(t)| = \\ |\vec{V}_{reference}(t)| + random(\ ) * SDR * \max\_speed \tag{1}$$

$$\theta_{member}(t) = \\ \theta_{reference}(t) + random(\ ) * ADR * \max\_angle \tag{2}$$

where *SDR* and *ADR* are the offset ratio of the speed and the angle that are used to control the offset degree between the RP and nodes, $0 \le SDR$, $ADR \le 1$; *max_speed* and *max_angle* are the maximum offset that the nodes in the group can obtain. In the RPGM, the RWP model is used to describe the movement of the RP.

3 MULTI-LAYER GROUP MOBILITY MODEL

RPGM is a single-layer group mobility model, in which each group has one logic center called the RP. While in Tactical Internet, the nodes' movement shows the characteristics of multi-layer group mobility according to the military establishment. In order to simulate the multi-layer group mobility of mobile nodes, we propose a Multi-layer Group Mobility model in this paper. In the MGM model, RPs belong to different layers. The top-layer RPs move according to the RWP, the speed and direction of the lower RP is a function of the upper RP, to which a random deviation is added. Similarly, the speed and direction of mobile nodes is a function of the lowest RP, to which a random deviation is added. Figure 1 shows an example of the three-layer group mobility model.

The speed and direction of the lower RP in the MGM model are expressed as follows:

$$|\vec{V}_{lower\_ref}| = \\ |\vec{V}_{upper\_ref}| + random(\ ) * SDR * \max\_speed \tag{3}$$

$$\theta_{lower\_ref} = \\ \theta_{upper\_ref} + random(\ ) * ADR * \max\_angle \tag{4}$$

where *SDR* and *ADR* are the offset ratio of the speed and the direction that are used to control the offset degree between the lower RP and the upper RP, $0 \le SDR$, $ADR \le 1$; *max_speed* and *max_angle* are the maximum offset that the lower RP can obtain.

In the MGM model, the top-layer RPs move based on the RWP model, and the lower RPs move with coordinates, as given by the following equations:

$$x\_LRP = \\ x\_URP + random(\ ) * XC * rp\_separation \tag{5}$$

$$y\_LRP = \\ y\_URP + random(\ ) * YC * rp\_separation \tag{6}$$

where *XC* and *YC* are the offset ratios of the X-axis and the Y-axis, $-1 \le XC$, $YC \le 1$; *rp_separation* is the maximum distance between the lower RP and the upper RP.

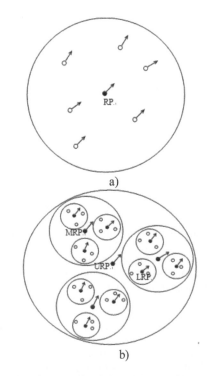

Figure 1. An example of the RPGM model (a) and the three-layer group mobility model (b).

4 EVALUATION CRITERION OF THE MOBILITY MODEL

MANET mobility models can be evaluated in both the physical link and routing protocol aspects. In the physical link aspect, by analyzing the position relationship among nodes, the average neighbor ratio and the average link maintained time can be calculated. In the routing protocol aspect, the average delivery ratio and the average end-to-end delay are analyzed according to RFC2501 and documented. Both of these aspects affect each other. By analyzing the average neighbor ratio and the average link maintained time, the difference in the routing protocol simulation results can be explained. Similarly, the routing protocol simulation results can reflect the influence of the node mobility model for the routing protocol.

Average Neighbor Ratio: Neighbor nodes are the nodes that locate in the transmission range of the specified node. $n_i(t)$ represents the neighbor number of the node i at the time of t, and N is the number of total nodes.

$$Average\ Neighbor\ Ratio = \frac{\sum_{i=1}^{N} \frac{n_i(t)}{N}}{N} \quad (7)$$

Average Link Maintained Time: Link maintained time is the time from one node entering the transmission range of another node to leave the transmission range. Average link maintained time is the average maintained time of all links within the simulation region. In the equation, t_i represents the link maintained time of the link i, and N is the number of total links.

$$Average\ Link\ Maintained\ Time = \sum_{i=1}^{N} t_i/N \quad (8)$$

Average Delivery Ratio: The average delivery ratio is the ratio of network packets received and all the packets sent, which can measure the packets' transmission situation in the network.

Average End-to-End Delay: The average end-to-end delay is the average interval between the source node sending the packets and the destination node receiving the packets, which can measure the real-time packets' transmission in the network.

5 SIMULATION ANALYSIS

The RWP and RPGM models are the most commonly used mobility models in mobile ad hoc network research. We compare the MGM model with the RWP and RPGM models in this paper.

GloMoSim is a scalable simulation library for wireless network systems developed by the UCLA. In this paper, GloMoSim is used for the simulation and performance evaluation of the AODV. All the simulation models have 135 mobile nodes placed within a 3000 m × 3000 m region. The radio propagation range for each node is 100 m with a channel capacity of 2Mbits/s. Each simulation is executed for 500 s. In each mobility model, the speed of all mobile nodes is set from 5 m/s to 50 m/s. In the RPGM model, we set 27 groups, and each group includes 5 nodes. In the MGM model, we set 1 regiment group that includes 3 battalion groups, each battalion group includes 3 company groups, each company group includes 3 platoon groups, and each platoon group includes 5 nodes. The value of *rp_separation* between the regiment RP and the battalion RP is set as 300 meters, and that of *rp_separation* between the battalion RP and the company RP is set as 200 meters, and the rp_separation is set as 150 meters between the company RP and the platoon RP. In each layer, we set 15 sender and receiver pairs, with each source generating a constant bit rate (CBR) traffic of 512 bytes data packet every 2 seconds.

Figure 2 and Figure 3 show that the average neighbor ratio and the average link maintained time can be obtained by analyzing the position relationship among mobile nodes. Because all nodes of the RWP model move randomly in the simulation region, mobile nodes are independent from each other, and the average neighbor ratio and the average link maintained time are low. In the RPGM model, dependency

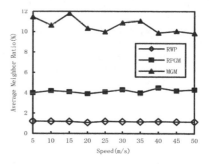

Figure 2.   Average neighbor ratio vs. speed.

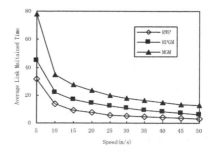

Figure 3.   Average link maintained time vs. speed.

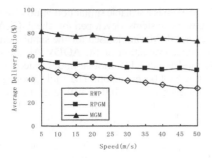

Figure 4.　Average delivery ratio vs. speed.

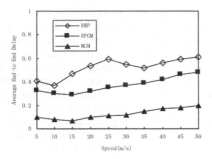

Figure 5.　Average end-to-end delay vs. speed.

among the mobile nodes of the same group leads to a higher average neighbor ratio and a longer average link maintained time compared with the RWP model. In the MGM model, there is a higher average neighbor ratio and a longer average link maintained time compared with the RPGM model because of the relationship among all the nodes.

Figure 4 and Figure 5 show the average delivery ratio and the average end-to-end delay vs. speed.

In Figure 4 and Figure 5, with the increase in the node moving speed, the network topology leads to a more frequent route failure, so the average delivery ratio decreases and the average end-to-end delay increases. In the MGM model, because the relativity of all nodes is larger than the RPGM and RWP models, the average delivery ratio is the largest and the average end-to-end delay is the smallest.

6　CONCLUSION

In order to simulate the performance of the MANET routing protocol in the Tactical Internet environment, we propose a group mobility model, namely the MGM model, in this paper, which can preferably simulate the nodes' movement characteristics of Tactical Internet. Compared with the RWM and RPGM models, the proposed model has an average end-to-end delay and has a larger average neighbor ratio and an average link maintained time and an average delivery ratio. The simulation

results show that different group mobility models have a profound effect in MANET routing protocols. In order to exactly simulate the performance of the MANET, we must establish a suitable mobility model according to different applications.

In this paper, we propose a group mobility model and simulate the performance of this model. In future work, we will emphasize the simulation of the performance of other MANET routing protocols using the MGM model, and propose a routing protocol fitting for Tactical Internet.

REFERENCES

Bai, F., N. Sadagopan, A Helmy. IMPORTANT: A framework to systematically analyze the Impact of Mobility on Performance of Routing protocols for Ad hoc Networks [C]. IEEE INFOCOM 2003.

Bettstetter, C., H. Hartenstein, P. Xavier, C. Rez. Stochastic properties of the random waypoint mobility model: epoch length, direction distribution, and cell change rate [C]. Proceedings of the 5th ACM MSWIN, Atlanta, Georgia, USA, 2002. ACM Press, 2002:7–14.

Campos, C.A.V., D.C. Otero, L.F.M. de Moraes. Realistic individual mobility Markovian models for mobile Ad Hoc networks [C]. Proceedings of IEEE Wireless Communications and Networking Conference (WCNC'04), Atlanta, GA, USA, March 2004. USA: IEEE, 2004, 4: 1980–1985.

Dong, C., P.L. Yang, C. Tian. Group Mobility Model for Ad Hoc Network [J]. Journal of System Simulation, 2006, 18(7): 1879–1883.

Hong, X., M. Gerla, G. Pei, and C.-C. Chiang. A Group Mobility Model for Ad Hoc Wireless Networks [C]. Proceedings of ACM/IEEE MSWIN'99, Seattle, WA. Aug 1999. USA: ACM, 1999:53–60.

Liang, B., Z.J. Haas. Virtual Backbone Generation and Maintenance in Ad hoc Network Mobility management [C]. IEEE INFOCOM 2000.

Michael, M. Stationary distributions of random walk mobility models for wireless ad hoc networks [C]. Proceedings of the 6th ACM Mobihoc, Urbana-Champaign, IL, USA, 2005. ACM Press, 2005:90–98.

Royer E.M., B. Santa, C. Toh. A review of current routing protocols for Ad hoc mobile wireless networks [J]. IEEE Personal Communications. 6(2), 46–55(1999).

Shi, C.J., Q.H. Ren, B. Zhen, Y.J. Liu. Design and Research a Double Rank Group Mobility Model in Mobile Ad Hoc Networks [J]. Journal of System Simulation, 2009, 21(22): 7139–7142.

Tracy Camp, Jeff Boleng, Vanessa Davies. A Survey of Mobility for Ad Hoc Network Research[R]. Dept. of Math. And Computer Sciences Colorado School of Mines, Golden, CO. April 12. 2002.

Xiang, Z., R. Bagrodia, M. Gerla. GloMoSim: a Library for Parallel Simulation of Large-scale Wireless Networks [C]. Proceedings of the 12th Workshop on Parallel and Distributed Simulations, Banff, Alberta, Canada, May 1998.

Xihong Zhang, Aizhen Liu; Chuanxin Gong. Research of Tactical internet Architecture and Its Application in Digital Battlefield [C]. Proceedings of 5th International Symposium on Test and Measurement (ISTM/2003), Atlanta USA, 2003:156–162.

Energy Science and Applied Technology – Fang (Ed.)
© 2016 Taylor & Francis Group, London, ISBN 978-1-138-02833-3

Real-time performance evaluation of Linux ARM virtualization

Feng Gu, Fei Hu & Haopeng Chen
Shanghai Jiao Tong University, Shanghai, China

ABSTRACT: In order to meet the time constraints of real-time tasks, embedded real-time systems are used to assign different real-time tasks to multiple processors. However, using multiple processors makes the system to consume more energy, harder to maintain and to grow in size and weight. In recent years, ARM CPUs have become increasingly common in embedded systems. With the improvement of its performance, more embedded systems tend to use only one high-performance processor to run all tasks. There is a growing demand for a solution to guarantee both isolation and time constraints of the tasks. We present our real-time virtual machine solution, which leverages the newest real-time scheduler of Linux and the hardware virtualization support of ARM CPUs. We take the maximum interrupt frequency of the virtual machine as one real-time performance metric. In addition, we consider the maximum execution frequency of a specific real-time task as the other metric. The results of our experiments demonstrate that the real-time performance of our virtual machine is roughly 23% of the native machine and the real-time performance of the virtual machine is linear with its CPU resource allocation ratio.

Keywords: Linux; ARM; Performance

1 INTRODUCTION

In order to meet the time constraints of real-time tasks, embedded real-time systems are used to assign different real-time tasks to multiple processors. Researchers have used two processors when they implement their flight control system for an unmanned helicopter. Their flight control software runs on an x86 CPU and the task to collect sensor data is assigned to an ARM CPU. The micro car uses an x86 CPU to process an image and an MCU to control the car speed. The multi-processor architecture provides a very strong temporal and spatial isolation, which guarantees that the tasks running on different processors will not preempt each other and their time constraints will be met. However, the multi-processor architecture will consume more energy and make the system more difficult to maintain. Besides, the size and weight of an embedded system will grow rapidly when an additional processor is added to the system. This creates a demand for a solution that can guarantee the isolation of tasks and meet the tasks' time constraints while keeping the system small, lightweight, energy-efficient and of low maintenance.

In recent years, ARM CPUs have become increasingly common in embedded systems because of their advantages in power efficiency. The ARM introduced hardware virtualization support in the latest ARMv7 and ARMv8 architectures, which improves the performance of the virtual machine on ARM CPUs. Although virtualization can provide spatial isolation between virtual machines, some extra effort is needed to ensure that the time constraints of the real-time tasks are met. This means that all the virtual machines and tasks have to be scheduled by real-time schedulers.

Our work makes three main contributions. First, we present a real-time virtual machine solution that uses an EDF scheduler to schedule the tasks and virtual machines. By using our solution, the tasks originally running on multiple processors will be migrated to one processor and will be assigned to several virtual machines to keep the spatial isolation. The scheduling parameters of the tasks and virtual machines will be set according to the time constraints of the tasks. Our implementation of the real-time virtual machine is based on the Linux SCHED_DEADLINE scheduler and the KVM/ARM hypervisor. SCHED_DEADLINE is the newly introduced real-time scheduler in the Linux kernel. It is an implementation of the EDF algorithm in the Linux scheduler. KVM/ARM is the ARM hypervisor in the mainline Linux kernel. It is the first hypervisor to leverage ARM hardware virtualization support to run unmodified guest operating systems on the ARM multicore hardware. Our experimental results demonstrate that KVM/ARM has modest virtualization performance and power costs. KVM/ARM was accepted as the ARM hypervisor in the mainline Linux kernel as of the Linux 3.9 kernel. Because of the dominance of Linux on ARM platforms, KVM/ARM and SCHED_DEADLINE can be easily adopted on ARM platforms. So, our real-time virtual machine solution is realized.

Second, we evaluate the real-time performance of our real-time virtual machine. We use two performance metrics to measure the real-time performance of our virtual machine. The maximum interrupt frequency of the virtual machine is one of the metrics. The maximum execution frequency of a specific real-time task is the other metric. The results of our experiments demonstrate that the real-time performance of our virtual machine is roughly 23% of the native machine and the real-time performance of the virtual machine is linear with its CPU resource allocation ratio.

Finally, we make several recommendations for developers who want to migrate the tasks into real-time virtual machines.

The rest of this paper is organized as follows. Section 2 provides the background of the techniques that we use in our implementation. In the next section, we describe the implementation of the real-time virtual machine. The real-time performance evaluation of our real-time virtual machine is presented in Section 3. Section 4 gives a few recommendations for developers who want to migrate the tasks into real-time virtual machines. Section 5 concludes this paper.

2 BACKGROUND

2.1 *Virtualization*

The virtual machine monitor (VMM) can be classified as type-I, type-II and hybrid VMM. The VMM that runs on bare hardware and performs a direct scheduling and resource allocation job is known as the type-I VMM. The VMM that runs on the host operating system is known as the type-II VMM. The hybrid VMM interprets all privileged instructions of the virtual machine. Meanwhile, the techniques of virtualization can be classified into two categories: full virtualization and para-virtualization. Full virtualization provides support for unmodified guest operating systems, while paravirtualization requires guest operating systems to be explicitly ported for the para-API.

2.2 *KVM/ARM*

ARM introduced a new CPU mode for running hypervisors, called the Hyp mode. However, the Hyp mode targets running a standalone hypervisor underneath the OS kernel, and is not designed to work well with a hosted hypervisor design. KVM/ARM introduces split-mode virtualization that splits the core hypervisor into two components,

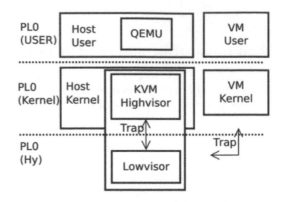

Figure 1. KVM/ARM system architecture (split-mode virtualization).

as shown in Figure 1: lowvisor and highvisor. By using split-mode virtualization, KVM/ARM leverages the existing OS mechanisms without any modification; at the same time, it leverages ARM hardware virtualization features.

2.3 *QEMU*

QEMU is a machine emulator. KVM uses a modified version of QEMU to emulate CPUs and other devices. KVM turns the single-threaded execution model of QEMU into a multi-threaded model. Normally, a QEMU process has two kinds of thread: main I/O thread and VCPU thread. While the I/O thread is used to manage emulated devices, the VCPU thread is used to run the guest code. An AIO thread is dynamically created to handle every AIO request and signal the completion to the I/O thread.

2.4 *SCHED_DEADLINE*

SCHED_DEADLINE is a new scheduling class for the Linux scheduler. It implements the Earliest Deadline First (EDF) real-time scheduling algorithm and uses the Constant Bandwidth Server (CBS) resource reservation scheduling technique to provide temporal isolation among non-interacting tasks. SCHED_DEADLINE can handle periodic tasks, sporadic tasks and aperiodic tasks. SCHED_DEADLINE ensures the temporal isolation of tasks.

2.5 *REAL-TIME COMPUTING*

A real-time system fails if its performance criteria are not met. Within this class of system, there are two broad categories: hard real-time systems and soft real-time systems. A hard real-time system is one that must meet its performance objectives every time and all the time. As soon as one of these systems does not meet one of its performance criteria, it fails. A soft real-time system is one that must meet its performance objectives on average only. This means that if every now and then a performance deadline is missed, the system does not fail. However, if the system repeatedly misses its performance deadlines, then it fails.

Real-time scheduling of virtual machines and real-time tasks.

We use the Linux SCHED_DEADLINE scheduler to schedule our real-time virtual machines. Meanwhile, the tasks running inside the virtual machines are scheduled by the same scheduler provided by the guest Linux kernel. Figure 2 shows the scheduling architecture of our real-time virtual machines.

2.6 Parameter definitions of tasks and virtual machines

SCHED_DEADLINE assigns each task a budget (sched_runtime) and a period (sched_period), considered to be equal to its deadline.

For simplicity, we use P_{vm} to represent the period of a virtual machine and B_{vm} to represent the budget of a virtual machine. We then define the frequency of a virtual machine as follows:

$$freq_{vm} = 1/P_{vm} \qquad (1)$$

The CPU resource allocation ratio of a virtual machine is given by

$$A_{vm} = B_{vm}/P_{vm} \times 100\% \qquad (2)$$

We use P_{task} and B_{task} to represent the period and budget of a task, respectively. The execution frequency of a task is given by

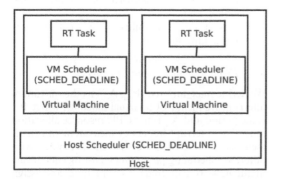

Figure 2. Scheduling architecture of real-time virtual machines. Both virtual machines and tasks are scheduled by the SCHED DEADLINE scheduler, which implements the EDF scheduling algorithm.

$$freq_{task} = 1/P_{task} \qquad (3)$$

The CPU resource allocation ratio of a task is given by

$$A_{task} = \frac{B_{task}}{P_{task}} \times 100\% \qquad (4)$$

Let R_{task} be the time that a real-time task has to run for in one period, then the utilization of a task is given by

$$U_{task} = \frac{R_{task}}{P_{task}} \times 100\% \qquad (5)$$

We use deadline_miss_count$_{task}$ to represent how many times the task missed its deadline during d seconds. Then, the deadline miss ratio of the task in d seconds is given by

$$
\begin{aligned}
deadline\_miss\_ratio_{task} = \\
\frac{deadline\_miss\_count_{task}}{d} \times 100\%
\end{aligned}
\qquad (6)
$$

or abbreviated as DMR_{task}.

A real-time task running in a virtual machine can be described as

$$
\begin{aligned}
W_{task,vm} = (P_{vm}, A_{vm}, freq_{task}, A_{task}, \\
U_{task}, DMR_{task}, d)
\end{aligned}
\qquad (7)
$$

or

$$
\begin{aligned}
W_{task,vm} = (freq_{vm}, A_{vm}, freq_{task}, A_{task}, \\
U_{task}, DMR_{task}, d)
\end{aligned}
\qquad (8)
$$

2.7 Real-time virtual machine

The Linux scheduler schedules the VCPU thread just as a normal thread. In order to make the SCHED_DEADLINE schedule our virtual machine, we make some modification to QEMU. First, we let QEMU accept EDF scheduling parameters. Our modified QEMU can now get its period and budget from command line arguments. Second, we make the VCPU thread set its scheduling parameter (P_{vm} and B_{vm}) at the moment it starts running. Besides, the mlock system call will be made during the start-up procedure of QEMU to keep the virtual machine from being swapped out to the disk.

2.8 Real-time task

The program that we use to simulate the behavior of a periodic real-time task in our real-time virtual machine is called the rt_task. The reason we write a new benchmark program is that we use a new performance metric, the maximum execution frequency of a real-time task, to evaluate the real-time performance of our virtual machine. rt_task gets its P_{rt_task}, B_{rt_task}, R_{rt_task}, and d from command line arguments. It sets its period and budget according to P_{rt_task} and B_{rt_task} when the program starts running. When rt_task is running, it continuously checks wall-clock time and thread cpu time by calling clock_gettime () function with different parameters to see whether it misses the deadline or its thread cpu time has reached R_{rt_task}. The rt_task will output the deadline miss ratio when the program exits after running d seconds. It should be noted that the rt_task has an inevitable overhead so that the actual running time of the rt_task in each period may be more than R_{rt_task}. However, the overhead is relatively small and can be ignored in most situations. This overhead has been excluded in the R_{rt_task} and U_{rt_task} that will be mentioned in the later part of this paper.

3 EVALUATION

In this section, we evaluate the real-time performance of our real-time virtual machine. We use two performance metrics to measure the real-time performance. One is the maximum interrupt frequency of the virtual machine, the other one is the maximum execution frequency of a real-time task. The results of our experiment demonstrate the usability of our real-time virtual machine solution.

3.1 *Experiment setup*

All our experiments are conducted on a Samsung ARM Chromebook. Table 1 lists the specification of the ARM Chromebook and the system information of the host and virtual machines. We refer the guide from Virtual Open Systems when we build the KVM/ARM virtualization environment. Since SCHED_DEADLINE was added to the Linux kernel mainline in version 3.14, we use Linux 3.14.10 to compile the kernel. We enable tracing in the kernel configuration for the sake of getting system scheduling information. We use the modified QEMU to run the real-time virtual machine. We pass "-nographic" argument to QEMU to disable graphical output in order to exclude the overhead brought by the X server. We assign 512 MiB memory and 2 GiB disk space to each virtual machine (Table 1). However, when we observe the system information of the host machine through the top command, we find that each virtual machine occupies nearly 700 MiB memory space.

Table 1. Specification of ARM Chromebook and the system information of the host and the virtual machine.

ARM Chromebook	
CPU	Samsung Exynos 5250
CPU SPEC	Cortex A15; 1.7GHz dual core
GPU	ARM Mali-T604 (Quad Core)
RAM	2 GiB DDR3
Disk	16 GiB SSD (connected to eMMC)
OS	Ubuntu 12.04
Kernel	Linux 3.14.10
Virtual machine	
OS	Ubuntu 12.04
Kernel	Linux 3.14.10
Memory	512 MiB
Disk	2 GiB

3.2 *Experiment purpose*

Real-time task always use clock interrupt to control its period. The maximum interrupt frequency is an important real-time performance metric. The higher clock interrupt rate leads to a better soft real-time performance. We write a simple program (int_freq) to determine the maximum interrupt frequency accordingly.

Most real-time tasks have sampling rate requirements. Only when the frequency of the task is higher than the required sampling rate, the input of the task is valid and the output of the task is useful. So, the maximum execution frequency of a real-time task can be considered as a real-time performance metric of a system.

The purpose of our experiments is to

1. Evaluate the real-time performance of the real-time virtual machine compared with the native machine;
2. Find out the scheduling parameters that affect the real-time performance of the virtual machine; and
3. Demonstrate the usability of our real-time virtual machine solution.

3.3 *Evaluation of interrupt frequency*

We define max_int_freq (P_{vm}, A_{vm}) as the maximum interrupt frequency of a virtual machine, where P_{vm} is the period and A_{vm} is the CPU resource allocation ratio. The interrupt frequency measurement experiment is conducted under virtual machines with different scheduling parameters. During each measurement, we run int_freq for several times and assign the maximum value to max_int_freq(P_{vm}, A_{vm}).

We suppose that both P_{vm} and A_{vm} have an influence on max_int_freq(P_{vm}, A_{vm}). As shown in Table 2, max_int_freq(P_{vm}, A_{vm}) grows very slow with the

Table 2. The relationship between the maximum interrupt frequency and the virtual machine period (P_{vm}) under the single virtual-machine environment ($A_{vm} = 50\%$).

P_{vm} (us)	$Freq_{vm}$ (Hz)	Max_int_freq(P_{vm}, A_{vm}) (Hz)	$Sched_{vm}$
1×10^8	1×10^{-2}	14705	39
1×10^7	0.1	14705	59
1×10^6	1	8474	87
1×10^5	10	8474	443
10000	100	8000	2729
1000	1000	7142	4395
100	10000	7092	3962
10	1×10^5	6849	3914

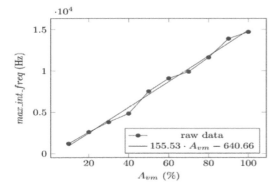

Figure 3. The relationship between the maximum interrupt frequency and the resource allocation of the virtual machine under the single virtual-machine environment ($P_{vm} = 1000$ us).

Table 3. Comparison of the maximum interrupt frequency between the native and virtual machines ($P_{vm} = 1000$ us, $A_{vm} = 100\%$).

Max_int_freq (Hz)		vm/native ratio (%)
vm	Native	
14705	62500	23.52

increase in P_{vm}. The $sched_{vm}$ column in Table 2 gives the time data that the virtual machine has been scheduled during the measurement. When P_{vm} is too large, the virtual machine almost occupies whole of the CPU resource so that max_int_freq(P_{vm}, A_{vm}) becomes abnormally large. As shown in Figure 3, max_int_freq(P_{vm}, A_{vm}) is linear with A_{vm}. Besides, we compare the max_int_freq(P_{vm}, A_{vm}) of our real-time virtual machine with that of the native machine. Table 3 summarizes the result of the comparison.

According to the result of the measurement experiment of max_int_freq(P_{vm}, A_{vm}), we can conclude that the virtual machine will cause a decrease in max_int_freq(P_{vm}, A_{vm}). max_int_freq(P_{vm}, A_{vm}) is linear with A_{vm}. P_{vm} has a minor influence on max_int_freq(P_{vm}, A_{vm}).

3.4 Evaluation of real-time task execution frequency

We use rt_task to simulate a real-time task. Before we evaluate the real-time performance of the virtual machine, we have to confirm a fact that DMR_{rt_task} will grow with $freq_{rt\_task}$ when we fix the values of U_{rt_task} and A_{rt_task}. As shown in Figure 4, DMR_{rt_task} does grow with $freq_{vm}$. We use $DMR_{threshold}$ to represent a threshold of DMR_{rt_task}. Based on the fact that we confirmed above, when a $DMR_{threshold}$ is given, we can find a maximum $freq_{rt\_task}$ that satisfies $DMR_{rt\_task} < DMR_{threshold}$. We use max_$freq_{rt\_task}$ to represent the maximum $freq_{rt\_task}$. A bisection method has been used to find max_$freq_{rt\_task}$. Algorithm 1 gives the bisection method. We set $period_{precision} = 1000$ns.

However, from Figure 4, we find that when $freq_{rt\_task} > freq_{vm}$, DMR_{rt_task} can still be very low (some are close to 0%). We guess that $freq_{vm}$ is not relevant to the actual scheduling frequency of the virtual machine. In Table 4, the scheduling frequency of the real-time task is compared with that of the virtual machine. We obtain the scheduling information by using ftrace. We can see that the scheduling frequency of the virtual machine just grows with the scheduling frequency of the rt_task as long as the DMR_{rt_task} is acceptably low. We can conclude that P_{vm} (or $freq_{vm}$) is not relevant to the actual scheduling frequency of the virtual machine. It is just the EDF scheduling parameter of the virtual machine.

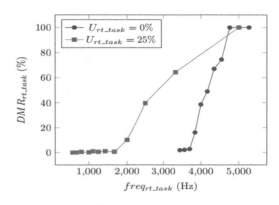

Figure 4. Task deadline miss ratio with different task frequencies under the single virtual-machine environment $P_{vm} = 1000$ us, $A_{vm} = 50\%$, $A_{rt\_task} = A_{vm}$).

Table 4 Comparison between the scheduling frequency of the virtual machine and the scheduling frequency of rt task. (P_{vm} = 1000 us, A_{vm} = 50%, A_{rt_task} = A_{vm}, U_{rt_task} = 0, d = 20 sec).

P_{rt_task} (us)	Scheduling Info of rt_task				Scheduling Info of vm		ratio
	$freq_{rt\_task}$ (Hz)	DMR_{rt_task} (%)	sched count	sched freq (Hz)	sched count	sched freq (Hz)	
$1 \cdot 10^5$	10,000	100	82,476	4,124	7,558	378	$9 \cdot 10^{-2}$
$2 \cdot 10^5$	5,000	100	94,700	4,735	27,354	1,363	0.29
$2.5 \cdot 10^5$	4,000	35.11	80,011	4,001	72,145	3,607	0.9
$3 \cdot 10^5$	3,333	3.51	66,676	3,334	$1.04 \cdot 10^5$	5,216	1.56
$4 \cdot 10^5$	2,500	0.38	50,003	2,500	93,928	4,696	1.88
$8 \cdot 10^5$	1,250	$1 \cdot 10^{-3}$	25,003	1,250	44,698	2,235	1.79
$1 \cdot 10^6$	1,000	0	20,003	1,000	38,651	1,933	1.93
$2 \cdot 10^6$	500	0	10,004	500	16,155	808	1.61

Algorithm 1 Bisection method to find max_freq$_{rt\_task}$.

Algorithm 1 Bisection method to find $max\_freq_{rt\_task}$

Require: Set P_{rt_task}, U_{rt_task}, $DMR_{threshold}$
Require: Set $DMR_{threshold}$, $period_{precision}$
1: *last_pass_period* $\Leftarrow$ null
2: *last_fail_period* $\Leftarrow$ 0
3: **repeat**
4:   run *rt_task* with P_{rt_task} and U_{rt_task}
5:   get DMR_{rt_task} form *rt_task* log
6:   **if** $DMR_{rt\_task} < DMR_{threshold}$ **then**
7:     *last_pass_period* $\Leftarrow$ P_{rt_task}
8:   **else**
9:     *last_fail_period* $\Leftarrow$ P_{rt_task}
10:   **end if**
11:   *last_period* $\Leftarrow$ P_{rt_task}
12:   **if** *last_pass_period* == null **then**
13:     $P_{rt\_task} \Leftarrow P_{rt\_task} \times 2$
14:   **else**
15:     $P_{rt\_task} \Leftarrow$ (*last_pass_period* + *last_fail_period*)/2
16:   **end if**
17: **until** *last_pass_period* $\neq$ null $\wedge$ $abs(P_{rt\_task}$ − *last_period*) < $period_{precision}$
18: $freq_{rt\_task} \Leftarrow \frac{1}{last\_pass\_period}$
19: $max\_freq_{rt\_task} \Leftarrow freq_{rt\_task}$
20: **return** $max\_freq_{rt\_task}$

3.5 Single VM environment

We suppose that max_freq$_{rt\_task}$ is relevant to P_{vm} and A_{vm}.

Figure 5 shows the relationship between max_freq$_{rt\_task}$ and P_{vm}. A very small P_{vm} does incur much more overhead. However, when P_{vm} varies in a normal range, max_freq$_{rt\_task}$ is very steady. This indicates that the influence of P_{vm} on max_freq$_{rt\_task}$ can be ignored when P_{vm} is set to a reasonable value. The sharp increase in max_freq$_{rt\_task}$ is caused by the extremely large P_{vm}, which makes the virtual machine occupy almost whole of the CPU resource during the experiment. Figure 6 shows the relationship between max_freq$_{rt\_task}$ and A_{vm}. We can clearly see that when U_{rt_task} $ A_{rt_task}$ $ A_{vm}$, max_freq$_{rt_task}$ is linear with A_{vm}. We then compare the max_freq$_{rt\_task}$ of the single VM environment

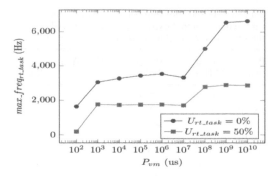

Figure 5. The relationship between the maximum task frequency and the period of the virtual machine under the single virtual-machine environment (A_{vm} = 50%, $DMR_{threshold}$ = 1%, A_{rt_task} = A_{vm}, d = 20 sec).

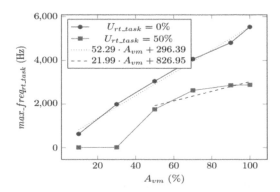

Figure 6. The relationship between the maximum task frequency and the virtual machine resource allocation under the single virtual-machine environment (P_{vm} = 1000 us, $DMR_{threshold}$ = 1%, A_{rt_task} = A_{vm}, d = 20 sec).

Table 5. Comparison of the maximum task frequency between the native environment and the single virtual-machine environment (P_{vm} = 1000 us, A_{vm} = 100%, $DMR_{threshold}$ = 1%, A_{rt_task} = A_{vm}, d = 20 sec).

U_{rt_task} (%)	Max_freq$_{rt\_task}$		vm/native ratio (%)
	vm	native	
0	5535.17	22756.2	24.32
50	2880.72	12642.1	22.79

with that of the native machine. Table 5 gives the result of the comparison. We can see that the performance of the real-time task is close to that of the interrupt. We can conclude that max_freq$_{rt\_task}$ is not relevant to P_{vm} as long as P_{vm} is set to a reasonable value, and max_freq$_{rt\_task}$ is linear with A_{vm} in the single virtual-machine environment.

4 RECOMMENDATIONS

According to our evaluation of the real-time performance of our real-time virtual machine, we offer a few recommendations for developers who want to migrate real-time tasks into real-time virtual machines.

We consider the decrease in real-time performance. The results of our experiment show that the decrease in real-time performance is quite significant. The real-time virtual machine only achieves roughly 23% performance of the native machine. We need to ensure that sampling rate requirements will still be met when migrating the tasks into the virtual machine.

We need to ensure that the maximum execution frequency of the task is affected by the CPU resource allocation ratio of the virtual machine. Not only the high-utilization tasks need more CPU resource, but also the high-frequency tasks need more CPU resource.

We set the period of the virtual machine to a reasonable value. If the period of the virtual machine is too small, it will incur much more overhead. Taking the ARM Chromebook as an example, the period of a virtual machine should be larger than 10^3 us.

5 CONCLUSIONS

In this paper, we present our real-time virtual machine solution to migrate real-time tasks from a multiple-processor architecture to a single-processor architecture. We use the KVM/ARM and Linux SCHED_DEADLINE scheduler to implement our real-time virtual machine. Our experimental results demonstrate that the real-time performance of the virtual machine is linear with its CPU resource allocation ratio. Meanwhile, the real-time performance of our real-time virtual machine is roughly 23% of that of the native machine. However, our solution is still usable and we offer several recommendations for developers who want to migrate the tasks into real-time virtual machines.

REFERENCES

A guide to setup KVM virtualization on samsung chromebook based on ARM cortex-a15. http://www.virtualopensystems.com/en/solutions/ guides/kvm-on-chromebook/.

Bellard, F. QEMU, a fast and portable dynamic translator. In USENIX Annual Technical Conference, FREENIX Track, pages 41–46, 2005.

Dall C. and J. Nieh. KVM/ARM: the design and implementation of the linux ARM hypervisor. In Proceedings of the 19th international conference on Architectural support for programming languages and operating systems, pages 333–348. ACM, 2014.

Etsion, Y., D. Tsafrir, and D.G. Feitelson. Effects of clock resolution on the scheduling of interactive and soft real-time processes. In ACM SIGMETRICS Performance Evaluation Review, volume 31, pages 172–183. ACM, 2003.

Faggioli, D., F. Checconi, M. Trimarchi, and C. Scordino. An EDF scheduling class for the linux kernel. In Proc. of the Real-Time Linux Workshop, 2009.

Goldberg, R.P. Architectural principles for virtual computer systems. Technical report, DTIC Document, 1973.

Hillary, N. Measuring performance for real-time systems. Freescale Semiconductor, November, 2005.

Kivity, A., Y. Kamay, D. Laor, U. Lublin, and A. Liguori. kvm: the linux virtual machine monitor. In Proceedings of the Linux Symposium, volume 1, pages 225–230, 2007.

Liu C.L. and J. W. Layland. Scheduling algorithms for multiprogramming in a hard-real-time environment. Journal of the ACM (JACM), 20(1):46– 61, 1973.

Paravirtualization. http://en.wikipedia.org/wiki/Paravirtualization, Aug. 2014. Page Version ID: 612470011.

Parri, A., J. Lelli, and M. Marinoni. Design and Implementation of the Multiprocessor Bandwidth Inheritance Protocol on Linux.

Samsung ARM chromebook -the chromium projects. http://www.chromium.org/chromium-os/developer-information-for-chrome-os-devices/samsung-arm-chromebook.

Topic: How to determine linux kernel timer interrupt frequency-ADVENAGE. http://www.advenage.com/topics/linux-timer-interrupt-frequency.php.

Wu, Y., F. Hu, and J. An. Design and implementation of flight control system software for unmanned helicopter. In Proceedings of 2011 International Conference on Computer Science and Network Technology, volume 4, pages 2196–2200, 2011.

Yang, M., Z. Lu, L. Guo, B. Wang, and C. Wang. Vision-based environmental perception and navigation of micro-intelligent vehicles. In Foundations and Applications of Intelligent Systems, pages 653–665. Springer, 2014.

Zhang, J., K. Chen, B. Zuo, R. Ma, Y. Dong, and H. Guan. Performance analysis towards a kvm-based embedded real-time virtualization architecture. In Computer Sciences and Convergence Information Technology (ICCIT), 2010 5th International Conference on, pages 421–426. IEEE, 2010.

Energy Science and Applied Technology – Fang (Ed.)
© 2016 Taylor & Francis Group, London, ISBN 978-1-138-02833-3

Cost optimization for green internet service in Content Distribution Networks

Haihang Zhou, Fei Hu & Haopeng Chen
Shanghai Jiao Tong University, Shanghai, China

ABSTRACT: Content service is an important type of Internet service to provide various contents to the end-users. To ensure the performance for content delivering, content service providers utilize a delivery framework: contents are collected or generated by multi-regional data centers and distributed by Content Distribution Networks (CDNs). The cost for content service mainly includes two components: electricity cost for data centers and usage cost for CDNs. As electricity prices vary within data centers and usage costs vary within CDNs, scheduling requests among data centers and CDNs may gain tremendous content service cost savings. In contrast, with the growing environmental issues (e.g. carbon emissions, global warming) and the depletion of traditional energy, the content service provider is under pressure (e.g. from environmental protection organizations) to improve the structure of energy consumed in data centers. In this paper, we optimize the content service cost with a real-life constraint and a renewable energy consumption constraint. The performance of the optimization problem is evaluated by using real-time electricity prices, real-life CDN usage prices and workload traces. The experimental results demonstrate that the proposed optimization method can effectively reduce the content service cost by more than 5.7% under the given green constraint.

Keywords: Design; Algorithms; Performance

1 INTRODUCTION

Many content service providers on the Internet such as Google and Facebook use Content Distribution Networks (CDNs) to deliver their content, such as video streaming. For instance, two of the most popular video service providers, Hulu and Netflix, adopt three CDNs, Limelight, Akamai and Level3, to deliver their contents. Besides CDNs, data center is also an important infrastructure used in content distribution. From the viewpoint of content service, a data center serves as the origin of content: it generates and stores content, and then serves and sends back the end-users' requests for the content; while the CDN acts like a content distributor: it replicates part of useful contents from a data center and serves the end-user's requests with the locally stored contents.

As the Internet service demand has grown rapidly in recent years, the consumption and cost of energy in data centers has increased dramatically. It has been reported that there is a total cost of about 5.6M dollars paid by large-scale data centers per year; data centers contribute to the fifth largest electricity consumption if we view them as a whole. There is much work focused on electricity cost minimization in data centers by distributing end-user's requests with dynamic electricity prices among multi-regional data centers or among different time slots. Besides electricity cost reduction in data centers, content service providers also focus on minimizing CDN usage costs. In addition, researchers gradually pay their attention on the problems of CDN selection for content delivery and minimization of the CDN usage cost.

Although content service cost is important in distributing requests among multi-regional data centers, the structure of energy consumption is also an essential problem for content service providers. With the growing negative environment implications (e.g. carbon emissions, global warming) and the depletion of brown energy, governments and environmental protection organizations exert more pressure on large-scale energy consumption companies. It has been reported that Apple aims to use more renewable energy in its data centers under the pressure of Greenpeace, which grades the Internet companies on filthiness of their cloud operations and Apple gains the lowest score in renewable energy in 2011. The US government provides up to 30% tax credit to promote renewable energy consumption. Improving the renewable energy consumption in data centers is an important problem. Researchers make the content service greener by scheduling renewable-aware workload in a single data center or among multi-regional data centers.

To reduce the content service cost, most recent works take the approach of either minimizing the electricity cost in data centers or minimizing the CDN usage costs. Few studies take both of the two approaches into consideration. The work has investigated the content service cost while taking both of the two approaches into account. However, renewable energy consumption of data centers has not been considered while scheduling requests among multi-regional data centers.

In this paper, we focus on how to choose CDNs and data centers on scheduling requests to minimize the content service cost while maintaining the renewable energy consumption ratio. We minimize the content service cost through a constrained optimization problem from the perspective of content service providers.

This paper makes two major contributions.

First, we optimize the sum of the electricity cost in data centers and CDN usage costs, while maintaining the renewable energy consumption ratio to a guaranteed proportion and satisfying other real-world constraints, such as processing capacity for CDNs and data centers. We generate the cost and green efficient request dispatching solution based on the proposed optimization problem.

Second, we evaluate the proposed optimization problem with real-time electricity prices, real-life CDN usage prices and workload traces. Our experimental results show that, compared with the traditional method in distributing requests among CDNs and data centers, the proposed optimization method can significantly reduce the content service cost with the given green constraint of renewable energy consumption ratio.

The rest of this paper is organized as follows. First, we formulate the content service cost minimization problem. Then, we discuss the simulation method and the experimental results. Finally, we conclude the paper.

2 PROBLEM

In this section, we model the content service entities and their relations. A cost optimization problem is also formulated to minimize the sum of data center cost and CDN usage cost while considering the ratio of the renewable energy consumption. The parameters referred to in the models and their corresponding meanings are listed in Table 1.

Data Center: A data center consists of a cluster of servers to serve the data requests from the CDN. In the modern power grid, electricity prices in data centers are time diversity and regional

diversity. The energy structure of data centers, which we define as the ratio of renewable energy consumption to total energy consumption, varies from region to region.

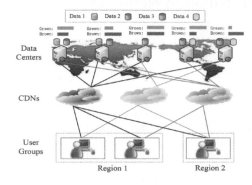

Figure 1. Relationships between user groups, data centers and CDNs.

Table 1. Notaions.

Notation	Definition
u_a	User group at area a
c_i	The i_{th} CDN
d_k	The k_{th} data center
S_q	The q_{th} type of data
$f(u_a, c_i, d_k, S_q)$	Number of requests from user group u_a through CDN c_i to data center d_k for data S_q
g	The given ratio of renewable energy consumption in the electricity bills of the content service provider
$G(d_k)$	The ratio of renewable energy consumption in each data center
A_i^l	Set of charging region l in the i_{th} CDN
$P_c(c_i, l)$	Pricing function of CDN c_i in charging region l
$P_d(d_k, S_q)$	Pricing function for processing per request for S_q in data center d_k
$N(d_k)$	Number of servers in data center d_k
$\mu(S_q)$	Number of requests a server could handle in one period (e.g. 1 second) of data S_q
$S(S_q)$	Average size of data S_q
$B(d_k, S_q)$	Boolean function; if data center d_k has the data S_q, the value is true, otherwise, false
$D(c_i, S_q)$	Boolean function; if CDN C_i can distribute the request to a data center which has the desired data S_q, then the value is true, otherwise, false
$r(u_a, S_q)$	Total requests for data S_q from user group u_a
$Q(c_i)$	The size of traffic a CDN could serve in one period (e.g. 1 second) with satisfactory performance for content delivering

410

Data: Data is modeled to represent a collection of content objects that the user groups request. There are always several content objects for the content service provider to serve, such as video streaming, a message or an email. Although there are both dynamic and static contents in a request, we only consider the dynamic one because the static content is always served by the replica servers in CDNs.

Cost Calculation. The content service providers need to pay the bills for the data center electricity cost and the CDN usage cost, which are the two main components of the cost function in our optimization problem. We will describe both of them in detail.

First, we calculate the CDN usage cost. The pricing of CDN usage is always related to its different charging regions (e.g. user group from Asia or Africa) and different traffic sizes that the user group sends to the CDN. The usage cost of the CDN paid by the content service provider can be formulated as follows:

$$P_{cdn} = \sum_{d_k} \sum_{c_i} \sum_{l} P_c(c_i, l) \sum_{s_q} S(s_q) \sum_{a \in A_i^l} f(u_a, c_i, d_k, s_q)$$
(1)

Second, we consider the data center electricity cost that depends on the electricity price and the number of requests and the data center electricity bills that the content service provider should pay, which can be formulated as follows:

$$P_{dc} = \sum_{d_k} \sum_{s_q} P_d(d_k, s_q) \sum_{c_i} \sum_{u_a} f(u_a, c_i, d_k, s_q).$$
(2)

Constraints. We have formulated the cost of content service in the above subsection. To satisfy the green demand of the content service provider and reflect the actual situations, we apply the constraints as follows.

For a data center d_k, the requests for data s_q will not be distributed to d_k if it does not have the data s_q, which means the corresponding number of requests $f(u_a, c_i, d_k, s_q) = 0$. The Boolean function $B(d_k, s_q)$ is used to represent the data set status of d_k, while the value equals 1 means the data center d_k has the data s_q and vice versa, which is given by

$$B(d_k, s_q) = \begin{cases} 1, & c_k \text{ has } s_q, \\ 0, & \text{otherwise}. \end{cases}$$
(3)

Then, the rule that the requests for data s_q will not be distributed to the data center d_k if the data center d_k does not have s_q can be formulated as follows:

$$\forall u_a, c_i, d_k, s_q, s.t. \ B(d_k, s_q) = 0, f(u_a, c_i, d_k, s_q) = 0$$
(4)

For a CDN c_i, the requests for the data s_q will not be allocated to c_i if c_i is not connected with a data center that has the data s_q. To further illustrate this situation, we define a Boolean function as

$$D(c_i, s_q) = \begin{cases} 1, & c_i \text{ can get } s_q, \\ 0, & \text{otherwise}. \end{cases}$$
(5)

The value of the Boolean function $D(c_i, s_q)$ equals 1, when the CDN c_i can distribute the requests for the data s_q to one or more of its connected data centers; otherwise, the value is 0. Thus, the situation that requests from the user group u_a will not allocate the data s_q through the CDN c_i can be represented as

$$\forall u_a, c_i, d_k, s_q, s.t. \ D(c_i, s_q) = 0, f(u_a, c_i, d_k, s_q) = 0.$$
(6)

To be environmental friendly, the content service provider focuses on improving the energy structure. We define the renewable energy consumption ratio as follows: the consumed electricity that is generated by renewable energy in the consumed electricity that is generated by all types of energy. The ratio in each data center d_k is represented as $G(d_k)$. With the given renewable energy consumption ratio g by the content service provider, the average ratio should be no less than g as follows:

$$\sum_{d_k} \sum_{s_q} \sum_{c_i} \sum_{u_a} f(u_a, c_i, d_k, s_q) G(d_k) / \sum_{u_a} \sum_{s_q} r(u_a, s_q) \geq g$$
(7)

The requests from each user group in one period should be handled. Thus, we formulate this constraint as

$$\forall u_a, s_q, \sum_{d_k} \sum_{c_i} f(u_a, c_i, d_k, s_q) = r(u_a, s_q).$$
(8)

It should be guaranteed that the data centers have the ability to handle the distributed requests, which means the performance for the content service should be guaranteed. The above constraint can be formulated as

$$\forall d_k, \sum_{s_q} \sum_{c_i} \sum_{u_a} f(u_a, c_i, d_k, s_q) / \mu(s_q) \leq N(d_k).$$
(9)

411

Similarly, the processing capacity of CDNs should also be guaranteed

$$\forall c_i, \sum_{s_q} S(s_q) \sum_{u_a} f(u_a, c_i, d_k, s_q) \leq Q(c_i). \quad (10)$$

The assigned requests must be non-negative:

$$\forall u_a, c_i, d_k, s_q, f(u_a, c_i, d_k, s_q) \geq 0. \quad (11)$$

Cost Optimization Problem Formulation. As discussed above, we model the cost minimization problem as a request distribution problem. The optimization problem can be formulated as follows:

$$\min_{f \in Z} P_{cdn} + P_{dc} \quad (12)$$

Subject to the constraints of (4), (6), (7), (8), (9), (10) and (11), it should be noted that the content service providers should pay for the total cost.

3 EVALUATIONS SETUP

In this section, we use the realistic parameters to setup our experiments to evaluate the performance of our proposed optimization method. First, we describe the real-life evaluation setup parameters, in particular the real-time electricity prices, the pricing function of CDNs, and the web request traces, as well as the ratio of renewable energy consumption. The experiment setup includes four main parts: data centers and workload data, the CDN usage pricing, the dynamic electricity price, and the ratio of the renewable energy consumption.

Data Centers and Workload Data. To evaluate the performance of our proposed optimization problem, we use the network configuration, as shown in Figure 1. We consider an evaluation environment that a content service provider runs five data centers in five different regions: Columbia District, North Dakota, Washington, Oregon and Arkansas. The number of running servers in the five data centers are 35000, 40000, 20000, 30000 and 40000, respectively.

To build our workloads in the experiment, we use the data of average number of requests per second on an EPA server (hereafter referred to as empirical workload) to represent the workload data. The workload data is generated 33 times for each request type $r(u_a, s_q)$ to evaluate the performance of the proposed optimization method in a massive workload.

Next, we consider the configuration of power consumption of each data center server. The processing speed of each server $\mu(s_q)$ in processing the four types of data is assumed to be 10, 6,

7 and 11 requests per second, respectively. In addition, the size of the four types of data is assumed to be 0.004MB, 0.010MB, 0.006MB and 0.011MB, respectively. We further assume that the servers in the data centers are power-proportional, which means the power consumption of the servers is linear to their workload. Accordingly, a server will consume 450 Watts and 270 Watts when running at the peak speed and idle, respectively. The power consumption of one server is assumed to be same in all of the data centers, which is true when homogeneous servers are configured in each data center.

CDN Usage Pricing. We assume that the content service provider can employ three CDNs. The real-life usage pricing function used by the three CDNs is as follows: MaxCDN, CDN77 and Cloud Front. The pricing function of CDNs indicate the following: different CDNs may have different charging regions or different charging fee structures; for a same CDN, the CDN usage price may be different while the user groups are in different charging regions or the monthly traffic varies. Thus, we assume there are three user groups, where the first two groups are located in France, and the third one is located in Hong Kong.

Renewable Energy Consumption Ratio. With the growing negative environment implication (e.g. global warming, carbon emission), the US government maintains the data of historical time series of energy production, consumption and prices for analysis and forecasting. To further illustrate, we define an index named the renewable energy consumption ratio g as follows:

$$g = \frac{\text{Renewable Energy Consumption}}{\text{Total Energy Consumption}} \quad (13)$$

This is an important index for public organizations (e.g. environment protection organization) or government to evaluate the energy consumption structure of the content service provider. The real-life data of renewable energy consumption and total energy consumption comes from the US Energy Information Administration (EIA) in 2012 of five states: Columbia District, North Dakota, Washington, Oregon, in which we assume that the five data centers are located. In our experiments, the energy structure of electricity consumed in the data center is assumed to be consistent with that in the home state of the data center.

Electricity Pricing Data. In the modern power grid, the electricity system in the USA can run time and location diversity mechanisms, which gives consumers' pricing information for the actual cost of electricity at any time. To calculate the electricity cost in data centers, we use the real-time electricity price trace from the New York Independent System Oper-

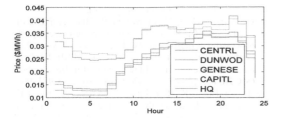

Figure 2.   Real-time electricity price.

Table 2.   Distance between user and the corresponding closest replica servers in each CDN.

Distance (k)	Max CDN	CDN77	Cloud front
Paris, France	380	370	440
Lyon, France	380	470	400
Hong Kong, China	210	180	190

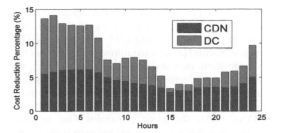

Figure 3.   Cost reduction under the same renewable energy consumption ratio.

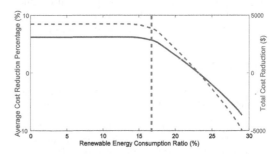

Figure 4. Relationship between cost reduction and renewable energy consumption ratio

ator (NYISO). The obtained day-head market price data is of 24 hours in June 1st, 2014 of five regions: CENTRL, DUNWOD, GENESE, CAPITL and HQ, which is used to simulate electricity price in the five data centers, respectively. Figure 2 shows the real-time electricity price in the five regions.

4   EVALUATION RESULTS

Compared with the traditional method, we evaluate the maximum content service cost reduction and the maximum renewable energy consumption ratio improvement gained by the proposed optimization method.

To evaluate the performance of the proposed optimization method in this paper, we compare the sum of the electricity cost and the CDN usage cost between the proposed optimization method and the traditional method. The traditional method adopts two main principles in distributing requests among the user groups, CDNs and data centers that (1) the user group sends its requests to a nearest replica server in CDNs to handle the requested data and (2) a CDN sends its requests equally to data centers with a desired data copy. The distance between user groups and the nearest replica servers in CDNs are listed in Table 2. The monthly traffic level is set to be the minimum level, and the processing capacity of each CDN is set to be 40GB.

First, we evaluate the reduction in the content service cost gained by the proposed optimization method under the constraint of the same renewable energy consumption ratio with the traditional method. According to the principles mentioned previously, the traditional method can achieve a renewable energy consumption ratio of about 16.84% in our experiments. Figure 3 shows the content service cost savings achieved by the proposed optimization method in twenty-four hours. Considering the regional diversity of the electricity price and the CDN usage cost, the user groups' requests are delivered to the most economical data centers and CDNs under the given green constraint. The achieved electricity cost reduction between 1 and 7H is more than that in the subse-

quent time of the day, which is caused by the electricity price difference among the five data centers that becomes closer to 8H, as shown in Figure 2. In average, the content service provider can gain more than 5.7% cost reduction in a day with the proposed optimization method.

In contrast, with the growing negative environment implications (e.g. carbon emissions, global warming), more and more content service providers take the renewable energy consumption into consideration. The goal of these companies is to improve the energy structure of the electricity consumed in their data centers. Thus, we evaluate the relationship between the content service cost reduction and the renewable energy consumption ratio, as shown in Figure 4. The red dash line represents the renewable energy consumption ratio achieved by the traditional method. It should be noted that with the growth of the renewable energy consumption ratio, the cost reduction may decrease due to the second-best economical request distribution under the green constraint. Further, when the ratio passes 23%, the content service is more expensive. The maximum ratio gained by the optimization method is 29%, which is more than 12% compared with the traditional method.

5 CONCLUSION

Three key questions faced by the content service provider are (1) how to reduce data center operating costs (e.g. electricity costs), (2) how to reduce CDN usage costs, and (3) how to improve renewable energy consumption used in data centers. In this paper, we tackle the problems by formulating an optimization problem with the cost function of the sum of electricity costs for data centers and usage costs for CDNs, under the green constraint. Our solution effectively delivers the requests between user groups, CDNs and data centers. With real-time electricity prices, real-life CDN usage prices and workload traces, our experimental results show that the proposed optimization method can significantly reduce the content service cost by more than 5.7% under the given renewable energy consumption ratio.

REFERENCES

"Apple aims to shrink its carbon footprint with new data centers." [Online]. Available: http://www.wired.com /2014/04/green-apple/.

Adhikari, V., Y. Guo, F. Hao, V. Hilt, and Z.-L. Zhang, "A tale of three cdns: An active measurement study of hulu and its cdns," in Proceedings of IEEE Conference on Computer Communications Workshops (INFOCOM WKSHPS), 2012, pp. 7–12.

Available: http://ita.ee.lbl.gov/.

Available: http://www.eia.gov/state/seds/.

Available: http://www.nyiso.com/.

Cook, G. and J. Van Horn, "How dirty is your data," A Look at the Energy Choices That Power Cloud Computing, 2011.

Fan, X., W.-D. Weber, and L. A. Barroso, "Power provisioning for a warehouse-sized computer," in Proceedings of the 34th Annual International Symposium on Computer Architecture (ISCA'07), San Diego, California, USA, 2007, pp. 13–23.

Goldenberg, D.K., L. Qiuy, H. Xie, Y.R. Yang, and Y. Zhang, "Optimizing cost and performance for multihoming," in Proceedings of the 2004 Conference on Applications, Technologies, Architectures, and Protocols for Computer Communications (SIGCOMM'04), New York, NY, USA, 2004, pp. 79–92.

Guo, Y. and Y. Fang, "Electricity cost saving strategy in data centers by using energy storage," IEEE Transactions on Parallel and Distributed Systems, vol. 24, no. 6, pp. 1149–1160, 2013.

Hamilton, J. "Cost of power in large-scale data centers." [Online]. Available: http://perspectives.mvdirona.com.

Information on: http://www.energystar.gov.

Kayaaslan, E., B.B. Cambazoglu, R. Blanco, F.P. Junqueira, and C. Aykanat, "Energy-price-driven query processing in multi-center web search engines," in Proceedings of the 34th International ACM SIGIR Conference on Research and Development in Information (SIGIR'11), New York, NY, USA, 2011, pp. 983–992.

Le, K., O. Bilgir, R. Bianchini, M. Martonosi, and T.D. Nguyen, "Managing the cost, energy consumption, and carbon footprint of internet services," in ACM SIGMETRICS Performance Evaluation Review, vol. 38, no. 1. ACM, 2010, pp. 357–358.

Li, C., W. Zhang, C.-B. Cho, and T. Li, "Solarcore: Solar energy driven multi-core architecture power management," in High Performance Computer Architecture (HPCA), 2011 IEEE 17th International Symposium on. IEEE, 2011, pp. 205–216.

Lin, M., A. Wierman, L.L.H. Andrew, and E. Thereska, "Dynamic right-sizing for power-proportional data centers," IEEE/ACM Trans. Netw., vol. 21, no. 5, pp. 1378–1391, Oct. 2013.

Liu, H.H., Y. Wang, Y.R. Yang, H. Wang, and C. Tian, "Optimizing cost and performance for content multihoming," SIGCOMM Comput. Commun. Rev., vol. 42, no. 4, pp. 371–382, Aug. 2012.

Liu, Z., M. Lin, A. Wierman, S. Low, and L. Andrew, "Greening geographical load balancing," in Proceedings of the International Conference on Measurement and Modeling of Computer Systems (ACM Sigmetrics'11), San Jose, California, USA, june 2011, pp. 7–11.

Luo, J., L. Rao, and X. Liu, "Temporal load balancing with service delay guarantees for data center energy cost optimization," IEEE Transactions on Parallel and Distributed Systems, vol. 99, no. PrePrints, 2013.

Meisner, D., B.T. Gold, and T.F. Wenisch, "Powernap: Eliminating server idle power," in Proceedings of the 14th International Conference on Architectural Support for Programming Languages and Operating Systems (ASPLOS'09), New York, NY, USA, 2009, pp. 205–216.

Rao, L., X. Liu, L. Xie, and W. Liu, "Minimizing electricity cost: Optimization of distributed internet data centers in a multi-electricity-market environment," in Proceedings of the 29th Annual Conference of the IEEE Communications Society (INFOCOM'10), San Diego, CA, USA, march 2010, pp. 1–9.

Ren, C., D. Wang, B. Urgaonkar, and A. Sivasubramaniam, "Carbon-aware energy capacity planning for datacenters," in Modeling, Analysis & Simulation of Computer and Telecommunication Systems (MASCOTS), 2012 IEEE 20th International Symposium on. IEEE, 2012, pp. 391–400.

Sharma, N., S. Barker, D. Irwin, and P. Shenoy, "Blink: managing server clusters on intermittent power," in ACM SIGPLAN Notices, vol. 47, no. 4. ACM, 2011, pp. 185–198.

Yao, J., H. Zhou, J. Luo, X. Liu, and h. Guan, "Comic: Cost optimization for internet content multihoming," Parallel and Distributed Systems, IEEE Transactions on, vol. PP, no. 99, pp. 1–1, 2014.

Yao, Y., L. Huang, A. Sharma, L. Golubchik, and M. Neely, "Data centers power reduction: A two time scale approach for delay tolerant workloads," in Proceedings of the 31st Annual Conference of the IEEE Communications Society (INFOCOM'12), march 2012, pp. 1431–1439.

Energy Science and Applied Technology – Fang (Ed.)
© 2016 Taylor & Francis Group, London, ISBN 978-1-138-02833-3

A new security scheme of trusted mobile platform based on MTM

Manzhi Yang & Huixiang Zhou
State Key Laboratory of Networking and Switching Technology, BUPT, Beijing, China

ABSTRACT: As a representative of the future development trend in the field of information security, the trusted computing technology is gaining increased attention, and has been widely used in the PC terminal. However, the more suitable environment for the application of trusted computing technology should be the new generation mobile communications network. Since the MTM (Mobile Trusted Module) has an incomparable superiority in performance, flexibility, scalability than the TPM (Trusted Platform Module), the mobile platform architecture design based on MTM can be more suitable for the needs of the next-generation mobile communication systems. This paper presents a secure boot process certified by direct measurement, which can not only meet the integrity requirement of the platform through the formal verification, but also avoid the trust losses caused by the traditional commission measure and verification. The results show that the proposed construction plan on the mobile platform and the secure boot process in this paper have better flexibility, security and credibility compared with the existing programs.

Keywords: mobile platform; security; trusted computing; Mobile communication; trusted module

1 INTRODUCTION

Because of the development of IP technology and the open network interface, the integration-oriented mobile network has more complex security threats. Traditional mobile network security technology cannot address these security threats. Meanwhile, due to the limited resources of mobile networks and other reasons, it is impossible to directly copy the IP network security technology. Therefore, we must first learn from the mature IP network security to defend technology and then explore a solution suitable for mobile network security.

As a representative of the future development trend in information security, the trusted computing technology is gaining increased attention, and has been widely used on the PC terminal. However, a more suitable environment for the trusted computing technology should be the new generation mobile communication network. Mobile networks have a relatively standardized management system and are easier to deploy and promote the trusted computing technology. In addition, the hardware-based trusted computing technology has an advantage of protecting mobile terminals that have limited storage and processing capacity. Therefore, this article applies the trusted computing technology to the mobile communication network, to achieve the UMTS network security defended by designing trusted mobile terminals.

2 RELATED TECHNOLOGY

2.1 *Trusted computing*

Trusted computing is a hot research field of information security, and the integration of mobile communications and the fixed network is an inevitable trend of future network development, while the design and implementation of a secure, reliable mobile platform is the basis for promoting the development of the integrated network. Using the trusted computing technology to protect the security of the mobile terminal is of great significance, and many research institutions at home and abroad and scholars have carried out studies on relevant aspects and achieved certain results.

1. Trusted Computing Group TCG established a Nokia-led working group committed to the extensions of related TCG specifications to solve the security issues of the mobile platform, and proposed the related specifications of the mobile trusted module (Mobile Trusted Module, MTM) in September 2006, but it did not propose specific solutions to build the mobile platform.
2. Intel, IBM, NTT and other companies launched the project of the Trusted Mobile Platform (TMP), and defined the different environment levels of the trusted execution, and used the idea of trusted computing to solve the security problem of mobile terminals. Based on the

Trusted Platform Module (TPM) of TCG, they proposed the architecture of software and hardware of the trusted mobile platform and protocol specifications.

3. Established in 2004, the organization of the Open Mobile Terminal OMTP has many important members including mobile operators, equipment vendors and content providers. The organization has especially established two research programs of applying security and trusted environment working on these security issues of trusted mobile devices, but there have been no published studies.

In summary, the current study of the trusted mobile platform is still in the exploratory step. TCG and OMTP have not yet put forward a specific plan to build a trusted mobile platform, and the present scheme of the trusted mobile platform is based on the TPM module, which is designed for the PC terminal, and cannot meet the unique attributes and applications of mobile platforms. On the basis of an in-depth study on the characteristics of the mobile platform, this paper presents a MTM-based construction plan for the mobile platform and the secure boot process. Compared with the traditional scheme, the proposed model has better flexibility, security and credibility.

2.2 Mobile trusted module

The main component of the trusted mobile platform is shown in Figure 1.

When applying security chip TPM suitable for the PC and server to the mobile platform, we encounter the following problems. First, the mobile platform is a restricted terminal with a limited computing power, storage capacity and limited energy and other properties, so the design of the security chip of the mobile platform and the

security policies of the mobile platform will differ from the PC. Second, given that the application environment of the mobile platform is more complex and has users, operators, content providers and equipment manufacturers as different owners, their requirements for the mobile platform are different; therefore, the design of the trusted mobile platform should meet these diverse needs.

3 SECURITY SCHEME OF TRUSTED MOBILE PLATFORM BASED ON MTM

Trusted mobile platform processor includes the baseband processor and the application processor. These two processors can be separated, and can also be one processor. Separating two processors can enhance the security of the system, because the baseband processor works in the mobile communication network environment with a certain security, which has the enclosed protocol and security safeguard, so safety problem is very small.

The application processor is located inside the mobile platform. This part has similar security threats to the PC, and attention on its security problem is relatively small, but this part does not have an effective security mechanism. Separating two processors to form two separate areas easily implements safety protection, so here we adopt a dual processor architecture.

Based on the MTM relevant specification proposed by the TCG and the mobile platform architecture proposed by the trusted mobile platform, we propose a trusted mobile platform system structure, as shown in Figure 2. It consists of the following security components constituting a reliable boundary:

1. The application processor: it controls the implementation of the system in guidance and after completion of guidance, it detaches the trusted application and its data from the untrusted application.
2. MTM: it provides a protected environment with a variety of security features including key algo-

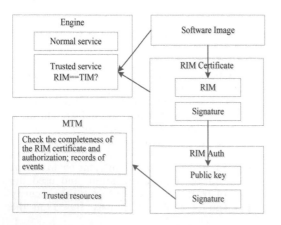

Figure 1. Components of the trusted mobile platform.

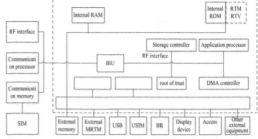

Figure 2. Architecture of the trusted mobile platform.

rithm, certification and reliable storage. The MTM can provide more flexible security and services, which satisfy the properties of most of the main mobile platforms. By the built-in MRTM, the equipment vendors ensure that the mobile platform reaches the trusted state after the secure guidance to provide a reliable guarantee for the upper application. Other upper applications, such as mobile operators, application providers and users, can realize credible services by other MRTM and MLTM.

3. RTV/RTM: RTM (Root of Trusted Measurement) measures the integrity of the subsequent components and RTV (Root of Trusted for Verification) verifies whether the field measurements and the RIM basic value are the same, and determines whether the integrity of the measured is destroyed or not. The MTM accesses the RTV/RTM by DMA.

4. DMA: in a high-level security system, the DMA is controlled by the trusted kernel, protecting the trusted application or the operating system access to physical memory.

4 SECURITY BOOT PROCESS

Assume that the generated safety boot process requires certificates of RIM, using the relevant specifications of the TCG, the direct measurement of the mobile platform security boot process is described as follows:

Step 1: MTM->TPM_init, TPM_startup. After call the TPM and TPM Startup, the MTM starts, all the PCR values initialize to 0;

Step 2: MTM->TPM_VertifyRIMCertAndAn Extend (e1).

Based on the certificates of RIM, the MTM check the integrity of the RTM/RTV and record the test results in the PCR;

Step 3: MTM->Measure (e), RTV-> Look UpRIMcert (e), MTM> TPM_ VertifyRIMCertAnd AnExtend (e), the RTM measures the integrity value of e2, the RTV loads the RIM certificate of e2 and verify whether the field measurements and the RIM basic value are the same. The MTM verifies the RIM certificate by TPM_VertifyRIMCertAndAnExtend and records the test results in the PCR, and e2 obtains the Executive power;

Step 4: MTM->Measure (e2), RTV-> LookUp RIMcert (e2), MTM> TPM_ VertifyRIMCertAndAn Extend (e2), the RTM measures the integrity value of e3, the RTV loads the RIM certificate of e3 and verifies whether the field measurements and the

RIM basic value are the same. The MTM verifies the RIM certificate by TPM_VertifyRIMCertAndAnExtend and records the test results in the PCR, and e3 obtains the Executive power;

Step 5: MTM->Measure (e3), RTV-> LookUp RIMcert (e3), MTM> TPM_ VertifyRIMCertAnd AnExtend(e3).

The RTM measures the integrity value of e4, the RTV loads the RIM certificate of e4 and verifies whether the field measurements and the RIM basic value are the same.

The MTM verifies the RIM certificate by TPM_Vertify RIM Cert and an Extend and records the test results in the PCR, and e4 obtains the Executive power. After the RTM and RTV measurement and verification, we ensure that all entities satisfy the integrity in the boot process, and safeguard the credibility of the entire platform. If any one of TPM_Verify RIM Cert and Extend calls the return value, which is an error, or cannot find the right certificate, then the boot process will be terminated.

5 CONCLUSIONS

The trusted mobile platform system structure is designed based on the TPM module, but these structures cannot meet the requirements of the mobile system's efficiency, size and support for many owners. The mobile platform based on the MTM has trusted storage, credible reports, trusted measurement and the ability to verify, which can effectively prevent tampering attack and keep the process of storage and conduct from the untrusted program intervention. The hardware technology ensures the system safety. In the aspect of flexibility, due to the introduction of the MTM, the process satisfies the security requirement of different owners of the mobile platform by generating the independent security engine and RIM certificate. In contrast to the existing safety boot process scheme, verification analysis indicated that the mobile platform construction scheme proposed in this paper has better flexibility, security and credibility, can meet the requirements of the integrity of the platform, avoid the traditional trust measurement and verify the trust loss.

ACKNOWLEDGMENTS

This work was supported by the NSFC (Grant Nos. 61300181, 61272057, 61202434, 61170270, 61100203, 61121061) and the Fundamental Research Funds for the Central Universities (Grant No. 2012RC0612, 2011YB01).

REFERENCES

Akram RN. Secure Smart Embedded Devices, Platforms and Applications [M]. Springer New York, 2014:71–93.

Bang J, Kim D. Effective operation and performance Improvement methods for OMTP BONDI-based mobile Web widget resources [J]. Journal of Zhejiang University SCIENCE C, 2011, 12(10):787–799.

Bang J, Kim D. Effective operation and performance improvement methods for OMTP BONDI-based mobile Web widget resources [J]. Journal of Zhejiang University-Science C (Computers & Electronics), 2011, (10).

Gilbert P. Toward trustworthy mobile sensing [C]. Proceedings of the Eleventh Workshop on Mobile Computing Systems & Applications. ACM, 2010:31–36.

Jøsang. Identity management and trusted interaction in internet and mobile computing [J]. Information Security, IET, 2014, 8(2):67–79.

Kang D. A study on migration scheme for a mobile trusted module [C]. Advanced Communication Technology, 2009. ICACT 2009. 11th International Conference on. IEEE, 2009:1672–1677.

Kim M, Ryou J. Information and Communications Security [M]. Springer Berlin Heidelberg, 2007: 375–385.

Energy Science and Applied Technology – Fang (Ed.)
© 2016 Taylor & Francis Group, London, ISBN 978-1-138-02833-3

Threshold-based distributed key management scheme for wireless sensor networks

Y. Jiang & Z.J. Lin
College of Computer Science and Technology, Chongqing University of Posts and Telecommunications, Chongqing, China

ABSTRACT: For the issues in wireless sensor networks, such as limited node computing and energy resources, and dynamic topology, this article designs a threshold mechanism-based key management scheme, which combines with the Lagrange interpolation and the hierarchical clustering structure. The scheme is based on secret sharing and uses a pre-shared key and distributed collaboration to realize dynamic cluster key management. In order to improve the flexibility of key management, the scheme introduces virtual nodes to help cluster key recovery when some nodes are invalid. Security analysis also shows that the scheme has a good anti-capture and data confidentiality.

Keywords: Wireless sensor networks; Lagrange interpolation; Threshold mechanism; Cluster key management; Virtual node

1 INTRODUCTION

Wireless Sensor Network (WSN) has many characteristics such as self-organizing, non-definite administrator or monitor, and open communication medium, which make it easily affected by a variety of security problems, such as key management, authentication, secure routing and attack response. An effective key management is the basis of security mechanisms such as secure routing, data fusion and the solutions for special attacks (Su Z).

Group key management schemes for the traditional network cost lots of resources. It cannot be used directly in WSNs. In recent years, much research has been conducted on group key management for WSNs. The existing key management schemes are usually based on the symmetric keys, and more suitable for WSNs than for the public key cryptosystem (LIU D, 2003). They use the symmetric binary polynomial in the key pre-distribution process, and propose a polynomial-based random key pre-distribution scheme. They also propose a group key update scheme, based on the key preset and local collaboration, significantly reducing the single node's overhead on storage and communication. In addition, they put forward a distributed group key management scheme based on cluster cooperation, which makes contribution to enhance the autonomy, security and scalability of WSNs. In addition, some asymmetric keys management schemes can be used in WSNs by improving the algorithm. They set servers and control centers in each area by using regional management model, and realize symmetric key distribution and agreement on nodes in a single region based on ECC and ElGamal. The above-mentioned schemes reduce the calculation and communication overhead, improve the scalability of the network by using the hierarchical structure in a collaborative way to manage keys. Some of these schemes need the location information of nodes; however, in the practical application of WSNs, the location information is usually unpredictable, and may be changed while working. So, there are many limitations in these schemes.

This paper designs a cluster key management scheme for WSNs. The scheme increases the flexibility of key management based on the key-sharing method (Shamir A, 1979), which was proposed by Shamir, clustering structure and threshold mechanism in the Lagrange polynomial interpolation method, and introducing virtual nodes (Deng SH, 2012).

2 KEY MANAGEMENT SCHEME

The best way to avoid the problems such as the big master key target or the accidentally damaged master key is to use a secret-sharing scheme. Shamir's secret-sharing scheme is a classic secret-sharing scheme, which can ensure the security of the number of collusion attack within the threshold.

In Zhou J, 2007, members in the group communication system are seen as a tree hierarchy, the construction of Lagrange interpolation polynomial and the distribution of group key are distributed to group members in order to complete, so that the group key distribution time can be greatly reduced. In You L, 2013, the Lagrange interpolation principle and one-way hash function are combined, the key pre-sharing and the local collaborative manner are used to update the group keys and polynomial shares, so that the system collusion-resistant capability is effectively improved.

2.1 Shamir's secret-sharing scheme

A shared key K is divided into n sub-keys (secret components) $k_1, k_2, ..., k_n$, and secretly assigned to n participants. The following two conditions need to be met:

1. It is easy to obtain K by the ki values that any s or more than s participants hold.
2. k_i values from $s-1$ or less than $s-1$ participants cannot obtain any information about K.

This method is called the (s, n) threshold scheme, where s is the threshold.

Since key reconstruction needs at least s sub-keys, the key would not be threatened when $\omega(\omega \leq s-1)$ sub-keys are exposed. So, collusion from less than s participants cannot get the key. In addition, if a sub-key is lost or destroyed accidentally, the key can still be recovered (as long as there is at least s valid sub-keys).

2.2 Network model and assumptions

WSNs usually adopt the hierarchical clustering architecture, as shown in Figure 1. The network hierarchy includes the base station, cluster head, and ordinary node, from top to bottom.

Base station (BS): the base station manages the entire network, and its computing resource, storage and energy are not limited. The base station receives all of the data sent by the nodes in the network. In certain circumstances, it also serves as the network control center, deploying the information carried for the nodes.

Cluster head (CH): the cluster head is selected from the ordinary nodes. The cluster head manages the cluster key.

Ordinary node (Node): the nodes collect and monitor the data, with very limited computing capability, storage and power.

We assume that the base station is trusted, and any node that has been captured or invalid could be recognized by its neighbors.

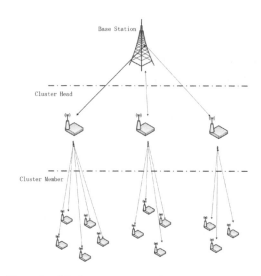

Figure 1. Wireless sensor network structure.

Table 1. Parameter description in this paper.

Parameter	Meaning	Parameter	Meaning
N	The number of sensor nodes in the network	$K_{i,BS}$	Key shared with the base station
BS	Base station	τ	Threshold of the key polynomial
ID_i	Unique Identifier of Node i	GK_i	The cluster key of Cluster i
M	The number of clusters in the network	k_i	Cluster key component of Node i
CH_i	The cluster head of Cluster i	r_i	Random number for Cluster i updating the cluster key
S_{ij}	Member j of Cluster i	P	The cluster key updating proportion parameter of the cluster members' number (can be set according to the actual situation)
n	The number of cluster members		

2.3 Threshold-based cluster key management scheme

Based on the dynamic characteristics of sensor networks, the key management scheme presented in this paper can be divided into three stages: initialization, cluster establishment, and cluster key establishment and update. When finish deploying the identity information and pre-shared key, the nodes cluster according to the energy they hold, and the cluster heads use Lagrange interpolation polynomial as the mathematical basis to assign cluster key components to the cluster members.

2.4 Related definitions

2.4.1 Initialization
Assume that the total number of sensor nodes in the network is N.

The unique identity identifier ID and a random key $K_{i,BS}$ shared with BS will be preset into each sensor node.

2.4.2 Cluster establishment
Nodes cluster according to the energy they hold at regular time by using the BPEC algorithm.

When clustering completed:

1. CH_i sends its cluster members' list to BS encrypted with $K_{i,BS}$.
2. The base station generates a random ID pool of $\sum_{i=1}^{M}(n_i - \tau_i)$ size for the use of virtual nodes.

2.4.3 Cluster key establishment
Cluster head CH_i re-establishes the cluster key when clustering is done.

In a n-member cluster, τ is the threshold, we choose a large prime number q, making $\tau < n < q$.

1. Choose any $\tau - 1$ distinct integers $a_1, a_2, ..., a_{\tau-1}$ from the finite field $GF(q)$, and select the cluster key GK_i, let $a_1 = GK_i$, and then construct a $(\tau - 1)$-degree randomized polynomial:
2. Calculate the key component $k_i = f(ID_i) \bmod q$ for each cluster member ID_i.
3. Assume that it is secure for a period of time after the completion of clustering, the cluster head assigned the cluster key component k_i to each member of the cluster, a MAC code is attached to the message for members to verify the validity of the message.

$$GK_i = \sum_{i=1}^{\tau} k_i \prod_{j=1, j \neq i}^{\tau} \frac{x_j}{x_i - x_j}$$

4. With any τ cluster key components k_i, cluster members can reconstruct the key-sharing polynomial $f(x)$ to recover the cluster key GK_i by using the Lagrange interpolation formula. Thus, each member of the cluster can be calculated

Cluster member S_{ij} needs to recover GK_i when it communicates with the cluster head or other members of the cluster. S_{ij} broadcasts a key component request to other nodes, and other members (e.g. S_{iu}) judge whether S_{ij} has been captured. If S_{ij} was verified as a legal node, then S_{iu} sends its key component k_u to S_{ij}, conversely, S_{iu} abandons the request from S_{ij}. When S_{ij} receives more than τ key components, it recovers GK_i and deletes the key components it just received.

Cluster key update. Due to circumstances that the node has been captured or energy depletion may occur in wireless sensor networks, the cluster key processing method under different conditions is discussed separately below.

2.1.4 Ordinary node leaves the cluster
a. Other members of the cluster can recognize the capture of S_{ij}, and send a message to CH_i. CH_i will remove S_{ij} from its member list and notice the base station. BS chooses a new random number r_i from $GF(q)$, and sends it to CH_i encrypted with $K_{i,BS}$. CH_i calculates $GK_i' = a_0 + r_i$ and $k_i = f(ID_i) \bmod q$, and sends the new key components to its cluster members encrypted with current GK_i.

b. Node failure or sleep due to the lack of energy

$$f(x) = a_0 + a_1 x + a_2 x^2 + ... + a_{\tau-1} x^{\tau-1}$$

First, CH_i judges whether the current number n of cluster members is less than the proportion parameter P of the initial number of cluster members. If not, then it judges whether the current number is less than the threshold τ. If it is not less than τ, GK_i will not be changed, else CH_i applies virtual nodes from BS[6]. BS chooses $n - \tau$ virtual nodes for the cluster head, then BS sends the encrypted virtual nodes' IDs to CH_i, and CH_i uses $f(x) = a_0 + a_1 x + a_2 x^2 + ... + a_{\tau-1} x^{\tau-1}$ to calculate the cluster key components of virtual nodes. For the rapid recovery of GK_i, key components of the virtual nodes are managed by CH_i.

When the current number of cluster members is less than P of the initial number, CH_i re-establishes GK_i based on the actual number of cluster members, and updates the key components for members.

2.1.5 Cluster head leaves the cluster
When the cluster head depletes energy or has been captured, the current cluster members elect ID_j as the cluster head CH_i' by energy, and CH_i' informs BS about the new cluster members.

The new cluster key components are established by CH_i'.

3 PERFORMANCE ANALYSIS

3.1 Computing and communication overhead comparison

WEN T, 2012 uses the key management scheme based on symmetric polynomial, which was proposed by Blundo, to propose the group key management scheme for the homogeneous network model. The scheme proposed in WEN T, 2012 is similar to this paper, which is called the (k, t, m) group polynomial scheme. The computing and communication overhead comparison between the scheme presented in this paper and the (k, t, m) group polynomial scheme is presented in Table 2.

L denotes the length of the node ID, multinomial coefficient, key, and random number; m denotes the number of polynomial (WEN T, 2012); δ denotes the number of the nodes in the dynamic cluster; δ' denotes the number of join/quit nodes in the dynamic cluster; t_{mac} denotes the message authentication computing time; t_h denotes the hash operation time; t_{ploy} denotes the (k, t, m) group polynomial operation time; t'_{ploy} denotes the operation time of the scheme presented in this paper; t_k denotes the time of encryption/decryption (symmetric encryption); and ①②③ denotes the case of node capture, the node energy depletion or hibernation, cluster head leaving the network, separately.

The time complexity of the (k, t, m) group polynomial operation is $O(k^t)$ and the time complexity of the scheme presented in this paper is $O(\tau^2)$. After normalizing the two parameters, the former can be expressed as $O(\tau^t)$, $t \geq \tau$. By comparison, it is concluded that time complexity of computing t_{ploy} is higher than t'_{ploy}, and computing time of t_h and t_{mac} can be ignored. As shown in Table 2, in instability sensor networks, when a node in the cluster changes, compared with the (k, t, m) group

polynomial operation scheme, the cluster head node overhead of the scheme presented in this paper is slightly higher; however, the computing and communication overhead generating by the normal node is smaller, so that the energy consumption is correspondingly reduced. The reduced energy consumption of numerous normal nodes can effectively prolong the overall life of the network. Furthermore, the polynomial computing of the (k, t, m) group polynomial operation scheme needs to calculate the k^tth power of ID_i, which is a great challenge for the computing and memory resources of the sensor node. To ensure the validity of the computing results (i.e. length of computing does not overflow), the upper limit t of nodes number in each cluster should be limited accordingly to a great extent. Thus, the scheme presented in this paper is energy efficient and the cost of computing is more suitable for wireless sensor networks.

3.2 Key anti-capturing

In the scheme presented in this paper, since $K_{i,BS}$, which are the nodes shared with the base station, are assigned before deployment, attackers cannot intercept the message during transmission. The key that each node shares with the base station is different from each other. If a node is captured, it only affects the direct participation part of the communication, but the security of other nodes even in the entire network is not affected. Moreover, in a single cluster, if the captured nodes are less than τ, the safety of the cluster keys is not affected; if the captured nodes are approximately τ or higher than the proportion number of cluster nodes, the key update can significantly increase the difficulty of attacker capturing.

3.3 Data confidentiality

When a node is captured or leaves (i.e. up to the threshold), the scheme updates the cluster key to ensure that only legitimate cluster members can obtain and generate the new cluster key, while the nodes that leave with the old cluster key component k_i are unable to recover the new polynomial $f'(x)$ to get the correct GK'_i, which cannot decrypt the message. When a new node i joins the network, its legitimacy $K_{i,BS}$ needs to be authenticated through the base station and only the legal one can be assigned to join the target cluster and get a key component of the cluster. Even if an attacker controls several nodes, since node numbers are less than the polynomial threshold τ, the attacker is unable to restore the polynomial and the right cluster keys through collusion, in order to ensure the confidentiality of the message.

Table 2. Computing and communication overhead comparison between the scheme presented in this paper and the (k, t, m) group polynomial scheme.

Group key operation	Overhead type	(k, t, m) group polynomial scheme		Scheme in this paper	
		Cluster head	Original node	Cluster head	Original node
Setup	Computing	$2t_k + t_{mac}$	$(t_{mac} + t_k + m \cdot t_{ploy}) \cdot \delta$	$t'_{ploy} \cdot \delta + t_k$	$t'_{ploy} \cdot \tau$
	Communication	$(3+\delta) \cdot L + 1$	$(3+\delta) \cdot L + 1) \cdot \delta$	$(2\delta+1) \cdot L$	$(\delta-1) \cdot L + 1$
Update	Computing	$t_k + t_{mac}$	$(t_{mac} + 2t_k + m \cdot t_{ploy}) \cdot \delta$	$t'_{ploy} \cdot \delta + t_k$	$t'_{ploy} \cdot \tau$
	Communication	$4L + 1$	$(4L+1) \cdot \delta$	$(2\delta+1) \cdot L$	$(\delta-1) \cdot L + 1$
Node leaving	Computing	$t_k + t_{mac}$	$(t_{mac} + 2t_k + m \cdot t_{ploy}) \cdot (\delta-\delta')$	①$(t'_{ploy}+t_k) \cdot (\delta-\delta')$ ②$\begin{cases}(\delta-\delta') \geq \tau : 0 \\ (\delta-\delta') < \tau : t'_{ploy} \\ (\delta-\delta') < \delta' : (\delta-\delta') \cdot t'_{mac}\end{cases}$	t_k
	Communication	$(3+\delta-\delta') \cdot L + 1$	$((3+\delta-\delta') \cdot L + 1) \cdot (\delta-\delta')$	①$(\delta-\delta') \cdot L + 2$ ②$\begin{cases}(\delta-\delta') \geq \tau : 0 \\ (\delta-\delta') < \tau : (\delta-\delta') \cdot L + 1\end{cases}$ ③$(2\delta+1) \cdot L + 2$	0

4 CONCLUSION

The nodes of this scheme only stores the cluster key components of clusters where the nodes are located and updated for the key can only focus on unsafe clusters, other than the whole network key. By this way, this scheme greatly reduces the storage cost and computation cost. At the same time, the scheme introduces assist computing of the base station's virtual nodes for cluster members in small number, which cannot reach the polynomial threshold and dynamically sets the cluster key update timing for a specific scene, so that the scheme has a high key management flexibility, ensuring the confidentiality of the message.

REFERENCES

Chai JG. Based on the fuzzy clustering sensor failure detection network node (J). Microelectronics & Computer, 2012, (7): 184–187.

Deng SH, Zhao ZM. A distributed group key management scheme of WSN (J).Information Security, 2012, (2): 18–20.

Liu D, Ning P. Location-based pairwise key establishments for static sensor networks(C). Proc. of the 1st ACM Workshop on Security of Ad Hoc and Sensor Networks. New York: ACM Press, 2003:72–82.

Shamir A. How to share a secret (J). Communication of the ACM, 1979, 24(11): 612–613.

Su Z, Lin C, Feng FJ, Ren FY, et al. Key management schemes and protocols for wireless sensor networks. Journal of Software, 2007, 18(5): 1218–1231.

Wen T, Zhang Y, Guo Q, et al. Dynamic group key management scheme for homogeneous wireless sensor networks (J). Journal on Communications, 2012, (6): 164–173.

Yang G, Yin GS, Yang W, et al. A reputation-based model for malicious node detection in WSNs (J). Journal of Harbin Institute of Technology, 2009, (10):158–162.

You L, Li ZG. A novel group key distribution via local collaboration (J).Computer Security, 2013, (1): 13–16.

Zeng P, Zhang L, Yang YT, et al. Lightweight and high security key management for wireless sensor networks (J). Application Research of Computers, 2014, (1):199–202.

Zeng WN, Lin YP, Lu QY. Distributed group key management scheme based on cluster collaboration in WSN (J). Journal of Computer Applications, 2009, (3): 638–642.

Zhang B, Wu JX. Key generation method based on Lagrange polynomial interpolation formula (J).Mini-micro System, 2001, (5): 583–585.

Zhang FT, Li JG, Wang XM, et al. Cryptography Tutorials (M). Wuhan: Wuhan University Press, 2006.9: 105.

Zhang W, Cao G. Group rekeying for filtering false data in sensor networks: a predistribution and local collaboration-based approach (C). INFOCOM 2005. 24th Annual Joint Conference of the IEEE Computer and Communications Societies. Proceedings IEEE. IEEE, 2005:503–514.

Zhou J, Zou SD, Li M, et al. A layered group key distribution scheme based on the Lagrange interpolating polynomial (J). Journal of Xiamen University (Natural Science), 2007:75–78.

Zhou XL, Wu M, Xu JB. BPEC: An energy-aware distributed clustering algorithm in WSNs (J). Journal of Computer Research and Development, 2009, (5):723–730.

Energy Science and Applied Technology – Fang (Ed.)
© 2016 Taylor & Francis Group, London, ISBN 978-1-138-02833-3

Ordered tree-based routing algorithm for Power Line Communication network

Huabing He, Tingting Yao, Yunfei Li & Juncheng Jia
School of Computer Science and Technology, Soochow University, Suzhou, China

ABSTRACT: Power Line Communication (PLC) is gaining increased attention because it only relies on the existing widespread power line facilities without additional wiring to transmit data. However, the current power line network still has the problems of low routing efficiency and high communication delay weakness. To overcome these issues, we propose a tree routing algorithm, namely PLC-TR, based on the characteristics of the tree topology of power line network. Specifically, PLC-TR first organizes the power line network into an ordered tree, and then selects routes by comparing the addresses. PLC-TR minimizes the network overhead successfully. The simulation results show that, compared with the shortest path routing (SPR), which is one of the traditional optimal algorithms, PLC-TR has a lower average packet transmission delay and a higher packet delivery rate, under the same disturbance degree.

Keywords: power line network, ordered tree, PLC-TR, tree routing algorithm, SPR

1 INTRODUCTION

In recent years, application and research in power line carrier communication has aroused people's attention (Slootweg H, 2009), such as high-speed broadband PLC Internet, intelligent home furnishing (Lin Y, 2002), large automation system and remote meter reading in power system (Bumiller G, 2010). As a method of data transmission, it uses power line carrier communication to transfer data, whose main advantage lies in the fact that it needs no additional wiring. Nevertheless, the electrical load environment for electricity distribution network is so complicated that there exist several harmful elements, such as over-damp and high noise. Regardless of whether dynamic trend for load in power line channel joins the Internet or not, it will lead to a strong signal decline. Consequently, Internet support is urgently needed to improve the reliability of power line communication, such as relaying and routing.

Recently, there have been few algorithms in electricity power line, most of which originated from the wireless network (Li H, 2009, Biagi M, 2012, Yoon S, 2014, Biagi M, 2010, Sanchez JA, 2009). Scholars at home and broad have performed more research on the Wireless Mobile AD Hoc Network. The general routing algorithm is based on location or topology. The wireless transmits signal around, while the power line transmits through the communication cable. Even communication nodes that are physically close cannot contact directly. Therefore, the position-based routing algorithm fails to apply the cable network. The shortest path routing (SPR) (Galli S, 2011) is used in the power line network to structure the shortest routing of networks, which proves to be the best traditional routing algorithm. As the SPR must exchange routing information with its neighbor node regularly, the Dijkstra algorithm applied in calculating the shortest route may inevitably increase network overheads.

The application of the ZIGBEE cluster tree algorithm (Kim T, 2014, Kim T, 2007, Ren Z, 201) in power line may give rise to the following problems. First, the fixed routing depth is difficult to adopt for the time-varying channel of power line, which has a strong interference and time variation. Another problem is the uneven distribution of nodes in the power line network. Large number of nodes is centralized in a certain branch, causing an overwhelming address distribution in branches with a small number of nodes.

Based on the characteristics of the above power line, this paper presents a routing algorithm in power line using the Ordered Tree (PLC-TR) that will eliminate the fixed routing depth and allocate the logical address according to the density distribution of the node. This algorithm will structure the whole network as an ordered tree. This paper is divided into four parts. The first part introduces the model of the power line network. The second part introduces the structure strategy of the ordered tree. The third part gives the routing strategy of PLC-TR. The fourth part carries out the simulation to prove its feasibility.

2 POWER-LINE NETWORK MODEL

There are three kinds of nodes in the network: gateway, router and ordinary node. Among these nodes, the gateway is the center and the hub node to net-com inquiry. The router not only functions as an ordinary node, but is also mainly responsible for the selection of routing and forwarding nodes. The ordinary node and the router are basic communication nodes. As shown in Figure 1, we represented the power-line network as a graph $G = \{N, E\}$.

Here, $N = \{n_1, n_2, n_3, ..., n_i, n_m\}$ represents the case of a single node i; M represents the number of nodes in the network; and E represents the edge between the node set which is empty in the initial state. We define a node set N_i as follows: $N_i = \{n_j \mid E_{ij} > E_{min} \ j = 0, 1, 2, 3 ...... M\}$. Here, n_j represents the node j; E_{ij} represents the energy value of the transmitting signals between n_i and n_j; E_{min} represents the minimum energy to guarantee a normal communication between two nodes.

Energy Value (EV): The EV refers to the energy of the data signal received by the receiving node of the signal. The nodes start to send the data signal with the same energy, but it gradually decays in the process of signal transmission on the power line. Therefore, the further the distance between the nodes, the smaller the received signal energy, and vice versa. The EV can be used to represent the maximum effective distance of point-to-point communication. Energy calculations can also rely on the power spectral density empirical formula. The power spectral density of the formula is as follows:

$$\mu(f, D) = (0.0034D + 1.0893)f + 0.1295D + 17.3481 \qquad (1)$$

where f is the frequency of the low voltage power line (KHZ) and D is the distance (m).

In order to reduce the routing maintenance overhead and enhance the efficiency of point-to-point communication, we build an ordered tree with the gateway as the head node, as shown in Figure 1. The tree must meet the following two conditions: first, it must have the shortest length of the average path, in order to reduce the time delay of communication and improve the pass rate; second, the tree must be sequential, in order to facilitate the routing. In the next section, we will show that the topology discovery process will make the tree satisfy the first condition, and the address allocation strategy will make the tree meet the second condition.

3 ORDERED TREE BUILDING STRATEGY

In the selection strategy of the router, we use the EV instead of the traditional algorithm to represent the distance between two nodes, which makes the algorithm more adaptable to the time-varying power line network. This article adopts tree breadth-first search algorithm to explore the unknown electric network topology. The algorithm will give priority to the distribution of the less energy node as the candidate router in order to detect a further distance. In the process of topology discovery, we use the carrier sense multipoint access collision avoidance (CSMA/CA) algorithm (Pinero-Escuer PJ, 2011) to improve the quality of the node data link layer communication. The process of EV determination and the candidate router selection process are as follows.

Step 1: the node n_i sends the topology discovery frame to explore the network. If a node n_k receives the explore frame that does not join the network, then the node n_k calculates the $EV(E_k)$ of the explore frame (formula (1)) and compares with the minimum EV (E_{min}) required for the normal communication.

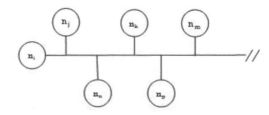

Figure 2.   Before the topology discovery.

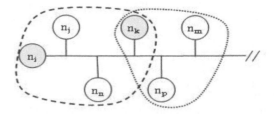

Figure 3.   After the topology discovery.

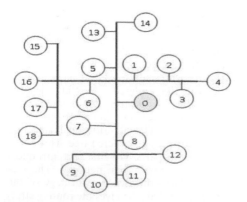

Figure 1.   Power-line network model.

If $E_{ik} > E_{min}$, then the node n_k will use its MAC address and E_{ik} to reply the node n_i. If the node n_i receives the reply from other nodes, then the node n_i will inform its parent node to assign it as the routing node. As shown in Figure 3, if the nodes n_j, n_n, n_k respond to the node n_i, then the node n_i will align the nodes n_j, n_n, n_k from the child node to arrange according to E_{ij}, E_{in}, E_{ik}.

Step 2: by using the tree breadth-first search strategy, the node n_i primarily chooses the minimum EV of the node set N_i, which means to select the farthest distance from the node n_i to the node n_k as the candidate routing node. Then, n_k repeats the process of n_i in the first step. If there is no reply, the nodes that do not join the network are assigned for the common nodes.

Step 3: it chooses the nodes whose EV ranks the second smallest in a node set Ni and then repeats the process of step 2, until all the nodes in the node set N_i are chosen as the candidate routing node, and executes the search process.

When the algorithm is complete, as shown in Figure 3, the node n_i searches out nodes n_j, n_n, n_k and the node n_k searches out the nodes n_m, n_p. Then, n_m, n_p become the router and get in a node set $N_i = \{n_j, n_n, n_k\}$ and $N_k = \{n_m, n_p\}$. Other nodes become the common nodes. After the completion of exploration, we logically construct the power-line network as a tree topology, as shown in Figure 3.

Hypothesis: the address field that the router n_s obtains is $[A_i, A_N]$; the number of routing nodes in the child node set Ns is k; the routing node set is $R = \{n_{s+1}, n_{s+2}, n_{s+3}, ..., n_{s+k}\}$; and the corresponding branch weight set is $W = \{W_{s+1}, W_{s+2}, W_{s+3}, ..., W_{s+k}\}$. The served percentage of the node ns address is p, where $0 < k < N$, $0 < p < 1$, $0 < i < k$, and then the distribution rules of the address field are as follows.

a. The node ns puts the minimum address in its address field as its own network address. The address node ns attained is A_1, where all the nodes set of routing nodes to a shared.

b. The address field that node n_s has assigned for n_{s+i} is $[A_{s+i}^{min}, A_{s+i}^{max}]$. A_{s+i}^{min} is the lower limit of the address field of the node n_{s+i} and A_{s+i}^{max} is the upper limit of the address field of the node n_{s+i}. The upper and lower computation formulas are as follows:

$$A_{s+i}^{max} = \begin{cases} A_{s+i-1}^{max} + 1 & i > 1 \\ A_2 & i = 1 \end{cases} \tag{3}$$

$$A_{s+i}^{max} = \frac{\sum_{j=1}^{i} W_{s+j}}{\left(\sum_{j=1}^{k} W_{s+j}\right)/(1-p)}(A_N - A_2) + A_2 \tag{4}$$

The space for the reserved address of the node ns is $[A_{s+k}^{max} + 1, A_N]$.

According to the way of allocation of the routing node address, it can be known that any routing node n_s receives an address field A_s^{min}, A_s^{max}.

The address field contains all addresses of all nodes in the sub-tree with the head node n_s. Therefore, the address assignment in such a way only needs to compare the address domain of head nodes.

4 ROUTING POLICY OF ORDERED TREE

Through the above elaboration of the address allocation method, it can be known that the network constitutes an orderly tree. In the tree, the address domain of any child routing is the refinement of that of its father routing node. Therefore, the routing process is the process of a combination between refinement and generalization. According to the direction of transmission, the tree network data frame can be divided into downward frames from the parent node to the child node (refining process) and the child node to the parent node of the uplink frame (generalization process). There are three kinds of communication model when any two nodes communicate.

The description of routing policy: the data frame received by the routing node is likely to be the downlink frame sent by the parent node or the uplink frame sent by the child node. For the downlink frame, if the destination network address is in the routing node, the node address domain will forward or it will not forward. For the uplink frame, if the destination node network address is in the domain, it turns the type of a data frame into the downlink frame and then it will forward, otherwise it will not forward. If the destination node network address is the same as routing node address, the uplink frame will jump within the MAC address to reach the destination node depending on the destination node. The routing algorithm of the pseudo code is given in Table 1.

Table 1. Tree routing algorithm (PLC-TR).

Function: Router n_s	START
forward data frames f	Receive data frames f.
by comparing itself	IF (D_ADDR = A_s^{min})
domain and the	data frames can reach
destination network	within one jump.
address.	IF (D_MAC = n_s_MAC)
Parameters:	Routing node n_s is the
D_ADDR: Destination	destination node.
network address.	ELSE: through D_MAC
D_MAC: Destination	direct send f to the
physical address.	destination node.
ns_MAC: n_s physical	ENDIF
address.	ELSE IF (D_ADDR >
A_s^{min}: n_s minimum	A_s^{min} AND D_ADDR A_s^{max})
addresses.	Routing nodes n_s forward
A_s^{max}: n_s maximum	f as downlink frame.
addresses.	ELSE IF (f is uplink frame)
	Routing nodes n_s forward
	f as uplink frame.
	ENDIF
	END

5 THE SIMULATION AND THE ANALYSIS OF SIMULATION RESULTS

In order to verify the PLC-TR algorithm performance, the simulation is carried out based on the NS2 environment. Due to the power line network time-varying interference in the actual situation, the nodes sent by the source node may not reach the forwarding node and the destination node correctly due to the interference problems. In order to simulate the situation, we set up the double states Markov model. The definition of the network model S_{error} state is that the router cannot receive the package, and that of the S_{run} state is that the packets can be sent and received normally. The line impact factors for p ($0 < p < 1$) and the line recovery factors for q ($0 < q < 1$). In the continuous time in seconds, the node in the S_{run} state comes into the S_{error} state for the possibility of p. The node in the S_{error} state comes out of the S_{run} state for the possibility of q. Without losing generality, this study chooses p = 0.02 and p = 0.2 to simulate under two kinds of disturbance degree, with q = 0.95. The simulation model is shown in Figure 1: the total length of power line is 6 km; the number of nodes is 40, 80, 120, 160, 200, 240, respectively and distribute along the power line uniformly; and the simulation time is 300 seconds. The effective transmission distance between nodes is 200 m, the data transmission speed is 20 KB/s; each packet size is 64byte; and the type of packets is CBR.

The simulation results in Figures 4 to 5 show that when p = 0.02, compared with the SPR, the average delivery rate of PLC-TR is increased by 7.9% and the average delay is reduced by 16.5%, and when p = 0.2, there is a 13.7% increase in the delivery rate and a 14.1% reduction in the delay.

As shown in Figure 4, the average end-to-end communication time delay increases gradually with the increasing node density, and it is significantly lower in the PLC-TR than in the SPR under the interference conditions p = 0.02 and p = 0.2. The reason is

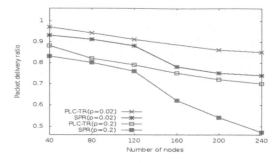

Figure 5. Relationships of the packet delivery rate and the quantitative nodes.

that each node in the SPR algorithm has to maintain the node-to-node path routing table; in this way, it needs to exchange routing tables with a neighbor node, with a certain network cost. With the increasing nodes, the costs grow rapidly, and it makes the end-to-end communication time delay much longer.

As shown in Figure 5, the speed of the packet delivery ratio increases with the increasing node density. It is more obvious when p = 0.2; under two kinds of interference, the PLC-TR is significantly higher than the SPR, because most of the nodes in the PLC-TR algorithm are the common node; the loss of them will not make other ordinary nodes inaccessible. All the nodes in the SPR algorithm may be on the shortest path of other source routers, so the loss in every node may result in the shortest path, and the PLC-TR needs to forward more than increase the probability of collision data. In addition, because the SPR's costs for maintenance of the shortest path network, which makes the packet collision retransmission times greater in the network, the package pass rate declines rapidly with the increase in nodes.

6 CONCLUSION

The PLC-TR proposed in this paper makes full use of the power network topology for network form and routing election, to solve the problem of inefficiency of the power line network routing to some extent, and reduce the network costs. The simulation experimental results show that in the process of line quality changes, the PLC-TR can still show a good performance in terms of the end-to-end delay, packet delivery ratio and other aspects.

REFERENCES

Biagi M, Greco S, Lampe L. Neighborhood-knowledge based geo-routing in PLC [C]. Power Line Communications and Its Applications, IEEE International Symposium on. IEEE, 2012: 7–12.

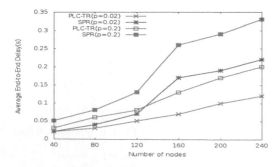

Figure 4. Relationships of the average end-to-end delay and the node number.

Biagi M, Lampe L. Location Assisted Routing Techniques for Power Line Communication in Smart Grids [C]. Smart Grid Communications, First IEEE International Conference on. IEEE, 2010: 274–278.

Bumiller G, Lampe L, Hrasnica H. Power line communication networks for large-scale control and automation systems [J]. Ommnaon Magazn, 2010, 48(4):106–113.

Galli S, Scaglione A, Wang Z. For the Grid and Through the Grid: The Role of Power Line Communications in the Smart Grid [J]. Rodng of H, 2011, 99: 998–1027.

Kim T, Kim S H, Yang J, et al. Neighbor Table Based Shortcut Tree Routing in ZigBee Wireless Networks [J]. Aralll and DrbdYmRanaon on, 2014, 25(3):706–716.

Kim T, Kim D, Park N. et al. Shortcut Tree Routing in ZigBee Networks [C]. Wireless Pervasive Computing, Iswpc 07, International Symposium on. IEEE, 2007.

Li H, Kensheng W, Li H, et al. Cluster Based Dynamic Routing on Powerline Carrier Network[C]. Wireless Communications, Networking and Mobile Computing, 2009. WiCom'09. 5th International Conference on. IEEE, 2009: 1–4.

Lin Y, Latchman H A, Lee M. et al. A power line communication network infrastructure for the smart home [J]. WrlOmmnaon, 2002, 9:104–111.

Pinero-Escuer PJ, Malgosa-Sanahuja J, Manzanares-Lopez P, et al. Homeplug-AV CSMA/CA Evaluation in a Real In-Building Scenario [J]. Communications Letters, IEEE, 2011, (6):683–685.

Ren Z, Tian L, Cao J, et al. An efficient hybrid routing algorithm for ZigBee networks[C]. Instrumentation and Measurement, Sensor Network and Automation, International Symposium on. IEEE, 2012: 415–418.

Slootweg H. Smart Grids—the future or fantasy? [C]. Smart Metering-making It Happen, Iet. IET, 2009:1–19.

Sanchez JA, Ruiz PM, Marin-Perez R. Beacon-less geographic routing made practical: challenges, design guidelines, and protocols [J]. Ommnaon Magazn, 2009, 47(8):85–91.

Yoon S, Jang S, Kim Y, et al. Opportunistic Routing for Smart Grid with Power Line Communication Access Networks [J]. Mar Grd Ranaon on, 2014, 5(1):303–311.

Energy Science and Applied Technology – Fang (Ed.)
© *2016 Taylor & Francis Group, London, ISBN 978-1-138-02833-3*

Research and realization of the indoor positioning system

Hongyun Guan & Hui Zhang

College of Information Science and Technology, Donghua University, Shanghai, China

ABSTRACT: Indoor positioning system has become one of the hot issues in the field of information technology. With the rapid development of networking and wireless communication technology, location-based service can apply to all works of life. This paper uses an android smartphone as a mobile terminal, using WiFi modules, orientation sensor and acceleration sensor on android to develop the corresponding client application program in order to get the signal strength (RSSI) values of the surrounding WiFi, the course of walking and the number of steps, and sends the information to the PC through the Socket Communication. Thus, this paper designs a hybrid indoor positioning system in which WiFi positioning is prioritized while the inertial sensor positioning ranks second. At the same time, this system has set up a simple PC experimental platform for testing the practicability of the whole hybrid indoor positioning system. The experimental results show that the whole hybrid indoor positioning system can meet the demand of the actual location.

Keywords: Indoor positioning; Android; WiFi module; Orientation sensor; Acceleration sensor; RSSI

1 INTRODUCTION

To date, the Global Positioning System (GPS) is being widely used, which is generally applied to all kinds of location-based service. However, the GPS fails to locate indoors, because this method of positioning requires the satellite to provide it with locating information, which is only suitable for the open and exposed outdoor environment. In the relatively closed indoor environment, the GPS cannot obtain information it needs to locate through the satellite. At this point, the WiFi Positioning Technology based on the Wireless Local Area Network and the Inertial Sensor Positioning Technology based on various sensors are developed rapidly. These two technologies are able to overcome the shortcoming of the GPS's indoor positioning, so that the location-based service can apply to all works of life.

The android operating system, based on Linux, mainly applies to mobile terminals such as smartphones and tablet computers. At present, the mobile terminal based on the android operating system is generally configured with the wireless module and all kinds of sensors, such as orientation sensor and acceleration sensor. On this basis, we can develop an indoor positioning system on the android mobile terminal.

This article, effectively combining WiFi positioning with the inertial sensor positioning, designs an original hybrid indoor positioning system on the android intelligent mobile terminal. These research findings can be widely applied in the fire rescue service, supermarket as well as shopping mall.

2 HYBRID INDOOR POSITIONING SYSTEM

Currently, the WiFi Positioning Technology and the Inertial Sensor Positioning Technology are the most widely used indoor positioning technologies. However, because the precision of the Inertial Sensor Positioning Technology depends on the pedestrian's initial position, and the course of walking and the distance traveled. However, there are errors when we measure the course of walking and distance during actual location. The longer the locating distance is, the bigger the errors are, so that locating errors become greater. Therefore, on the aspect of precision, the WiFi Positioning Technology is more dominating; however, there are some limitations for WiFi positioning. In places where WiFi signal intensity is weak or WiFi cannot cover, it is not available. Based on this limitation, this article designs a hybrid indoor positioning system in which WiFi positioning is prioritized, while the inertial sensor positioning ranks second.

This article targets the android smartphone as its client, and develops the relevant client application program on Eclipse, so that it can get RSSI values from surrounding WiFi, the course of walking and the number of steps that one takes. The locating information updates real time, and is sent to the PC through the Socket Communication. Every time the PC receives the information, it first selects three pre-established base stations and estimates whether the RSSI values are normal or not. If all RSSI values do not exceed the threshold value [–35~–95dBm], then RSSI values are received without error. Then, we can utilize the Triangle Centroid Localization Algorithm of WiFi Positioning Technology to locate; if at least one RSSI value exceeds the threshold value [–35~–95dBm] or RSSI values cannot receive (i.e. WiFi does not cover the area), then RSSI values are received abnormally. Then, we can utilize the course and the number of steps from the received information, and select the Dead Reckoning Method of Inertial Sensor Positioning Technology to locate.

3 TECHNICAL ROUTE

This system consists of the locating information acquisition module and the locating information processing module. The locating information acquisition module is the application program for the android client and the locating information processing module is the experimental platform on the PC. The information is sent from the acquisition module to the processing module through the Socket Communication.

4 THE REALIZATION OF APPLICATION PROGRAM FOR ANDROID CLIENT

The main interface of the application program has four functions: measuring course through the orientation sensor, measuring the number of steps through the acceleration sensor, obtaining WiFi name that is able to search and its RSSI values through the WiFi module and sending the information to the PC through the Socket Communication. The interface of the application program is shown in Figure 1.

The information on the top line of the interface is the mobile phone holder's course of walking, the number of steps that one takes and the reset steps (i.e. clear steps), which is used to locate by the Dead Reckoning Method; as for the WiFi list on the bottom, the left column shows all the WiFi signal names that the phone can search, and the right column shows the RSSI values of each WiFi at one point, which refreshes every five minutes, while the information of the list is used for Triangle Centroid Localization. The middle line of the interface is to realize the Socket Communication, input the IP address of the PC, as shown in Figure 3, and click on the send button to send the information in the whole page to the PC.

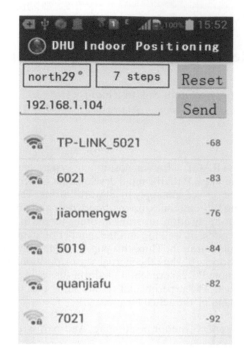

Figure 1. The main interface of the application program.

5 THE DESIGN AND CONSTRUCTION OF EXPERIMENTAL PLATFORM

5.1 *Design and function realization*

Design: first, the PC-side platform needs to display the PC's IP address and port number, so as to enter the IP address for the client and send the data to the corresponding port; second, this platform can load maps, to choose maps with area positioning, so we can find exactly the base station location for positioning. After loading the map, we select to start listening, and then we can get the WiFi list, the course of walking and the number of steps from the client application successfully and display them on the screen; Finally, select the positioning method based on the RSSI threshold value, if the RSSI values of three base stations that we have chosen did not exceed the threshold value, then locate the position of the three base stations on the map, and select triangle centroid localization. If at least one of the three base stations' RSSI value exceeds the threshold value, we mark the initial position on the map and select the dead reckoning method for positioning.

Function: the PC experimental platform designed in this topic can download different maps, making this platform useful in a variety of indoor environments. The IP address and port number will be displayed on the screen. When the Start Listening button is pressed, we can receive the data sent from the client application, and it will no longer receive the data when the Stop Listening button is pressed. First, we check three base stations in the WiFi list, and carry on the RSSI threshold value judgment: if all of them do not exceed the threshold value, press the Determine of Base Station button, and mark the location of the three stations on the map (red dot), and the coordinates of the three base stations will display at the same time, then press the Centroid Localization button after determining the base station, and we can see the anchor point (green dot) on the map. If at least one of the three base stations' RSSI value exceeds the threshold value, we should first press the Initial Position button and mark the original position on the map, (red dot), and then press the Dead Reckoning button, we will see the anchor point on the map (blue dot).

5.2 *Interface display of experimental platform*

Using the Microsoft Visual Studio 2010 design and built PC experimental platform, the loaded map is Donghua University the 28th dormitory, the 5th floor, corridor extends 2 m, and the top 24 m were to take as an experimental site. The PC's IP address and port number used for receiving the data are shown show. When the Start Listening button is pressed, the phone owners need to enter the IP address to locate, and click the Send button, the PC-side will display all of the positioning information received, as shown in Figure 2. The elements inside the red circle in the picture is the received orientation information.

Choosing three base stations: TP-LINK_5021, quanjiafu, jiaomengws, and all of the RSSI values of these three base stations do not exceed the threshold value, so press the Determination of Base Station button, we can mark the position of the three base stations on the map (red dots), and then press the Centroid Localization button, the anchor point will be obtained (green dot), as shown in Figure 3.

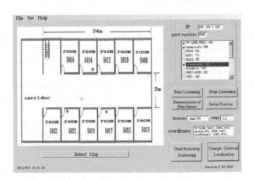

Figure 2.   PC receiving data interface.

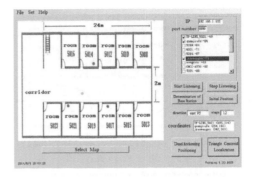

Figure 3.   Triangle centroid localization.

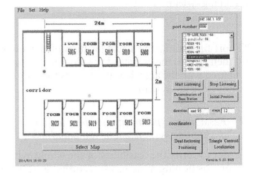

Figure 4.   Dead reckoning positioning.

Due to our choice of the three base stations, the RSSI value did not exceed the threshold of [−35—95dBm], so we choose centroid localization. Here, to illustrate how to locate through dead reckoning position, we uncheck the three base stations, first click on the Initial Position button, mark the start position on the map (red dot), and then click on the Dead Reckoning Positioning, we can position by the received course of walking and the number of steps, step is 0.75 meters set in the program, and the anchor point (blue dot) will be obtained, as shown in Figure 4.

6   EXPERIMENT AND DATA ANALYSIS

6.1 *Experimental conditions*

The fifth floor rooms and corridors of 28th dormitory building of Donghua University are selected to be experimental site, its corridor extends two meters wide, and we take the top 24 meters of the floor as the experiment site. The three base stations for positioning are the wireless router, WiFi names are as follows: TP—LINK_5021, quanjiafu and jiaomengws. The location are recorded by each experiment. The tester handles the mobile phone and the mobile at the same time, the data from the client application would send to the PC through the Socket Communication.

433

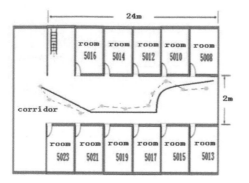

Figure 5. Feasibility experiment on the hybrid indoor positioning system.

PC port first marks the actual location of the tester, and then locates the estimated position of the tester, compared with the actual location, so as to verify the feasibility of the system.

6.2 Feasibility experiment on hybrid indoor positioning system and its data analysis

In order to verify the feasibility of the whole hybrid indoor positioning system, we recorded a walking path in the experiment, continuously positioning multiple points on the track. When each point is reached, the positioning information received from the mobile client is sent to the PC via the Socket Communication. The PC would test each information it has received, and choose the corresponding method to complete positioning and display the point on the map. After recording the positioning point, it continues to the next point positioning. Then, we can get the trajectory composed of multiple anchor points. As shown in Figure 5, the actual trajectory is a solid line (black), the positioning trajectory is a dotted line (black), the centroid location is represented by the green dot, and dead reckoning positioning is represented by the blue dot.

Data analysis: From Figure 6, we can see that although there is an error between the trajectory of the PC positioning and the actual path, the error is small and the trajectory of the general trend is consistent, so the hybrid indoor positioning system designed and implemented in this paper has strong feasibility, and it can meet the demand of the actual location.

7 CONCLUSION

Indoor positioning system has become one of the hot issues in the field of information technology. This paper has presented the research and implementation of the indoor positioning system based on android, from which the following conclusions can be drawn:

1. The development of the android client application can run normally;
   On the one hand, the application uses the corresponding WiFi module, gets all of the WiFi signal strength (RSSI) value for triangle centroid localization; on the other hand, it uses android's own orientation sensor and acceleration sensor with measurement of direction and steps for dead reckoning positioning. It also uses the Socket Communication to send all of the positioning information to the PC port experimental platform.
2. Strong feasibility on the indoor positioning system;
   This system has set up a simple PC port experimental platform for testing the practicability of the whole hybrid indoor positioning system. The experimental results show that the positioning can reach the expected goal and that the whole hybrid indoor positioning system can meet the demand of the actual location.

REFERENCES

Atia, M.M. 2013. Dynamic online-calibrated radio maps for indoor positioning in wireless local area networks. IEEE Transactions on Mobile Computing, 12(9): 1774–1787.

Atzori, L. 2012. Indoor navigation system using image and sensor data processing on a smartphone, In Proceedings of international conference on optimization of electrical and electronic equipment, 1158–1163.

Chen, F. 2010. Compressive sensing based positioning using RSS of WLAN access points [C]. IEEE INFOCOM, 1–9.

Doherty, L. 2011. Convex position estimation in wireless sensor networks [C]. Proc of the IEEE INFOCOM, 2001, 3, Anchorage: IEEE Computer and Communications Societies, 1655- 1663.

Harle, R. 2013. A survey of indoor inertial positioning systems for pedestrians. IEEE Communications Surveys and Tutorials, 15(3): 1281–1293.

Han Ning. 2010. Design of automatic forest fire positioning system based on video monitoring system [C]. Geoscience and Remote Sensing (IITA-GRS), 2010 Second IITA, 2:532–535.

Juang, S.Y. 2012. Real-time indoor surveillance based on smartphone and mobile robot, In Proceedings of IEEE international conference on industrial informatics, 475–480.

Li X. 2010. RSS-based location estimation with unknown path loss model [J]. IEEE Transactions on Wireless Communications, 5(12):3626–3633.

Wang, T.Q. 2013. Position accuracy of time-of-arrival based ranging using visible light with application in indoor localization systems. Journal of Light wave Technology, 31(20):3302–3308.

Zhang M.H. 2010. A new positioning method for location- based services in wireless LANs [J]. Chinese Journal of Electronics, 17(1):75–79.

Energy Science and Applied Technology – Fang (Ed.)
© 2016 Taylor & Francis Group, London, ISBN 978-1-138-02833-3

The design and implementation of remote monitoring terminal for electric vehicle based on Linux

Yang Xu & Zhaoxi Wang
Automobile Electronic Research Center, University of Posts and Telecommunications, Chongqing, China

ABSTRACT: The design and implementation of a remote monitoring terminal for the electric vehicle based on Linux is introduced in this paper. The terminal uses ARM920T as the hardware platform, which can be used for remote management of electric vehicles. In the hardware design of the terminal, the ARM, GPS, GPRS and CAN bus are combined to realize the function of remote monitoring and remote control. In the software design, the Linux operating system is used as the software platform. It realizes the transplantation of serial driver, SPI driver and other driver. In addition, the information frames gathered from CAN message and GPS message are analyzed and repackaged according to the predefined frame format. The experiments prove that this design can achieve the remote monitoring of electric vehicles.

Keywords: Vehicle Monitoring Terminal; Electric Vehicle; Linux; GPS

1 INTRODUCTION

With the environmental pollution increasingly becoming serious and the world energy crisis intensified, hybrid vehicles and pure electric vehicles step on the stage of history. The Twelfth Five Year Plan of China puts forward to the "three vertical and three horizontal three platform" as the development of layout, and continues to develop the new dynamic system and study the new dynamic key component, to promote the further development of electric vehicle industry (Lei Liangyu, 2010, Qi Weisu, 2012). The collection of vehicle running state data and fault information plays an important role in the development of electric vehicle technology. First, the running state data and fault information of the vehicle is the necessary reference for future development of electric vehicle technology. Second, if the running state of the vehicle can be monitored in real time and send some warning information about the potential danger of vehicles, it will greatly reduce the traffic accident caused by vehicle fault. Finally, the running state data and fault information of vehicles also play an important role in a traffic accident identified (Tian Ye, 2013). So, a stable and reliable remote monitoring terminal system is essential for the development of electric vehicles. In addition, the owner can also send control command to the remote monitoring via PC or an APP on mobile to control his vehicle. It can reduce vehicle theft to some extent.

2 THE VEHICLE MONITORING SYSTEM

The vehicle monitoring system consists of vehicle monitoring terminal, transmission network and monitoring center (Ma Teng, 2009). The system sends the running state data and fault information of vehicles to the monitoring center through the CAN and wireless data network, and then the data and information is processed and stored (Shao Yunin, 2008). At the same time, users can also send some control commands to vehicles to control it. The vehicle monitoring system is the base for vehicle anti-theft, vehicle anti-robbery, route monitoring of vehicles, running state information transmission of vehicles, traffic accident fast response, and it is also the foundation to solve the problems of the vehicles dynamic management. The whole structure of the vehicle monitoring system is shown in Figure 1.

3 THE HARDWARE IMPLEMENTATION OF THE MONITORING TERMINAL

The remote monitoring terminal of the vehicle designed in this paper is composed of a control module, storage module, GPS module, communication module, video capture module and power module. The communication module is composed of a CAN communication unit and the GPRS communication unit. The hardware structure of monitoring terminal is shown in Figure 2.

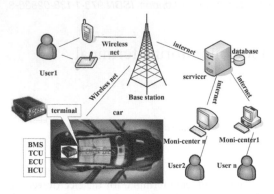

Figure 1. The vehicle monitoring system structure diagram.

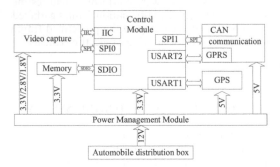

Figure 2. The hardware structure of monitoring terminal.

3.1 Control module

The main function of the control module is to collect and process the data. Considering the complexity of the algorithm and the development costs of the system, the S3C2440 is used as the main controller. The control unit consists of S3C2440, Nan Flash, SRAM and power supply components. During data acquisition, the MCU analyzes the latitude and longitude information of the vehicle and extracts the data part of the data frame, and then repackages the processed data to a new frame and sends it to the GPRS module. Meanwhile, it extracts the available information from CAN messages, and integrates different information from different CAN nodes, and then repackages the processed data and sends it to the GPRS module. In the remote control, the MCU handles the information read from the GPRS module and extracts the control command data. Then, it repackages the processed information in the CAN protocol format and sends it to the CAN bus to control vehicles.

3.2 Power modules

A stable, security, accurate power system is a prerequisite to ensure the stable operation of the vehicle monitoring terminal. The power supply is divided into two main parts in this paper; in other words, the power supply for the core part and for the peripheral parts is independent. In order to provide enough power for the MCU and peripheral circuit, the stable linear power supply module, K7805–2000, is used to change the voltage outputted from the automobile distribution box to 5V. The module can provide a stable output voltage and a higher power, and it can meet power demand of the GPS, GPRS and other large power device. In addition, the stable working voltage of 1.25V for the kernel of the CPU is generated by using the linear power supply MAX8860. In addition, to take into account the car work environment, the self-recovery fuse, high voltage capacitor and ESD is added in the design.

3.3 CAN module

Because the running state information of the vehicle is transmitted by the CAN bus, an efficient, fast, reliable CAN communication module is needed to collect car body information. The CAN controller, MCP2515, is used to achieve the communication between the vehicle and the monitoring terminal. MCP2515 provides a SPI interface and supports CAN specification of version 2.0 A/B and the data transmission rate up to 1Mbps. In addition, MCP2515 has an excellent EMC performance and low power consumption.

3.4 GPS and GPRS module

The GPS module is used to receive the signal from the GPS satellite, and calculates the location of the monitoring terminal. The ATK-NEO-6M is used in this design. The module has the characteristics of lower prices, GPS signal capture faster, lower power consumption, and high performance. It provides three styles of the serial port, IIC bus and SPI bus and uses the NMEA protocol to communicate with the MCU. The GPRS module is responsible for receiving control commands from the monitor center and transmitting vehicle information from the server via the TCP/IP protocol. The serial ports 1 and 2 are used for the GPS and the GPRS to communicate with the MCU, respectively.

3.5 Storage circuit

A SD card is used as the storage medium, the state information of vehicles from the CAN bus and the position information from the GPS is stored on it. The SD card is a swappable device; therefore, it adopts the ESD devices in the circuit to prevent the damage of the monitoring terminal in the process of plug.

4 THE SOFTWARE IMPLEMENTATION OF MONITORING TERMINAL

In order to ensure the system has a high extensibility and scalability, it uses the embedded Linux operating system as the bottom software platform. The Linux operating system has the characteristics of being safe, reliable, open source, extensive hardware support, and easy to develop, so it can shorten the development cycle to a large extent (Qi Weisu, 2012). The architecture of the system software based on Linux is shown in Figure 3.

4.1 The flow of system software

The GPS and GPRS modules communicate with the MCU via the serial port, and the SD card uses the SDIO interface and the CAN module uses the SPI in order to perform that. Therefore, the serial driver, SPI driver and SDIO driver should be transplanted to the Linux operating system. The application layer software of the terminal mainly includes the analysis of GPS data and CAN message, the repackaging of the processed CAN message and GPS data, and receiving and transmitting the data through the GPRS. Figure 4 shows the flow of the system software.

4.2 The software design of the GPS and GPRS module

The MCU communicates with the GPS using the NMEA-0183 protocol, which uses the ASCII code to transmit the information from the GPS positioning. The returned information from the GPS includes the current position, the number of currently available satellites, the number of visible satellites, ground speed, geodetic coordinate and the current time (UTC). The MCU configures the GPS and reads information from it through serial port 1.

The GPRS module has mainly three tasks: (1) to establish the communication link; (2) to establish the TCP/IP connection between the terminal

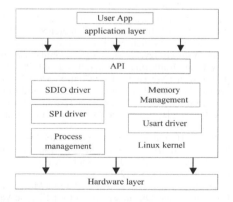

Figure 3.   The architecture of the system software.

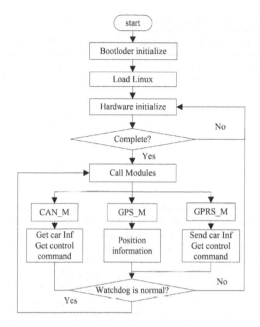

Figure 4.   The flowchart of the system software.

and the remote monitoring center; (3) to send and receive the data. The GPRS communicates with the MCU through the serial port to achieve the transmitted or received data.

4.3 The protocol design for repackaging

To ensure a reliable and secure data, this paper analyzes the message from the CAN bus and GPS module and repackages it. Then, the repackaged frames are sent to the GPRS module via the serial port, and the GPRS sends data to the server via the TCP/IP protocol. The format of the repackaged frames is presented in Table 1.

The first byte LEN indicates the total length in bytes of the command. The start byte is set to 0x55 and the car information part includes the specific information of the vehicle. The command byte shows the name of state information of vehicles (e.g. battery level and the vehicle speed) and the data (DT) part is the specific value of the state information control commands. The length of the checksum (CHK) is one byte, and it is calculated using equation (1), where the $oxFF$ is ASCII and X_{val} is the value that performed a bitwise OR on all the bytes before the checksum.

$$checksum = oxFF - X_{val} \qquad (1)$$

4.4 Information storage and pretreatment

When vehicles are in motion, the vehicle monitoring terminal acquires vehicle state information from the BMS, TCU, HCU and ECU every 10 ms

Table 1. The format of the repackaged frame.

LEN	STR	Car INF	Com byte	DT	CHK
1B	0 × 55	4B	1 B	n B	1 B

Table 2. The storage format of vehicle information.

Start	Time	Vehicle state information			
0 × 55	Time	TP	BS	MR	......

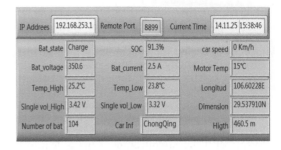

Figure 5. The test interface of the monitoring terminal.

by the CAN bus and then analyzes the information. The terminal control unit compares the processed information with the limit value of vehicles every time. The data are stored in the SD card and some warning information is sent to the server and the monitoring center if the value exceeds the threshold value, otherwise the data are stored in the SD card every 10 s. Table 2 presents the storage format.

The defaulted start byte of the stored information is 0x55, and the time information includes year, month, day, hour, minute, and second accounting for 10 bytes. The vehicle status information includes battery information, vehicle location information, electrical information, and tire pressure information.

5 TESTS FOR VEHICLE MONITORING TERMINAL

In the actual monitoring of the electric car, it does not require that the remote monitoring terminal should send relevant information to the server in real time, instead the information is stored in the SD card. In the following case, the monitoring terminal will send the vehicle state information stored in the SD card to the server. (1) When the vehicle data gathered by the terminal is abnormal, the terminal sends the vehicle data and alarming information to the server. (2) The terminal sends the information data stored in the SD card to the server and then removes the SD card data when the vehicle is charging. (3) When the control center sends commands to access the vehicle information, the current vehicle information will be sent to the server.

In order to test the stability and reliability of the vehicle monitoring terminal, this paper tests the terminal on a domestic electric vehicle platform. In order to test the readability data conveniently, a PC interface is designed by using Labview software to read and display the related information of the vehicle from the server. The information includes vehicle longitude and latitude, vehicle speed, motor speed, the current time, battery voltage, power, and temperature. The test interface is shown in Figure 5.

After the test, the vehicle monitoring terminal designed in this paper can be realized to monitor the vehicle location information and operating state. At the same time, it can control the vehicle remotely.

6 CONCLUSIONS

This paper introduces a design proposal of the remote monitoring terminal based on the embedded Linux operating system. It gives the realization of the hardware and software of the terminal in detail. The remote monitoring terminal designed in this paper can monitor and manage electric vehicles, and can reduce traffic accidents caused by the fault of the vehicle and curb the vehicle theft and robbery. So, it can be widely used in electric vehicles.

ACKNOWLEDGMENTS

This project was supported by the Natural Science Foundation Project of CQ CSTC cstc2012 jjA 60002".

REFERENCES

Lei Liangyu, Huang Jinpeng, Liu Jianjun. 2010. Research on Remote Monitoring System of Electric Vehicle [C]. Proceedings of International Conference on Power Electronics and Intelligent Transportation System.

Ma Teng, Yang Hongye. 2009. Design and implementation of Vehicle Monitoring Terminal Based on GPS/GPRS [J]. Electronic Measurement & Control, 32(4):71–74.

Qi Weisu, Zhao Yu, Pan Yong. 2012. Design and Implementation of Vehicle Monitoring Terminal Based on ARM-Linux [J]. Electronic Measurement & Control, 32(2):85–88.

Shao Yunin, Hu Gang, Chen Xueping. 2008. Design and implementation of Embedded Vehicle Monitoring Terminal Based on ARM [J]. Computer Measurement & Control, 16(8):1125–1128.

Tian Ye. 2013. The Development of Remote Monitoring, Calibration and Diagnostic System for Electric Vehicles [D]. Jilin University.

Energy Science and Applied Technology – Fang (Ed.)
© 2016 Taylor & Francis Group, London, ISBN 978-1-138-02833-3

A semantic-based cache management mechanism for MP2P

Yi Jiang & Feng Chen
School of Computer Science and Technology, Chongqing University of Posts and Telecommunications, Chongqing, China

ABSTRACT: With the data-interchange requirements growing continually, the MP2P network development has gained wide attention. The network resource storage procedure and the search process in the overlay network are inefficient resulting in the mobile Internet with high mobility and instability. A semantic-based cache management mechanism for the MP2P network is proposed, aiming at low accuracy and efficiency of searching, which means gathering a similar type of data that could improve the storage efficiency of the cache resource object in the method of semantic clustering. The simulations reveals that we can see a decent boost of searching resource efficiency in cache mechanism.

Keywords: mobile Internet; MP2P network; cache management mechanism; semantic clustering

1 INTRODUCTION

With the mobile Internet technology becoming more mature, the application of wireless Internet has developed rapidly. With effective resource sharing and nodes, the MP2P (Mobile Peer to Peer) network has attracted several researchers and become the research hotspot of the current wireless Internet. However, because of the constant movement of nodes, the randomness of entrance and exit nodes and the storage capacity difference of the terminal equipment in the MP2P network, there arise problems such as single point of failure, constant change of topological graph and low hit rate of Internet resource of the overlay network. The efficient resource distribution storage strategy can not only relieve the single point pressure effectively, but also improve the success rate of mobile data access and the network availability. Therefore, it becomes more important to design the suitable resource distribution storage strategy according to the characteristics of MP2P Internet in order to improve the access efficiency and performance of the MP2P network and then to improve the mobile terminal users' experience.

Currently, the resource storage of the MP2P network mostly applies the Gnutella mechanism, and resource orientation mainly adopts the flood pattern, but this pattern will result in network resource waste, network jam, and the constant node movement will also lead to difficulty in resource orientation. Based on the mobile Ad-hoc network (short in MANET), realized P2P overlay network (i.e. MP2P), to process resource sharing, this paper presents a MP2P cooperative caching mechanism, which can effectively remit network jam caused by resource hot issue, and at the same time, increase the efficiency of resource search and improve the performance of the whole network.

2 RELATED WORK

Nowadays, most of the literature takes P2P file sharing and data distribution on the mobile Ad-hoc network (MANET) as the mobile P2P problem for research.

In China, Zhou Xinxin et al. presented the mobile P2P network cooperative caching optimization strategy, which can comprehensively take into consideration the factors such as popularity of data in the network, enable the lowest data access cost of the entire network and decrease the data access delay. However, this method only uses the distributed idea but does not fully consider the dynamic change and node movement of the network. Zhang Guoyin et al. proposed the L-Tapestry network topology model and the cooperative cache algorithm based on geographical location information. Some improvement was made for the long delay and high energy consumption of data retrieval. The P2P network was deployed in the mobile campus network environment, and in-depth research has been made for the access control strategy, cache and replacement strategy and data consistency strategy, but how to divide into several areas according to the node interestingness for the topology of entire network was not taken into consideration. In the literature, Yan Yuanting used the virtual cluster of the network node caused by the different interestingness for the same resource to raise a mobile equivalent semantic community model. On this basis, for the self-interest of node under the MP2P network environment and the "free riding" phenomenon extensively exists, a task collaborative logic model oriented to the P2P community—TCLM-P2P based on the Agent theory has been put forward for resource discovery. In the literature, Song Benjie proposed the cache replacement algorithm, which solves the data demand problem under the circumstances of network short circuit or under heavy network load and improves the hit rate. He also presented the three-level cache architecture, which highly improves the cache utilization and solves the hot issues and the conflict of local demand.

In general, most of the current research mainly studies the resource search efficiency and network cost of resource search and does not put forward the thought of resource classification storage for the particular mobile environment. How to achieve the balance between the stability of the MP2P network and wireless resource consumption is the key issue needed to be solved.

3 RELEVANT DEFINITION

In contrast to the traditional MP2P network basically comparing with the resource content, the paper uses the semantic information to describe the resource target on the basis of the semantic MP2P cooperative caching control mechanism. It abstracts the center vector and position and the resource object storage according to semantic distance under the specific semantic overlay network topological structure. The relevant definitions are as follows.

3.1 Similarity definition

In the network, the resource nodes with close semantic can be organized to form SON (Semantic Overlay Network) in order to improve the access efficiency and the accuracy of resource access when responding to the resource access request.

Definition 1, Vector Similarity: $\cos\theta = \dfrac{W_1 \bullet W_2}{|W_1||W_2|}$

Here, W_1, W_2 are two vectors for the similarity to be calculated. Their similarity is judged through calculating the cosine value of the two vectors $\cos\theta$.

Definition 2, Vector Space Model (VSM): e paper shows the resource as the vectors composed of keywords and their weight. Each keyword is one dimensional of a vector. These keywords set constitute a multi-dimensional vector space. Each resource object is a vector in the multi-dimensional vector space. We take D as the vector space model of the resource object, which is given by

$$D = \{(X, key_1, W_1), (X, key_2, W_2), \dots (X, key_i, W_i)\}$$

where X is some resource object name; key_i is the resource object keyword; and W_i is the resource object keyword weight.

4 A SEMANTIC-BASED CACHE MANAGEMENT MECHANISM FOR MP2P

In the MP2P network, the resource stores on each node. Each node may store different resource objects, such as video, audio and text. We can use the internal relation that exists among the resources to organize and dispatch the resource objects in order to improve the search efficiency and ease the search pressure.

4.1 Basic idea

Based on the definition of the resource space vector under the MP2P network, we can provide the effective base for the storage and dispatching of resources on nodes by means of calculating the semantic similarity. The paper assumes the MP2P overlay network layered node structure, as shown in Figure 1. There are two kinds of nodes: weak-peer and super-peer. Both show the weak nodes and super nodes, respectively, in the overlay network. Weak-peer represents the nodes with strong mobility and weak stability. Within a certain time, it cannot form an interest accumulation point. However, super-peer has high stability and manages the entrance and exit of weak-peer. Also, it relates to the adjacent super-peer and represents a kind of resource interest node.

The proposed mechanism is shown in Figure 2.

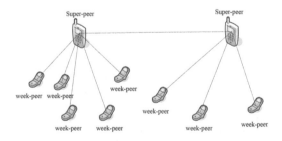

Figure 1. Network layer node structure.

Figure 2. A semantic-based cache management mechanism for MP2P.

When the cache object enters the MP2P overlay network, the Super-peer belonging to the uploading node calculates the semantic vector of the object and compares the center vectors of the nearby interest cluster. If the semantic distance is less than the threshold and the cache space is large enough, then it enters the interest cluster for storage. If the cache space is not large enough, the paper proposes the replacement mechanism of cache object. The new object will replace the object that is least used and with the distant semantic distance. For the objects not related to the interest cluster semantic, non-cached object storage mechanism will be used. Besides, the Super-peer will inspect the attribute set of non-cached objects on schedule to judge whether they can form a new interest cluster in order to enter the cache space.

4.2 Network initialization

In the paper, the MP2P network is composed of Weak-peer nodes. We choose the nodes with a long online time, weak mobile capacity and strong terminal processing ability as the Super-peer in all the Weak-peer nodes to form a semantic cluster managing the Weak-peer with similar semantic. The algorithm is as follows.

Algorithm 1: the algorithm if semantic cluster formation:
Input: t_{i0}, the time that the node enters the system; t_{in}, the current time of system; t_{ik}, the accumulative time that the node is online; V, the mobile data of the node; R, the direct communication radius of the node.
Output: if it will be a new cluster node.

1. Use formula (1) to calculate the node activity:

$$H_i = \frac{t_{ik}}{t_{in} - t_{i0}} \tag{1}$$

where t_{i0} is the time that the node enters the system; t_{in} is the current time of the system; and t_{ik} is the accumulative time that the node is online. The more the value of H_i is close to 1, the mode stable the node is, otherwise it shows that the offline time of the node is too long and not suitable to be taken as a semantic cluster.

2. Use formula (2) to calculate the node failure rate referring to papers:

$$L_i = 1 - (1 - V / R)^2, (R \geq V) \tag{2}$$

where V is the movement speed of the node and the direction is random and R is the direct communication radius of the node. The bigger the value of V/R, the bigger the failure rate L_i.

So, when the radius is specified, the faster the node moves, the higher the failure rate is and the stronger the dynamic is.
3. To calculate the average failure rate $\overline{L}$ and average activity $\overline{H}$, we repeat Steps 1 and 2 and calculate the average value to get the failure rate L and the average activity H of all the nodes.
4. To judge whether to become the new cluster node: $L_i > \overline{L}$ and $H_i > \overline{H}$.

We use the algorithm to calculate the nodes in the network to get the nodes with high stability as the Super-peer and others as the Weak-peer. Then, network initialization comes to the end.

4.3 The cache object placement strategy

The resource object placement strategy is composed of two algorithms. When the new cache resource object comes, we use Algorithm 1 to calculate the semantic distance between the new object and all the attribute center vectors in the peer-table, and we then use Algorithm 2 to calculate the semantic similarity weight as the base of whether the resource can enter the cache. The algorithm details are as follows:

Algorithm 2: the semantic distance algorithm of the center vector of the resource object:
Input: resource object $f_1, f_2, f_3, \ldots f_n$
Output: semantic distance $\cos\theta_1$, $\cos\theta_2$, $\cos\theta_3$, $\ldots, \cos\theta_n$

1. Assume a group of resource objects $f_1, f_2, f_3, \ldots f_n$ belonging to the same kind and use the TFLD method to extract the keyword $k_1, k_2, k_3, \ldots, k_i$ and calculate the weight.
2. Set up a two-dimensional vector about the keyword $K_i = \{(k_{i1}, w_{i1}), (k_{i2}, w_{i2}), \ldots, (k_{ii}, w_{ii}), \ldots (k_{in}, w_{in})\}$. Here, $k_{i1}, k_{i2}, \ldots, k_{in}$ is the i keyword name and $w_{i1}, w_{i2}, \ldots, w_{in}$ is the keyword weight. Then, choose the first n keywords according to the descending order of the weight.
3. Calculate the keyword eigenvalue p: take one *KUNION* as the set of all the keywords of all the files with the same attribute $KUNION = \{k \mid \exists(f_i, n), (k, w_{in}) \in K_i\}$. For each $k, k \in KUNION$. Then, calculate the eigenvalue $p = \sum_{\exists d_i, (k, w_{ij}) \in k_i} w_{ij}/n$. Also, w_{ij} is the weight of the j keyword and n is the number of resource objects.
4. Calculate the center vector of the resource: produce a feature pair (k, p) for each keyword k and produce the vector V according to the keyword features. Take it as $V = \{(k_1, p_1), \ldots, (k_i, p_i), \ldots, (k_n, p_n)\}$ and V is the equal of the center vector of the resource.
5. Extract the center vector of the cluster node: assume that there are k resource objects for the

cluster center node and each resource object keeps n keyword names and the corresponding weights. Sort the $k*n$ keyword weights and extract the first n biggest weights. Then, refer to steps 1~4 to produce the center vector of the node V_0.

6. Calculate the semantic distance between the resource object and the node: use the center vector obtained from 4 to transfer into distance to compare the angular deviation between two vectors and we can get $\cos\theta = \frac{V \cdot V_0}{|V||V_0|}$. Here, V is the center vector of the resource object from step 4 and V_0 is the center vector of the node from step 5.

7. Determine the semantic distance size:

$$\cos\theta = 1 \tag{3}$$
$$0 < \cos\theta < 1 \tag{4}$$
$$\cos\theta = 1 \tag{5}$$

Formula (3) shows that the two vectors coincide with each other, and they have the same semantic; formula (4) shows that the two vectors are similar but do not coincide with each other; formula (5) shows that the two vectors intersect but their semantics are quite different. Then, they can only be compared with the correlation with other clusters.

The above algorithm 1 shows the relation of the two vectors through the cosine value. If the vectors coincide, it means that the resource object is similar and can enter the cache; if the vectors intersect, then it cannot enter the cache. When the cosine value of the two vectors is within the value from 0 to 1, a further semantic similarity weight algorithm should be designed to judge whether it can enter some cluster.

Algorithm 3: the semantic similarity threshold:
Input: $p_1, p_2, \ldots, p_n$, the cosine similarity of the cluster node $\cos\theta_1, \cos\theta_2, \ldots, \cos\theta_n$
Output: *threshold*

1. Use formula (6) to calculate the standard deviation of each cosine similarity:

$$\partial(a) = \sqrt{\left(\sum_{i=1}^{n}(a_i - E(a)^2)\right)/n} \tag{6}$$

$E(a) = \sum_{i=1}^{n} a_i/n$, $E(a)$ is the mean value of all a and the value of a_i is equal to the value of $\cos\theta_i$.

2. Use formula (7) to calculate the threshold:

$$threshold = E(a) + k*\partial(a) \tag{7}$$

Here, k is a constant, which is used to adjust the threshold.

If the semantic distance is less than the weight, then the resource object enters the cache and chooses a semantic distance whose attribute is most close to the corresponding distance weight. If the

semantic distance is more than the corresponding distance weight in the peer-table, then the object is irrelevant to the common interest of the node and it does not enter the cache.

4.4 *Cache replacement strategy*

For a node, usually the interest is close in a period. So, we have reasons to think that the next access object is most likely to have the close semantic with this access object. Therefore, when an object enters the cache and the cache space is not large enough, we can compare the object and the semantic distance of the center vector and also the resource validation. The paper uses the method of combining the semantic distance with LRU, i.e. with the most distant semantic distance, the least to use the replacement principle. The algorithm is as following:

Algorithm 4: cache object replacement algorithm:
Input: resource center vector V_{i1}, V_{i1}, ..., V_{in}, node center vector V_{i0}
Output: if it will succeed to replace the resource

1. Calculate the semantic distance Dis: use Algorithm 1 to calculate the resource vector V_{ij}, which will join in and the node center vector V_{i0} and the semantic distance Dis.
2. Call over the resource stored in the LRU cache table and the semantic distance of the center vector Dis_1', Dis_2', Dis_3', ..., Dis_n'.
3. Compare each value of Dis_1', Dis_2', Dis_3', ..., Dis_n' with the value of Dis. If Dis is smaller than any one of them, then use the LRU algorithm to remove the least recently used cache data and at the same time move this resource into cache. If the value of Dis is bigger than any of them, then the cache gives up to store this resource.

5 EXPERIMENTAL ENVIRONMENT

This section shows the simulation of the algorithm by the experiment presented in the paper to compare with the flooding searching algorithm in the Gnutella network, and uses it to evaluate the MP2P cache mechanism after improvement. The experiment uses the search delay to measure the performance of the mechanism in the paper. The experiment assumes that the simulation is within the circle with the radius of 500 m. It divides into two situations. One is that the movement of the node is according to the Random Waypoint model for simulation, which means that the node is distributed in the network at random. The other is the two-layer structure based on semantic similarity in the paper, and then the Super-peer manages the Weak-peer with the similar interest. The network model is shown in Figure 3.

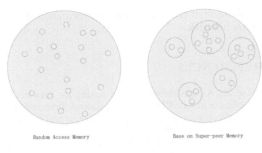

Random Access Memory   Base on Super-peer Memory

Figure 3.　Two storage strategy models.

The paper designs two experiments to compare the resource search efficiency of the above resources under different conditions, respectively.

5.1　*Average delay based on the change in network node*

The search delay means the hops from sending the request node to approving the request node, and the average delay is the average value of hops. It is one of the evaluation indices of network performance.

In the experiment, it is assumed that the node in the network cannot enter and exit the model range randomly, but can move at random within an effective range. When the node proposes the resource access request, the Gnutella mechanism searches the whole network by flooding searching. If it does not locate the resource within the hop threshold range T, then it will be considered that it does not hit the resource. In contrast to the flooding searching, the storage strategy after improvement requests to seek the Super-peer with the similar semantic as it puts the resources of the same kind into one Super-peer according to the semantic cluster and then it performs resource location.

The initial value of the number of nodes is 50, and this value will be increased each time (reorganize the network for each time adding the nodes in order to simulate the dynamically changing network condition). The hop weight of the resource access T is 5. In the mechanism presented in this paper, the low number setting of the Super-peer will result in insufficient clustering and the high number setting will violate the original intention of mobile P2P. So, the experiment assumes that the number of the Super-peer is 9. The simulation result is shown in Figure 4.

We can see from Figure 4 that the average delay of the resource searching of the storage mechanism after improvement and the Gnutella mechanism steadily rises with the increased addition of the number of nodes in the network; however, obviously, the average delay based on the semantic storage mechanism with the same number of nodes is lower than the relevant resources of the neighboring node search-

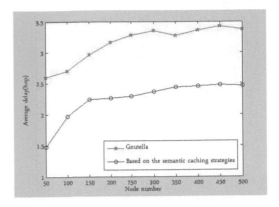

Figure 4.　Average delay comparison of adding nodes.

ing. If it does not jump to the next adjacent node, and the value of each hop threshold T decreases to 1, then it stops searching until T decreases to 0. To improve the mechanism, the search method finds the Super-peer with the similar semantic through the keyword vector and finds the resource address in the Super-peer cache to hit the resource quickly. The searching of resources is no longer aimless, and it can reduce searching significantly. We can also see from Figure 4 that when the number of nodes is less than 150, we can search the resources quickly within the hop threshold T and the curve rises quickly; when the number of nodes is more than 150, the two curves tend to flatten, and it shows that the average delay tends to be a stable value.

5.2　*Average delay based on the change in resource request*

The initial condition of the experiment is as follows: the initial value of the node is 100; the hop weight T is 10; the radius of the experiment area is 500 m; and the number of the Super-peer is 9. In order to simulate the instability of the dynamic network in the experiment, we add or reduce the number of nodes at random every once in a while in order to increase the complexity of the network. The simulation result is shown in Figure 5.

We can see from Figure 5 that the inquiry delay of the Gnutella mechanism increases with the increase in the number of requests. The increase in the entrance and exit nodes results in a large network fluctuation. The average delay of resource access in demand tends to rise quickly after a certain time. The mechanism proposed in the paper has an obvious advantage, with vertical comparison, and the average delay is lower than the average delay of flooding searching algorithm. Its increase is low and can better improve the performance of the whole network. Because of the stability of the

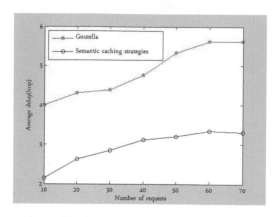

Figure 5. Comparison of the average delay of request increase.

Super-peer and the capacity of gathering the same kind of resources, the search request can locate the general direction of resources in demand quickly and greatly shorten the access delay.

In conclusion, the experimental results show that with its capacity of gathering the same kind of semantic resources and using the Super-peer to locate the resources quickly, the mechanism based on semantic grouping can greatly shorten the search delay and improve the search efficiency.

6 CONCLUSION

This paper proposes the MP2P cache management mechanism on the basis of the MP2P overlay network in the ad-hoc network. By improving the correlation of the same kind of resources, the hit rate and search efficiency is improved and then the problems of network congestion and the low hit rate of resources resulting from the mobility and instability of nodes are solved. The simulation results show that the storage strategy proposed in the paper is more effective and accurate than the current storage strategy in resource searching.

In addition, it can improve the performance of the entire network.

REFERENCES

Fikre, Z., A Mostefaoui. 2012. Caching for data availability in mobile P2P streaming systems. International Conference on Selected Topics in Mobile and Wireless Networking (iCOST). 48–53.

Lu, Y., K Sere, Z Xinrong. 2004. Towards an integrated architecture for peer-to-peer and ad hoc overlay network applications. 10th IEEE International Workshop on Future Trends of Distributed Computing Systems, FTDCS 312–318.

Mamei, M. 2004. Creating overlay data structures with the TOTA middleware to support content-based routing in mobile P2P networks. International Workshop on Hot Topics in Peer-to-Peer Systems. 74–79.

Song RenJie, Lu Zhe. 2014. P2P Network Caching Strategy. Science & Technology Vision, 16, 74–75.

Tang B, Gupta H, Das S. 2008. Benefit—based data caching in adhoc networks. IEEE Transactions on Mobile Computing, 3(7):289–304.

Tao Yuehua, et al. 2001. Vector-based Computing Scheme of similarity [J]. Journal of Yunnan Normal University, 21(5):18–19.

Wei Song, Weihua Zhuang. 2009. "Multi-Service load Sharing for Resource Management in the Cellular/WLAN Integrated Network," IEEE Transactions on Wireless Communications, vol. 8, no. 2, pp. 725–735.

Xian Youheng, Miao Fuyou, Xiong Yan, et al. 2010. Resource Discovery Algorithm Based on Super-peer in Mobile Peer-to-Peer Network. Journal of Chinese Computer Systems, 10(10):2066–2067.

Xinzheng, N., Z Dongmei, D Zhouhui. Energy adaptive cooperative data dissemination for mobile peer-to-peer network. 9th International Conference.

Yan Yuanting. 2012. Research on Resource Discovery in Mobile P2P Network based on Semantic Community Abstract. Anhui.

Zhang Guoying. 2013. Research on Distributed Storage and Transmission Mechanism Based on Mobile P2P. HaErB in.

Zhou Xinxin, Yu Zhenwei. 2013. An optimized M P2P cooperative caching method based on popularity and minimum access cost. Computer Engineering & Science, 35(8):33–35.

Energy Science and Applied Technology – Fang (Ed.)
© 2016 Taylor & Francis Group, London, ISBN 978-1-138-02833-3

Research on and realization of trusted boot mechanism for intelligent Set Top Box

Jian Zhang, Hao Zeng & Zhifang Zan
Chongqing University of Posts and Telecommunications, Chongqing, China

ABSTRACT: The trusted boot process is the basis of safe operation in an embedded intelligent terminal. In consideration of safety problems in current mobile intelligent terminals, this paper proposes the application of a digital signature-based method in the trusted boot of the embedded system without changing the hardware structure of existing mobile devices. Taking the 30 KAITIAN intelligent Set Top Box (STB) as an example, this paper analyzes the principle of digital signature to put the public key into SPI and OPT partitions and the private key is used to sign nand's partition. The function of verification for the boot partition of nand is added in the uboot of the intelligent STB. Finally, the safety test indicates that the mechanism can effectively protect the boot security of the embedded system.

Keywords: Digital signature; intelligent Set Top Box (STB); trusted boot; Android; embedded system

1 INTRODUCTION

With the development and progress of technology, mobile intelligent terminals become increasingly popular because of their open resources, portability and interactivity. Nowadays, embedded systems are widely used in industrial production, electronics, medical treatment, etc. With the increase in practicability, more and more manufacturers overemphasize the performance and neglect the fundamental security requirement. As a result, the security problems of mobile devices become much more serious, which creates an opportunity for hackers.

In recent years, scholars have conducted many studies on the lagging security of the existing intelligent STB and they attempted to transplant the trusted computing technology to the embedded device. However, the attempt is currently unavailable due to the hardware structure of intelligent TSB. This paper proposes a verification method based on digital signature to ensure the trusted boot of devices. Tests and verification indicate that the scheme is feasible and practicable.

2 DIGITAL SIGNATURE

2.1 Basic concepts

ISO7498-2 defines the digital signature as "some additional data in data units or change of codes for data units. A receiver can confirm the completeness and reliability of sources of data units with the data conversion and protect data from falsification and forgery".

2.2 Public key cryptography

The public key cryptography, also known as the asymmetric encryption technology, is the core and foundation of the digital signature. Both the encryption algorithm and the secret key for encryption are public, which is called the public key cryptosystem. Secret keys are divided into a pair of public keys and private keys and they are in one-to-one correspondence. Only the private key shall be encrypted. The public key can be public in any public place for use and downloading.

2.3 Realization of digital signature

This paper researches a method of digital signature based on the public key cryptography. There are a variety of methods for digital signature, such as RSA signature, DSA signature and digital elliptic curve signature algorithm As the RSA2048 algorithm is adopted for encryption in this paper, only the RSA signature is analyzed in detail below as shown in figure 1.

1. The sender generates a 256-bit hash value from the original data by sha256 algorithm which is also called the digest message.
2. The sender encrypts the hash value by RSA2048 and his own private key to generate a digest ciphertext, i.e. the digital signature of the sender.
3. Send the receiver the encrypted digital signature as a attachment of the original data and the original data.
4. The receiver works out the 56-bit value from the received original data by the sha256 algorithm.

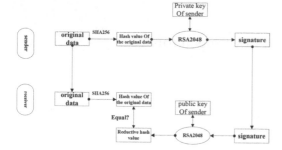

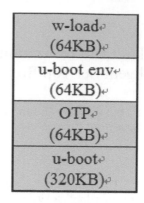

Figure 1. No security mechanism of RSA signature arithmetic.

Figure 2. 512 K SPI flash partitions.

5. The receiver inversely operates and decrypts the additional digital signature by RSA2048 algorithm and the public key and works out a hash value.

6. The receiver can confirm that the data are signed by the sender if the two hash values are the same.

3 OVERALL DESIGN OF TRUSTED BOOT MECHANISM

3.1 Boot process of intelligent STB

The intelligent TSB studied in this paper is an embedded device of Android system. Although Android is based on Linux kernel, the kernel of Android is expanded by several drives. For the boot of an Android intelligent TSB, the init process of Android replaces the init process of Linux to complete the initialization including initialization of devices, services and file systems. The virtual machine and programs are then started up.

As a type of embedded systems, the boot of the intelligent TSB consists of four parts: boot loader, kernel, filesystem and applications. The processor immediately executes a section of key codes in the fixed position of Flash in case of power-on and initial bootstrap of the system or system resetting. The boot loader is the first section of codes and is mainly used to initialize the processor and peripherals and call the kernel to execute in RAM. The Android kernel shall be mounted with the root filesystem, loaded with necessary kernel modules and start up the application. This is the whole process led by Linux during the boot of the Android intelligent TSB.

3.2 Flash of intelligent STB

After research on the boot process, the TSB used in this paper customizes Flash, i.e. the double Flash of SPI and NAND. The specific structure is as follows:

Nand Partitions

mtd4	logo
mtd5	boot
mtd6	recovery
mtd7	misc
mtd8	keydata
mtd9	system
mtd10	cache
mtd11	data

Signed Partition

Figure 3. Nand flash partitions.

1. SPI used in the device has a volume of 512k, and the detail storage method of SPI data is as fig. 2.

Note that items in blue can be flashed once only, i.e. read-only mode, thus ensuring the security of data in SPI. The Uboot-env partition stores some environment configuration parameters. OTP is similar to Uboot-env, but the partition option in this paper is to guarantee the security for the OTP partition stores the public key. The device chooses uboot as the boot loader. The uboot partition stores bootstrap and loading programs.

2. Nand data storage methods are as figure 3:

Partitions of Nand Flash are similar to that of most embedded devices, among which the boot partition is related to the boot process. Boot. img in the boot partition comprises kernel, ramdisk, filesystem, etc. Shadow sections are signed partitions.

3.3 Trusted boot principle of intelligent TSB

With analysis of the boot process of the device and the partition of the device Flash, this paper applies it to the trusted boot program of the device in combination with the digital signature algorithm.

1. Application in the signature process is shown in the figure 4:

Compute boot. Imp by sha256 to produce a boot. hash. Use the private key to sign boot. hash by RSA2048 algorithm to produce a signature file. Compress the signature file and boot. img together in boot, img. Signed and store it in the boot partition. Recovery. img is stored in the recovery partitions by using the same method to sign.

2. Application of the verification process is shown in the figure 5:

When uboot starts up after electricity connection, boot will use the public key to conduct a RSA inverse operation to the signature file and get a hash value. This hash value will be compared with the hash value directly computed by boot. img with sha256. Boot up the kernel in boot if the two values are identical. The theory of digital signature is applied to the process of trusted boot in this way.

3. Schematic of overall trusted boot scheme:

The figure 6 shows an overall schematic of the trusted boot. The W-load code in SPI runs first after power-on and normal startup of the STB, which is to guide the execution of loading programs of uboot in SPI. The loading program of uboot verifies the boot partition during RAM execution. The normal startup is realized with a correct signature; otherwise, the STB of the system is reset. Above is the overall design scheme of the trusted boot mechanism.

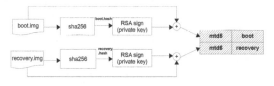

Figure 4.   Signature.

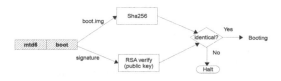

Figure 5.   Verify.

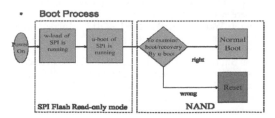

Figure 6.   Boot process.

4 SECURITY TESTING OF HIGHLY-SECURE (TRUSTED BOOT MECHANISM) STB

The scheme given in this paper can prevent the upgrade of unofficial upgrade packages. Even if an unofficial upgrade package is upgraded, the safety and reliability of check mechanism of uboot can be guaranteed by the uniqueness of public and private key pairs. Theoretically, any form of tampering upgrade package is invalid provided the private key without losing. The following is the security verification.

4.1 Verification of security upgrade

Program the SPI with a public key to make the board a device with check function. If a board with a public key can only program an installation package signed by a private key corresponding to the public key, the device is proved with a security upgrade mechanism.

1. Programming of mirror image package without a signature.

Directly put unsigned mirror images of system/data/boot/recovery/logo/wload/uboot/ubootenv into FirmwareInstal under the root directory of SD card and insert the newly finished highly-secure device. As shown in the figure 7, the upgrade process cannot be finished for the signature file failing to check "Unable to read "FirmwareInstall/boot. img. signed" from mmc 0:1". Similarly, the upgrade of other mirror images has such problem as well.

2. Programming of mirror images with different signature of private key.

Sign such unsigned mirror images as system/data/boot/recovery/logo/ wload /uboot/ ubootenv with another group of private keys. Put signed mirror images and signature files into Firmware Install under the root directory of SD card and insert them into the newly finished highly-secure device. Every mirror image with signature file can be read, but cannot be upgraded due to different hash values respectively inverse calculated by a signature file and boot. img, as the same to other mirror images. As shown in the figure 8.

Figure 7.   Update failed 1.

Figure 8.   Update failed 2.

4.2 Verification of trusted boot mechanism

As shown in the previous experiments, the highly-secure set top box has a mechanism of preventing flash, so an unofficial firmware package is hard to flash in. A special way to verify the actual effect of trusted boot mechanism is as follows.

1. Mirror image is complete while signature tampered

Upgrade a common STB by a firmware with the function of validation, then the STB will be equipped with the function of validation. Connect the STB by a cluster communication port (COM) and write the public key into spi and otp partitions, then the interior of STB has the function of validation and a public key while there is no a private key corresponding to the public key which is written into otp to signature the nand partition. Theoretically, the boot will fail. The verification result is shown as below. Reboot the STB after write in public key, SecureCRT will display invalid signature, as shown in the figure 9.

Thus, no corresponding private key signature in nand partition is verified. It fails to pass the verification.

2. Signature is complete while mirror image tampered

Taking a boot partition as an example to tamper that in an integrated device with high security, see the specific operation as follows:

Check Mtd where boot partition lies by cat/proc/mtd under command line to know that the boot partition is in mtd4. Then actively write a series

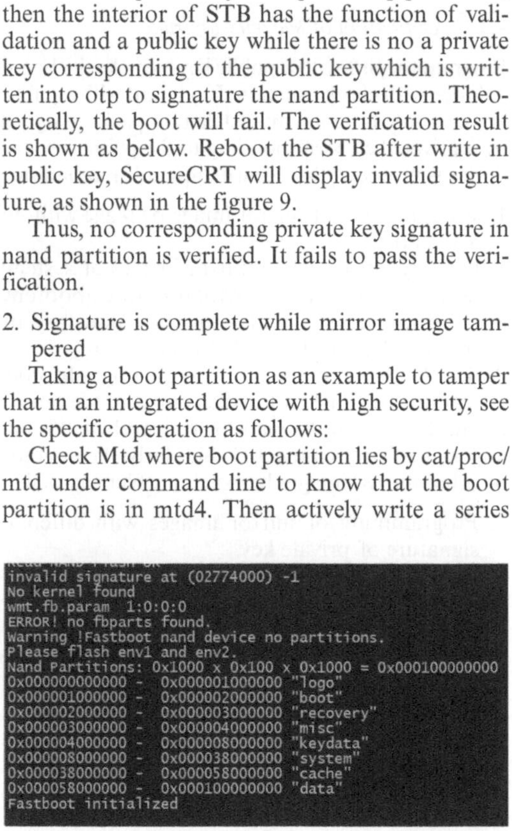

Figure 9. Boot failed 1.

Figure 10. Boot failed 2.

of 111111 word strings by redirection operator >> into the boot partition directly through printf 111111 >> /dev/block/mtdblock4, namely through printf way.

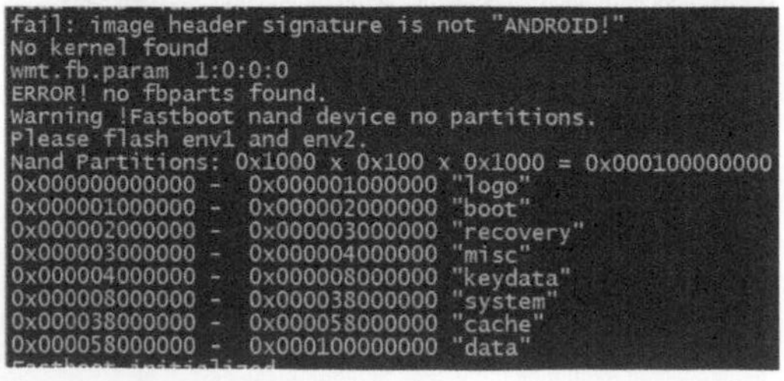

As shown in the figure 9:"image header signature is not "ANDROID!" NO kernel found" shown in the figure, I tampered the boot partition where stores kernel, so it cannot boot normally after being tampered.

5 CONCLUSION

Combined with the digital signature, this paper improves the flash partition scheme of intelligent set top box, adds the check function in uboot and provides a trusted boot scheme based on digital signature technology. After the test, this scheme will relatively guarantee the trusted boot of intelligent STB and can be widely used into other embedded devices.

REFERENCES

Fu Xiaofeng, 2013, touch stone across the river about OTT STB [J] Practical Audio-Visual Technique, 13–14, March 2013(3).

Jin Youli.2008, Several digital signature scheme [J] Network Security Technology & Application. 2008(02): 94–95.

Ju Hongwei, Li Fengyin, Yu Jiguo, Cao Baoxiang.2006,A Confirmer Signature Scheme Based on RSA [J] computer science engineering. 2006(07):154–156

Liu Peng. Research and Implementation of Unified Authentication System Based on Digital Signature [D]. North China Electric Power University, 2012.

Luo Jun, Jiang Jingqi, Min Zhisheng, Li Chengqing, 2012.

The research of trusted embedded system secure starup based on SHA-1 module Journal of Shandong University (Neo-confucianism edition) 2012, 47(9):1–6.

Ryan M.D. 2012, Automatic analysis of security properties of the TPM [M]//Trusted Systems. Berlin Heidelberg: Springer, 2012:1–4.

Shen xin, 2009, The Design and Realization of IPTV Set-Top Box's Software System [D], Beijing University of Posts and Telecommunications, 2009.

Wang Mingliang, the Key Technology Research of Set-up Boxes Based on the Android [D], 2013.

Wang Jianlei, 2013 Current situation and trend of development of the Internet TV set-top boxes [J]. Voice and Screen World, 2013.3(3): 61–63.

Yang J, Wang H.H, Wong F.F., 2012, Mobile trusted platform model for smart phone [J]. Computer Science, 2012(8):20–25.

Zhou Shufeng, Sun Yu Zhen, 2008, Software protection solution based on RSA Digital Signature [J] Computer Applications and Software. 2008(03):35–37.

Zhao Bo, Fei Yongkang, Xiang Shuang, et al. 2014, Research and implementation of secure boot mechanism for embedded systems. Computer Engineering and Applications, 2014.5(10):72–77.

System test, diagnosis, detection and monitoring

Energy Science and Applied Technology – Fang (Ed.)
© 2016 Taylor & Francis Group, London, ISBN 978-1-138-02833-3

Mid-long term load forecasting based on a new combined forecasting model

Yang Chao

School of Electrical Engineering, Tianjin University of Technology, Tianjin, China

ABSTRACT: A new combined forecasting model is presented in this paper, because mid-long term load forecasting has the problem of long time and low precision. First, this method utilizes the moving average method to adjust the original load data. Then, it combines the GM (1, 1) model and the BP neural network in a special way to forecast the mid-long term load. The validity of the new model has been evaluated by utilizing the actual data of a certain area. The experimental results indicate that the new combined model can improve the accuracy of load forecasting results, and it has an application value to mid-long term load forecasting.

Keywords: Mid-long term load forecasting; GM (1, 1) model; moving average method; BP neural network

1 INTRODUCTION

Mid-long term load forecasting provides data support for power system planning departments when they need to make decision. The accuracy of mid-long term load forecasting directly influences the rationality, economy of the power grid reformation and expansion, which has a profound significance to the development of power industry.

The grey prediction method needs fewer samples and has high precision in the short-term forecasting. The GM (1, 1) model is the classical model of the grey model. However, the grey model can only obtain a high accurate forecasting value of one or two sets of data in the future. This point limits the application of the grey model in the field of mid-long term load forecasting. Thus, the grey model needs to be improved.

The BP neural network has an advantage of dealing with nonlinear problems. Load forecasting itself is a nonlinear problem, so the BP neural network can fully show its advantage in the field of load forecasting. This paper combines the GM (1, 1) model and the BP neural network in an innovative way. In addition, the forecasting accuracy of the new model is improved.

2 MAIN THEORY AND MODEL

2.1 GM (1, 1) model

We use 1-AGO to generate the original data sequence $x^{(0)}$. Then, the new data sequence $x^{(1)}$

is obtained. The GM (1, 1) model is given as follows:

$$\begin{cases} \hat{x}^{(0)}(1) = \hat{x}^{(1)}(1) = x^{(0)}(1) \\ \hat{x}^{(0)}(k+1) = (1 - e^{\hat{a}})(x^{(0)}(1) - \dfrac{\hat{u}}{\hat{a}})e^{-\hat{a}k} \quad (k = 1, 2, \ldots) \end{cases}$$

$$(1)$$

2.2 Moving average method

The moving average method can avoid the large fluctuation of data. It is represented by the following formulas:

$$\begin{cases} x'^{(0)}(1) = \dfrac{3x^{(0)}(1) + x^{(0)}(2)}{4} & t = 1 \\ x'^{(0)}(t) = \dfrac{x^{(0)}(t-1) + 2x^{(0)}(t) + x^{(0)}(t+1)}{4} & t = 2,3,\ldots,n-1 \\ x'^{(0)}(n) = \dfrac{x^{(0)}(n-1) + 3x^{(0)}(n)}{4} & t = n \end{cases}$$

$$(2)$$

2.3 BP neural network

The BP neural network, which has a hidden layer, can improve the analysis ability of the network greatly, and it usually takes three layer network structures. The transfer function is significant to the neural network. It determines the ability and efficiency of the network to solve problems. Usually, Sigmoid function is selected as the transfer function:

$$f(x) = \frac{1}{1 + e^{-x}} \qquad (3)$$

3 COMBINED FORECASTING MODEL STEPS

The specific steps of the combined forecasting model are as follows:

Step 1: First, this paper uses 7 groups of load data as the original data, and then uses the moving average method to adjust the original data. In this way, the large fluctuation of the original data will be weakened. Then, the historical input data of the model are obtained.

Step 2: this paper uses the GM (1, 1) model to forecast the historical value and the future value of load.

Step 3: this paper uses the three-layer BP neural network structure and selects the sigmoid function as transfer function. It then calculates the number of neurons in the hidden layer by the following empirical formula:

$$t = \sqrt{p + q} + c \qquad (4)$$

where t is the number of nodes in the hidden layer; p is the number of nodes in the input layer; q is the number of nodes in the output layer; and c is a constant integer.

Step 4: this paper selects the GM (1, 1) model's forecasting results of historical year as neurons of the BP neural network's input layer. The actual historical load data are used as the expected value to adjust the BP neural network's weight. In order to convert the data to (0, 1), the data of the input layer should be normalized.

Step 5: After the BP neural network has been trained, this paper uses the forecasting results of the GM (1, 1) model in the future as the input layer's neurons of the BP neural network. After the BP neural network finishes its calculation, the output data can be obtained.

Step 6: Dealing with the output data by reduction treatment, and then the load forecasting value of future is obtained.

Step 7: The process is ended.

The flow chart of the combined model is shown in Figure 1.

4 EXPERIMENTAL RESULT ANALYSIS

In order to demonstrate the effectiveness of the combined model, the sample with a certain area's load data from the year 2000 to 2011 is selected. Data are shown in Table 1.

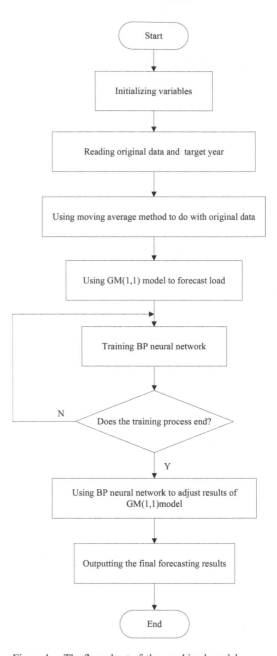

Figure 1.   The flow chart of the combined model.

The data from the year 2000 to 2006 in Table 1 are used as the original data. Then, this paper uses the moving average method to adjust the original data. The forecasting results of the GM (1, 1) model and the combined model are shown in Table 2.

In order to observe the model's forecasting results more intuitively, the curves of the actual load value and the model's forecasting results are shown in Figure 2.

Table 1. The actual load value of a certain area.

Year	Actual load value 10⁸ KWh
2000	384.43
2001	399.94
2002	439.96
2003	461.24
2004	510.11
2005	570.54
2006	611.57
2007	667.01
2008	689.72
2009	739.15
2010	809.9
2011	821.71

Table 2. Forecasting results.

Year	GM(1,1) 10⁸ KWh	Combined model 10⁸ KWh
2007	655.598	655.023
2008	711.179	710.738
2009	771.472	760.127
2010	836.877	799.736
2011	907.827	829.176

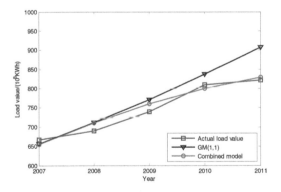

Figure 2. The curves of the actual load and the model's forecasting results.

From Table 2 and Figure 2, it can be seen that comparing with the forecasting curve of the GM (1, 1) model, the forecasting curve of the combined model is much closer to the curve of the actual load.

The relative error of the model's forecasting results is given in Table 3. The average absolute error of the model's forecasting results is given in Table 4.

In order to make the relative error of the model's forecasting results more intuitive, this paper uses

Table 3. The relative error of the model's forecasting results.

Year	2007	2008	2009	2010	2011
GM(1,1)	1.71%	−3.11%	−4.37%	−3.33%	−10.48%
Combined model	1.8%	−3.05%	−2.84%	1.25%	−0.91%

Table 4. The average absolute error of the model's forecasting results.

Model	MAE 10⁸ KWh
GM(1,1)	35.6575
Combined model	14.3223

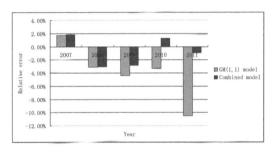

Figure 3. The relative error of the model's forecasting results.

the column chart. The column chart of the relative error is shown in Figure 3.

From Table 3, it can be found that the maximum absolute value of the relative error of the GM(1,1)'s forecasting results is 10.48%, but the maximum absolute value of the relative error of the combined model's forecasting results is 3.05%. Thus, the experimental results indicate that the error of the combined model is much smaller than that of the GM (1, 1) model. Form Figure 3, it can be seen that compared with the GM (1, 1) model, the combined model shows the advantage of the mid-long term load forecasting. The longer the time is, the more obvious advantage the combined model has.

In addition, from Table 4, it can be concluded that compared with the GM (1, 1) model, the MAE of the combined model's forecasting results reduces to 2133520000 KWh. So, the accuracy of forecasting results can be improved by the combined model.

5 CONCLUSIONS

A new combined forecasting model is proposed in this paper, which uses the moving average method to adjust the original data and then combines the

GM (1, 1) model with the BP neural network to forecast the load in the future. The new combined forecasting model has the advantage of the GM (1, 1) model and the BP neural network. It needs fewer samples and constructs model easily. It can also deal well with nonlinear problems. From the experimental results, it can be concluded that the accuracy of the forecasting results of the combined model can be improved obviously, so the method is feasible and effective.

REFERENCES

Ge Shaoyun, Jia Ousha, Liu Hong. 2012. A Gray Neural Network Model Improved by Genetic Algorithm for Short-Term Load Forecasting in Price-Sensitive Environment. Power System Technology, 36(1):224–229.

Li Wei, Yuan Ya-nan, Niu Dong-xiao. 2011. Long and medium term load forecasting based on grey model optimized by buffer operator and time response function. Power System Protection and Control, 39(10):59–63.

Li Yongzai, Fan Chengxian, Wang Yan, Wang Duan. 2012. The application of Improved GM (1, 1) Model in Long-Middle Term Load Forecasting. Computer Applications and Software, 29(5):13–19.

Lu Jia-liang, Zhang Zhen-gang. 2008. Error correcting Markov chains for medium /long term load rolling forecasting of BP models based on grey relational grade. East China Electric Power, 36(9):10–13.

Wang Dapeng, Wang Bingwen. 2013. Medium- and Long-Term Load Forecasting Based on Variable Weights Buffer Grey Model. Power System Technology, 37(1):167–171.

Wang Jie, Wu Guo-zhong, Li Yan-chang. 2009. Application of ant colony gray neural network combined forecasting model in load forecasting. Power System Protection and Control, 37(2):48–52.

Yang Ke, Tan Lun-nong. 2011. Analysis of Long-Term Load Forecasting Application in Power Market Environment. Proceedings of the CSU-EPSA, 23(3): 54–57.

Yuan Tie-jiang, Yuan Jian-dang, Chao Qin, et al. 2012. Study on the comprehensive model of mid-long term load forecasting. Power System Protection and Control, 40(14):143–146.

Zhang Cheng, Teng Huan, Fu Ting. 2013. Middle and long term power load forecasting based on grey discrete Verhulst model's theory. Power System Protection and Control, 41(4):45–49.

Zhou De-qiang, Wu Ben-ling. 2011. Optimization and power load forecasting of gray BP neural network model. Power System Protection and Control, 39(21): 65–69.

Energy Science and Applied Technology – Fang (Ed.)
© 2016 Taylor & Francis Group, London, ISBN 978-1-138-02833-3

Weighted-NMF based clustering method for the identification of cancer subtypes

Jia Peng & Ronghui Wu
Department of Information Science and Engineering, Hunan University, Changsha, China

ABSTRACT: It is well known that cancer is a kind of complex and heterogeneous disease, which consists of multiple molecular distinct subtypes. A reliable and accurate identification of cancer subtypes is crucial to the effective treatment of cancers. Gene expression data, which has been a powerful tool for the study of cancer subtype, typically contain thousands of genes on each chip. However, only a small part of genes are involved in the carcinogenic process. The inclusion of irrelevant or noisy variables may reduce the overall performance of various cancer identification methods, so feature selection is essential to the success of subtype recognition from high-dimensional data. Here we proposed a weighted nonnegative matrix factorization based clustering method for the cancer subtypes identification. We first rank each of the genes according to its importance to cancer subtypes identification, and then select the m top-ranking genes for the subsequent clustering. Another major contribution of our work is that we embed the weights of the selected genes directly into the clustering objective function.

Keywords: cancer subtype; clustering; gene expression data; non-negative matrix factorization

1 INTRODUCTION

Cancer is a kind of complex disease consisting of multiple molecular distinct subtypes, and different subtypes may respond differently to the same medical treatment. Therefore, a reliable and accurate identification of cancer subtypes is of great importance.

DNA microarray is a biotechnology that simultaneously monitors the expression of tens of thousands genes in cells, which makes the precise, objective, and systematic analysis of human complex disease possible (C.-H. Zheng, 2009). Non-negative matrix factorization (NMF) is a fast-developed machine learning algorithm based on decomposition by parts. Brunet et al. (JP B, 2004) first applied it to the problem of clustering cancer gene expression data, and demonstrated its ability to discover meaningful biological cancer subtypes. In application to three famous cancer datasets, they also concluded that NMF was more accurate than hierarchical clustering (HC) and more stable than self-organizing map (SOM.). Since then, various NMF extensions have been developed for gene expression data analysis. Gao and George enforced sparseness by combining the goal of minimizing reconstruction error with that of sparseness, and showed that the sparse NMF (SNMF) can improve clustering results to some extent (Y. Gao, 2005). Kong et al. also applied sparse constraints to original NMF, and they showed that clustering

performance of NMFSC was better than NMF as long as choosing appropriate degree of sparseness (X.Z. Kong, 2007).

In this work, we mainly discuss a kind of clustering approach to discover cancer subtypes, we call it weighted nonnegative matrix factorization based clustering method (WNMFC).Specially, we first assign a weight for each gene according to its importance for subtype identification, and then select the m top-ranking genes for the subsequent NMF based clustering. Gene selection in such context has the following two implications: on one hand, gene selection is actually a measure of dimensionality reduction, which will help improve the performance of NMF; on the other hand, only a small few of genes are involved in the carcinogenic process, most of the genes are irrelevant and redundant, and thus it is critical to get rid of these genes. Another main contribution of our work is that we embed the weights of the selected genes directly into the clustering objective function, and propose a weighted NMF method. We also compared the results by using NMF, SNMF, and our proposed WNMF with the selected genes respectively.

2 ASSIGNING WEIGHTS TO GENES

As a useful technique to decompose high-dimensional data matrix, independent component analysis (ICA) can also obtain mutually independent transformed coefficients. In the case of a $p \times n$ gene expression matrix X, which contains p gens and n samples, the ICA model $X = SA$ has the following implication: each row of X can be characterized by all the rows of A (eigengenes), and by the rows of S which correspond to their mutually independent coefficient on the eigengenes. In this paper, we utilize this ICA model to project the genes onto the desired independent directions, and obtain a ranking of the genes.

As previously mentioned, the rationale behind our proposed gene weight algorithm is that: genes that are statistically independent from others and showing bigger variations among samples should be assigned larger weights. Therefore, our weight-training approach is as follows:

$$w_l = (1-\partial)NMAD_l + \partial g_l, (1 \le l \le p) \qquad (2)$$

where w_l is the gene weight of the lth gene, NMAD is the normalized median absolute deviation(MAD) (Yiyi Liu, 2014):

$$NMAD_l = \frac{MAD_l}{\max(MAD)} \qquad (3)$$

where max (MAD) is the maximum of vector MAD. MAD can be used as a useful measure to express the expression variation of a gene among different samples (Yiyi Liu, 2014). We used the normalized MAD in order to make the weight-training mechanism stable and comparable with different overall expression levels. g_l is the rank of the lth gene obtained by ICA representing the statistical independence, interested readers can refer to. ∂ is a tuning parameter, here we choose a balanced value ($\partial = 0.5$). The final weight of each gene reflects both its dependence relations with other genes and its ability to separate the samples. Then all the genes are sorted in decreasing order according to their weights obtained above. For gene selection, we can select the first m genes, and reconstruct a $m \times n$ gene expression matrix Y.

Our ICA-MAD based weight-training method can be concluded as follows:

Step 1: Extracting z independent components $s_1, ..., s_z$ with zero mean and unit variance from the gene expression matrix using ICA. Obviously, z is an important parameter to be determined.

Step 2: For gene $l(l = 1, ..., p)$, the absolute score on each component $|s_{j,l}|$ is computed, and the ICA-based weight g_l for each gene can be denoted by $g_l = \max_j |s_{j,l}|$.

Step 3: For each gene $l (l = 1, ..., p)$, the normalized median absolute deviation $NMAD_l$ of gene l among different samples is computed.

Step 4: For each gene $l (l = 1, ..., p)$, gene weight w_l is computed according to formula (2).

3 CLUSTERING WITH NMF AND ITS EXTENSIONS

3.1 NMF method

Given a nonnegative matrix $Y_{m \times n}$, our goal is to find two new factor matrices (U and V) to approximate the original matrix:

$$Y_{ij} \approx (UV)_{ij} = \sum_{a=1}^{k} U_{ia}V_{aj}. \quad U \in R^{n \times k}, V \in R^{k \times m} \qquad (4)$$

where matrix U consists of k non-negative basis vectors, and column vectors of V represent the weights in the approximation of the columns in original matrix Y using the bases from U.

To obtain the two factor matrices in NMF, Lee and Seung proposed two algorithms: Euclidean distance-based algorithm and K-L divergence-based algorithm. As the divergence-based algorithm is better able to capture a subclass [1], we use the divergence-based method to implement NMF, and we mainly discuss the former one here.

The corresponding objective function that characterizes the approximation between original matrix Y and the product of two factor matrix U and V can be defined as the K-L divergence form:

$$
\begin{aligned}
F &= \sum_{i=1}^{m} \sum_{j=1}^{n} \left[Y_{ij} \log \frac{Y_{ij}}{(UV)_{ij}} - Y_{ij} + (UV)_{ij} \right] \\
&= \sum_{i=1}^{m} \sum_{j=1}^{n} \left[Y \circ \log \frac{[Y]}{[UV]} - Y + UV \right]_{ij}
\end{aligned} \qquad (5)
$$

where log(X) is the element-wise logarithm of X, $X \pm Y$ is the Hadamard product (or element by element product) of the matrices X and Y, and x//y is the Hadamard division (or element by element division) of the matrices X and Y . The iterative equations, which balances computational complexity and speed, has been derived as(V Blondel, 2001):

$$
\begin{aligned}
U &\leftarrow \frac{[U]}{[E_{m \times n}V^T]} \circ \left(\frac{[Y]}{[UV]} V^T \right), \\
H &\leftarrow \frac{[V]}{[U^T E_{m \times n}]} \circ \left(U^T \frac{[Y]}{[UV]} \right)
\end{aligned} \qquad (6)
$$

3.2 our proposed weighted NMF

Weighted measures are effective ways to improve performance of the basic NMF (V Blondel, 2001). In this work, we propose a weighted NMF method based on the weights of the selected genes in the gene expression data obtained by gene selection process described above. The aforementioned weights are presented in the diagonal matrix W, and we incorporate the "important indicator" of each gene into the clustering process by multiplying the weights to the row (genes) norms. Thus, the biological meaning of our WNMF method is obvious. The objective function with the weighted NMF is:

$$F_W = \sum_{i=1}^{m} \sum_{j=1}^{n} \left[W \circ \left(Y \circ \log \frac{[Y]}{[UV]} - Y + UV \right) \right]_{ij} \quad (7)$$

The weighted divergence F_W is local convergence under the following update rules:

$$F_W = \sum_{i=1}^{m} \sum_{j=1}^{n} \left[W \circ \left(Y \circ \log \frac{[Y]}{[UV]} - Y + UV \right) \right]_{ij} \quad (8)$$

The weighted divergence F_W is invariant under this updates if U and V are at a stationary point of the divergence. We don't describe the specific proof process here, the interested readers can refer to literature (Bin Gan, 2011) for details.

4 COMPUTATIONAL EXAMPLES AND ANALYSIS

We used the Leukemia Dataset to evaluate the performance of our method. The leukemia Dataset consists of 5000 genes in 38bone marrow samples, which were obtained from adult acute leukemia patients at the time of diagnosis and before chemotherapy. The total samples can be split into three subtypes: 19 cases of ALL-B, 8 cases of ALL-T, and 11 cases of AML.

When using matrix factorization algorithms to cluster samples, one key issue to be resolved is how to determine the number of clusters k, this is still an open problem. In our work, we utilize cophenetic correlation coefficients used in (Le L, 2008 Jin-long S, 2010) for this model selection problem. We first computed the cophenetic correlation coefficients p_k for each value of k ranging from 2 to 5 with all genes. As shown in Figure 1, clustering with k = 2 or k = 3 should be more stable than others. Therefore, we will select k = 2 and k = 3 for the subsequent comparison work. Then we applied our proposed ICA-MAD gene ranking method to the aforementioned dataset. Specifically, we calculated the gene ranking of all genes, and selected a sequence of m gene subsets for subsequent clustering process. We applied the selection results to NMF clustering and SNMF clustering and our WNMF clustering method respectively, and compared the identification accuracy rates with each other.

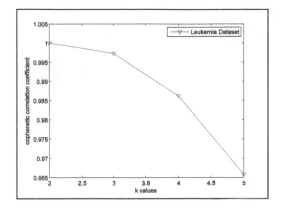

Figure 1. Corresponding cophenetic correlation coefficients.

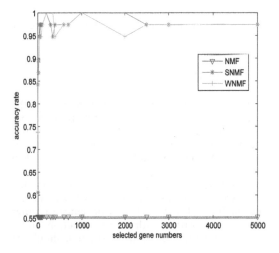

Figure 2. Clustering accuracy rate comparison of NMF and its extensions for k = 2.

Figure 2 demonstrates that, for k = 2, clustering performance of our WNMF is significantly better than the others when the number of selected genes is small. With the increase in the number of selected genes, the advantage of SNMF will gradually appear. As a whole, our WNMF method with ICA-MAD gene selection can achieve commensurate or better performance than NMF and SNMF.

Figure 3 showed that the advantages of our WNMF method are more obvious for k = 3. NMF method achieves its highest clustering accuracy rate 0.9734 when m = 65, and the accuracy rate essentially unchanged with the increase of selected genes. Thus, WNMF with ICA-MAD

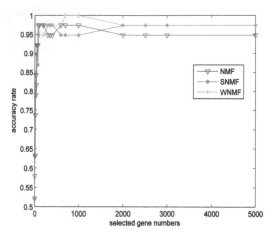

Figure 3. Clustering accuracy rate comparison of NMF and its extensions for k = 3.

gene selection can greatly improve the efficiency of clustering.

5 CONCLUSIONS

In this paper, we assigned a weight to each of the genes, embed it into the NMF objective function, and proposed our weighted nonnegative matrix factorization. Experiment results showed that our ICA-MAD gene ranking method is useful to emphasize informative genes which are useful for cancer gene expression data clustering, especially when coupled with our proposed WNMF clustering method. However, the performance of other clustering algorithms with our gene selection should be further explored. Another issue that should be resolved is the selection of the optimal parameters in WNMF clustering, including ∂ and m. Our next work is to develop better approaches to select the most appropriate parameters automatically.

REFERENCES

Bin Gan. Metasample Based Sparse Representation for Tumor Classification. IEEE/ACM Trans Comput Biol Bioinform, 8(5), pp. 1273–1282, 2011.

Blondel V. Algorithms for weighted matrix factorization for local representation, USA, Proc of Comp Vision and Pattern Recog, 2001.

Gao Y. & C. George. Improving molecular cancer class discovery through sparse non-negative matrix factorization. Bioinformatics, 21, pp. 3970–3975, 2005.

Jin-long S, Zhi-gang L. Research on the Advances of Nonnegative Matrix Factorization and Its Application in Bioinformatics (J). Computer Engineering & Science, 2010.

JP B. Metagenes and molecular pattern discovery using matrix factorization. Proc. Natl. Acad. Sci. USA, 101(12), pp. 4164–4169, 2004.

Kong X.Z. Molecular cancer class discovery using non-negative matrix factorization with sparseness constraint in Proc. Int. Conf. Intell. Comput., LNCS, 4681, pp. 792–802, 2007.

Le L & Yu-jin Z. A Survey on Algorithms of Non-Negative Matrix Factorization. Acta Electronica Sinica, 36(4), pp. 737–743, 2008.

Xue Yun. Clustering-based initialization for non-negative matrix factorization. Applied Mathematics and Computation, 205(2), pp. 525–536, 2008.

Yiyi Liu. A network-assisted co-clustering algorithm to discover cancer subtypes based on gene expression. BMC Bioinformatics, 15(1), pp. 2–23, 2014.

Zheng C.-H. Tumor Clustering Using Nonnegative Matrix Factorization With Gene Selection. Information Technology in Biomedicine, IEEE Transactions on, 13(4), pp. 599–607, 2009.

Energy Science and Applied Technology – Fang (Ed.)
© 2016 Taylor & Francis Group, London, ISBN 978-1-138-02833-3

Analysis of power system fault signals based on matrix pencil method

Z.Q. Xiao, Y.M. Liu & Q.H. Zhu
Liupanshui Power Supply Bureau of Guizhou Electric Power Company, Liupanshui, China

Y. Zhou & N. Wang
Guizhou Electric Power Institute, Guiyang, China

H.J. Wang
Department of Electric Engineering, North China Electric Power University, Baoding, China

ABSTRACT: A power system short-circuit fault signal contains decaying aperiodic and non-integer harmonics components, which are difficult to be accurately analyzed by discrete Fourier transform. In this paper, the noisy fault signal is filtered and analyzed by the matrix pencil method, and the simulation was carried out in the Matlab environment. The simulation results show that the fundamental component, decaying aperiodic and non-integer harmonics components can be accurately analyzed by the matrix pencil method in the case of a fault signal with noise. Collectively, this helps to better analyze the impact of the transient component on relay.

Keywords: Fault Signal Analysis; Discrete Fourier Transform; Matrix Pencil Method; Relay; Decaying Aperiodic Component

1   INTRODUCTION

Decaying a periodic and non-integer harmonics components will be in the fault signal when faults occur in the transmission line. The value and decay rate of decaying a periodic and non-integer harmonics components will influence the operation of the electromagnetic current relay, and cause the operation of current quick-breaking protection with a wrong protection zone or ultimately no zone. So, correctly and accurately analyzing the reasonable setting of the relay protection operating value has a great significance, and facilitates the improvement of the performance and the operation level of protection (Hu 2001, Li 2006 & Huang 2001).

The signal analysis of the traditional power system is mainly focused on Discrete Fourier Transform (DFT). When processing the discrete measured signal by DFT, and for a stable periodic signal, if the ratio of the sampling time to signal period is an integer, it can quickly obtain the amplitude, frequency and phase of the signal fundamental wave as well as each harmonic. However, for a transient fault signal, it is hard to use DFT or windowed interpolation DFT to analyze the elements of the signal (Zhang 1999, Niu 2012, Pang 2003 & Xue 2002). There are a vast number of harmonic analysis methods such as least squares algorithm (Meng 2012), wavelet transform method (Zeng 2012 & Zhang 2012), Prony algorithm (Ding 2005), and the algorithm combined FFT and wavelet package transform (Fang 2012). All of these methods are mainly focused on analyzing a harmonic wave and a simple harmonic wave, but ignore the decaying a periodic and non-integer harmonics components.

In order to analyze the elements of the fault status transmission line, this paper presents the matrix pencil method (MPM) (Sarkar 1995 & Suo 2010) to analyze the fault signal. This method can quickly obtain the fundamental component, decaying a periodic and non-integer harmonics components in the fault signal.

2   PROBLEM FORMULATION

The analyzed signal is given by

$$
\begin{aligned}
y(t) &= x(t) + n(t) \\
&= \sum_{i=1}^{M} R_i e^{(-\alpha_i + j\omega_i)t} + n(t) \quad 0 \le t \le T_s
\end{aligned} \tag{1}
$$

where $y(t)$ is the observed fault signal; $n(t)$ is the noise signal; $x(t)$ is the real fault signal; R_i, α_i and ω_i are the fault signal amplitude, decay factor and angular frequency of the element i, respectively; and Ts is the observation period.

Let Δt be the sampling interval, then the discrete sampling $y(t)$ and the sampling data sequence can be written as

$$y(k) = \sum_{i=1}^{M} R_i e^{(-\alpha_i + j\omega_i)k\Delta t} + n(k\Delta t)$$
$$k = 0, 1, ..., N-1 \tag{2}$$

Let $Z_i = e^{(-\alpha_i + j\omega_i)\Delta t}$. In this case, (2) can be reformulated as

$$y(k) = \sum_{i=1}^{M} R_i Z_i^{k} + n(k\Delta t)$$
$$k = 0, 1, ..., N-1 \tag{3}$$

The matrix Y can be defined as:

$$\mathbf{Y} = \begin{bmatrix} y(0) & y(1) & \cdots & y(L) \\ y(1) & y(2) & \cdots & y(L+1) \\ \vdots & \vdots & & \vdots \\ y(N-L-1) & y(N-L) & \cdots & y(N-1) \end{bmatrix}_{(N-L)\times(L+1)}$$
$$\tag{4}$$

In order to keep the better filtering effects, the value range of L is given by [$N/3$, $N/2$].

Then, the singular value decomposition is applied to the matrix Y:

$$\mathbf{Y} = \mathbf{U}\mathbf{\Sigma}\mathbf{V}^H \tag{5}$$

where U and V are unitary matrices and Σ is the diagonal matrix whose elements are the singular values of Y.

If there is no noise in the measurement data, the first M singular values of Y are non-zero, $\sigma_1 \geq \sigma_2 \geq ... \geq \sigma_M$, and the rest of the values are zero. At this time, the rank of Y is equal to M. However, during the practical measurement, the number of non-zero singular values will be greater than M. In this case, we keep the best first M singular values and treat others as zero. In this way, it could efficaciously reduce the negative effect of noise. Usually, we mark the satisfying condition ($\sigma_i/\sigma_1 < \mu$) singular value as the noise singular value, and the one as the useful signal. We put the maximum subscript i that satisfies the condition to replace M, and these M main singular vectors structure the filter matrix:

$$\mathbf{V'} = [v_1, v_2, ..., v_M] \tag{6}$$

According to (7) and (8), we can obtain matrices $\mathbf{Y}_1$ and $\mathbf{Y}_2$:

$$\mathbf{Y}_1 = \mathbf{U}\mathbf{\Sigma'}\mathbf{V}_1'^{H} \tag{7}$$
$$\mathbf{Y}_2 = \mathbf{U}\mathbf{\Sigma'}\mathbf{V}_2'^{H} \tag{8}$$

where $\mathbf{V'}_1$ and $\mathbf{V'}_2$ are the matrices without the first row and column of $\mathbf{V'}$ and $\mathbf{\Sigma'}$ is the matrix that is composed of M singular vectors corresponding to M main singular vectors from Σ.

According to the principle of MPM, we know that Z_i is the generalized eigenvalue of $[\mathbf{Y}_1, \mathbf{Y}_2]$:

$$\mathbf{Y}_2 R_i = Z_i \mathbf{Y}_1 R_i \tag{9}$$

where R_i is the generalized characteristic vector. Its equivalent form is given by

$$(\mathbf{Y}_1^{+}\mathbf{Y}_2 - Z_i\mathbf{I})R_i = [0] \tag{10}$$

where $\mathbf{Y}_1^{+}$ is the Moore-Penrose pseudo-inverse matrix of $\mathbf{Y}_1$.

Now, we find the eigenvalue Zi of the matrix Y1+Y2. According to (11), we can obtain the decay factor and the angular frequency from the signal as follows:

$$-\alpha_i + j\omega_i = \ln(Z_i)/T_s$$
$$i = 1, 2, ..., M \tag{11}$$

R_i satisfies the linear equations of (12):

$$\begin{bmatrix} y(0) \\ y(1) \\ \vdots \\ y()N-1 \end{bmatrix} = \begin{bmatrix} 1 & 1 & \cdots & 1 \\ z_1 & z_2 & \cdots & z_M \\ \vdots & \vdots & & \vdots \\ z_1^{N-1} & z_2^{N-1} & \cdots & z_M^{N-1} \end{bmatrix} \begin{bmatrix} R_1 \\ R_2 \\ \vdots \\ R_M \end{bmatrix} \tag{12}$$

Finally, this method uses the least squares algorithm to obtain R_i to determine the parameters of the signal.

3 CASE STUDY AND DISCUSSION

To study the effectiveness of this method, algorithms and simulations are accomplished using MATLAB.

The simulation signal, similar to formula (13), includes the fundamental sine signal, decaying aperiodic component, two decaying non-integer harmonic components and noise:

$$y(t) = 50\cos\left(2\pi f_1 t + \frac{\pi}{6}\right) + 4.5e^{-33t}$$
$$+ 3.8\cos\left(2\pi f_2 t + \frac{\pi}{3}\right)e^{-20t}$$
$$+ 0.8\cos\left(2\pi f_3 t + \frac{\pi}{4}\right)e^{-5t} + n(t) \tag{13}$$

where f_1 = 50 Hz (fundamental frequency), f_2 = 5.3*f_1 = 265 Hz (5.3 times larger than the fundamental frequency) and $f_3 = 11.7 * f_1 = 585$ Hz, and $n(t)$ is the zero-mean noise signal that obeys the normal distribution.

The sampling frequency is 6400 Hz. The method proposed needs 20 ms data from the simulation signal to analyze the discrete sampling data. Table 1 summarizes the analytic results of the single trial at different signal-to-noise ratios, and the amplitude errors are given in the form of relative error (percent), but the decaying constants, frequency and phase error are given in the absolute error form. A, B, C and D given in Table 1, respectively, represent the fundamental, decaying aperiodic component, decaying harmonic signal 1 and decaying harmonic signal 2.

It is hard to use the DFT signal to obtain the decaying constants and the amplitude from the decaying components. In addition, it is hard to use the 20 ms data to obtain the frequency of non-integer harmonic. Table 2 summarizes the analytic results and the error by using DFT in the discrete sampling of (13) at different signal-to-noise ratios.

From Table 2, we can know that it is hard to use DFT to accurately analyze the parameters of each element of the fault signal. The analytic results show a large error and amplitude of decaying aperiodic components. However, the optimal solution obtained by using the proposed method will only have a larger decaying harmonic amplitude error when the signal-to-noise ratio is low. So, the assessment of the result for Case A shows the effectiveness of the proposed method.

Table 1. The analytic results and the error by using this method at different signal-to-noise ratios.

parameters		40db calculation results	40db error	60db calculation results	60db error	80db calculation results	80db error
amplitude	A	50.06	0.120	49.981	-0.04	50.001	0.002
	B	4.52	0.441	4.538	0.844	4.498	-0.044
	C	3.76	-1.003	3.806	0.158	3.799	-0.026
	D	0.74	-7.632	0.805	0.625	0.801	0.125
Decay time constant	B	-34.15	-1.148	-32.888	0.112	-32.963	0.037
	C	-19.95	0.053	-20.211	-0.211	-19.989	0.011
	D	-48.15	2.846	-50.429	0.571	-51.042	-0.042
frequency /Hz	A	49.948	-0.0521	49.9996	-0.0004	49.9998	-0.0002
	B	265.067	0.0669	265.034	0.0341	264.997	-0.0030
	C	583.919	-1.0810	585.065	0.0650	585.003	0.0033
phase/°	A	59.915	-0.085	59.967	-0.033	60.001	0.001
	C	30.513	0.513	30.038	0.038	29.990	-0.010
	D	44.103	-0.897	45.245	0.245	45.027	0.027

Table 2. The analytic results and the error by using DFT at different signal-to-noise ratios.

parameters	40db calculation results	40db error (%)	60db calculation results	60db error (%)	80db calculation results	80db error (%)
Amplitude of fundamental	49.570	-0.86	49.630	-0.74	49.633	-0.73
amplitude of decaying DC component	3.230	-28.22	3.230	-28.22	3.230	-28.22

4 CONCLUSION

The decaying aperiodic components and decaying harmonic components of the power system short-circuit fault signal have some impact on the relay protection device. The traditional discrete FFT does not accurately analyze the parameters of the fault signals. In this paper, the fault signal is analyzed by the matrix pencil method, to achieve the accurate analysis of each component of the fault signal. On the whole, it is proved that this method has a greater significance for the analysis of the impact on the relay protection against transient component.

REFERENCES

Ding, Y.F. 2005. Spectrum Estimation of Harmonics and Interharmonics Based on Prony Algorithm. Transactions of China Electrotechnical Society, 25(10): 98–101.

Fang, G.Z. 2012. Detection of harmonic in power system based on FFT and wavelet packet. Power System Protection and Control, 30(5): 75–79.

Hu, Z.J. 2001. Study on protective algorithm for elimination of decaying aperiodic component. Power System Technology, 15(3): 7–11.

Hua, Y.B. & Sarker, T.K. 1990. Matrix pencil method for estimating parameters of exponentially damped/undamped sinusoids in noise. IEEE Transactions on Acoustics Speech Signal Process, 38(5): 814–824.

Huang, C. 2001. A new approach to eliminate non-period components from sampling data. Relay, 17(8): 10–12.

Li, Y.P. 2006. Influence of non-periodic components on transient saturation of current transformer. Electric Power Automation Equipment, 19(8): 15–18.

Li, W.C., Lin, G. & Yin, X. 2012. A detection method for non-integer harmonics measurement based on matrix pencil algorithm. Electrical Measurement & Instrumentation, 49(4): 24–26.

Meng, L.L., Sun, C.D. & Han, B.R. 2012. Algorithm for inter-harmonic detection based on least square method and ICA. Power System Protection and Control, 22(11):76–81.

Niu, S.S. 2012. An algorithm for electrical harmonic analysis based on triple-spectrum-line interpolation FFT. Proceedings of the CSEE, 32(16): 130–136.

Pang, H. 2003. An improved algorithm for harmonic analysis of power system using FFT techni-que. Proceedings of the CSEE, 23(6): 49–54.

Ritter, J. & Amdt, F. 1996. Efficient FDTD/matrix-pencil method for the full-wave scattering parameter analysis of wave guiding structures. Microwave Theory and Techniques, IEEE Transactions on, 44(12): 2450–2456.

Sarkar, T.K. & Pereira, O.1995. Using the matrix pencil method to estimate the parameters of a sum of complex exponentials. Antennas and Propagation Magazine, IEEE 37(1): 48–55.

Suo, Nan J.L. 2010. Harmonic Analysis of Fault Signal in UHV AC Transmission Line. High Voltage Engineering, 2010, 30(1): 37–43.

Xue, H. 2002. Precise algorithms for harmonic analysis based on FFT algorithm. Proceedings of the CSEE, 22(12): 106–110.

Yilmazer, N., Koh, J. & Sarkar T.K. 2006. Utilization of a unitary transform for efficient computation in the matrix pencil method to find the direction of arrival. *Antennas and Propagation, IEEE Transactions on,* 54(1):175–181.

Zeng, R.J., Yang, Z.B. & Liu, H.C. 2012. A method of power system harmonic detection based on wavelet transform. Power System Protection and Control, 20(15): 35–39.

Zhang, F.S. 1999. FFT algorithm with high accuracy for harmonic analysis in power system. *Proceedings of the CSEE,* 19(3): 63–66.

Zhang, P. & Li, H.B. 2012. A Novel Algorithm for Harmonic Analysis Based on Discrete Wavelet Transforms. *Transactions of China Electro technical Society,* 16(3): 252–259.

Energy Science and Applied Technology – Fang (Ed.)
© *2016 Taylor & Francis Group, London, ISBN 978-1-138-02833-3*

Mathematical modeling and simulation of VSC-HVDC system failure based on Matlab

Jianyuan Dong, Juanjuan Wang, Xue Zhang & Bin Lv
School of Mechanical and Electrical Engineering, Xi'an University of Architecture and Technology, Xi'an, China

ABSTRACT: Voltage-sourced converter-based HVDC (VSC-HVDC) is a new power transmission technology based on voltage-sourced converter and self-turn-off devices. The valves of VSC-HVDC are comparatively expensive. Therefore, the converter device must be protected from abnormal damage of over-voltage or over-current under a normal operation and condition of failure. It is necessary to simulate and analyze the fault of VSC-HVDC. An electromagnetic transient model and a control strategy for the VSC-HVDC system were proposed. Response is analyzed in a typical failure situation. SC-HVDC is simulated in MATLAB/Simulink, and rationality of parameter design method and validities of control strategies are verified by simulation results.

Keywords: VSC-HVDC; control strategy; transient stability; fault characteristic

1 INTRODUCTION

With the development and the progress of electric power technology, VSC-HVDC technology has become increasingly matured and widely used in the field of transmission. It has been adopted in the control systems as the grid interconnection of wind power station, a lonely island, and weak power supply and power supply project of the city.

In recent years, research institutions and scholars have conducted a further study on the VSC-HVDC model and control strategies. This paper mainly studies the mathematical model and control strategy of the VSC-HVDC system, ignoring the control strategy under the unbalance situation. This paper mainly studies the electromagnetic transient model under the unbalanced situation of the VSC-HVDC system, ignoring the system voltage and current stress under the fault condition. This paper only simulates the fault of the internal bus, and combines with the operating system under different application requirements, and put forwards the way for protection. Fault on the dc side and the change of load are not discussed in this paper.

When the VSC-HVDC system fails, it can cause damage to equipment or interrupt transmission. In view of this characteristic, the study focuses on the running condition of both the system under the three kinds of fault: dc line-to-earth fault, three-phase short-circuit fault on the inverter side and the load changes.

2 MATHEMATICAL MODEL VSC-HVDC

The system as a new kind of dc transmission technology is composed of converter station and dc transmission lines. This two-terminal system will be used as a case study, consisting of a single supply and a receiving end (VSC1 and VSC2 respectively), as shown in Figure 1. The DC capacitance C provides voltage support and has the function of filter. To reduce the higher-order harmonic that impacts the communication network, the system sets the filter on the ac side. We can get the mathematical model of the synchronous rotating coordinate system under a three-phase power grid voltage balance.

$$L\frac{di_{sd}}{dt} - \omega Li_{sq} + Ri_{sd} = u_{sd} - u_d \tag{1}$$

$$L\frac{di_{sq}}{dt} + \omega Li_{sd} + Ri_{sd} = u_{sq} - u_q \tag{2}$$

$$C\frac{du_{dc}}{dt} = \frac{3}{2}(s_d i_{sd} + s_q i_{sq}) - i_d \tag{3}$$

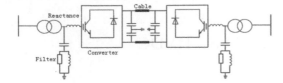

Figure 1. Proposed two-terminal multi-terminal DC system.

where u_{sd} and u_{sq} are the d and q axis components for the grid voltage, respectively; u_d and u_q are the d and q axis components for the AC voltage fundamental wave of VSC, respectively; i_{sd} and i_{sq} are the d and q axis components for the grid current, respectively.

In the $d - q$ synchronous rotating coordinate system, the active power p_s and the reactive power q_s can be represented as

$$\begin{cases} p_s = \dfrac{3}{2}(u_{sd}i_d + u_{sq}i_q) \\ q_s = \dfrac{3}{2}(u_{sd}i_q - u_{sq}i_d) \end{cases} \quad (4)$$

When $u_{sq} = 0$, type (4) can be written as

$$\begin{cases} p_s = \dfrac{3}{2}u_{sd}i_d \\ q_s = \dfrac{3}{2}u_{sd}i_q \end{cases} \quad (5)$$

By type (5), the system can regulate the active and reactive power through i_d and i_q.

3 CONTROL STRATEGY

The rectifier side of the VSC-HVDC system adopts the active power and reactive power control, and the control block diagram I, as shown in Figure 2. The inverter side uses the dc voltage and the reactive power control. In Figure 2, the controlling unit is divided into four parts as follows.

3.1 Measurement and calculation system

The phase lock 0f PLL: we measure the frequency f and the phase angle ωt of the three-phase voltage $U_{sa} \, U_{sb} \, U_{sc}$ and calculate the vector of A and B values, where $f = \omega/2\pi$.

The PCC voltage measurements: we measure the grid of the electromotive force of $U_{sa} \, U_{sb} \, U_{sc}$.

Current measurement: we measure the three phase current $i_a \, i_b \, i_c$ on the rectifier side.

Clark transformation: we introduce X_0 components, physical quantities i_a, i_b, i_c, U_{sa}, U_{sb}, U_{sc}

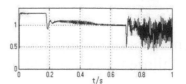

Dc voltage

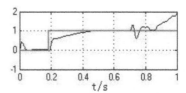

Active power

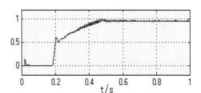

Direct-current power

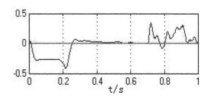

Reactive power

Figure 2. Simulation results of the rectifier during the dc line-to-earth fault.

under the three-phase static coordinate system of the ABC transform for physical quantities i_D, i_Q, i_0, U_{SD}, U_{SQ}, U_0 under the two-phase static coordinate system of DQ. The voltage's variation is similar to the current's, such as type (6):

$$\begin{bmatrix} i_D \\ i_Q \\ i_0 \end{bmatrix} = \sqrt{\frac{2}{3}} \begin{bmatrix} 1 & -\dfrac{1}{2} & -\dfrac{1}{2} \\ 0 & -\dfrac{\sqrt{3}}{2} & \dfrac{\sqrt{3}}{2} \\ \dfrac{1}{\sqrt{2}} & \dfrac{1}{\sqrt{2}} & \dfrac{1}{\sqrt{2}} \end{bmatrix} \begin{bmatrix} i_a \\ i_b \\ i_c \end{bmatrix} \quad (6)$$

Park transformation: physical quantities i_D, i_Q, i_0, U_{SD}, U_{SQ}, U_0 under the two-phase static coordinate system of the DQ transform for physical quantities i_D, i_Q, i_0^*, U_{sd}, U_{sq}, U_0^* under the two-phase rotating coordinate system of dq, the voltage's variation is similar to the current's, such as type (7):

$$\begin{bmatrix} i_d \\ i_q \\ i_0^* \end{bmatrix} = \begin{bmatrix} \sin(\omega t) & -\cos(\omega t) & 0 \\ \cos(\omega t) & \sin(\omega t) & 0 \\ 0 & 0 & 1 \end{bmatrix} \begin{bmatrix} i_D \\ i_Q \\ i_0 \end{bmatrix} \qquad (7)$$

Power calculation: we compute the active power P_s and the reactive power Q_s as the system actual output.

$$P_s = \frac{3}{2}(U_{sd}i_d + U_{sq}i_q) \qquad (8)$$

$$Q_s = \frac{3}{2}(U_{sd}i_q - U_{sq}i_d) \qquad (9)$$

3.2 Outer loop controller

Active power controller/reactive power controller/ dc voltage controller: outer loop controller tracking given reference signal and according to the VSC-HVDC system control target can realize dc voltage control, active power control, frequency control, reactive power control and voltage control.

Voltage—power slope module: according to slope characteristics and the size of dc transmission power, the system can adjust the voltage reference of the dc side.

3.3 Inner loop controller

Current decoupled: meeting the requirements of power factor control and making i_q to closely track the given value i_{qref}, where $i_{qref=0}$.

Feedforward compensation: using the open-loop control system compensates the measured disturbance signal;

Park inverse transformation: voltage wave U_d, U_q under the two-phase rotating coordinate system of the dq transform for voltage wave U_D, U_Q U_0 under the two-phase static coordinate system of DQ, such as type (10):

$$\begin{bmatrix} U_D \\ U_Q \\ U_0 \end{bmatrix} = \begin{bmatrix} \sin(\omega t) & \cos(\omega t) & 0 \\ -\cos(\omega t) & \sin(\omega t) & 0 \\ 0 & 0 & 1 \end{bmatrix} \begin{bmatrix} U_d \\ U_q \\ X_0 \end{bmatrix} \qquad (10)$$

Clark inverse transformation: the voltage U_D, U_Q U_0 under the two-phase static coordinate system of the DQ transform for voltage U_a, U_b U_c under the three-phase static coordinate system of ABC, such as type (11):

$$\begin{bmatrix} U_a \\ U_b \\ U_c \end{bmatrix} = \sqrt{\frac{2}{3}} \begin{bmatrix} 1 & 0 & \frac{1}{\sqrt{2}} \\ -\frac{1}{2} & -\frac{\sqrt{3}}{2} & \frac{1}{\sqrt{2}} \\ -\frac{1}{2} & \frac{\sqrt{3}}{2} & \frac{1}{\sqrt{2}} \end{bmatrix} \begin{bmatrix} U_D \\ U_Q \\ U_0 \end{bmatrix} \qquad (11)$$

3.4 DC side measurement calculation

Dc power calculation: $P_{dc} = U_{dc}i_{dc}$.

Dc current and voltage measurements: measurement of dc voltage U_{dc} and dc current i_{dc} on the dc side.

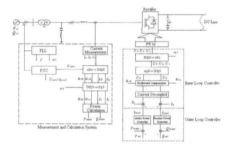

4 SIMULATION OF VSC-HVDC SYSTEM

Using the simulation software Mat lab/Simulink can simulate three kinds of fault state. The Dc line sends electricity power from AC System1 (230 KV, 2000 MV.A, 50 Hz) to another electricity power system of AC System2. The rectifier and the inverter are constructed by the antiparallel three levels of the IGBT/diode bridge model. The main parameters of the VSC-HVDC system are listed in Table 1.

4.1 DC line-to-earth fault

The cause of dc line-to-earth fault is cable grounding and cable insulation failure. When t = 0.7 s, the failure occurs. The simulation curve of failure is shown in Figure 3 and Figure 4.

By analyzing the simulation curve, the change in dc voltage presents three phases:

Phase 1: with the decrease in dc voltage, dc voltage controller increases the phase-shifting angle in order to recover dc voltage, and dc voltage drop can lead to a lower ac voltage amplitude.

Phase 2: the rectifier of dc voltage controller increases the voltage phase angle, improves the active power resulting in capacitor charging, and rises the voltage amplitude. With the decrease in the voltage phase-shifting angle, the active power of the inverter side is reduced.

Phase 3: after dc voltage recovery, voltage and current have an oscillation.

Table 1. The main technical parameters.

U/kV	f/Hz	S/MVA	L/km	X/H
220	50	200	75	0.15

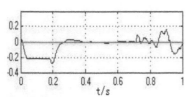

Active power

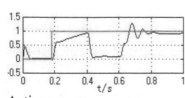

Reactive power

Figure 3. Simulation results of the inverter during the dc line-to-earth fault.

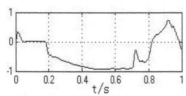

Active power

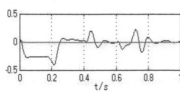

Reactive power

Figure 4. Simulation results of the rectifier during the three-phase short-circuit fault.

4.2 *Three-phase short-circuit fault*

When t = 0.4 s, the system has the three-phase short circuit fault on the inverter side, and this phenomenon exists 0.12 s. The simulation curve of failure is shown in Figure 5 and Figure 6.

Before the three-phase short circuit, the circuit is in a stable state. The system is divided into two loops when the failure occurs on the inverter side. A circuit is connected to the power supply and

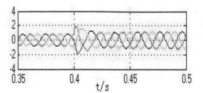

Three-phase voltage

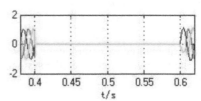

Three-phase current

Figure 5. Simulation results of the inverter during the three-phase short-circuit fault.

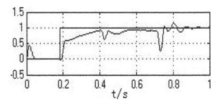

Active power

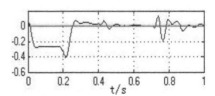

Reactive power

Figure 6. Simulation results of the rectifier during the load change.

the other is connected to the converter station. The power supply circuit of the current transient process consists of two components. The former belongs to the forced current, depending on the power supply voltage and short circuit loop impedance and the amplitude unchanged in the transient process. The latter belongs to the freedom of current, to make the flux and current in the inductive loop, and thus no mutation appeared.

4.3 *Load change*

Three-phase switch control: the three-phase load is 500 MW. We connect it to the simulation model of

466

the VSC-HVDC system. When t = 0.4 s, the three phase switch is closed. The simulation curve of failure is shown in Figure 7.

5 CONCLUSION

Through the analysis of the simulation curve, it can be concluded that when the system is subject to failure, on the dc side and ac side, there appears the phenomenon of over current and over voltage, and the failure will affect the size of transmission power. Thus, it effectively provides a theoretical basis for the research of failure protection in the future.

REFERENCES

Etxeberria-Otadui, I., U. Visarret, M. Caballer, A. Rufer. New optimized PWM VSC control structures and strategies under unbalanced voltage transients. IEEE Transactions on Industrial Electronics,54(5), p. 2902–2914, 2007.

Huang Fu-cheng, He Zhiyuan, Tang Guan-gfu, Ruan Jiang-jun, Wang Yan. Control Strategy for the Voltage Source Converter Based HVDC Transmission System Under Unbalanced AC Grid Conditions. Proceeding of the CSEE, 28(22), p. 144–151, 2008.

Ma Yu-ong, Xiao Xianning, Jiang Xu. Analysis of the impact of AC system single-phase earth fault on HVDC. Proceeding of the CSEE, 26(11), p.144–149, 2006.

Ni Jiajia. Studies. Transmission Line Transient Current Protection and Fault Location in HVDC System. Shang Hai: Shanghai Jiao Tong University, 2012.

Ren Xian-wen, Ma Yon-gtao, Wu Peng .Study on the commutation failure in HVDC transmission caused single-phase grounding at AC side. Proceeding of the CSEE, 35(5), p. 324–327, 2012.

Sanjay K, Remus Teodorescu, Pedro Rodriguez, Philip Carne Kjaer, Ani M. Gole .Negative Sequence Current in Wind Power Plants With VSC-HVDC Connection. IEEE Transactions on Sustainable Energy,3(3), p. 535–544. 2012.

Tang Guangfu. The technology of High Voltage Direct Current based on Voltage-Sourced Converter. Beijing: China Electric Power Press, 2010.

Wang Peng. The control strategy of VSC—HVDC system and experimental research. Baoding: North China Electric Power University, 2006.

Wang Yuxi. Analysis on Transient Characteristics of VSC-HVDC Power Supply System. Baoding: North China Electric Power University, 2012.

Wei Xiaoguang. Research on control strategies of voltage source converter based HVDC and its application in wind farm integration system. Beijing: China Electric Power Research Institute, 2007.

Ying, J., J. Fletcher, J.O' Reilly. Short-Circuit and Ground Fault Analyses and Location in VSC-based DC Network cables. IEEE Transactions on Industrial Electronics, 59(10), p. 182–197, 2012.

Energy Science and Applied Technology – Fang (Ed.)
© 2016 Taylor & Francis Group, London, ISBN 978-1-138-02833-3

Dynamic threshold Energy Detection based on spectrum prediction for cognitive radio

Q. Zhang, J.K. Guo, Z.Y. Yu & G.B. Liu
Xi'an Research Institute of High Technology, Xi'an, China

ABSTRACT: Energy Detection (ED) is a simple and popular method of spectrum sensing in cognitive radio systems. However, the performance of sensing techniques is largely affected when users experience fading and noise effects. To improve detection performance, a new spectrum sensing algorithm with spectrum prediction is proposed. The proposed algorithm adjusts the threshold of energy sensing detector according to the probability of the presence of primary users. The simulation results show that the proposed method outperforms the traditional method significantly in the low Signal-to-Noise Ratio (SNR) environment.

Keywords: Cognitive radio; Spectrum sensing; Spectrum prediction; Bayesian decision

1 INTRODUCTION

Spectrum sensing is important in cognitive radio to improve spectrum utilization (Vito 2013, Yücek & Arslan 2009). Energy detection (ED) (Urkowitz 1967) is the most popular sensing technique in spectrum sensing. However, its detection performance is influenced by the uncertainty of noise power and fading channel. By using the cooperative spectrum sensing scheme (Akyildiz et al. 2011), the above disadvantages can be overcome to a certain extent, but the cooperative spectrum technique needs more resources and it is not easy to implement in all the environments.

At present, the spectrum prediction technique has been extensively studied in the literature. The spectrum prediction accuracy is not affected by the noise uncertainty, since it only depends on the history knowledge and the prediction algorithm.

As a result, the performance of energy detection can be improved with the assistance of spectrum prediction. Therefore, the proposed method focuses on the combination of the spectrum sensing and the spectrum prediction.

2 SYSTEM MODEL

2.1 Energy detector

The cognitive user senses the primary user and makes decision according to the binary hypothesis test. H_1 and H_0 denote the presence and absence of the primary user, respectively. The received signal at the i-th cognitive user can be defined as follows:

$$x = \begin{cases} hs + \delta, & H_1 \\ \delta, & H_0 \end{cases} \tag{1}$$

where s is the primary user signal; h is the factor of the channel impairments of the sensing channel; δ is the zero-mean additive white Gaussian noise received by the cognitive user; s and δ are independent.

Energy detection is employed for local sensing. The measured statistic of energy detection Y is computed by

$$Y = \frac{1}{M} \sum_{t=1}^{t=M} x^2(t) \tag{2}$$

Here, M denotes the number of the sampling and $M = 2u = 2\tau_s W$, where W, τ_s, and u are the bandwidth of the measured frequency band, the measure duration and the time-bandwidth product, respectively.

The measured statistics for H_0 and H_1 satisfy the central and non-central chi-square distributions, respectively, each with "$2u$" degrees of freedom, i.e.,

$$Y \sim \begin{cases} \chi_{2u}^2(2\gamma), & H_1 \\ \chi_{2u}^2, & H_0 \end{cases} \tag{3}$$

where γ is the signal-to-noise ratio (SNR) in the sensing channel. To decide whether the primary user signal is present, we compare Y with a predefined energy threshold λ. When the sensing channel is the Rayleigh fading channel, the probability of false alarm p_f, detection probability p_d and misdetection probability p_m is given as follows (Digham et al. 2007):

$$p_f = \frac{\Gamma\left(u, \frac{\lambda}{2}\right)}{\Gamma(u)} = 1 - \frac{1}{\Gamma(u)} \int_0^{\lambda/2} e^{-t} t^{u-1} dt$$
$$\Gamma(\cdot) \text{ is Gamma function} \qquad (4)$$

$$p_d = \int_\gamma Q_u(\sqrt{2\gamma}, \sqrt{\lambda}) f(\gamma) d\gamma$$
$$= e^{-\lambda} \sum_{n=0}^{u-2} \frac{1}{n!} (\lambda/2)^2 + ((1+\overline{\gamma})/\overline{\gamma})^{u-1}$$
$$\times [e^{-\lambda/2} \sum_{n=0}^{u-2} \frac{1}{n!} (\lambda\overline{\gamma}/2(1+\overline{\gamma}))^n$$
$$- e^{-\lambda/2(1+\overline{\gamma})}] \qquad (5)$$

$$p_m = 1 - p_d \qquad (6)$$

where $\overline{\gamma}$ denotes the average SNR in the sensing channel, and $f(\gamma)$ is the probability density function (PDF) of γ. The derivate of p_f and p_d with respect to λ can be expressed as follows:

$$\frac{\partial p_f(\lambda)}{\partial \lambda} < 0 \qquad (7)$$

$$\frac{\partial p_d(\lambda)}{\partial \lambda} < 0 \qquad (8)$$

From Eqs. (4)–(8), in order to achieve a better detection probability, we can reduce threshold λ, but the p_f will raise simultaneously. In the case of a low SNR and deep fade, in order to achieve the system detection probability, p_f may be very high, which limits the application of the system in practice.

2.2 HMM predictor

The Hidden Markov Model (HMM) (Li et al. 2010) is widely used for spectrum prediction. In this paper, we use the HMM to predict the spectrum state.

A HMM can be described with a triple: $\lambda = [\pi, A, B]$, where π is a 1×2 vector, and A and B are 2×2 dimension matrices, respectively. Utilizing the spectrum state of past N time slots, the spectrum state at instant $N + 1$ can be predicted by training the HMM.

The training process of the HMM is to seek the optimal parameters $\lambda = [\pi, A, B]$ for maximizing the probability of generating observation sequence by the Expectation Maximum (EM) algorithm or the Baum-Welch algorithm. Then, the sequence O followed by a busy or idle slot at instant $N + 1$ can be calculated by the Viterbi algorithm. In other words, the joint probabilities $P(O, 1\backslash\lambda)$ and $P(O, 0\backslash\lambda)$ can be obtained, which can be considered as the prior probability of the primary user.

Sensing slot	Data slot	. . . . . .	Sensing slot	Data slot
τ	$T - \tau$			

Figure 1. Exhibition of sensing scheduling.

2.3 Sensing scheduling

We consider the period spectrum sensing and assume the sensing period as T. Cognitive users spend τ on sensing and decision. If the primary user signal is absent, cognitive users have $T - \tau$ on delivering data. Otherwise, the cognitive users wait for the next sensing period. The sensing scheduling is shown in Figure 1.

3 OPTIMAL ALGORITHM AND ANALYSIS

According to the signal detection theory, when the prior probability of the detection object is unknown, the N-P (Neyman-Pearson) rule is optimal. When the prior probability is available, the Bayesian decision will obtain the best trade-off between opportunity losing and interference probability. With the HMM predictor, the probability that the primary user signal is present can be obtained. We also define p_1 as the prior probability.

When the primary user signal is present, we assume that the cognitive user misses to detect it, and then the cognitive user signal will interfere with the primary user, so the interference probability p_{inter} can be defined as follows.

$$p_{inter} = p_1 p_m \qquad (9)$$

We define p_0 as the prior probability that the primary user signal is absent, which is given as follows:

$$p_0 = 1 - p_1; \qquad (10)$$

We assume and define the opportunity losing probability p_{los} as follows:

$$p_{los} = p_0 p_f \qquad (11)$$

Generally, p_{inter} is restricted to a certain upper bound $\overline{p}_{inter}$ in the cognitive radio system. Then, the problem of interest to us is formulated as follows:

$$\begin{cases} \min_\lambda f(\lambda) = p_1 p_m(\lambda) + p_0 p_f(\lambda) \\ \text{s.t. } p_1 p_m(\lambda) \le \overline{p}_{inter} \end{cases} \qquad (12)$$

where λ denotes the energy threshold. When λ is fixed, the values of p_m and p_f can be obtained by

Eqs (4)–(6). It should be noted that the objective function in Eq. (12) is actually the Bayesian minimum error rate decision formulation, which combines the effects including interference and opportunity losing probabilities. The constraint condition indicates that the interference probability is below a threshold, which makes Eq. (12) to be superior to the Bayesian minimum error rate decision formulation.

As shown in Figure 2a, the average error rate of the Bayesian minimum error rate decision is actually the sum of two error rates weighted with the corresponding prior probability (Zhang 2010). In the sense of posterior probability, the threshold $\lambda_{\text{Bayesian}}$ is fixed, which can gain a minimum error rate. However, the Bayesian decision cannot guarantee the protection of the primary user.

With the constraint condition, when $p_1 p_m > \overline{p}_{\text{inter}}$, the new algorithm will adjust the energy threshold to satisfy $p_1 p_m \leq \overline{p}_{\text{inter}}$. Hence, the new algorithm can guarantee the protection of the primary user. However, the error rate of the new algorithm is not the least, as shown in Figure 2b. Because the energy threshold of the new algorithm is dynamic along with p_1, so the algorithm is called the dynamic threshold energy detection (DT-ED) algorithm in this paper.

In formulation (12), the values of p_1 and p_0 can be known by the HMM predictor, and the values of p_m and p_f are determined by λ.

As we can see from Figure 2a, the objective function in Eq. (12) is strictly convex function, so it is easy to calculate the minimum point λ_0 without the constraint condition. If $p_1 p_m(\lambda_0) \leq \overline{p}_{\text{inter}}$, then λ_0 is satisfied. If $p_1 p_m(\lambda_0) > \overline{p}_{\text{inter}}$, then we can increase the value of λ_0 to meet the equation $p_1 p_m(\lambda_0) = \overline{p}_{\text{inter}}$. Finally, the value of λ_0 is the energy threshold, which can meet the requirement of Eq. (12).

4 SIMULATION AND DISCUSSION

In order to show the effectiveness of the DT-ED algorithm, we consider a system including a primary user and a cognitive user with the Rayleigh fading channel. We compare the DT-ED algorithm with the conventional ED detection algorithm, and the time-bandwidth time u of two algorithms is 5. For comparison, we choose the receiver operating characteristics (ROC) curves as baselines.

As can be seen from Figures 3–5, the ROC of the DT-ED algorithm is better than that of the conventional ED detection with the SNR of $\gamma = 0$ dB, $\gamma = -5$ dB, $\gamma = -10$ dB, respectively. Especially, when the SNR is -10dB, the performance of the ED algorithm is very poor, but the DT-ED algorithm still works well. Because the prior probability of the primary user is considered, the DT-ED algorithm received better sensing performance.

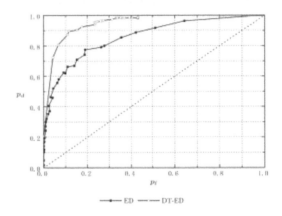

Figure 3.   ROC (SNR = 0dB).

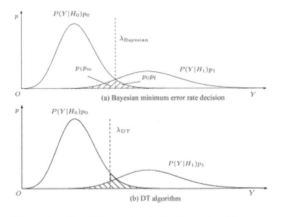

Figure 2.   The different threshold of two algorithms.

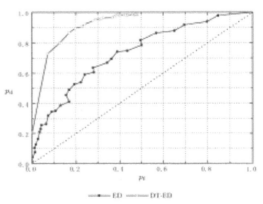

Figure 4.   ROC (SNR = –5dB).

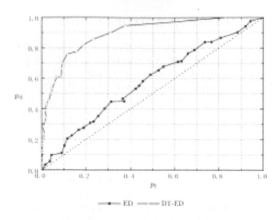

Figure 5.   ROC (SNR = −10dB).

5 CONCLUSIONS

A new energy detection method, which combines the sensing result and prediction result based on the Bayesian theory, has been proposed. The algorithm changes the energy threshold according to the prior probability of the primary user, and the prior probability is obtained by the HMM predictor. Furthermore, it was shown that the DT-ED algorithm improves the performance of the energy detector in the low SNR regime.

REFERENCES

Akyildiz et al. 2011. Cooperative spectrum sensing in cognitive radio networks: A survey. *Physical Communication* 4(1), 40–62.
Digham et al. 2007. On the energy detection of unknown signals over fading channels. *IEEE Transactions on Communications* 55(1), 21–24.
Li et al. 2010. Spectrum usage prediction based on high-order markov model for cognitive radio networks. In The 10th *IEEE International Conference on Computer and Information Technology*, Bradford, 29 Jun–1 July 2010.
Urkowitz, H. 1967. Energy detection of unknown deterministic signals. *Proceedings of the IEEE* 55(4), 523–531.
Vito, L.D. 2013. Methods and technologies for wideband spectrum sensing. *Measurement* (46), 3153–3165.
Yücek, T. & Arslan, H. 2009. A survey of spectrum sensing algorithms for cognitive radio applications. *IEEE Communications Surveys & Tutorials* 11(1), 116–130.
Zhang, X. 2010. *Pattern Recognition (Third Edition)*. Beijing: Tsinghua University press.

Energy Science and Applied Technology – Fang (Ed.)
© 2016 Taylor & Francis Group, London, ISBN 978-1-138-02833-3

Program fault localization based on improved program spectrum concept lattice

Dan Zhang, Minglei Gao & Baozhi Qiu
School of Information Engineering, Zhengzhou University, Zhengzhou, China

ABSTRACT: In program testing, various faults will overlap the failed tests, therefore decreasing the fault location efficiency. In this paper, an improved program spectrum concept lattice-based program fault localization (IPSCL-PFL) technique is proposed. Program spectrum is abstracted according to the test history, from which the program spectrum concept lattice is first constructed and then a novel suspiciousness computation method is defined by the IPSCL-PFL technique. The result of the experiment on Tetris in Eclipse shows that this technique has a higher efficiency than the state-of-art program spectrum-based technique.

Keywords: Program spectrum; Concept lattice; Suspiciousness; Program spectrum Concept lattice model; Fault Localization

1 INTRODUCTION

Automated software fault localization technique has become one of the hot research directions with a complex diversification and expansion of software. Software fault location technique mainly aims to narrow the search scope of fault to improve the efficiency of fault localization. Fault localization based on the program spectrum uses coverage information provided by test suites to compute likely faulty statements.

Compared with other fault localization techniques, the spectrum of program fault localization is simple with program information gathered on the spectrum, suitable for large-scale programs with a relatively high targeting efficiency, which can narrow down the possible locations of software faults and help save the developers' debugging time. However, this technique has two disadvantages: first, with the lack of application context information, the programmer has difficulty in determining the location of the fault according to the individual suspiciousness of program elements alone; second, with the increase in the number of faults in the program, different faults may cause the same statement test to fail, resulting in lower targeting efficiency. In view of the above limitations, a software fault localization technique based on the spectrum concept lattice (PSCL-SFL) was proposed by Wen. The concept lattice was introduced on the basis of the program spectrum to construct the suspiciousness calculation model. To a certain extent, it solved the problem of the low localization efficiency caused by the overlap of failed tests. However, the suspiciousness measurement interval (in) in this technique is narrow, the degree of differentiation is not obvious, especially for the statement of suspiciousness value is very close, and the programmer cannot locate the fault accurately and effectively. Thus, in this paper, fault localization based on the program spectrum concept lattice with a new suspiciousness calculation is proposed, which can locate the fault rapidly and accurately.

2 RELATED CONCEPTS

In this section, we present an overview of the related concepts in the fault localization. There are three steps in spectrum-based program fault localization. The first step involves the execution of the test cases to collect coverage information of program statements; the second one involves the use of the coverage information to calculate the suspiciousness of the covered statements; the last step involves locating of the fault with the descending order of suspiciousness.

Definition 1 (Program spectrum): For the given n statements in a program, the test suit $T = T_F \cup T_P$, $T_F = \{T_1, ..., T_s\}$ represents the failed test suits and $T_P = \{T_{s+1}, ..., T_m\}$ represents the past test suits. Then, the program spectrum can be shown as a two-dimensional matrix $M_{n,m}$, and the matrix element $b_{i,j}$ is given by

$$b_{i,j} \in M_{n,m} b_{i,j} = \begin{cases} \text{True, } T_j \text{ executes the statement of } s_i \\ \text{False, else} \end{cases}$$

Concept lattice is a hierarchical structure based on the formal concept of the generalization and specialization relationship. Formal context records the relationships between the form of objects and attributes. A formal concept is defined as a pair consisting of a set of objects (the "extent") and a set of attributes (the "intent"), such that the extent consists of all objects that share the given attributes, and the intent consists of all attributes shared by the given objects. Therefore, the program spectrum is taken as the formal context to create the concept lattice, which provides the node suspiciousness for fault localization.

Definition 2 (Concept Lattice): If $C(M)$ is the concept union in the concept context $M = (U, A, I)$, then the concept lattice of M is an ordered set $L(M) = (C(M), <=)$.

Definition 3 (Program Spectrum Concept Lattice): in the formal context $M = (U, A, I)$, the concept union is $C(M)$, where U is the program element union and A is the test case union, and $I \subseteq U \times A$ is the binary relation to be executed. Then, $L(M) = (C(M), <=)$ is the program spectrum concept lattice of $U \times A$.

Suspiciousness in software fault localization is closely related to the percentage of passed/failed test cases, which determines whether there is a fault in one statement. Wen counted the failed test cases of both lattice nodes and all nodes, and obtained the suspiciousness according to the calculation formula. The disadvantage is that the interval of suspiciousness obtained by the above formula is dense, which cannot locate fault statements rapidly. Therefore, the concept of entropy is introduced into the suspiciousness calculation model to enlarge the distribution range, and the suspiciousness calculation formula is as follows:

Equation 1 (Suspiciousness calculation formula):

entropy. Otherwise, the entropy will be low if there is a big gap between the number of fault and correct test cases. If the number of correct test cases is higher than that of fault ones, the numerator of the formula will be negative, which makes the suspiciousness value smaller than expected; otherwise, the numerator of the formula will be positive, which makes the suspiciousness value bigger than expected. The suspiciousness interval of the program fault can be enlarged according to the suspiciousness value, which makes it easier to locate the faulty statement.

3 IMPROVED PROGRAM SPECTRUM CONCEPT LATTICE-BASED FAULT LOCALIZATION

Fault localization based on the program spectrum concept lattice calculates the suspiciousness of each formal concept node, and then locates the fault according to the ordering set of that suspiciousness. In line with the localization theory, if there are more number of failed test cases than correct ones, the suspiciousness of the element is bigger. With the suspiciousness calculation formula based on entropy, we get the suspiciousness value of the lattice node and make a sort, and then the biggest node can be found out according to the sort result, which is also the likely faulty statement. Compared with PSCL-SFL, this method can shrink the fault search scope. For nodes with the same suspiciousness value, we again use Eq. 1 to look for the failed cases with different statements. After several rounds of tests, a higher localization can be achieved.

The following example in Table 1 presents the construction process of the program spectrum concept lattice. The program aims to get the biggest of four numbers, among which there are faults in s_6 and s_{11}.

$$\text{Sus}(s) = \frac{f(s) - p(s)}{(f(s) + p(s))\left(\dfrac{f(s)}{f(s) + p(s)}\log\dfrac{f(s)}{f(s) + p(s)} + \dfrac{p(s)}{f(s) + p(s)}\log\dfrac{p(s)}{f(s) + p(s)}\right)} \tag{1}$$

In Eq. 1, $f(s)$ is the number of failed test cases that executed statement s one or more times. Similarly, $p(s)$ is the number of passed test cases that executed statement s one or more times. In this definition, entropy is used to differentiate the suspiciousness degree, which means the higher the suspiciousness degree, the more likely the fault there is. When the number of fault test cases is more or less than that of correct test cases, the suspiciousness value is relatively small due to the high

Program spectrum formal context is generated according to the test cases of $T_1 \ldots T_8$. Table 2 presents the generated formal context.

In Table 2, $s_1, \ldots, s_{14}$ represent the corresponding statements of the program, where $T_1, \ldots, T_8$ are 8 test cases; * signifies that the test case is passed correctly for the statement, where the formal object in the row owns the formal attribute in the column.

According to the partial ordering relation in Definition 3, we can get the program spectrum

Table 1. The example program.

s1 int max = 0, a, b, c, d;	s2 Read (a, b, c, d);
s3 if(a > b)	s4 { max = a;

s5 if(max < c)	s6 max = a;
	Fault! Actually s6 ';
	s6 ' max = c;

s7 if(max < d)	s8 max = d; }else
s9 {max = b;	s10 if(max c)
	s11 max = c;
s11 max = b; Fault Indeed s11'	s12 if(max d)
s13 max = c; }	s14 print(max);

Table 2. Formal context of program spectrum.

Program	T1	T2	T3	T4	T5	T6	T7	T8
s1	*	*	*	*	*	*	*	*
s2	*	*	*	*	*	*	*	*
s3	*	*	*	*	*	*	*	*
s4			*	*	*			*
s5			*	*	*			*
s6			*	*				*
s7			*	*	*			*
s8				*				*
s9	*	*				*	*	
s10	*	*				*	*	
s11	*	*				*		
s12	*	*				*	*	
s13						*		
s14	*	*	*	*	*	*	*	*

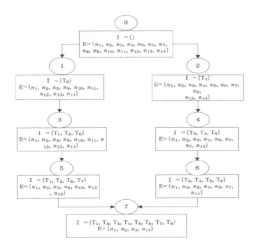

Figure 1. Program spectrum concept lattice.

concept lattice presented in Table 2. As shown in Figure 1, E and I are the extent and intent of the formal concept, respectively, with every node representing a formal concept. The attribute relationship

of program elements and test executions are shown in the figure.

On basis of the program spectrum formal context and the hierarchy relationships between formal objects and attributes, we generate the program concept lattice, as shown in Figure 1.

4 EXPERIMENT

In this section, we present an empirical evaluation to compare the efficiency of IPSCL-PFL with the other existing fault localization techniques that are described in Section 3.

4.1 Experimental subject programs and experimental setup

To evaluate the effectiveness of IPSCL-PFL techniques in the presence of multiple faults, this study leverages Tetris programs in Eclipse that are popular in IPSCL-PFL research. The programs implemented by object-oriented JAVA language programming contain 2397 line statements and 98 methods. In view of the keyboard input of Tetris, which are the left arrow, right arrow, down-arrow and space, one or two mistakes are manually and separately embedded into the corresponding source code methods. By the inspection of the IPSCL-PFL technique, which can correctly locate to the corresponding method, the effectiveness of the proposed technique can be verified. In the experiment, with 10 faulty versions in the same test inputs, the method of test coverage did not change, which led to the same program spectra well as the same formal context in the program spectrum concept lattice. For the sake of simplicity, a unified method sequence ( ) is used instead of the other methods.

There are 4 steps in the experiment process: (1) use code cover tools to collect the test coverage information and establish the program spectrum formal context; (2) construct the spectrum concept lattice figure; (3) calculate and count up the number of failed and successful tests according to the program spectrum concept lattice diagram and work out the suspiciousness of each statement; (4) order and locate the fault according to the degree of suspiciousness.

Table 3 illustrates the application spectrum formal context established by the IPSCL-PFL technique. In Table 3, the first column represents the corresponding methods in Tetris game programs, the columns T_1 to T_8 represent the eight test executions in view of the above methods, and * represents the corresponding line method covered by the corresponding columns.

Table 3. The formal context in the Tetris programs.

Program	T1	T2	T3	T4	T5	T6	T7	T8
Move left()	*		*	*				
Move right()		*	*					
Move down()	*	*	*		*		*	*
Rotate clockwise ()	*	*	*	*	*	*		*
Sequence ()	*	*	*	*	*	*	*	*

4.2 Experimental results and analysis

Table 4 presents the suspiciousness calculation results for the proposed approach and the PSCL-SFL technique in the above example programs. Among them, the first column indicates the grid nodes in the program spectrum concept lattice in Figure 2, the second column represents the extents of the grid nodes in the program spectrum concept lattice, the third column indicates the suspiciousness calculated by the PSCL-SFL technique, and the last column represents the suspiciousness calculated by the proposed technique.

The results shown in Figure 2 depict the data in Table 4. Points and connecting lines are drawn for the two techniques. The legend to the right shows how to interpret the lines representing each technique. The labels in the legend are abbreviated for space. The horizontal axis represents the grid nodes in the formal context of the program spectrum concept lattice. The vertical axis represents the suspiciousness by the fault localization technique. Compared with the suspiciousness calculated by the two different suspiciousness model in Table 4 and Figure 2, the distribution of the suspiciousness in the PSCL-SFL technique is dense, the variable scope arrangement is from 0 to 1, and recognition degree is not high, which cannot accurately pinpoint the location of the fault. While the distribution of the suspiciousness in the IPSCL-PFL technique is a relative dispersion and the variable scope arrangement is bigger, resulting in the fact that the fault localization is more accurate and rapid, which improves the efficiency of the localization to a certain extent.

Table 5 presents the suspiciousness calculation results for the proposed approach and the PSCL-SFL technique in Tetris programs, which consists of some faulty methods in 10 versions. The first column indicates the whole manually embedded faulty methods in 10 versions of the Tetris programs. The second to sixth columns, respectively, indicate the suspiciousness and the number of failed tests of the grid nodes in the manually embedded methods. The suspiciousness in each column consists of two parts separated by commas: the first part indicates the suspiciousness or the number of the failed tests in the PSCL-SFL technique and the value in parentheses indicates

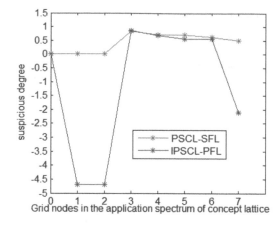

Figure 2. Comparison of fault localization techniques.

Table 4. Comparison of fault localization techniques.

Grid node	Extent	Suspiciousness 1	Suspiciousness 2
0	Φ	0	0
1	s13	0	−4.69
2	s8	0	−4.69
3	s11	0.83	0.86
4	s6	0.71	0.69
5	s9, s10, s12	0.71	0.55
6	s4, s5, s7	0.63	0.55
7	s1, s2, s3, s14	0.50	−2.1

Table 5. Comparison of fault localization techniques in Tetris.

Fault methods	move left	move right	move down	rotatecloc kwise	sequence
moveleft	1, 3.5	0.83, 2.7	0.56, 1.4	0.56, 1.4	0.5, 1.1
moveright	0.75, 2.3	1, 3.2	0.6, 1.8	0.55, 1.4	0.5, 1.1
movedown	0.44, 1.3	1(2), -1.2	1(7),- 4.6	0.44, 1.3	0.5, 1.1
rotate-clock wise	1(3), -2.7	1(2), -1.2	0.44, 1.3	1(7), -4.6	0.5, 1.1
moveleft, moveright	1(3), -2.7	1(2), -1.2	0.57, 1.4	0.57, 1.4	0.5, 1.1
moveleft, movedown	1(3), -2.7	1(2), -1.2	1(7), -4.6	0.45, 1.3	0.5, 1.1
moveleft,rota- teclockwise	1(3), -2.7	1(2), -1.2	0.44, 1.3	1(7), -4.6	0.5, 1.1
moveright, movedown	0.55, 1.4	1(2), -1.2	1(7), -4.6	0.44, 1.3	0.5, 1.1
moveright,rot- ateclockwise	1(3), -2.7	1(2), -1.2	0.44, 1.3	1(7), -4.6	0.5, 1.1
movedown,ro- tateclockwise	1(3), -2.7	1(2), -1.2	1(7),- 4.6	1(7), -4.6	1(8), -4.6

the number of failed tests in the PSCL-SFL technique; the second part indicates the suspiciousness in the method proposed in this paper. For example, at the first line and third column of Table 5, the suspiciousness in the method move right () in the PSCL-SFL technique is 0.83, while the suspiciousness in the IPSCL-PFL technique is 2.7. We can see from Table 5 that the IPSCL-PFL technique expands the measurement range of the suspiciousness, which is able to guide the programmer to the fault more accurately and improve the efficiency of the fault localization effectively.

5 CONCLUSIONS

This paper proposes a new concept and technique, namely the improved program spectrum concept lattice-based program fault localization (IPSCL-PFL) technique, based on a new suspiciousness calculation formula, which can expand the suspiciousness measurement range. In comparison with other existing prioritization techniques in Tetris programs, we have shown that the technique proposed makes the suspiciousness more obvious, which may be more effective in reducing the debugging cost.

REFERENCES

Abreu R, Zoeteweij P, van Gemund AJC. On the accuracy of spectrum-based fault localization [C]. Proceedings of Testing: Academic and Industrial Conference-Practice and Research Techniques. Windsor: IEEE Computer Society, 2007: 89–98.

Abreu R, Zoeteweij P, Gemund AV. Spectrum-based multiple fault localization[C].Proceedings of the IEEE/ACM International Conference on Automated Software Engineering. Auckland: IEEE Computer Society, 2009:88–99.

Bandyopadhyay A. Improving spectrum-based fault localization using proximity-based weighting of test cases [C]. In Proc. of the 26th IEEE on Automated Software Engineering. Lawrence: IEEE Computer Society, 2011:660–664.

Jones JA, Harrold MJ. Empirical evaluation of the tarantula automatic fault localization technique [C]. Proceedings of the 20th IEEE/ACM International Conference on Automated Software Engineering. Long Beach: IEEE Computer Society, 2005:273–282.

Liu Yong-po, Wu Ji. Experiment study of BBN-Based fault localization. Journal of Computer Research and Development. 2010, 47(4):707–715.

Wen Wan-zhi, Chen Xiang, Sun Xiao-bing. Software fault localization based on program spectrum lattice (J), Journal of Sichuan University (Engineering science edition), 2014, 46(2):87–94.

Wen Wan-zhi, Li Bi-xin, Sun Xiaobing. Technique of Software fault localization based on hierarchical slicing spectrum (J), Journal of software, 2013, 24(5):977–992.

Wen Wan-hi, Li Bi-xin, Sun Xiao-bing. A technique of multiple fault localization based on conditional execution slicing spectrum (J), Journal of Computer Research and development, 2013, 50(5):1030–1043.

Energy Science and Applied Technology – Fang (Ed.)
© 2016 Taylor & Francis Group, London, ISBN 978-1-138-02833-3

A novel configuration system for FPGA test based on algorithm and ping-pong operation

Xianjian Zheng, Zhiping Wen, Fan Zhang, Lei Chen & Changlei Feng
Beijing Microelectronics Technology Institute, Beijing, China

ABSTRACT: Recently, partial reconfiguration technology and external configuration circuit have been proposed to reduce the configuration time in the FPGA manufacturing test. However, there is still room for improvement to reduce the configuration time. This paper proposed a novel configuration system to reduce the total configuration time based on the algorithm and ping-pong operation. This system is implemented on a series of SRAM-based FPGA. The result of the experiment shows that the system proposed to reduce at least 56% of the total configuration time. Thus, this system is effective in reducing the total configuration time in the FPGA manufacturing test, and can be generally applied to all current SRAM-based FPGA.

Keywords: FPGA manufacturing test; algorithm; and ping-pong operation

1   INTRODUCTION

FPGA test is important and its challenge is the increasingly high cost. In contrast to the ASIC test, the FPGA test usually needs hundreds of configurations to achieve a reasonable fault coverage. FPGA is usually tested by an automatic test equipment (ATE) in the process of manufacture. After FPGA under test is configured and energized, ATE analyses the response of FPGA and reports the result. However, the memory of ATE cannot meet the demand of saving a large number of configuration files (called bitfile) when ATE tests an FPGA. In addition, using the memory card provided by equipment manufacturers would be a huge cost. Thus, an external circuit for configuration was proposed to meet the demand of saving bitfiles and reducing the cost. However, the configuration speed of the external circuit is usually slower than ATE's.

Conversely, due to cost considerations, the test time of FPGA should be as short as possible. The FPGA test requires many iterations: it programs the device with a test configuration and applies test stimuli and observes the response. However, the time spent on programming the device is much larger than the time required to apply the test stimuli. Thus, people endeavour to reduce the number of configurations and the time for a single configuration to minimize the test costs. It is a challenge to complete a large number of bitfiles in a short time for the FPGA volume production test. The FPGA test requires many iterations: it programs the device with a test configuration and applies test stimuli and observes

the response. However, the time spent on programming the device is much longer than the time required to apply the test stimuli (Tahoori, M. B, 2003). Thus, most people endeavour to reduce the number of configurations to minimize the test time, as reported previously (Dixon, B). The methods mentioned above have to sacrifice some of the fault coverage, and few people consider the issue from the perspective of reducing the single configuration time.

In the case without reducing the fault coverage and for meeting the demand of saving a large number of bitfiles, this paper proposed a novel configuration system based on software and ping-pong operation to reduce the configuration time.

The rest of this paper is organized as follows. Section II introduces the ping-pong operation and some techniques for reducing the configuration time in the FPGA manufacturing test. Section III gives the specific algorithm used in the software section and the details of a lower machine that achieves the ping-pong operation. The monolithic construction of the novel configuration system is shown in Section IV. The experimental results are also discussed in Section IV. Finally, conclusions are drawn in Section V.

2   BACKGROUND

Ping-pong operation is a technique used in controlling the data flow. The typical ping-pong operation is shown in Figure 1.

The input data flow is distributed to two different data buffers by the multiplexer of input data

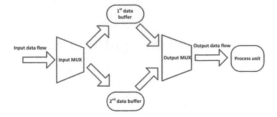

Figure 1. The typical ping-pong operation.

flow. The data buffers usually consist of DPRAM or FIFO. The input data flow is saved in the 1st data buffer during the first cycle. In the second cycle, the input data flow is saved in the 2nd data buffer by controlling the multiplexer of input data flow, and the data saved in the 1st data buffer is sent to the data processing unit by controlling the multiplexer of output data flow. In the third cycle, the data saved in the 2nd data buffer is sent to the data processing unit, and at the same time, the input data flow is saved in the 1st data buffer. By repeating these cycles, the input data flow is processed continuously and most efficiently. Thus, a low-speed processing unit is able to process high-speed data flow.

SRAM-based FPGA architectures have configuration memory arranged in frames that are tiled about the device (Xilinx, 2009). These frames are the smallest addressable segments of the device configuration memory space, and all operations must therefore act upon whole configuration frames. The length of the frame is determined by the size of the device. Partial reconfiguration (PR) technology allows the modification of an operating FPGA design by loading a partial reconfiguration file (Xilinx, 2011). After a full configuration file configures the FPGA, partial reconfiguration files can be downloaded to modify reconfigurable regions in FPGA without compromising the integrity of the applications running on those parts of the device that are not being reconfigured. In fact, the minimum unit could be reconfigured by frame-based partial reconfiguration technology. In Abramovici, M, 1993, partial reconfiguration technology was used to detect and diagnose the fault, and in Xilinx, 2000, it was used to construct BIST. In fact, partial reconfiguration technology was applied to reduce the configuration time in the FPGA manufacturing test in Wang L.

3 THE DETAIL OF THE PROPOSED SYSTEM

In this section, the specific algorithm used in the software is proposed. In addition, the achievement of the lower machine, which is based on the ping-pong operation, is described in detail.

3.1 Optimized test partial reconfiguration based on the algorithm

Here, the most extreme case is considered: there is only one same frame between two configuration files. In this case, a command header is substituted for that same frame. Usually, the length of the header is much shorter than a frame. It means that the partial reconfiguration technology can be used to reduce the configuration time on any situation.

However, it is not the most optimized way to configure by simply using the partial reconfiguration technology. Reasonably arranging the order of bitfiles can further decrease the size of total configuration files, as the similarity between configuration files is different. An algorithm is proposed to sort bitfiles to ensure that any two sequential configuration files have a minimum discrepancy in this paper.

The algorithm for sorting is shown in Algorithm 1.

The smaller the discrepancy ratio between two sequential bitfiles is, the smaller the reconfiguration file needed. Consequently, the sequence of total bitfiles impacts the size of total bitfiles using partial reconfiguration technology. The formula for calculating the discrepancy ratio between two sequential bitfiles (labeled A and B) is as follows:

$$Discrepancy\,(A,\,B) = \frac{number\ of\ different\ frames\ between\ A\ and\ \boldsymbol{B}}{total\ number\ of\ frames\ in\ a\ bitfile}$$

(1)

The formula for calculating the total discrepancy ratio of a list is given in formula (3);

Algorithms 1. Algorithms for sorting bitfiles.

Create a new list named Source-List which includes
  all of the bitfiles
Create a new empty list named Result-List
Repeat
Pick the first bitfile (labeled A) from Source-List
Compare A with all the bitfiles in Result-List
Find the bitfile (labeled B) which is the most similar
  to A in Result-List
Suppose to place A behind B
Count total discrepancy ratio (sign as x)
  in Result-List
Suppose to place A in front of B
Count total discrepancy ratio (sign as y)
  in Result-List
If x is greater than y
Place A in front of B
Else place A behind B
Remove A from Result-List
Until Result-List is empty

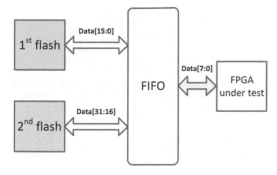

Figure 2. Connected relationship between the flash memory array and the FIFO.

$$List = \{List[0],\ List[1],\ List[2],\ \dots,$$
$$List[N-1],\ List[N]\} \qquad (2)$$

$$Discrepancy\ (List) =$$
$$\sum_{i=1}^{N} Discrepancy(List[i-1],\ List[i]) \qquad (3)$$

3.2 Lower machine based on the ping-pong operation

It needs hundreds of bitfiles to complete the test in FPGA manufacturing. These bitfiles require a large amount of storage capacity. Thus, the configuration circuit with high storage capacity was proposed to configure the FPGA under test. However, the reading speed of large-capacity memory such as flash is about 30MB/s, and the highest configuration speed of FPGA under test is 50MB/s. It is obviously that the reading speed of flash is much lower than the configuration speed of FPGA. A dual-port FIFO is used as the data buffer to operate two flash circuits, with the reading speed of 30MB/s in the configuration system proposed in this paper. Thus, the highest reading speed of the flash memory array is 60MB/s, which could meet the demand of the configuration of FPGA. The schematic is shown in Figure 2.

The master FPGA transforms 8-bits data into 32-bits data, and then the FIFO saves the 16 low-order bits data into the 1st flash and the 16 high-order bits data into the 2nd flash. This design equally splits the bitfiles automatically when the bitfiles are written in the flash memory array. When the FPGA under test needs to be configured, the FIFO would read 32 bits data from the flash memory array during one clock period and output the data in the 8-bits form. To ensure an efficient and continuous configuration, the clock frequency of the reading flash should be greater than 84% of the clock frequency of configuration.

4 IMPLEMENTATION AND RESULT

The system is implemented on a series of FPGA. The software of PC screens and sorts the configuration files, and generates reconfiguration files. The lower machine based on the ping-pong operation and flash memory is designed to save the reconfiguration files and configure the FPGA under test.

4.1 Software implementation

According to Algorithm 1, a list that includes bitfiles is exported. This list is processed to export the optimized reconfiguration files according to the flow-process diagram, as shown in Figure 3.

As shown in Figure 3, a series of optimized reconfiguration files are exported. These reconfiguration files must be configured according to the order arranged by Algorithm 1 and the flow-process diagram in the test, as shown in Figure 3.

The software based on Visual Studio is proposed to process the configuration files according to Algorithm 1 and the flow-process diagram, as shown in Figure 3.

4.2 Hardware implementation

The structure of the configuration system proposed in this paper is shown in Figure 4.

The configuration system includes the programming mode and the test mode. Partial reconfiguration files are programed to the flash memory array with the software of PC in order when the configuration system is in the programming mode. The system would identify the handshake signal from ATE and configure the FPGA under test when the system connects to ATE. The specific process is shown in Figure 5.

Figure 6 is the actual picture of the lower machine. The lower machine is controlled by a

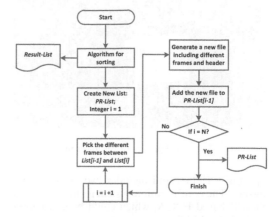

Figure 3. Flow-process diagram for exporting PR files.

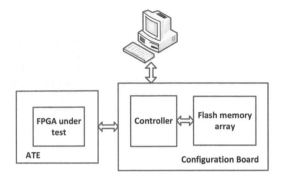

Figure 4. The structure of the configuration system.

Figure 5. Process of the configuration system.

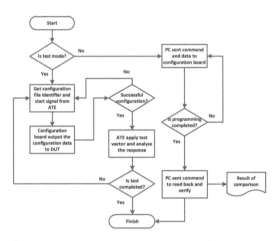

Figure 6. The actual picture of the lower machine.

Spartan-6 FPGA and includes two 2Gbit NOR-flashes. The configuration system is suitable for all SRAM-based FPGA, which could be configured in the SelectMAP mode.

Table 1. The result of the experiment.

Device	Size of bitfile (Mb)	Total size of bitfiles for test (Mb)		Total configuration time (s)	
		Normal system	Proposed system	Normal system	Proposed system
XC2V1000	3.9	612.3	428	2.55	1.07
XC3P5000	12.7	2311.4	1664	9.63	4.16
XC5VLX85	20.8	4784	3205	19.93	8.01

4.3 *Result*

This system was implemented on a series of SRAM-based FPGA with an analogy circuit that simulates ATE. The result of the experiment is shown in Table 1.

The result of the experiment shows that the configuration system proposed in this paper reduces at least 56% of the total configuration time.

5 CONCLUSION

This paper proposed a novel configuration system to reduce the total configuration time based on the algorithm and the ping-pong operation. This system was implemented on a series of FPGA device. The result of the experiment shows that this system reduces at least 56% of the total configuration time. It proves that this system is effective in reducing the total configuration time in the FPGA manufacturing test, and this system can be generally applied to all current SRAM-based FPGA.

REFERENCES

Abramovici, M. BIST-based Test and Diagnosis of FPGA Logic Blocks, IEEE Trans. Very Large Scale Integration Systems, Vol. 9, No. 1, 1993, 73–82.

Dixon, B. Built-In Self-Test of the Programmable Interconnect in Field Programmable Gate Arrays. Auburn: Auburn University.

Tahoori, M.B. Automatic Configuration Generation for FPGA Interconnect Testing. In Proceedings of the 21st IEEE VLSI Test Symposium, 2003.

Xilinx, Inc. Virtex-4 FPGA Configuration User Guide, UG071 (v1.1) (June 9, 2009).

Xilinx, Inc. Partial Reconfiguration User Guide, UG702 (v13.3) (October 19, 2011).

Xilinx, Inc. Correcting Single-Event Upsets through Virtex Partial Configuration, XAPP216 (v1.0) (June 1, 2000).

Wang, L. & Stroud, C. System on Chip Test Architectures. Amsterdam: Elsevier.

Energy Science and Applied Technology – Fang (Ed.)
© 2016 Taylor & Francis Group, London, ISBN 978-1-138-02833-3

Hadoop real-time monitoring system based on Ganglia, Nagios and MongoDB

Qiao Zhu & Li Miao
Department of Computer Science and Engineering, Hunan University, Hunan, China

ABSTRACT: Currently, Hadoop is widely used in a growing number of network applications. Based on the advantages of its open source framework for distributed systems, it implements Hadoop distributed file system and Map Re-duce distributed computing framework. Therefore, how to conduct performance monitoring nodes in the cluster to ensure the normal and efficient operation of the whole cluster is one of the key studies on Hadoop. Although there are a large number of third-party monitoring tools to achieve monitoring for Hadoop cluster, limitations of monitoring indicators lead to inadequate monitoring granularity. This article is a supplement of Ganglia, which is a real-time monitoring system. Using JMX interface which Hadoop supports to get more monitoring indicators, we can achieve full performance monitoring of Hadoop cluster when running tasks. Meanwhile, it makes use of the advantages of Nagios and MongoDB, and resolves the problem that Ganglia cannot store data permanently.

Keywords: Hadoop; Ganglia; Nagios; MongoDB; real-time monitoring

1   INTRODUCTION

Today's society has already entered the era of big data. Whether it is the businesses, government agencies or research institutions, the amount of data that they need to deal with is growing daily. Along with its development is the emergence of the Map Reduce framework, which can be distributed to a large data efficiently and run on thousands of nodes in order to solve the performance problems of data processing effectively. As one of its open source implementation, Hadoop Map Reduce has been used in the actual work by companies and organizations such as Facebook and Yahoo.

With the extensive application of the Map Reduce framework, the amount of data to be processed is increasing rapidly and the number of nodes in the cluster becomes more extensive. Today, many companies are deployed Map Reduce nodes and the number is more than one thousand, even thousands. Faced with so many nodes, how to manage Hadoop cluster becomes a rather tricky problem, especially when performance problems occur. Therefore, it is a significant job to build a Hadoop cluster monitoring system to monitor their operation condition and performance in time to ensure the normal operation of the cluster system.

There are three main problems that need to be solved by the Hadoop real-time monitoring system.

The first one is about collecting task indicator data: how to approach Name Node and Data Node performance indicators from Hadoop cluster, and get Map Reduce tasks execution state from Job Tracker and Task Tracker, including start time, running time, schedule strategy and the number of times the task is killed. The second one is the collection of system indicator data: system performance indicators from the cluster, such as memory usage, JVM heap usage, cpuIdel, network traffic and IO speed. The third is the visual analysis, which is about how to display these data in the graphical form in order to provide a good UI interface to users who can make appropriate and timely treatment when problems occur.

This paper starts from these three aspects, and completes a real-time monitoring system for Hadoop. It can accomplish a full range of indicator data collection and presentation during the task. It is based on Ganglia, which is a distributing monitoring system under the high-performance computing system. At the same time, we take advantage of Nagios, which is an open source monitoring system working on Linux/Unix platforms as another functional module of Hadoop real-time monitoring system. In addition to keep long-term storage of monitoring data in order to generate reports and optimization decision of log analysis, we integrate MongoDB as the database of the Hadoop real-time monitoring system.

2 INTRODUCTION OF HADOOP

Hadoop is an Apache Software Foundation open source and Java-based implementation of the Map/Reduce framework, which is a computing framework that the users can make full use of clusters to carry out high-speed progressing and storage without knowing about the details of the distribution. With HDFS (Hadoop Distributed File System) and Map Reduce (open source of Google Map Reduce) as its core, Hadoop aims at constructing a distributing system with good reliability and expansibility. Hadoop can run application on a cluster composed of lots of cheap hardware devices with a group of steady interface.

As an open source implementation of Google Map Reduce framework, Map Reduce is able to calculate a huge amount of work down into roughly equal work and data block corresponding to it, and then distributes among the nodes in the cluster for parallel processing. There are two characters in Hadoop for carrying out Map Reduce task: Job Tracker and Task Tracker. There is only one Job Tracker in each Hadoop cluster, and it is a manager to schedule work that runs on the main code. Task Tracker is used to perform the work running on codes of every cluster. They communicate with each other by heartbeat. Each Map Reduce task is initialized to a job in Hadoop. The Map Reduce framework introduces a way of processing data that consists of two main phases: Map phase and Reduce phase, which are represented by two functions, namely the Map function and the Reduce function. Each function that the user defined is implemented by the mapper and the reducer.

The Map phase: during the map phase, each mapper reads the input record by record, converts it into a key/value pair and propagates the pair to the user-defined map function. The latter transforms the input pair (based on the user's implementation) and outputs a new key/value pair. The map function's output pairs are further propagated to subsequent phases where they are partitioned/grouped and sorted by their keys. Grouping is achieved using a partitioner that partitions the key space by applying a hash function over the keys. The total number of partitions is the same as the number of reduce tasks (reducers) for the job. Finally, these intermediate pairs are stored locally on the mappers.

The Reduce phase: during the reduce phase, the Data Nodes that implement the reducers retrieve the intermediate data from the mappers. Specifically, each reducer fetches a corresponding partition from each mapper node in a fashion that all values with the same key are passed to the same reducer. The fetched map's output pairs are merged constructing pairs of <key; list (values)> based on the same key. The newly structured pairs of key/list are propagated to the user-defined reduce function. The latter receives an iterator over each list, combines these values together and returns a new key/value pair. The final output pairs are written to output files and stored in the HDFS. The whole Map/Reduce procedure is depicted in Figure 1.

HDFS is based on the GFS basic framework developed by Google Inc., which is a distributed file system with a high fault tolerance, suitable for deployment on a cheap machine. In contrast, it provides a high-throughput data access, for the reason that it is ideal for applications with large data sets. HDFS uses master/slave structure model and a cluster of it is made up of a Name Node and several Data Nodes. Figure 2 shows the system of HDFS.

Name Node is responsible for managing the HDFS directory tree and files related to metadata information, including namespace of file and block, correspondence between files and blocks, and locations of each copy. The information is stored in the local disk in the form of two documents: editlog and fsimage, which will be reconstructed when HDFS starts. Still, it is responsible for monitoring the health status of each Data Node. Once a Data Node is shut down, it will remove the Data Node from HDFS and backup the data again. Data Node, which accomplishes the actual data storage, is installed on each node, while the files are divided into several data blocks and stored in a group of Data Nodes. What's more, Data Node also creates, deletes and copies data blocks in the control of Name Node. It organizes file contents in fixed-size (default size 64M) blocks as a basic unit. When uploading a huge file to HDFS, it will be divided into several blocks and stored in different

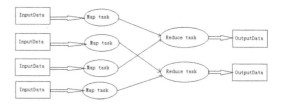

Figure 1.   Execution overview.

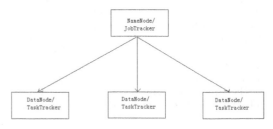

Figure 2.   HDFS architecture.

nodes. In order to ensure data reliability, the same block will be written into several different Data Nodes in pipelined manner. In addition, HDFS has its specific allocation principle: the default number of copies is three, one copy is saved on a machine in the same rack, the second one is saved on a machine in a different rack, and the third one is saved on a machine that is in the same rack of the second. In each cluster, there is a Secondary Name Node, the most important task of which is not to be a hot metadata backup of Name Node, but to merge the fsimage and editlog logs regularly and transfer to Name Node in order to reduce the pressure.

3 DESIGN OF HADOOP REAL-TIME MONITORING SYSTEM

3.1 *Java Management Extensions*

JMX (Java Management Extensions) is a framework for managing functional implants for applications, devices and systems, which can span a range of heterogeneous operating system platforms, system architectures and network transport protocols in order to develop a seamless integrated system, network and service management applications deftly.

The bottom-up of the JMX architecture's three levels is the instrumentation level, agent level and distributed service level. Instrumentation level defines how to manage resources to achieve JMX specification. The resources can be a Java application, a service or a device that can use Java to develop, or at least use Java for packaging, and can be placed in the JMX framework, thus becoming a JMX management component MBean. Agent level is a management entity run on the JVM, which is active in between management resources and the manager, used to directly manage resources, and these resources can be controlled by a remote management program. Moreover, it is composed of an MBean server and a series of servers that process the management resources. All management members are required to register to the MBean Server, which is the core part of MBean before being managed. Distributed service level mainly provides the application management platform to achieve JMX interface, defines the management interfaces and components that can operate on the agent layer so that the manager can operate agents.

3.2 *Ganglia*

Ganglia is an open source cluster monitoring project launched by UC Berkeley, including gmond, gmetad and a web end. Ganglia is used to monitor the performance of the system, such as cpu, mem, hard disk utilization, I/O load and network traffic situation. Through the curve, it is easy to figure out the working state of each node, which plays an important role in reasonable adjustment, allocation of system resources and improving the overall system performance.

There is a daemon process gmond running on each node to collect and send measurement data. It works just like an agent in the traditional monitoring system, which is responsible for interaction with the operating system to obtain indicator data, such as processor speed and memory usage. Gmetad can be deployed on any node in the cluster or an independent host that can connect to the cluster through a network. It is a simple poller that polls each node in the cluster to collect indicator data and writes into RRD. RRD Tool deals with them and displays them in the graphical way in a web for managers to view.

3.3 *Nagios*

Nagios is a popular open source network monitoring tool, which can effectively monitor Windows, Linux and UNIX host status, printers and routers, switches and other network settings. Once the system or service status becomes abnormal, it will send out an e-mail or SMS alarm notification to site operation and maintenance personnel at the first time, and issue a regular notification about the status of the recovery.

After starting Nagios up, it will automatically call the plug periodically to detect the server status and maintain a queue for keeping all plug-in returns to the status information. Nagios begins to read the information from the beginning of the queue each time, and displays the results of states by the web after processing. Nagios can tell four different status return information: 0(OK) indicates the status of normal/green; 1(WARNING) indicates a warning/yellow; 2(CRITICAL) indicates a serious mistake/red; 3 (UNKNOWN) indicates an unknown error/ deep yellow. Users can judge the state of the monitored object by the return information of plug-in, and notice abnormal in time once the problem is found.

3.4 *MongoDB*

NoSQL Database is produced in order to solve the challenges of large-scale data collection and multiple data types, especially large data application problems. In addition, document database can be seen as an upgraded version of the key database and can be allowed to nest the key between, whose query efficiency is higher than the key database.

Here, MongoDB is the most typical one, whose document model is free and flexible. Besides, the built-in horizontal expansion mechanism provides data from 1 million to 1 billion level processing capabilities; it not only fully meets the Web2.0 and mobile Internet data storage needs, but also greatly reduces the cost of operation and maintenance for small and medium websites.

3.5 *Hadoop real-time monitoring model*

Hadoop real-time monitoring system is based on the distributing monitoring system Ganglia. After the daemons including Name Node, Secondary Name Node, Data Node, Job Tracker and Task Tracker in Hadoop start, corresponding metric will be registered to the local MBean Server in order to become resources, which can be monitored by JConsole. JMX takes the initiative to write the relevant data in the container during running tasks. In addition, gmetad of Ganglia will also pull JMX monitoring data into RRD and display them in the two-dimensional graphical way on the main node. Therefore, in this system, we expand the indicators Ganglia monitors from system indicators to task indicators of Hadoop cluster, such as performance indicator of Name Node, HDFS information of Data Node, performance data of JobTacker and Task Tracker. RRDtool, a database tool used in Ganglia, except for storage data, is still a powerful graphics engine that is able to complete the graphical display of data.

Meanwhile, we integrate Nagios as the alert alarm module of Hadoop real-time monitoring system in order to monitor all the nodes and services and provide exception notification in case of abnormal situations.

We use MongoDB as our database for two reasons. One is the RRD Tool as the default database of Ganglia, which has a huge loophole except for obvious advantages in monitoring data. Its database RRD is a ring one and the size of the file is fixed so that the way to store new data is added behind the existing data. Once to the end of the file, the new data will cover the former, which results in losing the overdue data. In order to keep long-term data, we need the non-coverage database to accomplish the job.

The other reason is that developers use Web Server as Hadoop in-built Jetty, after Name Node starts, it will call class Http Server to start Jetty up, in which there is a JMX Json Servlet used to capture Name Node monitoring data by Rest as Json format. MongoDB's storage format is Bson, and it is one of the extensions of Json. So far, MongoDB changes from the candidate to the best choice of Hadoop real-time monitoring system. Figure 3 shows the architecture of the system.

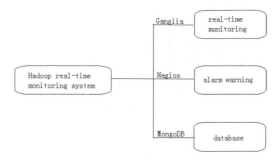

Figure 3. Hadoop real-time monitoring system.

4 IMPLEMENTATION

After the daemons in Hadoop cluster start, including Name Node, Secondary Name Node, Data Node, Job Tracker and Task Tracker, corresponding metric will be registered to the local MBean Server. The configuration of interface to access its operational indicator is in hadoop-env.sh, such as jmxremote port. It should be noted that each configuration corresponds to a Java process, each port must be different and can only be bound to one process. So, we can obtain performance indices of each daemon in runtime by JMX.

4.1 *Ganglia and Nagios integration*

Ganglia installation and configuration: Ganglia installation includes the installation of the control node and each monitoring node. Each monitoring node needs to install ganglia-monitor to collect local information. Control node needs to install gmetad to collect data and RRD Tool to store the data and display operation status in the graphical way. Control node needs to modify the configuration file /etc/ganglia/gmond. conf to name a cluster name, which is consistent with that of the data source of /etc/ganglia/gmetad. conf. Still, we need to modify /etc/ganglia to change the configuration of the cluster.

Nagios installation and configuration: It includes the installation of both Nagios and Nagios-Plugin. It is necessary to install the web side and set an authentication code for Nagios. After modifying the configuration file {nagios_home}/etc/ to configure each file, it is vital to check each configuration file against nagios. cfg to ensure the accuracy.

Deploy Hadoop cluster: Modify the configuration file {hadoop_home}/conf/hadoop_env.sh, so that we can approach performance indicators of each daemon. Under normal circumstances, we can use 8004–8009 as default ports to monitor Name Node, Secondary Name Node, Data Node, Blancer, Job Tracker and Task Tracker successively. Any unused port can be used according to the need of users. We modify the configuration file /etc./hadoop/hadoop-metrics2.properties to configure Ganglia. As we known, 239.2.11.71 is the broadcast address of Ganglia. In our experiment, we change it to the address of gmetad because of the unicast mode. Remote terminal can get monitoring data by linking to this port, which is monitored by TCP and the default is 8679. In addition, through the port, gmetad collects the XML data, and then sends the configured files to each node.

We restart the whole hadoop cluster to refresh the configuration. We start daemon ganglia-monitor, gmetad and nagios daemons.

Users can access http://namenode/ganglia to see the home of Ganglia monitoring platform on the master node, as shown in Figure 4. We can observe the state of load, CPU operating status, memory usage of cluster, network traffic monitoring and Hadoop cluster running index by Ganglia. In contrast, after logging in the monitoring page of Nagios Web http://namenode/nagios/, we can check the related information about the whole Hadoop cluster, as shown in Figure 5.

4.2 Data storage of monitoring system

Data obtained by Ganglia is stored in the RRD file format and each file of each machine generates a RRD file, which is not suitable for long-term preser-

```
{
  "beans" : [ {
    "name" : "java.lang:type=MemoryPool,name=PS Eden Space",
    "modelerType" : "sun.management.MemoryPoolImpl",
    "Name" : "PS Eden Space",
    "Type" : "HEAP",
    "Valid" : true,
    "CollectionUsage" : {
      "committed" : 16449536,
      "init" : 16449536,
      "max" : 344129536,
      "used" : 0
    },
    "CollectionUsageThreshold" : 0,
    "CollectionUsageThresholdCount" : 0,
    "MemoryManagerNames" : [ "PS MarkSweep", "PS Scavenge" ],
    "PeakUsage" : {
      "committed" : 16449536,
      "init" : 16449536,
      "max" : 344129536,
      "used" : 16449536
    },
    "Usage" : {
      "committed" : 16449536,
      "init" : 16449536,
      "max" : 344129536,
      "used" : 5627664
    },
    "CollectionUsageThresholdExceeded" : true,
    "CollectionUsageThresholdSupported" : true,
    "UsageThresholdSupported" : false
  }, {
    "name" : "java.lang:type=Memory",
    "modelerType" : "sun.management.MemoryImpl",
    "Verbose" : false,
    "HeapMemoryUsage" : {
      "committed" : 62914560,
      "init" : 65584832,
      "max" : 932118528,
      "used" : 7397800
    },
    "NonHeapMemoryUsage" : {
      "committed" : 19136512,
      "init" : 19136512,
      "max" : 117440512,
      "used" : 14420304
    },
    "ObjectPendingFinalizationCount" : 0
  }, {
    "name" : "Hadoop:service=TaskTracker,name=MetricsSystem,sub=Control",
    "modelerType" : "org.apache.hadoop.metrics2.impl.MetricsSystemImpl"
```

Figure 6. Json data from JMX.

vation of data. However, some monitor information of our system requires long-term storage for log analysis to predict the formation of decision-making and reporting. In order to acquire a long-term storage of data, we use MongoDB instead of RRD as the database of the Hadoop real-time monitoring system.

Later than Hadoop 1.0 version, it treats own monitoring interface as an invisible property, which is JMX. Through this, we can obtain a more detailed cluster monitoring. By accessing specific URL, users can monitor the Web page data, where monitoring data are presented in Json format. As shown in Figure 6, we can obtain part of the data about Name Node in Json by accessing the address http://namenode/jmx?get = hadoop:service = NameNode.

We use the web crawler Web Magic, to which the URL is passed as a parameter. According to the URL, it gets Json data from the corresponding page and delivers to MongoDB to complete the collection of monitoring data.

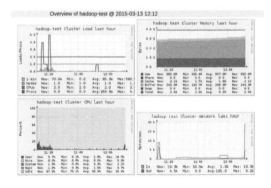

Figure 4. Web interface of Ganglia.

Figure 5. Web interface of Nagios.

5 CONCLUSION

Hadoop is increasingly used to process big data, so the real-time performance monitoring of Hadoop cluster is of great importance. We integrate these two distributed systems Ganglia and Nagios, which can not only monitor the hardware resources and other operational indicator of Hadoop, but also inform the managers in case of emergency, so that managers can respond in time. MongoDB is integrated as the database of this system in order to analyze expired data for reporting and optimization decision.

REFERENCES

Cai, Liuqing. 2011. Design and Implementation of Cloud Monitoring with MongoDB [D]. Beijing: Beijing Jiaotong University.

Dean, J & Ghemawat, S. 2004. Map Reduce: Simplified data processing on large clusters[C]. In USENIX Symposium on Operating Systems Design and Implementation: 137–150.

Ganglia monitoring Hadoop cluster. 2012. http://www.congci.comlitemlganglia-jiankong-hadoop-jiqun.

Pavlo, A & Paulson, E. 2009. A comparison of approaches to large-scale data analysis, Proceedings of the 2009 ACM SIGMOD International Conference on Management of data: 165–178.

White, Tom. 2010. Hadoop: The Definitive Guide [M].

Energy Science and Applied Technology – Fang (Ed.)
© 2016 Taylor & Francis Group, London, ISBN 978-1-138-02833-3

Research on on-line monitoring and fault diagnosis of wind turbine

Q.H. Zhu & T.Y. Liu
Shanghai Dianji University, Shanghai, China

S.G. Zhu
Shandong Agriculture and Engineering University, Shandong, China

ABSTRACT: With the continuous development of society and the increase of human demand for energy, wind energy as a kind of clean renewable energy, receives the increasing attention from all over the world. As the industry of wind power has developed rapidly in recent years, the global capacity is enlarged constantly, the structure of wind power generation is becoming more and more complicated, at the same time, the accident of wind turbines also occur frequently. Therefore, it is necessary to carry out the research on the wind turbines in the real-time online monitoring, found the failure of the unit earlier and maintenance in time, this can improve the reliability of wind turbines, which is of great significance to the study of the condition monitoring of wind turbine and maintenance costs and other issues.

Keywords: wind energy; the global installed capacity; wind turbine; on-line monitoring system; fault diagnosis technology

1 INTRODUCTION

As the problems of energy crisis and environmental pollution are becoming increasingly serious, wind energy which belongs to the clean and renewable energy resource has been a high degree of concern and attention of the countries all over the world, and it has become an important part of national strategy of sustainable development. In 2013, the global installed capacity of wind power reached 325.75GW, growing 12.5% compared to the same period. The rapid development of wind power in China continues to play a leading role in the wind power industry, and now China is a major national leading wind power development in Asia. In 2013, the new installed capacity of wind power in China is 16.01GW; compared with 2012 it grows 3.12GW. At the same time, China's offshore wind power project construction has made break through progress, planning construct 5GW offshore wind power by the year of 2020 and 30GW by 2015 (GWEC, 2014).

With the rapid development of wind power, the high costs of the operating maintenance for the wind turbines will affect the economic benefits of wind farms. Due to geographical conditions and the limitation of wind energy resources, the wind farms generally distributes in the place which is far from the city and the wind power company; the cabin is usually located on the altitude over 50~80 m, and this increases the difficulty of operating maintenance. The expensive cost of the operating maintenance will reduce the economic benefits of wind power and increase the operation cost. Therefore, it's necessary to carry on the real-time online monitoring, to find the failure earlier and maintenance in time. This can improve the reliability of wind turbines, to ensure the normal safe operation of the wind turbine.

2 STRUCTURE OF WIND TURBINE

Wind turbine, which includes both wind turbines and generator, is a device for converting wind energy into electrical energy, there are two energy conversion process in fact, namely wind energy, mechanical energy, electricity, its working principle is shown in Fig. 1.

Wind turbine mainly consists of the wind wheels, transmission system, device (yaw system of wind), hydraulic system, braking system, control and security system, engine room, tower and other components. The wind wheel will rotate by wind turbine through wind, and then through the transmission system of growth to achieve the speed of the generator and then drives generators to generate power, effectively converting wind energy into electrical energy. The structure of the wind turbine is shown in Fig. 2.

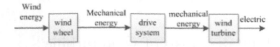

Figure 1. The working principle of wind power.

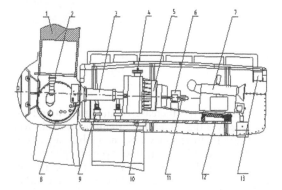

Figure 2. The structure of wind power generation.

(1. blade 2. electric pitch system bearings 3. principal axis 4. cabin 5. gearbox 6. the high speed shaft brake 7. generator 8. axial flow fan 9. mount 10. slip ring 11. yaw bearing 12. yawing driven 13. hub)

Wind turbine is the device which can capture of wind energy, and convert wind energy into mechanical energy, including leaf blade, hub, and variable propeller system. Wind drives the impeller to rotate, implements the transformation between wind energy and mechanical energy; Wheel hub will convey the mechanical energy to drive system; pitch system is used to adjust the pitch Angle of blade, so that the wind wheel has a constant speed and max output power. The drive system will transfer mechanical energy which is produced by the wind wheel to generator, generally by the high-speed shaft, low speed shaft, gear box, coupling, and mechanical brake institutions. Gear box is used to increase the impeller rotating speed so as to drive the generator; coupling is a kind of soft structure connecting the gear box to the generator; the mechanical brake mechanism is composed of a fixed to the low speed shaft or the brake disc and arranged on the high speed shaft of the hydraulic clamp formed around the yaw system, which makes the wind wheel always in the upwind state in order to improve the efficiency(Song, H.H, 2009). Wind turbine will transport rotating mechanical energy into electrical energy, its main type includes cage asynchronous generator, brushless doubly-fed generator, AC excitation generator, synchronous generator etc.

Wind technology has been developed from fixed pitch wind turbine to variable pitch wind turbine; the wind turbine has been developed from the constant speed control to the variable speed constant frequency control; the variable speed constant frequency power system have appeared in the variable speed constant frequency doubly-fed induction machine electrical system, synchronous motor variable speed constant frequency electric system and permanent magnet synchronous motor variable speed constant frequency system, etc. (Guo, X. 2013) Doubly-fed induction generator is mostly used presently.

3 ON-LINE MONITORING SYSTEM OF WIND TURBINE

The on-line monitoring and diagnosis system of wind turbine is the collection of signal acquisition, on-line monitoring and signal analysis, which integrates multi-functional on-line monitoring and diagnosis analysis system the on-line monitoring of the wind turbine of the vibration, the temperature, the pressure and electric parameters, comparing the result of the on-line monitoring with the set and diagnosis system can timely found abnormal running and alarm, can undertake all kinds of analysis of the collected data processing, which can accurately determine the equipment failure. The most commonly used is the SCADA system for real-time monitoring of wind turbines.

SCADA system is the system of data acquisition and monitoring control, production process control and dispatching automation system based on computer, can be used to on-site equipment monitoring and control, achieve the function of data acquisition, equipment control, measure and parameters adjustment and all kinds of signal alarm and other functions. The structure diagram of megawatt wind power SCADA system is shown in Fig. 3. Firstly, the annular communication link is made up by communication cable or optical cable between fans, take one or more units are connected as a host and data switches the fan, you can also connect directly to the monitoring center. When receiving the data, the monitoring center would analysis and processing to complete the whole process of communication.

Usually, the hardware structure of the monitoring system is mainly composed of the signal detection part, data acquisition part, site server and center server. Signal detection module includes the sensor (vibration sensor is the most commonly used), the signal transmitter, the signal preprocessing plate. The signal transmitter will make the signals which is collected by the sensor do amplifying and filtering preliminarily, in order to make its have anti-interference and stability, and then translate the signal to the signal preprocessing

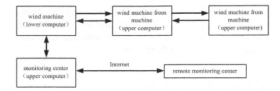

Figure 3. Structure diagram of the SCADA system.

board to complete the function of amplifying, filtering, transferring and transforming on the signal. Data acquisition is the equipment which is located in the host computer that can translate the vibration, temperature, pressure and other analog conversion to digital quantity in order to facilitate the follow-up data processing device. The site server is the communications equipment which is connected to the acquisition instrument installed in wind turbine cabin, mainly for data recording, processing, storage and management, sending data to the diagnosis of central server through Internet based on the Internet. The center server is used to receive and store the frontend server sends over the various data and parameters.

4 THE FAILURE DIAGNOSIS OF WIND TURBINE

The fault diagnosis technology is that monitoring the wind turbines in time, determining the unit as a whole or partial is normal or not, and can find fault and its' reason in the early; forecasting the development trend of fault. There is a lot of diagnosing methods for fault diagnosis, such as vibration testing, oil analyzing, technologies of infrared temperature measurement technology; and wavelet analysis, fault tree method, the pattern recognition method and the diagnostic methods of mathematical analysis and genetic algorithm, neural network and expert system etc.

4.1 *The fault mode*

The fault mode is a kind of wind turbine fault phenomenon forms, and it also can be determine the cause of the fault of wind turbine through the fault model, and take corresponding measures to reduce the loss. The fault mode usually can be divided into the following seven types: 1) damage type fault model, the phenomenon is that for the open circuit and short circuit, dislocation, crack, fracture; 2) aging failure model, the phenomenon is of corrosion, abrasion, aging, discoloration and deterioration; 3) fall type fault modes, the phenomenon is loose, torn off; 4) imbalance fault pattern, the phenomenon is as fault clearance discomfort, stroke,

improper voltage, flow, pressure and improper; 5) of blockage or leak age fault model, the phenomenon is for the blockage, oil leakage, water seepage, leakage, water leakage, leakage; 6) functional fault model, the phenomenon is for the indicator the parameters of output function is not normal, unstable performance degradation, motion lag, motion interference, the flow is not smooth, disorder, high temperature, drift and vibration, poor contact; 7) other fault modes, the phenomenon is poor lubrication, water supply, lack of oil.

4.2 *The common faults on wind turbine*

Wind farms usually built in remote areas far from the city or more offshore areas, the maintenance of wind turbines is very difficult, because traffic inconvenience and the unit in high altitude, so the reliability of wind turbines and the failure rate is one of the focuses of the wind power technical factors. Large fault of megawatt wind power unit are mainly concentrated in the gearbox, generator, high speed shaft, low speed shaft, blade, electrical system, yaw system, control system, such as key components. This paper mainly discusses the blade, gearbox, wind turbine for failure analysis.

The fault of blade. The blade is the most easily damaged parts of the wind turbine, and is also the most expensive part. As China's onshore wind farms basic locate in the northern area, the icy wind turbine blade will cause a great impact, at the same time in changing the blade aerodynamic shape and low efficiency caused by the rotation of the fan blade is not balanced; leaf blade during rotation, load by centrifugal force, with the speed changing alternating, aggravate the fatigue failure of blade. Climate and weather factor such as lightning, bad will cause great influence on the wind turbine blade, such as leaf damage, cracking of blades, the blade root screw loosening, and blade material aging. The external climate factors will rotate the blade is not balanced, so that the load variation of blade, blade fatigue failure of intensified, and even make the blade stall flutter damage, corrosion, damage and fracture. The running state can be monitored by the wind turbine blades of the wind turbine output power characteristics of inspection, so as to realize the fault diagnosis of the blade.

The fault of the gear box. The gear box often includes gear, rolling bearing, axis, the fault generated from these components affect each other. The internal parts of the gear box are very complicated, which requires the gear box manufacturing is accurate, skilled operation, reducing the fault after using. Gearbox in the long-term variable speed load harsh conditions of operation, normal faults of the gear (gear pitting damage, broken gear), bearing damage, the tooth error, tooth bonding, gear eccentricity, shaft misalignment, shaft bending, axial, oil leakage, oil temperature too high, bad lubrication of rolling bearing failure, and a box body resonance. Doing math analysis on the data signal which is collected from the monitoring system, we can get the characteristics of the fault. The fault of the gear box has its unique signal feature, so we can do confirm the type and the location to realize the fault diagnosis of the gear box.

The fault of the wind turbine. The wind turbine usually works in the electromagnetic environment long time; there will inevitably be the following problems: the excessive vibration and noise generators, generators bearing overheating, noise are not normal and insulation damage etc…The common faults of wind turbine generator are mainly after the bearing cage fracture, front bearing wear fault, rotor broken bar fault, air gap eccentricity, lubrication of bearings and other(Shan, G.K, 2012). Wind power generator failure, still can continue to operate, if not detected early enough to maintain, hidden faults developed to a very serious when can cause major accidents. Inside of the generator winding fault can be diagnosed by temperature rise of stator turn-to-turn short circuit, rotor broken bars, air-gap eccentric of the three common faults can be carried out on the stator current sampling analysis for diagnosis.

5 CONCLUSIONS

The wind farm is generally located in more remote areas, the regular maintenance and repair can greatly reduce the cost of wind energy. On-line monitoring system will evaluate the operation of the unit before damaging, effectively avoid the on-site maintenance personnel to take measure to be blind, timely, comprehensive understanding of the operation condition of the wind farm, has an important significance in the maintenance and management of the field unit.

ACKNOWLEDGMENTS

This work was supported by Shanghai Municipal Natural Science Foundation (No. 14ZR1417200), Innovation Program of Shanghai Municipal Education Commission (No. 14YZ157), and National Natural Science Foundation of China (No. 61374136).

REFERENCES

Guo, X. 2013.The study on the no-line monitoring system of wind generating set, Yangzhou University: pp. 18–21.
GWEC. 2014. Global Wind Report. GWEC.
Shan, G.K. 2012. Megawatt wind turbine condition monitoring and fault diagnosis research, Shenyang University of Technology: pp. 25–31.
Song, H.H. 2009. The wind power generation technology and engineering, China Water & Power Press, Beijing: pp. 57–59.
Wang, R.C. & Lin, F.H. 2009.Research on On-line Monitoring and Fault Diagnosis of Wind Turbine, East China Electric Power, (1):190–193.
Wang, Z.J., Xu, Y.F., Liu, S.M. & Sun, X. 2013. Large wind turbine condition monitoring and intelligent fault diagnosis, Shanghai Jiao Tong University Press, Shanghai: pp. 34–37.
Xie, Y. & Jiao, B. 2010. State of the art of condition monitoring system and fault diagnosis methods of wind turbine, Journal of Shanghai Dianji University, 13(6): pp. 328–333.
Yang, M.M. 2009. Fault mode statistic& analysis and failure diagnosis of large-scale wind turbines, North China Electric Power University: pp. 8–16.

Energy Science and Applied Technology – Fang (Ed.)
© 2016 Taylor & Francis Group, London, ISBN 978-1-138-02833-3

The design of transformer on-line monitoring system based on DSP

Tengfei Ding & Chunyuan Yin

Shenyang Institute of Automation, Chinese Academy of Sciences, Shenyang, China
University of Chinese Academy of Sciences, Beijing, China
Key Laboratory of Networked Control System, CAS, Shenyang, China

Yunxia Cao

Shenyang Institute of Automation, Chinese Academy of Sciences, Shenyang, China
Key Laboratory of Networked Control System, CAS, Shenyang, China

ABSTRACT: The security of the power transformer operation directly affects the whole power system. Thus, this paper aims to design the system for "real-time" monitoring the parameters of the transformer based on the TMS320F28335 DSP designed by TI Company. First, the system makes use of different sensors' signal acquisition. Then, the inconsistent analog signal can be converted into the standard signal for AD conversion through the analog signal photoelectric isolation amplifier. Finally, the DSP will process them, and the processed data will be transferred to the host computer for storage and display. The test results show that the system can monitor the equipment running status accurately, stably and timely.

Keywords: Transformer; DSP; on-line monitoring

1 INTRODUCTION

With the rapid development of economy, the increasing demand for electricity and improvement of the power grid capacity and voltage level make power delivery more important, so it needs to ensure not only the supply of electricity, but also the quality of power supply. As the indispensable equipment in the power system, the state of the transformer is related to the security of the entire grid. Therefore, "real-time" monitoring of transformer is of great significance.

With the development of information and intelligence, the transformer on-line monitoring system has been able to reflect the state of the transformer in real time, and warn timely when transformer exception occurs. At the same time, it can also have a quick check of the corresponding point of failure. In the stable situation, the monitoring of various operating parameters can provide an important reference for improving the quality and capacity of power supply, and guarantee the security of the power system. The system first collects the semaphore by the modern sensor, then the signal is processed by high-performance DSP, and finally the data is displayed in the host computer. The system can monitor the data accurately, completely and reliably; therefore, it can fully ensure a safe and reliable operation of the transformer.

2 OVERALL STRUCTURE AND FUNCTION

The monitoring system is mainly divided into three parts: the first part is composed of various sensors, which are used to collect and transmit the important monitoring data of the transformer; the second part is mainly the analog-to-digital conversion and data processing based on DSP; the third part is the display and storage of the data in the host computer. The system is powerful, comprehensive and cost-effective, and can run a more comprehensive monitoring of the state. The overall system structure diagram is shown in Figure 1.

The system is mainly used for monitoring the important operating parameters of the transformer and the fault-prone parts, which operates under a strong voltage and a complex electromagnetic environment. Its main functions are as follows.

Sensors are mainly responsible for obtaining the transformer primary side and secondary side voltage current effective values, core grounding current, oil temperature, hot temperature, and oil level. These parameters can fully reflect the running state of the transformer, so the faults can be found in time according to the data changes. The performance of sensors will affect the accuracy of the final results, so the sensors must have high stability and good sensitivity.

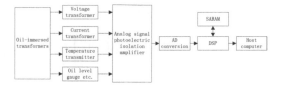

Figure 1. The system structure.

Host computer obtains the processed data from the DSP through CAN communication to display and store in a large-capacity storage device, which can be used for the subsequent development of intelligent algorithms and expert systems.

3 SYSTEM DESIGN

3.1 *Signal standardization*

The format of output data through sensors is not consistent, so it will make the circuit design and data process complicated. For the convenience of unified processing of the signal, it is necessary to convert the inconsistent electrical signal to the standard voltage signal. So, the initial signal obtained by different sensors needs to be converted to 0V to 5V voltage signal through the analog signal photoelectric isolation amplifier, and then adjust the circuit voltage signal to convert it to a suitable range that can make AD conversion. It can enhance the generality of system hardware and reduce the complexity of data processing.

3.2 *Data acquisition and processing*

Unified voltage signal cannot be directly converted through the AD conversion chip, because it needs to convert the electrical signals to the appropriate voltage input range that the AD chip accepts through the voltage regulation circuit. This system adopts ADS7844, which has an 8-channel, 12-bit sampling Analog-to-Digital Converter (ADC), with a conversion rate of up to 200 KHz. DSP can control AD chip sampling through a 32-bit CPU timer sampling, and then process the sample and control ADS7844 to send and receive the data through the Mc BSP communication mode. After the serial AD chip collecting a set of measured data, DSP can process them.

In order to monitor the change in the dynamic voltage current signal timely, the computation speed and access speed need to be improved. DSP used in this system is a high-speed floating-point processor. Its sampling frequency is 4096 Hz, which can meet the requirement of high speed, and a large amount of data is stored in RAM. At the same time, in order to meet the requirements of

"real time", the ping-pong operation is applied for data processing, and the pipeline parallel computing is used to calculate the data flow, thus making the buffering and processing of data more perfect.

In view of the core grounding current, oil temperature, oil level and the temperature signal change smoothly, so the sampling frequency (1000 Hz) can meet the requirements. As long as the average of a set of data is calculated, it can obtain the relevant information accurately.

3.3 *Data display*

F28335DSP has the enhanced CAN module that is a completely CAN controller, having 32 fully configurable mailboxes and real-time mail functions. The data processed by DSP will be transmitted to the host computer through the CAN bus, and then monitoring software of the host computer can complete some tasks, such as real-time displaying of the transformer's running state and diagnosing the faults of transformer analysis. Monitoring equipment should be used ancillary with the host computer's software, i.e. the system needs both start and referring to the IEC61850 communication standard communication interface.

4 DATA CALCULATION

4.1 *Oil temperature and oil level of transformer*

Data about oil temperature, oil level and temperature can be calculated by the linear relationship between the measured signal and signals from the sensor. For example, the range of the temperature controller can be measured as −20°C–120°C, and the corresponding output current as 4 mA–20 mA.

Abnormal situation may arise when the oil temperature is too high: when the temperature rises above 55 K, it can give an alarm signal; once more than 65 K, it will give a trip warning signals. Top oil temperature rise = top oil temperature—environment temperature bottom oil temperature rise = bottom oil temperature—environmental temperature.

The oil level information is used to judge whether to send out alarm signals or not: when the oil level is higher than 10 or less than 0, which, respectively, corresponding to 20 mA or 4 mA, it will give an alarm signal.

4.2 *Sampling frequency calculation*

Fast Fourier Transformation (FFT) can convert a signal to the frequency domain and find the characteristics of the signal in it, so the analog signal sampled through the AD can be converted into the digital signal to make FFT transformation to catch the corresponding sampling point's results. N sampling points corresponds to the N point FFT results. In order to reduce the effects of frequency spectrum leakage and fence, before carrying out the spectrum analysis, the original FFT algorithm needs a windowing optimization to minimize the errors.

Monitoring the harmonic signal needs to set up a suitable sampling frequency, the frequency of harmonic that needs to be monitored reaches 11 times the fundamental frequency, and the fundamental frequency is 50 Hz, so it can calculate the frequency of 11 times harmonic: 50 Hz × 11 = 550 Hz. According to the Shannon's sampling theorem, the sampling frequency must be greater than twice the highest frequency of the monitoring signal, so the minimum sampling frequency is 1100 Hz. However, the sampling frequency is more than twice the signal frequency in practice. At the same time, for the convenience of FFT computation, the sampling frequency usually takes N power of 2. Considering the accuracy and speed of the processor, the sampling frequency is set as 4096 Hz. Thus, we can both meet the requirements of the sampling theorem and improve accuracy. Besides, in order to achieve "real time", setting sampling points as 4096 points, the corresponding resolution will be 1 Hz; in this case, the AD chip transmits the data to DSP one time per second.

4.3 *Electricity parameter calculation*

The fundamental frequency signal cycle is set to T and instantaneous voltage for u(t), so it can calculate the effective voltage value according to the following formula: $u = \sqrt{\frac{1}{T}\int_0^T u^2(t)dt}$. The fundamental frequency is f = 50 Hz, and the sampling frequency is f1 = 4096 Hz, f1 >> f. The Ux (n) and Ix (n) are the cycle discrete sampling sequence from N points sampled by the voltage and the current. The sampling is the process of electrical signals' discretization, and in line with the integral principle. Therefore, the continuous voltage current value can be replaced with the discrete sampling sequence. According to the principle of integral, it can get the voltage effective value within a cycle: $U_i = \sqrt{\frac{1}{T}\sum_{i=0}^{N-1} u_i^2 \Delta T}$, where u_i is the voltage sample values and ΔT is the sampling period.

Because the uniform sampling is under the control of the CPU timer, the sampling interval is therefore constant: N = T/ΔT, and the effective voltage value is given by

$$U = \sqrt{\frac{1}{N}\sum_{i=0}^{N-1} u_x^2(n)}$$

In order to improve the calculation accuracy, the sampling cycle can be set as 50 to calculate the voltage effective value of each phase, and then calculate the average to get a more accurate voltage effective value: $\overline{U} = \frac{Ua+Ub+Uc}{3}$.

So, the current effective value is $I = \sqrt{\frac{1}{N}\sum_{i=0}^{N-1} u_x^2(n)}$ and the average current effective of three-phase is $\overline{I} = \frac{Ia+Ib+Ic}{3}$.

The active power is $P = \frac{1}{N}\sum_{i=0}^{N-1} U_x^2(n)I_x^2(n)$. Because the Ux(n) and Ix(n) are sampling values of the voltage and the current at the same time, the active power can be obtained after discretization:

$$P = \frac{1}{N}\sum_{i=0}^{N-1} U_x^2(n)I_x^2(n).$$

According to the current and voltage effective value and the relationship between them, it can get the apparent power $S = \overline{U}\overline{I}$, the reactive power $Q = \sqrt{S^2 - P^2}$ and the power factor $\cos\varphi = P/S$.

5 CONCLUSION

The transformer on-line monitoring system is based on DSP, which makes a full use of the precision sensor, the serial AD conversion chip, the internal resources and floating-point operation function of DSP to monitor the basic semaphore and the higher harmonic. A variety of optimization methods are used to improve the processing speed in the design process. Both the experiment test and the enterprise transformer test show that the system can monitor the operation parameters accurately and operate stably. This system has been successfully applied in the enterprise, and provides a powerful guarantee for the modern power system.

REFERENCES

Lin Guang-min, Huang Yi-feng, Ouyang Sen, et al. Design of a power quality monitoring device based on DSP and CPLD [J]. Power System Protection and Control, 2009, 37(18):97–101.

Liu Xian-yong. Research on Photoacoustic Sensing of Transformer Fault Gases[C]. Proceedings of Photonics Asia 2004, Sensors and Imaging 5633 volume, November 2004, Beijing, China.

Shi, H. X., Z. H. Li, and Y. X. Bi. An on-line cavitation monitoring system for large Kaplan turbines. In Proc. of 2007 IEEE PES General Meeting, USA, 2007, 1–6.

Energy Science and Applied Technology – Fang (Ed.)
© *2016 Taylor & Francis Group, London, ISBN 978-1-138-02833-3*

Duplicate Address Detection process with variable-length prefix

Guangjia Song & Zhenzhou Ji
School of Computer Science and Technology, Harbin Institute of Technology, Harbin, China

ABSTRACT: Duplicate Address Detection (DAD) is an important component of the address resolution protocols, which determines whether an Internet Protocol (IP) address can be used. In the traditional DAD process, critical information is broadcast through the network. This is a vulnerability that malicious nodes can exploit to mount attacks, resulting in failure to configure IP addresses. To resolve the issue, a new DAD process with variable-length prefix, DAD-vP, was proposed. With DAD-vP, prefix is used to indicate a specific range rather than the entire destination address being tested. This significantly increases the difficulty of mounting an attack. Simulation results show that when under attack, address configuration using the DAD-vP process has a much higher success rate compared to that using the traditional DAD process.

Keywords: network security; variable prefix; address resolution; neighbor discovery; duplicate address detection

1 INTRODUCTION

One of the main functions of a computer network is the exchange of data between nodes. In this process, Internet Protocol (IP) packets (formatted units of data) are transferred from the source to destination nodes via various layers of intermediate devices (such as routers or switches).

There are two modes by which IP packets are delivered: direct and indirect. The former occurs when both source and destination nodes are on the same link (also referred to as being in the same local area network, LAN). In this situation, the switch uses its own <Port, MAC> mapping table to locate the corresponding port of the destination media access control (MAC) address. The IP packets are then forwarded directly to the port of the destination node.

When the source and destination nodes are on different links, the packets must be delivered indirectly. Packets must be sent through routers until they reach the network in which the destination host is located. The local switch (or router) of the destination host then carries out a direct delivery.

Regardless of the delivery mode, packets can only be delivered to the corresponding host of the destination IP according to its MAC address. The process by which a destination IP address obtains its corresponding MAC address is known as address resolution, which is dependent upon the address resolution protocols (ARP and NDP) for completion. When an attack occurs during the address resolution process, the consequences are usually dire. For example, a typical man-in-the-middle attack can result in data interceptions or falsifications, or even network interruptions. Attacks that target the address resolution process are major threats to LAN security.

In the IPv4 system, address resolution is mainly done using ARP. However, in the IPv6 protocol system, Neighbor Discovery Protocol (NDP) is used instead. NDP is an important basic protocol for IPv6. It combines various enhanced IPv4 protocols, including ARP, Internet Control Message Protocol (ICMP) routing discovery, and ICMP routing redirection. In addition, as the basic protocol for IPv6, the NDP also has other various functions, such as prefix discovery, Neighbor Unreachability Detection (NUD), Duplicate Address Detection (DAD), and stateless address auto-configuration (SLAAC).

The Internet Engineering Task Force (IETF) has since proposed the secure neighbor discovery (SEND) protocol to address various security and other concerns. When analyzed in terms of the framework of protocols, ARP, NDP, and SEND all minimally contain the following three main links:

The word "data" is plural, not singular. Process to obtain the <IP, MAC> mapping of the destination node;
DAD process;
Data structure maintenance phase.

2 DEVELOPMENT SITUATION OF DAD

The basic format of ARP packets is shown in Fig. 1. The MAC and IP addresses of the source node are indicated by Src MAC and Src IP, respectively. Those of the destination node are indicated by Dest MAC and Dest IP, respectively.

The ARP protocol initially depends on gratuitous ARP for DAD. The gratuitous ARP process can be described as follows: Host A decides to use IPx, but before doing so, it must broadcast an ARP request. This is different from other general ARP requests because during the broadcast, the address of the node (i.e., IPx) is stated in the Dest IP field, which is the destination address to be resolved. The purpose of doing so is to check whether IPx has already been used. If the node receives a response to that broadcast, it means that there is a conflict with that address.

The Network Working Group request for comments RFC 5227 proposed a new approach to Address Conflict Detection (ACD) to carry out DAD more effectively. Two new packets are added in this method: 1) ARP probe and 2) ARP announcement. An ARP probe is similar to an ARP request, but its Src IP field is filled with 0.0.0.0 to reduce pollution. This is because ARP contains a mechanism for passively obtaining <IP, MAC> mapping, such that upon receipt of the broadcasted ARP request, the other hosts will update their own caches based on the IP and MAC data contained therein. For the ACD process, when a host uses a new IP address, it first broadcasts an ARP probe packet to confirm that there is no conflict, after which it will send out an ARP announcement packet, usually thrice. At this time, the Src IP and Dest IP in the announcement packet

will be its own IPx. This is a declaration that it will be using IPx.

For NDP to be realized, the detection process relies mainly on the Neighbor Solicitation (NS) and Neighbor Advertisement (NA) packets. The format of an NDP message is shown in Fig. 2. The target address is usually the destination address to be resolved. The meaning of the "option" field varies depending on the message type contained therein. In most cases, the link-layer address is provided. The "type" field represents the message type, where 135 and 136 represent the type values of NS and NA messages, respectively.

To illustrate, let us assume that a network contains two hosts, A and B, with MAC addresses of 00E0-FC00-0001 and 00E0-FC00-0002, respectively. The local link-layer address of Host B is 1::2:B. When Host A wishes to configure the same address, it will broadcast an NS message. Host B will reply with an NA message indicating a configuration conflict in the address. An example of an NS and an NA message is shown in Fig. 3.

Currently, security research is mainly focused on the first and third stages, with the former being a way to ensure that IP and MAC mapping are

Ethernet Header	Dest MAC
	Src MAC
	Type
IPv6 Header	Src IP
	Dest IP
	Next header
IPv6 Data	Flags
	Type
	Target address
	option

Figure 2. Format of an NDP message.

Ethernet Header	Dest MAC
	Src MAC
	Type
IPv4 Header	Hard type
	Port type
	Hard size
	Port size
	op
IPv4 Data	Src MAC
	Src IP
	Dest MAC
	Dest IP

Figure 1. Packet format of ARP.

NS

Ethernet Header	3333-FF02-000B 00E0-FC00-0001 0x0806
IPv6 Header	:: FF02::1:FF02:B 0x3A
IPv6 Data	135 1::2:B 00E0-FC00-0001

(a)

NA

Ethernet Header	3333-0000-0001 00E0-FC00-0002 0x0806
IPv6 Header	1::2:B FF02::1 0x3A
IPv6 Data	S=1 136 1::2:B 00E0-FC00-0002

(b)

Figure 3. NS and NA messages.

obtained, while the latter acts to prevent the cache from becoming polluted. There has been relatively little research on the second stage because DAD is used less often during the communication process compared to the other two stages. DAD is usually carried out when a host starts up, or when there are changes to the IP or MAC addresses. Furthermore, each host generally only has one IP address within an IPv4 environment, and the IP address of the host rarely changes.

However, that situation is changing with the increasing popularity of IPv6. In this system, a node can have multiple IP addresses, such as a link-local address, a globally unique address, or other network addresses derived according to the previous prefixes of broadcasts by various routers. One of the main features of IPv6 is SLAAC, which allows a node to combine its own MAC address based on the extended unique identifier EUI-64 regulations to generate a link-local address automatically; artificial intervention or manual operation is not required. Unlike IPv4, the DAD process occurs much more frequently in IPv6. Consequently, attacks on the DAD process will have a much greater effect on an IPv6 network.

The DAD process is deemed to have failed as long as any node claims that the address has already been taken and therefore is not available. The source node must then reconfigure another address for testing in a new DAD process. This is the same situation regardless if the system is IPv4 or IPv6. The attacker can take advantage of this characteristic and respond repeatedly to all random DAD packets received that the address to be detected has been taken. Eventually, the source node will not have any available addresses for use.

3 DAD PROCESS WITH VARIABLE-LENGTH PREFIX

The main reasons for the existing DAD process to be vulnerable to attacks are:

1. Making the wrong assumption that all nodes on the network can be trusted; and
2. Broadcasting an important message to the network (the destination address being tested) during the DAD process. This makes it convenient for an attacker to launch an attack simply by forging a packet in response to the destination address being tested, thereby causing the DAD process to fail.

Therefore, a new DAD process, DAD-vP, with variable-length prefix, is proposed to address the two aforementioned issues. First, two new types of packets are added: 1) NS with variable-length prefix (NS-vP) and 2) NA with variable-length prefix

Ethernet Header	Dest MAC
	Src MAC
	Type
IPv6 Header	SRC IP
	DST IP
	Next header
IPv6 Data	Flags
	Prefix
	Type
	Target address
	option

Figure 4.   Message format of DAD-vP.

(NA-vP). The format of these two packets is similar to the NS and NA of NDP. The only difference is the addition of the prefix field. The specific format of the new packets is shown in Fig. 4.

The DAD-vP process can be described as follows: After Host A has configured an address, it will first wait for a random length of time before broadcasting the NS-vP packets. Unlike the traditional DAD process, the destination address being tested is not specified in the target address field in the NS-vP packet. Instead, the prefix field contains only the prefix information of the destination address being detected. When the other nodes receive the NS-vP, they will check the prefix field contained therein and see if any of their existing addresses match that prefix. If yes, the node will unicast an NA-vP in response. Within that NA-vP, the prefix field is the same as that in the NS-vP. The target address field will be the address that matches the prefix information. Host A will check the target address field of all the NA-vPs received within the stipulated time. If a NA-vP's target address field match with the detection address, then the DAD-vP will fail, otherwise DAD-vP success. The flowchart of the DAD-vP process is shown in Fig. 5.

Using the previous example of Hosts A and B again, if Host A has configured 1::2:B during the DAP-vP process, it must send out an NS-vP for DAD prior to using that address. In the NS-vP, the target address field is null, while the prefix field is 1:0:0:0 (i.e., a 64-bit prefix was issued). The details are shown in Fig. 6(a). The length of the prefix is variable and can range from 0–128 bits. The suggested length is less than 100 bits. This is because if the prefix is too long, a malicious node can combine the prefix and the multicast address to deduce the destination address being detected.

After Host B receives the NS-vP, it will check the prefix field contained therein. If any of its IP

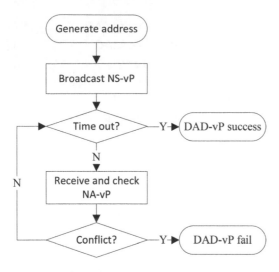

Figure 5.   Message format of DAD-vP.

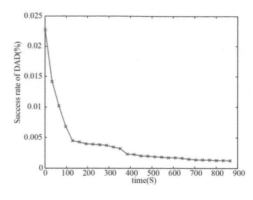

Figure 7.   Address configuration success rate of DAD.

	NS-vP			NA-vP
Ethernet Header	3333-FF02-000B 00E0-FC00-0001 0x0806		Ethernet Header	3333-0000-0001 00E0-FC00-0002 0x0806
IPv6 Header	:: FF02::1:FF02:B 0x3A		IPv6 Header	1::2:B FF02::1 0x3A
IPv6 Data	135 1:0:0:0 Null 00E0-FC00-0001		IPv6 Data	S=1 1:0:0:0 136 1::2:B 00E0-FC00-0002
	(a)			(b)

Figure 6.   NS-vP and NA-vP.

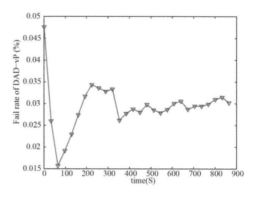

Figure 8.   Address configuration failure rate of DAD-vP.

addresses match that prefix, it will send out an NA-vP for which the prefix field is similar to the content of the NS-vP. The target address field will include the IP address that matches the prefix. The details are shown in Fig. 6(b). Upon receiving the NA-vP and checking the target address field, Host A will discover that the address is similar to the one being detected. Given this conflict in the address, Host A will realize that it should configure other addresses.

4   SIMULATIONS

The software OPNET (OPNET Technologies, Inc., Bethesda, MD, USA) was used to simulate a LAN that contains one switch node and eight peripheral nodes (of which 7 are normal and 1 is malicious). Each normal node contains two packets transmission sources. The first is used to generate DAD packet which has a random detection address at uniform distribution intervals (the average value

was 1). The other source is used to generate background traffic, which was based on the 30-day average traffic experienced by the core switch of a particular university. The data acquisition software used was Orion (SolarWinds, Austin, TX, USA).

The simulation comprised two scenarios: the normal DAD and DAD-vP processes. For the DAD process, the malicious node forged the NA based on the target address stated in the NS. For the DAD-vP process, the malicious node responded with a randomly selected address based on the prefix message in the NS. The statistics for the simulation measured the failure rate for address configuration. The results of the two scenarios with the normal DAD and DAD-vP processes are shown in Figs. 7 and 8, respectively.

In the situation under the traditional DAD process where a malicious node exists, the destination address that is being detected is exposed during the broadcast. The malicious node was able to send false replies in a targeted manner, resulting in the success rate of address configuration by the normal nodes to be very low and close to the link loss rate (Fig.7).

However, because the address being detected is concealed under the DAD-vP process, the malicious node could only send random replies based on the prefix information, making it difficult to mount successful attacks. Therefore, the failure rate of address configuration by the normal nodes was very low (Fig. 8).

5 CONCLUSIONS

With the increase in the number of network nodes and the increasing popularity and application of IPv6, the effects of attacks targeting DAD will gradually become greater. The traditional DAD process is vulnerable and can hardly withstand the attacks of malicious nodes. By hiding the destination address to be detected with the help of prefix, DAD-vP makes it much more difficult for malicious nodes to mount attacks. To be compatible with technologies such as source address validation improvement (SAVI), a traditional DAD process can be executed after the DAD-vP process. This will facilitate the binding of the source address by the switches. Although DAD-vP results in increased communication volumes compared with traditional DAD, the communication overhead is justifiable because of the increase in the success rates of DAD.

REFERENCES

Arkko, Jari, et al. Secure neighbor discovery (SEND). RFC 3971, March, 2005.
Bagnulo M, García-Martínez A. SAVI: The IETF standard in address validation. Communications Magazine, IEEE, 2013, 51(4): 66–73.
Bi J, Yao G, Halpern J, et al. SAVI for Mixed Address Assignment Methods Scenario. Working in progress, 2011.
Cheshire, Stuart. "IPv4 Address conflict detection." (2008).
Fall, Kevin R., and W. Richard Stevens. TCP/IP illustrated, volume 1: The protocols. addison-Wesley, 2011.
Jinhua, Gao, and Xia Kejian. "ARP spoofing detection algorithm using ICMP protocol." Computer Communication and Informatics (ICCCI), 2013 International Conference on. IEEE, 2013.
Narten, Thomas, et al. "RFC4861: Neighbor Discovery for IP version 6 (IPv6)." Standards Track, http://www.ietf. org/rfc/rfc4861. txt (2007).
Nikander, Pekka, James Kempf, and Erik Nordmark. "IPv6 neighbor discovery (ND) trust models and threats." RFC3756, Internet Engineering Task Force 99 (2004).
Pandey, Poonam. "Prevention of ARP spoofing: a probe packet based technique." Advance Computing Conference (IACC), 2013 IEEE 3rd International. IEEE, 2013.
Wu J, Bi J, Li X, et al. Rfc5210: A source address validation architecture (sava) testbed and deployment experience. 2008.
Yang, Xinyu, Ting Ma, and Yi Shi. "Typical dos/ddos threats under ipv6." Computing in the Global Information Technology, 2007. ICCGI 2007. International Multi-Conference on. IEEE, 2007.

Energy Science and Applied Technology – Fang (Ed.)
© 2016 Taylor & Francis Group, London, ISBN 978-1-138-02833-3

Research on developing a miniature temperature measurement system based on uncooled infrared detector

Leizi Jiao, Yun Lang & Daming Dong
National Engineering Research Center for Information Technology in Agriculture,
Beijing Academy of Agriculture and Forestry Sciences, Beijing, China

Kun Zhang
College of Information Engineering, Taiyuan University of Technology, Taiyuan, China

ABSTRACT: This paper introduces the process of research and development of an uncooled temperature measurement system, which transforms the temperature distributed on the surface of the object to the gray-scale image, and then displays the image in real time on the monitor, and finally gets the temperature field of the object's surface. Starting from the purpose of miniaturization, real time and high speed, this system exploits the uncooled long-wavelength infrared micro bolometer camera TAU336 as its imaging system and single FPGA (Field Programmable Gate Array) EP4CE115F29 processor to capture and compute the picture. Meanwhile, it employs the cameralink protocol to work out the speed matching problem between the high-speed display and the relative low-speed capture of the picture. In this paper, a whole system design scheme is proposed, including the project of the hardware and software, and the particular description of the video capture plus the VGA (Video Graphics Array) display system, respectively. The result of the experiment demonstrates that the system can display the object's surface temperature field in real time and has the characteristics of low power consumption and miniaturization.

Keywords: Uncooled infrared detector; Cameralink; FPGA; VGA

1 INTRODUCTION

Since the first real-time infrared thermal imager equipped in the field of military in the early 1960s, the research and development of the infrared detector has never ceased (KocerArslan & Besikci, 2012; Xing, 2005) . The original infrared detectors are all cooled type, with its dissatisfied natural traits such as expensive, poor reliability and heavy volume, with all of these disadvantages seriously limiting its promotion in the field of military, industry and commerce and having a strict control on the cost (Fièque, Tissot, Trouilleau, Crastes, & Legras, 2007; Yang, 2013). It is meaningless if the progress in science and technology cannot benefit the human society, but fortunately, after decades of rapid development, especially benefits from the progress of MEMS technology and materials, uncooled infrared detectors have made a great progress in its key technical performances including the NETD (Noise Equivalent Temperature Difference) and the ability of response (Huang & Shen, 2014; Liu, 2011). Indubitably, uncooled infrared detectors, relying on its advantage of high reliability, small volume, low power consumption and low cost, have greatly expanded its applications across all fields (Cheng, Paradis, Bui, & Almasri, 2011).

It is the significant success of uncooled infrared thermal imaging technology in the field of temperature measurement that embodies its application value tremendously. Pan took advantage of the infrared imaging technology to measure the temperature of gas and low density plasma in order to diagnose the existence of natural gas (PanBarker & Meschanov, 2002). Kargel made use of this technology to detect the local heating that affects people's ear, in order to reduce its electromagnet radiation and try to change our habits of using cell phones (Kargel, 2005). Fokaides measured the temperature distribution of the building's surface with the help of an uncooled thermal infrared imager, and analyzed the inconsistent sensitivity factors between the theory and the result (Fokaides & Kalogirou, 2011). Also, Mechantd C.J established relevant models of Sahara Desert based on an uncooled thermal infrared imager to explain the influence of the bright temperature caused by the sand (Merchant, Embury, Le Borgne, & Bellec, 2006). Besides, Zhang and his colleague studied the application of infrared thermal imaging temperature measurement technology in the field of metal mechanical processing (Zhang & Huang, 2001). In this paper, we exploit the FILR Corporation's uncooled long-wave infrared thermal imaging camera as the detector and the cameralink protocol to transmit the video data, combined with FPGA's high-speed data processing ability to design a kind of miniature real-time temperature measurement system, finally displaying a series of continuous pictures on the monitor by the VGA connector. The system designed in this paper has the merits of vivacity, small scale and real time. The framework of this paper is as follows: project of the system, design of the software system, analysis of the image and summary of the paper.

2 THE PROJECT OF SYSTEM

2.1 The uncooled infrared detector

When the detector receives IR (Infrared Radiation) from the object, it converts the radiation power to electrical signals and processes these signals through the electronic system so as to obtain the infrared thermal image or gray image of the object's surface. Therefore, undoubtedly, the detector is the key part of the temperature that measures the system. The popular photoelectric conversion materials of the uncooled detector are BST (Barium Strontium Titanate), $\partial - Si$ (amorphous silicon) and VO_x (vanadium oxide), and because of its outstanding TCR (Temperature Coefficient of Resistance), VO_x has been widely used by manufacturers (Gokhale & Rais-Zadeh, 2014; Wang, 2012). Currently, the uncooled IR detector market is shared by these powerful companies such as FLIR, Fluke and Raytek. In this project, we employ the TAU 336 uncooled long-wave infrared thermal imaging camera manufactured by FILR as our detector (Systems, 2010), as shown in Figure 1.

The operating temperature of this detector is −40°C to +80°C external, the resolution is 336×256 and the spectral band is 7.5–13.5 μ–m.

It has an MDR26 interface itself, which can transmits the data swiftly to the main board, and also supports the cameralink communication protocol by the special connector on it, totally meets some extraordinary requirements in a high-speed way, outputs an 8-bit or 14-bit serial LVDS (Low Voltage Differential Signaling) digital video, and PAL (Phase Alternating Line) and NTSC (National Television System Committee) video formats are supported. In this project, we choose the PAL format to mate our video output, and the detector obtains a 5DC voltage from the main board through its mini-USB interface. It is proved in our experiment that the 8-bit digital video input to the FPGA processor is feasible in a normal temperature measurement.

2.2 Design of FPGA hardware system

We aim to design a miniature temperature measurement system, which could get gray-scale map processed by the FPGA chip from the detector, and finally display the video in real time on the monitor. In the general design, also two processors are employed: FPGA and DSP (Digital Signal Processor). The former uses to control the signal and the latter uses to process the image. Although it is easy to implement relatively, this scheme has a high loss of power and volume (Wang, 2013; Yang & Zhou, 2013). By considering comprehensively, one FPGA processor is sufficient in this project in order to economize the volume and power consumption of the system. Figure 2 shows the principle of the hardware system.

In this project, FPGA reads the digital video signal from the uncooled infrared detector transmitted by the cameralink connector. The outside SDRAM (Synchronous Dynamic random access memory) fulfills the cache of data, and the data is outputted through the VGA interface. Detailed explanations for the hardware system are as follows.

The FPGA chip of this system employs EP4CE115F29 (Altera, 2011), which owns a large

Figure 1.   Appearance of TAU 336.

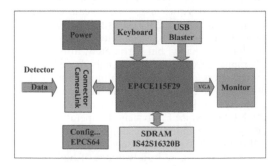

Figure 2.   The principle of the hardware system.

capacity, includes more than 114,480 logical units and 432M9 K memory modules at the same time, and satisfies the requirement of the design and further development. The configuration chip employs EPCS64, which supports the JTAG (Joint Test Action Group) and the AS (Active Serial Configuration) programming modes via the USB-blaster interface, making the program to debug and download easily. In addition, the FLASH memory chip is used to store the program.

One pair of SDRAM IS4216320B has been planned, 32M × 16Bit, to cache the images read in; compared with SRAM (Static Random Access Memory), it has a higher integration, less power consumption, and what's more, a smaller size at the same capacity.

The power system provides the voltage for the whole system, which includes 5VDC, 3.3VDC, 2.5VDC and 1.2VDC for the detector, SDRAM and FPGA, respectively. Meanwhile, a keyboard electric circuit is designed to provide OFF/ON and RESET buttons for the system.

2.3 Design of video acquisition system

Cameralink is a kind of special serial communication protocol for machine vision application, it applies LVDS to accomplish data transmission, which solves the image rate matching problem between outputting and gathering effectively (Huang & LIU, 2013; Wu, 2013). The cameralink protocol includes three parts: the video data signal is based on channel link technology, together with the video control signal and the serial communication signal, of which consist the cameralink signal. The connector designed is shown in Figure 3.

A brief description for the above picture is as follows.

The chip DS90CR288 A is in charge of a 28-bit video data series-parallel conversion

The chip DS90 LV047 is in charge of the camera contravention

The DS90 LV019 chip is in charge of the serial communication between the FPGA and the camera

Three video transmission modules are supported by the cameralink connector, BASE, MEDIUM and FULL. We choose an 8-bit digital video signal to accomplish our design in this project; therefore, the BASE module is appropriate. When the signals of the frame, line and data are valid, as well as the camera control signals at a high level, the infrared camera begins to work. Apparently, it is easy to configure the capture of data and control of camera signal according to the handbook, in order to make sure that the FPGA reads the data (transmitted by the MDR26 interface) via the cameralink connector correctly.

2.4 Design of VGA display

In this project, we adopt the standard VGA video transmission, which is commonly used and supports the multiple frame frequency and video formats. It is a 15 pin D-type connector, which is responsible for the transmission of signals. Among these pins, the very important are the three primary color signals R (Red), G (Green) and B (Blue), HS (horizontal synchronization) signal plus VS (vertical synchronization) signals (Zhang & Liu, 2010). The primary color signal is the analog signal; however, the FPGA outputs is the digital signal. It is necessary to use a D/A (Digital to Analog) conversion chip between the FPGA chip and the video output. Therefore, we employ a 10-bit chip ADV7123, which is sufficient since the data used in this project is 8-bit. The principle of the VGA display is shown in Figure 4.

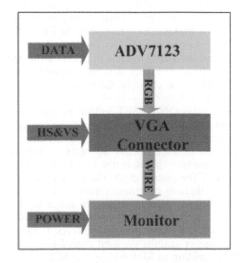

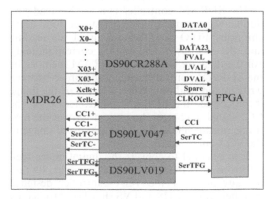

Figure 3.   The principle of the cameralink connector.

Figure 4.   The principle of the VGA display.

The chip ADV7123 is responsible for the conversion of the 8-bit digital signal from the FPGA to the 8-bit analog signal. We exploit only one channel of the three RGB channels because the picture is a gray-scale map, meanwhile keeping the HS and VS signals controlled by the FPGA at a high level to guarantee the gray-map display on the screen. The whole design of this system is introduced in this section: we mainly introduce the uncooled infrared detector TAU336, and discuss the principle of the video capture and the VGA display, and also the type of the main chips that we have selected, which lays a good hardware foundation on the subsequent image storage and display in real time.

3 THE DESIGN OF SOFTWARE SYSTEM

From the analysis and discussion in Section 2, we have confirmed the system design and the type of components. It is the timing of the interfaces, the process and cache of the image data that determine the design of the software system. On the one hand, there are the 50 MHz, 60 MHz and 100 MHz clock frequencies from the cameralink protocol, VGA-timing and SDRAM controller that separately need to be combined. We solve this problem by relying on the inner chip asynchronous FIFO (First In First Out) module, whereby the system can accomplish the matching of the clock and the write-read of the data successfully. On the other hand, in order to realize the real-time displaying, the data from the upper grade will be stored in SDRAM and the data stored in SDRAM before should be sent to the next grade at the same time (Tong, 2012; Wu, 2013).

It is convenient to use the inner resources of the FPGA by means of the software Quartus II (an integrated development environment provided by Altera). It makes it easy to design the PLL (Phase Locking Loop) module to produce the system clock and the FIFO module to generate the width and depth of the data that the software system needs. The flow chart of the system is shown in Figure 5.

When the ON/RESET button is pressed, the system uploads the program from the configuration chip EPCS64 to the FPGA chip, and completes the initialization of all parts of the chips and connectors subsequently. During the effective period of the system clock, the FPGA will detect whether there is continuous data reading via the cameralink connector or not; in other words, it detects whether the FVAL and LVAL (discussed in Section 2.3) signals are at a high level. If the data is reading in exactly, the 8-bit digital video signal will be sending to SDRAM by the SDRAM controller, or the system will be waiting, as shown in

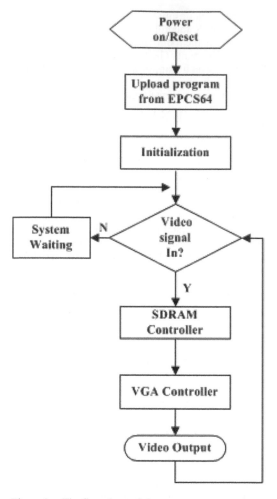

Figure 5. The flow chart of the system.

the chart. It is unnecessary to make a color space conversion since the screen displayed is a grayscale map, hence only one channel is employed.

The SDRAM controller is a double-read and double-write port, which consists of four asynchronous FIFO modules: two used for reading and the rest used for writing correspondingly. All the data are buffered in SDRAM memory by the SDRAM controller, giving the order between write in and read out. When a frame of the image is processed, the SDRAM controller sends the CC1 signal to control the camera to read the next frame; meanwhile, it sends the processed image to the VGA controller.

It is essential to convert the digital image read out from the FPGA to the analog image. By taking advantage of the VGA controller designed together with the HS and VS signals controlled by the FPGA, we can accomplish the display of one frame of the video. The system output adopts PAL 60 Hz video format, thanks to the 800x600 normal VGA-timing, and we then make an appropriate cutting of it so that to obtain the effective timing signal of this project, from 448 to 783 for the active time of horizontal timing and 199 to 454 for that of the vertical timing. Finally, the resolution of the output grayscale video is 336×256..

In this section, the design of the system software is discussed above, we give a particular introduction of the SDRAM controller: what does it consists of and how does it work, and illustrate how to get the non-standard VGA timing of this project.

4 ANALYSIS OF THE RESULTS

In the first two sections, we have accomplished the whole design of the system, including the hardware and software design together with the type of components. Nevertheless, the completive design is not secure for the unbroken output of the image for two reasons: first, the non-uniformity caused by the manufacturing process, material and the disturbance from the circumstance; second, the radiated noise and current noise. All the unfavorable factors together make the image have a worse quality, edge blur, low intensity and low resolution, and it is not easy to distinguish the target from the background (Liu, 2013; Wang, 2013). Owing to these disadvantages, it is important to process the original image because of its non-uniformity and make its enhancement. In this project, we utilize the one-point calibration algorithm for the non-uniformity correction and the histogram equalization to improve the quality of the image. After programming and debugging by means of Quartus II, we take two typical grayscale maps indoor and outdoor for an illustration, as shown in Figure 6 and Figure 7.

Figure 6 and Figure 7 show the indoor and outdoor images, respectively. From these figures, we

Figure 6.   The indoor grayscale image.

Figure 7.   The outdoor grayscale image.

can clearly see the outline of the portrait in the two grayscale maps and that the facial brightness are significantly higher than the surroundings, which suggests that the temperature of this part is obviously higher than the background; what's more, it is because of wearing the glass that the temperature of this region is extremely lower than the other and reflected deep gray on the maps. Besides, the contrast of Figure 7 is higher than that of Figure 6, which is caused by the high temperature difference between each other.

5 CONCLUSION

In this scheme, we make use of the uncooled infrared detector TAU336 to design and implement a normal temperature measurement system based on the FPGA processor. By designing the cameralink interface, the system realizes the velocity matching of the image between the input and the output. The design of the hardware platform mainly consists of one FPGA processor, one outside SDRAM memory and the digital to analog chip ADV7123, together with other hardware resources. In the design of the software system, we introduce the working flow of the system and illustrate specific parameters such as the match of the system clock, the cache of data and the algorithm for processing the grayscale map. It is reasonable that the system designed in this paper satisfies the measurement requirement through the contrast of the picture displayed. In addition, the system has the characteristics of small size, real time and vivacity, and also has its application in the engineering practice to some extent.

ACKNOWLEDGMENTS

This work was supported by the Beijing Natural Science Foundation (No. 4131002).

REFERENCES

Altera. (2011). Cyclone IV_Handbook.

Cheng, Q., Paradis, S., Bui, T., & Almasri, M. (2011). Design of Dual-Band Uncooled Infrared Microbolometer. IEEE Sensors Journal, 11(1), 167–175.

Fièque, B., Tissot, J. L., Trouilleau, C., Crastes, A., & Legras, O. (2007). Uncooled Microbolometer Detector: Recent Developments at Ulis (Vol. 49, pp. 187–191).

Fokaides, P. A., & Kalogirou, S. A. (2011). Application of Infrared Thermography for the Determination of the Overall Heat Transfer Coefficient (U-Value) in Building Envelopes. Applied Energy, 88(12), 4358–4365.

Gokhale, V. J., & Rais-Zadeh, M. (2014). Uncooled Infrared Detectors Using Gallium Nitride on Silicon Micromechanical Resonators. Journal of Microelectromechanical Systems, 23(4), 803–810.

Huang, D., & LIU, X. (2013). Design of High-speed Image Acquisition and Processing System Based on CameraLink. Journal of Jilin University (Engineering and Technology Edition) (S1), 309–312.

Huang, Z., & Shen, N. (2014). A Low-cost Infrared Absorbing Structure for An Uncooled Infrared Detector in a Standard CMOS Process. Journal of Semiconductors (3), 97–101.

Kargel, C. (2005). Infrared thermal imaging to measure local temperature rises caused by handheld mobile phones. IEEE Transactions on Instrumentation and Measurement, 54(4SI), 1513–1519.

Kocer, H., Arslan, Y., & Besikci, C. (2012). Numerical Analysis of Long Wavelength Infrared HgCdTe Photodiodes. Infrared Physics & Technology, 55(1), 49–55.

Liu, J. (2011). Overview of the MEMS technology in uncooled infrared detectors, 2011International Conference on Commercialization of Sensor & MEMS.

Liu, Z. (2013). Research on Infrared Image Nonuniformity Correction and Image Enhancement Technology, Harbin Engineering University.

Merchant, C. J., Embury, O., Le Borgne, P., & Bellec, B. (2006). Saharan dust in nighttime thermal imagery: Detection and reduction of related biases in retrieved sea surface temperature. Remote Sensing of Environment, 104(1), 15–30.

Pan, X., Barker, P., & Meschanov, A. (2002). Temperature measurements by coherent Rayleigh scattering. Optical SOC AMER.

Systems, F. C. (2010). Tau 640 Slow Video Camera User's Manual.

Tong, X. (2012). Study of Video Image Processing System Based on FPGA., Xidian University.

Wang, H. (2013). Research on the Key Technologies of Temperature Field Measurement Based on Thermal Infrared Imager., University of Chinese Academy of Sciences (Xi'an Institute of Optics & Precision Mechanics).

Wang, M. (2012). Research on the room temperature measurement system based on uncooled infrared thermal imager., Guilin University of Electronic Technology.

Wu, Z. (2013). FPGA-Based Technology Research on Camera's Image Acquisition and Processing with Camera Link Interface., Harbin Institute of Technology.

Xing, S. (2005). Study of Uncooled Infrared Thermal Imaging System., Nanjing University of Science and Technology.

Yang, P. (2013). Design and Implementation of the Miniaturization Infrared Image Signal Processor., Changchun University of Science and Technology.

Yang, Q., & Zhou, Y. (2013). Design of Video Image Acquisition and Display System Based on FPGA. Computer Engineering and Design (06), 1988–1992.

Zhang, G., & Huang, W. (2001). Applications of the Infrared Thermography on the Metal Mechanical Processing. Infrared and Laser Engineering, P74-P78.

Zhang, H., & Liu, X. (2010). Digital Video Collection and Conversion System Design Based on FPGA. Computer & Digital Engineering(06), 77–81.

Energy Science and Applied Technology – Fang (Ed.)
© 2016 Taylor & Francis Group, London, ISBN 978-1-138-02833-3

The design and implementation of mobile application security detection system

Congming Wei, Qi Li & Yanhui Guo
Beijing University of Posts and Telecommunications, Beijing, China

ABSTRACT: With the increasing frequency of using mobile intelligent terminal, mobile application software security is particularly important. So, it is necessary to design and achieve an efficient, comprehensive and rapid mobile application detection system. This paper puts forward a kind of improved method combined with static application detection and machine learning methods. The key of the improved method is to determine the characteristics of the study object and machine learning algorithms. This paper proposes the use of support vector machine to realize the automation application safety assessment. For the system functional requirements, the mobile application security testing system consists of application analysis subsystem, machine learning subsystem, and the processing subsystem. Finally, the test results show that the mobile application security testing system designed in this paper can well meet the functional requirements of the application security testing, and ensures a better rate of detection and recognition, and thus achieves the expected goal.

Keywords: Static analysis; Machine learning; Support vector machine; Feature mapping

1 INTRODUCTION

In recent years, with the gradual improvement of mobile intelligent terminal computing ability, mobile intelligent terminal functions are becoming more complex, mobile application security situation are becoming more severe, and mobile malicious applications have become one of the most important factors threatening the development of mobile Internet. To this end, efforts are made to effectively prevent and control mobile intelligent terminal malicious programs related to China's mobile Internet industry's healthy development and the vital interests of the user of the mobile terminal. However, there are many differences between the traditional mobile security environment and the security environment. Compared with the traditional terminal, mobile intelligent terminal has the characteristics of mobility and other characteristics such as preserving personal privacy data. No traditional method of software safety analysis and protection provided an effective support for mobile platforms, so the intelligent mobile terminal software security analysis and protection technology research has important significance. In recent years, many researchers have studied the detection of malicious applications. Wang Rui proposed a kind of malicious code behavior feature extraction and detection method based on semantics, using the behavior graph of semantic abstract to describe the behavior characteristics of malicious codes. Sathyanarayan extracted API call features from known malicious samples using the static analysis method for the characteristics of the malicious code family, as an alternative to the feature extraction of a single application of malicious samples. Igor Santos and others proposed a method to analyze malicious applications based on the sequence of instructions, for the inspection of unknown malicious applications of the known malicious family. Mamoun Alazab used the API call as a starting point of malicious feature detection. They analyzed the features of the structure and behavior, and developed a set of automatic disassembly and extraction system, which can identify a malicious software and a normal software. A. H. Sung J. proposed a kind of malicious code detection method based on the behavior characteristics of system call sequences.

However these algorithms have some shortcomings. First, the existing methods cannot detect unknown viruses and variants of known viruses, as the basis of feature analysis is the malicious code in the library, if the malware samples are yet to be analyzed, the traditional analysis method based on the feature analysis cannot identify the virus, resulting in false negative. For the known viruses after deformation and confusion, the characteristic analysis is incapable. Second, the accuracy of the method depends on the update of the feature library. Despite the new virus appearing continuously, the characteristics of scanning tools must constantly be upgrade and the virus database must be updated. If the virus database is not updated on time, there will be the risk of acquiring new virus infection.

According to the analysis of mobile application behavior sequence and classification problem, this paper proposes a malicious application detection method based on support vector machine. The method combines the signature information and static detection feature, and establishes the support vector machine model to detect malicious applications. The experiment shows that the method has a high analysis accuracy, and can be used to detect an abnormal behavior in the application.

2 FRAMEWORK REVIEW

The ultimate purpose of this detection system is to provide a simple, efficient mobile application security testing tool, which classifies Apps on the basis of features by using the machine learning methods, and protects the users from security risks. The framework consists of application analysis subsystem, machine learning subsystem and processing subsystem. Application analysis subsystem is mainly responsible for decompiling Android application, scanning source code and related documents and extracting the application feature. Machine learning subsystem, the core of the whole detection system, is responsible for learning the characteristics of the software detected, so as to judge the security of the software. The processing subsystem is responsible for the output of test results. The overall framework is shown in Figure 1.

Figure 1 shows the interaction between the main function module of the whole detection system, the user layer and the resource layer. The arrow indicates the relationship between the modules and the subsystem with a call to the called.

Application analysis subsystem includes application receiving module, signature extraction module, preprocessing module and feature extraction module. Preprocessing module is mainly responsible for the decompression of application, classes. It decks binary files decompilation and the decompilation of AndroidManifest.xml. Feature extraction

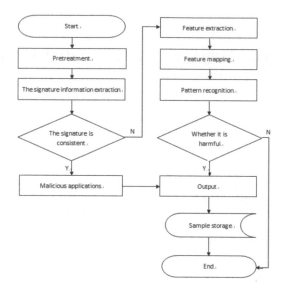

Figure 2. Implementation process.

module extracts the application features from the source code, AndroidManifest.xml, resource file according to the result of application decompilation, and then submits them to the machine learning module.

Machine learning subsystem consists of feature mapping module and the pattern recognition module. Feature mapping module is responsible for mapping the feature to the vector space through 0, 1 processing. The pattern recognition module completes two classification treatments of the processed feature, so as to judge the security of the application.

According to the output results of the machine learning subsystem, the result processing subsystem automatically generates the test document, and stores the application of the sample and documents in the resource layer.

The method combining the sample analysis and machine learning can improve the detection accuracy of malicious software. The whole detection system implementation process is shown in Figure 2.

3 DESIGN OF MACHINE LEARNING MODULE

3.1 *Support vector machine*

The core idea of support vector machine is to construct one or more hyperplanes to make a distance from the data need to classify to the decision plane are the farthest. The basic principle of support vector machine can take a simple two-

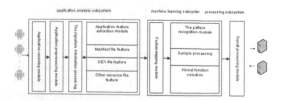

Figure 1. System framework.

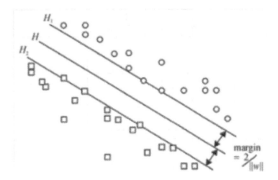

Figure 3. Optimal hyperplane under the condition of linear reparability.

dimensional classification problem as an example. As shown in Figure 3, we can use a straight line to separate these two categories of data: the line is the hyperplane H, assuming that y is −1 for the circular datapoints on a H hyperplane side, and 1 for the circular datapoints on the other side. The solid line in Figure 3 is the Optimal HyperPlane, whose distances to the two edges are equal. This distance is also called classification intervals. The points on H1 and H2 are the "support vector".

From the above figure, we assume that the training sample set is as follows: x_i in R^n, y_i in {−1, 1}, $l = 1, ..., n$. The corresponding classification function can be expressed as

$$y_i[(w \cdot x_i) + b] - 1 \ni 0, i = 1, ..., n \qquad (3-1)$$

where x_i, y_i is the training sample, R^n is expressed as the training sample space, the n value is the number of training samples, the b value is the bias term, $w \cdot x_i$ is the vector inner product, sample classification intervals can be expressed as $2b\|w\|$; therefore, to make the maximum classification intervals, $(\|w\|)/2$ should be minimum. So, the face both meeting the classification function formula (3–1) and letting $(\|w\|)/2$ minimize is the optimal classification face. In order to solve the above problem, we can use Lagrange's theorem to turn it into a dual problem, which satisfies the following constraint conditions.

In the condition of $\sum_{i=1}^{n} y_i a_i = 0$ and $a_i \geq 0$, $i = 1, ..., n$, solution function (3-2) of the maximum value is

$$Q(a) = \sum_{i=1}^{n} a_i - \frac{1}{2} \sum_{i,j=1}^{n} a_i a_j y_i y_j (x_i \cdot x_j) \qquad (3-2)$$

where A is the Lagrange multipliers corresponding to each sample, and the sample solved is the support vector, and then we can obtain the optimal classification function:

$$f(x) = \text{sgn}\{(w^* * x) + b^*\}$$
$$= \text{sgn}\left\{\sum_{i=1}^{n} a_i * y_i \times (a_i \cdot y_i) + b^*\right\} \qquad (3-3)$$

In Formula (4-4), w* is a weight vector, if the problem is linear, it turns into a linear separable problem in high-dimensional space, then through the nonlinear transformation, the problem eventually becomes the optimal classification plane search in high-dimensional space. In fact, selecting the appropriate kernel function K $(x_i * x_j)$ can achieve this purpose, and the objective function for the corresponding transformation is

$$Q(a) = \sum_{i=1}^{n} a_i - \frac{1}{2} \sum_{i,j=1}^{n} a_i a_j y_i y_j K(x_i \cdot x_j) \qquad (3-4)$$

The classification function becomes

$$f(x) = \text{sgn}\{(w^* * x) + b^*\}$$
$$= \text{sgn}\left\{\sum_{i=1}^{n} a_i * y_i \times K(a_i \cdot y_i) + b^*\right\} \qquad (3-5)$$

The formula (3-5) shows that, for the SVM non-linear classification problem, the kernel function is the key. This paper uses the kernel function based on polynomial $K(x_i, x_j) = ((x_i \cdot x_j) + 1)^d$.

For the sample weighted processing method, this paper takes the importance of different samples as basis, references the thought of the class distance between the same type of samples in KNN classification algorithm, and designs an empirical method to evaluate the sample weight distribution. The representation of m_i is the mean vector for class i samples:

$$m_i = \frac{1}{n_i} \sum_{k=1}^{n_k} X_k^{(i)} \qquad (3-6)$$

In Formula (3-6): n_i is defined as the number of class i of sample, i is defined as the category of the sample. If the total number of samples is 2, then $2 = n_1 = n_2$, and note that $n_{min} = \min(n_1, n_2)$; in the sample concentration, $x^{(i)}$ means all belong to the category i sample collection. The mean vector of the m_i sample class is m_j, then the Euclidean distance samples x_i and m_j can be expressed as

$$d(x_i, m_j) = (x_i - m_j)^T (x_i - m_j)$$
$$= \sum_{k=1}^{n} (x_i^k - m_j^k)^2 \qquad (3-7)$$

In the formula (3–7): The k dimension component of x_i is x_i^k, and the k dimension component of m_j is m_j^k, the dimension eigenvector of sample is n. Then, the estimation sample weights can be expressed accordingly as

$$d(x_i, m_j) = (x_i - m_j)^T (x_i - m_j)$$
$$= \sum_{k=1}^{n} (x_i^k - m_j^k)^2 \qquad (3\text{-}8)$$

3.2 The machine learning module

3.2.1 Feature mapping module

Usually, the characteristics of malicious applications will reflect the specific mode and combination, our goal is to express dependencies among features with the concept of using machine learning. First of all, we need to put these extracted features values mapped to the corresponding feature space. Therefore, this paper defines a set of S, by using the set S, and we define a |S| dimensional vector, and each element is 1 or 0, then the application of x can be mapped to the $\varphi(x)$ vector, so each character s is extracted from x whose corresponding value is 1 or 0. Therefore, the feature vector function of application $\varphi(x)$ can be defined as follows:

$$\varphi: x \rightarrow \{0,1\}^{|S|}, \ \varphi(x) \mapsto (I(x,s))_{s \subset S} \qquad (3\text{-}9)$$

The function $I(x, s)$ can be defined as follows

$$I(x, s) = \begin{cases} 1 & \text{if application x concludes characteristic s} \\ 0 & \text{if x doesn't conclude characteristic s} \end{cases}$$
$$(3\text{-}10)$$

In the vector space, the distance between applications that share the similar characteristics is very near, and the distance between applications that share inconsistent characteristics is far. Therefore, we can judge whether the application is harmful by determining its position.

3.2.2 Pattern recognition

Before describing the construction method of pattern recognition module, first we build a verification method of pattern recognition module to metric the error value of classification function. This paper chooses the most common verification method, namely the cross validation method, to achieve. It is often used to determine the accuracy of the model in the case of small samples. Training pattern recognition module based on the support vector machine method, the pattern recognition module achieves according to the process shown in Figure 4. The specific steps are as follows:

A. raining sample processing is mainly the sample feature extraction;
B. Map extraction samples feature to a vector space.
C. Train pattern recognition module based on the cross validation method on the condition of different kernel functions, until the optimal.

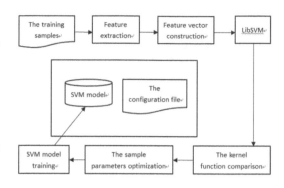

Figure 4. The machine learning module processing flow.

D. Based on the sample processing method, we calculate sample weights and optimize the parameters according to sample weights.
E. Based on the above steps, we train the SVM model to get the final model.

4 EXPERIMENTAL VERIFICATION

4.1 Sample selection

In order to ensure the diversity of samples, this paper collected from multiple channels to download the application sample of 131611, including hazardous and normal samples, of which 96150 applications are from the Google Play official store, 19545 applications are from three stores, and 15916 applications are from the forum and website.

4.2 Performance verification

According to the research on machine learning algorithm in Section 3.1, this section tested and optimized the model proposed in this paper based on the above sample, which proved the performance of the algorithm and the model.

The performance of the algorithm. As the decision tree classification algorithm is not suitable for this environment, only the performances of bias algorithm, KNN algorithm and the algorithm of support vector machine are compared. The experimental results are presented in the table: the average accuracy of support vector machines is 88.5% higher than the accuracy of the Bayesian algorithm and its improvement. Compared with the K nearest neighbor, K-NN, the average accuracy rate of the support vector machine algorithm is slightly higher, but the error is 0.32% far less than the other two methods, showing that the method has good stability.

Comparison with other detection methods. In this test, we compare the detection system with other detection tools. We compared our tools with the methods proposed by Kirin, RCP Peng et al.

Table 1. Application sample distribution.

Store	Google play	Three party stores	Website and forum
The number of samples	96150	19545	15916

Table 2. The classification accuracy of different algorithms.

The data mining method	The accuracy rate %	Error %
NB	67.81	9.20
NB + FCBF	85.02	2.32
NBK	86.54	1.93
NBK + FCBF	87.02	3.6
KNN	87.88	2.20
SVM	88.50	0.32

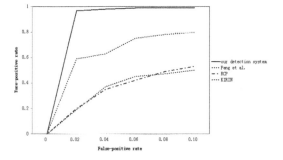

Figure 5. Comparison with other detection methods.

The test results are shown as the ROC curve in Figure 5. As can be seen from the figure, the detection rate of this system is much higher than that of other methods. The correct detection rate reaches up to 94%, and the error rate is only 1%, and the error rate of other methods is 10%~50%.

5 CONCLUSION

This paper mainly completed the following tasks. It introduces the Android platform security threat situation, and illustrates the necessity and importance of the design and realization of the application security detection system. It summarizes the principle and problems of existing malware detection technology, and puts forward a detection method of the signature information detection, static analysis method and machine learning method combining.

Starting from the overall system architecture, system flow, this paper designs the application security detection system, and discusses the key function of the application security detection system module design.

This paper introduces the method of machine learning, and puts forward the research method based on support vector machine as the application security classification algorithm. In addition, in order to avoid the deficiency of existing algorithms, the paper puts forward the corresponding optimization strategy.

In this paper, the main functions of the system and the performance of the algorithm is tested and analyzed. First, we collect a large number of samples used to provide good data to support the validation of the model optimization and performance. Then, we compare the detection and recognition rate of different data mining methods, verifying the accuracy of the support vector machine algorithm. Finally, based on the real experimental data, we compare our tool with other detection tools, to verify the availability and effectiveness of the system.

ACKNOWLEDGMENTS

This work was supported by the National Natural Science Foundation of China Project (61302087, 61401038).

REFERENCES

Alazab M, et al. Malware detection based on structural and behavioural features of api calls. 2010.

Cristianini N, et al. An introduction to support vector machines [M]. Publishing House of electronics industry, 2004.

Ding Changfu, Wang Liang. Neural network based on BP cross validation method to fault diagnosis of steam turbine [J]. Electric power science and Engineering, 2008, 24(3): 31–34.

Li Panchi, et al. Application of support vector machine in pattern classification [J]. Journal of Daqing Petroleum Institute, 2003, 27(2): 59–61.

Mirzaei N, et al. 2012.Testing Android Apps through Symbolic Execution. SIGSOFT Soft. Eng. Notes, November, 2012, 37(6):1–5.

Peng H, et al. Using probabilistic generative models for ranking risks of android apps [C]. Proceedings of the 2012 ACM conference on Computer and communications security. ACM, 2012: 241–252.

Santos I, et al. Idea: Opcode-Sequence-Based Malware Detection. Massacci F, Wallach D, Zannone N. Springer Berlin Heidelberg, 2010:35–43.

Sarma BP, et al. Android permissions: a perspective combining risks and benefits [C]. Proceedings of the 17th ACM symposium on Access Control Models and Technologies. ACM, 2012: 13–22.

Sathyanarayan VS, et al. Signature Generation and Detection of Malware Families: ACISP '08, Berlin, Heidelberg, 2008[C]. Springer-Verlag.

Sung A. H., et al. Static analyzer of vicious executables (save), 2004 [C].

Wang Rui, et al. The behavior of malicious code detection and feature extraction method based on semantic. Journal of software, 2012(02):378–393.

Energy Science and Applied Technology – Fang (Ed.)
© 2016 Taylor & Francis Group, London, ISBN 978-1-138-02833-3

Study on high reliability urban distribution network based on a multiple-looped network

Yuebin Xu, Weiyong Han & Zheng Gao
Energy China Tianjin Electric Power Design Institute, Tianjin, China

ABSTRACT: Based on a study on the typical electric distribution network structure, this article proposes a new distribution network structure on the basis of the existing distribution automation, which can suit the developmental power grid, improve the distribution network's power supply reliability and lay a solid foundation for the access of distributed energy.

Keywords: electric distribution network structure; distribution automation; multiple-looped network

1 INTRODUCTION

With the sustained and rapid economic development, requirements of residents and users for electricity and power quality have improved significantly. Electric distribution network to users directly is to ensure the power reliability and quality, and to improve the efficiency of the power grid and innovative service. It is also a network frame foundation for realizing urban power distribution network automation and distributed power access. Therefore, improving the power supply reliability of a 10kv distribution network of power grid has an important significance for the development of new energy (Zhang, 2005, Zheng, 1999, Wang, 2014). Based on the domestic space truss structure and the abroad advanced space truss structure of the urban power supply network, this article considers the adaptability of new energy access for the electrical network structure and the related distribution automation technology, and puts forward a new type of distribution network that is suitable for the urban power grid structure.

2 COMPARATIVE ANALYSIS FOR TYPICAL CONNECTION OF DISTRIBUTION NETWORK

Domestic distribution network connection is mainly divided into the aerial area network connection, cable area network and both composite structure. Most of the urban distribution network uses the cable network frame as a result of the limitation of terrain and spatial (Sun, 2009, Chen *et al*, 2000). The 10kv cable line network connection mode mainly has the following four ways: single loop network of cable backbone, double-loop network of cable backbone, two actives plus one standby, and triple-double network (Zhu, 2006, Distribution, 2007).

2.1 Single loop network of cable backbone

In the single loop network, two lines of electricity is derived from the same medium-voltage bus bar of two substations or different medium-voltage busbars of the same substation in the power supply area, and it feeds out the single loop network, by using an open-loop operation mode. The load rate of this connection mode can reach up to 50%, which is suitable for instances where users are concentrated or the power supply reliability requirements are general. It can be converted to a double-loop network in the later process of power grid construction.

2.2 Double-loop network of cable backbone

Double-loop network connection of the power supply is derived from different substations or 4 of the 10kv lines led by different 10kv substation buss, by using the open-loop operation mode. The load rate of this connection mode can also reach up to 50%. However, the double-loop network can provide two power supplies for users, which can significantly improve the power supply reliability. It also has a stable space truss structure, suitable for the urban core area and higher requirement areas.

2.3 Two actives plus one standby

Two overheads or cable lines from two 10kv bus bars of different substations or the same substation form a loop. Then, they increase one other line on the basis of a public spare line, as shown in Figure 3. If two main power lines are derived from bus bars of different substations, the standby power should also be from different substations, and the two main lines should be able to meet the power supply of the normal line when the loss of electrical power supply occurs. The load rate of the main power supply loop can reach up to 67%. In this way, it becomes mainly suitable for the linear distribution and load capacity of more than 3000kva single power supply load or one side of the multi-power supply load.

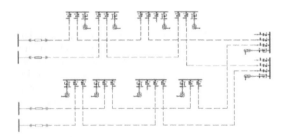

Figure 1. Connection mode of the double-loop network of cable backbone.

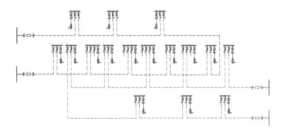

Figure 2. Connection mode of the double-loop network of cable backbone.

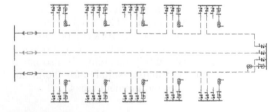

Figure 3. The connection mode of two actives plus one standby.

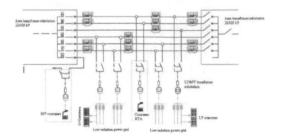

Figure 4. The schematic diagram of a 20kV triple-double network in Paris.

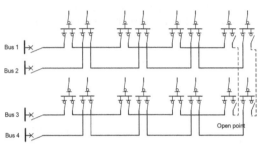

Figure 5. The typical triple-double connection mode of the cable line in Zhejiang Province.

2.4 Triple-double network

"Triple-double network" refer to "double power sources, double lines and double accesses". Double power resources refer to two superior 110kv substations. Double lines refer to the two medium-voltage cables connecting the "double power sources". Double accesses refer to on-off switches connected to the "double lines" connecting to the public. In this mode, the three key modules of the power supply system have their own "backup" at the same time. When failure occurs in a module, the intelligent device will switch to the standby module, in order to ensure continuous power supply by the user. The triple-double network structure has been put into operation in Paris and Zhejiang, as shown in Figure 4 and Figure 5. Figure 6 shows the double accesses of the triple-double network in Zhejiang. Compared with common loop methods, the triple-double network makes two power accesses in a single match, which causes a higher power supply reliability but at a higher cost and a complex operation process in the site construction.

The analysis shows that no method can meet the demand of power supply reliability, load rate, power failure, network component failure, large capacity user access, utilization rate of equipment, and important load applicability at the same time. Therefore, this paper puts forward a new type of

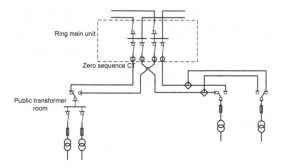

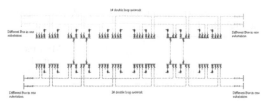

Figure 7. The connection mode of the multiple-looped network (double H network).

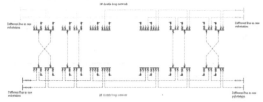

Figure 8. The connection mode of the multiple-looped network (binocular network).

Figure 9. The connection mode of the multiple-looped network (monocular network).

Figure 6. The access connection mode of the double power public distribution transform in Zhejiang Province.

power distribution network structure, namely the multiple-looped network.

3 NEW DISTRIBUTION NETWORK

New distribution network development, which is based on the existing traditional network frame, should be according to the distributed power supply, distribution automation and the requirement in the future (Wang, 2011, Zhang, 2012). Now, the urban distribution is done mainly by the single-loop network of cable backbone, double-loop network of cable backbone and two actives plus one standby. With the development of smart grids and the progress in the pilot and scale of the distribution automation in the medium voltage network application, under the condition of the large distribution master station, the ability and advanced strategy to solve network complexity breakthrough are still ongoing (Zhao et al., 2013, Li, 2014). The single-loop network, the double-loop network and the other traditional network frame can be further improved in the multiple-looped network. The network self-healing can be more flexible, when multiple powers and distributed power insert can be used as a reserve capacity in harmony with the focus on the intelligent and distributed intelligent application strategy, enhancing the stability and economic operation of the traditional network frame.

3.1 Multiple-looped network structure

Based on a single-loop network, there is an appropriate increase in the contact between rings, or the connection between them on the basis of the regional road on both sides, which forms the multiple-looped network. According to different access power supplies and communications, the network can be classified as the double H network, the binocular network and the monocular network, as shown in Figures 7, 8 and 9. The line load rate of

this connection mode can reach up to 50%-70%. It can provide users with 4 lines or below power supply, and the reliability safety level can be up to N−2. Making the power limitation according to the network access can significantly improve the level of access to the large capacity. This mode is better used in areas covered by distribution automation, and should be consistent with the city construction step. It is suitable for the high reliability requirements of important areas, users, and the general area of the double power supply.

3.2 Multiple-looped network compared with common methods

Power failure. It is one of the important indicators in the urban distribution network. Without considering differences under the premise of equipment, corresponding indicators by the calculation and analysis of different network structures are listed in Table 1. The analysis shows that the multiple-looped network can meet the "N-2" or "N-2" principle when the "quasi N-2" achieves up to "N-2" on the basis of manual intervention.

Network component failure. It is formed by evaluating the satisfaction assessment based on the continuous power supply and the lateral load

Table 1. Statistical table of the network power supply failure's effect.

Connection mode reliability	Typical primary system				Multiple-looped network	
	Single loop	Double loop	Two actives plus one standby	Triple-double	Double H	Binocular
N-1	√	√	√	√	√	√
Reliable N-2	×	√	√	√	√	√
N-2	×	×	×	√	×	√

Note: × for dissatisfy; √ for satisfy.

Table 2. The impact assessment of the network component failure on the load side uninterrupted power supply.

Transformer room access mode	Public or dedicated transformer room Double(over) power source device						
Access networks component failure	Single loop	Double loop	Two actives plus one standby	Triple-double		Binocular	Double H
				Zhejiang	Paris		
10kV access line	×/×	√	√	√	√	√	√
10kV bus	×/×	√	√	√	√	√	√
10kV bus to transformer feeder line	×/√	√	×	√	√	√	√
Distribution transformer	×/√	√	×	√	√	√	√

Note: × for cannot supply power uninterruptible; √ for can supply power uninterruptible; o for cannot ensure uninterruptible power supply.

Table 3. Capacity utilization rate under different network connection modes.

Connection mode	Typical primary system				Multiple-looped network	
	Single loop	Double loop	Two actives plus one standby	Triple-double	Double H	Binocular
Equipment utilization	50%	50%	67%	50%	75%	75%

Table 4. The applicability of different load levels and capacities for different distribution networks.

Connection mode Important Load applicability	Typical primary system				Multiple-looped network	
	Single loop	Double loop	Two actives plus one standby	Triple-double	Double H	Binocular
Extra first order load	×	√	×	×	√	√
First order load	×	√	√	×	√	√
Second order load	o	√	√	√	√	√
Third order load	√	√	√	√	√	√
Great capacity user	1× limit	2× limit	2× limit	2× limit	4× limit	4× limit

Note: × for not applicable; o for partly applicable; √ for applicable.

electrical link consists of the power supply network and the system associated with electrical components. System analyses are presented in Table 2.

Equipment utilization analysis. Under the condition of safety standards of the distribution network, equipment utilization analysis can be defined as the maximum load rate of distribution equipment. Besides, there is a certain connection between the maximum load rate of distribution equipment and the line, as given in Table 3.

Analysis of important load applicability. Determining the corresponding power supply strategy should be according to different load types, different user types, and different sizes in order to ensure the safety, reliability, progressiveness and rationality of the power distribution system. Taking Tianjin electric power company as an example,

Table 5. Assessment form of various constraint conditions.

Connection mode Influence condition	Typical primary system				Multiple-looped network	
	Singleble loop	Dou-ble loop	Two actives plus one standby	Triple-double	Dou-ble H	Bino-cular
Power failure	C	B	B	B	A	A
Network component failure	C	C	C	C	A	A
Equipment utiliza-tion	B	B	B	B	A	A
Important load applica-bility	C	B	B	C	A	A

Note: C for poor; B for ordinary; A for excellent.

the applicability in different network connection modes and different power supply load levels are given in Tables 4. The limitation is the largest capacity, according to analysis of the power grid structure, reserve capacity and overload ratio in different areas.

Analysis of important load applicability. By the analysis presented in Tables 5, we can be seen that no matter which connection mode of the distribution network or distribution transformer is selected, the limitation and applicable conditions still remain. Of the different kinds of connection mode, the double H network and the binocular network are more helpful to improve power supply reliability and promote the optimization and upgrading of existing distribution networks. The level of equipment utilization rate, power supply redundancy, power supply reliability, large capacity and user access in the multiple-looped network can be further increased.

4 CONCLUSION

From the above analysis, the double H network and the binocular network of the multiple-looped network perform better in power supply at a maximum load in order to meet the need for reliability and flexibility of self-healing. The harmonious development of the multiple-looped network and distribution automation lays the foundation for increasing the rationality and optimization of distributed power access, which can raise the stability of the traditional network frame.

REFERENCES

2007. Distribution network planning for Yizhuang economic and technological development zone in Beijing.
Chen Tingji, Cheng Haozhong, He Ming. 2000. Research on Connection Modes of Urban Middle Voltage Distribution Networks. *Power system Technology*, 24(9):35–28.
Li Xiaoliang. 2014. Research for Improving Power Quality Based on Distributed Generation in aMicro. Grid. *Beijing Jiaotong University.*
Sun Lifeng. 2009. Research on Connection Mode of 10 kV Distribution Network in City Power Network Layout. *Electric Power Science and Engineering*, 25(12):69–72.
Wang Chengshan. 2014. Research on Key Technologies of Microgrid. *Transactions of China Electrotechnical Society*, 29(2):1–12.
Wang Kai. 2011. Research on Complexity and Vulnerability of Power Network Structure Based on Complex Network Theory. *Huazhong University of Science and Technology.*
Zhang Huawei. 2005. Research on the Optimizational Planning of Distribution Network in Urban Network Renovation. *North China Electric Power University (Beijing).*
Zhang Si. 2012. Optimizational Planning for transmission network considering the short circuit current. *Zhejiang University.*
Zhao Jiaqing, Wang Ke, Yang Shengchun, etal. 2013. Analysis for evaluation indexes of complex structural security in transmission networks. *Power System Protection and Control*, 14:48–53.
Zheng Meite. 1999. Study on the Principle of Power Network Configuration Planning. *China Power*, 6:11–14.
Zhu Yong. 2006. Research on Connection Modes of Urban Middle Voltage Distribution Network in Zheng dong new district. *North China Electric Power University (Beijing).*

Energy Science and Applied Technology – Fang (Ed.)
© 2016 Taylor & Francis Group, London, ISBN 978-1-138-02833-3

Information fusion framework for feature classification in machine fault diagnosis

Li Li & Songlin Wu

Faculty of Mechanical Engineering, Xijing University, Xijing, China

ABSTRACT: On the basis of information fusion theory and machine Maintenance Information Fusion System, the process of information fusion and experimental results of various classifiers for equipment fault diagnoses are studied in detail. A new and general information fusion framework for feature extraction is suggested for the diagnosis, using MLP, RBFNN and KNN classifiers and Dempster-Shafer theory. The fault feature extraction and the application of the information fusion framework in fault diagnosis are carried out, respectively. It has been shown that the performance of the proposed framework is efficient in the computational experiment.

Keywords: Information Fusion; Fault Diagnosis; Feature Extraction

1 INTRODUCTION

Information fusion definitely is a process with multi-levels and multi-aspects, combining multi-source information and data to deal with detection, relation and estimation. Its purpose is to obtain the correct station and identity estimation, and obtain the results on machine condition completely and timely. It has been used widely in the machine maintenance system, such as maintenance process monitoring, optimization, decision-making support and performance condition estimation. Essentially, machine fault diagnosis is to comprehensively evaluate the information by making use of knowledge and data corresponding to machine running conditions. To date, the study and research work have been mainly focused on the information fusion algorithm and its application to solve some special problems in machine fault diagnosis. For example, the Intelligent Fusion Diagnosis System for Aviation Engine developed by the Nanjing University of Aeronautics and Astronautics, the Information Fusion System for Engine Diagnostics and Health Management developed by NASA Glenn GRC, neuron networks, pattern recognition, statistic parameters estimation and fuzzy logical have usually been used in these applications to implement the typical fault diagnosis and prognosis for engines.

Information fusion still gets its primary level in machine fault diagnosis. For instance, a fault diagnosis framework based on information fusion was proposed by Ma, in which condition parameters and signals were optimally selected and processed to extract the features. Wang and Wan presented another machine fault diagnosis framework based on the multi-sensor information fusion. In addition, multi-diagnostic methods are fused into the system, by which the much more precision diagnosis results were achieved than by single information. As the single information is of higher accuracy, the multi-methods and multi-sensors methods mean nothing.

Practically, a large amount of information-related machine would be used during the repair maintenance process. To efficiently and reasonably make use of these data and improve the correct rate of machine fault diagnosis and the versatility of the diagnosis system, on the basis of the machine maintenance information system, several key problems in information fusion-based diagnosis system are discussed, and a general information fusion framework for fault feature detection is presented in this paper. The corresponding experimental study is performed.

2 INFORMATION FUSION FRAMEWORK FOR FAULT FEATURE DETECTION

Machine maintenance information fusion system can be broken up into three classifications in its structure: original data collection or acquisition, feature detection and decision-making, as shown in Figure 1. At the data acquisition level, original data collected directly from sensors are analyzed comprehensively, including fault information collection, arranging and running parameter recording, statistical analysis, and trending prediction. The purpose of fault feature detection is to imple-

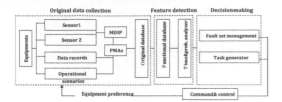

Figure 1. Machine maintenance information fusion system.

ment the data compression. The compressed data are then summarized and analyzed comprehensively according to the application. The fault features are extracted to implement the process of mapping from the fault evidence space into the fault space, i.e. the general inspection and diagnosis for machine faults. At the decision-making level, the reasoning for trouble shooting and positioning and the parameters' reliability estimation and prognosis, are carried out generally according to the decision-making target.

In the information fusion of fault feature detection, the process of machine fault diagnosis is to carry out the pattern recognition for representing machine running conditions, using multi-sensors information. The information is generally non-linear separable and non-parameters. The fault diagnosis can be viewed as a process to classify the symptom data. During the process of implementing the various information mapping from the fault symptom space into the fault space, f:S→E, some classifiers often have different accuracies for a part of categories, sometimes higher and at other time lower. Therefore, an efficient method for the pattern classification should be set up. Various classifiers are considered to be used as supplementary. In this paper, some classifiers widely used in fault diagnosis, such as multi-layer-perception (MLP), RBFNN and kNN, are analyzed experimentally for fault diagnosis. The results are presented in Table 1.

Clearly, the performances of the classifiers are different. MLP has a higher correct rate for conditions N and C2. RBPNN has a higher correct rate for N and C1 than for C2, and kNN is better for C1 and C2 than for N. To improve diagnostic performance, the three classifiers are concurrently used in the proposed information fusion framework.

From the above analysis, a general information fusion framework in fault feature detection level and used for machine fault diagnosis is proposed, as shown in Figure 2. Fusion calculation for feature classification is carried out to accurately implement the condition classified on the basis of multi-source data. An expert system is then used to verify the diagnosis results.

Table 1. Diagnostic performance of three classifiers for testing samples.

Input	Classifier	Classifications of conditions		
		N	C2	C1
N	MLP	91.5		8.5
	RBPNN	96	1.5	2.5
	kNN	68.5	14.5	17
C1	MLP	14.5	85.5	
	RBPNN	4.5	93.5	2
	kNN	2	98	
C2	MLP	6	3	91
	RBPNN		9	91
	kNN	1	1.5	97.5

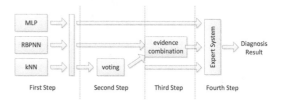

Figure 2. Information fusion frameworks for fault feature detection.

3 EXPERIMENTS FOR ROTATING MACHINE FAULT DIAGNOSIS

The fusion framework shown in Figure 2 is used in the fault diagnosis of rotating machine to verify its performance. A total of 300 groups of data sampled from the normal condition, oil whirling and mass unbalance are used as training samples to train each single classifier. Another 700 groups of data are used as testing samples for the final verification. Each single classifier's output is then set as $[1\ 0\ 0]^T$, the normal condition as $[0\ 1\ 0]^T$, oil whirling as $[0\ 0\ 1]^T$ for the mass unbalance.

In the first step, BP algorithm is applied for the MLP training, during which the kinetic correction and the variable learning rate are determined. The mean square error (MSE) is set as 0.03 and the kinetic factor as 0.85. The structure of MLP with two layers is designed as $4 \times 11 \times 3$. The networks reach their exception value after 5482 times training. The RBPNN networks' SPREAD constant is set as 0.1. The kNN is applied when k = 1.

The testing sample is then used to verify the classifier's performance. The results are presented in the CM form in Table 2.

From Table 2, it is clear that each classifier has different performances from the others. With respect to the oil whirling fault, MLP and RBPNN

Table 2. Diagnosis result of the first step.

Input	Classifiers	Classified conditions		
		Normal	Oil whirling	Mass-unbalance
Normal	MLP	74.4	9	16.6
	RBPNN	86.1	9.2	4.7
	kNN	91.3	7	1.7
Oil Whirling	MLP	5.3	91.4	3.3
	RBPNN		89.3	10.7
	kNN	18	67.9	14.1
Mass-Unbalance	MLP	3	11.4	85.6
	RBPNN	14.7	7	78.3
	kNN		7.6	92.4

Table 3. Classification results after the voting.

Input	Classified conditions			
	Normal	Oil whirling	Mass-unbalance	No result
Normal	91.9		8.1	
Oil whirling		89.9		10.1
Mass-unbalance		3	88.3	8.7

have a higher accuracy, while kNN has a lower accuracy.

To reduce the difference and improve the diagnostic performance, a voting rule to the three classifiers is used in the second step. If two of the three support a condition, then the result is determined as this condition. Otherwise, the result is rejected when the three outputs are totally different. The result of the second step is presented in Table 3.

The fusion framework's performance has been improved to some extent after the second voting process. Still, there appears situations with no result, except for the two conditions of oil whirling and unbalance. These situations are inputted into the third process, namely the evidence-combination. The situations with no result will be classified by evidence-combination into applicable classifications, as presented in Table 4.

From Table 4, we can see that there are some unclear classified situations in each condition. The final process, the expert system, will determine the end result. As shown in tables, an expert system with the flowing sample rules is designed.

If the condition is determined as mass-unbalance after the second step, THEN if the first step's result is mass-unbalance by kNN THEN.

The condition is mass-unbalance. ELSE the condition is whiting END. END.

Table 4. Classification results through evidence combination.

Input	Classified conditions		
	Normal	Oil whirling	Mass-unbalance
Normal	91.9		8.1
Oil whirling		93.6	6.4
Mass-unbalance		9.9	90.1

Table 5. Final diagnosis results.

Input	Classified conditions		
	Normal	Oil whirling	Mass-unbalance
Normal	91.9		8.1
Oil whirling		93.6	6.4
Mass-unbalance		6.8	93.2

After the fourth step, the final diagnosis results are achieved through the expert system, as presented in Table 5.

With respect to the rotating machine fault, the average diagnostic accuracy reaches up to 92.9%. It is shown that the information fusion framework for fault feature detection has a promising performance.

4 CONCLUSIONS

On the basis of the principle of information fusion and machine maintenance information fusion system, a general fault diagnosis information fusion framework for fault feature detection is proposed. Various sensors' information is fully applied to carry out the machine condition diagnosis. It aims at the common experiments for rotating machine fault diagnosis, the package of two layers of decomposition and reconstruction of the measured signal using db20 wavelet, and extraction of the normal and fault condition of the information as the feature vector of a single classifier. The fusion frame is checked by using extraction and the training sample data of the same type. The results of the experiments for rotating machine fault diagnosis indicate that the fusion frame has an ideal result, showing a diagnostic accuracy of 92.9%.

REFERENCES

Allan J. Volponi, Tom Brotherton, Robert Luppold, etc, 2004. Development of an Information Fusion System for Engine Diagnostics and Health Management [EB/OL]. TM-2004-212924.

Chen G, 2005. Intelligent Fusion Diagnosis for Wear Fault of Aviation Engine [J]. Mechanical Engineering of China, 16(4):299~302.

Lawrence A. Klein, 1999. Sensor and Data Fusion Concepts and Applications [M]. SPIE Optical Engineering Press.

Ma L., Sun Y., 2002 Information Fusion based Fault Diagnosis for Control System [J]. Infrared and Laser Engineering, 31(1):36~40.

Wu S. L. Han J.T., 2001 Integrated Maintenance Information System based Data Fusion and its Applications [J], Journal of Air Force Engineering University (Natural Science Version) Vol.2 No. 4. 4:4–7.

Wang M., Wang W. J., 2001 Multi-Sensors based Data Fusion for Fault Diagnostic Techniques [J]. Journal of Huazhong University of Science and Technology, 29(2):96~98

Wu Songlin, 2004. Dynamic Diagnostic Model for Aviation Maintenance [J]. Trans. of VSB- Technical University of Ostrava (Czech Rep.), Mechanical Series, Vol. L, No.1, p. 219–225.

Wald, 1999. Definitions and Terms of Reference in Data Fusion [J]. International Archives of Photogrammetry and Remote Sensing (IAPRS), 32(6):2~6.

Recognition, video and image processing

Energy Science and Applied Technology – Fang (Ed.)
© 2016 Taylor & Francis Group, London, ISBN 978-1-138-02833-3

2DPCA and 2DLDA algorithm on hyperspectral image classification

Ju Cai & Li Ma

Faculty of Mechanical and Electronic and Information, China University of Geosciences (Wuhan),
Wuhan, China

ABSTRACT: Based on hyperspectral image data, windowing on each sample data enables us to obtain the image patch, and achieve the two-dimensional Principal Component Analysis algorithm and two-dimensional Linear Discriminant Analysis algorithm based on its statistical covariance. This methodology can effectively take advantages of the spatial-spectral information of the hyperspectral image data. The experimental results in this paper illustrate that it outperforms the one-dimensional feature extraction algorithm in terms of the classification of hyperspectral image data.

Keywords: hyperspectral image; dimensionality reduction; classification; two-dimensional; spatial information

1 INTRODUCTION

Hyperspectral imaging has becoming one of the most promising techniques such as ecological science, geological science, precision agriculture, and military applications. Consisting of two spatial dimensions and one spectral dimension data cube provides valuable observations of contiguous and very narrow spectral bands (H. Pu 2014, A. Mohan 2007). Just as the high-dimensional real-world data, hyperspectral images usually contain many significant large proportions of pixels that are highly redundant. The redundancy coming from spectral and spectral and spatial domains makes it difficult to improve classification accuracy for hyperspectral imagery. Meanwhile, the high dimension and the high dimensionality and complexity in the data structure of the hyperspectral images not only increase computational complexity but also may result in the Hughes phenomenon, leading to challenges in classification. Therefore, dimensionality reduction techniques are often performed prior to the classification process, which reduce the redundancy in the original subspace by preserving important information about objects of interest for classification purposes. Principal Component Analysis (PCA) and Linear Discriminant Analysis (LDA) are most widely used methods among DR, and these two algorithms are also used in the DR of hyperspectral image processing (J. Li 2014, L. Ma 2010). However, these two algorithms make use of the sample covariance only, while they do not take advantage of the spatial information round the sample.

When PCA and LDA are used on the human face recognition, we first reshape the two-dimension human face image data to vectors, then calculate the two order statistics. In the frame of PCA, we calculate the covariance matrix of the samples. The LDA calculates the between-class scatter matrix and within-class scatter matrix. Compared with PCA and LDA, two dimension principal component analysis (L. Sirovich 1987, Jian Yang 2004) (2DPCA) and two dimension linear discriminant analysis (2DLDA) (S. Kongsontana 2005, F.E. Alsaqre 2006) get the covariance matrix using image matrix straightforwardly. These two algorithms can get the local image texture information when extracting features.

To apply the algorithms 2DPCA and 2DLDA on the feature extract on hyperspectral image, we window on the each pixel on the image. Then, each pixel could be turn into a small image patch, which could be treated as a two dimension matrix or an image. The advantages of 2DPCA and 2DLDA over other algorithms are to make use of both the spatial information and the spectral information of each pixel in the process of feature extracting of hyperspectral image.

2 PROPOSED ALGORITHM

2DPCA and 2DLDA are proposed in the domain of feature extract of image and image face recognition. Besides, they can do better when compared with PCA and LDA. After the feature extract of the image, the classification of the algorithm can be the K-nearest-neighbor. Each pixel is a vector, when we window on the pixel, each pixel is expanded into a matrix. Then, the matrix is in the form that can be used by 2DPCA and 2DLDA.

2.1 Spatial information

In the hyperspectral image, the spectral information of the two pixels on the image is similar when the two pixels are closer, and they could get more probability in the same class. Some paper has taken some method of windowing on each pixel on the hyperspectral image, and they get some good performance in the algorithm. In this paper, we adopt same method, taken by others, and window on each pixel, use the spatial information of each pixel. Then, a simple pixel can be turned into an image patch.

Given some pixels, $x_1, x_2, ..., x_n \in \Re^{d \times 1}$, window on some pixel x, e.g. using a 3×3 window. Then, the $x \subset \Re^{d \times 1}$ is turned into an image patch $x \subset \Re^{d \times 9}$. Among the pixels, X_5 is the pixel x, and $X = [X_1, X_2, ..., X_9]$, as shown in Figure 1.

We take the same method on the all pixels, and all pixels are turned into image patches.

2.2 2DPCA and 2DLDA

After the processing of the windowing on each pixel, the hyperspectral image data could be in the mode, which could be used in the algorithm 2DPCA and 2DLDA.

2DPCA uses the image patch to get the covariance matrix. 2DLDA uses the image patch to calculate the between-class scatter matrix and within-class scatter matrix. The following steps of the algorithm 2DPCA and 2DLDA are described as follows.

2DPCA Algorithm:
data set: x_i, $i = 1, ..., N$ windowed to get x_i, $i = 1, ..., N$;
get the mean of all X_i, $\bar{X}$;
get the covariance matrix

$$C = \sum_i^N (X_i - \bar{X})(X_i - \bar{X})^T$$

Calculate the projection vectors of the covariance matrix C, $U = [U_1, ..., U_d]$

To get the projection of all image patch, which are centralized, $X_i - \bar{X}$, the projection, $U^T(X_i - \bar{X})$

2DLDA Algorithm:
Data set: x_i, $i = 1, ..., N$ windowed to get x_i, $i = 1, ..., N$;
Calculate the within-class scatter matrix

$$C_w = \sum_c^C \sum_i^{N_c} (X_i - \bar{X_c})(X_i - \bar{X_c})^T$$

Calculate the between-class scatter matrix

$$C_b = \sum_c^C N_c (\bar{X_c} - \bar{X})(\bar{X_c} - \bar{X})^T$$

Calculate the projection vectors of the matrix $C_w^{-1} C_b$, $U = [U_1, ..., U_d]$

To get the projection, $U^T X_i$

2.3 Classification

Given two feature matrix,

$$X_i = [X_{i1}, ..., X_{in}], X_j = [X_{j1}, ..., X_{jn}]$$

And the similarity is defined as

$$d[X_i, X_j] = \sum_{k=1}^n \|X_{ik} - X_{jk}\|_2$$

where $\|X_{ik} - X_{jk}\|_2$ denotes the Euclidean distance between the feature vectors X_{ik}, X_{jk}.

3 EXPERIMENTS AND ANALYSIS

In this section, we demonstrate the effectiveness of the proposed 2DPCA and 2DLDA algorithms on some hyperspectral image, AVIRIS Indian Pine (INP). Indian Pine hyperspectral image data set in our experiment is commonly used. The AVIRIS sensor generates 220 bands across the spectral range from 0.2 to 2.4 um. In the experiments, the number of bands is reduced to 200 by removing 20 water absorbing bands. This image has spatial resolution of 20 m per pixel and spatial dimension 145×145. It contains 16 ground-truth classes. The specific information about these three hyperspectral image data set is given in Table 1.

3.1 Indian pine data set

In the experiments, the proposed algorithm in this paper 2DPCA and 2DLDA makes use of spatial and spectral information of the hyperspectral image, and extracts the feature of the image. So, the algorithms proposed are linear feature extract algorithm, PCA and LDA, and a simple spatial-spectral-information-used algorithm, Spatially Coherence Distance (SCD). The classification is the 1 NN neighbor algorithm. Above all, the five algorithms

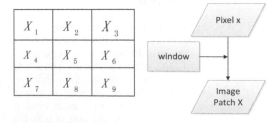

Figure 1.   Pixel to image patch.

Table 1. Ground truth of Indian pine.

	Class name	Number
1	Alfalfa	54
2	Corn-notill	1434
3	Corn-min	834
4	Corn	234
5	Grass/Pasture	497
6	Grass/Trees	747
7	Grass/Pasture-mowed	26
8	Hay-windrowed	489
9	Oats	20
10	Soybeans-notill	968
11	Soybeans-min	2468
12	Soybean-clean	614
13	Wheat	212
14	Woods	1294
15	Building-Grass-Tree-Drives	380
16	Stone-steel Towers	95

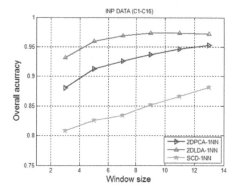

Figure 3. Effect of spatial window size for INP.

4 CONCLUSION

Combining the spatial and spectral information on the hyperspectral image data set, the proposed algorithm 2DPCA and 2DLDA can extract more information from the data set. Going forward, we will try to apply the kernel function in the feature extract algorithm.

REFERENCES

Alsaqre F.E. & R. Qiuqi. Face Recognition Using Diagonal 2D Linear Discriminant Analysis[C]. Proc. IEEE Int. Conf. ICSP. 2006.

Boser B.E. & I.M. Guyon. A Training Algorithm for Optimal Margin Classifiers[C]. 5th Annu. Workshop Compute Learn Theory, 1992.

Jian Yang, David Zhang. Two-Dimensional PCA: A New Approach to Appearance-Based Face Representation and Recognition [J]. IEEE Trans. Pattern Analysis and Machine Intelligence, 2004, 26(1):131–137.

Kongsontana S. & Y. Rangsanseri. Face Recognition Using 2DLDA Algorithm [J]. IEEE International Symposium on Signal Processing and Information Technology, 2005, 5(2):675–678.

Li J. & H. Zhang. Hyperspectral Image Classification by Nonlocal Joint Collaborative Representation With a Locally Adaptive Dictionary [J]. IEEE Transactions on Geoscience and Remote Sensing, 2014, 52(6):3707–3719.

Ma L. & M.M. Crawford. Local Manifold Learning-Basedk–Nearest-Neighbor for Hyperspectral Image Classification [J]. IEEE Transactions on Geoscience and Remote Sensing, 2010, 48(11):4099–4199.

Mohan A. & G. Sapiro. Spatially Coherent Nonlinear Dimensionality Reduction and Segmentation of Hyperspectral Images [J]. IEEE Geoscience and Remote Sensing Letters, 2007, 4(2):206–210.

Pu H. & Z. Chen. A Novel Spatial-Spectral Similarity Measure for Dimensionality Reduction and Classification of Hyperspectral Imagery [J]. IEEE Transactions on Geoscience and Remote Sensing, 2014, 52(11):7008–7022.

Sirovich L. Low-dimensional procedure for the characterization of human faces [J]. Optical Society of America, 1987, 4(3):519–524.

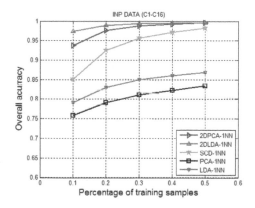

Figure 2. Effect of the number of training samples for INP.

are as follows: ① 2DPCA+1 NN, ② 2DLDA+1 NN, ③ SCD+1 NN, ④ PCA+1 NN, ⑤ LDA+1 NN.

The overall accuracy of the classification of the five algorithms is shown in Figure 2. The x-axis is the percent of the data, which are chosen as training data, from 0.1 to 0.5. The y-axis is the overall accuracy of the classification. In Figure 2, it is obvious that the proposed algorithms outperform other algorithms on comparison. And the window size is chosen on the window of 9×9. The overall accuracy of the window size as a factor from 3×3 to 9×9 is shown in Figure 3. 10% of the data are randomly chosen as training samples, and the remaining 90% are used for testing. In Figure 3, it can be seen that even if the window size used is 3×3, the proposed algorithm, 2DPCA and 2DLDA, can achieve a very high accuracy in comparison with other algorithms.

Energy Science and Applied Technology – Fang (Ed.)
© 2016 Taylor & Francis Group, London, ISBN 978-1-138-02833-3

Design and implementation of indoor wireless video surveillance system based on ODMA-WiFi

Z.C. Wang
Department of Electronics and Information Engineering, Hebei University, Baoding, China

J.W. Cui & Z. Li
Hebei University, Baoding, China

ABSTRACT: Given that the traditional wireless video surveillance system has many drawbacks such as limited coverage, poor network robustness and low data rate, this article introduces how to construct a new type of wireless video surveillance system based on the ODMA-WiFi network. The network combines the advantages of both techniques: the advantages of ODMA, e.g. opportunity-driven multiple access, multi-hop communication and being flexible in building network, have been fully applied, and its defects, i.e. poor scalability and not being widely used, have been remedied. A modular design with three communication sublayers is used in the hardware, which makes it more flexible to deploy a network, and the LRMS matching the system can intelligently manage all ODMA network nodes. The successful construction of the ODMA-WiFi wireless video surveillance system further proves that ODMA-WiFi has significant technical advantages in terms of transmission efficiency and communication distance compared with traditional wireless networks.

Keywords: ODMA-WiFi; Local Resource Management (LRMS); Communication Sublayer

1 INTRODUCTION

In general, wireless video surveillance refers to a type of monitoring system, which uses radio waves instead of generic cabling to the transmit video, audio and data signal. In some sense, it is the perfect combination of video surveillance technology and wireless network technology.

Wireless transmission technology is essential to the wireless video surveillance system and determines its reliability, availability and scalability. Currently, the popular technologies in this domain include Wireless Fidelity (WiFi), Zigbee, Wimax and Ad-Hoc, which have their respective characteristics and various application fields. Because of its high bandwidth capacity and good compatibility, WiFi technology can be used for a host of applications. However, it also has some shortcomings such as limited coverage, poor network security and robustness. The advantages of Zigbee lie in its low-cost networking and low device consumption, but the transfer-rate is less than satisfactory. Though the data rate of Wimax is high, its deployment costs are enormous because a large number of mobile base stations are required to be constructed. The interconnection between various ODMA nodes can be achieved through automatic configuration and a user node communication with the backbone network by means of multi-hop based on the relay-communication function of the neighboring nodes.

2 ODMA-WIFI VIDEO SURVEILLANCE SYSTEM DESIGN

The video surveillance system based on ODMA-WiFi is being applied to the laboratory building of College of Electronic and Information Engineering of Hebei University. Due to the complexity of the internal structure of the building and lots of solid obstacles inside, it is difficult for a common wireless network to achieve the ideal monitoring effect. After several tests and comparison, the ODMA-WIFI network has ultimately been chosen to set up this indoor monitoring system, which can real-time monitor the internal environment of the building and precision instrument in laboratory. Meanwhile, by using this system, the building also achieves the full coverage of wireless hotspots by which teachers and students from any place in the building can get access to the campus network and the Internet, providing a great convenience for research and teaching.

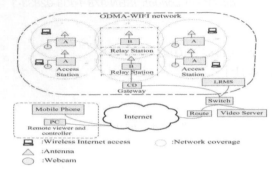

Figure 1. Communication structure of the ODMA-WiFi network.

ODMA-WiFi communication network can be divided into three sub-layers, as shown in Figure 1.

The first one is the access layer composed of device A that is connected by a wire or radio to the network camera, PC or other smart terminal equipment. Terminals connected to different devices that have the function of both access points and relay points can communicate with each other in the local ODMA network, regardless of the failure of the upper network. The second one is the relay layer consisting of B devices, which have high transmitting power and strong penetration and is responsible for relaying and convergence of the network. ODMA multi-hop transmission can be created between A and B or B and B. The last one is the CD device that makes up the gateway layer that links up with Ethernet as well as B devices, and is responsible for data exchange with an external network node, reporting the performance of network nodes to the LRMS (network management system), and distributing configuration data to the destination node, in order that intelligent centralized management of the ODMA network can be achieved.

In other words, the design and division of the three communication sub-layers of ODMA-WiFi can give full coverage of the advantages of ODMA technology, e.g. being flexible in building network and wisdom path selection. Combined with the function of routing forwarding, the ODMA network seamlessly integrated with the Internet could makes it possible that remote users can view and control ODMA nodes through the intelligent terminal.

3 ODMA-WIFI WIRELESS VIDEO SURVEILLANCE SYSTEM IN HEBEI UNIVERSITY

Depending on the layout of the building and the key monitoring area that have been designated, overall fifty webcams and twenty-six ODMA devices are used in the network-building process

on this occasion. Each floor builds a single ODMA network and every network can communicate with each other by means of wired Ethernet. While each CD is placed indoors connected with switch, B and A are fixed onto the wall under the necessity of networking. Webcams can be linked to a nearby A device by the network cable or wirelessly and other intelligent terminals in the building can get access to the campus network through the A's function of wireless hot-spot.

The monitoring center, a convergence spot for the entire network in the building, is located in Data Communications Laboratory on the second floor, to which the ODMA-WiFi network is linked with wired Ethernet, as shown in Figure 2.

As a centralized monitoring unit, the video server in monitoring center is responsible for incessantly compressing and storing video stream sent by acquisition equipment. Meanwhile, the server can also be used as the "transit station" and the "springboard", making it possible for remote users to view and manage the internal monitoring network through the Internet.

LRMS (Local Resource Management System) is an intelligent management system designed specifically for the ODMA network. In addition to monitoring the real-time running information of each node in the network, the system mainly achieves two major functions.

The first one is the division of ODMA subnet. ODMA-WiFi equipment occupies the 5.8G spectrum, which is divided into 13 channels. In the process of communication, the channels do not interfere with each other, which makes it possible to achieve the corresponding division of subnet according to different channels. The divided ODMA network can not only utilize the Spectrum resources effectively, but reduce data collisions as well, thereby enhancing the overall stability and reliability of the network. In accordance with the advance planning, the same data stream (distinguished by the IP address or port) could be assigned to a particular channel (subnet) by LRMS in order to acquire better transmission quality and to achieve the division of subnet intuitively and easily, as shown in Figure 3.

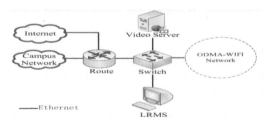

Figure 2. Sketch map of the monitoring center.

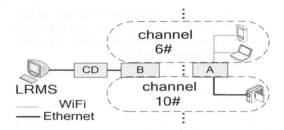

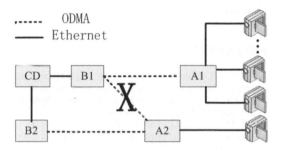

Figure 3. Schematic diagram of the subnet division.

Figure 5. Schematic diagram of ODMA-WiFi equipment and webcam.

ODMA
Ethernet

Figure 4. Schematic diagram of route designation.

The second one is the route-designating. Although "wisdom path selection" is one of the advantages of the ODMA network, the targeted route-intervention can make the distribution of the cyber source more reasonable for an imbalanced local network. As shown in Figure 4, according to the "shortest route principle", B1 should be the next hop of A2, but the load of B1 is very large, so B2, which is relatively idle, can be designated as the prior next-hop of A2 by the remote configuration of LRMS, eventually achieving load balancing of each link.

4 EQUIPMENT INTRODUCTION AND INSTALLATION CONFIGURATION

Although the internal chip is different, the packaging appearance of each kind of equipment (A, B, CD) is the same (Figure 5, left). The power supply mode of the device is flexible (AC/DC transformation or direct powering by the USB port) and the flat design is conducive to heat dissipation and installation.

Four LAN ports of a device are used for connecting the camera, PC or other Ethernet terminals, and the highest rate is up to 1000Mbps. The LAN port of B equipment mainly has two functions. On the one hand, it is responsible for wired communications with the superior device of CD. On the other hand, more than one B can form a cascade connection through the port by means of the wire, thus expanding the coverage of the relay

layer. There are totally five standard Ethernet ports on the CD device: one WAN port is linked with the local area network and four LAN ports are connected downward to the B device. The IP address of the WAN port and that of the LAN port must be in the same network segment.

The webcam (Figure 5, right) as a video capture terminal adopts the compression format of H.264 Main Profile@Level3.0 for easy network transmitting and saving storage space. Video streams can be transmitted to the specified video server for processing and storing because of the support for remote storage. Remote monitoring through the Internet can be realized with the router owing to the function of port mapping and UPNP.

The CD device on each floor is linked through parallel lines with Ethernet switch in the laboratory. The static IP address of the CD's WAN port can be set up depending on the configuration information of the local area network in the building.

B devices just need to be fixed in place according to the planning without software settings. To ensure the quality of signal transmission, the length of the cable between B and CD or B and B should be within 80M.

Video signal can be transmitted between a device and the webcam via the network cable (less than 10M) or wireless WIFI. Each A can be connected at most to five cameras, whose IP address should be in the same subnet segment as that of the CD. Under the condition of ample bandwidth, there is no limit on the number of other terminals that can get access to a device via WiFi connection.

Based on the embedded IP address of each ODMA device (usually without modification), the LRMS is able to visit and manage every node in the network. The devices are required to write the IP address of the LRMS to the software system of itself in order to give all of its management authority to the LRMS.

5 NETWORK TESTING

After the installation and configuration of the device is completed, a test for overall network performance is launched in a gradual and graded way. The test method is as follows.

Each ODMA equipment is linked through a twisted pair cable with a laptop, whose IP address should be in the same network segment as the

ping	trace

Figure 6.   Schematic diagram of ping and trace.

embedded address of the device. The user can log on to the equipment management interface of each piece of equipment through the browser, then using the built-in tools such as "trace" and "ping" to evaluate the connectivity of a certain network node. If the average return value is less than 3 ms (Figure 6, left) after executing the Ping command or the CD value related to certain multi-hop paths is less than 50 (Figure 6, right) after tracing the gateway address, it can be concluded that the communication quality between the source and the destination is relatively poor and that there are some issues in the local area network. Aiming at the existing problems, certain measures, e.g. adjusting device location, adding relay node and expanding wired bandwidth, could be taken into consideration to solve the faults.

6   CONCLUSIONS

As a burgeoning Ad-Hoc communication technology, ODMA has a great potential that is definitely worth exploring. On the basis of an in-depth study on the ODMA technology, this paper designs an indoor wireless video monitoring system based on the ODMA-WiFi network, which can not only effectively avoid the typical problems of the traditional monitoring system, such as complex cabling and high construction costs, but also successfully resolve certain problems that the existing wireless systems have encountered in the indoor environment. After having been tested and applied in different application scenarios for a long time, there is no doubt that the whole system with certain distinctive features, such as fast data transmission, good stability and easy management, has achieved the anticipated goal. Of course, the contents described in this article are just some of the basic application of ODMA-WiFi technology. On this basis, more new applications of the Internet of things, such as information acquisition, indoor positioning and regional rights management, can be developed.

Communication operators could also take advantage of this technology to enrich the application of the existing broadband network and develop a new data communication market.

ACKNOWLEDGMENTS

This work was supported by the Natural Science Foundation of Hebei F2014201168. The project name is "Research on Cooperative Communication Routing Algorithm for Heterogeneous Networks".

REFERENCES

Awerbuch B, Holmer D, Rubens H. High throughput route selection in multi-rate ad hoc wireless networks [M]//Wireless on-demand network systems. Springer Berlin Heidelberg, 2004: 253–270.

Conti M, Maselli G, Turi G, et al. Cross-layering in mobile ad hoc network design [J]. Computer, 2004, 37(2): 48–51.

Fan Z. High throughput reactive routing in multi-rate ad hoc networks [J]. Electronics letters, 2004, 40(25): 1591–1592.

Farahani S. ZigBee wireless networks and transceivers [M]. Access Online via Elsevier, 2011.

Hart M, Vadgama S. Factor that affect performance of a mobile multihop relay system [J]. IEEE C802. 16 mmr-05–017r1, 2005.

Hart M J, Vadgama S K. Link and system-level analysis of structured multi-hop networks[C]//Proceedings of the International Workshop on Wireless Ad-hoc Networks. 2005: 234–239.

Liu G R, Wang L L, Zhou S. The research and design of embedded wireless video monitoring system [C]//E-Business and E-Government (ICEE), 2011 International Conference on. IEEE, 2011: 1–3.

Sauter M. Beyond 3G-Bringing networks, terminals and the web together: LTE, WiMAX, IMS, 4G devices and the mobile Web 2.0 [M]. Wiley. Com, 2011.

Wang Z C, Wang L, Tian X Y, et al. A campus safety monitoring system based on Ad-Hoc network [J]. Advanced Materials Research, 2012, 562: 1792–1795.

Yuen W H, Lee H, Andersen T D. A simple and effective cross layer networking system for mobile ad hoc networks[C]//Personal, Indoor and Mobile Radio Communications, 2002. The 13th IEEE International Symposium on. IEEE, 2002, 4: 1952–1956.

Yang G, Xiao M, Chen H, et al. A novel cross-layer routing scheme of ad hoc networks with multi-rate mechanism[C]//Wireless Communications, Networking and Mobile Computing, 2005. Proceedings. 2005 International Conference on. 2005, 2: 701–704.

Energy Science and Applied Technology – Fang (Ed.)
© 2016 Taylor & Francis Group, London, ISBN 978-1-138-02833-3

The study of the printed Data Matrix (DM) code identification

Chengxiang Yin

School of Information and Electronics, Beijing Institute of Technology, Beijing, China

ABSTRACT: With the characteristics of high information capacity, simple and small size, the Data Matrix (DM) code is playing an increasingly important role in the area of the portable equipment identification. This paper mainly proposes a DM code identification scheme based on the image processing for the printed DM code. This identification scheme has been proved to be feasible and simple according to the experimental results.

Keywords: Data Matrix (DM) code; DM code identification; printed DM code

1 INTRODUCTION

Because of the improving requirements of the information storage capacity and the identification accuracy of the barcode, the one-dimensional barcode, which has great limitations in information density and error correction capability, is no longer trailed. To overcome these limitations, two-dimensional barcode came into being. Two-dimensional barcode is an applied technology, which can record information on the two-dimensional plane in a pattern of black and white using a particular geometry. It can be mainly divided into stacked barcode represented by the PDF417 code and the matrix code represented by QR and DM codes. With respect to the one-dimensional barcode, the two-dimensional barcode has advantages of smaller area, larger amount of data, stronger resistance to damage and the ability to express the content of text and images.

Due to the lack of the professional identification equipment in the area of the product security, online shopping and merchandising electronic discount coupons, and the portable camera identification are more suitable for these areas, for example, the phone camera identification, hence there is a need for a simple two-dimensional code. The DM code has a high density and a small size compared with the other kinds of two-dimensional code, and makes the mobile phone identification possible. Relative to the QR code, the most popular code currently, the information capacity of the DM code is comparable and the application of it is simpler. It is known as the "simple code" and has a low demand of identification terminal. Because the DM code can provide labels with small size, it is very suitable for identification of small parts and can be printed directly on the entity for anti-counterfeit labels, such as the heritage counterfeiting, circuit identification and the management of the archaeological heritage. Meanwhile, the DM code adopts sophisticated error correction techniques and has a superior dirt capacity. Even if the coding part is partially damaged, the information identification can be carried out successfully.

According to the superiorities of the DM code, it can be printed directly on the entity for security identification. In this paper, the research object is the printed DM code. The current research on the two-dimensional code is mainly focused on the QR code and the PDF417 code; however, the research on the DM code is rare, especially on the platform of the mobile phone. Given the portability of the java platform, this paper proposes a DM code identification scheme based on the image processing for the printed DM code.

2 THE IDENTIFICATION OF THE DM CODE

2.1 *The composition of the DM code*

The DM code can be divided into two versions: ECC000-140 and ECC200.

ECC000-140 has a variety of different levels of error correction functions. However, ECC200 uses the Reed-Solomon algorithm as the error correction algorithm. ECC200 is universally applied for its easy algorithm and size elasticity. On the image, the difference between the above two versions is that the upper right corner module of the ECC200 is light colored, but for ECC000–140, it is dark colored.

The graph of the DM code is composed of the data area and the detection area. The data area is formed by square modules that are regularly arranged. In Figure 1, the data area is the 9×9 matrix in the central. The function of the data area is to store the useful information. Around the data area is the detection area and contains two adjacent sides formed by dark solid lines to form an L-shaped border. The other two adjacent sides are composed of alternating dark and light modules. The function of the detection area is limiting the physical size, locating border and handling the distortion.

2.2 *The identification process*

In this paper, the identification process of the printed DM code includes three stages: image pre-processing stage, border determination stage and image segmentation stage. The process diagram is shown in Figure 2.

The image preprocessing stage. When an image of a non-standard printed DM code is obtained by the camera, the first thing that must be done is to extract the symbol of the DM code, as there exist problems caused by the image capturing and the printing, such as unevenness of the image brightness, the stains on the image and the interference of the Gaussian noise. The captured image of the printed DM code is shown in Figure 3.

To extract the symbol of the DM code, the critical operation is to do image binarization for the captured image. The image binarization modifies the gray value of each pixel according to the threshold value T of the image. If the gray value of a pixel is smaller than T, we set its gray value to 0; otherwise we set its gray value to 255.

As there exists a large difference between the gray value of the light module and the dark module in the captured image as the contrast ratio of the image is high, we use the average value of all the pixels' gray value as the threshold value T. The threshold value T can be expressed as

Figure 1.   An example of the ECC200 DM code.

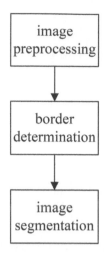

Figure 2.   Identification process of the printed DM code.

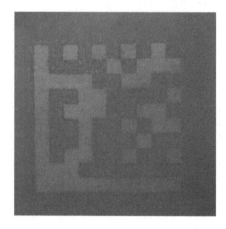

Figure 3.   Captured image of the printed DM code.

$$T = \frac{S}{w \times h} \qquad (1)$$

where S represents the sum of the gray value of each pixel; w represents the width of the image; and h represents the height of the image. The gray value Y can be expressed as

$$Y = a \times red + b \times green + c \times blue \qquad (2)$$

where a, b and c represent the weights of the three primary colors *red*, *green* and *blue* and it must be guaranteed that $a + b + c = 1$.

Then, we reset the gray value of each pixel sequentially. If the gray value of a pixel is smaller than T, then we set its gray value to 0; otherwise we set its gray value to 255. Finally, we get the binarized image.

2.2.1 *The border determination stage*

After the image preprocessing stage, the captured image of the printed DM code is converted into its binarized form. However, we still cannot locate the actual position of the symbol of the DM code, as there exists the blank area outside the detection area. We then determine the border of the DM code, which is the key point to solve the problem.

In this paper, the strategy of the border determination is to locate the two adjacent sides which form an L-shaped border of the DM code, and determine the location of the other two sides based on the L-shaped border. The flow chart of the algorithm is shown in Figure 4.

The value of $(countymax\text{-}50)$ and $(countxmax\text{-}50)$ is determined through the experiments.

At the step of determining the top border, the algorithm adopts the method of traversal detection, i.e. traversing all the possible location $x2$ of the top border. The $x2$ can be expressed as

$$x2 = x1 - x + 1 \tag{3}$$

where x represents the distance between the top border and the bottom border. The x has a minimum value of $counti$, which is the number of the black pixels of the left border $y1$, and the x will have an incremental of 1 after each detection. The detection criterion is that the four pixels above the top border must be all white pixels.

At the step of determining the right border, the method is also the traversal detection, i.e. traversing all the possible location $y2$ of the top border. The $y2$ can be expressed as

$$y2 = y1 + y - 1 \tag{4}$$

where y represents the distance between the left border and the right border. The y has a minimum value of $countj$, which is the number of the black pixels of the bottom border $x1$, and the y will have an incremental of 1 after each detection. The detection criterion is that the four pixels right to the right border must be all white pixels.

The schematic diagram of the border determination stage is shown in Figure 5.

After the above two steps, the effective width of the binarized image is just the value of y, and the effective height of the binarized image is just the value of x. Then, the effective pixels of the binarized image can be intercepted, and finally the border determined image can be obtained.

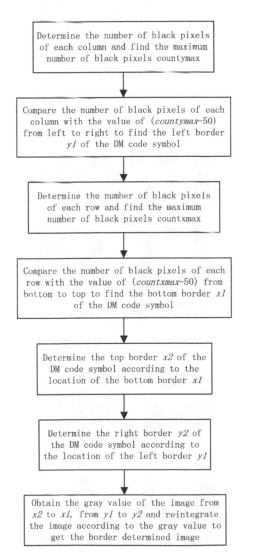

Figure 4. Flow chart of the border determination stage.

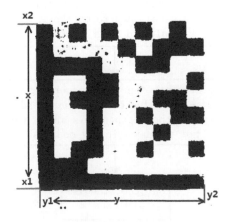

Figure 5. Schematic diagram of the border determination stage.

2.2.2 *The image segmentation stage*

After the border determination stage, the border determined image of the DM code can be obtained. The last thing needs to be finished is to convert the non-standard DM code to its standard form.

In this paper, the key point of the image segmentation stage is to divide the border determined image to a hundred modules with the origin, which is the vertex of the L border, and make a decision for each module to judge whether it is black or white. The flow chart of the algorithm is shown in Figure 6.

The threshold value 175 is determined through the experiments. The standard DM code means that the size of the whole image is 224×224, and the size of the DM code symbol is 160×160, i.e.

the side length ratio of the whole image and the DM code symbol is 1.4.

3 RESULTS

3.1 *The results of the image preprocessing stage*

Through the preprocessing stage, the captured image is converted to the binarized image. The problems of the brightness unevenness and the interference caused by the Gaussian noise have been solved. The binarized image is after the preprocessing stage, as shown in Figure 7.

3.2 *The results of the border determination stage*

Through the border determination stage, the blank area outside the detection area in the binarized image is clipped. The image obtained is just the section of the DM code, which is ultimately required, and this operation is conducive to the subsequent image segmentation operation. The image after the border determination stage is as shown in Figure 8.

```
┌─────────────────────────────────────┐
│ Divide the border determined image   │
│ To 100 modules with the same size    │
└─────────────────────────────────────┘
                   │
                   ▼
┌─────────────────────────────────────┐
│ Calculate the average value of the   │
│ gray value for each module           │
└─────────────────────────────────────┘
                   │
                   ▼
┌─────────────────────────────────────┐
│ Compare the average gray value of    │
│ each module with the value of 175 to │
│ judge whether it is black or white   │
└─────────────────────────────────────┘
                   │
                   ▼
┌─────────────────────────────────────┐
│ Traverse each pixel of the image in  │
│ module unit and reset the gray       │
│ value of each pixel                  │
└─────────────────────────────────────┘
                   │
                   ▼
┌─────────────────────────────────────┐
│ Reintegrate the image according to   │
│ the new value of each pixel and      │
│ discard the extra pixels which are   │
│ not included in any modules          │
└─────────────────────────────────────┘
                   │
                   ▼
          ┌──────────────────┐
          │ Get the standard │
          │ DM code          │
          └──────────────────┘
```

Figure 6. Flow chart of the image segmentation stage.

Figure 7. Binarized image.

Figure 8. Border determined image.

Figure 9.   Standard DM code.

3.3 *The results of the image segmentation stage*

Through the image segmentation stage, the standard DM code image can be obtained as shown in Figure 9.

According to the results obtained above, this method of the DM code identification is feasible and has a good effect. It is simple enough and can be realized easily.

4   CONCLUSIONS

There exist many available identification algorithms of the DM code; however, most of these algorithms are cumbersome when they are targeted at some simple DM code, for example, the printed DM code with a relatively low capacity. Especially when these complicated algorithms are realized in the platform of mobile phone with lower computing capability, they may result in redundant processing time. This paper, fully combined with the own characteristics of the printed DM code, such as simple, high contrast ratio and the L-shaped border, adopts simple algorithms in each stage, and finally realizes the complete automatic identification function. It lays a foundation for the application of the DM code to mobile phones.

REFERENCES

Du C. The design and the identification of the DataMatrix barcode [J]. Computer CD Software and Applications, 2012(12):149–149.

Fröschle H K, Gonzales-Barron U, McDonnell K, et al. Investigation of the potential use of e-tracking and tracing of poultry using linear and 2D barcodes[J]. Computers and Electronics in Agriculture, 2009, 66(2): 126–132.

Farina S, Paganucci S, Landini W. Data Matrix Codes: Experimental use in a Museum Exhibition [J].

Hu X, He J. The Technology Research of Decoding Data Matrix Code [J]. Journal of Hangzhou Dianzi University, 2008, 28(5):124–126.

Martínez-Moreno J, Marcén P G, Torcal R M. Data matrix (DM) codes: A technological process for the management of the archaeological record [J]. Journal of Cultural Heritage, 2011, 12(2): 134–139.

Mc Inerney B, Corkery G, Ayalew G, et al. A preliminary in vivo study on the potential application of e-tracking in poultry using ink printed 2D barcodes [J]. Computers and electronics in agriculture, 2010, 73(2): 112–117.

Plain-Jones C. Data matrix identification [J]. Sensor Review, 1995, 15(1): 12–15.

Peng F. The applications of the two-dimensional matrix code (DM) in archaeological specimen's management [J]. Acta Anthropologica Sinica, 2011, 30(3): 343–344.

Yuan Y, Zhao X, Yang D. The correction of DataMatrix barcode distortion [J]. Application of the computer system, 2008(10): 47–50.

Energy Science and Applied Technology – Fang (Ed.)
© 2016 Taylor & Francis Group, London, ISBN 978-1-138-02833-3

Research on automatic code generation technology on the basis of UML and visualization of templates

Peishun Liu, Maochun Zheng & Yongquan Yang
Computer Science and Technology Department, Ocean University of China, China

ABSTRACT: In recent years, the demand of data management systems for various businesses is growing, resulting in an increasingly large system size and amount of codes which costs more time and manpower. Therefore, in order to improve the efficiency of software development, the development of an automatic code generator is imperative. Through the study of previous code generating theories, it is obvious that a few issues such as "automatic generation", "operation visualization", and "multi-tables association" remain unresolved. First, through the research of MDA theory, this paper puts forward an automatic code generating project based on the combination of UML data modeling, HTML5 visual controls, template file definitions and template engines, and finally validates the use and functionality of the code generator through a specific test.

Keywords: Automatic code generation; UML; Visualization; Customizable templates; Metadata

1 INTRODUCTION

In today's world with highly developed software engineering, most software developers still use manual coding. The development cycle of this traditional software development method is longer, and there is a lot of duplication of effort. In this case, issues such as delay of large software projects, cost overrun plans and performance below expectations usually appear. These issues compel software developers to think how to reduce software development and maintenance costs and improve development efficiency (Dong Yuming, 2012).

Then, in the traditional development approach, these problems are always difficult to solve perfectly. This requires a new way of software development that can avoid low-level duplication of development and achieve standardized production of softwares (Anitha Rani Marneni, 2010), in order to help businesses cope with the rapidly changing market environment. Automatic code generation technology solves this problem to a certain degree.

This paper performs an in-depth research on code generation technology used in the current software development technology, puts forward a solution making use of visual UML modeling for automatic code generation, and achieves an auto-matic code generator based on the combination of free marker, spring and Hibernate.

2 RESEARCH ON STATUS OF AUTOMATIC CODE GENERATION TECHNOLOGY

In recent years, domestic and foreign researchers have proposed a number of ideas and solutions on the research and implementation of automatic code generation technology (Abid Mehmood 2013). Currently, the common solutions of code generation programs are based on metadata-driven, design patterns (Hamad I. Alsawalqah 2014), UML models, XML and templates. The backgrounds that these programs get involved with vary, but they all have a certain academic or practical value as well as a certain influence on the development of automatic code generation technology.

Among these, the code generation scheme based on the UML models can generate directly multilayer system, when generating codes, and retain the relationship embodied in the model hierarchy. However, the program uses XML for the description of modeling information, indicating that the business model is not easy to manage and therefore operation cannot be visualized. Template-based code generation scheme has characteristics such as relatively easy to achieve, code templates that are easy to maintain and flexibility and strong and easy to follow expansion (Guan Taiyang, 2007). However, it lacks corresponding template rules when handling multiple tables.

3 SOLUTIONS OF STATUS OF AUTOMATIC CODE GENERATION TECHNOLOGY

Automatic code generation techniques presented in this paper is based on UML visual modeling and template techniques. The overall flow of the system is shown in Figure 1.

Users abstract the universal framework of all template layers according to the maturity framework, and input these templates into the code generator. Users will conduct the model design of the associated business according to the actual needs of projects. Backend code generators generate objects that are used to encapsulate metadata based on user-designed business models. Handler will combine the imported data objects with templates and use the free marker template engine to generate project code, database files generated by Hibernate. After code generation, it prompts the user to download the code, the user-generated code and mature business integration framework underlying code, while it inputs the file into the target database to complete the initialization project system.

The current code generation technology can only deal with ordinary adding, deleting, modifying, and searching business logics. This program focuses on solving business process of the relationship between

data tables, developing relevant template rules, and improves the efficiency and formativeness of the code written by automatically generating the code.

4 KEY TECHNOLOGIES OF AUTOMATIC CODE GENERATION TECHNOLOGY

Business models design. This paper proposes a set of mechanism for business models based on database design by studying UMl data modeling as well as combining with the design principle of the database. This mechanism is divided into three levels, as shown in Figure 2: Project, DynaClass and DynaAttribute. Project refers to the basic environmental information that is required for the project to establish the basic objectives. After the establishment of the project information, UML modeling on the project starts, which include dynamic class and static class data modeling. Business objects are described through these attributes, and the generator combines with logic modules based on these attributes to generate a database table file and specific business handling code.

Business model relationship treatment. In the process of code generating, the association between business objects costs more effort, but currently most automatic code generators cannot handle multi-table associated with them, so this paper proposes a multi-table associated with handling mechanism.

Multi-table association refers to one to one, one to many, many to one and many to many. By studying the multi-table association database as well as query methods, many too many and one to one relationships will be expressed using the combination of two-way association mapping. In this builder, we set mtmRelations in List <MT Relation> type for Dyna class to mark bidirectional association and achieve association mapping. Among these, MT Relation's set properties are as follows: master class, master class properties, controlled class, con-

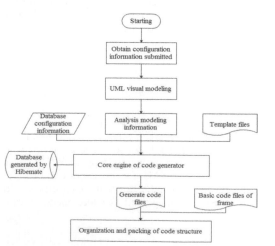

Figure 1. The design process of the code generator.

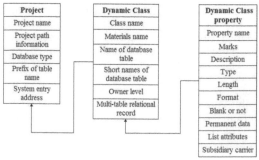

Figure 2. Business model mechanism designs.

trolled class attributes, and associated logo (one to one or many to many). For the one to many and many to one relationship, means to control transfer and mapping the column are used to achieve unidirectional association mapping. Dynamic class (DynaClass) set of property will be Scope Class, which is the controlling side of one to many relationships. Class attribute (DynaAttribute) sets the Boolean type attribute is Ref Class that indicates whether it is a list of many, indicates that the attribute points to a specific dynamic class, sets the type of property DynaClass ref Class, and indicates the specific dynamic class points.

Free marker template technique is simple with lightweight and excellent performance (Yang Xingtao 2009). Therefore, this paper adopts free marker as the template language. The template example is shown in Figure 3, indicating that code generation templates include fixed code and variable code. $ {} is the interpolation part of free marker, thus completing the combination of data model and template files (John Grundy 2013) by the definition of interpolation. In order to deal with the characteristics of the business models, the template takes full advantage of the logic syntax and the operators' built-in free marker, thereby easily generating personalized configuration code in accordance with the requirements.

UML visual modeling technology. JsUML2 library is a HTML5/JavaScript database that supports UML2 diagrams and provides Web. Developers can easily visually edit UML models in their own websites. By studying the JsMUL2 modeling database, according to JsMUL2 rule, this paper uses xml (Figure 4) as the interaction channel between the visual model and the database model, thus completing the combination between it and the business model design mechanisms. Users can switch the view, to edit on the surface of the UML modeling the database, which are free conversions between UML modeling and database design interfacial pages.

```
<%--many to many--%>
[#if (clazz.mtmRelations?size>0)]
    [#list clazz.mtmRelations as mtm]
    [#assign masterClassName]$[@lib.simpleClassName clazz=mtm.master/][/#assign]
    [#assign masterId]$(masterClassName?uncap_first}Id[/#assign]//Setting the master
    [#assign masterIdPath]$(masterClassName?lower_case)[/#assign]
    $(mtm.slave.name)
@ResponseBody
public String getOuter$(slaveClassName)ListBy$(masterClassName)(
    Map<String, Object> model,[#if clazz.scopeClass??]$(cls.myScopeIdPathVar) [/#if]
    @PathVariable("$(masterId)") int $(masterId),BindingResult result){
    page.setExt(String.valueOf($(masterId)));
    [#if clazz.scopeClass??]page.setScope(getScope("$(cls.myPathVariable)"));[/#if]
    return service.getOuter$(slaveClassName)ListBy$(masterClassName)(page); }
    [/#list]
[/#if]
```

Figure 3. Sample code of template files.

```
<umldiagrams>
<UMLClassDiagram name="Class diagram" >//Class Designer
  <UMLClass id="0.7:UML.Class_2"  abstract="false"/>//Class 1
    <superitem id="stereotypes" visibleSubComponents="true"/>
    <item id="name" value="Class1"/>//Class 1 member information
    <superitem id="attributes" visibleSubComponents="true">
    <item value="«ID»+/ID:int"/></superitem>
    <superitem id="operations" visibleSubComponents="true"/>
  </UMLClass>
  <UMLClass id="0.7:UML.Class_1" abstract="false">//Class 2
    <superitem id="stereotypes" visibleSubComponents="true"/>
    <item id="name" value="Class2"/>
    <superitem id="attributes" visibleSubComponents="true"/>
    <superitem id="operations" visibleSubComponents="true"/>
  </UMLClass>
  <UMLGeneralization id="0.7:UML.Generalization_0"
    side_A="0.7:UML.Class_2" side_B="0.7:UMLClass_1"/>//Class 1 and Class 2 relationship
    <point x="353.281045751634" y="150"/>
    <point x="322.954248366013 1" y="182"/>
    <superitem id="stereotype" visibleSubComponents="true"/>
    <item id="name" value=""/>
  </UMLGeneralization>
</UMLClassDiagram>
</umldiagrams>
```

Figure 4. Sample codes in UML view of the XML schema.

```
@RequestMapping(value="/classOne-{classId}", method=RequestMethod.GET)
public String toclassPage(Map<String, Object> model,  @PathVariable("classID") int masterId){
    model.put("classID", masterId);
    return getMyView("classOne.ftl");
}
@RequestMapping(value="/outer/masterIdPath-{masterId}", method=RequestMethod.POST)
@ResponseBody
public String getOuterslaveClassNameListBymasterClassName(Map<String, Object> model,
    int masterId         BindingResult result){
    page.setExt(String.valueOf(masterId));
    return service.getOuterslaveClassNameListBymasterClassName(page);
}
```

Figure 5. Samples of code generation.

5 EXPERIMENTAL RESULTS

In a running program, input the above XML examples and templates file as import, and the test results obtained are directly generated codes related to templates as output, as shown in Figure 5. Watching from the output results, there are relevant properties and related functions, the system achieves the purpose of the original design from the aspect of functions.

6 CONCLUSION

This paper presents a solution to realize Java EE code generator, which functions relatively well, and not only proposes solutions for multi-table related issues, but also realizes UML-based visual design. Of course, there is a large gap between the systems generated by the code generator in this paper away from mature enterprise-class applications, such as the lack of IDE's support for the code level, the generated code cannot be directly imported by Eclipse and other IDE. On the basis of this paper, IDE-based projects and integration of automatic codes deploy, thus we can achieve a more comprehensive information system software solutions for automatic code generation.

ACKNOWLEDGEMENTS

Scientific research item: Marine public welfare industry research special funds for the project (201105033) National Key Technology R&D Program under Grant No. 2012BAH17F03.

REFERENCES

Anitha Rani Marneni, 2010. Automatic generation of object-oriented class implementations from behavioral specifications [D].The University of Texas at San Antonio.

Abid Mehmood, Dayang N.A. Jawawi, 2013. Aspect-oriented model-driven code generation: A systematic mapping study [J]. Information and Software Technology, 552:

Dong Yuming, 2012. Research and Application of Code Generation Technology in Development of Management Information System [D]. Jilin University: Jilin University.

Guan Taiyang, 2007. Research on automatic code generation based on templates [D]. University of Electronic Science and Technology.

Hamad I. Alsawalqah, Sung on Kang, Jihyun Lee, 2014. A method to optimize the scope of a software product platform based on end-user features [J]. The Journal of Systems & Software, 98.

John Grundy, John Hosking, 2013. Guest editor's introduction: special issue on innovative automated software engineering tools [J]. Automated Software Engineering, 20(2).

Yang Xingtao, Su Guiping, Wang Ruifang and Wang Xiaofang, 2009. Research and implementation of domain-specific modeling and code generation [J]. Computer System, 04:100–103.

Energy Science and Applied Technology – Fang (Ed.)
© 2016 Taylor & Francis Group, London, ISBN 978-1-138-02833-3

Research on digital watermarking based on colorful two-dimension code

H.Z. Lu, F.H. Lan & H.C. Li
School of the Engineering University of CAPF, Xi'an, China

ABSTRACT: Digital Watermarking, as an important branch of information hiding, is widely used to protect the security of digital media. At the same time, the application of two-dimension code has become widely popular; however, security, counterfeiting and forgery have become intensely serious. The birth of the colorful two-dimension code makes the traditional black and white one to become more beautiful without changing its essence. In view of the characteristic, actual application and security status of the two-dimension code combines with the characteristics of a colorful two-dimension code. For the first time, the application of digital watermark in the colorful two-dimension code is studied. A colorful two-dimension code watermarking algorithm to resist geometric attacks is designed, which makes the colorful two-dimension code to have a more practical function and substantial significance.

Keywords: digital watermarking, colorful two-dimension code, anti-fake, resist geometric attacks

1 INTRODUCTION

With the rapid development of electronic information technology, two-dimension code generation tool on the Internet is easy to get. Many criminals spread the virus, Trojan horse programs, and fish websites with the two-dimension code. In addition, the two-dimension code is also widely used in the food, pharmaceutical and other living consumption goods and in anti-counterfeiting and verification. In order to make more attractive appearance, the colorful two-dimension code emerges as the times require. However, the difference between the colorful and the black and white two-dimension code is only to make it more beautiful, without code scanning tools changing and the amount of information increasing. However, from the perspective of digital watermarking, the colorful image compared with the two value and the gray images has a greater watermark embedding capacity and a higher invisibility in the watermark embedding strength under the same conditions.

At present, study on two-dimension code digital watermarking is very scarce, and the research object and direction are more focused on the black and white image in two-dimension code of each monochrome block of airspace processing to embed watermark information, and these methods' watermark capacity is small. Therefore, after the analysis and study of the current situation of the development of the two-dimension code and the existing two-dimension code watermark tech-

nology, this paper proposes a digital watermarking technology based on the colorful two-dimension code.

2 TWO-DIMENSION CODE

The two-dimension code can encode information in the horizontal and vertical directions simultaneously. It has the characteristics of large information capacity, high reliability, low cost and strong robustness. In the late 80's, America and Japan began to study the two-dimension code technology, and developed a variety of molding ways of coding. The use of two kinds of coding method is the QR (Quickly Response) code and the PDF417 (Portable Data File) code. The two methods differ as follows: QR codes is superior to the PDF417 code in the reading direction (360°a full range of reading) and reading speed. So, this paper focuses on the research on the watermarking of QR code.

The QR code symbol, as shown in Figure 1.1, is a square array consisting of a square module that contains the function of graphics and the coding region. Among them, the function of the graphic is position detection, location and calibration, and that of the coding region is data coding. The sides of the symbol leave at least four modules of blank area with width.

Colorful two-dimension code is a special kind of two-dimension code. It not only has all the features of the common black-and-white two-dimension

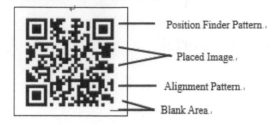

Figure 1.1.　QR code symbol.

Position Finder Pattern.

Placed Image.

Alignment Pattern.

Blank Area.

Figure 1.2.　Colorful two-dimension code.

codes, but also shows colorful appearance. However, the colorful two-dimension code must satisfy the foreground colorful image of dark and background colorful image of light, and the colorful image of the anchor point is similar to the colorful image of the other module, otherwise it will affect the reading of scanning software. Figure 1.2 is a colorful two-dimension code image, the foreground for multi-colorful, background colorful is pure white. In addition, some colorful two-dimension codes add no more than 1/10 of the QR code size colorful images at the center part. This is the application of two-dimension code that its error correcting ability maximum is no more than 30%.

3　TWO-DIMENSION CODE WATERMARK

The two-dimension code always appears on print or electronic screen. In printing and scanning yards, shooting will inevitably cause the two-dimension code image rotation, scaling and a certain amount of colorful fluctuation error. Many watermark embedding algorithm cannot to be used directly. The two-dimension code itself has characteristics of the rotational correction function, so the watermarking algorithms cannot destroy the two-dimension code to be read, and has the ability of anti-scaling attack. In addition, in practice, two-

Figure 2.1.　DCT partial coefficient recovery image.

dimension code watermark needs to realize blind extraction (does not need the original image while extracting the watermark).

Figure 1.2 shows after discrete cosine transform (DCT). Taking the top $2/5 \times 2/5(16\%)$ in the low frequency coefficients and DCT inverse transform the recovered image, as shown as Figure 2.1. We can see that the partial recovery coefficient diagram (Figure 2.1) and the original colorful of two-dimension code image (Figure 1.2) have a obvious difference, but the image can still be correctly read by the two-dimension code detector. This indicates that in a certain intensity modified the remaining 84% high frequency coefficient in DCT does not affect the correct reading two-dimension code, so that the watermark has a large number of embedded space and large embedding strength in DCT domain. Therefore, in theory, embedding the watermark in the DCT domain of two-dimension code is feasible.

Due to the blank area size of two-dimension code is variable, and changes in this area will affect the normal reading two-dimension code, so before embedding watermark should make the blank part cut off, retaining only the middle part as the watermark embedding area. Two-dimensional code can be generated artificially or by the system according to the amount of data and error correction rate to determine a single module occupying a pixel matrix size. For example, in Figure 1.2, a single module to the pixel size is 4×4. In order to make the watermark to resist scaling attack, before the watermark embedded in the two-dimension code needs to be unified format standard. First by scaling adjust the two-dimension code module size for 4×4 (other reasonable standard can be), then embedded the watermark information. When extracting the watermark, using the two-dimension code positioning and correction function to rotate and scale the correction to make the module size adjustment is 4×4, thus restoring the size of the embedding, and then the watermark can be extracted. Figure 2.2 shows

Figure 2.2.   Blank to standardize.

the core part of the two-dimension code removed the blank area and module standardized. In order to reduce the influence of colorful fluctuation error to the watermark, we can select multiple position redundant embedding, to ensure that the watermark can be extracted correctly.

4   COLORFUL TWO-DIMENSION CODE WATERMARK ALGORITHM

Because the characteristics of two-dimension code, the common two-dimension code image is not too large, so the two-dimension code watermark capacity is limited. Usually, it will have a certain identity information of binary sequence as the watermark, and the length can be adjusted according to the image. In order to directly, may use 32×32 binary image as watermark, expand it into one-dimensional binary sequence length of 1024. The carrier image is a colorful two-dimension code image generated by a text, as shown in Figure 1.2.

4.1   *Watermarking pretreatment*

In order to destroy the spatial correlation of the original watermark, first Arnold scrambled the watermark image W, and then make it chaotic encryption after putting it into one-dimensional sequences, and get the watermark sequence Wk (if the watermark itself is a binary sequence, it can be directly used for chaotic encryption). The scrambling times n and the initial value of chaotic generator as the key to save.

4.2   *Watermark embedding*

The process of watermark embedding are as follows:

1. Remove the blank area of colorful two-dimension code image I0, standardized module and get the 4×4 size of image I.

2. RGB colorful separate I, and get grayscale, three components of the R, G, B.
3. Divided R component into the size of 8×8 non-overlapping block (considering the characteristics of the QR code, the block size should be equal to the integer times of the size of the module, to resist scaling attacks).
4. Do DCT transform of the block, and get the DC coefficient recorded as dij.
5. For a block, the embedding strength is α, do the following transformation to dij of the block: s = round(dij/α), that is four to five homes in rounding,
   If the mod(s +Wk,2) = 1, dij' = (s −0.5) × α;
   If the mod(s +Wk,2) = 0, dij' = (s+0.5) × α.
6. For each block DCT transform, and get the R component image of the watermarked recorded as R.
7. 90° rotating G component and cycle spinning to the right to half of the image size. 180° rotation B component and cycle spinning upward to half of the image size. Repeat steps 2 to 5 as R component on the component after transformation separately. Then, inverse rotation and cycle spinning, respectively, to get G' and B'.
8. Merge R', G', B 'three components and get watermarked colorful image, and then add the blank area to make it into the standard two-dimension code image I'.

4.3   *Watermark extraction*

The watermark extraction basically is the inverse process of watermark embedding without the original carrier image participating, and then realize the blind watermark extraction.

1. Remove the blank area after correcting the colorful two-dimension code image with watermark, and standardized module to get the 4×4 size of image Ic.
2. RGB colorful separates Ic, and get grayscale with three components.
3. Divide R component into the size of 8×8 non-overlapping block.
4. Do DCT transform of the block, and get the DC coefficient recorded as dij.
5. For a block, the embedding strength is α, do the following detection processing to dij of the block: s = floor(dij/α), that is round numbers,
   If mod(s,2) = 1, W1(k) = W1(k)+1;
   If mod(s,2) = 0, W0(k) = W0(k)+1;
   When all the block DC coefficient testing is completed, if W1 (k) > W0(k), extracted the watermark WR(k) = 1; otherwise, WR(k) = 0.
6. 90° rotating G component and cycle spinning to the right to half of the image size. 180° rotation B component and cycle spinning upward to half of

the image size. And the extraction process is the same as R component, then get WG and WB.

7. Merge WR, WG, WB each in accordance with the principle of majority and get the one-dimensional watermark sequence W'.

8. Chaos decrypt W', then do Arnold inverse transformation after the reduction to a two-dimension matrix, and get the watermarking image Wc.

5 SIMULATION EXPERIMENT AND ANALYSIS

Choose the image of the size of 326×326, Figure 1.2, as the original carrier image. Choose the image of the size of 32×32, Figure 4.1, as the watermark image. Use the software Matlab7.0 as the experimental platform.

Figures 4.2 and 4.3 respectively for the carrier image after embedding watermark and extracted watermark image. PSNR = 37.1891, NC = 1. The watermarked carrier image can be read correctly through the two-dimension code detection scanner, which means that the two-dimension code watermark is invisible. Thus, the watermark algorithm is effective.

Scaling attack to the embedding watermark carrier image, as shown in Figure 4.4 (a) and 4.5 (a). Zoom in and out two times, then extract the watermark, as shown in Figure 4.4 (b) and 4.5 (b).

Figure 4.1. Original carrier.

Figure 4.2. Embedding watermark.

Figure 4.3. Extracted watermark.

a Attacked image b Extracted watermark

Figure 4.4. Magnified 2 times.

a Attacked image b Extracted watermark

Figure 4.5. Reduced 1/2.

Because of the standardized scale correction when extracting the watermark, the NC value can reach to 1 and 0.9916, respectively, which proved that the algorithm has good ability to resist scaling attack. So, this algorithm meets the requirements of the two-dimension code watermarking.

Cutting 1/5 of the central part of watermark image, as shown in Figure 4.6 (a). Watermark extracted from different component and the final watermark are shown in Figure 4.6 (b), NC = 0.9859, which means that algorithm has strong resistance to cut smudge attacks. For the common attack not more than 1/10 picture of the two-dimension code center, the watermark information can be accurately extracted out. At the same time, it's directly to see watermark extracted from different components. Through the rotation and translation of different components of two-dimension code vector image when embedding the watermark can distract the influence of local daub continuous attack on the same position of watermark information. In addition, because the NC value from the single component extracted watermark can, respectively, reach 0.7142, 0.8824, and 0.8685, which basically meet the watermark effectiveness index. If there is no special requirements to anti-cut smudge attacks, different components can be embedded different watermarking information, which can greatly increase the watermark capacity.

Figure 4.6(a).   Center clipping 1/5 image.

Figure 4.6(b).   Extracted watermark.

6   CONCLUSION

Through theoretical analysis and simulation experiment analysis, it is feasible to embed watermark sequence in the colorful two-dimension code. Colorful two-dimension code provides more embedding position of the watermark and higher visibility, which has an obvious effect on improving the ability against the watermark attack and enhances the practicability of the two-dimension code watermarking.

REFERENCES

Chen, Z et al. Digital watermarking technique based on two-dimension code, Computer Application, 2006, 26(8): 1998–2000.

Guo, Z.H et al. An Algorithm Based on Double-Fragile Digital Watermarking of Distance Education's Electronic Seal. Proceedings of 2010 3rd IEEE International Conference on Computer Science and Information Technology VOL.4.

Li, P. Research and design based on the secret two-dimension code and the security of digital watermarking Figure code, Jilin university degree thesis, 2013.

Neelesh, M. & Madhu, S. Dual Watermarking Scheme For Secure Buyer-Seller Watermarking Protocol, Proceedings of 2011 3rd IEEE International Conference on Information Management and Engineering (ICIME 2011) VOL.06.

Zhu, B.W et al. Research on Digital Watermarking Algorithm Based on LSB for QR Code, Journal of Chengdu Information Engineering College, 2012, 27(6): 542–547.

Author index